FIFTH EDITION

COLLEGE ALGEBRA AND TRIGONOMETRY

Richard N. Aufmann

Vernon C. Barker

Richard D. Nation

Palomar College

Houghton Mifflin Company

Boston   New York

Publisher: Jack Shira
Senior Sponsoring Editor: Lynn Cox
Senior Development Editor: Dawn Nuttall
Associate Editor: Jennifer King
Assistant Editor: Melissa Parkin
Senior Project Editor: Tamela Ambush
Editorial Assistant: Lisa Sullivan
Senior Production/Design Coordinator: Carol Merrigan
Manufacturing Manager: Florence Cadran
Marketing Manager: Danielle Potvin
Marketing Associate: Nicole Mollica

Cover photograph: © David Norton / Getty Images

PHOTO CREDITS

Chapter P: *p. 1* NASA, N. Benitez, T. Broadhurst (The Hebrew University), H. Ford (JHU), M. Clampin (STScI), G. Hartig (STScI), G. Illingworth (UCO/Link Observatory), the ACS Science Team and ESA; *p. 2* AP/Wide World Photos; *p. 64* NASA, ESA, and the Hubble Heritage Team (STScI/AURA). **Chapter 1:** *p. 81* Bettmann /CORBIS; *p. 81* Spencer Grant / PhotoEdit, Inc.; *p. 93* Mona Lisa by Leonardo da Vinci. © Gianni Dagli Orti/CORBIS; *p. 98* © The British Museum / Topham - HIP / The Image Works; *p. 108* The Granger Collection; *p. 112* Jeff Greenberg / PhotoEdit, Inc.; *p. 139* AP/Wide Wolrd Photos; *p. 148* Danielle Austen / Syracuse Newspapers / The Image Works. **Chapter 2:** *p. 161* Charles O'rear / CORBIS; *p. 224* CORBIS. **Chapter 3:** *p. 277* Sonda Dawes / The Image Works; *p. 278* David Young-Wolff / PhotoEdit, Inc.; *p. 288* Syndicated Features Limited / The Image Works; *p. 309* The Granger Collection; *p. 319* Bettmann/CORBIS; *p. 321* Bettmann / CORBIS; *p. 342* Richard T. Nowitz / CORBIS. **Chapter 4:** *p. 353* Chris McLaughlin / CORBIS; *p. 373* Bettmann / CORBIS; *p. 379* Charles O'rear / CORBIS; *p. 379* David James / Getty Images; *p. 383* Bettmann / CORBIS; *p. 428* Tom Brakefield / CORBIS; *p. 435* Bettmann / CORBIS; *p. 440* AP / Wide World Photos. **Chapter 5:** *p. 459* Art; *p. 474* Courtesy of NASA and JPL; *p. 488* Art; *p. 488* Tony Craddock/Getty Images; *p. 507* Reuters/NewMedia Inc./CORBIS. **Chapter 6:** *p. 591* Courtesy of Richard Nation; *p. 611* Art. **Chapter 7:** *p. 627* Massimo Listr / CORBIS; *p. 643* MacDuff Everton / CORBIS; *p. 672* The Granger Collection. **Chapter 8:** *p. 683* John Gay, US Navy /Getty Images; *p. 692* 1998 International Conference on Quality Control by Artificial Vision-QCAV'98. Kagawa Convention Center, Takamatsu, Kagawa, Japan, November 10-12, 1998, pp. 521-528; *p. 693* Reprinted with permission of Ian Morison, Jodrell Bank Conservatory; *p. 709* Hugh Rooney / Eye Ubiquitous / CORBIS; *p. 717* The Granger Collection. **Chapter 9:** n/a. **Chapter 10:** *p. 831* Boyd Norton / The Image Works; *p. 863* Comstock images. **Chapter 11:** *p. 902* Stephen Johnson / Getty Images; *p. 902* Stephen Johnson / Getty Images.

Printed in the U.S.A.

Library of Congress Control Number: 2003110128

ISBNs:
Student's Edition: 0-618-38680-7
Instructor's Annotated Edition: 0-618-38681-5

2 3 4 5 6 7 8 9-DOW-08 07 06 05 04

CONTENTS

5 TRIGONOMETRIC FUNCTIONS   *459*

6 TRIGONOMETRIC IDENTITIES AND EQUATIONS   *553*

7   APPLICATIONS OF TRIGONOMETRY   683

8   TOPICS IN ANALYTIC GEOMETRY   683

9 SYSTEMS OF EQUATIONS 767

10 MATRICES 831

11 SEQUENCES, SERIES, AND PROBABILITY 901

PREFACE

With each successive edition of *College Algebra and Trigonometry* we strive to enhance and refine our instructional materials. In this edition we have continued our emphasis on *doing* mathematics rather than duplicating mathematics through extensive drill. Students are urged to investigate concepts, apply those concepts, and then present their findings. We are ever cognizant of the motivating influence that contemporary, relevant applications have on students. As a result, we have added many new application exercises and deleted those that are of little student interest.

Technology is introduced naturally to illustrate or enhance the current topic. We integrate technology into a discussion when it can be used to foster and promote a better understanding of a concept. The optional *Integrating Technology* boxes and graphing calculator exercises are designed to instill in students an appreciation of both the power and the limitations of technology.

In this edition, we have retained our basic philosophy, which is to deliver a comprehensive and mathematically sound treatment of the topics considered essential for a college algebra and trigonometry course. To help students master these concepts, we have tried to maintain a balance among theory, application, modeling, and drill. Carefully developed mathematics is complemented by abundant, creative applications that are both contemporary and representative of a wide range of disciplines. Many application exercises are accompanied by a diagram that helps the student visualize the mathematics of the application.

Changes for the Fifth Edition

Many new student support features have been added to this edition:

A *Focus on Problem Solving* feature has been added at the beginning of each chapter. This feature reviews and demonstrates the strategies used by successful problem-solvers.

Review Notes, identified by ⇆, direct students to the page where a previously discussed concept can be reviewed. Students needing to review this concept can do so before proceeding with new material.

Prepare for Next Section Exercises, found in each section's exercise set (except the last section of a chapter), allow students to practice prerequisite skills and concepts needed to be successful in the next section. Each exercise references the section of the text that contains the concepts related to the question for students to easily review. All answers are provided in the student answer section.

Cumulative Review Exercises have been added to every chapter, after Chapter P. These 20 Cumulative Review Exercises allow students to review material from previous chapters. The answers for all of these exercises are included in the student answer section. Next to each answer is a section reference that directs the student to the section of the text from which the exercise was taken.

Answers to All Chapter Review Exercises   are now included in the student answer section. Next to each answer is a section reference that directs the student to the section of the text from which the exercise was taken.

Some specific chapter changes are indicated below.

Chapter P, *Preliminary Concepts*, has been reorganized and now includes a review of the basic operations on complex numbers. This chapter is a review of many topics that are a prerequisite to success in college algebra. Students who have successfully completed an intermediate algebra course may not need to cover this chapter.

Chapter 1, *Equations and Inequalities*, now includes several new Examples that illustrate contemporary applications of mathematics. Also, each exercise set includes some new application exercises. For instance, in Section 1.2 formulas are provided for assessing the reading grade level of written material and for assessing the passing performance of quarterbacks.

Chapter 2, *Functions and Graphs*, includes a completely new discussion on composition of functions that is motivated through an application problem. Concise reviews of vertical and horizontal translations have been added. Many new application exercises were added.

Chapter 3, *Polynomial and Rational Functions*, starts with a more detailed discussion of polynomial division. Several new worked examples and contemporary application exercises have been included in this chapter.

Chapter 4, *Exponential and Logarithmic Functions*, now includes additional contemporary application exercises. For instance, some exercises illustrate an exponential relationship between various water temperatures and the time required for a scuba diver to reach hypothermia.

In Chapter 5, *Trigonometric Functions*, new applications from astronomy, aviation, and physics have been included.

Chapter 7, *Applications of Trigonometry*, now includes a discussion of how one of Mollweide's formulas can be used to check whether a triangle has been solved correctly.

In Chapter 8, *Topics in Analytic Geometry*, we have included several new applications of a more contemporary nature. We have continued the use of graphing utilities to illustrate important concepts.

Chapter 10, *Matrices*, now includes some application exercises involving heat transfer that can be solved using matrices.

CHAPTER OPENER FEATURES

CHAPTER OPENER

Each chapter begins with a **Chapter Opener** that illustrates a specific application of a concept from the chapter. There is a reference to a particular exercise within the chapter that asks the student to solve a problem related to the chapter opener topic.

The icons at the bottom of the page let students know of additional resources available on CD, video/DVD, in the *Student Study Guide*, and online at math.college.hmco.com/students.

Page 354

Page 353

CHAPTER **4**

EXPONENTIAL AND LOGARITHMIC FUNCTIONS

4.1 INVERSE FUNCTIONS
4.2 EXPONENTIAL FUNCTIONS AND THEIR APPLICATIONS
4.3 LOGARITHMIC FUNCTIONS AND THEIR APPLICATIONS
4.4 LOGARITHMS AND LOGARITHMIC SCALES
4.5 EXPONENTIAL AND LOGARITHMIC EQUATIONS
4.6 EXPONENTIAL GROWTH AND DECAY
4.7 MODELING DATA WITH EXPONENTIAL AND LOGARITHMIC FUNCTIONS

Modeling Data with an Exponential Function

The following table shows the time, in hours, before the body of a scuba diver, wearing a 5-millimeter-thick wet suit, reaches hypothermia (95°F) for various water temperatures.

Water Temperature, °F	Time, hours
36	1.5
41	1.8
46	2.6
50	3.1
55	4.9

Source: Data as scanned from the American Journal of Physics, vol. 71, no. 4 (April 2003), Fig. 3, p. 336.

The following function, which is an example of an exponential function, closely models the data in the table:

$$T(F) = 0.1509(1.0639)^F$$

In this function F represents the Fahrenheit temperature of the water, and T represents the time in hours. A diver can use the function to determine the time it takes to reach hypothermia for water temperatures that are not included in the table. See Exercise 21, page 445. The function $T(F)$ was determined by using exponential regression, which is one of the topics in Section 4.7.

---

21. HYPOTHERMIA The following table shows the time T, in hours, before a scuba diver wearing a 3-millimeter-thick wet suit reaches hypothermia (95°F) for various water temperatures F, in degrees Fahrenheit.

Water temperature, °F	Time T, hours
41	1.1
46	1.4
50	1.8
59	3.7

a. Find an exponential regression model for the data. Round the constants a and b to the nearest hundred thousandth.

b. Use the model from part **a.** to estimate the time it takes for the diver to reach hypothermia in water that has a temperature of 65°F. Round to the nearest tenth of an hour.

Page 445

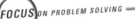
FOCUS ON PROBLEM SOLVING

Use Two Methods to Solve and Compare Results

Sometimes it is possible to solve a problem in two or more ways. In such situations it is recommended that you use at least two methods to solve the problem, and compare your results. Here is an example of an application that can be solved in more than one way.

Example

In a league of eight basketball teams, each team plays every other team in the league exactly once. How many league games will take place?

Solution

Method 1: *Use an analytic approach.* Each of the eight teams must play the other seven teams. Using this information, you might be tempted to conclude that there will be $8 \cdot 7 = 56$ games, but this result is too large because it counts each game between two individual teams as two different games. Thus the number of league games will be

$$\frac{8 \cdot 7}{2} = \frac{56}{2} = 28$$

Method 2: *Make an organized list.* Use the letters A, B, C, D, E, F, G, and H to represent the eight teams. Use the notation AB to represent the game between team A and team B. Do not include BA in your list because it represents the same game between team A and team B.

AB	AC	AD	AE	AF	AG	AH
BC	BD	BE	BF	BG	BH	
	CD	CE	CF	CG	CH	
		DE	DF	DG	DH	
			EF	EG	EH	
				FG	FH	
					GH	

The list shows that there will be 28 league games.

The procedure of using two different solution methods and comparing results is employed often in this chapter. For instance, see Example 2, page 409. In this example, a solution is found by applying algebraic procedures and also by graphing. Notice that both methods produce the same result.

NEW! *FOCUS ON PROBLEM SOLVING*

A **Focus on Problem Solving** follows the Chapter Opener. This feature highlights and demonstrates a problem-solving strategy that may be used to successfully solve some of the problems presented in the chapter.

— AUFMANN INTERACTIVE METHOD (AIM) —

INTERACTIVE PRESENTATION

College Algebra and Trigonometry is written in a style that encourages the student to interact with the textbook.

EXAMPLES

Each section contains a variety of worked examples. Examples are **titled** so that the student can see at a glance the type of problem being illustrated, often accompanied by **annotations** that assist the student in moving from step to step; and offers the **final answer in color** so that it is readily identifiable.

TRY EXERCISES

Following every example is a suggested **Try Exercise** from that section's exercise set for the student to work. The exercises are color coded by number in the exercise set and the *complete solution* of that exercise can be found in an appendix to the text.

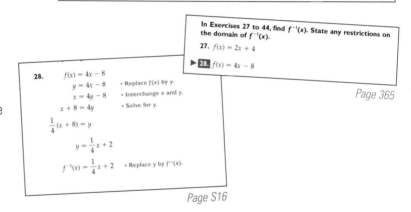

Page 360

EXAMPLE 3 Find the Inverse of a Function

Find the inverse of $f(x) = 3x + 8$.

Solution

$$f(x) = 3x + 8$$
$$y = 3x + 8 \qquad \bullet \text{ Replace } f(x) \text{ by } y.$$
$$x = 3y + 8 \qquad \bullet \text{ Interchange } x \text{ and } y.$$
$$x - 8 = 3y \qquad \bullet \text{ Solve for } y.$$
$$\frac{x - 8}{3} = y$$
$$\frac{1}{3}x - \frac{8}{3} = f^{-1}(x) \qquad \bullet \text{ Replace } y \text{ by } f^{-1}(x).$$

The inverse function is given by $f^{-1}(x) = \dfrac{1}{3}x - \dfrac{8}{3}$.

▶ **Try Exercise 28, page 365**

In Exercises 27 to 44, find $f^{-1}(x)$. State any restrictions on the domain of $f^{-1}(x)$.

27. $f(x) = 2x + 4$

▶ **28.** $f(x) = 4x - 8$

Page 365

28.
$$f(x) = 4x - 8$$
$$y = 4x - 8 \qquad \bullet \text{ Replace } f(x) \text{ by } y.$$
$$x = 4y - 8 \qquad \bullet \text{ Interchange } x \text{ and } y.$$
$$x + 8 = 4y \qquad \bullet \text{ Solve for } y.$$
$$\frac{1}{4}(x + 8) = y$$
$$y = \frac{1}{4}x + 2$$
$$f^{-1}(x) = \frac{1}{4}x + 2 \qquad \bullet \text{ Replace } y \text{ by } f^{-1}(x).$$

Page S16

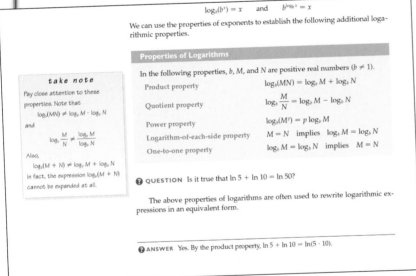

$$\log_b(b^x) = x \qquad \text{and} \qquad b^{\log_b x} = x$$

We can use the properties of exponents to establish the following additional logarithmic properties.

Properties of Logarithms

In the following properties, b, M, and N are positive real numbers ($b \neq 1$).

Product property	$\log_b(MN) = \log_b M + \log_b N$
Quotient property	$\log_b \dfrac{M}{N} = \log_b M - \log_b N$
Power property	$\log_b(M^p) = p \log_b M$
Logarithm-of-each-side property	$M = N$ implies $\log_b M = \log_b N$
One-to-one property	$\log_b M = \log_b N$ implies $M = N$

take note

Pay close attention to these properties. Note that
$$\log_b(MN) \neq \log_b M \cdot \log_b N$$
and
$$\log_b \frac{M}{N} \neq \frac{\log_b M}{\log_b N}$$
Also,
$$\log_b(M + N) \neq \log_b M + \log_b N$$
In fact, the expression $\log_b(M + N)$ cannot be expanded at all.

❓ **QUESTION** Is it true that $\ln 5 + \ln 10 = \ln 50$?

The above properties of logarithms are often used to rewrite logarithmic expressions in an equivalent form.

❓ **ANSWER** Yes. By the product property, $\ln 5 + \ln 10 = \ln(5 \cdot 10)$.

Page 395

QUESTION/ANSWER

In every section, we pose at least one **Question** to the student about the material being read. This question encourages the reader to pause and think about the current discussion and to answer the question. To make sure the student does not miss important information, the **Answer** to the question is provided as a footnote on the same page.

REAL DATA AND APPLICATIONS

APPLICATIONS

One way to motivate an interest in mathematics is through applications. Applications require the student to use problem-solving strategies, along with the skills covered in a section, to solve practical problems. This careful integration of applications generates student awareness of the value of algebra as a real-life tool.

Applications are taken from many disciplines including agriculture, business, chemistry, construction, Earth science, education, economics, manufacturing, nutrition, real estate, and sociology.

Page 428

428 Chapter 4 Exponential and Logarithmic Functions

In the following example we determine a logistic growth model for a coyote population.

EXAMPLE 8 **Find and Use a Logistic Model**

At the beginning of 2002, the coyote population in a wilderness area was estimated at 200. By the beginning of 2004, the coyote population had increased to 250. A park ranger estimates that the carrying capacity of the wilderness area is 500 coyotes.

a. Use the given data to determine the growth rate constant for the logistic model of this coyote population.

b. Use the logistic model determined in part **a.** to predict the year in which the coyote population will first reach 400.

Solution

a. If we represent the beginning of the year 2002 by $t = 0$, then the beginning of the year 2004 will be represented by $t = 2$. In the logistic model, make the following substitutions: $P(2) = 250$, $c = 500$, and

$$a = \frac{c - P_0}{P_0} = \frac{500 - 200}{200} = 1.5.$$

$$P(t) = \frac{c}{1 + ae^{-bt}}$$

$$P(2) = \frac{500}{1 + 1.5e^{-b \cdot 2}} \qquad \text{• Substitute the given values.}$$

$$250 = \frac{500}{1 + 1.5e^{-b \cdot 2}}$$

$$250(1 + 1.5e^{-b \cdot 2}) = 500 \qquad \text{• Solve for the growth rate constant } b.$$

$$1 + 1.5e^{-b \cdot 2} = \frac{500}{250}$$

$$1.5e^{-b \cdot 2} = 2 - 1$$

$$e^{-b \cdot 2} = \frac{1}{1.5}$$

$$-2b = \ln\left(\frac{1}{1.5}\right)$$

$$b = -\frac{1}{2}\ln\left(\frac{1}{1.5}\right)$$

$$b \approx 0.20273255$$

Page 448

448 Chapter 4 Exponential and Logarithmic Functions

Time t (minutes)	0	5	10	15	20	25
Coffee temp. T (°F)	165°	140°	121°	107°	97°	89°
$T - 70°$	95°	70°	51°	37°	27°	19°

a. Use a graphing utility to find an exponential model for the difference $T - 70$ as a function of t.

b. Use the model to predict how long it will take (to the nearest minute) for the coffee to cool to 80°F.

35. OLYMPIC DISTANCES The following table shows the winning Olympic distances for the men's shot put for the years 1948 to 2000.

Men's Olympic Shot Put, 1948 to 2000

Year	Distance	Year	Distance
1948	56 ft 2 in.	1976	69 ft $\frac{3}{4}$ in.
1952	57 ft 1$\frac{1}{2}$ in.	1980	70 ft 1$\frac{1}{2}$ in.
1956	60 ft 11 in.	1984	69 ft 9 in.
1960	64 ft 6$\frac{3}{4}$ in.	1988	73 ft 8$\frac{3}{4}$ in.
1964	66 ft 8$\frac{1}{4}$ in.	1992	71 ft 2$\frac{1}{2}$ in.
1968	67 ft 4$\frac{3}{4}$ in.	1996	70 ft 11$\frac{1}{4}$ in.
1972	69 ft 6 in.	2000	69 ft 10$\frac{1}{4}$ in.

Source: Time Almanac 2002

Represent the year 1948 by $t = 48$.

a. Use the regression features of a graphing utility to determine a logistic growth model and a logarithmic model for the data.

b. Use graphs of the models in part **a.** to determine which model provides the better fit for the data.

c. Use the model you selected in part **b.** to predict the men's shot put distance for the year 2008. Round to the nearest hundredth of a foot.

36. WORLD POPULATION The following table lists the years in which the world's population first reached 3, 4, 5, and 6 billion.

World Population Milestones

Year	Population
1960	3 billion
1974	4 billion
1987	5 billion
1999	6 billion

Source: Time Almanac 2002, p. 708.

a. Find a logistic growth model, $P(t)$, for the data in the table. Let t represent the number of years after 1960 ($t = 0$ represents the year 1960).

b. According to the logistic growth model, what will the world's population approach as $t \to \infty$? Round to the nearest billion.

37. DESALINATION The following table shows the amount of fresh water w (in cubic yards) produced from saltwater after t hours of a desalination process.

t	1	2.5	3.5	4.0	5.1	6.5
w	18.2	46.6	57.4	61.5	68.7	76.2

a. Use a graphing utility to find a linear model and a logarithmic model for the data.

b. Examine the correlation coefficients of the two regression models to determine which model provides the better fit for the data. State the correlation coefficient r for each model.

c. Use the model you selected in part **b.** to predict the amount of fresh water that will be produced after 10 hours of the desalination process. Round to the nearest tenth of a cubic yard.

38. A CORRELATION COEFFICIENT OF 1 A scientist uses a graphing utility to model the data set {(2, 5), (4, 6)} with a logarithmic function. The following display shows the results.

What is the significance of the fact that the correlation coefficient for the regression equation is $r = 1$?

```
LnReg
y=a+blnx
a=4
b=1.442695041
r²=1
r=1
```

REAL DATA

Real data examples and exercises, identified by , ask students to analyze and create mathematical models from actual situations. Students are often required to work with tables, graphs, and charts drawn from a variety of disciplines.

TECHNOLOGY

INTEGRATING TECHNOLOGY

The **Integrating Technology** feature contains optional discussions that can be used to further explore a concept using technology. Some introduce technology as an alternative way to solve certain problems and others provide suggestions for using a calculator to solve certain problems and applications. Additionally, optional graphing calculator examples and exercises (identified by) are presented throughout the text.

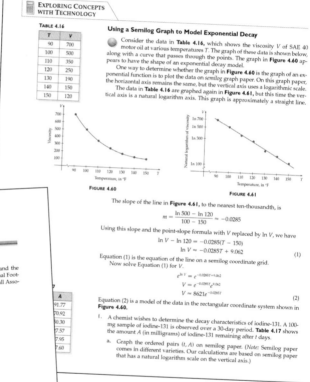

INTEGRATING TECHNOLOGY

The graph of $y^2 > 2\,8x$ is shown in **Figure 8.5.** Note that the graph is not the graph of a function. To graph $y^2 > 2\,8x$ with a graphing utility, we first solve for y to produce $y > 6$ ˇ $2\,8x$. From this equation we can see that for any x , 0, there are two values of y. For example, when $x > 2\,2$,

$$y > 6 \text{ ˇ } \text{s2 8d's2 2d}> 6 \text{ ˇ } 16 > 6\,4$$

The graph of $y^2 > 2\,8x$ in **Figure 8.5** was drawn by graphing both $y_1 > $ ˇ $2\,8x$ and $y_2 > 2$ ˇ $2\,8x$ in the same window.

FIGURE 8.5

Page 678

EXPLORING CONCEPTS WITH TECHNOLOGY

A special end-of-chapter feature, **Exploring Concepts with Technology**, extends ideas introduced in the text by using technology (graphing calculator, CAS, etc.) to investigate extended applications or mathematical topics. These explorations can serve as group projects, class discussions, or extra-credit assignments.

Page 256

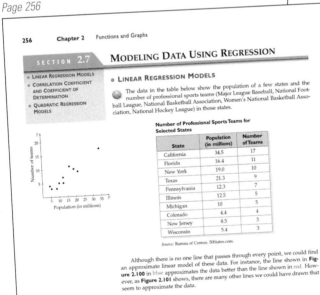

450   Chapter 4   Exponential and Logarithmic Functions

EXPLORING CONCEPTS WITH TECHNOLOGY

Using a Semilog Graph to Model Exponential Decay

Consider the data in **Table 4.16,** which shows the viscosity V of SAE 40 motor oil at various temperatures T. The graph of these data is shown below, along with a curve that passes through the points. The graph in **Figure 4.60** appears to have the shape of an exponential decay model.

One way to determine whether the graph in **Figure 4.60** is the graph of an exponential function is to plot the data on *semilog* graph paper. On this graph paper, the horizontal axis remains the same, but the vertical axis uses a logarithmic scale.

The data in **Table 4.16** are graphed again in **Figure 4.61,** but this time the vertical axis is a natural logarithm axis. This graph is approximately a straight line.

TABLE 4.16

T	V
90	700
100	500
110	350
120	250
130	190
140	150
150	120

FIGURE 4.60

FIGURE 4.61

The slope of the line in **Figure 4.61,** to the nearest ten-thousandth, is

$$m = \frac{\ln 500 - \ln 120}{100 - 150} \approx -0.0285$$

Using this slope and the point-slope formula with V replaced by $\ln V$, we have

$$\ln V - \ln 120 = -0.0285(T - 150)$$
$$\ln V \approx -0.0285T + 9.062 \qquad (1)$$

Equation (1) is the equation of the line on a semilog coordinate grid. Now solve Equation (1) for V.

$$e^{\ln V} = e^{-0.0285T + 9.062}$$
$$V = e^{-0.0285T}e^{9.062}$$
$$V \approx 8621e^{-0.0285T} \qquad (2)$$

Equation (2) is a model of the data in the rectangular coordinate system shown in **Figure 4.60.**

1.   A chemist wishes to determine the decay characteristics of iodine-131. A 100-mg sample of iodine-131 is observed over a 30-day period. **Table 4.17** shows the amount A (in milligrams) of iodine-131 remaining after t days.

a.   Graph the ordered pairs (t, A) on semilog paper. (*Note:* Semilog paper comes in different varieties. Our calculations are based on semilog paper that has a natural logarithm scale on the vertical axis.)

Page 450

MODELING

Special modeling sections, which rely heavily on the use of a graphing calculator, are incorporated throughout the text. These optional sections introduce the idea of a mathematical model using various real-world data sets that further motivate students and help them see the relevance of mathematics.

---

256   Chapter 2   Functions and Graphs

SECTION 2.7   **MODELING DATA USING REGRESSION**

- LINEAR REGRESSION MODELS
- CORRELATION COEFFICIENT AND COEFFICIENT OF DETERMINATION
- QUADRATIC REGRESSION MODELS

LINEAR REGRESSION MODELS

The data in the table below show the population of a few states and the number of professional sports teams (Major League Baseball, National Football League, National Basketball Association, Women's National Basketball Association, National Hockey League) in those states.

Number of Professional Sports Teams for Selected States

State	Population (in millions)	Number of Teams
California	34.5	17
Florida	16.4	11
New York	19.0	10
Texas	21.3	9
Pennsylvania	12.3	7
Illinois	12.5	5
Michigan	10	5
Colorado	4.4	4
New Jersey	8.5	3
Wisconsin	5.4	3

Source: Bureau of Census, 50States.com.

Although there is no one line that passes through every point, we could find an approximate linear model of these data. For instance, the line shown in **Figure 2.100** in blue approximates the data better than the line shown in red. However, as **Figure 2.101** shows, there are many other lines we could have drawn that seem to approximate the data.

FIGURE 2.100

FIGURE 2.101

STUDENT PEDAGOGY

TOPIC LIST

At the beginning of each section is a list of the major topics covered in the section.

KEY TERMS AND CONCEPTS

Key terms, in bold, emphasize important terms. Key concepts are presented in blue boxes in order to highlight these important concepts and to provide for easy reference.

MATH MATTERS

These margin notes contain interesting sidelights about mathematics, its history, or its application.

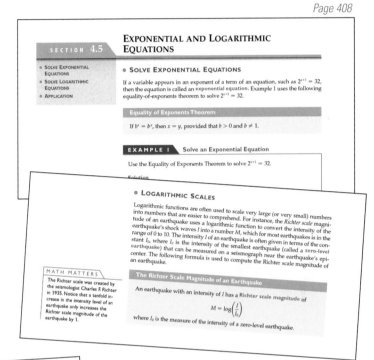

Page 408

SECTION **4.5**

EXPONENTIAL AND LOGARITHMIC EQUATIONS

● SOLVE EXPONENTIAL EQUATIONS
● SOLVE LOGARITHMIC EQUATIONS
● APPLICATION

● **SOLVE EXPONENTIAL EQUATIONS**

If a variable appears in an exponent of a term of an equation, such as $2^{x+1} = 32$, then the equation is called an exponential equation. Example 1 uses the following equality-of-exponents theorem to solve $2^{x+1} = 32$.

Equality of Exponents Theorem

If $b^x = b^y$, then $x = y$, provided that $b > 0$ and $b \neq 1$.

EXAMPLE 1   Solve an Exponential Equation

Use the Equality of Exponents Theorem to solve $2^{x+1} = 32$.

● **LOGARITHMIC SCALES**

Logarithmic functions are often used to scale very large (or very small) numbers into numbers that are easier to comprehend. For instance, the *Richter scale* magnitude of an earthquake uses a logarithmic function to convert the intensity of the earthquake's shock waves I into a number M, which for most earthquakes is in the range of 0 to 10. The intensity I of an earthquake is often given in terms of the constant I_0, where I_0 is the intensity of the smallest earthquake (called a *zero-level earthquake*) that can be measured on a seismograph near the earthquake's epicenter. The following formula is used to compute the Richter scale magnitude of an earthquake.

MATH MATTERS

The Richter scale was created by the seismologist Charles F. Richter in 1935. Notice that a tenfold increase in the intensity level of an earthquake only increases the Richter scale magnitude of the earthquake by 1.

The Richter Scale Magnitude of an Earthquake

An earthquake with an intensity of I has a Richter scale magnitude of

$$M = \log\left(\frac{I}{I_0}\right)$$

where I_0 is the measure of the intensity of a zero-level earthquake.

Page 398

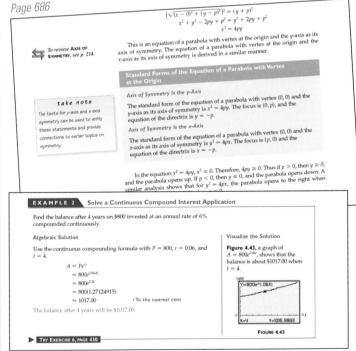

Page 686

$$(\sqrt{(x-0)^2 + (y-p)^2})^2 = (y+p)^2$$
$$x^2 + y^2 - 2py + p^2 = y^2 + 2py + p^2$$
$$x^2 = 4py$$

To review AXIS OF SYMMETRY, *see p. 214.*

This is an equation of a parabola with vertex at the origin and the y-axis as its axis of symmetry. The equation of a parabola with vertex at the origin and the x-axis as its axis of symmetry is derived in a similar manner.

Standard Forms of the Equation of a Parabola with Vertex at the Origin

Axis of Symmetry Is the y-Axis

The standard form of the equation of a parabola with vertex $(0, 0)$ and the y-axis as its axis of symmetry is $x^2 = 4py$. The focus is $(0, p)$, and the equation of the directrix is $y = -p$.

Axis of Symmetry Is the x-Axis

The standard form of the equation of a parabola with vertex $(0, 0)$ and the x-axis as its axis of symmetry is $y^2 = 4px$. The focus is $(p, 0)$, and the equation of the directrix is $x = -p$.

take note

The tests for y-axis and x-axis symmetry can be used to verify these statements and provide connections to earlier topics on symmetry.

In the equation $x^2 = 4py$, $x^2 \geq 0$. Therefore, $4py \geq 0$. Thus if $p > 0$, then $y \geq 0$, and the parabola opens up. If $p < 0$, then $y \leq 0$, and the parabola opens down. A similar analysis shows that for $y^2 = 4px$, the parabola opens to the right when

EXAMPLE 2   Solve a Continuous Compound Interest Application

Find the balance after 4 years on \$800 invested at an annual rate of 6% compounded continuously.

Algebraic Solution

Use the continuous compounding formula with $P = 800$, $r = 0.06$, and $t = 4$.

$$A = Pe^{rt}$$
$$= 800e^{0.06(4)}$$
$$= 800e^{0.24}$$
$$= 800(1.27124915)$$
$$\approx 1017.00 \qquad \text{• To the nearest cent}$$

The balance after 4 years will be \$1017.00.

Visualize the Solution

Figure 4.43, a graph of $A = 800e^{0.06t}$, shows that the balance is about \$1017.00 when $t = 4$.

Y=800e^(.06X)

X=4   Y=1016.9993

FIGURE 4.43

▶ TRY EXERCISE 6, PAGE 430

Page 422

NEW! *REVIEW NOTES*

A ⇄ directs the student to the place in the text where the student can review a concept that was previously discussed.

TAKE NOTE

These margin notes alert students to a point requiring special attention or are used to amplify the concept under discussion.

VISUALIZE THE SOLUTION

For appropriate examples within the text, we have provided both an algebraic solution and a graphical representation of the solution. This approach creates a link between the algebraic and visual components of a solution.

EXERCISES

TOPICS FOR DISCUSSION

These special exercises provide questions related to key concepts in the section. Instructors can use these to initiate class discussions or to ask students to write about concepts presented.

EXERCISES

The exercise sets in *College Algebra and Trigonometry* were carefully developed to provide a wide variety of exercises. The exercises range from drill and practice to interesting challenges. They were chosen to illustrate the many facets of topics discussed in the text. Each exercise set emphasizes skill building, skill maintenance, and, as appropriate, applications. **Icons** identify appropriate writing , group , data analysis ●, web www,

and graphing calculator 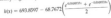 exercises.

Page 429

TOPICS FOR DISCUSSION

1. Explain the difference between compound interest and simple interest.

2. What is an exponential growth model? Give an example of an application for which the exponential growth model might be appropriate.

3. What is an exponential decay model? Give an example of an application for which the exponential decay model might be appropriate.

4. Consider the exponential model $P(t) = P_0 e^{kt}$ and the logistic model $P(t) = \dfrac{c}{1 + ae^{-bt}}$. Explain the similarities and differences between the two models.

Functions   **447**

mp, 1968 to 2000

Year	Distance
2988	6 ft 8 in.
992	6 ft 7$\frac{1}{2}$ in.
996	6 ft 8$\frac{3}{4}$ in.
2000	6 ft 7 in.

Year	Distance
1980	6 ft 5$\frac{1}{2}$ in.
1984	6 ft 7$\frac{1}{2}$ in.

Source: *Time Almanac* 2002.

b. Use the model you chose in part **a.** to find the q-value associated with a pH of 8.2. Round to the nearest tenth.

30. WORLD POPULATION The following table lists the years in which the world's population first reached 3, 4, 5, and 6 billion.

World Population Milestones

Year	Population
1960	3 billion
1974	4 billion
1987	5 billion
1999	6 billion

Source: *Time Almanac* 2002, p. 708.

a. Find an exponential model for the data in the table. Let $x = 0$ represent the year 1960.

b. Use the model to predict the year in which the world's population will first reach 7 billion.

31. PANDA POPULATION One estimate gives the world panda population as 3200 in 1980 and 590 in 2000.

a. Find an exponential model for the data and use the model to predict the year in which the panda population p will be reduced to 200. (Let $t = 0$ represent the year 1980.)

b. Because the exponential model in part **a.** fits the data perfectly, does this mean that the model will accurately predict future panda populations? Explain.

32. OLYMPIC RECORDS The following table shows the Olympic gold medal distances for the women's high jump from 1968 to 2000.

Represent the year 1968 by 68.

a. Use a graphing utility to determine a linear model and a logarithmic model for the data, with the distance measured in inches. State the correlation coefficient r for each model.

b. Use the correlation coefficient for each of the models in part **a.** to determine which model provides the better fit for the data.

c. Use the model you selected in part **b.** to predict the women's Olympic gold medal high jump distance in 2012. Round to the nearest tenth of an inch.

33. NUMBER OF AUTOMOBILES In 1900, the number of automobiles in the United States was around 8000. By 2000, the number of automobiles in United States had reached 200 million.

a. Find an exponential model for the data and use the model to predict the number of automobiles, to the nearest 100,000, in the United States in 2010. Use $t = 0$ to represent the year 1900.

b. According to the model, in what year will the number of automobiles in the United States first reach 300 million?

34. TEMPERATURE OF COFFEE A cup of coffee is placed in a room that maintains a constant temperature of 70°F. The following table shows both the coffee temperature T after t minutes and the difference between the coffee temperature and the room temperature after t minutes.

Page 447

Included in each exercise set are **Connecting Concepts** exercises. These exercises extend some of the concepts discussed in the section and require students to connect ideas studied earlier with new concepts.

NEW! *EXERCISES TO PREPARE FOR THE NEXT SECTION*

Every section's exercise set (except for the last section of a chapter) contains exercises that allow students to practice the previously-learned skills they will need to be successful in the next section. Next to each question, in brackets, is a reference to the section of the text that contains the concepts related to the question for students to easily review. All answers are provided in the Answer Appendix.

PROJECTS

Projects are provided at the end of each exercise set. They are designed to encourage students to do research and write about what they have learned. These Projects generally emphasize critical thinking skills and can be used as collaborative learning exercises or as extra-credit assignments.

Page 381

4.2   Exponential Functions and Their Applications   **381**

CONNECTING CONCEPTS

60. Verify that the hyperbolic cosine function $\cosh(x) = \dfrac{e^x + e^{-x}}{2}$ is an even function.

61. Verify that the hyperbolic sine function $\sinh(x) = \dfrac{e^x - e^{-x}}{2}$ is an odd function.

62. Graph $g(x) = 10^x$, and then sketch the graph of g reflected across the line given by $y = x$.

63. Graph $f(x) = e^x$, and then sketch the graph of f reflected across the line given by $y = x$.

In Exercises 64 to 67, determine the domain of the given function. Write the domain using interval notation.

64. $f(x) = \dfrac{e^x - e^{-x}}{e^x + e^{-x}}$

65. $f(x) = \dfrac{e^{|x|}}{1 + e^x}$

66. $f(x) = \sqrt{1 - e^x}$

67. $f(x) = \sqrt{e^x - e^{-x}}$

PREPARE FOR SECTION 4.3

68. If $2^x = 16$, determine the value of x. [4.2]

69. If $3^{-x} = \dfrac{1}{27}$, determine the value of x. [4.2]

70. If $x^4 = 625$, determine the value of x. [4.2]

71. Find the inverse of $f(x) = \dfrac{2x}{x + 3}$. [4.1]

72. State the domain of $g(x) = \sqrt{x - 2}$. [2.2]

73. If the range of $h(x)$ is the set of all positive real numbers, then what is the domain of $h^{-1}(x)$? [4.2]

PROJECTS

1. THE SAINT LOUIS GATEWAY ARCH The Gateway Arch in Saint Louis was designed in the shape of an inverted catenary, as shown by the red curve in the drawing at the right. The Gateway Arch is one of the largest optical illusions ever created. As you look at the arch (and its basic shape defined by the catenary curve), it appears to be much taller than it is wide. However, this is not the case. The height of the catenary is given by

$$h(x) = 693.8597 - 68.7672\left(\dfrac{e^{0.0100333x} + e^{-0.0100333x}}{2}\right)$$

where x and $h(x)$ are measured in feet and $x = 0$ represents the position at ground level that is directly below the highest point of the catenary.

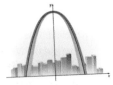

a. Use a graphing utility to graph $h(x)$.

b. Use your graph to find the height of the catenary for $x = 0$, 100, 200, and 299 feet. Round each result to the nearest tenth of a foot.

— END OF CHAPTER —

Page 451

CHAPTER SUMMARY

At the end of each chapter there is a Chapter Summary that provides a concise section-by-section review of the chapter topics.

TRUE/FALSE EXERCISES

Following each chapter summary are true/false exercises. These exercises are intended to help students understand concepts and can be used to initiate class discussions.

CHAPTER 4 SUMMARY

4.1 Inverse Functions

- If f is a one-to-one function with domain X and range Y, and g is a function with domain Y and range X, then g is the inverse function of f if and only if $(f \circ g)(x) = x$ for all x in the domain of g and $(g \circ f)(x) = x$ for all x in the domain of f.

- A function f has an inverse function if and only if it is a one-to-one function. The graph of a function f and the graph of the inverse function f^{-1} are symmetric with respect to the line given by $y = x$.

4.2 Exponential Functions and Their Applications

- For all positive real numbers b, $b \neq 1$, the exponential function defined by $f(x) = b^x$ has the following properties:
  1. f has the set of real numbers as its domain.
  2. f has the set of positive real numbers as its range.
  3. f has a graph with a y-intercept of $(0, 1)$.
  4. f has a graph asymptotic to the x-axis.
  5. f is a one-to-one function.

CHAPTER 4 TRUE/FALSE EXERCISES

In Exercises 1 to 16, answer true or false. If the statement is false, give an example or state a reason to demonstrate that the statement is false.

1. Every function has an inverse function.

2. If $(f \circ g)(a) = a$ and $(g \circ f)(a) = a$ for some constant a, then f and g are inverse functions.

3. If $7^x = 40$, then $\log_7 40 = x$.

4. If $\log_4 x = 3.1$, then $4^{3.1} = x$.

5. If $f(x) = \log x$ and $g(x) = 10^x$, then $f[g(x)] = x$ for all real numbers x.

6. If $f(x) = \log x$ and $g(x) = 10^x$, then $g[f(x)] = x$ for all real numbers x.

7. The exponential function $h(x) = b^x$ is an increasing function.

8. The logarithmic function $j(x) = \log_b x$ is an increasing function.

9. The exponential function $h(x) = b^x$ is a one-to-one function.

10. The logarithmic function $j(x) = \log_b x$ is a one-to-one function.

11. The graph of $f(x) = \dfrac{2^x + 2^{-x}}{2}$ is symmetric with respect to the y-axis.

12. The graph of $f(x) = \dfrac{2^x - 2^{-x}}{2}$ is symmetric with respect to the origin.

13. If $x > 0$ and $y > 0$, then $\log(x + y) = \log x + \log y$.

14. If $x > 0$, then $\log x^2 = 2 \log x$.

15. If M and N are positive real numbers, then
$$\ln \frac{M}{N} = \ln M - \ln N$$

16. For all $p > 0$, $e^{\ln p} = p$.

Page 453

Page 454

CHAPTER 4 REVIEW EXERCISES

In Exercises 1 to 4, determine whether the given functions are inverses.

1. $F(x) = 2x - 5 \qquad G(x) = \dfrac{x + 5}{2}$

2. $h(x) = \sqrt{x} \qquad k(x) = x^2, \; x \geq 0$

3. $l(x) = \dfrac{x + 3}{x} \qquad m(x) = \dfrac{3}{x - 1}$

4. $p(x) = \dfrac{x - 5}{2x} \qquad q(x) = \dfrac{2x}{x - 5}$

In Exercises 5 to 8, find the inverse of the function. Sketch the graph of the function and its inverse on the same set of coordinate axes.

5. $f(x) = 3x - 4$

6. $g(x) = -2x + 3$

7. $h(x) = -\dfrac{1}{2}x - 2$

8. $k(x) = \dfrac{1}{x}$

In Exercises 31 and 32, use a graphing utility [to graph] each function.

31. $f(x) = \dfrac{4^x + 4^{-x}}{2}$

32. $f(x) = \dfrac{3^x - 3^{-x}}{2}$

In Exercises 33 to 36, change each logarithmic equation to its exponential form.

33. $\log_4 64 = 3$

34. $\log_{1/2} 8 = -3$

35. $\log_{\sqrt{5}} 4 = 4$

36. $\ln 1 = 0$

In Exercises 37 to 40, change each exponential equation to its logarithmic form.

37. $5^3 = 125$

38. $2^{10} = 1024$

39. $10^0 = 1$

40. $8^{1/2} = 2\sqrt{2}$

In Exercises 41 to 44, write the given logarithm in terms of logarithms of x, y, and z.

CHAPTER REVIEW EXERCISES

Review exercises are found at the end of each chapter. These exercises are selected to help the student integrate all of the topics presented in the chapter.

CHAPTER TEST

The Chapter Test exercises are designed to simulate a possible test of the material in the chapter.

NEW! CUMULATIVE REVIEW EXERCISES

Cumulative Review Exercises, which appear at the end of each chapter (except Chapter P), help students maintain skills learned in previous chapters.

NEW! The answers to all **Chapter Review Exercises,** all **Chapter Test Exercises,** and all **Cumulative Review Exercises** are given in the Answer Section. Along with the answer, there is a reference to the section that pertains to each exercise.

CHAPTER 4 TEST

1. Find the inverse of $f(x) = 2x - 3$. Graph f and f^{-1} on the same coordinate axes.

2. Find the inverse of $f(x) = \dfrac{x}{4x - 8}$. State the domain and the range of f^{-1}.

3. **a.** Write $\log_b(5x - 3) = c$ in exponential form.
   b. Write $3^{1/2} = y$ in logarithmic form.

4. Write $\log_b \dfrac{z^2}{y^3 \sqrt{x}}$ in terms of logarithms of x, y, and z.

5. Write $\log(2x + 3) - 3\log(x - 2)$ as a single logarithm with a coefficient of 1.

6. Use the change-of-base formula and a calculator to approximate $\log_4 12$. Round your result to the nearest ten thousandth.

7. Graph: $f(x) = 3^{-x/2}$

8. Graph: $f(x) = -\ln(x + 1)$

9. Solve: $5^x = 22$. Round your solution to the nearest ten thousandth.

10. Find the exact solution of $4^{5-x} = 7^x$.

11. Solve: $\log(x + 99) - \log(3x - 2) = 2$

12. Solve: $\ln(2 - x) + \ln(5 - x) = \ln(37 - x)$

Page 456

CUMULATIVE REVIEW EXERCISES

1. Solve $|x - 4| \leq 2$. Write the solution set using interval notation.

2. Solve $\dfrac{x}{2x - 6} \geq 1$. Write the solution set using set-builder notation.

3. Find, to the nearest tenth, the distance between the points $(5, 2)$ and $(11, 7)$.

4. The height, in feet, of a ball released with an initial upward velocity of 44 feet per second and at an initial height of 8 feet is given by $h(t) = -16t^2 + 44t + 8$, where t is the time in seconds after the ball is released. Find the maximum height the ball will reach.

5. Given $f(x) = 2x + 1$ and $g(x) = x^2 - 5$, find $(g \circ f)$.

6. Find the inverse of $f(x) = 3x - 5$.

7. The load that a horizontal beam can safely support varies jointly as the width and the square of the depth of the beam. It has been determined that a beam with a width of 4 inches and a depth of 8 inches can safely support a load of 1500 pounds. How many pounds can a beam of the same material and the same length safely support if it has

Page 457

Instructor Resources

College Algebra and Trigonometry has a complete set of support materials for the instructor.

Instructor's Annotated Edition   This edition contains a replica of the student text with additional resources for the instructor. These include: *Instructor Notes, Alternative Example* notes, *PowerPoint* icons, *Suggested Assignments,* and answers to all exercises.

Instructor's Solutions Manual   The *Instructor's Solutions Manual* contains worked-out solutions for all exercises in the text.

Instructor's Resource Manual with Testing   This resource includes six ready-to-use printed *Chapter Tests* per chapter, and a *Printed Test Bank* providing a printout of one example of each of the algorithmic items on the *HM Testing* CD-ROM program.

HM ClassPrep w/ HM Testing CD-ROM   *HM ClassPrep* contains a multitude of text-specific resources for instructors to use to enhance the classroom experience. These resources can be easily accessed by chapter or resource type and also can link you to the text's website. *HM Testing* is our computerized test generator and contains a database of algorithmic test items, as well as providing **online testing** and **gradebook** functions.

Instructor Text-specific Website   The resources available on the *ClassPrep CD* are also available on the instructor website at math.college.hmco.com/instructors. Appropriate items are password protected. Instructors also have access to the student part of the text's website.

Student Resources

Student Study Guide   The *Student Study Guide* contains complete solutions to all odd-numbered exercises in the text, as well as study tips and a practice test for each chapter.

Math Study Skills Workbook by Paul D. Nolting   This workbook is designed to reinforce skills and minimize frustration for students in any math class, lab, or study skills course. It offers a wealth of study tips and sound advice on note taking, time management, and reducing math anxiety. In addition, numerous opportunities for self assessment enable students to track their own progress.

HM Eduspace® Online Learning Environment   *Eduspace* is a text-specific, web-based learning environment that combines an algorithmic tutorial program with homework capabilities. Specific content is available 24 hours a day to help you further understand your textbook.

HM mathSpace® Tutorial CD-ROM   This tutorial CD-ROM allows students to practice skills and review concepts as many times as necessary by providing algorithmically-generated exercises and step-by-step solutions for practice.

SMARTHINKING™ Live, Online Tutoring   Houghton Mifflin has partnered with SMARTHINKING to provide an easy-to-use and effective online tutorial service.

Whiteboard Simulations and **Practice Area** promote real-time visual interaction. Three levels of service are offered.

- **Text-specific Tutoring** provides real-time, one-on-one instruction with a specially qualified 'e-structor.'
- **Questions Any Time** allows students to submit questions to the tutor outside the scheduled hours and receive a reply within 24 hours.
- **Independent Study Resources** connect students with around-the-clock access to additional educational services, including interactive websites, diagnostic tests, and Frequently Asked Questions posed to SMARTHINKING e-structors.

Houghton Mifflin Instructional Videos and DVDs   Text-specific videos and DVDs, hosted by Dana Mosely, cover all sections of the text and provide a valuable resource for further instruction and review.

Student Text-specific Website   Online student resources can be found at this text's website at math.college.hmco.com/students.

Acknowledgments

The authors would like to thank the people who have reviewed this manuscript and provided many valuable suggestions.

Ioannis K. Argyros, *Cameron University, OK*
Peter Arvanites, *Rockland Community College, NY*
Linda Berg, *University of Great Falls, MT*
Paul Bialek, *Trinity College, IL*
Zhixiong Cai, *Barton College, NC*
Cheryl F. Cavaliero, *Butler County Community College, PA*
Jennie Cox, *Central Georgia Technical College, GA*
Marilyn Danchanko, *Cambria County Area Community College, PA*
Laura Davis, *Garland County Community College, AR*
Sylvia Dorminey, *Savannah Technical College, GA*
Gay Grubbs, *Griffin Technical College, GA*
Kathryn Hodge, *Midland College, TX*
Clement S. Lam, *Mission College, CA*
Dr. Helen Medley
Carla Monticelli, *Camden County College, NJ*
J. Reid Mowrer, *University of New Mexico–Valencia Campus, NM*
Sue Neal, *Wichita State University, KS*
Georgie O'Leary, *Warner Southern College, FL*
Suzanne Pauly, *Delaware Technical and Community College, DE*
Lauri Semarne
Mike Shirazi, *Germanna Community College, VA*
Anthony Tongen, *Trinity International University, IL*
Hanson Umoh, *Delaware State University, DE*
Rebecca Wells, *Henderson Community College, KY*

Special thanks to Christi Verity for her diligent preparation of the solutions manuals and to Sandy Doerfel for her contribution to this project.

PRELIMINARY CONCEPTS

The Red Shift

You may have noticed that the sound of an approaching siren has a higher pitch (frequency) than that of a receding siren. Like sound, when a light source moves toward or away from us, the frequency of the light changes. Astronomers use this fact to measure the distance of a galaxy from us.

The photograph at the left was taken by the Hubble Space Telescope. Many of the bright objects in this photograph are galaxies. When a galaxy is moving away from us, the characteristic frequencies of certain light waves, the black lines in the spectrum below the photo, are shifted to the red side of the spectrum. The amount of this *red shift*, as astronomers call it, can be used to determine the speed at which the galaxy is receding from Earth. The formula used to compute the speed involves a rational expression, which is one of the topics of this chapter. **Exercise 67 on page 64** is an example of this formula.

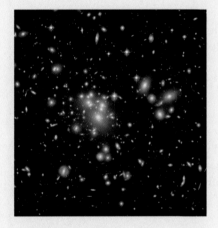

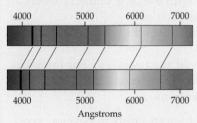

4000   5000   6000   7000
Angstroms

Source: Original art at
http://www.astro.ucla.edu/~wright/
doppler.htm.

VIDEO & DVD

SSG

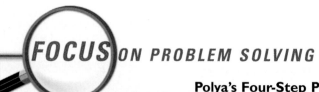

FOCUS ON PROBLEM SOLVING

Polya's Four-Step Process

Your success in mathematics and your success in the workplace are heavily dependent on your ability to solve problems. George Polya (1887–1985) was one of the foremost mathematicians to study problem solving. The basic structure that Polya advocated for problem solving has four steps, as outlined below.

1. Understand the problem.
   - Can you restate the problem in your own words?
   - Can you determine what is known about this type of problem?
   - Is there missing information that you need in order to solve the problem?
   - Is there information given that is not needed?
   - What is the goal?

2. Devise a plan.
   - Make a list of the known information.
   - Make a list of information that is needed to solve the problem.
   - Make a table or draw a diagram.
   - Work backwards.
   - Try to solve a similar but simpler problem.
   - Research the problem to determine whether there are known techniques for solving problems of its kind.
   - Try to determine whether some pattern exists.
   - Write an equation.

3. Carry out the plan.
   - Work carefully.
   - Keep an accurate and neat record of all your attempts.
   - Realize that some of your initial plans will not work and that you may have to return to step 2 and devise another plan or modify your existing plan.

4. Review your solution.
   - Make sure that the solution is consistent with the facts of the problem.
   - Interpret the solution in the context of the problem.
   - Ask yourself whether there are generalizations of the solution that could apply to other problems.
   - Determine the strengths and weaknesses of your solution. For instance, is your solution only an approximation to the actual solution?
   - Consider the possibility of alternative solutions.

As you go through this course, make a conscious effort to develop good problem-solving skills. One way to do this is to create problems and then solve them. Do not focus only on math problems. Think about your major or strategy games and create problems in those areas and then solve them. Here is one to get you started.

> Three containers A, B, and C can hold, respectively, 8, 5, and 3 gallons of water. Initially container A is filled completely, and the other two containers are empty. Without using any measuring devices other than these containers, divide the water into two equal parts by pouring from one container to another.

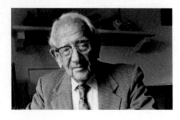

SECTION **P.1**

THE REAL NUMBER SYSTEM

- SETS
- UNION AND INTERSECTION OF SETS
- ABSOLUTE VALUE AND DISTANCE
- INTERVAL NOTATION
- ORDER OF OPERATIONS AGREEMENT
- SIMPLIFYING VARIABLE EXPRESSIONS

MATH MATTERS

Archimedes (c. 287–212 B.C.) was the first to calculate π with any degree of precision. He was able to show that

$$3\frac{10}{71} < \pi < 3\frac{1}{7}$$

from which we get the approximation

$$3\frac{1}{7} = \frac{22}{7} \approx \pi.$$

The use of the symbol π for this quantity was introduced by Leonhard Euler (1707–1783) in 1739, approximately 2000 years after Archimedes.

SETS

Human beings share the desire to organize and classify. Ancient astronomers classified stars into groups called *constellations*. Modern astronomers continue to classify stars by such characteristics as color, mass, size, temperature, and distance from Earth. In mathematics it is useful to place numbers with similar characteristics into **sets**. The following sets of numbers are used extensively in the study of algebra:

Integers	$\{\ldots, -3, -2, -1, 0, 1, 2, 3, \ldots\}$
Rational numbers	{all terminating or repeating decimals}
Irrational numbers	{all nonterminating, nonrepeating decimals}
Real numbers	{all rational or irrational numbers}

If a number in decimal form terminates or repeats a block of digits, then the number is a rational number. Here are two examples of rational numbers.

0.75 is a terminating decimal.

$0.2\overline{45}$ is a repeating decimal. The bar over the 45 means that the digits 45 repeat without end. That is, $0.2\overline{45} = 0.24545454\ldots$.

Rational numbers also can be written in the form $\frac{p}{q}$, where p and q are integers and $q \neq 0$. Examples of rational numbers written in this form are

$$\frac{3}{4} \qquad \frac{27}{110} \qquad -\frac{5}{2} \qquad \frac{7}{1} \qquad \frac{-4}{3}$$

Note that $\frac{7}{1} = 7$, and in general, $\frac{n}{1} = n$ for any integer n. Therefore, all integers are rational numbers.

When a rational number is written in the form $\frac{p}{q}$, the decimal form of the rational number can be found by dividing the numerator by the denominator.

$$\frac{3}{4} = 0.75 \qquad \frac{27}{110} = 0.2\overline{45}$$

In its decimal form, an irrational number neither terminates nor repeats. For example, 0.272272227… is a nonterminating, nonrepeating decimal and thus is an irrational number. One of the best-known irrational numbers is pi, denoted by the Greek symbol π. The number π is defined as the ratio of the circumference of a circle to its diameter. Often in applications the rational number 3.14 or the rational number $\frac{22}{7}$ is used as an approximation of the irrational number π.

Every real number is either a rational number or an irrational number. If a real number is written in decimal form, it is a terminating decimal, a repeating decimal, or a nonterminating and nonrepeating decimal.

The relationship between the various sets of numbers is shown in **Figure P.1.**

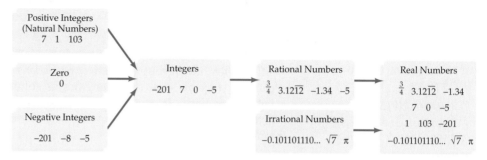

FIGURE P.1

Prime numbers and *composite numbers* play an important role in almost every branch of mathematics. A **prime number** is a positive integer other than 1 that has no positive-integer factors[1] other than itself and 1. The 10 smallest prime numbers are 2, 3, 5, 7, 11, 13, 17, 19, 23, and 29. Each of these numbers has only itself and 1 as factors.

A **composite number** is a positive integer greater than 1 that is not a prime number. For example, 10 is a composite number because 10 has both 2 and 5 as factors. The 10 smallest composite numbers are 4, 6, 8, 9, 10, 12, 14, 15, 16, and 18.

EXAMPLE 1   **Classify Real Numbers**

Determine which of the following numbers are

a. integers   b. rational numbers   c. irrational numbers
d. real numbers   e. prime numbers   f. composite numbers

$$-0.2, \quad 0, \quad 0.\overline{3}, \quad \pi, \quad 6, \quad 7, \quad 41, \quad 51, \quad 0.71771777177771\ldots$$

Solution

a. Integers: $0, 6, 7, 41, 51$

b. Rational numbers: $-0.2, 0, 0.\overline{3}, 6, 7, 41, 51$

c. Irrational numbers: $0.71771777177771\ldots, \pi$

d. Real numbers: $-0.2, 0, 0.\overline{3}, \pi, 6, 7, 41, 51, 0.71771777177771\ldots$

e. Prime numbers: $7, 41$

f. Composite numbers: $6, 51$

▶ **TRY EXERCISE 2, PAGE 15**

---

[1] Recall that a factor of a number divides the number evenly. For instance, 3 and 7 are factors of 21; 5 is not a factor of 21.

Each member of a set is called an **element** of the set. For instance, if $C = \{2, 3, 5\}$, then the elements of C are 2, 3, and 5. The notation $2 \in C$ is read "2 is an element of C." Set A is a **subset** of set B if every element of A is also an element of B, and we write $A \subseteq B$. For instance, the set of **negative integers** $\{-1, -2, -3, -4, \ldots\}$ is a subset of the set of integers. The set of **positive integers** $\{1, 2, 3, 4, \ldots\}$ (also known as the set of **natural numbers**) is also a subset of the set of integers.

❷ QUESTION Are the integers a subset of the rational numbers?

take note

The order of the elements of a set is not important. For instance, the set of natural numbers less than 6 given at the right could have been written $\{3, 5, 2, 1, 4\}$. It is customary, however, to list elements of a set in numerical order.

The **empty set**, or **null set**, is the set that contains no elements. The symbol $\varnothing$ is used to represent the empty set. The set of people who have run a two-minute mile is the empty set.

The set of natural numbers less than 6 is $\{1, 2, 3, 4, 5\}$. This is an example of a **finite set**; all the elements of the set can be listed. The set of all natural numbers is an example of an **infinite set**. There is no largest natural number, so all the elements of the set of natural numbers cannot be listed.

Sets are often written using **set-builder notation**. Set-builder notation can be used to describe almost any set, but it is especially useful when writing infinite sets. For instance, the set

$$\{2n \,|\, n \in \text{natural numbers}\}$$

is read as "the set of elements $2n$ such that n is a natural number." By replacing n by each of the natural numbers, this is the set of positive even integers: $\{2, 4, 6, 8, \ldots\}$.

The set of real numbers greater than 2 is written:

$$\{x \,|\, x > 2, x \in \text{real numbers}\}$$

and is read "the set of x such that x is greater than 2 and x is an element of the real numbers."

Much of the work we do in this text uses the real numbers. With this in mind, we will frequently write, for instance, $\{x \,|\, x > 2, x \in \text{real numbers}\}$ in a shortened form as $\{x \,|\, x > 2\}$, where we assume that x is a real number.

MATH MATTERS

A **fuzzy set** is one in which each element is given a "degree" of membership. The concepts behind fuzzy sets are used in a wide variety of applications such as traffic lights, washing machines, and computer speech recognition programs.

EXAMPLE 2   **Use Set-Builder Notation**

List the four smallest elements in $\{n^3 \,|\, n \in \text{natural numbers}\}$.

Solution
Because we want the four *smallest* elements, we choose the four smallest natural numbers. Thus $n = 1, 2, 3,$ and 4. Therefore, the four smallest elements of $\{n^3 \,|\, n \in \text{natural numbers}\}$ are 1, 8, 27, and 64.

▶ **TRY EXERCISE 6, PAGE 15**

❷ ANSWER Yes.

● UNION AND INTERSECTION OF SETS

Just as operations such as addition and multiplication are performed on real numbers, operations are performed on sets. Two operations performed on sets are union and intersection. The union of two sets A and B is the set of elements that belong to A or to B or to both A and B.

Union of Two Sets

The **union** of two sets, written $A \cup B$, is the set of all elements that belong to either A or B. In set-builder notation, this is written

$$A \cup B = \{x \,|\, x \in A \text{ or } x \in B\}$$

For instance, given $A = \{2, 3, 4\}$ and $B = \{0, 1, 2, 3\}$, then $A \cup B = \{0, 1, 2, 3, 4\}$. Note that an element that belongs to both sets is listed only once.

The intersection of the two sets A and B is the set of elements that belong to both A and B.

Intersection of Two Sets

The **intersection** of two sets, written $A \cap B$, is the set of all elements that are common to both A and B. In set-builder notation, this is written

$$A \cap B = \{x \,|\, x \in A \text{ and } x \in B\}$$

For instance, given $A = \{2, 3, 4\}$ and $B = \{0, 1, 2, 3\}$, $A \cap B = \{2, 3\}$.

If the intersection of two sets is the empty set, the two sets are said to be **disjoint.** For example, if $A = \{2, 3, 4\}$ and $B = \{7, 8\}$, then $A \cap B = \varnothing$ and A and B are disjoint sets.

EXAMPLE 3   Find the Union and Intersection of Sets

Find each intersection or union given $A = \{0, 2, 4, 6, 10, 12\}$,
$B = \{0, 3, 6, 12, 15\}$, and $C = \{1, 2, 3, 4, 5, 6, 7\}$.

a. $A \cup C$
b. $B \cap C$
c. $A \cap (B \cup C)$
d. $B \cup (A \cap C)$

Solution

a. $A \cup C = \{0, 1, 2, 3, 4, 5, 6, 7, 10, 12\}$ • The elements that belong to **A** or **C**

b. $B \cap C = \{3, 6\}$ • The elements that belong to **B** and **C**

c.   First determine $B \cup C = \{0, 1, 2, 3, 4, 5, 6, 7, 12, 15\}$. Then

$$A \cap (B \cup C) = \{0, 2, 4, 6, 12\}$$ • **The elements that belong to A and (B ∪ C)**

d.   First determine $A \cap C = \{2, 4, 6\}$. Then

$$B \cup (A \cap C) = \{0, 2, 3, 4, 6, 12, 15\}$$ • **The elements that belong to B or (A ∩ C)**

▶ **TRY EXERCISE 16, PAGE 15**

● ABSOLUTE VALUE AND DISTANCE

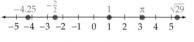

FIGURE P.2

The real numbers can be represented geometrically by a **coordinate axis** called a **real number line. Figure P.2** shows a portion of a real number line. The number associated with a point on a real number line is called the **coordinate** of the point. The point corresponding to zero is called the **origin.** Every real number corresponds to a point on the number line, and every point on the number line corresponds to a real number.

The *absolute value* of a real number a, denoted $|a|$, is the distance between a and 0 on the number line. For instance, $|3| = 3$ and $|-3| = 3$ because both 3 and -3 are 3 units from zero. See **Figure P.3.**

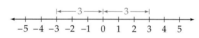

FIGURE P.3

In general, if $a \geq 0$, then $|a| = a$; however, if $a < 0$, then $|a| = -a$ because $-a$ is positive when $a < 0$. This leads to the following definition.

Definition of Absolute Value
The **absolute value** of the real number a is defined by $$|a| = \begin{cases} a & \text{if } a \geq 0 \\ -a & \text{if } a < 0 \end{cases}$$

The definition of *distance* between two points on a real number line makes use of absolute value.

Distance Between Points on a Real Number Line
If a and b are the coordinates of two points on a number line, the **distance** between the graph of a and the graph of b, denoted by $d(a, b)$, is given by $d(a, b) = |a - b|$.

As an example of th[...] [...]se coordinate is −2 and the poi[...]

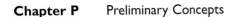

FIGURE P.4

Note from **Figure P.4** tha[...] [...]ber line. Also note that the order [...]

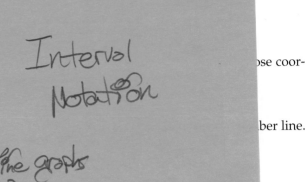

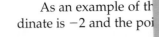

EXAMPLE 4

Express the distance between a and -3 on the number line using absolute value.

Solution

$$d(a, -3) = |a - (-3)| = |a + 3|$$

▶ **TRY EXERCISE 48, PAGE 16**

● INTERVAL NOTATION

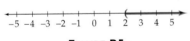

FIGURE P.5

The graph of $\{x \mid x > 2\}$ is shown in **Figure P.5.** The set is the real numbers greater than 2. The parenthesis at 2 indicates that 2 is not included in the set. Rather than write this set of real numbers using set-builder notation, we frequently write the set in **interval notation** as $(2, \infty)$.

In general, the interval notation

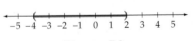

FIGURE P.6

(a, b) represents all real numbers between a and b, not including a and b. This is an **open interval.** In set-builder notation, we write $\{x \mid a < x < b\}$. For instance, the graph of $(-4, 2)$ is shown in **Figure P.6.**

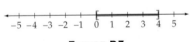

FIGURE P.7

$[a, b]$ represents all real numbers between a and b, including a and b. This is a **closed interval.** In set-builder notation, we write $\{x \mid a \leq x \leq b\}$. For instance, the graph of $[0, 4]$ is shown in **Figure P.7.** The brackets at 0 and 4 indicate that those numbers are included in the graph.

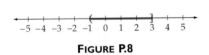

FIGURE P.8

$(a, b]$ represents all real numbers between a and b, not including a but including b. This is a **half-open interval.** In set-builder notation, we write $\{x \mid a < x \leq b\}$. For instance, the graph of $(-1, 3]$ is shown in **Figure P.8.**

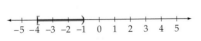

FIGURE P.9

$[a, b)$ represents all real numbers between a and b, including a but not including b. This is a **half-open interval.** In set-builder notation, we write $\{x \mid a \leq x < b\}$. For instance, the graph of $[-4, -1)$ is shown in **Figure P.9.**

Subsets of the real numbers whose graphs extend forever in one or both directions can be represented by interval notation using the **infinity symbol** ∞ or the **negative infinity symbol** −∞.

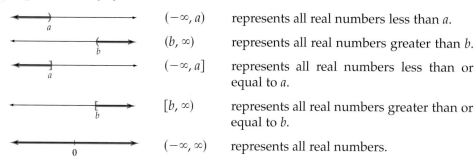

$(-\infty, a)$ represents all real numbers less than a.

(b, ∞) represents all real numbers greater than b.

$(-\infty, a]$ represents all real numbers less than or equal to a.

$[b, \infty)$ represents all real numbers greater than or equal to b.

$(-\infty, \infty)$ represents all real numbers.

EXAMPLE 5   Graph a Set in Interval Notation

Graph $(-\infty, 3]$. Write the interval in set-builder notation.

Solution

The set is the real numbers less than or equal to 3. In set-builder notation, this is the set $\{x \mid x \leq 3\}$. Draw a right bracket at 3, and darken the number line to the left of 3, as shown in **Figure P.10.**

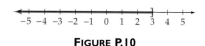

FIGURE P.10

▶ **TRY EXERCISE 54, PAGE 16**

---

take note

It is never correct to use a bracket when using the infinity symbol. For instance, $[-\infty, 3]$ is not correct. Nor is $[2, \infty]$ correct. Neither negative infinity nor positive infinity is a real number and therefore cannot be contained in an interval.

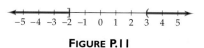

FIGURE P.11

FIGURE P.12

The set $\{x \mid x \leq -2\} \cup \{x \mid x > 3\}$ is the set of real numbers that are either less than or equal to -2 or greater than 3. We also could write this in interval notation as $(-\infty, -2] \cup (3, \infty)$. The graph of the set is shown in **Figure P.11.**

The set $\{x \mid x > -4\} \cap \{x \mid x < 1\}$ is the set of real numbers that are greater than -4 *and* less than 1. Note from **Figure P.12** that this set is the interval $(-4, 1)$, which can be written in set-builder notation as $\{x \mid -4 < x < 1\}$.

EXAMPLE 6   Graph Intervals

Graph the following. Write **a.** and **b.** using interval notation. Write **c.** and **d.** using set-builder notation.

a. $\{x \mid x \leq -1\} \cup \{x \mid x \geq 2\}$ **b.** $\{x \mid x \geq -1\} \cap \{x \mid x < 5\}$
c. $(-\infty, 0) \cup [1, 3]$ **d.** $[-1, 3] \cap (1, 5)$

Solution

a. $(-\infty, -1] \cup [2, \infty)$

b. $[-1, 5)$

Continued ▶

c. 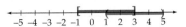 $\{x \mid x < 0\} \cup \{x \mid 1 \le x \le 3\}$

d. The graphs of $[-1, 3]$, in red, and $(1, 5)$, in blue, are shown below.

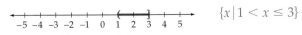

Note that the intersection of the sets occurs where the graphs intersect. Although $1 \in [-1, 3]$, $1 \notin (1, 5)$. Therefore, 1 does not belong to the intersection of the sets. On the other hand, $3 \in [-1, 3]$ and $3 \in (1, 5)$. Therefore, 3 belongs to the intersection of the sets. Thus we have the following.

$\{x \mid 1 < x \le 3\}$

▶ **TRY EXERCISE 64, PAGE 16**

● ORDER OF OPERATIONS AGREEMENT

The approximate pressure p, in pounds per square inch, on a scuba diver x feet below the water's surface is given by

$$p = 15 + 0.5x$$

The pressure on the diver at various depths is given below.

10 feet	$15 + 0.5(10) = 15 + 5 = 20$ pounds
20 feet	$15 + 0.5(20) = 15 + 10 = 25$ pounds
40 feet	$15 + 0.5(40) = 15 + 20 = 35$ pounds
70 feet	$15 + 0.5(70) = 15 + 35 = 50$ pounds

Note that the expression $15 + 0.5(70)$ has two operations, addition and multiplication. When an expression contains more than one operation, the operations must be performed in a specified order, as listed below in the Order of Operations Agreement.

The Order of Operations Agreement

If grouping symbols are present, evaluate by performing the operations within the grouping symbols, innermost grouping symbols first, while observing the order given in steps 1 to 3.

Step 1 Evaluate exponential expressions.
Step 2 Do multiplication and division as they occur from left to right.
Step 3 Do addition and subtraction as they occur from left to right.

Therefore, the expression $15 + 0.5(70)$ is simplified by first performing the multiplication and then performing the addition, as we did above.

take note

Recall that subtraction can be rewritten as addition of the opposite. Therefore,

$3x^2 - 4xy + 5x - y - 7$
$= 3x^2 + (-4)xy + 5x + (-y) + (-7)$

In this form, we can see that the terms (addends) are $3x^2$, $-4xy$, $5x$, $-y$, and -7.

One of the ways the Order of Operations Agreement is used is to evaluate variable expressions. The addends of a variable expression are called **terms.** The terms for the expression at the right are $3x^2$, $-4xy$, $5x$, $-y$, and -7. Observe that the sign of a term is the sign that immediately precedes it.

$$3x^2 - 4xy + 5x - y - 7$$

The terms $3x^2$, $-4xy$, $5x$, and $-y$ are **variable terms.** The term -7 is a **constant term.** Each variable term has a **numerical coefficient** and a **variable part.** The numerical coefficient for the term $3x^2$ is 3; the numerical coefficient for the term $-4xy$ is -4; the numerical coefficient for the term $5x$ is 5; and the numerical coefficient for the term $-y$ is -1. When the numerical coefficient is 1 or -1 (as in x and $-x$), the 1 is usually not written.

To **evaluate** a variable expression, replace the variables by their given values and then use the Order of Operations Agreement to simplify the result.

EXAMPLE 7   **Evaluate a Variable Expression**

a. Evaluate $\dfrac{x^3 - y^3}{x^2 + xy + y^2}$ when $x = 2$ and $y = -3$.

b. Evaluate $(x + 2y)^2 - 4z$ when $x = 3$, $y = -2$, and $z = -4$.

Solution

a. $\dfrac{x^3 - y^3}{x^2 + xy + y^2}$

$\dfrac{2^3 - (-3)^3}{2^2 + 2(-3) + (-3)^2} = \dfrac{8 - (-27)}{4 - 6 + 9} = \dfrac{35}{7} = 5$

b. $(x + 2y)^2 - 4z$

$(3 + 2(-2))^2 - 4(-4) = (3 + (-4))^2 - 4(-4)$

$= (-1)^2 - 4(-4) = 1 - 4(-4)$

$= 1 + 16 = 17$

▶ **TRY EXERCISE 74, PAGE 16**

● SIMPLIFYING VARIABLE EXPRESSIONS

Addition, multiplication, subtraction, and *division* are the operations of arithmetic. **Addition** of the two real numbers a and b is designated by $a + b$. If $a + b = c$, then c is the **sum** and the real numbers a and b are called **terms.**

Multiplication of the real numbers a and b is designated by ab or $a \cdot b$. If $ab = c$, then c is the **product** and the real numbers a and b are called **factors** of c.

The number $-b$ is referred to as the **additive inverse** of b. **Subtraction** of the real numbers a and b is designated by $a - b$ and is defined as the sum of a and the additive inverse of b. That is,

$$a - b = a + (-b)$$

If $a - b = c$, then c is called the **difference** of a and b.

The **multiplicative inverse** or **reciprocal** of the nonzero number b is $1/b$. The **division** of a and b, designated by $a \div b$ with $b \neq 0$, is defined as the product of a and the reciprocal of b. That is,

$$a \div b = a\left(\frac{1}{b}\right) \qquad \text{provided that } b \neq 0$$

If $a \div b = c$, then c is called the **quotient** of a and b.

The notation $a \div b$ is often represented by the fractional notation a/b or $\dfrac{a}{b}$.

The real number a is the **numerator,** and the nonzero real number b is the **denominator** of the fraction.

Properties of Real Numbers

Let a, b, and c be real numbers.

	Addition Properties	**Multiplication Properties**
Closure	$a + b$ is a unique real number.	ab is a unique real number.
Commutative	$a + b = b + a$	$ab = ba$
Associative	$(a + b) + c = a + (b + c)$	$(ab)c = a(bc)$
Identity	There exists a unique real number 0 such that $a + 0 = 0 + a = a.$	There exists a unique real number 1 such that $a \cdot 1 = 1 \cdot a = a.$
Inverse	For each real number a, there is a unique real number $-a$ such that $a + (-a) = (-a) + a = 0.$	For each *nonzero* real number a, there is a unique real number $1/a$ such that $a \cdot \dfrac{1}{a} = \dfrac{1}{a} \cdot a = 1.$
Distributive		$a(b + c) = ab + ac$

EXAMPLE 8   Identify Properties of Real Numbers

Identify the property of real numbers illustrated in each statement.

a. $(2a)b = 2(ab)$

b. $\left(\dfrac{1}{5}\right)11$ is a real number.

c. $4(x + 3) = 4x + 12$

d. $(a + 5b) + 7c = (5b + a) + 7c$

e. $\left(\dfrac{1}{2} \cdot 2\right)a = 1 \cdot a$

f. $1 \cdot a = a$

Solution

a.   Associative property of multiplication

b.   Closure property of multiplication of real numbers

c.   Distributive property

d.   Commutative property of addition

e.   Inverse property of multiplication

f.   Identity property of multiplication

▶ **TRY EXERCISE 86, PAGE 16**

We can identify which property of real numbers has been used to rewrite expressions by closely comparing the expressions and noting any changes. For instance, to simplify $(6x)2$, both the commutative property and associative property of multiplication are used.

$$(6x)2 = 2(6x) \qquad \text{• Commutative property of multiplication}$$
$$= (2 \cdot 6)x \qquad \text{• Associative property of multiplication}$$
$$= 12x$$

To simplify $3(4p + 5)$, use the distributive property.

$$3(4p + 5) = 3(4p) + 3(5) \qquad \text{• Distributive property}$$
$$= 12p + 15$$

Terms that have the same variable part are called **like terms.** The distributive property is also used to simplify an expression with like terms such as $3x^2 + 9x^2$.

$$3x^2 + 9x^2 = (3 + 9)x^2 \qquad \text{• Distributive property}$$
$$= 12x^2$$

Note from this example that like terms are combined by adding the coefficients of the like terms.

> *take note*
>
> Normally, we will not show, as we did at the right, all the steps in the simplification of a variable expression. For instance, we will just write $(6x)2 = 12x$, $3(4p + 5) = 12p + 15$, and $3x^2 + 9x^2 = 12x^2$. It is important to know, however, that every step in the simplification depends on one of the properties of real numbers.

❓ **QUESTION** Are the terms $2x^2$ and $3x$ like terms?

EXAMPLE 9   **Simplify Variable Expressions**

a.   Simplify $5 + 3(2x - 6)$.

b.   Simplify $4x - 2[7 - 5(2x - 3)]$.

Continued ▶

❓ **ANSWER** No. The variable parts are not the same. The variable part of $2x^2$ is $x \cdot x$. The variable part of $3x$ is x.

Solution

a. $5 + 3(2x - 6) = 5 + 6x - 18$ • Use the distributive property.

$= 6x - 13$ • Add the constant terms.

b. $4x - 2[7 - 5(2x - 3)]$

$= 4x - 2[7 - 10x + 15]$ • Use the distributive property to remove the inner parentheses.

$= 4x - 2[-10x + 22]$ • Simplify.

$= 4x + 20x - 44$ • Use the distributive property to remove the brackets.

$= 24x - 44$ • Simplify.

▶ **TRY EXERCISE 106, PAGE 17**

An **equation** is a statement of equality between two numbers or two expressions. There are four basic properties of equality that relate to equations.

Properties of Equality

Let a, b, and c be real numbers.

Reflexive	$a = a$
Symmetric	If $a = b$, then $b = a$.
Transitive	If $a = b$ and $b = c$, then $a = c$.
Substitution	If $a = b$, then a may be replaced by b in any expression that involves a.

EXAMPLE 10   **Identify Properties of Equality**

Identify the property of equality illustrated in each statement.

a.   If $3a + b = c$, then $c = 3a + b$.

b. $5(x + y) = 5(x + y)$

c.   If $4a - 1 = 7b$ and $7b = 5c + 2$, then $4a - 1 = 5c + 2$.

d.   If $a = 5$ and $b(a + c) = 72$, then $b(5 + c) = 72$.

Solution

a. Symmetric   **b.** Reflexive   **c.** Transitive   **d.** Substitution

▶ **TRY EXERCISE 90, PAGE 16**

TOPICS FOR DISCUSSION

1. Archimedes determined that $\dfrac{223}{71} < \pi < \dfrac{22}{7}$. Is it possible to find an exact representation for π of the form $\dfrac{a}{b}$, where a and b are integers?

2. If $I = \{\text{irrational numbers}\}$ and $Q = \{\text{rational numbers}\}$, name the sets $I \cup Q$ and $I \cap Q$.

3. If the proposed simplification shown at the right is correct, so state. If it is incorrect, show a correct simplification.
$$2 \cdot 3^2 = 6^2 = 36$$

4. Are there any even prime numbers? If so, name them.

5. Does every real number have an additive inverse? Does every real number have a multiplicative inverse?

6. What is the difference between an open interval and a closed interval?

EXERCISE SET P.1

In Exercises 1 and 2, determine whether each number is an integer, a rational number, an irrational number, a prime number, or a real number.

1. $-\dfrac{1}{5}, 0, -44, \pi, 3.14, 5.05005000500005\ldots, \sqrt{81}, 53$

▶ 2. $\dfrac{5}{\sqrt{7}}, \dfrac{5}{7}, 31, -2\dfrac{1}{2}, 4.235653907493, 51, 0.888\ldots$

In Exercises 3 to 8, list the four smallest elements of each set.

3. $\{2x \,|\, x \in \text{positive integers}\}$

4. $\{|x| \,|\, x \in \text{integers}\}$

5. $\{y \,|\, y = 2x + 1, x \in \text{natural numbers}\}$

▶ 6. $\{y \,|\, y = x^2 - 1, x \in \text{integers}\}$

7. $\{z \,|\, z = |x|, x \in \text{integers}\}$

8. $\{z \,|\, z = |x| - x, x \in \text{negative integers}\}$

In Exercises 9 to 18, perform the operations given
$A = \{-3, -2, -1, 0, 1, 2, 3\}$, $B = \{-2, 0, 2, 4, 6\}$,
$C = \{0, 1, 2, 3, 4, 5, 6\}$, and $D = \{-3, -1, 1, 3\}$.

9. $A \cup B$
10. $C \cup D$

11. $A \cap C$
12. $C \cap D$

13. $B \cap D$
14. $B \cup (A \cap C)$

15. $D \cap (B \cup C)$
▶ 16. $(A \cap B) \cup (A \cap C)$

17. $(B \cup C) \cap (B \cup D)$
18. $(A \cap C) \cup (B \cap D)$

In Exercises 19 to 30, graph each set. Write sets given in interval notation in set-builder notation and write sets given in set-builder notation in interval notation.

19. $(-2, 3)$
20. $[1, 5]$

21. $[-5, -1]$
22. $(-3, 3)$

23. $[2, \infty)$
24. $(-\infty, 4)$

25. $\{x \,|\, 3 < x < 5\}$
26. $\{x \,|\, x < -1\}$

27. $\{x \,|\, x \geq -2\}$

28. $\{x \,|\, -1 \leq x < 5\}$

29. $\{x \,|\, 0 \leq x \leq 1\}$

30. $\{x \,|\, -4 < x \leq 5\}$

In Exercises 31 to 40, write each expression without absolute value symbols.

31. $-|-5|$

32. $-|-4|^2$

33. $|3| \cdot |-4|$

34. $|3| - |-7|$

35. $|\pi^2 + 10|$

36. $|\pi^2 - 10|$

37. $|x - 4| + |x + 5|$, given $0 < x < 1$

38. $|x + 6| + |x - 2|$, given $2 < x < 3$

39. $|2x| - |x - 1|$, given $0 < x < 1$

40. $|x + 1| + |x - 3|$, given $x > 3$

In Exercises 41 to 50, use absolute value notation to describe the given situation.

41. Distance between x and 3

42. Distance between a and -2

43. The distance between x and -2 is 4.

44. The distance between z and 5 is 1.

45. $d(m, n)$

46. $d(p, 8)$

47. The distance between a and 4 is less than 5.

▶ **48.** The distance between z and 5 is greater than 7.

49. The distance between x and -2 is greater than 4.

50. The distance between y and -3 is greater than 6.

In Exercises 51 to 66, graph each set.

51. $(-\infty, 0) \cup [2, 4]$

52. $(-3, 1) \cup (3, 5)$

53. $(-4, 0) \cap [-2, 5]$

▶ **54.** $(-\infty, 3] \cap (2, 6)$

55. $(1, \infty) \cup (-2, \infty)$

56. $(-4, \infty) \cup (0, \infty)$

57. $(1, \infty) \cap (-2, \infty)$

58. $(-4, \infty) \cap (0, \infty)$

59. $[-2, 4] \cap [4, 5]$

60. $(-\infty, 1] \cap [1, \infty)$

61. $(-2, 4) \cap (4, 5)$

62. $(-\infty, 1) \cap (1, \infty)$

63. $\{x \,|\, x < -3\} \cup \{x \,|\, 1 < x < 2\}$

▶ **64.** $\{x \,|\, -3 \leq x < 0\} \cup \{x \,|\, x \geq 2\}$

65. $\{x \,|\, x < -3\} \cup \{x \,|\, x < 2\}$

66. $\{x \,|\, x < -3\} \cap \{x \,|\, x < 2\}$

In Exercises 67 to 78, evaluate the variable expression for $x = 3$, $y = -2$, and $z = -1$.

67. $-y^3$

68. $-y^2$

69. $2xyz$

70. $-3xz$

71. $-2x^2y^2$

72. $2y^3z^2$

73. $xy - z(x - y)^2$

▶ **74.** $(z - 2y)^2 - 3z^3$

75. $\dfrac{x^2 + y^2}{x + y}$

76. $\dfrac{2xy^2z^4}{(y - z)^4}$

77. $\dfrac{3y}{x} - \dfrac{2z}{y}$

78. $(x - z)^2(x + z)^2$

In Exercises 79 to 92, state the property of real numbers or the property of equality that is used. P. 12

79. $(ab^2)c = a(b^2c)$

80. $2x - 3y = -3y + 2x$

81. $4(2a - b) = 8a - 4b$

82. $6 + (7 + a) = 6 + (a + 7)$

83. $(3x)y = y(3x)$

84. $4ab + 0 = 4ab$

85. $1 \cdot (4x) = 4x$

▶ **86.** $7(a + b) = 7(b + a)$

87. $x^2 + 1 = x^2 + 1$

88. If $a + b = 2$, then $2 = a + b$.

89. If $2x + 1 = y$ and $3x - 2 = y$, then $2x + 1 = 3x - 2$.

▶ **90.** If $4x + 2y = 7$ and $x = 3$, then $4(3) + 2y = 7$.

91. $4 \cdot \dfrac{1}{4} = 1$

92. $ab + (-ab) = 0$

In Exercises 93 to 106, simplify the variable expression.

93. $3(2x)$

94. $-2(4y)$

95. $3(2 + x)$

96. $-2(4 + y)$

97. $\dfrac{2}{3}a + \dfrac{5}{6}a$

98. $\dfrac{3}{4}x - \dfrac{1}{2}x$

99. $2 + 3(2x - 5)$

100. $4 + 2(2a - 3)$

101. $5 - 3(4x - 2y)$

102. $7 - 2(5n - 8m)$

103. $3(2a - 4b) - 4(a - 3b)$

104. $5(4r - 7t) - 2(10r + 3t)$

105. $5a - 2[3 - 2(4a + 3)]$

▶ **106.** $6 + 3[2x - 4(3x - 2)]$

107. AREA OF A TRIANGLE The area of a triangle is given by area $= \dfrac{1}{2}bh$, where b is the base of the triangle and h is its height. Find the area of a triangle whose base is 3 inches and whose height is 4 inches.

108. VOLUME OF A BOX The volume of a rectangular box is given by

volume $= lwh$

where l is the length, w is the width, and h is the height of the box. Find the volume of a classroom that is 40 feet long, 30 feet wide, and 12 feet high.

109. PROFIT FROM SALES The profit, in dollars, a company earns from the sale of x bicycles is given by

profit $= -0.5x^2 + 120x - 2000$

Find the profit the company earns from selling 110 bicycles.

110. MAGAZINE CIRCULATION The circulation, in thousands of subscriptions, of a new magazine n months after its introduction can be approximated by

circulation $= \sqrt{n^2 - n + 1}$

Find, to the nearest hundred, the circulation of the magazine after 12 months.

111. HEART RATE The heart rate, in beats per minute, of a certain runner during a cool-down period can be approximated by

$$\text{Heart rate} = 65 + \frac{53}{4t + 1}$$

where t is the number of minutes after the start of cool-down. Find the runner's heart rate in 10 minutes. Round to the nearest whole number.

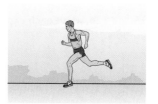

112. BODY MASS INDEX According to the National Institutes of Health, body mass index (BMI) is measure of body fat based on height and weight that applies to both adult men and women, with values between 18.5 and 24.9 considered healthy. BMI is calculated as BMI $= \dfrac{705w}{h^2}$, where w is the weight of the person in pounds and h is the person's height in inches. Find the BMI for a person who weighs 160 pounds and is 5 feet 10 inches tall. Round to the nearest whole number.

113. PHYSICS The height, in feet, of a ball t seconds after it is thrown upward is given by height $= -16t^2 + 80t + 4$. Find the height of the ball 2 seconds after it has been released.

114. CHEMISTRY Salt is being added to water in such a way that the concentration, in grams per liter, is given by concentration $= \dfrac{50t}{t + 1}$, where t is the time in minutes after the introduction of the salt. Find the concentration of salt after 24 minutes.

CONNECTING CONCEPTS

In Exercises 115 to 118, perform the given operation given A is any set.

115. $A \cup A$

116. $A \cap A$

117. $A \cap \varnothing$

118. $A \cup \varnothing$

119. If A and B are two sets and $A \cup B = A$, what can be said about B?

120. If A and B are two sets and $A \cap B = B$, what can be said about B?

121. Is division of real numbers an associative operation? Give a reason for your answer.

122. Is subtraction of real numbers a commutative operation? Give a reason for your answer.

123. Which of the properties of real numbers are satisfied by the integers?

124. Which of the properties of real numbers are satisfied by the rational numbers?

In Exercises 125 and 126, write each expression without absolute value symbols.

125. $\left| \dfrac{x + 7}{|x| + |x - 1|} \right|$, given $0 < x < 1$.

126. $\left| \dfrac{x + 3}{\left| x - \dfrac{1}{2} \right| + \left| x + \dfrac{1}{2} \right|} \right|$, given $0 < x < 0.2$.

In Exercises 127 to 132, use absolute value notation to describe the given statement.

127. x is closer to 2 than it is to 6.

128. x is closer to a than it is to b.

129. x is farther from 3 than it is from -7.

130. x is farther from 0 than it is from 5.

131. x is more than 2 units from 4 but less than 7 units from 4.

132. x is more than b units from a but less than c units from a.

PREPARE FOR SECTION P.2

133. Simplify: $2^2 \cdot 2^3$ [P.1]

134. Simplify: $\dfrac{4^3}{4^5}$ [P.1]

135. Simplify: $(2^3)^2$ [P.1]

136. Simplify: $3.14(10^5)$ [P.1]

137. True or false: $3^4 \cdot 3^2 = 9^6$ [P.1]

138. True or false: $(3 + 4)^2 = 3^2 + 4^2$ [P.1]

PROJECTS

1. NUMBER PUZZLE A number n has the following properties:

When n is divided by 6, the remainder is 5.
When n is divided by 5, the remainder is 4.
When n is divided by 4, the remainder is 3.
When n is divided by 3, the remainder is 2.
When n is divided by 2, the remainder is 1.

What is the smallest value of n?

2. OPERATIONS ON INTERVALS Besides finding unions and intersections of intervals, it is possible to apply other operations to intervals. For instance, $(-1, 2)^2$ is the interval that results from squaring every number in the interval $(-1, 2)$. This gives $[0, 4)$. Thus $(-1, 2)^2 = [0, 4)$.

a. Find $(-4, 2)^2$.

b. Find $ABS(-4, 5)$, the absolute value of every number in $(-4, 5)$.

c. Find $\sqrt{(0, 9)}$, the square root of every number in $(0, 9)$.

d. Find $\dfrac{1}{(0, 1)}$, the reciprocal of every number in $(0, 1)$.

3. FACTORS OF A NUMBER Explain why the square of a natural number always has an odd number of natural number factors.

To change a number from scientific notation to its decimal form, reverse the procedure. That is, if the exponent is positive, move the decimal point to the right the same number of places as the exponent. For example,

$$3.5 \times 10^5 = 350,000$$

5 places

If the exponent is negative, move the decimal point to the left the same number of places as the absolute value of the exponent. For example,

$$2.51 \times 10^{-8} = 0.0000000251$$

8 places

Most scientific calculators display very large and very small numbers in scientific notation. The number $450,000^2$ is displayed as $\boxed{\textbf{2.025} \quad \textbf{E 11}}$. This means $450,000^2 = 2.025 \times 10^{11}$.

EXAMPLE 3   Simplify an Expression Using Scientific Notation

The Andromeda galaxy is approximately 1.4×10^{19} miles from Earth. If a spacecraft could travel 2.8×10^{12} miles in 1 year (about one-half the speed of light), how many years would it take to reach the Andromeda galaxy?

Solution

To find the time, divide the distance by the speed.

$$t = \frac{1.4 \times 10^{19}}{2.8 \times 10^{12}} = \frac{1.4}{2.8} \times 10^{19-12} = 0.5 \times 10^7 = 5.0 \times 10^6$$

It would take 5.0×10^6 (or 5,000,000) years to reach the Andromeda galaxy.

▶ **TRY EXERCISE 46, PAGE 32**

● **RATIONAL EXPONENTS AND RADICALS**

To this point, the expression b^n has been defined for real numbers b and integers n. Now we wish to extend the definition of exponents to include rational numbers so that expressions such as $2^{1/2}$ will be meaningful. Not just any definition will do. We want a definition of rational exponents for which the properties of integer exponents are true. The following example shows the direction we can take to accomplish our goal.

If the product property for exponential expressions is to hold for rational exponents, then for rational numbers p and q, $b^p b^q = b^{p+q}$. For example,

$$9^{1/2} \cdot 9^{1/2} \quad \text{must equal} \quad 9^{1/2+1/2} = 9^1 = 9$$

Thus $9^{1/2}$ must be a square root of 9. That is, $9^{1/2} = 3$.

The example suggests that $b^{1/n}$ can be defined in terms of roots according to the following definition.

Definition of $b^{1/n}$

If n is an even positive integer and $b \geq 0$, then $b^{1/n}$ is the nonnegative real number such that $(b^{1/n})^n = b$.

If n is an odd positive integer, then $b^{1/n}$ is the real number such that $(b^{1/n})^n = b$.

As examples,

- $25^{1/2} = 5$ because $5^2 = 25$.
- $(-64)^{1/3} = -4$ because $(-4)^3 = -64$.
- $16^{1/2} = 4$ because $4^2 = 16$.
- $-16^{1/2} = -(16^{1/2}) = -4$.
- $(-16)^{1/2}$ is not a real number.
- $(-32)^{1/5} = -2$ because $(-2)^5 = -32$.

If n is an even positive integer and $b < 0$, then $b^{1/n}$ is a *complex number*. Complex numbers are discussed in Section P.6.

To define expressions such as $8^{2/3}$, we will extend our definition of exponents even further. Because we want the power property $(b^p)^q = b^{pq}$ to be true for rational exponents also, we must have $(b^{1/n})^m = b^{m/n}$. With this in mind, we make the following definition.

Definition of $b^{m/n}$

For all positive integers m and n such that m/n is in simplest form, and for all real numbers b for which $b^{1/n}$ is a real number,

$$b^{m/n} = (b^{1/n})^m = (b^m)^{1/n}$$

Because $b^{m/n}$ is defined as $(b^{1/n})^m$ and also as $(b^m)^{1/n}$, we can evaluate expressions such as $8^{4/3}$ in more than one way. For example, because $8^{1/3}$ is a real number, $8^{4/3}$ can be evaluated in either of the following ways:

$$8^{4/3} = (8^{1/3})^4 = 2^4 = 16$$
$$8^{4/3} = (8^4)^{1/3} = 4096^{1/3} = 16$$

Of the two methods, the $b^{m/n} = (b^{1/n})^m$ method is usually easier to apply, provided you can evaluate $b^{1/n}$.

Here are some additional examples.

$$64^{2/3} = (64^{1/3})^2 = 4^2 = 16$$

$$32^{-6/5} = \frac{1}{32^{6/5}} = \frac{1}{(32^{1/5})^6} = \frac{1}{2^6} = \frac{1}{64}$$

$$81^{0.75} = 81^{3/4} = (81^{1/4})^3 = 3^3 = 27$$

☐ INTEGRATING
☐ TECHNOLOGY

For the examples on page 24, the base of the exponential expression was an integer power of the denominator of the exponent.

$$64 = 4^3 \qquad 32 = 2^5 \qquad 81 = 3^4$$

If the base of the exponential expression is not a power of the denominator of the exponent, a calculator is used to evaluate the expression. Here are some examples.

```
16^(3/8)
              2.828427125
25^(-1/5)
              .5253055609
42^(.14)
              1.687543205
```

In each of these examples, the value of the exponential expression is an irrational number. The decimal display is only an approximation of the actual result.

The following exponent properties were stated earlier, but they are restated here to remind you that they have now been extended to apply to rational exponents.

Properties of Rational Exponents

If p, q, and r represent rational numbers and a and b are positive real numbers, then

Product $\qquad b^p \cdot b^q = b^{p+q}$

Quotient $\qquad \dfrac{b^p}{b^q} = b^{p-q}$

Power $\qquad (b^p)^q = b^{pq} \qquad (a^p b^q)^r = a^{pr} b^{qr}$

$$\left(\dfrac{a^p}{b^q}\right)^r = \dfrac{a^{pr}}{b^{qr}} \qquad b^{-p} = \dfrac{1}{b^p}$$

Recall that an exponential expression is in simplest form when no powers of powers or negative exponents appear and each base occurs at most once.

EXAMP

Simplify: (

Solution

$$= \dfrac{x^{5/2}}{y}$$

▶ TRY EXE

● SIMPLIF

Radicals, expressed by the notation $\sqrt[n]{b}$, are also used to denote roots. The number b is the **radicand**, and the positive integer n is the **index** of the radical.

Definition of $\sqrt[n]{b}$

If n is a positive integer and b is a real number such that $b^{1/n}$ is a real number, then $\sqrt[n]{b} = b^{1/n}$.

If the index n equals 2, then the radical $\sqrt[2]{b}$ is written as simply $\sqrt{b}$, and it is referred to as the **principal square root of b** or simply the **square root of b.**

The symbol $\sqrt{b}$ is reserved to represent the nonnegative square root of b. To represent the negative square root of b, write $-\sqrt{b}$. For example, $\sqrt{25} = 5$, whereas $-\sqrt{25} = -5$.

Definition of $(\sqrt[n]{b})^m$

For all positive integers n, all integers m, and all real numbers b such that $\sqrt[n]{b}$ is a real number, $(\sqrt[n]{b})^m = \sqrt[n]{b^m} = b^{m/n}$.

When $\sqrt[n]{b}$ is a real number, the equations

$$b^{m/n} = \sqrt[n]{b^m} \qquad \text{and} \qquad b^{m/n} = (\sqrt[n]{b})^m$$

can be used to write exponential expressions such as $b^{m/n}$ in radical form. Use the denominator n as the index of the radical and the numerator m as the power of the radicand or as the power of the radical. For example,

$$(5xy)^{2/3} = (\sqrt[3]{5xy})^2 = \sqrt[3]{25x^2y^2}$$

• Use the denominator 3 as the index of the radical and the numerator 2 as the power of the radical.

The equations

$$b^{m/n} = \sqrt[n]{b^m} \qquad \text{and} \qquad b^{m/n} = (\sqrt[n]{b})^m$$

also can be used to write radical expressions in exponential form. For example,

$$\sqrt{(2ab)^3} = (2ab)^{3/2}$$ • Use the index 2 as the denominator of the power and the exponent 3 as the numerator of the power.

The definition of $\left(\sqrt[n]{b}\right)^m$ often can be used to evaluate radical expressions. For instance,

$$(\sqrt[3]{8})^4 = 8^{4/3} = (8^{1/3})^4 = 2^4 = 16$$

Care must be exercised when simplifying even roots (square roots, fourth roots, sixth roots,…) of variable expressions. Consider $\sqrt{x^2}$ when $x = 5$ and when $x = -5$.

Case 1   If $x = 5$, then $\sqrt{x^2} = \sqrt{5^2} = \sqrt{25} = 5 = x$.

Case 2   If $x = -5$, then $\sqrt{x^2} = \sqrt{(-5)^2} = \sqrt{25} = 5 = -x$.

These two cases suggest that

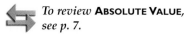
To review **ABSOLUTE VALUE**, *see p. 7.*

$$\sqrt{x^2} = \begin{cases} x, & \text{if } x \geq 0 \\ -x, & \text{if } x < 0 \end{cases}$$

Recalling the definition of absolute value, we can write this more compactly as $\sqrt{x^2} = |x|$.

Simplifying odd roots of a variable expression does not require using the absolute value symbol. Consider $\sqrt[3]{x^3}$ when $x = 5$ and when $x = -5$.

Case 1   If $x = 5$, then $\sqrt[3]{x^3} = \sqrt[3]{5^3} = \sqrt[3]{125} = 5 = x$.

Case 2   If $x = -5$, then $\sqrt[3]{x^3} = \sqrt[3]{(-5)^3} = \sqrt[3]{-125} = -5 = x$.

Thus $\sqrt[3]{x^3} = x$.

Although we have illustrated this principle only for square roots and cube roots, the same reasoning can be applied to other cases. The general result is given below.

Definition of $\sqrt[n]{b^n}$

If n is an even natural number and b is a real number, then

$$\sqrt[n]{b^n} = |b|$$

If n is an odd natural number and b is a real number, then

$$\sqrt[n]{b^n} = b$$

Here are some examples of these properties.

$$\sqrt[4]{16z^4} = 2|z| \qquad \sqrt[5]{32a^5} = 2a$$

Because radicals are defined in terms of rational powers, the properties of radicals are similar to those of exponential expressions.

Properties of Radicals

If m and n are natural numbers and a and b are nonnegative real numbers, then

Product $\sqrt[n]{a} \cdot \sqrt[n]{b} = \sqrt[n]{ab}$

Quotient $\dfrac{\sqrt[n]{a}}{\sqrt[n]{b}} = \sqrt[n]{\dfrac{a}{b}}$

Index $\sqrt[m]{\sqrt[n]{a}} = \sqrt[mn]{a}$

A radical is in **simplest form** if it meets all of the following criteria.

1. The radicand contains only powers less than the index. ($\sqrt{x^5}$ does not satisfy this requirement because 5, the exponent, is greater than 2, the index.)

2. The index of the radical is as small as possible. ($\sqrt[9]{x^3}$ does not satisfy this requirement because $\sqrt[9]{x^3} = x^{3/9} = x^{1/3} = \sqrt[3]{x}$.)

3. The denominator has been rationalized. That is, no radicals appear in the denominator. ($1/\sqrt{2}$ does not satisfy this requirement.)

4. No fractions appear under the radical sign. ($\sqrt[4]{2/x^3}$ does not satisfy this requirement.)

Radical expressions are simplified by using the properties of radicals. Here are some examples.

EXAMPLE 5 Simplify Radical Expressions

Simplify.

a. $\sqrt[4]{32x^3y^4}$ b. $\sqrt[3]{162x^4y^6}$

Solution

a. $\sqrt[4]{32x^3y^4} = \sqrt[4]{2^5x^3y^4} = \sqrt[4]{(2^4y^4) \cdot (2x^3)}$ • Factor and group factors that can be written as a power of the index.

$= \sqrt[4]{2^4y^4} \cdot \sqrt[4]{2x^3}$ • Use the product property of radicals.

$= 2|y|\sqrt[4]{2x^3}$ • Recall that for n even, $\sqrt[n]{b^n} = |b|$.

b. $\sqrt[3]{162x^4y^6} = \sqrt[3]{(2 \cdot 3^4)x^4y^6}$ • Factor and group factors that can be written as a power of the index.

$= \sqrt[3]{(3xy^2)^3 \cdot (2 \cdot 3x)}$

$= \sqrt[3]{(3xy^2)^3} \cdot \sqrt[3]{6x}$ • Use the product property of radicals.

$= 3xy^2\sqrt[3]{6x}$ • Recall that for n odd, $\sqrt[n]{b^n} = b$.

▶ **TRY EXERCISE 78, PAGE 32**

Like radicals have the same radicand and the same index. For instance,

$$3\sqrt[3]{5xy^2} \qquad \text{and} \qquad -4\sqrt[3]{5xy^2}$$

are like radicals. Addition and subtraction of like radicals are accomplished by using the distributive property. For example,

$$4\sqrt{3x} - 9\sqrt{3x} = (4 - 9)\sqrt{3x} = -5\sqrt{3x}$$
$$2\sqrt[3]{y^2} + 4\sqrt[3]{y^2} - \sqrt[3]{y^2} = (2 + 4 - 1)\sqrt[3]{y^2} = 5\sqrt[3]{y^2}$$

The sum $2\sqrt{3} + 6\sqrt{5}$ cannot be simplified further because the radicands are not the same. The sum $3\sqrt[3]{x} + 5\sqrt[4]{x}$ cannot be simplified because the indices are not the same.

Sometimes it is possible to simplify radical expressions that do not appear to be like radicals by simplifying each radical expression.

EXAMPLE 6 **Combine Radical Expressions**

Simplify: $5x\sqrt[3]{16x^4} - \sqrt[3]{128x^7}$

Solution

$5x\sqrt[3]{16x^4} - \sqrt[3]{128x^7}$

$= 5x\sqrt[3]{2^4x^4} - \sqrt[3]{2^7x^7}$ • Factor.

$= 5x\sqrt[3]{2^3x^3} \cdot \sqrt[3]{2x} - \sqrt[3]{2^6x^6} \cdot \sqrt[3]{2x}$ • Group factors that can be written as a power of the index.

$= 5x(2x\sqrt[3]{2x}) - 2^2x^2 \cdot \sqrt[3]{2x}$ • Use the product property of radicals.

$= 10x^2\sqrt[3]{2x} - 4x^2\sqrt[3]{2x}$ • Simplify.

$= 6x^2\sqrt[3]{2x}$

▶ **TRY EXERCISE 86, PAGE 32**

Multiplication of radical expressions is accomplished by using the distributive property. For instance,

$$\sqrt{5}(\sqrt{20} - 3\sqrt{15}) = \sqrt{5}(\sqrt{20}) - \sqrt{5}(3\sqrt{15}) \qquad \text{• Use the distributive property.}$$

$$= \sqrt{100} - 3\sqrt{75} \qquad \text{• Multiply the radicals.}$$
$$= 10 - 3 \cdot 5\sqrt{3} \qquad \text{• Simplify.}$$
$$= 10 - 15\sqrt{3}$$

The product of more complicated radical expressions may require repeated use of the distributive property.

EXAMPLE 7 **Multiply Radical Expressions**

Perform the indicated operation:

$$(\sqrt{3} + 5)(\sqrt{3} - 2)$$

Solution

$$
\begin{aligned}
(\sqrt{3} + 5)(\sqrt{3} - 2) \\
&= (\sqrt{3} + 5)\sqrt{3} - (\sqrt{3} + 5)2 && \bullet \text{ Use the distributive property.} \\
&= (\sqrt{3}\sqrt{3} + 5\sqrt{3}) - (2\sqrt{3} + 2 \cdot 5) && \bullet \text{ Use the distributive property.} \\
&= 3 + 5\sqrt{3} - 2\sqrt{3} - 10 \\
&= -7 + 3\sqrt{3}
\end{aligned}
$$

▶ **TRY EXERCISE 96, PAGE 33**

To **rationalize the denominator** of a fraction means to write it in an equivalent form that does not involve any radicals in its denominator.

EXAMPLE 8 **Rationalize the Denominator**

Rationalize the denominator. **a.** $\dfrac{5}{\sqrt[3]{a}}$ **b.** $\sqrt{\dfrac{3}{32y}}$

Solution

a. $\dfrac{5}{\sqrt[3]{a}} = \dfrac{5}{\sqrt[3]{a}} \cdot \dfrac{\sqrt[3]{a^2}}{\sqrt[3]{a^2}} = \dfrac{5\sqrt[3]{a^2}}{\sqrt[3]{a^3}} = \dfrac{5\sqrt[3]{a^2}}{a}$ • Use $\sqrt[3]{a} \cdot \sqrt[3]{a^2} = \sqrt[3]{a^3} = a.$

b. $\sqrt{\dfrac{3}{32y}} = \dfrac{\sqrt{3}}{\sqrt{32y}} = \dfrac{\sqrt{3}}{4\sqrt{2y}} = \dfrac{\sqrt{3}}{4\sqrt{2y}} \cdot \dfrac{\sqrt{2y}}{\sqrt{2y}} = \dfrac{\sqrt{6y}}{8y}$

▶ **TRY EXERCISE 106, PAGE 33**

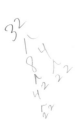

To rationalize the denominator of a fractional expression such as

$$\frac{1}{\sqrt{m} + \sqrt{n}}$$

we make use of the conjugate of $\sqrt{m} + \sqrt{n}$, which is $\sqrt{m} - \sqrt{n}$. The product of these conjugate pairs does not involve a radical.

$$(\sqrt{m} + \sqrt{n})(\sqrt{m} - \sqrt{n}) = m - n$$

In Example 9 we use the conjugate of the denominator to rationalize the denominator.

EXAMPLE 9   **Rationalize the Denominator**

Rationalize the denominator.

a. $\dfrac{2}{\sqrt{3} + \sqrt{2}}$ b. $\dfrac{a + \sqrt{5}}{a - \sqrt{5}}$

Solution

a. $\dfrac{2}{\sqrt{3} + \sqrt{2}} = \dfrac{2}{\sqrt{3} + \sqrt{2}} \cdot \dfrac{\sqrt{3} - \sqrt{2}}{\sqrt{3} - \sqrt{2}} = \dfrac{2\sqrt{3} - 2\sqrt{2}}{3 - 2} = 2\sqrt{3} - 2\sqrt{2}$

b. $\dfrac{a + \sqrt{5}}{a - \sqrt{5}} = \dfrac{a + \sqrt{5}}{a - \sqrt{5}} \cdot \dfrac{a + \sqrt{5}}{a + \sqrt{5}} = \dfrac{a^2 + 2a\sqrt{5} + 5}{a^2 - 5}$

▶ **TRY EXERCISE 110, PAGE 33**

 TOPICS FOR DISCUSSION

1. Given that a is a real number, discuss when the expression $a^{p/q}$ represents a real number.

2. The expressions $-a^n$ and $(-a)^n$ do not always represent the same number. Discuss the situations in which the two expressions are equal and those in which they are not equal.

3. The following calculator screen shows the value of the quotient of two radical expressions. Is the answer correct? Explain what happened.

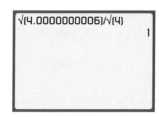

4. If you enter the expression for $\sqrt{5}$ on your calculator, the calculator will respond with 2.236067977 or some number close to that. Is this the exact value of $\sqrt{5}$? Is it possible to find the exact decimal value of $\sqrt{5}$ with a calculator? with a computer?

EXERCISE SET P.2

In Exercises 1 to 12, evaluate each expression.

1. -5^3

2. $(-5)^3$

3. $\left(\dfrac{2}{3}\right)^0$

4. -6^0

5. 4^{-2}

6. 3^{-4}

7. $\dfrac{1}{2^{-5}}$

8. $\dfrac{1}{3^{-3}}$

9. $\dfrac{2^{-3}}{6^{-3}}$

▶ **10.** $\dfrac{4^{-2}}{2^{-3}}$

11. $-2x^0$

12. $\dfrac{x^0}{4}$

In Exercises 13 to 32, write the exponential expression in simplest form.

13. $2x^{-4}$

14. $3y^{-2}$

15. $(-2ab^4)(-3a^2b^4)$

16. $(9xy^2)(-2x^2y^2)$

17. $\dfrac{6a^4}{8a^8}$

18. $\dfrac{12x^3}{16x^4}$

19. $\dfrac{12x^3y^4}{18x^5y^2}$

20. $\dfrac{5v^4w^{-3}}{10v^8}$

21. $\dfrac{36a^{-2}b^3}{3ab^4}$

22. $\dfrac{-48ab^{10}}{-32a^4b^3}$

23. $(-2m^3n^2)(-3mn^2)^2$

24. $(2a^3b^2)^3(-4a^4b^2)$

25. $(x^{-2}y)^2(xy)^{-2}$

26. $(x^{-1}y^2)^{-3}(x^2y^{-4})^{-3}$

27. $\left(\dfrac{3a^2b^3}{6a^4b^4}\right)^2$

28. $\left(\dfrac{2ab^2c^3}{5ab^2}\right)^3$

29. $\dfrac{(-4x^2y^3)^2}{(2xy^2)^3}$

▶ **30.** $\dfrac{(-3a^2b^3)^2}{(-2ab^4)^3}$

31. $\left(\dfrac{a^{-2}b}{a^3b^{-4}}\right)^2$

32. $\left(\dfrac{x^{-3}y^{-4}}{x^{-2}y}\right)^{-2}$

In Exercises 33 to 36, write the number in scientific notation.

33. 2,011,000,000,000

34. 49,100,000,000

35. 0.000000000562

36. 0.000000402

In Exercises 37 to 40, change the number from scientific notation to decimal notation.

37. 3.14×10^7

38. 4.03×10^9

39. -2.3×10^{-6}

40. 6.14×10^{-8}

In Exercises 41 to 48, perform the indicated operation and write the answer in scientific notation.

41. $(3 \times 10^{12})(9 \times 10^{-5})$

42. $(8.9 \times 10^{-5})(3.4 \times 10^{-6})$

43. $\dfrac{9 \times 10^{-3}}{6 \times 10^8}$

44. $\dfrac{2.5 \times 10^8}{5 \times 10^{10}}$

45. $\dfrac{(3.2 \times 10^{-11})(2.7 \times 10^{18})}{1.2 \times 10^{-5}}$

▶ **46.** $\dfrac{(6.9 \times 10^{27})(8.2 \times 10^{-13})}{4.1 \times 10^{15}}$

47. $\dfrac{(4.0 \times 10^{-9})(8.4 \times 10^5)}{(3.0 \times 10^{-6})(1.4 \times 10^{18})}$

48. $\dfrac{(7.2 \times 10^8)(3.9 \times 10^{-7})}{(2.6 \times 10^{-10})(1.8 \times 10^{-8})}$

In Exercises 49 to 70, simplify each exponential expression.

49. $4^{3/2}$

50. $-16^{3/2}$

51. $-64^{2/3}$

52. $125^{4/3}$

53. $9^{-3/2}$

54. $32^{-3/5}$

55. $\left(\dfrac{4}{9}\right)^{1/2}$

56. $\left(\dfrac{16}{25}\right)^{3/2}$

57. $\left(\dfrac{1}{8}\right)^{-4/3}$

58. $\left(\dfrac{8}{27}\right)^{-2/3}$

59. $(4a^{2/3}b^{1/2})(2a^{1/3}b^{3/2})$

60. $(6a^{3/5}b^{1/4})(-3a^{1/5}b^{3/4})$

61. $(-3x^{2/3})(4x^{1/4})$

▶ **62.** $(-5x^{1/3})(-4x^{1/2})$

63. $(81x^8y^{12})^{1/4}$

64. $(27x^3y^6)^{2/3}$

65. $\dfrac{16z^{3/5}}{12z^{1/5}}$

66. $\dfrac{6a^{2/3}}{9a^{1/3}}$

67. $(2x^{2/3}y^{1/2})(3x^{1/6}y^{1/3})$

68. $\dfrac{x^{1/3}y^{5/6}}{x^{2/3}y^{1/6}}$

69. $\dfrac{9a^{3/4}b}{3a^{2/3}b^2}$

70. $\dfrac{12x^{1/6}y^{1/4}}{16x^{3/4}y^{1/2}}$

In Exercises 71 to 80, simplify each radical expression.

71. $\sqrt{45}$

72. $\sqrt{75}$

73. $\sqrt[3]{24}$

74. $\sqrt[3]{135}$

75. $\sqrt[3]{-135}$

76. $\sqrt[3]{-250}$

77. $\sqrt{24x^2y^3}$

▶ **78.** $\sqrt{18x^2y^5}$

79. $\sqrt[3]{16a^3y^7}$

80. $\sqrt[3]{54m^2n^7}$

In Exercises 81 to 88, simplify each radical and then combine like radicals.

81. $2\sqrt{32} - 3\sqrt{98}$

82. $5\sqrt[3]{32} + 2\sqrt[3]{108}$

83. $-8\sqrt[4]{48} + 2\sqrt[4]{243}$

84. $2\sqrt[3]{40} - 3\sqrt[3]{135}$

85. $4\sqrt[3]{32y^4} + 3y\sqrt[3]{108y}$

▶ **86.** $-3x\sqrt[3]{54x^4} + 2\sqrt[3]{16x^7}$

87. $x\sqrt[3]{8x^3y^4} - 4y\sqrt[3]{64x^6y}$

88. $4\sqrt{a^5b} - a^2\sqrt{ab}$

In Exercises 89 to 98, find the indicated products and express each result in simplest form.

89. $\left(\sqrt{5} + 3\right)\left(\sqrt{5} + 4\right)$ **90.** $\left(\sqrt{7} + 2\right)\left(\sqrt{7} - 5\right)$

91. $\left(\sqrt{2} - 3\right)\left(\sqrt{2} + 3\right)$ **92.** $\left(2\sqrt{7} + 3\right)\left(2\sqrt{7} - 3\right)$

93. $\left(3\sqrt{z} - 2\right)\left(4\sqrt{z} + 3\right)$

94. $\left(4\sqrt{a} - \sqrt{b}\right)\left(3\sqrt{a} + 2\sqrt{b}\right)$

95. $\left(\sqrt{x} + 2\right)^2$ ▶ **96.** $\left(3\sqrt{5y} - 4\right)^2$

97. $\left(\sqrt{x - 3} + 2\right)^2$ **98.** $\left(\sqrt{2x + 1} - 3\right)^2$

In Exercises 99 to 112, simplify each expression by rationalizing the denominator. Write the result in simplest form.

99. $\dfrac{2}{\sqrt{2}}$ **100.** $\dfrac{3x}{\sqrt{3}}$

101. $\sqrt{\dfrac{5}{18}}$ **102.** $\sqrt{\dfrac{7}{40}}$

103. $\dfrac{3}{\sqrt[3]{2}}$ **104.** $\dfrac{2}{\sqrt[3]{4}}$

105. $\dfrac{4}{\sqrt[3]{8x^2}}$ ▶ **106.** $\dfrac{2}{\sqrt[4]{4y}}$

107. $\dfrac{3}{\sqrt{3} + 4}$ **108.** $\dfrac{2}{\sqrt{5} - 2}$

109. $\dfrac{6}{2\sqrt{5} + 2}$ ▶ **110.** $\dfrac{-7}{3\sqrt{2} - 5}$

111. $\dfrac{3}{\sqrt{5} + \sqrt{x}}$ **112.** $\dfrac{5}{\sqrt{y} - \sqrt{3}}$

113. **NATIONAL DEBT** In February of 2003, the U.S. national debt was approximately 6.4×10^{12} dollars. At that time, the population of the U.S. was 2.89×10^8 people. In February of 2003, what was the U.S. debt per person?

114. COLOR MONITORS A color monitor for a computer can display 2^{32} colors. A physiologist estimates that the human eye can detect approximately 36,000 different colors. How many colors, to the nearest thousand, would go undetected by a human eye using this monitor?

115. **WEIGHT OF AN ORCHID SEED** An orchid seed weighs approximately 3.2×10^{-8} ounce. If a package of seeds contains 1 ounce of orchid seeds, how many seeds are in the package?

116. LASER WAVELENGTH The wavelength of a certain helium-neon laser is 800 nanometers. (1 nanometer is 1×10^{-9} meter.) The frequency, in cycles per second, of this wave is $\dfrac{1}{\text{wavelength}}$. What is the frequency of this laser?

117. **DOPPLER EFFECT** Astronomers can approximate the distance to a galaxy by measuring its *red shift*, which is a shift in the wavelength of light due to the velocity of the galaxy. This is similar to the way the sound of a siren coming toward you seems to have a higher pitch than when the siren is moving away from you. A formula for red shift is red shift $= \dfrac{\lambda_r - \lambda_s}{\lambda_s}$, where λ_r and λ_s are wavelengths of a certain frequency of light. Calculate the red shift for a galaxy for which $\lambda_r = 5.13 \times 10^{-7}$ meter and $\lambda_s = 5.06 \times 10^{-7}$ meter.

118. **ASTRONOMICAL UNIT** Earth's mean distance from the sun is 9.3×10^7 miles. This distance is called the *astronomical unit* (AU). Jupiter is 5.2 AU from the sun. Find the distance in miles from Jupiter to the sun.

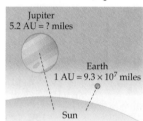

Jupiter
5.2 AU = ? miles

Earth
1 AU = 9.3×10^7 miles

Sun

119. **ASTRONOMY** The sun is approximately 1.44×10^{11} meters from Earth. If light travels 3×10^8 meters per second, how many minutes does it take light from the sun to reach Earth?

120. **MASS OF AN ATOM** One gram of hydrogen contains 6.023×10^{23} atoms. Find the mass of one hydrogen atom.

121. CELLULAR PHONE PRODUCTION An electronics firm estimates that the revenue R it will receive from the sale of x cell phones (in thousands) can be approximated by $R = 1250x(2^{-0.007x})$. What is the estimated revenue when the company sells 20,000 cell phones? Round to the nearest dollar.

122. DRUG POTENCY The amount A (in milligrams) of digoxin, a drug taken by cardiac patients, remaining in the blood t hours after a patient takes a 2-milligram dose is given by $A = 2(10^{-0.0078t})$.

a. How much digoxin remains in the blood of a patient 4 hours after taking a 2-milligram dose?

b. Suppose that a patient takes a 2-milligram dose of digoxin at 1:00 P.M. and another 2-milligram dose at 5:00 P.M. How much digoxin remains in the patient's blood at 6:00 P.M.?

123. **WORLD POPULATION** An estimate of the world's future population P is given by $P = 5.9(2^{0.0119n})$, where n is the number of years after 2000 and P is in billions. Using this estimate, what will the world's population be in 2050?

124. **LEARNING THEORY** In a psychology experiment, students were given a nine-digit number to memorize. The

percent P of students who remembered the number t minutes after it was read to them can be given by $P = 90 - 3t^{2/3}$. What percent of the students remembered the number after 1 hour?

125. **OCEANOGRAPHY** The percent P of light that will pass to a depth d, in meters, at a certain place in the ocean is given by $P = 10^{2 - (d/40)}$. Find, to the nearest percent, the amount of light that will pass to a depth of **a.** 10 meters and **b.** 25 meters below the surface of the ocean.

CONNECTING CONCEPTS

126. If $2^x = y$, then find 2^{x-4} in terms of y.

127. If a and b are nonzero numbers and $a < b$, is the statement $a^{-1} < b^{-1}$ a true statement? Give a reason for your answer.

128. How many digits are in the product $4^{50} \cdot 5^{100}$?

In Exercises 129 to 132, find the value of p for which the statement is true.

129. $a^{2/5}a^p = a^2$

130. $b^{-3/4}b^{2p} = b^3$

131. $\dfrac{x^{-3/4}}{x^{3p}} = x^4$

132. $(x^4x^{2p})^{1/2} = x$

In Exercises 133 to 136, rationalize the numerator.

133. $\dfrac{\sqrt{4 + h} - 2}{h}$

134. $\dfrac{\sqrt{9 + h} - 3}{h}$

135. $\sqrt{n^2 + 1} - n \left(Hint: \sqrt{n^2 + 1} - n = \dfrac{\sqrt{n^2 + 1} - n}{1} \right)$

136. $\sqrt{n^2 + n} - n \left(Hint: \sqrt{n^2 + n} - n = \dfrac{\sqrt{n^2 + n} - n}{1} \right)$

137. Evaluate: $\left(\sqrt{2^{\sqrt{2}}} \right)^{\sqrt{2}}$

PREPARE FOR SECTION P.3

138. Simplify: $-3(2a - 4b)$ [P.1]

139. Simplify: $5 - 2(2x - 7)$ [P.1]

140. Simplify: $2x^2 + 3x - 5 + x^2 - 6x - 1$ [P.1]

141. Simplify: $4x^2 - 6x - 1 - 5x^2 + x$ [P.1]

142. True or false: $4 - 3x - 2x^2 = 2x^2 - 3x + 4$ [P.1]

143. True or false: $\dfrac{12 + 15}{4} = \dfrac{\overset{3}{\cancel{12}} + 15}{\cancel{4}} = 18$ [P.1]

PROJECTS

1. **RELATIVITY THEORY** A moving object has energy, called *kinetic energy*, by virtue of its motion. As mentioned earlier in this chapter, the theory of relativity uses the formula at the right for kinetic energy.

$$K.E._r = mc^2 \left(\dfrac{1}{\sqrt{1 - \dfrac{v^2}{c^2}}} - 1 \right)$$

When the speed of an object is much less than the speed of light (3.0×10^8 meters per second) the formula

$$\text{K.E.}_n = \frac{1}{2}mv^2$$

is used. In each formula, v is the velocity of the object in meters per second, m is its rest mass in kilograms, and c is the speed of light given above. Calculate the percent error (in **a.** through **e.**) for each of the given velocities. The formula for percent error is

$$\% \text{ error} = \frac{|\text{K.E.}_r - \text{K.E.}_n|}{\text{K.E.}_r} \times 100$$

a. $v = 30$ meters per second (speeding car on an expressway)

b. $v = 240$ meters per second (speed of a commercial jet)

c. $v = 3.0 \times 10^7$ meters per second (10% of the speed of light)

d. $v = 1.5 \times 10^8$ meters per second (50% of the speed of light)

e. $v = 2.7 \times 10^8$ meters per second (90% of the speed of light)

f. Use your answers from **a.** through **e.** to give a reason why the formula for kinetic energy given by K.E._n is adequate for most of our common experiences involving motion (walking, running, bicycle, car, plane).

g. According to Relativity Theory, the mass, m, of an object changes as its velocity according to

$$m = \frac{m_0}{\sqrt{1 - \dfrac{v^2}{c_2}}}$$

where m_0 is the rest mass of the object. The approximate rest mass of an electron is 9.11×10^{-31} kilogram. What is the percent change, from its rest mass, in the mass of an electron that is traveling at $0.99c$ (99% of the speed of light)?

h. According to the Theory of Relativity, a particle (such as an electron or a spacecraft) cannot exceed the speed of light. Explain why the equation for K.E._r suggests such a conclusion.

SECTION P.3 POLYNOMIALS

- OPERATIONS ON POLYNOMIALS
- APPLICATIONS OF POLYNOMIALS

● OPERATIONS ON POLYNOMIALS

A **monomial** is a constant, a variable, or a product of a constant and one or more variables, with the variables having only nonnegative integer exponents. The constant is called the **numerical coefficient** or simply the **coefficient** of the monomial. The **degree of a monomial** is the sum of the exponents of the variables. For example, $-5xy^2$ is a monomial with coefficient -5 and degree 3.

The algebraic expression $3x^{-2}$ is not a monomial because it cannot be written as a product of a constant and a variable with a *nonnegative* integer exponent.

A sum of a finite number of monomials is called a **polynomial.** Each monomial is called a **term** of the polynomial. The **degree of a polynomial** is the largest degree of the terms in the polynomial.

Terms that have exactly the same variables raised to the same powers are called **like terms.** For example, $14x^2$ and $-31x^2$ are like terms; however, $2x^3y$ and $7xy$ are not like terms because x^3y and xy are not identical.

A polynomial is said to be simplified if all its like terms have been combined. For example, the simplified form of $4x^2 + 3x + 5x$ is $4x^2 + 8x$. A simplified polynomial that has two terms is a **binomial,** and a simplified polynomial that has three terms is a **trinomial.** For example, $4x + 7$ is a binomial, and $2x^3 - 7x^2 + 11$ is a trinomial.

A nonzero constant, such as 5, is called a **constant polynomial**. It has degree zero because $5 = 5x^0$. The number 0 is defined to be a polynomial with no degree.

Standard Form of a Polynomial

The **standard form of a polynomial** of degree n in the variable x is

$$a_n x^n + a_{n-1} x^{n-1} + \cdots + a_2 x^2 + a_1 x + a_0$$

where $a_n \neq 0$ and n is a nonnegative integer. The coefficient a_n is the **leading coefficient**, and a_0 is the **constant term**.

If a polynomial in the variable x is written with decreasing powers of x, then it is in **standard form**. For example, the polynomial

$$3x^2 - 4x^3 + 7x^4 - 1$$

is written in standard form as

$$7x^4 - 4x^3 + 3x^2 - 1$$

The following table shows the leading coefficient, degree, terms, and coefficients of the given polynomials.

Polynomial	Leading Coefficient	Degree	Terms	Coefficients
$9x^2 - x + 5$	9	2	$9x^2, -x, 5$	$9, -1, 5$
$11 - 2x$	-2	1	$-2x, 11$	$-2, 11$
$x^3 + 5x - 3$	1	3	$x^3, 5x, -3$	$1, 5, -3$

EXAMPLE 1   Identify Terms Related to a Polynomial

Write the polynomial $6x^3 - x + 5 - 2x^4$ in standard form. Identify the degree, terms, constant term, leading coefficient, and coefficients of the polynomial.

Solution

A polynomial is in standard form when the terms are written in decreasing powers of the variable. The standard form of the polynomial is $-2x^4 + 6x^3 - x + 5$. In this form, the degree is 4; the terms are $-2x^4$, $6x^3$, $-x$, and 5; the constant term is 5. The leading coefficient is -2; the coefficients are $-2, 6, -1$, and 5.

▶ **TRY EXERCISE 12, PAGE 41**

To add polynomials, add the coefficients of the like terms.

EXAMPLE 2   **Add Polynomials**

Add: $(3x^3 - 2x^2 - 6) + (4x^2 - 6x - 7)$

Solution

$(3x^3 - 2x^2 - 6) + (4x^2 - 6x - 7)$
$$= 3x^3 + (-2x^2 + 4x^2) + (-6x) + [(-6) + (-7)]$$
$$= 3x^3 + 2x^2 - 6x - 13$$

▶ **TRY EXERCISE 24, PAGE 41**

The **additive inverse of the polynomial** $3x - 7$ is

$$-(3x - 7) = -3x + 7$$

❓ QUESTION What is the additive inverse of $3x^2 - 8x + 7$?

To subtract a polynomial, we add its additive inverse. For example,

$$(2x - 5) - (3x - 7) = (2x - 5) + (-3x + 7)$$
$$= [2x + (-3x)] + [(-5) + 7]$$
$$= -x + 2$$

The distributive property is used to find the product of polynomials. For instance, to find the product of $(3x - 4)$ and $(2x^2 + 5x + 1)$, we treat $3x - 4$ as a *single* quantity and *distribute it* over the trinomial $2x^2 + 5x + 1$, as shown in Example 3.

EXAMPLE 3   **Multiply Polynomials**

Simplify: $(3x - 4)(2x^2 + 5x + 1)$

Solution

$(3x - 4)(2x^2 + 5x + 1)$
$$= (3x - 4)(2x^2) + (3x - 4)(5x) + (3x - 4)(1)$$
$$= (3x)(2x^2) - 4(2x^2) + (3x)(5x) - 4(5x) + (3x)(1) - 4(1)$$
$$= 6x^3 - 8x^2 + 15x^2 - 20x + 3x - 4$$
$$= 6x^3 + 7x^2 - 17x - 4$$

▶ **TRY EXERCISE 32, PAGE 41**

In the following calculation, a vertical format has been used to find the product of $(x^2 + 6x - 7)$ and $(5x - 2)$. Note that like terms are arranged in the same vertical column.

❓ ANSWER The additive inverse is $-3x^2 + 8x - 7$.

$$
\begin{array}{r}
x^2 + 6x - 7 \\
5x - 2 \\
\hline
\end{array}
$$

$$
\begin{array}{ll}
-\ 2x^2 - 12x + 14 & = -2(x^2 + 6x - 7) \\
5x^3 + 30x^2 - 35x & = 5x(x^2 + 6x - 7) \\
\hline
5x^3 + 28x^2 - 47x + 14 &
\end{array}
$$

If the terms of the binomials $(a + b)$ and $(c + d)$ are labeled as shown below, then the product of the two binomials can be computed mentally by the **FOIL** method.

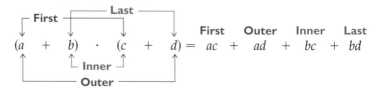

In the following illustration, we find the product of $(7x - 2)$ and $(5x + 4)$ by the FOIL method.

$$
\begin{array}{ccccc}
& \textbf{First} & \textbf{Outer} & \textbf{Inner} & \textbf{Last} \\
(7x - 2)(5x + 4) = & (7x)(5x) & + (7x)(4) & + (-2)(5x) & + (-2)(4) \\
= & 35x^2 & + 28x & - 10x & - 8 \\
= & \multicolumn{4}{l}{35x^2 + 18x - 8}
\end{array}
$$

Certain products occur so frequently in algebra that they deserve special attention.

Special Product Formulas

Special Form	Formula(s)
(Sum)(Difference)	$(x + y)(x - y) = x^2 - y^2$
(Binomial)^2	$(x + y)^2 = x^2 + 2xy + y^2$ $(x - y)^2 = x^2 - 2xy + y^2$

The variables x and y in these special product formulas can be replaced by other algebraic expressions, as shown in Example 4.

EXAMPLE 4   Use the Special Product Formulas

Find each special product.   **a.** $(7x + 10)(7x - 10)$ **b.** $(2y^2 + 11z)^2$

Solution

a. $(7x + 10)(7x - 10) = (7x)^2 - (10)^2 = 49x^2 - 100$

b. $(2y^2 + 11z)^2 = (2y^2)^2 + 2[(2y^2)(11z)] + (11z)^2 = 4y^4 + 44y^2z + 121z^2$

▶ **TRY EXERCISE 56, PAGE 41**

Many application problems require you to *evaluate polynomials*. To **evaluate a polynomial**, substitute the given value(s) for the variable(s) and then perform the indicated operations using the Order of Operations Agreement.

EXAMPLE 5   Evaluate a Polynomial

Evaluate the polynomial $2x^3 - 6x^2 + 7$ for $x = -4$.

Solution

$2x^3 - 6x^2 + 7$

$2(-4)^3 - 6(-4)^2 + 7 = 2(-64) - 6(16) + 7$ • Substitute -4 for x. Evaluate the powers.

$= -128 - 96 + 7$ • Perform the multiplications.

$= -217$ • Perform the additions and subtractions.

▶ Try Exercise 66, page 41

● APPLICATIONS OF POLYNOMIALS

EXAMPLE 6   Solve an Application

The number of singles tennis matches that can be played among n tennis players is given by the polynomial $\frac{1}{2}n^2 - \frac{1}{2}n$. Find the number of singles tennis matches that can be played among 4 tennis players.

Solution

$\frac{1}{2}n^2 - \frac{1}{2}n$

$\frac{1}{2}(4)^2 - \frac{1}{2}(4) = \frac{1}{2}(16) - \frac{1}{2}(4) = 8 - 2 = 6$ • Substitute 4 for n. Then simplify.

Therefore, 4 tennis players can play a total of 6 singles matches. See **Figure P.13.**

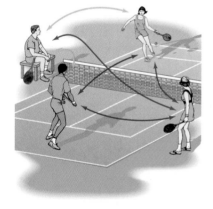

FIGURE P.13

4 tennis players can play a total of 6 singles matches.

▶ Try Exercise 76, page 42

EXAMPLE 7   Solve an Application

A scientist determines that the average time in seconds that it takes a particular computer to determine whether an n-digit natural number is prime or composite is given by

$$0.002n^2 + 0.002n + 0.009, \quad 20 \le n \le 40$$

The average time in seconds that it takes the computer to factor an n-digit number is given by

$$0.00032(1.7)^n, \quad 20 \le n \le 40$$

Continued ▶

Estimate the average time it takes the computer to

a. determine whether a 30-digit number is prime or composite

b. factor a 30-digit number

Solution

a. $0.002n^2 + 0.002n + 0.009$

$0.002(30)^2 + 0.002(30) + 0.009 = 1.8 + 0.06 + 0.009 = 1.869 \approx 2$ seconds

b. $0.00032(1.7)^n$

$0.00032(1.7)^{30} \approx 0.00032(8{,}193{,}465.726)$

≈ 2600 seconds

▶ **TRY EXERCISE 78, PAGE 42**

TOPICS FOR DISCUSSION

1. Discuss the definition of the term *polynomial*. Give some examples of expressions that are polynomials and some examples of expressions that are not polynomials.

2. Suppose that P and Q are both polynomials of degree n. Discuss the degrees of $P + Q, P - Q, PQ, P + P$, and $P - P$.

3. Suppose that you evaluate a polynomial P of degree n for larger and larger values of x (for instance, when $x = 1, 2, 3, 4, \ldots$). Discuss whether the value of the polynomial would eventually (for very large values of x) continually increase, decrease, or fluctuate between increasing and decreasing.

4. Discuss the similarities and differences among monomials, binomials, trinomials, and polynomials.

EXERCISE SET P.3

In Exercises 1 to 10, match the descriptions, labeled A, B, C,...., J, with the appropriate examples.

A. $x^3y + xy$ B. $7x^2 + 5x - 11$

C. $\dfrac{1}{2}x^2 + xy + y^2$ D. $4xy$

E. $8x^3 - 1$ F. $3 - 4x^2$

G. 8 H. $3x^5 - 4x^2 + 7x - 11$

I. $8x^4 - \sqrt{5}x^3 + 7$ J. 0

1. A monomial of degree 2.

2. A binomial of degree 3.

3. A polynomial of degree 5.

4. A binomial with leading coefficient of -4.

5. A zero-degree polynomial.

6. A fourth-degree polynomial that has a third-degree term.

7. A trinomial with integer coefficients.

8. A trinomial in x and y.

9. A polynomial with no degree.

10. A fourth-degree binomial.

In Exercises 11 to 16, for each polynomial determine its *a.* standard form, *b.* degree, *c.* coefficients, *d.* leading coefficient, *e.* terms.

11. $2x + x^2 - 7$

▶ **12.** $-3x^2 - 11 - 12x^4$

13. $x^3 - 1$

14. $4x^2 - 2x + 7$

15. $2x^4 + 3x^3 + 5 + 4x^2$ **16.** $3x^2 - 5x^3 + 7x - 1$

In Exercises 17 to 22, determine the degree of the given polynomial.

17. $3xy^2 - 2xy + 7x$ **18.** $x^3 + 3x^2y + 3xy^2 + y^3$

19. $4x^2y^2 - 5x^3y^2 + 17xy^3$ **20.** $-9x^5y + 10xy^4 - 11x^2y^2$

21. xy **22.** $5x^2y - y^4 + 6xy$

In Exercises 23 to 34, perform the indicated operations and simplify if possible by combining like terms. Write the result in standard form.

23. $(3x^2 + 4x + 5) + (2x^2 + 7x - 2)$

▶ **24.** $(5y^2 - 7y + 3) + (2y^2 + 8y + 1)$

25. $(4w^3 - 2w + 7) + (5w^3 + 8w^2 - 1)$

26. $(5x^4 - 3x^2 + 9) + (3x^3 - 2x^2 - 7x + 3)$

27. $(r^2 - 2r - 5) - (3r^2 - 5r + 7)$

28. $(7s^2 - 4s + 11) - (-2s^2 + 11s - 9)$

29. $(u^3 - 3u^2 - 4u + 8) - (u^3 - 2u + 4)$

30. $(5v^4 - 3v^2 + 9) - (6v^4 + 11v^2 - 10)$

31. $(4x - 5)(2x^2 + 7x - 8)$

▶ **32.** $(5x - 7)(3x^2 - 8x - 5)$

33. $(3x^2 - 2x + 5)(2x^2 - 5x + 2)$

34. $(2y^3 - 3y + 4)(2y^2 - 5y + 7)$

In Exercises 35 to 48, use the **FOIL** method to find the indicated product.

35. $(2x + 4)(5x + 1)$ **36.** $(5x - 3)(2x + 7)$

37. $(y + 2)(y + 1)$ **38.** $(y + 5)(y + 3)$

39. $(4z - 3)(z - 4)$ **40.** $(5z - 6)(z - 1)$

41. $(a + 6)(a - 3)$ **42.** $(a - 10)(a + 4)$

43. $(5x - 11y)(2x - 7y)$ **44.** $(3a - 5b)(4a - 7b)$

45. $(9x + 5y)(2x + 5y)$ **46.** $(3x - 7z)(5x - 7z)$

47. $(3p + 5q)(2p - 7q)$ **48.** $(2r - 11s)(5r + 8s)$

In Exercises 49 to 54, perform the indicated operations and simplify.

49. $(4d - 1)^2 - (2d - 3)^2$ **50.** $(5c - 8)^2 - (2c - 5)^2$

51. $(r + s)(r^2 - rs + s^2)$ **52.** $(r - s)(r^2 + rs + s^2)$

53. $(3c - 2)(4c + 1)(5c - 2)$

54. $(4d - 5)(2d - 1)(3d - 4)$

In Exercises 55 to 62, use the special product formulas to perform the indicated operation.

55. $(3x + 5)(3x - 5)$ ▶ **56.** $(4x^2 - 3y)(4x^2 + 3y)$

57. $(3x^2 - y)^2$ **58.** $(6x + 7y)^2$

59. $(4w + z)^2$ **60.** $(3x - 5y^2)^2$

61. $[(x + 5) + y][(x + 5) - y]$

62. $[(x - 2y) + 7][(x - 2y) - 7]$

In Exercises 63 to 70, evaluate the given polynomial for the indicated value of the variable.

63. $x^2 + 7x - 1$, for $x = 3$

64. $x^2 - 8x + 2$, for $x = 4$

65. $-x^2 + 5x - 3$, for $x = -2$

▶ **66.** $-x^2 - 5x + 4$, for $x = -5$

67. $3x^3 - 2x^2 - x + 3$, for $x = -1$

68. $5x^3 - x^2 + 5x - 3$, for $x = -1$

69. $1 - x^5$, for $x = -2$

70. $1 - x^3 - x^5$, for $x = 2$

71. RECREATION The air resistance (in pounds) on a cyclist riding a bicycle in an upright position can be given by $0.016v^2$, where v is the speed of the cyclist in miles per hour. Find the air resistance on a cyclist when

   a. $v = 10$ mph   **b.** $v = 15$ mph

72. HIGHWAY ENGINEERING On an expressway, the recommended *safe distance* between cars in feet is given by $0.015v^2 + v + 10$, where v is the speed of the car in miles per hour. Find the safe distance when

   a. $v = 30$ mph   **b.** $v = 55$ mph

73. GEOMETRY The volume of a right circular cylinder (as shown below) is given by $\pi r^2 h$, where r is the radius of the base and h is the height of the cylinder. Find the volume when

   a. $r = 3$ inches, $h = 8$ inches

   b. $r = 5$ cm, $h = 12$ cm

74. AUTOMOTIVE ENGINEERING The fuel efficiency (in miles per gallon of gas) of a car is given by the expression $-0.02v^2 + 1.5v + 2$, where v is the speed of the car in miles per hour. Find the fuel efficiency when

   a. $v = 45$ mph   **b.** $v = 60$ mph

75. PSYCHOLOGY Based on data from one experiment, the reaction time, in hundredths of a second, of a person to visual stimulus varies according to age and is given by $0.005x^2 - 0.32x + 12$, where x is the age of the person. Find the reaction time to the stimulus for a person when

   a. $x = 20$ years old   **b.** $x = 50$ years old

▶ **76. COMMITTEE MEMBERSHIP** The number of committees consisting of exactly 3 people that can be formed from a group of n people is given by the polynomial

$$\frac{1}{6}n^3 - \frac{1}{2}n^2 + \frac{1}{3}n$$

Find the number of committees consisting of exactly 3 people that can be formed from a group of 21 people.

77. CHESS MATCHES Find the number of chess matches that can be played between the members of a group of 150 people. Use the formula from Example 6.

▶ **78. COMPUTER SCIENCE** A computer scientist determines that the time in seconds it takes a particular computer to calculate n digits of π is given by the polynomial

$$4.3 \times 10^{-6}n^2 - 2.1 \times 10^{-4}n$$

where $1000 \le n \le 10{,}000$. Estimate the time it takes the computer to calculate π to

   a. 1000 digits   **b.** 5000 digits   **c.** 10,000 digits

79. COMPUTER SCIENCE If n is a positive integer, then $n!$, which is read "n factorial," is given by

$$n(n - 1)(n - 2) \cdots 2 \cdot 1$$

For example, $4! = 4 \cdot 3 \cdot 2 \cdot 1 = 24$. A computer scientist determines that each time a program is run on a particular computer, the time in seconds required to compute $n!$ is given by the polynomial

$$1.9 \times 10^{-6}n^2 - 3.9 \times 10^{-3}n$$

where $1000 \le n \le 10{,}000$. Using this polynomial, estimate the time it takes this computer to calculate 4000! and 8000!.

80. AIR VELOCITY OF A COUGH The velocity, in meters per second, of the air that is expelled during a cough is given by velocity $= 6r^2 - 10r^3$, where r is the radius of the trachea in centimeters.

   a. Find the velocity as a polynomial in standard form.

   b. Find the velocity of the air in a cough when the radius of the trachea is 0.35 cm. Round to the nearest hundredth.

81. SPORTS The height, in feet, of a baseball released by a pitcher t seconds after it is released is given by (ignoring air resistance)

$$\text{Height} = -16t^2 + 4.7881t + 6$$

For the pitch to be a strike, it must be at least 2 feet high when it crosses home plate and must be no higher than 5 feet high. If it takes 0.5 second for the ball to reach home plate, will the ball be high enough to be a strike?

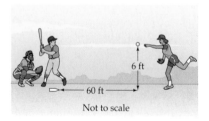

82. MEDICINE The temperature, in degrees Fahrenheit, of a patient after receiving a medication is given by

$$\text{Temperature} = 0.0002t^3 - 0.0114t^2 + 0.0158t + 104$$

where t is the number of minutes after receiving the medication.

   a. What was the patient's temperature just before the medication was taken?

b. What was the patient's temperature 25 minutes after taking the medication?

83. PRIME FACTORIZATION Find n given

$$n! = 2^{11} \cdot 3^6 \cdot 5^3 \cdot 7^2 \cdot 11 \cdot 13$$

CONNECTING CONCEPTS

The following special product formulas can be used to find the cube of a binomial.

$$(x + y)^3 = x^3 + 3x^2y + 3xy^2 + y^3$$
$$(x - y)^3 = x^3 - 3x^2y + 3xy^2 - y^3$$

In Exercises 84 to 89, make use of the given special product formulas to find the indicated products.

84. $(a + b)^3$ **85.** $(a - b)^3$ **86.** $(x - 1)^3$

87. $(y + 2)^3$ **88.** $(2x - 3y)^3$ **89.** $(3x + 5y)^3$

PREPARE FOR SECTION P.4

90. Simplify: $\dfrac{6x^3}{2x}$ [P.2]

91. Simplify: $(-12x^4)3x^2$ [P.2]

92. Express x^6 as a power of **a.** x^2 and **b.** x^3. [P.2]

In Exercises 93 to 95, replace the question mark to make a true statement.

93. $6a^3b^4 \cdot ? = 18a^3b^7$ [P.2]

94. $-3(5a - ?) = -15a + 21$ [P.1]

95. $2x(3x - ?) = 6x^2 - 2x$ [P.1]

PROJECTS

1. **ODD NUMBERS** Every odd number can be written in the form $2n - 1$, and every even number can be expressed as $2n$, where n is a natural number. Explain, by writing a few paragraphs and giving the supporting mathematics, why the product of two odd numbers is an odd number, the product of two even numbers is an even number, and the product of an even number and an odd number is an even number.

2. **PRIME NUMBERS** Fermat's Little Theorem states, "If n is a prime number and a is *any* natural number, then $a^n - a$ is divisible by n." For instance, for $n = 11$ and $a = 14$, $\dfrac{14^{11} - 14}{11} = 368,142,288,150$. The important aspect of this theorem is that no matter what natural number is chosen for a, $a^{11} - a$ is evenly divisible by 11.

Knowing whether a number is prime plays a central role in the security of computer systems. A restatement (called the *contrapositive*) of Fermat's Little Theorem is "If n is a number and a is some number for which $a^n - a$ is *not* divisible by n, then n is *not* a prime number" is used to determine when a number is *not* prime. For example, if $n = 14$, then $\dfrac{2^{14} - 2}{14} = \dfrac{8191}{7}$, and thus there is some number ($a = 2$) for which $2^{14} - 2$ is not evenly divisible by 14. Therefore, 14 is not prime.

a. Explain the meaning of the *contrapositive* (used above) of a theorem. Use your explanation to write the contrapositive of "If two triangles are congruent, then they are similar."

b. $7^{14} - 7$ is divisible by 14. Explain why this does not contradict the fact that 14 is not a prime.

c. Explain the meaning of the *converse* of a theorem. State the converse of Fermat's Little Theorem.

d. The number 561 has the property that $a^{561} - a$ is divisible by 561 for all natural numbers a. Can you use

Fermat's Little Theorem to conclude that 561 is a prime number? Explain.

e. Suppose that $a^n - a$ is divisible by n for all values of a. Can you conclude that n is a prime number? Explain your answer.

f. Find a definition of a Carmichael number. What do Carmichael numbers have to do with the information in part **e.**?

SECTION P.4   # FACTORING

- GREATEST COMMON FACTOR
- FACTORING TRINOMIALS
- SPECIAL FACTORING
- FACTOR BY GROUPING
- GENERAL FACTORING

Writing a polynomial as a product of polynomials of lower degree is called **factoring**. Factoring is an important procedure that is often used to simplify fractional expressions and to solve equations.

In this section we consider only the factorization of polynomials that have integer coefficients. Also, we are concerned only with **factoring over the integers**. That is, we search only for polynomial factors that have integer coefficients.

● GREATEST COMMON FACTOR

The first step in any factorization of a polynomial is to use the distributive property to factor out the **greatest common factor (GCF)** of the terms of the polynomial. Given two or more exponential expressions with the same prime number base or the same variable base, the GCF is the exponential expression with the smallest exponent. For example,

$$2^3 \text{ is the GCF of } 2^3, 2^5, \text{ and } 2^8 \quad \text{and} \quad a \text{ is the GCF of } a^4 \text{ and } a$$

The GCF of two or more monomials is the product of the GCFs of all the *common* bases. For example, to find the GCF of $27a^3b^4$ and $18b^3c$, factor the coefficients into prime factors and then write each common base with its smallest exponent.

$$27a^3b^4 = 3^3 \cdot a^3 \cdot b^4 \qquad 18b^3c = 2 \cdot 3^2 \cdot b^3 \cdot c$$

The only common bases are 3 and b. The product of these common bases with their smallest exponents is 3^2b^3. The GCF of $27a^3b^4$ and $18b^3c$ is $9b^3$.

The expressions $3x(2x + 5)$ and $4(2x + 5)$ have a common *binomial* factor that is $2x + 5$. Thus the GCF of $3x(2x + 5)$ and $4(2x + 5)$ is $2x + 5$.

EXAMPLE I   **Factor Out the Greatest Common Factor**

Factor out the GCF.

a. $10x^3 + 6x$ **b.** $15x^{2n} + 9x^{n+1} - 3x^n$ (where n is a positive integer)

c. $(m + 5)(x + 3) + (m + 5)(x - 10)$

Solution

a. $10x^3 + 6x = (2x)(5x^2) + (2x)(3)$ • The GCF is 2x.

$= 2x(5x^2 + 3)$ • Factor out the GCF.

b. $15x^{2n} + 9x^{n+1} - 3x^n$

$= (3x^n)(5x^n) + (3x^n)(3x) - (3x^n)(1)$ • The GCF is $3x^n$.

$= 3x^n(5x^n + 3x - 1)$ • Factor out the GCF.

c. $(m + 5)(x + 3) + (m + 5)(x - 10)$ • Use the distributive property to factor out $(m + 5)$.

$= (m + 5)[(x + 3) + (x - 10)]$

$= (m + 5)(2x - 7)$ • Simplify.

▶ **TRY EXERCISE 6, PAGE 53**

● FACTORING TRINOMIALS

Some trinomials of the form $x^2 + bx + c$ can be factored by a trial procedure. This method makes use of the FOIL method in reverse. For example, consider the following products:

$(x + 3)(x + 5) = x^2 + 5x + 3x + (3)(5) = x^2 + 8x + 15$
$(x - 2)(x - 7) = x^2 - 7x - 2x + (-2)(-7) = x^2 - 9x + 14$
$(x + 4)(x - 9) = x^2 - 9x + 4x + (4)(-9) = x^2 - 5x - 36$

The coefficient of *x* is the sum of the constant terms of the binomials.

The constant term of the trinomial is the product of the constant terms of the binomials.

❓ **QUESTION** Is $(x - 2)(x + 7)$ the correct factorization of $x^2 - 5x - 14$?

Points to Remember to Factor $x^2 + bx + c$

1. The constant term c of the trinomial is the product of the constant terms of the binomials.

2. The coefficient b in the trinomial is the sum of the constant terms of the binomials.

3. If the constant term c of the trinomial is positive, the constant terms of the binomials have the same sign as the coefficient b of the trinomial.

4. If the constant term c of the trinomial is negative, the constant terms of the binomials have opposite signs.

❓ **ANSWER** No. $(x - 2)(x + 7) = x^2 + 5x - 14$.

EXAMPLE 2 Factor a Trinomial of the Form $x^2 + bx + c$

Factor: $x^2 + 7x - 18$

Solution

We must find two binomials whose first terms have a product of x^2 and whose last terms have a product of -18; also, the sum of the product of the outer terms and the product of the inner terms must be $7x$. Begin by listing the possible integer factorizations of -18 and the sum of those factors.

Factors of -18	Sum of the Factors
$1 \cdot (-18)$	$1 + (-18) = -17$
$(-1) \cdot 18$	$(-1) + 18 = 17$
$2 \cdot (-9)$	$2 + (-9) = -7$
$(-2) \cdot 9$	$(-2) + 9 = 7$

• **Stop. This is the desired sum.**

Thus -2 and 9 are the numbers whose sum is 7 and whose product is -18. Therefore,

$$x^2 + 7x - 18 = (x - 2)(x + 9)$$

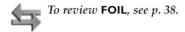 *To review* **FOIL**, *see p. 38.*

The FOIL method can be used to verify that the factorization is correct.

▶ **TRY EXERCISE 12, PAGE 53**

The trial method sometimes can be used to factor trinomials of the form $ax^2 + bx + c$, which do not have a leading coefficient of 1. We use the factors of a and c to form trial binomial factors. Factoring trinomials of this type may require testing many factors. To reduce the number of trial factors, make use of the following points.

Points to Remember to Factor $ax^2 + bx + c$, $a > 0$

1. If the constant term of the trinomial is positive, the constant terms of the binomials have the same sign as the coefficient b in the trinomial.

2. If the constant term of the trinomial is negative, the constant terms of the binomials have opposite signs.

3. If the terms of the trinomial do not have a common factor, then neither binomial will have a common factor.

EXAMPLE 3 Factor a Trinomial of the Form $ax^2 + bx + c$

Factor: $6x^2 - 11x + 4$

Solution

Because the constant term of the trinomial is positive and the coefficient of the x term is negative, the constant terms of the binomials will both be negative. This time we find factors of the first term as well as factors of the constant term.

Factors of $6x^2$	Factors of 4 (both negative)
$x, 6x$	$-1, -4$
$2x, 3x$	$-2, -2$

Use these factors to write trial factors. Use the FOIL method to see whether any of the trial factors produce the correct middle term. If the terms of a trinomial do not have a common factor, then a binomial factor cannot have a common factor (point 3). Such trial factors need not be checked.

Trial Factors	Middle Term	
$(x - 1)(6x - 4)$	Common factor	• 6x and 4 have a common factor.
$(x - 4)(6x - 1)$	$-1x - 24x = -25x$	
$(x - 2)(6x - 2)$	Common factor	• 6x and 2 have a common factor.
$(2x - 1)(3x - 4)$	$-8x - 3x = -11x$	• This is the correct middle term.

Thus $6x^2 - 11x + 4 = (2x - 1)(3x - 4)$.

▶ **TRY EXERCISE 16, PAGE 53**

Sometimes it is impossible to factor a polynomial into the product of two polynomials having integer coefficients. Such polynomials are said to be **nonfactorable over the integers**. For example, $x^2 + 3x + 7$ is nonfactorable over the integers because there are no integers whose product is 7 and whose sum or difference is 3.

If you have difficulty factoring a trinomial, you may wish to use the following theorem. It will indicate whether the trinomial is factorable over the integers.

Factorization Theorem

The trinomial $ax^2 + bx + c$, with integer coefficients a, b, and c, can be factored as the product of two binomials with integer coefficients if and only if $b^2 - 4ac$ is a perfect square.

EXAMPLE 4   Apply the Factorization Theorem

Determine whether each trinomial is factorable over the integers.

a. $4x^2 + 8x - 7$ b. $6x^2 - 5x - 4$

Continued ▶

Solution

a. The coefficients of $4x^2 + 8x - 7$ are $a = 4$, $b = 8$, and $c = -7$. Applying the factorization theorem yields

$$b^2 - 4ac = 8^2 - 4(4)(-7) = 176$$

Because 176 is not a perfect square, the trinomial is nonfactorable over the integers.

b. The coefficients of $6x^2 - 5x - 4$ are $a = 6$, $b = -5$, and $c = -4$. Thus

$$b^2 - 4ac = (-5)^2 - 4(6)(-4) = 121$$

Because 121 is a perfect square, the trinomial is factorable over the integers. Using the methods we have developed, we find

$$6x^2 - 5x - 4 = (3x - 4)(2x + 1)$$

▶ **TRY EXERCISE 24, PAGE 53**

Certain trinomials can be expressed as quadratic trinomials by making suitable variable substitutions. A trinomial is **quadratic in form** if it can be written as

$$au^2 + bu + c$$

If we let $x^2 = u$, the trinomial $x^4 + 5x^2 + 6$ can be written as shown at the right.

$$\begin{aligned} x^4 + 5x^2 + 6 \\ = (x^2)^2 + 5(x^2) + 6 \end{aligned}$$

The trinomial is quadratic in form.

$$= u^2 + 5u + 6$$

If we let $xy = u$, the trinomial $2x^2y^2 + 3xy - 9$ can be written as shown at the right.

$$\begin{aligned} 2x^2y^2 + 3xy - 9 \\ = 2(xy)^2 + 3(xy) - 9 \end{aligned}$$

The trinomial is quadratic in form.

$$= 2u^2 + 3u - 9$$

When a trinomial that is quadratic in form is factored, the variable part of the first term in each binomial factor will be u. For example, because $x^4 + 5x^2 + 6$ is quadratic in form when $x^2 = u$, the first term in each binomial factor will be x^2.

$$\begin{aligned} x^4 + 5x^2 + 6 &= (x^2)^2 + 5(x^2) + 6 \\ &= (x^2 + 2)(x^2 + 3) \end{aligned}$$

The trinomial $x^2y^2 - 2xy - 15$ is quadratic in form when $xy = u$. The first term in each binomial factor will be xy.

$$\begin{aligned} x^2y^2 - 2xy - 15 &= (xy)^2 - 2(xy) - 15 \\ &= (xy + 3)(xy - 5) \end{aligned}$$

EXAMPLE 5

Factor: **a.** $6x^2y^2 - xy - 12$ **b.** $2x^4 + 5x^2 - 12$

Solution

a. $6x^2y^2 - xy - 12$

$$= (3xy + 4)(2xy - 3)$$

• **The trinomial is quadratic in form when** $xy = u.$

b. $2x^4 + 5x^2 - 12$

   $= (x^2 + 4)(2x^2 - 3)$

   • The trinomial is quadratic in form when $x^2 = u$.

▶ **TRY EXERCISE 36, PAGE 53**

● SPECIAL FACTORING

The product of a term and itself is called a **perfect square.** The exponents on variables of perfect squares are always even numbers. The **square root of a perfect square** is one of the two equal factors of the perfect square. To find the square root of a perfect square variable term, divide the exponent by 2. For the examples below, assume the variables represent positive numbers.

Term		Perfect Square	Square Root
7	$7 \cdot 7 =$	49	$\sqrt{49} = 7$
y	$y \cdot y =$	y^2	$\sqrt{y^2} = y$
$2x^3$	$2x^3 \cdot 2x^3 =$	$4x^6$	$\sqrt{4x^6} = 2x^3$
x^n	$x^n \cdot x^n =$	x^{2n}	$\sqrt{x^{2n}} = x^n$

take note

The **sum** of two squares does not factor over the integers. For instance, $49x^2 + 144$ does not factor over the integers.

The factors of the difference of two perfect squares are the sum and difference of the square roots of the perfect squares.

Factors of the Difference of Two Perfect Squares

$$a^2 - b^2 = (a + b)(a - b)$$

EXAMPLE 6   Factor the Difference of Squares

Factor: $49x^2 - 144$

Solution

$49x^2 - 144 = (7x)^2 - (12)^2$ • Recognize the difference-of-squares form.

   $= (7x + 12)(7x - 12)$ • The binomial factors are the sum and the difference of the square roots of the squares.

▶ **TRY EXERCISE 40, PAGE 53**

   A **perfect-square trinomial** is a trinomial that is the square of a binomial. For example, $x^2 + 6x + 9$ is a perfect-square trinomial because

$$(x + 3)^2 = x^2 + 6x + 9$$

Every perfect-square trinomial can be factored by the trial method, but it generally is faster to factor perfect-square trinomials by using the following factoring formulas.

> ### Factors of a Perfect-Square Trinomial
>
> $$a^2 + 2ab + b^2 = (a + b)^2$$
> $$a^2 - 2ab + b^2 = (a - b)^2$$

EXAMPLE 7   Factor a Perfect-Square Trinomial

Factor: $16m^2 - 40mn + 25n^2$

Solution

Because $16m^2 = (4m)^2$ and $25n^2 = (5n)^2$, try factoring $16m^2 - 40mn + 25n^2$ as the square of a binomial.

$$16m^2 - 40mn + 25n^2 \stackrel{?}{=} (4m - 5n)^2$$

Check:

$$
\begin{aligned}
(4m - 5n)^2 &= (4m - 5n)(4m - 5n) \\
&= 16m^2 - 20mn - 20mn + 25n^2 \\
&= 16m^2 - 40mn + 25n^2
\end{aligned}
$$

The factorization checks. Therefore, $16m^2 - 40mn + 25n^2 = (4m - 5n)^2$.

▶ **TRY EXERCISE 50, PAGE 54**

> **take note**
>
> It is important to check the proposed factorization. For instance, consider $x^2 + 13x + 36$. Because x^2 is the square of x and 36 is the square of 6, it is tempting to factor, using the perfect-square trinomial formulas, as $x^2 + 13x + 36 \stackrel{?}{=} (x + 6)^2$. Note, however, that $(x + 6)^2 = x^2 + 12x + 36$, which is not the original trinomial. The correct factorization is
> $$x^2 + 13x + 36 = (x + 4)(x + 9)$$

The product of the same three terms is called a **perfect cube**. The exponents on variables of perfect cubes are always divisible by 3. The **cube root of a perfect cube** is one of the three equal factors of the perfect cube. To find the cube root of a perfect cube variable term, divide the exponent by 3.

Term		Perfect Cube	Cube Root
5	$5 \cdot 5 \cdot 5 =$	125	$\sqrt[3]{125} = 5$
z	$z \cdot z \cdot z =$	z^3	$\sqrt[3]{z^3} = z$
$3x^2$	$3x^2 \cdot 3x^2 \cdot 3x^2 =$	$27x^6$	$\sqrt[3]{27x^6} = 3x^2$
x^n	$x^n \cdot x^n \cdot x^n =$	x^{3n}	$\sqrt[3]{x^{3n}} = x^n$

The following factoring formulas are used to factor the sum or difference of two perfect cubes.

> **take note**
>
> Note the pattern of the signs when factoring the sum or difference of two perfect cubes.
>
> $$a^3 + b^3 = (a + b)(a^2 - ab + b^2)$$
> Same sign, Opposite signs
>
> $$a^3 - b^3 = (a - b)(a^2 + ab + b^2)$$
> Same sign, Opposite signs

> ### Factors of the Sum or Difference of Two Perfect Cubes
>
> $$a^3 + b^3 = (a + b)(a^2 - ab + b^2)$$
> $$a^3 - b^3 = (a - b)(a^2 + ab + b^2)$$

EXAMPLE 8   Factor the Sum or Difference of Cubes

Factor: **a.** $8a^3 + b^3$ **b.** $a^3 - 64$

Solution

a. $8a^3 + b^3 = (2a)^3 + b^3$ • Recognize the sum-of-cubes form.

$$= (2a + b)(4a^2 - 2ab + b^2)$$ • Factor.

b. $a^3 - 64 = a^3 - 4^3$ • Recognize the difference-of-cubes form.

$$= (a - 4)(a^2 + 4a + 16)$$ • Factor.

▶ **TRY EXERCISE 56, PAGE 54**

● FACTOR BY GROUPING

take note

$-a + b = -(a - b)$. Thus, $-4y + 14 = -(4y - 14)$.

Some polynomials can be **factored by grouping**. Pairs of terms that have a common factor are first grouped together. The process makes repeated use of the distributive property, as shown in the following factorization of $6y^3 - 21y^2 - 4y + 14$.

$$6y^3 - 21y^2 - 4y + 14$$

$$= (6y^3 - 21y^2) - (4y - 14)$$ • Group the first two terms and the last two terms.

$$= 3y^2(2y - 7) - 2(2y - 7)$$ • Factor out the GCF from each of the groups.

$$= (2y - 7)(3y^2 - 2)$$ • Factor out the common binomial factor.

When you factor by grouping, some experimentation may be necessary to find a grouping that is of the form of one of the special factoring formulas.

EXAMPLE 9   **Factor by Grouping**

Factor by grouping.   a. $a^2 + 10ab + 25b^2 - c^2$ b. $p^2 + p - q - q^2$

Solution

a. $a^2 + 10ab + 25b^2 - c^2$

$$= (a^2 + 10ab + 25b^2) - c^2$$ • Group the terms of the perfect-square trinomial.

$$= (a + 5b)^2 - c^2$$ • Factor the trinomial.

$$= [(a + 5b) + c][(a + 5b) - c]$$ • Factor the difference of squares.

$$= (a + 5b + c)(a + 5b - c)$$ • Simplify.

b. $p^2 + p - q - q^2$

$$= p^2 - q^2 + p - q$$ • Rearrange the terms.

$$= (p^2 - q^2) + (p - q)$$ • Regroup.

$$= (p + q)(p - q) + (p - q)$$ • Factor the difference of squares.

$$= (p - q)(p + q + 1)$$ • Factor out the common factor $(p - q)$.

▶ **TRY EXERCISE 66, PAGE 54**

● GENERAL FACTORING

Here is a general factoring strategy for polynomials:

General Factoring Strategy

1. Factor out the GCF of all terms.

2. Try to factor a binomial as

  a. the difference of two squares

  b. the sum or difference of two cubes

3. Try to factor a trinomial

  a. as a perfect-square trinomial

  b. using the trial method

4. Try to factor a polynomial with more than three terms by grouping.

5. After each factorization, examine the new factors to see whether they can be factored.

EXAMPLE 10   **Factor Using the General Factoring Strategy**

Completely factor: $x^6 + 7x^3 - 8$

Solution
Factor $x^6 + 7x^3 - 8$ as the product of two binomials.

$$x^6 + 7x^3 - 8 = (x^3 + 8)(x^3 - 1)$$

Now factor $x^3 + 8$, which is the sum of two cubes, and factor $x^3 - 1$, which is the difference of two cubes.

$$x^6 + 7x^3 - 8 = (x + 2)(x^2 - 2x + 4)(x - 1)(x^2 + x + 1)$$

▶ **TRY EXERCISE 72, PAGE 54**

 TOPICS FOR DISCUSSION

1. Discuss the meaning of the phrase *nonfactorable over the integers*.

2. You know that if $ab = 0$, then $a = 0$ or $b = 0$. Suppose a polynomial is written in factored form and then set equal to zero. For instance, suppose

$$x^2 - 2x - 15 = (x - 5)(x + 3) = 0$$

Discuss what implications this has for the values of x. Do not answer this question only for the polynomial above, but also for any polynomial written as a product of linear factors and then set equal to zero.

3. Let P be a polynomial of degree n. Discuss the number of possible distinct linear polynomials that can be a factor of P.

4. A method of evaluating polynomials, sometimes called Horner's method, involves factoring a polynomial in a certain manner. For instance,

$$4x^3 - 2x^2 + 5x - 3 = [(4x - 2)x + 5]x - 3$$
$$5x^4 - 2x^3 + 4x^2 + x - 6 = \{[(5x - 2)x + 4]x + 1\}x - 6$$

To evaluate the polynomial, the factored form is evaluated. Discuss the advantages and disadvantages of using this method to evaluate a polynomial.

5. If n is a natural number, $n! = n(n - 1)(n - 2) \cdots 3 \cdot 2 \cdot 1$. Explain why none of the following consecutive integers is a prime number.

$$5! + 2 \qquad 5! + 3 \qquad 5! + 4 \qquad 5! + 5$$

How many numbers are in the following list of consecutive integers? How many of those numbers are prime numbers?

$$k! + 2, \quad k! + 3, \quad k! + 4, \quad k! + 5, \quad \ldots, \quad k! + k$$

Explain why this result means that there are arbitrarily long sequences of consecutive natural numbers that do not contain a prime number.

EXERCISE SET P.4

In Exercises 1 to 8, factor out the GCF from each polynomial.

1. $5x + 20$

2. $8x^2 + 12x - 40$

3. $-15x^2 - 12x$

4. $-6y^2 - 54y$

5. $10x^2y + 6xy - 14xy^2$

▶ 6. $6a^3b^2 - 12a^2b + 72ab^3$

7. $(x - 3)(a + b) + (x - 3)(a + 2b)$

8. $(x - 4)(2a - b) + (x + 4)(2a - b)$

In Exercises 9 to 22, factor each trinomial over the integers.

9. $x^2 + 7x + 12$

10. $x^2 + 9x + 20$

11. $a^2 - 10a - 24$

▶ 12. $b^2 + 12b - 28$

13. $6x^2 + 25x + 4$

14. $8a^2 - 26a + 15$

15. $51x^2 - 5x - 4$

▶ 16. $57y^2 + y - 6$

17. $6x^2 + xy - 40y^2$

18. $8x^2 + 10xy - 25y^2$

19. $x^4 + 6x^2 + 5$

20. $x^4 + 11x^2 + 18$

21. $6x^4 + 23x^2 + 15$

22. $9x^4 + 10x^2 + 1$

In Exercises 23 to 28, use the factorization theorem to determine whether each trinomial is factorable over the integers.

23. $8x^2 + 26x + 15$

▶ 24. $16x^2 + 8x - 35$

25. $4x^2 - 5x + 6$

26. $6x^2 + 8x - 3$

27. $6x^2 - 14x + 5$

28. $10x^2 - 4x - 5$

In Exercises 29 to 36, factor over the integers.

29. $x^4 - x^2 - 6$

30. $x^4 + 3x^2 + 2$

31. $x^2y^2 - 2xy - 8$

32. $2x^2y^2 + xy - 1$

33. $3x^4 + 11x^2 - 4$

34. $2x^4 + 3x^2 - 9$

35. $3x^6 + 2x^3 - 8$

▶ 36. $8x^6 - 10x^3 - 3$

In Exercises 37 to 46, factor each difference of squares over the integers.

37. $x^2 - 9$

38. $x^2 - 64$

39. $4a^2 - 49$

▶ 40. $81b^2 - 16c^2$

41. $1 - 100x^2$

42. $1 - 121y^2$

43. $x^4 - 9$

44. $y^4 - 196$

45. $(x + 5)^2 - 4$

46. $(x - 3)^2 - 16$

In Exercises 47 to 54, factor each perfect-square trinomial.

47. $x^2 + 10x + 25$

48. $y^2 + 6y + 9$

49. $a^2 - 14a + 49$

▶ **50.** $b^2 - 24b + 144$

51. $4x^2 + 12x + 9$

52. $25y^2 + 40y + 16$

53. $z^4 + 4z^2w^2 + 4w^4$

54. $9x^4 - 30x^2y^2 + 25y^4$

In Exercises 55 to 62, factor each sum or difference of cubes over the integers.

55. $x^3 - 8$

▶ **56.** $b^3 + 64$

57. $8x^3 - 27y^3$

58. $64u^3 - 27v^3$

59. $8 - x^6$

60. $1 + y^{12}$

61. $(x - 2)^3 - 1$

62. $(y + 3)^3 + 8$

In Exercises 63 to 68, factor (over the integers) by grouping in pairs.

63. $3x^3 + x^2 + 6x + 2$

64. $18w^3 + 15w^2 + 12w + 10$

65. $ax^2 - ax + bx - b$

▶ **66.** $a^2y^2 - ay^3 + ac - cy$

67. $6w^3 + 4w^2 - 15w - 10$

68. $10z^3 - 15z^2 - 4z + 6$

In Exercises 69 to 88, use the general factoring strategy to completely factor each polynomial. If the polynomial does not factor, then state that it is nonfactorable over the integers.

69. $18x^2 - 2$

70. $4bx^3 + 32b$

71. $16x^4 - 1$

▶ **72.** $81y^4 - 16$

73. $12ax^2 - 23axy + 10ay^2$

74. $6ax^2 - 19axy - 20ay^2$

75. $3bx^3 + 4bx^2 - 3bx - 4b$

76. $2x^6 - 2$

77. $72bx^2 + 24bxy + 2by^2$

78. $64y^3 - 16y^2z + yz^2$

79. $(w - 5)^3 + 8$

80. $5xy + 20y - 15x - 60$

81. $x^2 + 6xy + 9y^2 - 1$

82. $4y^2 - 4yz + z^2 - 9$

83. $8x^2 + 3x - 4$

84. $16x^2 + 81$

85. $5x(2x - 5)^2 - (2x - 5)^3$

86. $6x(3x + 1)^3 - (3x + 1)^4$

87. $4x^2 + 2x - y - y^2$

88. $a^2 + a + b - b^2$

CONNECTING CONCEPTS

In Exercises 89 and 90, find all positive values of k such that the trinomial is a perfect-square trinomial.

89. $x^2 + kx + 16$

90. $36x^2 + kxy + 100y^2$

In Exercises 91 and 92, find k such that the trinomial is a perfect-square trinomial.

91. $x^2 + 16x + k$

92. $x^2 - 14xy + ky^2$

In Exercises 93 and 94, use the general strategy to factor each polynomial. In each exercise n represents a positive integer.

93. $x^{4n} - 1$

94. $x^{4n} - 2x^{2n} + 1$

In Exercises 95 to 98, write, in its factored form, the area of the shaded portion of each geometric figure.

95.

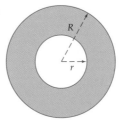

96.

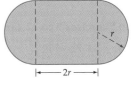

97.

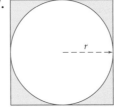

98.

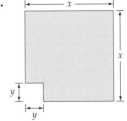

have an LCD of $(x + 3)(2x - 1)$. The rational expressions

$$\frac{5x}{(x + 5)(x - 7)^3} \quad \text{and} \quad \frac{7}{x(x + 5)^2(x - 7)}$$

have an LCD of $x(x + 5)^2(x - 7)^3$.

EXAMPLE 3   Add and Subtract Rational Expressions

Perform the indicated operation and then simplify if possible.

a. $\dfrac{5x}{48} + \dfrac{x}{15}$ b. $\dfrac{x}{x^2 - 4} - \dfrac{2x - 1}{x^2 - 3x - 10}$

Solution

a. Determine the prime factorization of the denominators.

$$48 = 2^4 \cdot 3 \quad \text{and} \quad 15 = 3 \cdot 5$$

The desired common denominator is the product of each of the prime factors raised to its largest power. Thus the common denominator is $2^4 \cdot 3 \cdot 5 = 240$. Write each rational expression as an equivalent rational expression with a denominator of 240.

$$\frac{5x}{48} + \frac{x}{15} = \frac{5x \cdot 5}{48 \cdot 5} + \frac{x \cdot 16}{15 \cdot 16} = \frac{25x}{240} + \frac{16x}{240} = \frac{41x}{240}$$

b. Factor each denominator to determine the LCD of the rational expressions.

$$x^2 - 4 = (x + 2)(x - 2)$$
$$x^2 - 3x - 10 = (x + 2)(x - 5)$$

The LCD is $(x + 2)(x - 2)(x - 5)$. Forming equivalent rational expressions that have the LCD, we have

$$\frac{x}{x^2 - 4} - \frac{2x - 1}{x^2 - 3x - 10}$$

$$= \frac{x(x - 5)}{(x + 2)(x - 2)(x - 5)} - \frac{(2x - 1)(x - 2)}{(x + 2)(x - 5)(x - 2)}$$

$$= \frac{x^2 - 5x - (2x^2 - 5x + 2)}{(x + 2)(x - 2)(x - 5)} = \frac{x^2 - 5x - 2x^2 + 5x - 2}{(x + 2)(x - 2)(x - 5)}$$

$$= \frac{-x^2 - 2}{(x + 2)(x - 2)(x - 5)} = -\frac{x^2 + 2}{(x + 2)(x - 2)(x - 5)}$$

▶ **TRY EXERCISE 30, PAGE 63**

● COMPLEX FRACTIONS

A **complex fraction** is a fraction whose numerator or denominator contains one or more fractions. Complex fractions can be simplified by using one of the following two methods.

Methods for Simplifying Complex Fractions

Method 1: Multiply by 1 in the form $\dfrac{LCD}{LCD}$.

1. Determine the LCD of all the fractions in the complex fraction.

2. Multiply both the numerator and the denominator of the complex fraction by the LCD.

3. If possible, simplify the resulting rational expression.

Method 2: Multiply the numerator by the reciprocal of the denominator.

1. Simplify the numerator to a single fraction and the denominator to a single fraction.

2. Using the definition for dividing fractions, multiply the numerator by the reciprocal of the denominator.

3. If possible, simplify the resulting rational expression.

EXAMPLE 4   **Simplify Complex Fractions**

Simplify: $\dfrac{\dfrac{2}{x-2}+\dfrac{1}{x}}{\dfrac{3x}{x-5}-\dfrac{2}{x-5}}$

Solution

First simplify the numerator to a single fraction and then simplify the denominator to a single fraction.

$$\dfrac{\dfrac{2}{x-2}+\dfrac{1}{x}}{\dfrac{3x}{x-5}-\dfrac{2}{x-5}} = \dfrac{\dfrac{2\cdot x}{(x-2)\cdot x}+\dfrac{1\cdot(x-2)}{x\cdot(x-2)}}{\dfrac{3x-2}{x-5}}$$

• Simplify numerator and denominator.

$$= \dfrac{\dfrac{2x+(x-2)}{x(x-2)}}{\dfrac{3x-2}{x-5}} = \dfrac{\dfrac{3x-2}{x(x-2)}}{\dfrac{3x-2}{x-5}}$$

$$= \dfrac{3x-2}{x(x-2)}\cdot\dfrac{x-5}{3x-2}$$

• Multiply the numerator by the reciprocal of the denominator.

$$= \dfrac{x-5}{x(x-2)}$$

▶ **TRY EXERCISE 42, PAGE 63**

> **EXAMPLE 5**   Simplify a Fraction

Simplify the fraction $\dfrac{c^{-1}}{a^{-1} + b^{-1}}$.

Solution

The fraction written without negative exponents becomes

$$\frac{c^{-1}}{a^{-1} + b^{-1}} = \frac{\dfrac{1}{c}}{\dfrac{1}{a} + \dfrac{1}{b}}$$ • Using $x^{-n} = \dfrac{1}{x^n}$

$$= \frac{\dfrac{1}{c} \cdot abc}{\left(\dfrac{1}{a} + \dfrac{1}{b}\right) abc}$$ • Multiply the numerator and the denominator by abc, which is the LCD of the fraction in the numerator and the fraction in the denominator.

$$= \frac{ab}{bc + ac}$$

> TRY EXERCISE 60, PAGE 64

take note

It is a mistake to write

$$\frac{c^{-1}}{a^{-1} + b^{-1}} \quad \text{as} \quad \frac{a + b}{c}$$

because a^{-1} and b^{-1} are terms and cannot be treated as factors.

● APPLICATION OF RATIONAL EXPRESSIONS

> **EXAMPLE 6**   Solve an Application

The *average speed* for a round trip is given by the complex fraction

$$\frac{2}{\dfrac{1}{v_1} + \dfrac{1}{v_2}}$$

where v_1 is the average speed on the way to your destination and v_2 is the average speed on your return trip. Find the average speed for a round trip if $v_1 = 50$ mph and $v_2 = 40$ mph.

Solution

Evaluate the complex fraction with $v_1 = 50$ and $v_2 = 40$.

$$\frac{2}{\dfrac{1}{v_1} + \dfrac{1}{v_2}} = \frac{2}{\dfrac{1}{50} + \dfrac{1}{40}} = \frac{2}{\dfrac{1 \cdot 4}{50 \cdot 4} + \dfrac{1 \cdot 5}{40 \cdot 5}}$$ • Substitute and simplify the denominator.

$$= \frac{2}{\dfrac{4}{200} + \dfrac{5}{200}} = \frac{2}{\dfrac{9}{200}}$$

$$= 2 \cdot \frac{200}{9} = \frac{400}{9} = 44\frac{4}{9}$$

The average speed for the round trip is $44\dfrac{4}{9}$ mph.

> TRY EXERCISE 64, PAGE 64

? QUESTION In Example 6, why is the speed for the round trip *not* the average of v_1 and v_2?

 TOPICS FOR DISCUSSION

1. Discuss the meaning of the phrase *rational expression*. Is a rational expression the same as a fraction? If not, give some examples of fractions that are not rational expressions.

2. What is the domain of a rational expression?

3. Explain why the following is *not* correct.

$$\frac{2x^2 + 5}{x^2} = 2 + 5 = 7$$

4. Consider the rational expression $\dfrac{x^2 - 3x - 10}{x^2 + x - 30}$. By simplifying this expression, we have

$$\frac{x^2 - 3x - 10}{x^2 + x - 30} = \frac{(x - 5)(x + 2)}{(x - 5)(x + 6)} = \frac{x + 2}{x + 6}$$

Does this really mean that $\dfrac{x^2 - 3x - 10}{x^2 + x - 30} = \dfrac{x + 2}{x + 6}$ for every value of x? If not, for what values of x are the two expressions equal?

? ANSWER Because you were traveling slower on the return trip, the return trip took longer than the time spent going to your destination. More time was spent traveling at the slower speed. Thus the average speed is less than the average of v_1 and v_2.

EXERCISE SET P.5

In Exercises 1 to 10, simplify each rational expression.

1. $\dfrac{x^2 - x - 20}{3x - 15}$

▶ 2. $\dfrac{2x^2 - 5x - 12}{2x^2 + 5x + 3}$

3. $\dfrac{x^3 - 9x}{x^3 + x^2 - 6x}$

4. $\dfrac{x^3 + 125}{2x^3 - 50x}$

5. $\dfrac{a^3 + 8}{a^2 - 4}$

6. $\dfrac{y^3 - 27}{-y^2 + 11y - 24}$

7. $\dfrac{x^2 + 3x - 40}{-x^2 + 3x + 10}$

8. $\dfrac{2x^3 - 6x^2 + 5x - 15}{9 - x^2}$

9. $\dfrac{4y^3 - 8y^2 + 7y - 14}{-y^2 - 5y + 14}$

10. $\dfrac{x^3 - x^2 + x}{x^3 + 1}$

In Exercises 11 to 40, simplify each expression.

11. $\left(-\dfrac{4a}{3b^2}\right)\left(\dfrac{6b}{a^4}\right)$

12. $\left(\dfrac{12x^2y}{5z^4}\right)\left(-\dfrac{25x^2z^3}{15y^2}\right)$

13. $\left(\dfrac{6p^2}{5q^2}\right)^{-1}\left(\dfrac{2p}{3q^2}\right)^2$

14. $\left(\dfrac{4r^2s}{3t^3}\right)^{-1}\left(\dfrac{6rs^3}{5t^2}\right)$

15. $\dfrac{x^2 + x}{2x + 3} \cdot \dfrac{3x^2 + 19x + 28}{x^2 + 5x + 4}$

? QUESTION What is the real part and the imaginary part of $3 - 5i$?

Note from the following diagram that the real numbers are a subset of the complex numbers, and imaginary numbers are a subset of the complex numbers. The real numbers and imaginary numbers are disjoint sets.

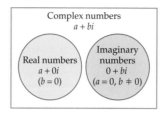

Complex numbers
$a + bi$

Real numbers
$a + 0i$
$(b = 0)$

Imaginary numbers
$0 + bi$
$(a = 0, b \neq 0)$

Example 1 illustrates writing a complex number in the standard form $a + bi$.

EXAMPLE 1 **Write a Complex Number in Standard Form**

Write $7 + \sqrt{-45}$ in the form $a + bi$.

Solution

$$
\begin{aligned}
7 + \sqrt{-45} &= 7 + i\sqrt{45} \\
&= 7 + i\sqrt{9} \cdot \sqrt{5} \\
&= 7 + 3i\sqrt{5}
\end{aligned}
$$

▶ **TRY EXERCISE 8, PAGE 71**

● **ADDITION AND SUBTRACTION OF COMPLEX NUMBERS**

All the standard arithmetic operations that are applied to real numbers can be applied to complex numbers.

Definition of Addition and Subtraction of Complex Numbers

If $a + bi$ and $c + di$ are complex numbers, then

Addition	$(a + bi) + (c + di) = (a + c) + (b + d)i$
Subtraction	$(a + bi) - (c + di) = (a - c) + (b - d)i$

Basically, these rules say that to add two complex numbers, add the real parts and add the imaginary parts. To subtract two complex numbers, subtract the real parts and subtract the imaginary parts.

? ANSWER Real part: 3; imaginary part: -5

EXAMPLE 2   Add or Subtract Complex Numbers

Simplify.

a. $(7 - 2i) + (-2 + 4i)$ **b.** $(-9 + 4i) - (2 - 6i)$

Solution

a. $(7 - 2i) + (-2 + 4i) = (7 + (-2)) + (-2 + 4)i = 5 + 2i$

b. $(-9 + 4i) - (2 - 6i) = (-9 - 2) + (4 - (-6))i = -11 + 10i$

▶ **TRY EXERCISE 18, PAGE 72**

● **MULTIPLICATION OF COMPLEX NUMBERS**

When multiplying complex numbers, the term i^2 is frequently a part of the product. Recall that $i^2 = -1$. Therefore,

$$3i(5i) = 15i^2 = 15(-1) = -15$$
$$-2i(6i) = -12i^2 = -12(-1) = 12$$
$$4i(3 - 2i) = 12i - 8i^2 = 12i - 8(-1) = 8 + 12i$$

When multiplying square roots of negative numbers, first rewrite the radical expressions using i. For instance,

$$\sqrt{-6} \cdot \sqrt{-24} = i\sqrt{6} \cdot i\sqrt{24} \qquad \bullet \sqrt{-6} = i\sqrt{6}, \sqrt{-24} = i\sqrt{24}$$
$$= i^2\sqrt{144} = -1 \cdot 12$$
$$= -12$$

Note from this example that it would have been incorrect to multiply the radicands of the two radical expressions. To illustrate:

$$\sqrt{-6} \cdot \sqrt{-24} \neq \sqrt{(-6)(-24)}$$

❓ QUESTION What is the product of $\sqrt{-2}$ and $\sqrt{-8}$?

To multiply two complex numbers, we use the following definition.

Definition of Multiplication of Complex Numbers

If $a + bi$ and $c + di$ are complex numbers, then
$$(a + bi)(c + di) = (ac - bd) + (ad + bc)i$$

Because every complex number can be written as a sum of two terms, it is natural to perform multiplication on complex numbers in a manner consistent

take note

Recall that the definition of the product of radical expressions required that the radicand be a positive number. Therefore, when multiplying expressions containing negative radicands, we must first rewrite the expression using i and a positive radicand.

❓ ANSWER $\sqrt{-2} \cdot \sqrt{-8} = i\sqrt{2} \cdot i\sqrt{8} = i^2\sqrt{16} = -1 \cdot 4 = -4$

with the operation defined on binomials and the definition $i^2 = -1$. By using this analogy, you can multiply complex numbers without memorizing the definition.

EXAMPLE 3 — Multiply Complex Numbers

Simplify.   **a.** $(3 - 4i)(2 + 5i)$ **b.** $\left(2 + \sqrt{-3}\right)\left(4 - 5\sqrt{-3}\right)$

Solution

a.
$$
\begin{aligned}
(3 - 4i)(2 + 5i) &= 6 + 15i - 8i - 20i^2 \\
&= 6 + 15i - 8i - 20(-1) && \text{• Replace } i^2 \text{ by } -1. \\
&= 6 + 15i - 8i + 20 && \text{• Simplify.} \\
&= 26 + 7i
\end{aligned}
$$

b.
$$
\begin{aligned}
\left(2 + \sqrt{-3}\right)\left(4 - 5\sqrt{-3}\right) &= \left(2 + i\sqrt{3}\right)\left(4 - 5i\sqrt{3}\right) \\
&= 8 - 10i\sqrt{3} + 4i\sqrt{3} - 5i^2\sqrt{9} \\
&= 8 - 10i\sqrt{3} + 4i\sqrt{3} - 5(-1)(3) \\
&= 8 - 10i\sqrt{3} + 4i\sqrt{3} + 15 = 23 - 6i\sqrt{3}
\end{aligned}
$$

▶ **TRY EXERCISE 34, PAGE 72**

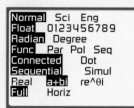

INTEGRATING TECHNOLOGY

Some graphing calculators can be used to perform operations on complex numbers. Here are some typical screens for a TI-83 Plus.

Press $\boxed{\text{MODE}}$. Use the down arrow key to highlight $a + bi$.

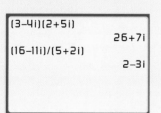

Press $\boxed{\text{ENTER}}$ $\boxed{\text{2nd}}$ [QUIT].
Here are two examples of computations on complex numbers. To enter an i, use $\boxed{\text{2nd}}$ [i], which is above the decimal point key.

```
(3-4i)(2+5i)
                26+7i
(16-11i)/(5+2i)
                 2-3i
```

● DIVISION OF COMPLEX NUMBERS

Recall that the number $\dfrac{3}{\sqrt{2}}$ is not in simplest form because there is a radical expression in the denominator. Similarly, $\dfrac{3}{i}$ is not in simplest form because $i = \sqrt{-1}$. To write this expression in simplest form, multiply the numerator and denominator by i.

$$
\frac{3}{i} \cdot \frac{i}{i} = \frac{3i}{i^2} = \frac{3i}{-1} = -3i
$$

Here is another example.

$$
\frac{3 - 6i}{2i} = \frac{3 - 6i}{2i} \cdot \frac{i}{i} = \frac{3i - 6i^2}{2i^2} = \frac{3i - 6(-1)}{2(-1)}
$$
$$
= \frac{3i + 6}{-2} = -3 - \frac{3}{2}i
$$

Recall that to simplify the quotient $\dfrac{2 + \sqrt{3}}{5 + 2\sqrt{3}}$, we multiply the numerator and denominator by the conjugate of $5 + 2\sqrt{3}$, which is $5 - 2\sqrt{3}$. In a similar manner, to find the quotient of two complex numbers, we multiply the numerator and denominator by the conjugate of the denominator.

The complex numbers $a + bi$ and $a - bi$ are called **complex conjugates** or **conjugates** of each other. The conjugate of the complex number z is denoted by $\bar{z}$. For instance,

$$
\overline{2 + 5i} = 2 - 5i \qquad \text{and} \qquad \overline{3 - 4i} = 3 + 4i
$$

Consider the product of a complex number and its conjugate. For instance,

$$(2 + 5i)(2 - 5i) = 4 - 10i + 10i - 25i^2$$
$$= 4 - 25(-1) = 4 + 25$$
$$= 29$$

Note that the product is a *real* number. This is always true.

Product of Complex Conjugates

The product of a complex number and its conjugate is a real number. That is, $(a + bi)(a - bi) = a^2 + b^2$.

For instance, $(5 + 3i)(5 - 3i) = 5^2 + 3^2 = 25 + 9 = 34$.

The next example shows how the quotient of two complex numbers is determined by using conjugates.

EXAMPLE 4   Divide Complex Numbers

Simplify: $\dfrac{16 - 11i}{5 + 2i}$

Solution

$$\frac{16 - 11i}{5 + 2i} = \frac{16 - 11i}{5 + 2i} \cdot \frac{5 - 2i}{5 - 2i}$$

• Multiply numerator and denominator by the conjugates of the denominator.

$$= \frac{80 - 32i - 55i + 22i^2}{5^2 + 2^2}$$

$$= \frac{80 - 32i - 55i + 22(-1)}{25 + 4}$$

$$= \frac{80 - 87i - 22}{29}$$

$$= \frac{58 - 87i}{29}$$

$$= \frac{29(2 - 3i)}{29} = 2 - 3i$$

▶ **TRY EXERCISE 48, PAGE 72**

● POWERS OF *i*

The following powers of *i* illustrate a pattern:

$$i^1 = i \qquad\qquad i^5 = i^4 \cdot i = 1 \cdot i = i$$
$$i^2 = -1 \qquad\qquad i^6 = i^4 \cdot i^2 = 1(-1) = -1$$
$$i^3 = i^2 \cdot i = (-1)i = -i \qquad\qquad i^7 = i^4 \cdot i^3 = 1(-i) = -i$$
$$i^4 = i^2 \cdot i^2 = (-1)(-1) = 1 \qquad\qquad i^8 = (i^4)^2 = 1^2 = 1$$

Because 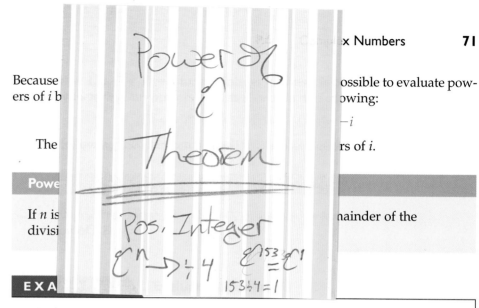 ossible to evaluate powers of i b   owing:

$-i$

The   rs of i.

Powe

If n is   nainder of the
divisi

E X A

Evaluate: i^{153}

Solution

Use the powers of i theorem.

$$i^{153} = i^1 = i \qquad \bullet \text{ Remainder of } 153 \div 4 \text{ is } 1.$$

▶ **TRY EXERCISE 60, PAGE 72**

TOPICS FOR DISCUSSION

1. What is an imaginary number? What is a complex number?

2. How are the real numbers related to the complex numbers?

3. Is zero a complex number?

4. What is the conjugate of a complex number?

5. If a and b are real numbers and $ab = 0$, then $a = 0$ or $b = 0$. Is the same true for complex numbers? That is, if u and v are complex numbers and $uv = 0$, must one of the numbers be zero?

EXERCISE SET P.6

In Exercises 1 to 10, write the complex number in standard form.

1. $\sqrt{-81}$

2. $\sqrt{-64}$

3. $\sqrt{-98}$

4. $\sqrt{-27}$

5. $\sqrt{16} + \sqrt{-81}$

6. $\sqrt{25} + \sqrt{-9}$

7. $5 + \sqrt{-49}$

▶ **8.** $6 - \sqrt{-1}$

9. $8 - \sqrt{-18}$

10. $11 + \sqrt{-48}$

In Exercises 11 to 36, simplify and write the complex number in standard form.

11. $(5 + 2i) + (6 - 7i)$

12. $(4 - 8i) + (5 + 3i)$

13. $(-2 - 4i) - (5 - 8i)$

14. $(3 - 5i) - (8 - 2i)$

15. $(1 - 3i) + (7 - 2i)$

16. $(2 - 6i) + (4 - 7i)$

17. $(-3 - 5i) - (7 - 5i)$

▶ **18.** $(5 - 3i) - (2 + 9i)$

19. $8i - (2 - 8i)$

20. $3 - (4 - 5i)$

21. $5i \cdot 8i$

22. $(-3i)(2i)$

23. $\sqrt{-50} \cdot \sqrt{-2}$

24. $\sqrt{-12} \cdot \sqrt{-27}$

25. $3(2 + 5i) - 2(3 - 2i)$

26. $3i(2 + 5i) + 2i(3 - 4i)$

27. $(4 + 2i)(3 - 4i)$

28. $(6 + 5i)(2 - 5i)$

29. $(-3 - 4i)(2 + 7i)$

30. $(-5 - i)(2 + 3i)$

31. $(4 - 5i)(4 + 5i)$

32. $(3 + 7i)(3 - 7i)$

33. $\left(3 + \sqrt{-4}\right)\left(2 - \sqrt{-9}\right)$

▶ **34.** $\left(5 + 2\sqrt{-16}\right)\left(1 - \sqrt{-25}\right)$

35. $\left(3 + 2\sqrt{-18}\right)\left(2 + 2\sqrt{-50}\right)$

36. $\left(5 - 3\sqrt{-48}\right)\left(2 - 4\sqrt{-27}\right)$

In Exercises 37 to 54, write each expression as a complex number in standard form.

37. $\dfrac{6}{i}$

38. $\dfrac{-8}{2i}$

39. $\dfrac{6 + 3i}{i}$

40. $\dfrac{4 - 8i}{4i}$

41. $\dfrac{1}{7 + 2i}$

42. $\dfrac{5}{3 + 4i}$

43. $\dfrac{2i}{1 + i}$

44. $\dfrac{5i}{2 - 3i}$

45. $\dfrac{5 - i}{4 + 5i}$

46. $\dfrac{4 + i}{3 + 5i}$

47. $\dfrac{3 + 2i}{3 - 2i}$

▶ **48.** $\dfrac{8 - i}{2 + 3i}$

49. $\dfrac{-7 + 26i}{4 + 3i}$

50. $\dfrac{-4 - 39i}{5 - 2i}$

51. $(3 - 5i)^2$

52. $(2 + 4i)^2$

53. $(1 + 2i)^3$

54. $(2 - i)^3$

In Exercises 55 to 62, evaluate the power of i.

55. i^{15} **56.** i^{66} **57.** $-i^{40}$ **58.** $-i^{51}$

59. $\dfrac{1}{i^{25}}$ ▶ **60.** $\dfrac{1}{i^{83}}$ **61.** i^{-34} **62.** i^{-52}

In Exercises 63 to 68, evaluate $\dfrac{-b + \sqrt{b^2 - 4ac}}{2a}$ **for the given values of a, b, and c. Write your answer as a complex number in standard form.**

63. $a = 3, b = -3, c = 3$

64. $a = 2, b = 4, c = 4$

65. $a = 2, b = 6, c = 6$

66. $a = 2, b = 1, c = 3$

67. $a = 4, b = -4, c = 2$

68. $a = 3, b = -2, c = 4$

CONNECTING CONCEPTS

The property that the product of conjugates of the form $(a + bi)(a - bi)$ is equal to $a^2 + b^2$ can be used to factor the sum of two perfect squares over the set of complex numbers. For example, $x^2 + y^2 = (x + yi)(x - yi)$. In Exercises 69 to 74, factor the binomial over the set of complex numbers.

69. $x^2 + 16$

70. $x^2 + 9$

71. $z^2 + 25$

72. $z^2 + 64$

73. $4x^2 + 81$

74. $9x^2 + 1$

75. Show that if $x = 1 + 2i$, then $x^2 - 2x + 5 = 0$.

76. Show that if $x = 1 - 2i$, then $x^2 - 2x + 5 = 0$.

77. When we think of the cube root of 8, $\sqrt[3]{8}$, we normally mean the *real* cube root of 8 and write $\sqrt[3]{8} = 2$. However, there are two other cube roots of 8 that are complex numbers. Verify that $-1 + i\sqrt{3}$ and $-1 - i\sqrt{3}$ are cube roots of 8 by showing that $\left(-1 + i\sqrt{3}\right)^3 = 8$ and $\left(-1 - i\sqrt{3}\right)^3 = 8$.

78. It is possible to find the square root of a complex number.

Verify that $\sqrt{i} = \dfrac{\sqrt{2}}{2}(1 + i)$ by showing that

$$\left[\frac{\sqrt{2}}{2}(1 + i)\right]^2 = i.$$

79. Simplify $i + i^2 + i^3 + i^4 + \cdots + i^{28}$.

80. Simplify $i + i^2 + i^3 + i^4 + \cdots + i^{100}$.

PROJECTS

ARGAND DIAGRAM Just as we can graph a real number on a real number line, we can graph a complex number. This is accomplished by using one number line for the real part of the complex number and one number line for the imaginary part of the complex number. These two number lines are drawn perpendicular to each other and pass through their respective origins, as shown below.

The result is called the *complex plane* or an *Argand diagram* after Jean-Robert Argand (1768–1822), an accountant and amateur mathematician. Although he is given credit for this representation of complex numbers, Caspar Wessel (1745–1818) actually conceived the idea before Argand.

To graph the complex number $3 + 4i$, start at 3 on the real axis. Now move 4 units up (for positive numbers move up, for negative numbers move down) and place a dot at that point, as shown in the diagram. Graphs of several other complex numbers are also shown.

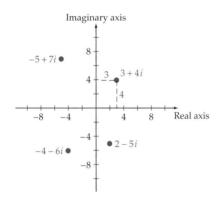

In Exercises 1 to 8, graph the complex number.

1. $2 + 5i$ **2.** $4 - 3i$ **3.** $-2 + 6i$ **4.** $-3 - 5i$

5. 4 **6.** $-2i$ **7.** $3i$ **8.** -5

The absolute value of a complex number is given by $|a + bi| = \sqrt{a^2 + b^2}$. In Exercises 9 to 12, find the absolute value of the complex number.

9. $2 + 5i$ **10.** $4 - 3i$ **11.** $-2 + 6i$ **12.** $-3 - 5i$

13. The additive inverse of $a + bi$ is $-a - bi$. Show that the absolute value of a complex number and the absolute value of its additive inverse are equal.

14. A *real* number and its additive inverse are the same distance from zero but on opposite sides of zero on a real number line. Describe the relationship between the graphs of a complex number and its additive inverse.

EXPLORING CONCEPTS WITH TECHNOLOGY

Can You Trust Your Calculator?

You may think that your calculator always produces correct results in a *predictable* manner. However, the following experiment may change your opinion.

First note that the algebraic expression

$$p + 3p(1 - p)$$

is equal to the expression

$$4p - 3p^2$$

Use a graphing calculator to evaluate both expressions with $p = 0.05$. You should find that both expressions equal 0.1925. So far we do not observe any unexpected results. Now replace p in each expression with the current value of that expression (0.1925 in this case). This is called *feedback* because we are feeding our outputs back into each expression as inputs. Each new evaluation is referred to as an *iteration*. This time each expression takes on the value 0.65883125. Still no surprises. Continue the feedback process. That is, replace p in each expression *with the current value* of that expression. Now each expression takes on the value 1.33314915207, as shown in the following table. The iterations were performed on a *TI-85* calculator.

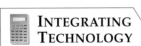

INTEGRATING TECHNOLOGY

To perform these iterations with a *TI* graphing calculator, first store 0.05 in p and then store $p + 3p(1 - p)$ in p as shown below.

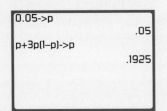

Each time you press ENTER, the expression $p + 3p(1 - p)$ will be evaluated with p equal to the previous result.

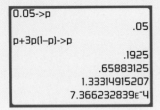

Iteration	$p + 3p(1 - p)$	$4p - 3p^2$
1	0.1925	0.1925
2	0.65883125	0.65883125
3	1.33314915207	1.33314915207

The following table shows that if we continue this feedback process on a calculator, the expressions $p + 3p(1 - p)$ and $4p - 3p^2$ will start to take on different values starting with the fourth iteration. By the 37th iteration, the values do not even agree to two decimal places.

Iteration	$p + 3p(1 - p)$	$4p - 3p^2$
4	7.366232839E-4	7.366232838E-4
5	0.002944865294	0.002944865294
6	0.011753444481	0.0117534448
7	0.046599347553	0.046599347547
20	1.12135618652	1.12135608405
30	0.947163304835	0.947033128433
37	0.285727963839	0.300943417861

1. Use a calculator to find the first 20 iterations of $p + 3p(1 - p)$ and $4p - 3p^2$, with the initial value of $p = 0.5$.

2. Write a report on chaos and fractals. Include information on the "butterfly effect." An excellent source is *Chaos and Fractals, New Frontiers of Science* by Heinz-Otto Peitgen, Hartmut Jurgens, and Dietmar Saupe (New York: Springer-Verlag, 1992).

3. Equations of the form $p_{n+1} = p_n + rp_n(1 - p_n)$ are called Verhulst population models. Write a report on Verhulst population models.

CHAPTER P SUMMARY

P.1 The Real Number System

- The following sets of numbers are used extensively in algebra:

Integers	$\{\ldots, -3, -2, -1, 0, 1, 2, 3, \ldots\}$
Rational numbers	{all terminating or repeating decimals}
Irrational numbers	{all nonterminating, nonrepeating decimals}
Real numbers	{all rational and irrational numbers}

- Set-builder notation is a method of writing sets and has the form {variable | condition on the variable}.

- Union and intersection are two operations on sets.

- The absolute value of a real number a is given by

$$|a| = \begin{cases} a, & a \geq 0 \\ -a, & a < 0 \end{cases}$$

- For any two real numbers a and b, the distance between the graph of a and the graph of b is given by $d(a, b) = |a - b|$.

- The Order of Operations Agreement is used to simplify expressions.

P.2 Integer and Rational Number Exponents

- If b is any real number and n is any natural number, then

$$b^n = \underbrace{b \cdot b \cdot b \cdot \cdots \cdot b}_{n \text{ factors of } b}$$

- For any nonzero real number b, $b^0 = 1$.

- If $b \neq 0$ and n is any natural number, then $b^{-n} = \dfrac{1}{b^n}$ and $\dfrac{1}{b^{-n}} = b^n$.

- **Properties of Rational Exponents**
  If p, q, and r represent rational numbers, and a and b are positive real numbers, then

  Product $\quad b^p \cdot b^q = b^{p+q}$

  Quotient $\quad \dfrac{b^p}{b^q} = b^{p-q}$

  Power $\quad (b^p)^q = b^{pq} \quad (a^p b^q)^r = a^{pr} b^{qr}$

  $\quad\quad\quad\ \left(\dfrac{a^p}{b^q}\right)^r = \dfrac{a^{pr}}{b^{qr}} \quad b^{-p} = \dfrac{1}{b^p}$

- **Properties of Radicals**
  If m and n are natural numbers, and a and b are positive real numbers, then

  Product $\quad \sqrt[n]{a} \cdot \sqrt[n]{b} = \sqrt[n]{ab}$

  Quotient $\quad \dfrac{\sqrt[n]{a}}{\sqrt[n]{b}} = \sqrt[n]{\dfrac{a}{b}}$

  Index $\quad \sqrt[m]{\sqrt[n]{b}} = \sqrt[mn]{b}$

P.3 Polynomials

- A polynomial is an expression of the form

$$a_n x^n + a_{n-1} x^{n-1} + \cdots + a_2 x^2 + a_1 x + a_0$$

- Special product formulas are as follows:

Special Form	Formula(s)
(Sum)(Difference)	$(x + y)(x - y) = x^2 - y^2$
(Binomial)^2	$(x + y)^2 = x^2 + 2xy + y^2$ $(x - y)^2 = x^2 - 2xy + y^2$

P.4 Factoring

- Factoring formulas are as follows:

Special Form	Formula(s)
Difference of two squares	$x^2 - y^2 = (x + y)(x - y)$
Perfect-square trinomials	$x^2 + 2xy + y^2 = (x + y)^2$ $x^2 - 2xy + y^2 = (x - y)^2$
Sum of cubes	$x^3 + y^3 = (x + y)(x^2 - xy + y^2)$
Difference of cubes	$x^3 - y^3 = (x - y)(x^2 + xy + y^2)$

- To factor a polynomial, use the general factoring strategy.

P.5 Rational Expressions

- A rational expression is a fraction in which the numerator and denominator are polynomials. The properties of rational expressions are used to simplify a rational expression and to find the sum, difference, product, and quotient of two rational expressions.

- Complex fractions can be simplified in either of the following ways:

  Method 1: Multiply both the numerator and the denominator by the LCD of all the fractions in the complex fraction.

  Method 2: Simplify the numerator to a single fraction and the denominator to a single fraction. Multiply the numerator by the reciprocal of the denominator.

P.6 Complex Numbers

- The number i, called the *imaginary unit*, is the number such that $i^2 = -1$.

- If a is a positive real number, then $\sqrt{-a} = i\sqrt{a}$. The number $i\sqrt{a}$ is called an *imaginary number*.

- A *complex number* is a number of the form $a + bi$, where a and b are real numbers and $i = \sqrt{-1}$. The number a is the *real part* of $a + bi$, and b is the *imaginary part*.

- The complex numbers $a + bi$ and $a - bi$ are called *complex conjugates* or *conjugates* of each other.

- **Operations on Complex Numbers**

  $(a + bi) + (c + di) = (a + c) + (b + d)i$

  $(a + bi) - (c + di) = (a - c) + (b - d)i$

  $(a + bi)(c + di) = (ac - bd) + (ad + bc)i$

  $\dfrac{a + bi}{c + di} = \dfrac{a + bi}{c + di} \cdot \dfrac{c - di}{c - di}$ • **Multiply numerator and denominator by the conjugate of the denominator.**

CHAPTER P TRUE/FALSE EXERCISES

In Exercises 1 to 10, answer true or false. If the statement is false, give an example or a reason to show that the statement is false.

1. If a and b are real numbers, then $|a - b| = |b - a|$.

2. If a is a real number, then $a^2 \geq a$.

3. The set of rational numbers is closed under the operation of addition.

4. The set of irrational numbers is closed under the operation of addition.

5. Let $x \oplus y$ denote the average of the two real numbers x and y. That is,

$$x \oplus y = \frac{x + y}{2}$$

The operation $\oplus$ is an associative operation because $(x \oplus y) \oplus z = x \oplus (y \oplus z)$ for all real numbers x, y, and z.

6. Using interval notation, we write the inequality $x > a$ as $[a, \infty)$.

7. If n is a real number, then $\sqrt{n^2} = n$.

8. $(a + b)^2 = a^2 + b^2$

9. $\sqrt[3]{a^3 + b^3} = a + b$

10. $\sqrt{-2}\sqrt{-8} = 4$

CHAPTER P REVIEW EXERCISES

In Exercises 1 to 4, classify each number as one or more of the following: integer, rational number, irrational number, real number, prime number, composite number.

1. 3   2. $\sqrt{7}$ 3. $-\dfrac{1}{2}$ 4. $0.\overline{5}$

In Exercises 5 and 6, use $A = \{1, 5, 7\}$ and $B = \{2, 3, 5, 11\}$ to find the indicated intersection or union.

5. $A \cup B$ 6. $A \cap B$

In Exercises 7 to 14, identify the real number property or the property of equality that is illustrated.

7. $5(x + 3) = 5x + 15$

8. $a(3 + b) = a(b + 3)$

9. $(6c)d = 6(cd)$

10. $\sqrt{2} + 3$ is a real number.

11. $7 + 0 = 7$

12. $1x = x$

13. If $7 = x$, then $x = 7$.

14. If $3x + 4 = y$, and $y = 5z$, then $3x + 4 = 5z$.

In Exercises 15 and 16, graph each inequality and write the inequality using interval notation.

15. $-4 < x \le 2$

16. $x \le -1$ or $x > 3$

In Exercises 17 and 18, graph each interval and write each interval as an inequality.

17. $[-3, 2)$

18. $(-1, \infty)$

In Exercises 19 to 22, write each real number without absolute value symbols.

19. $|7|$ **20.** $|2 - \pi|$ **21.** $|4 - \pi|$ **22.** $|-11|$

In Exercises 23 and 24, find the distance on the real number line between the points whose coordinates are given.

23. $-3, 14$

24. $\sqrt{5}, -\sqrt{2}$

In Exercises 25 and 26, evaluate each expression.

25. $-5^2 + (-11)$

26. $\dfrac{(2^2 \cdot 3^{-2})^2}{3^{-1} \cdot 2^3}$

In Exercises 27 and 28, simplify each expression.

27. $(3x^2y)(2x^3y)^2$

28. $\left(\dfrac{2a^2b^3c^{-2}}{3ab^{-1}} \right)^2$

In Exercises 29 and 30, evaluate each exponential expression.

29. $25^{1/2}$

30. $-27^{2/3}$

In Exercises 31 to 34, simplify each expression.

31. $x^{2/3} \cdot x^{3/4}$

32. $\left(\dfrac{8x^{5/4}}{x^{1/2}} \right)^{2/3}$

33. $\left(\dfrac{x^2y}{x^{1/2}y^{-3}} \right)^{1/2}$

34. $(x^{1/2} - y^{1/2})(x^{1/2} + y^{1/2})$

In Exercises 35 to 44, simplify each radical expression. Assume the variables are positive real numbers.

35. $\sqrt{48a^2b^7}$

36. $\sqrt{12a^3b}$

37. $\sqrt{72x^2y}$

38. $\sqrt{18x^3y^5}$

39. $\sqrt{\dfrac{54xy^3}{10x}}$

40. $-\sqrt{\dfrac{24xyz^3}{15z^6}}$

41. $\dfrac{7x}{\sqrt[3]{2x^2}}$

42. $\dfrac{5y}{\sqrt[3]{9y}}$

43. $\sqrt[3]{-135x^2y^7}$

44. $\sqrt[3]{-250xy^6}$

In Exercises 45 and 46, write each number in scientific notation.

45. $620,000$

46. 0.0000017

In Exercises 47 and 48, change each number from scientific notation to decimal form.

47. 3.5×10^4

48. 4.31×10^{-7}

In Exercises 49 to 52, perform the indicated operation and express each result as a polynomial in standard form.

49. $(2a^2 + 3a - 7) + (-3a^2 - 5a + 6)$

50. $(5b^2 - 11) - (3b^2 - 8b - 3)$

51. $(2x^2 + 3x - 5)(3x^2 - 2x + 4)$

52. $(3y - 5)^3$

In Exercises 53 to 56, completely factor each polynomial over the integers.

53. $3x^2 + 30x + 75$

54. $25x^2 - 30xy + 9y^2$

55. $20a^2 - 4b^2$

56. $16a^3 + 250$

In Exercises 57 and 58, simplify each rational expression.

57. $\dfrac{6x^2 - 19x + 10}{2x^2 + 3x - 20}$

58. $\dfrac{4x^3 - 25x}{8x^4 + 125x}$

In Exercises 59 to 62, perform the indicated operation and simplify if possible.

59. $\dfrac{10x^2 + 13x - 3}{6x^2 - 13x - 5} \cdot \dfrac{6x^2 + 5x + 1}{10x^2 + 3x - 1}$

60. $\dfrac{15x^2 + 11x - 12}{25x^2 - 9} \div \dfrac{3x^2 + 13x + 12}{10x^2 + 11x + 3}$

61. $\dfrac{x}{x^2 - 9} + \dfrac{2x}{x^2 + x - 12}$

62. $\dfrac{3x}{x^2 + 7x + 12} - \dfrac{x}{2x^2 + 5x - 3}$

In Exercises 63 and 64, simplify each complex fraction.

63. $\dfrac{2 + \dfrac{1}{x - 5}}{3 - \dfrac{2}{x - 5}}$

64. $\dfrac{1}{2 + \dfrac{3}{1 + \dfrac{4}{x}}}$

In Exercises 65 and 66, write the complex number in standard form.

65. $5 + \sqrt{-64}$

66. $2 - \sqrt{-18}$

In Exercises 67 to 74, perform the indicated operation and write the answer in simplest form.

67. $(2 - 3i) + 4 + 2i$

68. $(4 + 7i) - (6 - 3i)$

69. $2i(3 - 4i)$

70. $(4 - 3i)(2 + 7i)$

71. $(3 + i)^2$

72. i^{345}

73. $\dfrac{4 - 6i}{2i}$

74. $\dfrac{2 - 5i}{3 + 4i}$

CHAPTER P TEST

1. For real numbers a, b, and c, identify the property that is illustrated by $(a + b)c = ac + bc$.

2. Given $A = \{0, 2, 4, 6, 8\}$ and $B = \{1, 3, 5, 7, 9\}$, find $A \cup B$.

3. Find the distance between the points -12 and -5 on the number line.

4. Simplify: $(-2x^0y^{-2})^2(-3x^2y^{-1})^{-2}$

5. Simplify: $\dfrac{(2a^{-1}bc^{-2})^2}{(3^{-1}b)(2^{-1}ac^{-2})^3}$

6. Write 0.00137 in scientific notation.

7. Simplify: $\dfrac{x^{1/3}y^{-3/4}}{x^{-1/2}y^{3/2}}$

8. Simplify: $3x\sqrt[3]{81xy^4} - 2y\sqrt[3]{3x^4y}$

9. Simplify: $\dfrac{x}{\sqrt[4]{2x^3}}$

10. Simplify: $\dfrac{3}{\sqrt{x} + 2}$

11. Simplify: $(x - 2y)(x^2 - 2x + y)$

12. Evaluate the polynomial $3y^3 - 2y^2 - y + 2$ for $y = -3$.

13. Factor: $7x^2 + 34x - 5$

14. Factor: $3ax - 12bx - 2a + 8b$

15. Factor: $16x^4 - 2xy^3$

16. Simplify: $\dfrac{x^2 - 2x - 15}{25 - x^2}$

17. Simplify: $\dfrac{x}{x^2 + x - 6} - \dfrac{2}{x^2 - 5x + 6}$

18. Simplify: $\dfrac{2x^2 + 3x - 2}{x^2 - 3x} \div \dfrac{2x^2 - 7x + 3}{x^3 - 3x^2}$

19. Simplify: $\dfrac{3}{a + b} \cdot \dfrac{a^2 - b^2}{2a - b} - \dfrac{5}{a}$

20. Simplify: $x - \dfrac{x}{x + \dfrac{1}{2}}$

❓ QUESTION If $ax + b = c$, does $x = \dfrac{c}{a} - b$?

● APPLICATIONS

Linear equations emerge in a variety of application problems. In solving such problems, it generally helps to apply specific techniques in a series of small steps. The following general strategies should prove to be helpful in the remaining portion of this section.

Strategies for Solving Application Problems

1. Read the problem carefully. If necessary, reread the problem several times.

2. When appropriate, draw a sketch and label parts of the drawing with the specific information given in the problem.

3. Determine the unknown quantities, and label them with variables. Write down any equation that relates the variables.

4. Use the information from step 3, along with a known formula or some additional information given in the problem, to write an equation.

5. Solve the equation obtained in step 4, and check to see whether these results satisfy all the conditions of the original problem.

EXAMPLE 3   Dimensions of a Painting

One of the best known paintings is *Mona Lisa* by Leonardo da Vinci. It is on display at the Musee du Louvre, in Paris. The length (or height) of this rectangular-shaped painting is 24 centimeters more than its width. The perimeter of the painting is 260 centimeters. Find the width and the length of the painting.

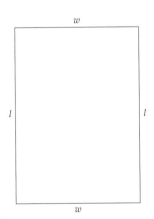

Solution

1. Read the problem carefully.

2. Draw a rectangle. See **Figure 1.3**.

3. Label the rectangle. We have used w for its width and l for its length. The problem states that the length is 24 centimeters more than the width. Thus l and w are related by the equation

$$l = w + 24$$

Continued ▶

FIGURE 1.3

❓ ANSWER No. $x = \dfrac{c - b}{a}$, provided $a \neq 0$.

4. The perimeter of a rectangle is given by the formula $P = 2l + 2w$. To produce an equation that involves only constants and a single variable (say, w), substitute 260 for P and $w + 24$ for l.

$$P = 2l + 2w$$
$$260 = 2(w + 24) + 2w$$

5. Solve for w.

$$260 = 2w + 48 + 2w$$
$$260 = 4w + 48$$
$$212 = 4w$$
$$w = 53$$

The length is 24 centimeters more than the width. Thus $l = 53 + 24 = 77$.

A check verifies that 77 is 24 more than 53 and that twice the length (77) plus twice the width (53) gives the perimeter (260). The width of the painting is 53 centimeters and its length is 77 centimeters.

▶ **TRY EXERCISE 20, PAGE 99**

Many *uniform motion* problems can be solved by using the formula $d = rt$, where d is the distance traveled, r is the rate of speed, and t is the time.

EXAMPLE 4 **A Uniform Motion Problem**

A runner runs a course at a constant speed of 6 mph. One hour after the runner begins, a cyclist starts on the same course at a constant speed of 15 mph. How long after the runner starts does the cyclist overtake the runner?

Solution

If we represent the time the runner has spent on the course by t, then the time the cyclist takes to overtake the runner is $t - 1$. The following table organizes the information and helps us determine how to write the distances each person travels.

	rate r	·	time t	=	distance d
Runner	6	·	t	=	$6t$
Cyclist	15	·	$t - 1$	=	$15(t - 1)$

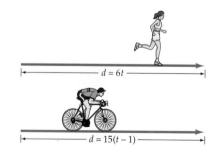

$d = 6t$

$d = 15(t - 1)$

FIGURE 1.4

Figure 1.4 indicates that the runner and the cyclist cover the same distance. Thus

$$6t = 15(t - 1)$$
$$6t = 15t - 15$$
$$-9t = -15$$
$$t = 1\frac{2}{3}$$

The cyclist overtakes the runner $1\frac{2}{3}$ hours after the runner starts.

▶ **TRY EXERCISE 24, PAGE 99**

Many business applications can be solved by using the equation

$$\text{Profit} = \text{revenue} - \text{cost}$$

EXAMPLE 5 A Business Application

It costs a tennis shoe manufacturer $26.55 to produce a pair of tennis shoes that sells for $49.95. How many pairs of tennis shoes must the manufacturer sell to make a profit of $14,274.00?

Solution

The *profit* is equal to the *revenue* minus the *cost*. If x equals the number of pairs of tennis shoes to be sold, then the revenue will be $49.95x$ and the cost will be $26.55x$. Therefore,

$$\text{Profit} = \text{revenue} - \text{cost}$$
$$14{,}274.00 = 49.95x - 26.55x$$
$$14{,}274.00 = 23.40x$$
$$610 = x$$

The manufacturer must sell 610 pairs of tennis shoes to make the desired profit.

▶ **TRY EXERCISE 32, PAGE 100**

Simple interest problems can be solved by using the formula $I = Prt$, where I is the interest, P is the principal, r is the simple interest rate per period, and t is the number of periods.

EXAMPLE 6 An Investment Problem

An accountant invests part of a $6000 bonus in a 5% simple interest account and the remainder of the money is invested at 8.5% simple interest. Together the investments earn $370 per year. Find the amount invested at each rate.

Solution

Let x be the amount invested at 5%. The remainder of the money is $6000 - x$, which will be the amount invested at 8.5%. Using the simple interest formula, $I = Prt$, with $t = 1$ year, yields

$$\text{Interest at 5\%} = x \cdot 0.05 = 0.05x$$
$$\text{Interest at 8.5\%} = (6000 - x) \cdot (0.085) = 510 - 0.085x$$

The interest earned on the two accounts equals $370.

$$0.05x + (510 - 0.085x) = 370$$
$$-0.035x + 510 = 370$$
$$-0.035x = -140$$
$$x = 4000$$

The accountant invested $4000 at 5% and the remaining $2000 at 8.5%. Check as before.

▶ **TRY EXERCISE 36, PAGE 100**

Percent mixture problems involve combining solutions or alloys that have different concentrations of a common substance. Percent mixture problems can be solved by using the formula $pA = Q$, where p is the percent of concentration, A is the amount of the solution or alloy, and Q is the quantity of a substance in the solution or alloy. For example, in 4 liters of a 25% acid solution, p is the percent of acid (25%), A is the amount of solution (4 liters), and Q is the amount of acid in the solution, which equals $(0.25) \cdot (4)$ liters $= 1$ liter.

EXAMPLE 7   A Percent Mixture Problem

A chemist mixes an 11% hydrochloric acid solution with a 6% hydrochloric acid solution. How many milliliters (ml) of each solution should the chemist use to make a 600-milliliter solution that is 8% hydrochloric acid?

Solution

Let x be the number of milliliters of the 11% solution. Because the final solution will have a total of 600 milliliters of fluid, $600 - x$ is the number of milliliters of the 6% solution. See **Figure 1.5.**

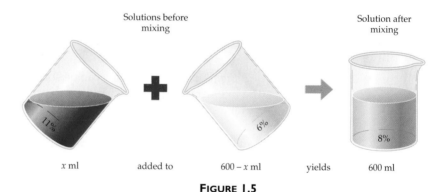

Solutions before mixing		Solution after mixing

| x ml | added to | $600 - x$ ml | yields | 600 ml |

FIGURE 1.5

Because all the hydrochloric acid in the final solution comes from either the 11% solution or the 6% solution, the number of milliliters of hydrochloric acid in the 11% solution added to the number of milliliters of hydrochloric acid in the 6% solution must equal the number of milliliters of hydrochloric acid in the 8% solution.

$$\left(\begin{array}{c} \text{ml of acid in} \\ \text{11\% solution} \end{array} \right) + \left(\begin{array}{c} \text{ml of acid in} \\ \text{6\% solution} \end{array} \right) = \left(\begin{array}{c} \text{ml of acid in} \\ \text{8\% solution} \end{array} \right)$$

$$0.11x \quad + \quad 0.06(600 - x) \quad = \quad 0.08(600)$$

$$0.11x + 36 - 0.06x = 48$$

$$0.05x + 36 = 48$$

$$0.05x = 12$$

$$x = 240$$

The chemist should use 240 milliliters of the 11% solution and 360 milliliters of the 6% solution to make a 600-milliliter solution that is 8% hydrochloric acid.

▶ **TRY EXERCISE 40, PAGE 100**

To solve a *work problem*, use the equation

Rate of work × time worked = part of task completed

For example, if a painter can paint a wall in 15 minutes, then the painter can paint 1/15 of the wall in 1 minute. The painter's *rate of work* is 1/15 of the wall each minute. In general, if a task can be completed in x minutes, then the rate of work is $1/x$ of the task each minute.

EXAMPLE 8   A Work Problem

Pump A can fill a pool in 6 hours, and pump B can fill the same pool in 3 hours. How long will it take to fill the pool if both pumps are used?

Solution

Because pump A fills the pool in 6 hours, $\dfrac{1}{6}$ represents the part of the pool filled by pump A in 1 hour. Because pump B fills the pool in 3 hours, $\dfrac{1}{3}$ represents the part of the pool filled by pump B in 1 hour.

Let t = the number of hours to fill the pool together. Then

$$t \cdot \frac{1}{6} = \frac{t}{6}$$ • **Part of the pool filled by pump A**

$$t \cdot \frac{1}{3} = \frac{t}{3}$$ • **Part of the pool filled by pump B**

$$\begin{pmatrix} \text{Part filled} \\ \text{by pump A} \end{pmatrix} + \begin{pmatrix} \text{Part filled} \\ \text{by pump B} \end{pmatrix} = \begin{pmatrix} 1 \text{ filled} \\ \text{pool} \end{pmatrix}$$

$$\frac{t}{6} \qquad + \qquad \frac{t}{3} \qquad = \qquad 1$$

Multiplying each side of the equation by 6 produces

$$t + 2t = 6$$
$$3t = 6$$
$$t = 2$$

Check: Pump A fills $\dfrac{2}{6}$, or $\dfrac{1}{3}$, of the pool in 2 hours and pump B fills $\dfrac{2}{3}$ of the pool in 2 hours, so 2 hours is the time required to fill the pool if both pumps are used.

▶ **TRY EXERCISE 50, PAGE 100**

 TOPICS FOR DISCUSSION

1. A student solves the formula $A = P + Prt$ for the variable P. The student's answer is $P = A - Prt$. Is this a correct response? Explain.

2. A student takes reciprocals of each term to write the formula

$$\frac{1}{f} = \frac{1}{d_0} + \frac{1}{d_i}$$

as $f = d_0 + d_i$. Did this technique produce a valid formula? Explain.

3. In the formula $S = \dfrac{a_1}{(1 - r)}$, what restrictions are placed on the variable r?

4. The formula $A = \dfrac{1}{2}bh$ can also be expressed as $A = \dfrac{bh}{2}$. Do you agree?

EXERCISE SET 1.2

In Exercises 1 to 10, solve the formula for the specified variable.

1. $V = \dfrac{1}{3}\pi r^2 h$; h (geometry)

2. $P = S - Sdt$; t (business)

3. $I = Prt$; t (business)

▶ **4.** $A = P + Prt$; P (business)

5. $F = \dfrac{Gm_1 m_2}{d^2}$; m_1 (physics)

6. $A = \dfrac{1}{2}h(b_1 + b_2)$; b_1 (geometry)

7. $a_n = a_1 + (n - 1)d$; d (mathematics)

8. $y - y_1 = m(x - x_1)$; x (mathematics)

9. $S = \dfrac{a_1}{1 - r}$; r (mathematics)

10. $\dfrac{P_1 V_1}{T_1} = \dfrac{P_2 V_2}{T_2}$; V_2 (chemistry)

11. **QUARTERBACK RATING** During the 2002 season, Peyton Manning, the quarterback of the Indianapolis Colts, completed 66.33% of his passes. He averaged 7.11 yards per pass attempt, 4.57% of his passes were for touchdowns, and 3.21% of his passes were intercepted. Determine Manning's quarterback rating for the 2002 season. Round to the nearest tenth. (*Hint:* See Example 2, page 92.)

12. **QUARTERBACK RATING** During the 2002 season, Drew Bledsoe, the quarterback of the Buffalo Bills, completed 61.48% of his passes. He averaged 7.15 yards per pass attempt, 3.93% of his passes were for touchdowns, and 2.46% of his passes were intercepted. Determine Bledsoe's quarterback rating for the 2002 season. Round to the nearest tenth. (*Hint:* See Example 2, page 92.)

The SMOG (Simplified Measure of Gobbledygook) readability formula is often used to estimate the reading grade level required by a person if he or she is to *fully* understand the written material being assessed. The formula is given by

SMOG reading grade level = $\sqrt{w} + 3$

where w is the number of words that have three or more syllables in a sample of 30 sentences.

13. **ASSESSING A READING LEVEL** A sample of 30 sentences from *Alice's Adventures in Wonderland*, by Lewis Carroll, shows a total of 42 words that have three or more syllables. Use the SMOG reading grade level formula to estimate the reading grade level required to fully understand this novel. Round the reading grade level to the nearest tenth.

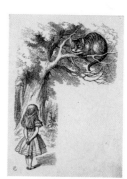

▶ **14.** **ASSESSING A READING LEVEL** A sample of 30 sentences from *A Tale of Two Cities*, by Charles Dickens, shows a total of 105 words that have three or more syllables. Use the SMOG reading grade level formula to estimate the reading grade level required to fully understand this novel. Round the reading grade level to the nearest tenth.

Another popular readability formula is the Gunning Fog Index. Here is the formula:

Gunning Fog Index = $0.4(A + W)$

where A is the average number of words per sentence and W is the percentage of words that have four or more syllables. The Gunning Fog Index is defined as the minimum grade level at which the writing is *easily* read.

15. **ASSESSING A READING LEVEL** In a sample of sentences from the novel *Bridget Jones's Diary*, by Helen Fielding, the average number of words per sentence is

23.0, and five words have four or more syllables. Use the Gunning Fog Index to estimate the reading grade level required to easily read this material.

16. **ASSESSING A READING LEVEL** In a sample of sentences from the book *Winnie-the-Pooh Meets Tigger*, by A. A. Milne, the average number of words per sentence is 11.47, and only one word has four or more syllables. Use the Gunning Fog Index to estimate the reading grade level required to easily read this material. Round the reading grade level to the nearest tenth.

In Exercises 17 to 52, solve by using the Strategies for Solving Application Problems (see page 93).

17. One-fifth of a number plus one-fourth of the number is 5 less than one-half the number. What is the number?

18. The numerator of a fraction is 4 less than the denominator. If the numerator is increased by 14 and the denominator is decreased by 10, the resulting number is 5. What is the original fraction?

19. **GEOMETRY** The length of a rectangle is 3 feet less than twice the width of the rectangle. If the perimeter of the rectangle is 174 feet, find the width and the length.

▶ 20. **GEOMETRY** The width of a rectangle is 1 meter more than half the length of the rectangle. If the perimeter of the rectangle is 110 meters, find the width and the length.

21. **GEOMETRY** A triangle has a perimeter of 84 centimeters. Each of the two longer sides of the triangle is three times as long as the shortest side. Find the length of each side of the triangle.

22. **GEOMETRY** A triangle has a perimeter of 161 miles. Each of the two smaller sides of the triangle is two-thirds the length of the longest side. Find the length of each side of the triangle.

23. **UNIFORM MOTION** Running at an average rate of 6 meters per second, a sprinter ran to the end of a track and then jogged back to the starting point at an average rate of 2 meters per second. The total time for the sprint and the jog back was 2 minutes 40 seconds. Find the length of the track.

▶ 24. **UNIFORM MOTION** A motorboat left a harbor and traveled to an island at an average rate of 15 knots. The average speed on the return trip was 10 knots. If the total trip took 7.5 hours, how far is the harbor from the island?

25. **UNIFORM MOTION** A plane leaves an airport traveling at an average speed of 240 kilometers per hour. How long will it take a second plane traveling the same route at an

average speed of 600 kilometers per hour to catch up with the first plane if it leaves 3 hours later?

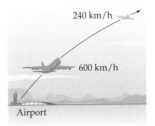

26. **UNIFORM MOTION** A plane leaves Chicago headed for Los Angeles at 540 mph. One hour later, a second plane leaves Los Angeles headed for Chicago at 660 mph. If the air route from Chicago to Los Angeles is 1800 miles, how long will it take for the first plane to pass the second plane? How far from Chicago will they be at that time?

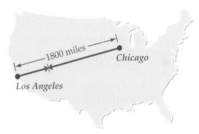

27. **UNIFORM MOTION** Marlene rides her bicycle to her friend Jon's house and returns home by the same route. Marlene rides her bike at constant speeds of 6 mph on level ground, 4 mph when going uphill, and 12 mph when going downhill. If her total time riding was 1 hour, how far is it to Jon's house?

28. **UNIFORM MOTION** A car traveling at 80 kilometers per hour is passed by a second car going in the same direction at a constant speed. After 30 seconds, the two cars are 500 meters apart. Find the speed of the second car.

29. **FINDING AN AVERAGE** A student has test scores of 80, 82, 94, and 71. What score does the student need on the next test to produce an average score of 85?

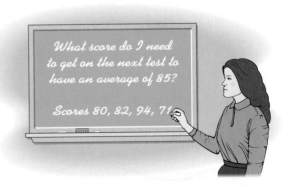

30. FINDING AN AVERAGE A student has test scores of 90, 74, 82, and 90. The next examination is the final examination, which will count as two tests. What score does the student need on the final examination to produce an average score of 85?

31. BUSINESS It costs a manufacturer of sunglasses $8.95 to produce sunglasses that sell for $29.99. How many sunglasses must the manufacturer sell to make a profit of $17,884?

▶ **32. BUSINESS** It costs a restaurant owner 18 cents per glass for orange juice, which is sold for 75 cents per glass. How many glasses of orange juice must the restaurant owner sell to make a profit of $2337?

33. BUSINESS The price of a computer fell 20% this year. If the computer now costs $750, how much did it cost last year?

34. BUSINESS The price of a magazine subscription rose 4% this year. If the subscription now costs $26, how much did it cost last year?

35. INVESTMENT An investment adviser invested $14,000 in two accounts. One investment earned 8% annual simple interest, and the other investment earned 6.5% annual simple interest. The amount of interest earned for 1 year was $1024. How much was invested in each account?

▶ **36. INVESTMENT** A total of $7500 is deposited into two simple interest accounts. On one account the annual simple interest rate is 5%, and on the second account the annual simple interest rate is 7%. The amount of interest earned for 1 year was $405. How much was invested in each account?

37. INVESTMENT An investment of $2500 is made at an annual simple interest rate of 5.5%. How much additional money must be invested at an annual simple interest rate of 8% so that the total interest earned is 7% of the total investment?

38. INVESTMENT An investment of $4600 is made at an annual simple interest rate of 6.8%. How much additional money must be invested at an annual simple interest rate of 9% so that the total interest earned is 8% of the total investment?

39. METALLURGY How many grams of pure silver must a silversmith mix with a 45% silver alloy to produce 200 grams of a 50% alloy?

▶ **40. CHEMISTRY** How many liters of a 40% sulfuric acid solution should be mixed with 4 liters of a 24% sulfuric acid solution to produce a 30% solution?

41. CHEMISTRY How many liters of water should be evaporated from 160 liters of a 12% saline solution so that the solution that remains is a 20% saline solution?

42. AUTOMOTIVE A radiator contains 6 liters of a 25% antifreeze solution. How much should be drained and replaced with pure antifreeze to produce a 33% antifreeze solution?

43. COMMERCE A ballet performance brought in $61,800 on the sale of 3000 tickets. If the tickets sold for $14 and $25, how many of each were sold?

44. COMMERCE A vending machine contains $41.25. The machine contains 255 coins, which consist only of nickels, dimes, and quarters. If the machine contains twice as many dimes as nickels, how many of each type of coin does the machine contain?

45. COMMERCE A coffee shop decides to blend a coffee that sells for $12 per pound with a coffee that sells for $9 per pound to produce a blend that will sell for $10 per pound. How much of each should be used to yield 20 pounds of the new blend?

46. DETERMINE NUMBER OF COINS A bag contains 42 coins, with a total weight of 246 grams. If the bag contains only gold coins that weigh 8 grams each and silver coins that weigh 5 grams each, how many gold and how many silver coins are in the bag?

47. METALLURGY How much pure gold should be melted with 15 grams of 14-karat gold to produce 18-karat gold? (*Hint:* A karat is a measure of the purity of gold in an alloy. Pure gold measures 24 karats. An alloy that measures x karats is $\frac{x}{24}$ gold. For example, 18-karat gold is $\frac{18}{24} = \frac{3}{4}$ gold.)

48. METALLURGY How much 14-karat gold should be melted with 4 ounces of pure gold to produce 18-karat gold? (*Hint:* See Exercise 47.)

49. INSTALL ELECTRICAL WIRES An electrician can install the electric wires in a house in 14 hours. A second electrician requires 18 hours. How long would it take both electricians, working together, to install the wires?

▶ **50. PRINT A REPORT** Printer A can print a report in 3 hours. Printer B can print the same report in 4 hours. How long would it take both printers, working together, to print the report?

51. DETERMINE INDIVIDUAL PRICES A book and a bookmark together sell for $10.10. If the price of the book is $10.00 more than the price of the bookmark, find the price of the book and the price of the bookmark.

52. SHARE AN EXPENSE Three people decide to share the cost of a yacht. By bringing in an additional partner, they can reduce the cost for each person by $4000. What is the total cost of the yacht?

CONNECTING CONCEPTS

The *Archimedean law of the lever* **states that for a lever to be in a state of balance with respect to a point called the** *fulcrum,* **the sum of the downward forces times their respective distances from the fulcrum on one side of the fulcrum must equal the sum of the downward forces times their respective distances from the fulcrum on the other side of the fulcrum. The accompanying figure shows this relationship.**

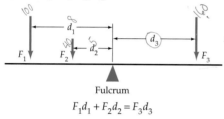

Fulcrum

$$F_1 d_1 + F_2 d_2 = F_3 d_3$$

53. LOCATE THE FULCRUM A 100-pound person 8 feet to the left of the fulcrum and a 40-pound person 5 feet to the left of the fulcrum balance with a 160-pound person on a teeter-totter. How far from the fulcrum is the 160-pound person?

54. LOCATE THE FULCRUM A lever 21 feet long has a force of 117 pounds applied to one end of the lever and a force of 156 pounds applied to the other end. Where should the fulcrum be located to produce a state of balance?

55. DETERMINE A FORCE How much force applied 5 feet from the fulcrum is needed to lift a 400-pound weight that is located on the other side, 0.5 foot from the fulcrum?

56. DETERMINE A FORCE Two workers need to lift a 1440-pound rock. They use a 6-foot steel bar with the fulcrum 1 foot from the rock, as the accompanying figure shows. One worker applies 180 pounds of force to the other end of the lever. How much force will the second worker need to apply 1 foot from that end to lift the rock?

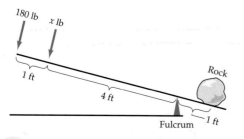

57. SPEED OF SOUND IN AIR Two seconds after firing a rifle at a target, the shooter hears the impact of the bullet. Sound travels at 1100 feet per second and the bullet at 1865 feet per second. Determine the distance to the target (to the nearest foot).

58. SPEED OF SOUND IN WATER Sound travels through sea water 4.62 times faster than through air. The sound of an exploding mine on the surface of the water and partially submerged reaches a ship through the water 4 seconds before it reaches the ship through the air. How far is the ship from the explosion (to the nearest foot)? Use 1100 feet per second as the speed of sound through the air.

59. AGE OF DIOPHANTUS The work of the ancient Greek mathematician Diophantus had great influence on later European number theorists. Nothing is known about his personal life except for the information given in the following epigram. "Diophantus passed $\frac{1}{6}$ of his life in childhood, $\frac{1}{12}$ in youth, and $\frac{1}{7}$ more as a bachelor. Five years after his marriage was born a son who died four years before his father, at $\frac{1}{2}$ his father's (final) age." How old was Diophantus when he died?

60. EQUIVALENT TEMPERATURES The relationship between the Fahrenheit temperature (*F*) and the Celsius temperature (*C*) is given by the formula

$$F = \frac{9}{5}C + 32$$

At what temperature will a Fahrenheit thermometer and a Celsius thermometer read the same?

PREPARE FOR SECTION 1.3

---

61. Factor: $x^2 - x - 42$ [P.4]

62. Factor: $6x^2 - x - 15$ [P.4]

63. Write $3 + \sqrt{-16}$ in $a + bi$ form. [P.6]

64. If $a = -3$, $b = -2$, and $c = 5$, evaluate
$$\frac{-b - \sqrt{b^2 - 4ac}}{2a} \text{ [P.1/P.2]}$$

65. If $a = 2$, $b = -3$, and $c = 1$, evaluate
$$\frac{-b + \sqrt{b^2 - 4ac}}{2a} \text{ [P.1/P.2]}$$

66. If $x = 3 - i$, evaluate $x^2 - 6x + 10$. [P.6]

PROJECTS

---

1. A WORK PROBLEM AND ITS EXTENSIONS If a pump can fill a pool in A hours, and a second pump can fill the same pool in B hours, then the total time T, in hours, needed to fill the pool with both pumps working together is given by

$$T = \frac{AB}{A + B}$$

a. Verify this formula.

b. Consider the case where a pool is to be filled by three pumps. One can fill the pool in A hours, a second in B hours, and a third in C hours. Derive a formula in terms of A, B, and C for the total time T needed to fill the pool.

c. Consider the case where a pool is to be filled by n pumps. One pump can fill the pool in A_1 hours, a second in A_2 hours, a third in A_3 hours,..., and the nth pump can fill the pool in A_n hours. Write a formula in terms of $A_1, A_2, A_3, \ldots, A_n$ for the total time T needed to fill the pool.

The following chart is called an *alignment chart* or a *nomogram*. If you know any two of the values A, B, and T, then you can use the alignment chart to determine the unknown value. For example, the straight line segment that connects 3 on the A-axis with 6 on the B-axis crosses the T-axis at 2. Thus the total time required for a pump that takes 3 hours to fill the pool and a pump that takes 6 hours to fill the pool is 2 hours when they work together.

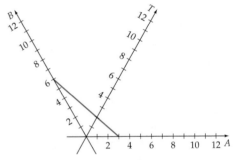

Alignment Chart for $T = \dfrac{AB}{A + B}$

d. Consider the case where a pool is to be filled by three pumps. One can fill the pool in $A = 6$ hours, a second in $B = 8$ hours, and a third in $C = 12$ hours. Write a few sentences explaining how you could make use of the alignment chart above to show that it takes about 2.7 hours for the three pumps to fill the pool when they work together.

2. RESISTANCE OF PARALLEL CIRCUITS The alignment chart shown in Project 1 can be used to solve some problems in electronics that concern the total resistance of a *parallel* circuit. Read an electronics text, and write a paragraph or two that explain this problem and how it is related to the problem of filling a pool with two pumps.

QUADRATIC EQUATIONS

SOLVE QUADRATIC EQUATIONS BY FACTORING

In Section 1.1 you solved linear equations. In this section you will learn to solve a type of equation that is referred to as a *quadratic equation*.

Definition of a Quadratic Equation

A quadratic equation in x is an equation that can be written in the standard quadratic form

$$ax^2 + bx + c = 0$$

where a, b, and c are real numbers and $a \neq 0$.

MATH MATTERS

The term *quadratic* is derived from the Latin word *quadrare*, which means "to make square." Because the area of a square that measures x units on each side is x^2, we refer to equations that can be written in the form $ax^2 + bx + c = 0$ as equations that are quadratic in x.

Several methods can be used to solve a quadratic equation. For instance, if you can factor $ax^2 + bx + c$ into linear factors, then $ax^2 + bx + c = 0$ can be solved by applying the following property.

The Zero Product Principle

If A and B are algebraic expressions such that $AB = 0$, then $A = 0$ or $B = 0$.

The zero product principle states that if the product of two factors is zero, then at least one of the factors must be zero. In Example 1, the zero product principle is used to solve a quadratic equation.

EXAMPLE 1   Solve by Factoring

Solve each quadratic equation by factoring.

a. $x^2 + 2x - 15 = 0$ **b.** $2x^2 - 5x = 12$

Solution

a. $x^2 + 2x - 15 = 0$

$(x - 3)(x + 5) = 0$ • **Factor.**

$x - 3 = 0$ $x + 5 = 0$ • **Set each factor equal to zero.**

$x = 3$ $x = -5$ • **Solve each linear equation.**

A check shows that 3 and -5 are both solutions of $x^2 + 2x - 15 = 0$.

Continued ▶

b. $2x^2 - 5x = 12$

$2x^2 - 5x - 12 = 0$ • **Write in standard quadratic form.**

$(x - 4)(2x + 3) = 0$ • **Factor.**

$x - 4 = 0 \qquad 2x + 3 = 0$ • **Set each factor equal to zero.**

$x = 4 \qquad\quad 2x = -3$ • **Solve each linear equation.**

$$x = -\frac{3}{2}$$

A check shows that 4 and $-\dfrac{3}{2}$ are both solutions of $2x^2 - 5x = 12$.

▶ **TRY EXERCISE 6, PAGE 113**

Some quadratic equations have a solution that is called a *double root*. For instance, consider $x^2 - 8x + 16 = 0$. Solving this equation by factoring, we have

$x^2 - 8x + 16 = 0$

$(x - 4)(x - 4) = 0$ • **Factor.**

$x - 4 = 0 \qquad x - 4 = 0$ • **Set each factor equal to zero.**

$x = 4 \qquad\quad x = 4$ • **Solve each linear equation.**

The only solution of $x^2 - 8x + 16 = 0$ is 4. In this situation, the single solution 4 is called a **double solution** or **double root** because it was produced by solving the two identical equations $x - 4 = 0$, both of which have 4 as a solution.

● SOLVE QUADRATIC EQUATIONS BY TAKING SQUARE ROOTS

In the following example we solve $x^2 = 25$ by factoring.

$x^2 = 25$

$x^2 - 25 = 0$

$(x - 5)(x + 5) = 0$ • **Factor.**

$x - 5 = 0 \qquad x + 5 = 0$ • **Set each factor equal to zero.**

$x = 5 \qquad\quad x = -5$ • **Solve each linear equation.**

The solutions of $x^2 = 25$ also can be found by taking the square root of each side of the equation. In the following work a plus or minus sign is placed in front of the square root of 25 to produce both solutions. The notation $x = \pm 5$ means $x = 5$ or $x = -5$.

$x^2 = 25$

$x = \pm\sqrt{25}$ • **Take the square root of each side of the equation. Insert a plus or minus sign in front of the radical on the right.**

$x = \pm 5$

$x = 5 \quad$ or $\quad x = -5$

We will refer to the preceding method of solving a quadratic equation as the *square root procedure*.

The Square Root Procedure

If $x^2 = c$, then $x = \sqrt{c}$ or $x = -\sqrt{c}$, which can also be written as $x = \pm\sqrt{c}$.

EXAMPLE 2   **Solve by Using the Square Root Procedure**

Use the square root procedure to solve each equation.

a. $3x^2 + 12 = 0$ b. $(x + 1)^2 = 49$

Solution

a. $3x^2 + 12 = 0$

$\qquad\qquad 3x^2 = -12$ • Solve for x^2.

$\qquad\qquad\; x^2 = -4$

$\qquad\qquad\;\; x = \pm\sqrt{-4}$ • Take the square root of each side of the equation and insert a plus or minus sign in front of the radical.

$\qquad\qquad\;\; x = \pm 2i$

b. $(x + 1)^2 = 49$

$\qquad\quad x + 1 = \pm\sqrt{49}$ • Take the square root of each side of the equation and insert a plus or minus sign in front of the radical.

$\qquad\qquad\; x = -1 \pm 7$ • Simplify.

$\qquad\qquad\; x = 6$ or -8

▶ **TRY EXERCISE 20, PAGE 113**

● SOLVE QUADRATIC EQUATIONS BY COMPLETING THE SQUARE

Consider the following binomial squares and their perfect-square trinomial products.

Square of a Binomial	=	Perfect-Square Trinomial
$(x + 5)^2$	=	$x^2 + 10x + 25$
$(x - 3)^2$	=	$x^2 - 6x + 9$

In each of the preceding perfect-square trinomials, the coefficient of x^2 is 1, and the constant term is the square of half the coefficient of the x term.

$$x^2 + 10x + 25, \qquad \left(\frac{1}{2}\cdot 10\right)^2 = 25$$

$$x^2 - 6x + 9, \qquad \left(\frac{1}{2}\cdot(-6)\right)^2 = 9$$

Adding to a binomial of the form $x^2 + bx$ the constant term that makes the binomial a perfect-square trinomial is called **completing the square**. For example, to complete the square of $x^2 + 8x$, add

$$\left(\frac{1}{2} \cdot 8\right)^2 = 16$$

to produce the perfect-square trinomial $x^2 + 8x + 16$.

Completing the square is a powerful procedure because it can be used to solve *any* quadratic equation. For instance, to solve $x^2 - 6x + 13 = 0$, begin by writing the variable terms on one side of the equation and the constant term on the other side.

$x^2 - 6x = -13$ • Subtract 13 from each side of the equation.

$x^2 - 6x + 9 = -13 + 9$ • Complete the square by adding $\left[\frac{1}{2}(-6)\right]^2 = 9$ to each side of the equation.

$(x - 3)^2 = -4$ • Factor and solve by the square root procedure.

$x - 3 = \pm\sqrt{-4}$

$x - 3 = \pm 2i$

$x = 3 \pm 2i$

The solutions of $x^2 - 6x + 13 = 0$ are $3 - 2i$ and $3 + 2i$. You can check these solutions by substituting each solution into the original equation. For instance, the following check shows that $3 - 2i$ does satisfy the original equation.

$x^2 - 6x + 13 = 0$

$(3 - 2i)^2 - 6(3 - 2i) + 13 = 0$ • Substitute $3 - 2i$ for x.

$9 - 12i + 4i^2 - 18 + 12i + 13 = 0$ • Simplify.

$4 + 4(-1) = 0$

$0 = 0$ • The left side equals the right side, so $3 - 2i$ checks.

EXAMPLE 3   **Solve by Completing the Square**

Solve $x^2 = 2x + 6$ by completing the square.

Solution

$x^2 = 2x + 6$

$x^2 - 2x = 6$ • Isolate the constant term.

$x^2 - 2x + 1 = 6 + 1$ • Complete the square.

$(x - 1)^2 = 7$ • Factor and simplify.

$x - 1 = \pm\sqrt{7}$ • Apply the square root procedure.

$x = 1 \pm \sqrt{7}$ • Solve for x.

The exact solutions of $x^2 = 2x + 6$ are $x = 1 - \sqrt{7}$ and $x = 1 + \sqrt{7}$. A calculator can be used to show that $1 - \sqrt{7} \approx -1.646$ and $1 + \sqrt{7} \approx 3.646$. The decimals -1.646 and 3.646 are approximate solutions of $x^2 = 2x + 6$.

▶ **TRY EXERCISE 26, PAGE 113**

Completing the square by adding the square of half the coefficient of the x term requires that the coefficient of the x^2 term be 1. If the coefficient of the x^2 term is not 1, then first multiply each term on each side of the equation by the reciprocal of the coefficient of x^2 to produce a coefficient of 1 for the x^2 term.

EXAMPLE 4 — Solve by Completing the Square

Solve $2x^2 + 8x - 1 = 0$ by completing the square.

Solution

$2x^2 + 8x - 1 = 0$

$2x^2 + 8x = 1$ • Isolate the constant term.

$\dfrac{1}{2}(2x^2 + 8x) = \dfrac{1}{2}(1)$ • Multiply both sides of the equation by the reciprocal of the coefficient of x^2.

$x^2 + 4x = \dfrac{1}{2}$

$x^2 + 4x + 4 = \dfrac{1}{2} + 4$ • Complete the square.

$(x + 2)^2 = \dfrac{9}{2}$ • Factor and simplify.

$x + 2 = \pm\sqrt{\dfrac{9}{2}}$ • Apply the square root procedure.

$x = -2 \pm 3\sqrt{\dfrac{1}{2}}$ • Solve for x.

$x = -2 \pm 3\dfrac{\sqrt{2}}{2}$ • Simplify.

$x = \dfrac{-4 \pm 3\sqrt{2}}{2}$

The solutions of $2x^2 + 8x - 1 = 0$ are $x = \dfrac{-4 + 3\sqrt{2}}{2}$ and $x = \dfrac{-4 - 3\sqrt{2}}{2}$.

▶ **TRY EXERCISE 30, PAGE 113**

• SOLVE QUADRATIC EQUATIONS BY USING THE QUADRATIC FORMULA

Completing the square on $ax^2 + bx + c = 0$ $(a \neq 0)$ produces a formula for x in terms of the coefficients a, b, and c. The formula is known as the *quadratic formula*, and it can be used to solve *any* quadratic equation.

The Quadratic Formula

If $ax^2 + bx + c = 0$, $a \neq 0$, then

$$x = \frac{-b \pm \sqrt{b^2 - 4ac}}{2a}$$

MATH MATTERS

Ancient mathematicians thought of "completing the square" in a geometric manner. For instance, to complete the square of $x^2 + 8x$, draw a square that measures x units on each side, and add four rectangles that measure 1 unit by x units to the right side and the bottom of the square.

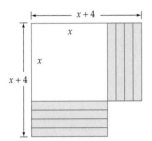

Each of the rectangles has an area of x square units, so the total area of the figure is $x^2 + 8x$. To make this figure a complete square, we must add 16 squares that measure 1 unit by 1 unit, as shown below.

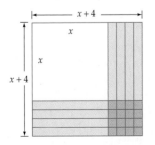

This figure is a *complete square* whose area is

$(x + 4)^2 = x^2 + 8x + 16$

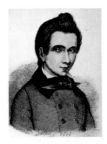
Proof We assume a is a positive real number. If a were a negative real number, then we could multiply each side of the equation by -1 to make it positive.

$$ax^2 + bx + c = 0 \quad (a \neq 0)$$
• Given.

$$ax^2 + bx = -c$$
• Isolate the constant term.

$$x^2 + \frac{b}{a}x = -\frac{c}{a}$$
• Multiply each term on each side of the equation by $\frac{1}{a}$.

$$x^2 + \frac{b}{a}x + \left(\frac{b}{2a}\right)^2 = \left(\frac{b}{2a}\right)^2 - \frac{c}{a}$$
• Complete the square.

$$\left(x + \frac{b}{2a}\right)^2 = \frac{b^2}{4a^2} - \frac{c}{a}$$
• Factor the left side. Simplify the powers on the right side.

$$\left(x + \frac{b}{2a}\right)^2 = \frac{b^2}{4a^2} - \frac{4a}{4a} \cdot \frac{c}{a}$$
• Use a common denominator to simplify the right side.

$$x + \frac{b}{2a} = \pm\sqrt{\frac{b^2 - 4ac}{4a^2}}$$
• Apply the square root procedure.

$$x + \frac{b}{2a} = \pm\frac{\sqrt{b^2 - 4ac}}{2a}$$
• Because $a > 0$, $\sqrt{4a^2} = 2a$.

$$x = -\frac{b}{2a} \pm \frac{\sqrt{b^2 - 4ac}}{2a}$$
• Add $-\frac{b}{2a}$ to each side.

$$x = \frac{-b \pm \sqrt{b^2 - 4ac}}{2a}$$
◆

As a general rule, you should first try to solve quadratic equations by factoring. If the factoring process proves difficult, then solve by using the quadratic formula.

EXAMPLE 5   Solve by Using the Quadratic Formula

Use the quadratic formula to solve each of the following.

a. $4x^2 - 4x - 3 = 0$ **b.** $x^2 = 3x + 5$

Solution

a.   For the equation $4x^2 - 4x - 3 = 0$, we have $a = 4$, $b = -4$, and $c = -3$. Substituting in the quadratic formula produces

$$x = \frac{-b \pm \sqrt{b^2 - 4ac}}{2a}$$

$$= \frac{-(-4) \pm \sqrt{(-4)^2 - 4(4)(-3)}}{2(4)}$$

$$= \frac{4 \pm \sqrt{64}}{8} = \frac{4 \pm 8}{8} = \frac{1 \pm 2}{2} = \frac{3}{2} \text{ or } -\frac{1}{2}$$

The solutions of $4x^2 - 4x - 3 = 0$ are $x = \frac{3}{2}$ and $x = -\frac{1}{2}$.

To review **COMPLEX CONJUGATES**, *see p. 69.*

TRY EXERCISE 38, PAGE 113

take note

Although the equation in Example **5a** can be solved by factoring, we have solved it by using the quadratic formula to illustrate the procedures involved in applying the quadratic formula.

b. The standard form of $x^2 = 3x + 5$ is $x^2 - 3x - 5 = 0$. Substituting $a = 1$, $b = -3$, and $c = -5$ in the quadratic formula produces

$$x = \frac{-(-3) \pm \sqrt{(-3)^2 - 4(1)(-5)}}{2(1)}$$

$$= \frac{3 \pm \sqrt{29}}{2}$$

The solutions of $x^2 = 3x + 5$ are $x = \dfrac{3 + \sqrt{29}}{2}$ and $x = \dfrac{3 - \sqrt{29}}{2}$.

? QUESTION Can the quadratic formula be used to solve any quadratic equation $ax^2 + bx + c = 0$ with real coefficients and $a \neq 0$?

THE DISCRIMINANT OF A QUADRATIC EQUATION

The solutions of $ax^2 + bx + c = 0$, $a \neq 0$, are given by

$$x = \frac{-b \pm \sqrt{b^2 - 4ac}}{2a}$$

The expression under the radical, $b^2 - 4ac$, is called the **discriminant** of the equation $ax^2 + bx + c = 0$. If $b^2 - 4ac \geq 0$, then $\sqrt{b^2 - 4ac}$ is a real number. If $b^2 - 4ac < 0$, then $\sqrt{b^2 - 4ac}$ is not a real number. Thus the sign of the discriminant can be used to determine whether the solutions of a quadratic equation are real numbers.

The Discriminant and the Solutions of a Quadratic Equation

The equation $ax^2 + bx + c = 0$, with real coefficients and $a \neq 0$, has as its discriminant $b^2 - 4ac$.

- If $b^2 - 4ac > 0$, then $ax^2 + bx + c = 0$ has *two distinct real solutions.*
- If $b^2 - 4ac = 0$, then $ax^2 + bx + c = 0$ has *one real solution.* The solution is a double solution.
- If $b^2 - 4ac < 0$, then $ax^2 + bx + c = 0$ has *two distinct nonreal complex solutions.* The solutions are conjugates of each other.

EXAMPLE 6   Use the Discriminant to Determine the Number of Real Solutions

For each equation, determine the discriminant and state the number of real solutions.

a. $2x^2 - 5x + 1 = 0$ **b.** $3x^2 + 6x + 7 = 0$ **c.** $x^2 + 6x + 9 = 0$

Continued ▶

? ANSWER Yes. However, it is sometimes easier to find the solutions by factoring, by the square root procedure, or by completing the square.

Solution

a. The discriminant of $2x^2 - 5x + 1 = 0$ is $b^2 - 4ac = (-5)^2 - 4(2)(1) = 17$. Because the discriminant is positive, $2x^2 - 5x + 1 = 0$ has two distinct real solutions.

b. The discriminant of $3x^2 + 6x + 7 = 0$ is $b^2 - 4ac = 6^2 - 4(3)(7) = -48$. Because the discriminant is negative, $3x^2 + 6x + 7 = 0$ has no real solutions.

c. The discriminant of $x^2 + 6x + 9 = 0$ is $b^2 - 4ac = 6^2 - 4(1)(9) = 0$. Because the discriminant is 0, $x^2 + 6x + 9 = 0$ has one real solution.

▶ **TRY EXERCISE 48, PAGE 113**

● APPLICATIONS OF QUADRATIC EQUATIONS

A **right triangle** contains one 90° angle. The side opposite the 90° angle is called the **hypotenuse.** The other two sides are called **legs.** The lengths of the sides of a right triangle are related by a theorem known as the Pythagorean Theorem.

The Pythagorean Theorem

If a and b denote the lengths of the legs of a right triangle and c the length of the hypotenuse, then $c^2 = a^2 + b^2$.

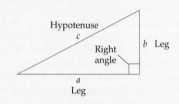

The Pythagorean Theorem states that the square of the length of the hypotenuse of a right triangle is equal to the sum of the squares of the lengths of the legs. This theorem is often used to solve applications that involve right triangles.

EXAMPLE 7   Determine the Dimensions of a Television Screen

A television screen measures 60 inches diagonally, and its *aspect ratio* is 16 to 9. This means that the ratio of the width of the screen to the height of the screen is 16 to 9. Find the width and height of the screen.

take note

Many movies are designed to be shown on a screen that has a 16 to 9 aspect ratio.

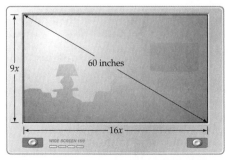

A 60-inch television screen with a 16:9 aspect ratio.

Solution

Let $16x$ represent the width of the screen and let $9x$ represent the height of the screen. Applying the Pythagorean Theorem gives us

$$(16x)^2 + (9x)^2 = 60^2$$

$$256x^2 + 81x^2 = 3600 \qquad \text{• Solve for } x.$$

$$337x^2 = 3600$$

$$x^2 = \frac{3600}{337} \qquad \text{• Apply the square root procedure.}$$

$$x = \sqrt{\frac{3600}{337}} \approx 3.268 \text{ inches} \qquad \text{• The plus or minus sign is not used in this application because we know } x \text{ is positive.}$$

The height of the screen is about $9(3.268) \approx 29.4$ inches and the width of the screen is about $16(3.268) \approx 52.3$ inches.

▶ **TRY EXERCISE 58, PAGE 114**

EXAMPLE 8   **Determine the Dimensions of a Candy Bar**

At the present time, a company makes rectangular solid candy bars that measure 5 inches by 2 inches by 0.5 inch. Due to difficult financial times, the company has decided to keep the price of the candy bar fixed and reduce the volume of the bar by 20%. What should be the dimensions of the new candy bar if it is decided to keep the height at 0.5 inch and to keep the length of the candy bar 3 inches longer than the width?

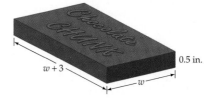

Solution

The volume of a rectangular solid is given by $V = lwh$. The original candy bar had a volume of $5 \cdot 2 \cdot 0.5 = 5$ cubic inches. The new candy bar will have a volume of $80\%(5) = 0.80(5) = 4$ cubic inches.

Let w represent the width and $w + 3$ represent the length of the new candy bar. For the new candy bar we have:

$$lwh = V$$

$$(w + 3)(w)(0.5) = 4 \qquad \text{• Substitute in the volume formula.}$$

$$(w + 3)(w) = 8 \qquad \text{• Multiply each side by 2.}$$

$$w^2 + 3w = 8$$

$$w^2 + 3w - 8 = 0 \qquad \text{• Write in } ax^2 + bx + c = 0 \text{ form.}$$

Continued ▶

INTEGRATING TECHNOLOGY

In many application problems it is helpful to use a calculator to estimate the solutions of a quadratic equation by applying the quadratic formula. For instance, the following figure shows the use of a graphing calculator to estimate the solutions of $w^2 + 3w - 8 = 0$.

```
(-3+√(3²-4*1*(-8)))/2
                1.701562119
(-3-√(3²-4*1*(-8)))/2
               -4.701562119
```

$$w = \frac{-(3) \pm \sqrt{(3)^2 - 4(1)(-8)}}{2(1)}$$

• Use the quadratic formula.

$$= \frac{-3 \pm \sqrt{41}}{2}$$

$$\approx 1.7 \quad \text{or} \quad -4.7$$

We can disregard the negative value because the width must be positive. The width of the new candy bar should be 1.7 inches, to the nearest tenth of an inch. The length should be 3 inches longer, which is 4.7 inches.

▶ **TRY EXERCISE 70, PAGE 115**

Quadratic equations are often used to determine the height (position) of an object that has been dropped or projected. For instance, the *position equation* $s = -16t^2 + v_0t + s_0$ can be used to estimate the height of a projected object near the surface of the earth at a given time t, in seconds. In this equation, v_0 is the initial velocity of the object in feet per second, and s_0 is the initial height of the object in feet.

EXAMPLE 9   Determine the Time of Descent

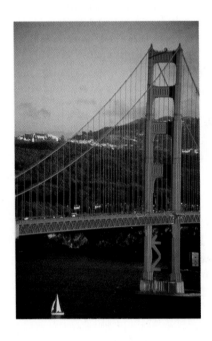

A ball is thrown downward with an initial velocity of 5 feet per second from the Golden Gate Bridge, which is 220 feet above the water. How long will it take for the ball to hit the water? Round your answer to the nearest hundreth of a second.

Solution

The distance s, in feet, of the ball above the water after t seconds is given by $s = -16t^2 - 5t + 220$. We have replaced v_0 with -5 because the ball is thrown downward. (If the ball had been thrown upward, we would use $v_0 = 5$). To determine the time it takes the ball to hit the water, substitute 0 for s in the equation $s = -16t^2 - 5t + 220$ and solve for t. In the following work, we have solved by using the quadratic formula.

$$0 = -16t^2 - 5t + 220$$

$$t = \frac{-(-5) \pm \sqrt{(-5)^2 - 4(-16)(220)}}{2(-16)}$$

• Use the quadratic formula.

$$= \frac{5 \pm \sqrt{14,105}}{-32}$$

• Use a calculator to estimate t.

$$\approx -3.87 \quad \text{or} \quad 3.56$$

Because the time must be positive, we disregard the negative value. The ball will hit the water in about 3.56 seconds.

▶ **TRY EXERCISE 72, PAGE 115**

 TOPICS FOR DISCUSSION

1. Name the four methods of solving a quadratic equation that have been discussed in this section. What are the advantages and disadvantages of each?

2. If x and y are real numbers and $xy = 0$, then $x = 0$ or $y = 0$. Do you agree with this statement? Explain.

3. If x and y are real numbers and $xy = 1$, then $x = 1$ or $y = 1$. Do you agree with this statement? Explain.

4. Explain how to complete the square on $x^2 + bx$.

5. If the discriminant of $ax^2 + bx + c = 0$ with real coefficients and $a \neq 0$ is negative, then what can be said concerning the solutions of the equation?

EXERCISE SET 1.3

In Exercises 1 to 10, solve each quadratic equation by factoring and applying the zero product principle.

1. $x^2 - 2x - 15 = 0$

2. $x^2 + 3x - 10 = 0$

3. $2x^2 - x = 1$

4. $2x^2 + 5x = 3$

5. $8x^2 + 189x - 72 = 0$

▶ 6. $12x^2 - 41x + 24 = 0$

7. $3x^2 - 7x = 0$

8. $5x^2 = -8x$

9. $(x - 5)^2 - 9 = 0$

10. $(3x + 4)^2 - 16 = 0$

In Exercises 11 to 20, use the square root procedure to solve each quadratic equation.

11. $x^2 = 81$

12. $x^2 = 225$

13. $2x^2 = 48$

14. $3x^2 = 144$

15. $3x^2 + 12 = 0$

16. $4x^2 + 20 = 0$

17. $(x - 5)^2 = 36$

18. $(x + 4)^2 = 121$

19. $(x - 3)^2 + 16 = 0$

▶ 20. $(x + 2)^2 + 28 = 0$

In Exercises 21 to 32, solve each quadratic equation by completing the square.

21. $x^2 + 6x + 1 = 0$

22. $x^2 + 8x - 10 = 0$

23. $x^2 - 2x - 15 = 0$

24. $x^2 + 2x - 8 = 0$

25. $x^2 + 4x + 5 = 0$

▶ 26. $x^2 - 6x + 10 = 0$

27. $x^2 + 3x - 1 = 0$

28. $x^2 + 7x - 2 = 0$

29. $2x^2 + 4x - 1 = 0$

▶ 30. $2x^2 + 10x - 3 = 0$

31. $3x^2 - 8x = -1$

32. $4x^2 - 4x = -15$

In Exercises 33 to 46, solve each quadratic equation by using the quadratic formula.

33. $x^2 - 2x = 15$

34. $x^2 - 5x = 24$

35. $x^2 = -x + 1$

36. $x^2 = -x - 1$

37. $2x^2 + 4x = -1$

▶ 38. $2x^2 + 4x = 1$

39. $3x^2 - 5x + 3 = 0$

40. $3x^2 - 5x + 4 = 0$

41. $\dfrac{1}{2}x^2 + \dfrac{3}{4}x - 1 = 0$

42. $\dfrac{2}{3}x^2 - 5x + \dfrac{1}{2} = 0$

43. $24x^2 = 22x + 35$

44. $72x^2 + 13x = 15$

45. $0.5x^2 + 0.6x = 0.8$

46. $1.2x^2 + 0.4x - 0.5 = 0$

In Exercises 47 to 56, determine the discriminant of the quadratic equation, and then state the number of real solutions of the equation. Do not solve the equation.

47. $2x^2 - 5x - 7 = 0$

▶ 48. $x^2 + 3x - 11 = 0$

49. $3x^2 - 2x + 10 = 0$

50. $x^2 + 3x + 3 = 0$

51. $x^2 - 20x + 100 = 0$

52. $4x^2 + 12x + 9 = 0$

53. $24x^2 = -10x + 21$

54. $32x^2 - 44x = -15$

55. $12x^2 + 15x = -7$

56. $8x^2 = 5x - 3$

57. **GEOMETRY** The length of each side of an equilateral triangle is 31 centimeters. Find the altitude of the triangle. Round to the nearest tenth of a centimeter.

▶ **58.** DIMENSIONS OF A BASEBALL DIAMOND How far, to the nearest tenth of a foot, is it from home plate to second base on a baseball diamond? (*Hint:* The bases in a baseball diamond form a square that measures 90 feet on each side.)

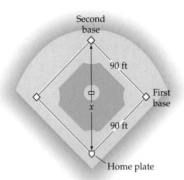

59. DIMENSIONS OF A TELEVISION SCREEN A television screen measures 54 inches diagonally, and its aspect ratio is 4 to 3. Find the width and the height of the screen.

60. PUBLISHING COSTS The cost, in dollars, of publishing x books is $C(x) = 40{,}000 + 20x + 0.0001x^2$. How many books can be published for $250{,}000?

61. COST OF A WEDDING The average cost of a wedding, in dollars, is modeled by

$$C(t) = 38t^2 + 291t + 15{,}208$$

where $t = 0$ represents the year 1990 and $0 \le t \le 14$. Use the model to determine the year during which the average cost of a wedding first reached $19{,}000.

62. REVENUE The demand for a certain product is given by $p = 26 - 0.01x$, where x is the number of units sold per month and p is the price, in dollars, at which each item is sold. The monthly revenue is given by $R = xp$. What number of items sold produces a monthly revenue of $16{,}500?

63. PROFIT A company has determined that the profit, in dollars, it can expect from the manufacture and sale of x tennis racquets is given by

$$P = -0.01x^2 + 168x - 120{,}000$$

How many racquets should the company manufacture and sell to earn a profit of $518{,}000?

64. QUADRATIC GROWTH A plant's ability to create food through the process of photosynthesis depends on the surface area of its leaves. A biologist has determined that the surface area A of a maple leaf can be closely approximated by the formula $A = 0.72(1.28)h^2$, where h is the height of the leaf in inches.

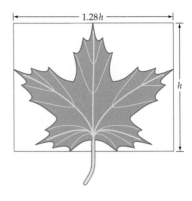

a. Find the surface area of a maple leaf with a height of 7 inches. Round to the nearest tenth of a square inch.

b. Find the height of a maple leaf with an area of 92 square inches. Round to the nearest tenth of an inch.

65. DIMENSIONS OF AN ANIMAL ENCLOSURE A veterinarian wishes to use 132 feet of chain-link fencing to enclose a rectangular region and subdivide the region into two smaller rectangular regions, as shown in the following figure. If the total enclosed area is 576 square feet, find the dimensions of the enclosed region.

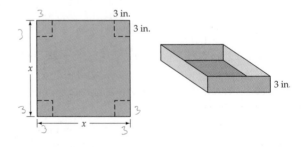

66. CONSTRUCTION OF A BOX A square piece of cardboard is formed into a box by cutting out 3-inch squares from each of the corners and folding up the sides, as shown in the following figure. If the volume of the box needs to be 126.75 cubic inches, what size square piece of cardboard is needed?

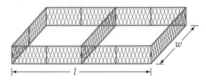

67. POPULATION DENSITY OF A CITY The population density D (in people per square mile) of a city is related to the horizontal distance x, in miles, from the center of the city by $D = -45x^2 + 190x + 200$, $0 < x < 5$. At what distances from the center of the city does the population density equal 250 people per square mile? Round each result to the nearest tenth of a mile.

68. TRAFFIC CONTROL Traffic engineers install "flow lights" at the entrances of freeways to control the number of cars entering the freeway during times of heavy traffic. For a particular freeway entrance, the number of cars N waiting to enter the freeway during the morning hours can be approximated by $N = -5t^2 + 80t - 280$, where t is the time of the day and $6 \leq t \leq 10.5$. According to this model, when will there be 35 cars waiting to enter the freeway?

69. 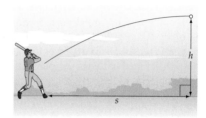 **DAREDEVIL MOTORCYCLE JUMP** In March of 2000, Doug Danger made a successful motorcycle jump over an L-1011 jumbo jet. The horizontal distance of his jump was 160 feet, and his height, in feet, during the jump was approximated by $h = -16t^2 + 25.3t + 20$, $t \geq 0$. He left the takeoff ramp at a height of 20 feet, and he landed on the landing ramp at a height of about 17 feet. How long, to the nearest tenth of a second, was he in the air?

▶ **70. DIMENSIONS OF A CANDY BAR** At the present time a company makes rectangular solid candy bars that measure 5 inches by 2 inches by 0.5 inch. Due to difficult financial times, the company has decided to keep the price of the candy bar fixed and reduce the volume of the bar by 20%. What should be the dimensions, to the nearest tenth of an inch, of the new candy bar if it is decided to keep the height at 0.5 inch and to make the length of the new candy bar 2.5 times longer than its width?

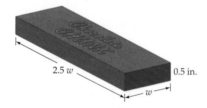

2.5 w   0.5 in.
w

71. HEIGHT OF A ROCKET A model rocket is launched upward with an initial velocity of 220 feet per second. The height, in feet, of the rocket t seconds after the launch is given by $h = -16t^2 + 220t$. How many seconds after the launch will the rocket be 350 feet above the ground? Round to the nearest tenth of a second.

▶ **72. BASEBALL** The height h, in feet, of a baseball above the ground t seconds after it is hit is given by $h = -16t^2 + 52t + 4.5$. Use this equation to determine the number of seconds, to the nearest tenth of a second, from the time the ball is hit until the ball hits the ground.

73. BASEBALL Two equations can be used to track the position of a baseball t seconds after it is hit. For instance, suppose $h = -16t^2 + 50t + 4.5$ gives the height, in feet, of a baseball t seconds after it is hit, and $s = 103.9t$ gives the horizontal distance, in feet, the ball is from home plate t seconds after it is hit. See the figure at the top of the next column. Use these equations to determine whether this particular baseball will clear a 10-foot fence positioned 360 feet from home plate.

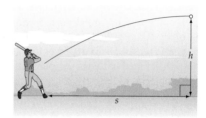

74. BASKETBALL Michael Jordan was known for his "hang time," which is the amount of time a player is in the air when making a jump toward the basket. An equation that approximates the height s, in inches, of one of Jordan's jumps is given by $s = -16t^2 + 26.6t$, where t is time in seconds. Use this equation to determine Michael Jordan's hang time, to the nearest tenth of a second, for this jump.

75. NUMBER OF HANDSHAKES If everyone in a group of n people shakes hands with everyone other than themselves, then the total number of handshakes h is given by

$$h = \frac{1}{2}n(n - 1)$$

The total number of handshakes that are exchanged by a group of people is 36. How many people are in the group?

76. 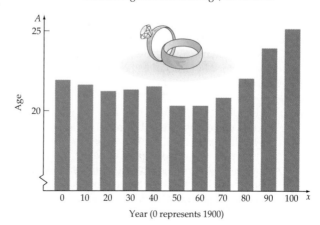 **MEDIAN AGE AT FIRST MARRIAGE** During the first 60 years of the 20th century, couples tended to marry at younger and younger ages. During the last 40 years, that trend was reversed. The median age A, in years, at first marriage for women can be modeled by

$$A = 0.0013x^2 - 0.1048x + 22.5256$$

where $x = 0$ represents the year 1900 and $x = 100$ represents the year 2000. Use the model to predict in what year in the future the median age at first marriage for women will first reach 26 years. (*Source:* U.S. Census Bureau, www.Census.gov.)

Median Age at First Marriage, for Women

77. 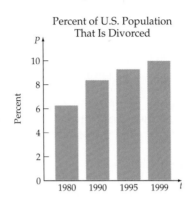 **PERCENT OF DIVORCED CITIZENS** The percent P of U.S. citizens who are divorced can be closely approximated by $P = -0.0016t^2 + 0.225t + 6.201$, where t is time in years, with $t = 0$ representing 1980. Use this model to predict in what year the percent of U.S. citizens who are divorced will first reach 11.0%. (*Source:* U.S. Census Bureau, www.Census.gov.)

Percent of U.S. Population That Is Divorced

78. **AUTOMOTIVE ENGINEERING** The number of feet N that a car needs to stop on a certain road surface is given by

$N = -0.015v^2 + 3v$, $0 \leq v \leq 90$, where v is the speed of the car in miles per hour when the driver applies the brakes. What is the maximum speed, to the nearest mile per hour, that a motorist can be traveling and stop the car within 100 feet?

79. **ORBITAL DEBRIS** The amount of space debris orbiting Earth has been increasing at an alarming rate. In 1995, there were about 14 million pounds of debris orbiting Earth, and by the year 2000, the amount of debris had increased to over 25 million pounds. (*Source:* http://orbitaldebris.jsc.nasa.gov/.)

The equation $A = 0.05t^2 + 2.25t + 14$ closely models the amount of debris orbiting Earth, where A is the amount of debris in millions of pounds and t is the time in years, with $t = 0$ representing the year 1995. Use the equation to

a. estimate the amount of orbital debris we can expect in the year 2006

b. estimate in what year the amount of orbital debris will first reach 50 million pounds

---

CONNECTING CONCEPTS

80. a. Show that the equation $x^2 + bx - 4 = 0$ always has two distinct real number solutions, regardless of the value of b.

b. For what values of k does $x^2 - 6x + k = 0$ have two distinct real number solutions?

81. **GOLDEN RECTANGLES** A rectangle is called a *golden rectangle* provided its length l and its width w satisfy the equation

$$\frac{l}{w} = \frac{l + w}{l}$$

a. Solve this formula for l. (*Hint:* Multiply both sides of the equation by wl, and then use the quadratic formula to solve for l in terms of w. Because l must be positive, state only the positive solution.)

b. If the width of a golden rectangle measures 101 feet, what is the length of the rectangle? Round to the nearest tenth of a foot.

c. Measure the width and the length of a credit card. Would you say that the credit card closely approximates a golden rectangle?

The following theorem is known as the *sum and product of the roots theorem.*

> Let $ax^2 + bx + c = 0$, $a \neq 0$, be a quadratic equation. Then r_1 and r_2 are roots of the equation if and only if
>
> $$r_1 + r_2 = -\frac{b}{a} \quad \text{and} \quad r_1 r_2 = \frac{c}{a}$$

In Exercises 82 to 86, use the sum and product of the roots theorem to determine whether the given numbers are roots of the quadratic equation.

82. $x^2 + 4x - 21 = 0$; $-7, 3$

83. $2x^2 - 7x - 30 = 0$; $-\dfrac{5}{2}, 6$

84. $9x^2 - 12x - 1 = 0$; $\dfrac{2 + \sqrt{5}}{3}, \dfrac{2 - \sqrt{5}}{3}$

85. $x^2 - 2x + 2 = 0$; $1 + i, 1 - i$

86. $x^2 - 4x + 12 = 0$; $2 + 3i, 2 - 3i$

───── **PREPARE FOR SECTION 1.4** ─────

87. Factor: $x^3 - 16x$ [P.4]

88. Factor: $x^4 - 36x^2$ [P.4]

89. Evaluate: $8^{2/3}$ [P.2]

90. Evaluate: $16^{3/2}$ [P.2]

91. Find $\left(1 + \sqrt{x-5}\right)^2$, $x > 5$ [P.2/P.3]

92. Find $\left(2 - \sqrt{x+3}\right)^2$, $x > -3$ [P.2/P.3]

───── **PROJECTS** ─────

1. THE SUM AND PRODUCT OF THE ROOTS THEOREM Use the quadratic formula to prove the sum and product of the roots theorem stated just before Exercise 82.

2. VISUAL INSIGHT

President James A. Garfield is credited with the following proof of the Pythagorean Theorem. Write the supporting reasons for each of the steps in this proof.

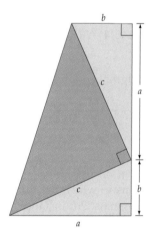

$$\text{Area} = \frac{1}{2}c^2 + 2\left(\frac{1}{2}ab\right) = \frac{1}{2}(\text{height})(\text{sum of bases})$$

$$\frac{1}{2}c^2 + ab = \frac{1}{2}(a+b)(a+b)$$

$$\frac{1}{2}c^2 = \frac{1}{2}a^2 + \frac{1}{2}b^2$$

$$c^2 = a^2 + b^2$$

OTHER TYPES OF EQUATIONS

POLYNOMIAL EQUATIONS

Some polynomial equations that are neither linear nor quadratic can be solved by the various techniques presented in this section. For instance, the **third-degree equation**, or **cubic equation**, in Example 1 can be solved by factoring the polynomial and using the zero product property.

EXAMPLE 1   Solve a Polynomial Equation by Factoring

Solve: $x^3 - 16x = 0$

Solution

$$x^3 - 16x = 0$$
$$x(x^2 - 16) = 0 \qquad \text{• Factor out the GCF, } x.$$
$$x(x + 4)(x - 4) = 0 \qquad \text{• Factor the difference of squares.}$$

Set each factor equal to zero.

$$x = 0 \quad \text{or} \quad x + 4 = 0 \quad \text{or} \quad x - 4 = 0$$
$$x = 0 \quad \text{or} \quad x = -4 \quad \text{or} \quad x = 4$$

A check will show that $-4, 0$, and 4 are roots of the original equation.

▶ **TRY EXERCISE 6, PAGE 125**

take note

If you attempt to solve Example 1 by dividing each side by x, you will produce the equation $x^2 - 16 = 0$, which has roots of only -4 and 4. In this case the division of each side of the equation by the variable x does not produce an equivalent equation. To avoid this common mistake, factor out any variable factors that are common to each term instead of dividing each side of the equation by the factor.

RATIONAL EQUATIONS

A **rational equation** is an equation that involves fractions in which the numerators and/or the denominators of the fractions are polynomials. For instance,

$$\frac{x}{x - 3} = \frac{9}{x - 3} - 5$$

is a rational equation.

Many rational equations can be solved by multiplying each side of the equation by a variable expression to produce a polynomial equation. When we multiply each side of an equation by a variable expression, we restrict the variable so that the expression is not equal to zero. Example 2b illustrates the fact that you may produce incorrect results if you fail to restrict the variable.

EXAMPLE 2   Solve Rational Equations

Solve each equation.

a. $\dfrac{x}{x - 3} = \dfrac{9}{x - 3} - 5$ **b.** $1 + \dfrac{x}{x - 5} = \dfrac{5}{x - 5}$

Solution

a. First, note that the denominator $x - 3$ would equal zero if x were 3. To produce a simpler equivalent equation, multiply each side by $x - 3$, with the restriction that $x \neq 3$.

$$(x - 3)\left(\frac{x}{x - 3}\right) = (x - 3)\left(\frac{9}{x - 3} - 5\right) \qquad \bullet \, x \neq 3.$$

$$x = (x - 3)\left(\frac{9}{x - 3}\right) - (x - 3)5$$

$$x = 9 - 5x + 15$$

$$6x = 24$$

$$x = 4$$

Substituting 4 for x in the original equation establishes that 4 is the solution.

b. To produce a simpler equivalent equation, multiply each side of the equation by $x - 5$, with the restriction that $x \neq 5$.

$$(x - 5)\left(1 + \frac{x}{x - 5}\right) = (x - 5)\left(\frac{5}{x - 5}\right) \qquad \bullet \, x \neq 5.$$

$$(x - 5)1 + (x - 5)\left(\frac{x}{x - 5}\right) = 5$$

$$x - 5 + x = 5$$

$$2x = 10$$

$$x = 5$$

Although we have obtained 5 as a proposed solution, 5 is *not* a solution of the original equation because it contradicts our restriction $x \neq 5$. Substitution of 5 for x in the original equation results in denominators of 0. In this case the original equation has no solution.

▶ **TRY EXERCISE 14, PAGE 125**

TRY EXERCISE 14, PAGE 125

---

take note

When we multiply both sides of an equation by $x - a$, we assume that $x \neq a$.

MATH MATTERS

Determine the incorrect step in the following "proof" that $2 = 1$.

$a = b$	Given.
$a^2 = ab$	Multiply by a.
$a^2 - b^2 = ab - b^2$	Subtract b^2.
$(a + b)(a - b) = b(a - b)$	Factor.
$\dfrac{(a + b)(a - b)}{(a - b)} = \dfrac{b(a - b)}{(a - b)}$	Divide by $a - b$.
$a + b = b$	Simplify.
$b + b = b$	Substitute b for a.
$2b = b$	Simplify.
$2 = 1$	Divide by b.

EXAMPLE 3   A Medical Application

Young's rule is often used by physicians to determine what portion of the recommended adult dosage of a medication should be administered to a child. In equation form, Young's rule is given by

$$\text{Portion of an adult dosage} = \frac{x}{x + 12}$$

where x represents the age, in years, of the child. Using Young's rule, a physician has determined that Elizabeth should receive only $\dfrac{1}{7}$ the recommended adult dosage of a medication. Determine Elizabeth's age.

Continued ▶

Solution

Elizabeth is to receive $\dfrac{1}{7}$ of an adult dosage of a particular medication. Thus we need to solve the following rational equation for x.

$$\frac{1}{7} = \frac{x}{x + 12}$$

$$7(x + 12)\frac{1}{7} = 7(x + 12)\left(\frac{x}{x + 12}\right) \qquad \text{• Multiply each side by } 7(x + 12).$$

$$x + 12 = 7x \qquad \text{• Simplify.}$$

$$12 = 6x \qquad \text{• Solve for } x.$$

$$x = 2$$

Elizabeth is 2 years old.

▶ **TRY EXERCISE 60, PAGE 126**

● RADICAL EQUATIONS

Some equations that involve radical expressions can be solved by using the following result.

The Power Principle

If P and Q are algebraic expressions and n is a positive integer, then every solution of $P = Q$ is a solution of $P^n = Q^n$.

EXAMPLE 4 **Solve a Radical Equation**

Use the power principle to solve $\sqrt{x + 4} = 3$.

Solution

$$\sqrt{x + 4} = 3$$

$$\left(\sqrt{x + 4}\right)^2 = 3^2 \qquad \text{• Square each side of the equation. (Apply the power principle with } n = 2.)$$

$$x + 4 = 9$$

$$x = 5$$

$$\text{Check: } \sqrt{x + 4} = 3$$

$$\sqrt{5 + 4} \stackrel{?}{=} 3 \qquad \text{• Substitute 5 for } x.$$

$$\sqrt{9} \stackrel{?}{=} 3$$

$$3 = 3 \qquad \text{• 5 checks.}$$

The only solution is 5.

▶ **TRY EXERCISE 28, PAGE 126**

It is possible to solve equations that are quadratic in form without making a formal substitution. For example, to solve $x^4 + 5x^2 - 36 = 0$, factor the equation and apply the zero product property.

$$x^4 + 5x^2 - 36 = 0$$
$$(x^2 + 9)(x^2 - 4) = 0$$

$$x^2 + 9 = 0 \quad \text{or} \quad x^2 - 4 = 0$$
$$x^2 = -9 \quad \text{or} \quad x^2 = 4$$
$$x = \pm 3i \quad \text{or} \quad x = \pm 2$$

  TOPICS FOR DISCUSSION

1. If P and Q are algebraic expressions and n is a positive integer, then the equation $P^n = Q^n$ is equivalent to the equation $P = Q$. Do you agree? Explain.

2. Consider the equation $(x^2 - 1)(x - 2) = 3(x - 2)$. Dividing each side of the equation by $x - 2$ yields $x^2 - 1 = 3$. Is this second equation equivalent to the first equation?

3. A tutor claims that cubing each side of $(4x - 1)^{1/3} = -2$ will not introduce any extraneous solutions. Do you agree?

4. What would be an appropriate substitution that would enable you to write $x^{-2} - \dfrac{2}{x} = 15$ as a quadratic equation?

5. A classmate solves the equation $x^2 + y^2 = 25$ for y and produces the equation $y = \sqrt{25 - x^2}$. Do you agree with this result?

EXERCISE SET 1.4

In Exercises 1 to 12, solve each polynomial equation by factoring and using the zero product principle.

1. $x^3 - 25x = 0$

2. $x^3 - x = 0$

3. $x^3 - 2x^2 - x + 2 = 0$

4. $x^3 - 4x^2 - 2x + 8 = 0$

5. $2x^5 - 18x^3 = 0$

▶ **6.** $x^4 - 36x^2 = 0$

7. $x^4 - 3x^3 - 40x^2 = 0$

8. $x^4 + 3x^3 - 8x - 24 = 0$

9. $x^4 - 16x^2 = 0$

10. $x^4 - 16 = 0$

11. $x^3 - 8 = 0$

12. $x^3 + 8 = 0$

In Exercises 13 to 26, solve each rational equation and check your solution(s).

13. $\dfrac{3}{x + 2} = \dfrac{5}{2x - 7}$

▶ **14.** $\dfrac{4}{y + 2} = \dfrac{7}{y - 4}$

15. $\dfrac{30}{10 + x} = \dfrac{20}{10 - x}$

16. $\dfrac{6}{8 + x} = \dfrac{4}{8 - x}$

17. $\dfrac{3x}{x + 4} = 2 - \dfrac{12}{x + 4}$

18. $\dfrac{8}{2m + 1} - \dfrac{1}{m - 2} = \dfrac{5}{2m + 1}$

19. $2 + \dfrac{9}{r - 3} = \dfrac{3r}{r - 3}$

20. $\dfrac{t}{t - 4} + 3 = \dfrac{4}{t - 4}$

21. $\dfrac{5}{x - 3} - \dfrac{3}{x - 2} = \dfrac{4}{x - 3}$

22. $\dfrac{4}{x - 1} + \dfrac{7}{x + 7} = \dfrac{5}{x - 1}$

23. $\dfrac{x}{x - 3} = \dfrac{x + 4}{x + 2}$

24. $\dfrac{x}{x - 5} = \dfrac{x + 7}{x + 1}$

25. $\dfrac{x + 3}{x + 5} = \dfrac{x - 3}{x - 4}$

26. $\dfrac{x - 6}{x + 4} = \dfrac{x - 1}{x + 2}$

In Exercises 27 to 40, use the power principle to solve each radical equation. Check all proposed solutions.

27. $\sqrt{x - 4} - 6 = 0$

▶ 28. $\sqrt{10 - x} = 4$

29. $x = 3 + \sqrt{3 - x}$

▶ 30. $x = \sqrt{5 - x} + 5$

31. $\sqrt{3x - 5} - \sqrt{x + 2} = 1$

32. $\sqrt{6 - x} + \sqrt{5x + 6} = 6$

33. $\sqrt{2x + 11} - \sqrt{2x - 5} = 2$

▶ 34. $\sqrt{x + 7} - 2 = \sqrt{x - 9}$

35. $\sqrt{x + 7} + \sqrt{x - 5} = 6$

36. $x = \sqrt{12x - 35}$

37. $2x = \sqrt{4x + 15}$

38. $\sqrt[3]{7x - 3} = \sqrt[3]{2x + 7}$

39. $\sqrt[3]{2x^2 + 5x - 3} = \sqrt[3]{x^2 + 3}$

40. $\sqrt[4]{x^2 + 20} = \sqrt[4]{9x}$

In Exercises 41 to 56, find all the real solutions of each equation by first rewriting each equation as a quadratic equation.

41. $x^4 - 9x^2 + 14 = 0$

▶ 42. $x^4 - 10x^2 + 9 = 0$

43. $2x^4 - 11x^2 + 12 = 0$

44. $6x^4 - 7x^2 + 2 = 0$

45. $x^6 + x^3 - 6 = 0$

46. $6x^6 + x^3 - 15 = 0$

47. $x^{1/2} - 3x^{1/4} + 2 = 0$

48. $2x^{1/2} - 5x^{1/4} - 3 = 0$

49. $3x^{2/3} - 11x^{1/3} - 4 = 0$

▶ 50. $6x^{2/3} - 7x^{1/3} - 20 = 0$

51. $9x^4 = 30x^2 - 25$

52. $4x^4 - 28x^2 = -49$

53. $x^{2/5} - 1 = 0$

54. $2x^{2/5} - x^{1/5} = 6$

55. $9x - 52\sqrt{x} + 64 = 0$

56. $8x - 38\sqrt{x} + 9 = 0$

57. FENCE CONSTRUCTION A worker can build a fence in 8 hours. With the help of an assistant, the fence can be built in 5 hours. How long should it take the assistant, working alone, to build the fence?

58. ROOF REPAIR A roofer and an assistant can repair a roof together in 6 hours. Working alone the assistant can complete the repair in 14 hours. If both the roofer and the assistant work together for 2 hours and then the assistant is left alone to finish the job, how much longer should the assistant need to finish the repairs?

59. AVERAGE GOLF SCORE Renee has played four rounds of golf this season. Her average score is 92. If she can score 86 on each round she plays in the future, how many more rounds will she need to play to bring her average down to 88? (*Hint:* A player's average golf score is equal to the total number of strokes divided by the total number of rounds played.)

▶ 60. MEDICAL DOSAGE FOR A CHILD A physician has used Young's rule (see Example 3) to determine that Sandy should receive $\dfrac{1}{2}$ of an adult dose of a medication. How old is Sandy?

61. WRITING FOR A PARTICULAR READING LEVEL A writer of books for adolescents has decided to write a book that can be fully understood by adolescents at the sixth grade level. According to the SMOG reading grade level formula, what is the maximum number of words with three or more syllables that should appear in any sample of 30 sentences of the book? (*Hint:* See Example 5.)

▶ 62. WRITING FOR A PARTICULAR READING LEVEL A writer of books for children has decided to write a book that can be fully understood by children at the fourth grade level. According to the SMOG reading grade level formula, what is the maximum number of words with three or more syllables that should appear in any sample of 30 sentences of the book? (*Hint:* See Example 5.)

63. RADIUS OF A CONE A conical funnel has a height h of 4 inches and a lateral surface area L of 15π square inches. Find the radius r of the cone. (*Hint:* Use the formula $L = \pi r \sqrt{r^2 + h^2}$.)

64. DIAMETER OF A CONE As flour is poured onto a table, it forms a right circular cone whose height is one-third the diameter of the base. What is the diameter of the base when the cone has a volume of 192 cubic inches? Round to the nearest tenth of an inch.

65. PRECIOUS METALS A solid silver sphere has a diameter of 8 millimeters, and a second silver sphere has a diameter of 12 millimeters. The spheres are melted down and recast to form a single cube. What is the length s of each edge of the cube? Round your answer to the nearest tenth of a millimeter.

66. PENDULUM The period T of a pendulum is the time it takes the pendulum to complete one swing from left to right and back. For a pendulum near the surface of the earth,

$$T = 2\pi\sqrt{\frac{L}{32}}$$

where T is measured in seconds and L is the length of the pendulum in feet. Find the length of a pendulum that has a period of 4 seconds. Round to the nearest tenth of a foot.

67. DISTANCE TO THE HORIZON On a ship, the distance d that you can see to the horizon is given by $d = 1.5\sqrt{h}$, where h is the height of your eye measured in feet above sea level and d is measured in miles. How high is the eye level of a navigator who can see 14 miles to the horizon? Round to the nearest foot.

CONNECTING CONCEPTS

68. RADIUS OF A CIRCLE The radius r of a circle inscribed in a triangle with sides of lengths a, b, and c is given by

$$r = \sqrt{\frac{(s-a)(s-b)(s-c)}{s}}$$

where $s = \frac{1}{2}(a + b + c)$.

a. Find the length of the radius of a circle inscribed in a triangle with sides of 5 inches, 6 inches, and 7 inches. Round to the nearest hundredth of an inch.

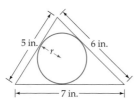

b. The radius of a circle inscribed in an equilateral triangle measures 2 inches. What is the exact length of each side of the equilateral triangle?

69. RADIUS OF A CIRCLE The radius r of a circle that is circumscribed about a triangle with sides of lengths a, b, and c is given by

$$r = \frac{abc}{4\sqrt{s(s-a)(s-b)(s-c)}}$$

where $s = \frac{1}{2}(a + b + c)$.

a. Find the radius of a circle that is circumscribed about a triangle with sides of 7 inches, 10 inches, and 15 inches. Round to the nearest hundredth of an inch.

b. A circle with radius 5 inches is circumscribed about an equilateral triangle (see the following figure). What is the exact length of each side of the equilateral triangle?

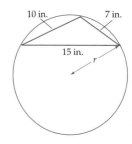

FIGURE FOR EXERCISE 69(b)

In Exercises 70 and 71, the depth s from the opening of a well to the water can be determined by measuring the total time between the instant you drop a stone and the moment you hear it hit the water. The time (in seconds) it takes the stone to hit the water is given by $\sqrt{s}/4$, where s is measured in feet. The time (also in seconds) required for the sound of the impact to travel up to your ears is given by $s/1100$. Thus the total time T (in seconds) between the instant you drop the stone and the moment you hear its impact is

$$T = \frac{\sqrt{s}}{4} + \frac{s}{1100}$$

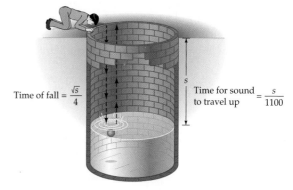

Time of fall $= \dfrac{\sqrt{s}}{4}$ Time for sound to travel up $= \dfrac{s}{1100}$

70. TIME OF FALL One of the world's deepest water wells is 7320 feet deep. Find the time between the instant you drop a stone and the time you hear it hit the water if the surface of the water is 7100 feet below the opening of the well. Round your answer to the nearest tenth of a second.

71. Solve $T = \dfrac{\sqrt{s}}{4} + \dfrac{s}{1100}$ for s.

72. DEPTH OF A WELL Use the result of Exercise 71 to determine the depth from the opening of a well to the water level if the time between the instant you drop a stone and the moment you hear its impact is 3 seconds. Round your answer to the nearest foot.

PREPARE FOR SECTION 1.5

73. Find: $\{x \mid x > 2\} \cap \{x \mid x > 5\}$ [P.1]

74. Evaluate $3x^2 - 2x + 5$ for $x = -3$. [P.1]

75. Evaluate $\dfrac{x + 3}{x - 2}$ for $x = 7$. [P.1/P.5]

76. Factor: $10x^2 + 9x - 9$ [P.4]

77. For what value of x is $\dfrac{x - 3}{2x - 7}$ undefined? [P.1/P.5]

78. Solve: $2x^2 - 11x + 15 = 0$ [1.3]

PROJECTS

1. **THE REDUCED CUBIC** The mathematician Francois Vieta knew a method of solving the "reduced cubic" $x^3 + mx + n = 0$ by using the substitution $x = \dfrac{m}{3z} - z$.

a. Show that this substitution results in the equation
$$z^6 - nz^3 - \frac{m^3}{27} = 0.$$

b. Show that the equation in **a.** is quadratic in form.

c. Solve the equation in **a.** for z.

d. Use your solution from **c.** to find the real solution of the equation $x^3 + 3x = 14$.

2. **FERMAT'S LAST THEOREM** One of the most famous theorems is known as *Fermat's Last Theorem*. Write an essay on Fermat's Last Theorem. Include information about

- the history of Fermat's Last Theorem.

- the relationship between Fermat's Last Theorem and the Pythagorean Theorem.

- Dr. Andrew Wiles's proof of Fermat's Last Theorem.

The following list includes a few of the sources you may wish to consult.

- *Fermat's Enigma,* by Simon Singh. Walker and Company, New York, 1997.

- *The Last Problem,* by Eric Temple Bell. The Mathematical Association of America, 1990.

- "Andrew Wiles: A Math Whiz Battles 350-Year-Old Puzzle," by Gina Kolata, *Math Horizons,* Winter 1993, pp. 8–11. The Mathematical Association of America.

- "Introduction to Fermat's Last Theorem," by David A. Cox, *The American Mathematical Monthly,* vol. 101, no. 1 (January 1994), pp. 3–14.

INEQUALITIES

● PROPERTIES OF INEQUALITIES

In Section P.1 we used inequalities to describe the order of real numbers and to represent subsets of real numbers. In this section we consider inequalities that involve a variable. In particular, we consider how to determine which real numbers make an inequality a true statement.

The **solution set** of an inequality is the set of all real numbers for which the inequality is a true statement. For instance, the solution set of $x + 1 > 4$ is the set of all real numbers greater than 3. Two inequalities are **equivalent inequalities** if they have the same solution set. We can solve many inequalities by producing *simpler* but equivalent inequalities until the solutions are readily apparent. To produce these simpler but equivalent inequalities, we often apply the following properties.

MATH MATTERS

Another property of inequalities, called the *transitive property*, states that for real numbers a, b, and c, if $a > b$ and $b > c$, then $a > c$. We say that the relationship "is greater than" is a transitive relationship.

Not all relationships are transitive relationships. For instance, consider the game of scissors, paper, rock. In this game, scissors wins over paper, paper wins over rock, but scissors does not win over rock!

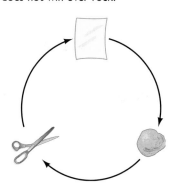

Properties of Inequalities

Let a, b, and c be real numbers.

1. *Addition-Subtraction Property*   If the same real number is added to or subtracted from each side of an inequality, the resulting inequality is equivalent to the original inequality.

   $a < b$ and $a + c < b + c$ are equivalent inequalities.

2. *Multiplication-Division Property*

   a. Multiplying or dividing each side of an inequality by the same *positive* real number produces an equivalent inequality.

   If $c > 0$, then $a < b$ and $ac < bc$ are equivalent inequalities.

   b. Multiplying or dividing each side of an inequality by the same *negative* real number produces an equivalent inequality provided the direction of the inequality symbol is *reversed*.

   If $c < 0$, then $a < b$ and $ac > bc$ are equivalent inequalities.

Note the difference between Property 2a and Property 2b. Property 2a states that an equivalent inequality is produced when each side of a given inequality is multiplied (divided) by the same *positive* real number and the inequality symbol is not changed. By contrast, Property 2b states that when each side of a given inequality is multiplied (divided) by a *negative* real number, we must *reverse* the direction of the inequality symbol to produce an equivalent inequality. For instance, multiplying both sides of $-b < 4$ by -1 produces the equivalent inequality $b > -4$. (We multiplied both sides of the first inequality by -1, and we changed the less than symbol to a greater than symbol.)

EXAMPLE 1   **Solve Linear Inequalities**

Solve each of the following inequalities.

a. $2x + 1 < 7$ **b.** $-3x - 2 \leq 10$

Solution

a. $2x + 1 < 7$

$\qquad 2x < 6$ • **Add -1 to each side and keep the inequality symbol as is.**

$\qquad x < 3$ • **Divide each side by 2 and keep the inequality symbol as is.**

The inequality $2x + 1 < 7$ is true for all real numbers less than 3. In set-builder notation the solution set is given by $\{x \mid x < 3\}$. In interval notation the solution set is $(-\infty, 3)$. See the following figure.

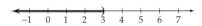

b. $-3x - 2 \leq 10$

$\qquad -3x \leq 12$ • **Add 2 to each side and keep the inequality symbol as is.**

$\qquad x \geq -4$ • **Divide each side by -3 and reverse the direction of the inequality symbol.**

The inequality $-3x - 2 \leq 10$ is true for all real numbers greater than or equal to -4. In set-builder notation the solution set is given by $\{x \mid x \geq -4\}$. In interval notation the solution set is $[-4, \infty)$. See the following figure.

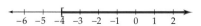

▶ **TRY EXERCISE 6, PAGE 140**

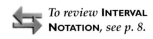

To review **INTERVAL NOTATION,** *see p. 8.*

---

take note

Solutions of inequalities are often stated using set-builder notation or interval notation. For instance, the solutions of $2x + 1 < 7$ can be written in set-builder notation as $\{x \mid x < 3\}$ or in interval notation as $(-\infty, 3)$.

● **COMPOUND INEQUALITIES**

A **compound inequality** is formed by joining two inequalities with the connective word *and* or *or*. The inequalities shown below are compound inequalities.

$$x + 1 > 3 \quad \text{and} \quad 2x - 11 < 7$$
$$x + 3 > 5 \quad \text{or} \quad x - 1 < 9$$

The solution set of a compound inequality with the connective word *or* is the *union* of the solution sets of the two inequalities. The solution set of a compound inequality with the connective word *and* is the *intersection* of the solution sets of the two inequalities.

EXAMPLE 2   **Solve Compound Inequalities**

Solve each compound inequality. Write each solution in set-builder notation.

a. $2x < 10$ or $x + 1 > 9$ **b.** $x + 3 > 4$ and $2x + 1 > 15$

Solution

a. $2x < 10 \qquad$ or $\qquad x + 1 > 9$

$\qquad x < 5 \qquad\qquad\qquad x > 8$ • **Solve each inequality.**

$\qquad \{x \,|\, x < 5\} \qquad\qquad \{x \,|\, x > 8\}$ • **Write each solution as a set.**

$\qquad \{x \,|\, x < 5\} \cup \{x \,|\, x > 8\} = \{x \,|\, x < 5 \text{ or } x > 8\}$ • **Write the union of the solution sets.**

b. $x + 3 > 4 \qquad$ and $\qquad 2x + 1 > 15$

$\qquad x > 1 \qquad\qquad\qquad 2x > 14$ • **Solve each inequality.**

$\qquad\qquad\qquad\qquad\qquad\qquad x > 7$

$\qquad \{x \,|\, x > 1\} \qquad\qquad \{x \,|\, x > 7\}$ • **Write each solution as a set.**

$\qquad \{x \,|\, x > 1\} \cap \{x \,|\, x > 7\} = \{x \,|\, x > 7\}$ • **Write the intersection of the solution sets.**

▶ **TRY EXERCISE 10, PAGE 140**

❓ **QUESTION** What is the solution set of the compound inequality $x > 1$ or $x < 3$?

take note

We reserve the notation $a < b < c$ to mean $a < b$ and $b < c$. Thus the solution set of $2 > x > 5$ is the empty set, because there are no numbers less than 2 and greater than 5.

The inequality given by

$$12 < x + 5 < 19$$

is equivalent to the compound inequality $12 < x + 5$ *and* $x + 5 < 19$. You can solve $12 < x + 5 < 19$ by either of the following methods.

Method 1 Find the intersection of the solution sets of the inequalities $12 < x + 5$ and $x + 5 < 19$.

$$12 < x + 5 \qquad \text{and} \qquad x + 5 < 19$$
$$7 < x \qquad \text{and} \qquad x < 14$$

The solution set is $\{x \,|\, x > 7\} \cap \{x \,|\, x < 14\} = \{x \,|\, 7 < x < 14\}$.

take note

The compound inequality $a < b$ and $b < c$ can be written in the compact form $a < b < c$. However, the compound inequality $a < b$ or $b > c$ cannot be expressed in a compact form.

Method 2 Subtract 5 from each of the three parts of the inequality.

$$12 < \quad x + 5 \quad < 19$$
$$12 - 5 < x + 5 - 5 < 19 - 5$$
$$7 < \quad x \quad < 14$$

The solution set is $\{x \,|\, 7 < x < 14\}$.

● **ABSOLUTE VALUE INEQUALITIES**

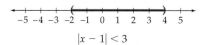

$|x - 1| < 3$

FIGURE 1.6

The solution set of the absolute value inequality $|x - 1| < 3$ is the set of all real numbers whose distance from 1 is *less than* 3. Therefore, the solution set consists of all numbers between -2 and 4. See **Figure 1.6.** In interval notation, the solution set is $(-2, 4)$.

❓ **ANSWER** The set of all real numbers. Using interval notation, the solution set is written as $(-\infty, \infty)$.

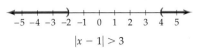

$|x - 1| > 3$

FIGURE 1.7

The solution set of the absolute value inequality $|x - 1| > 3$ is the set of all real numbers whose distance from 1 is *greater than* 3. Therefore, the solution set consists of all real numbers less than -2 *or* greater than 4. See **Figure 1.7**. In interval notation, the solution set is $(-\infty, -2) \cup (4, \infty)$.

The following properties are used to solve absolute value inequalities.

Properties of Absolute Value Inequalities

For any variable expression E and any nonnegative real number k,

$$|E| \le k \quad \text{if and only if} \quad -k \le E \le k$$
$$|E| \ge k \quad \text{if and only if} \quad E \le -k \quad \text{or} \quad E \ge k$$

These properties also hold true when the $<$ symbol is substituted for the $\le$ symbol and when the $>$ symbol is substituted for the $\ge$ symbol.

> ***take note***
>
> Some inequalities have a solution set that consists of all real numbers. For example, $|x + 9| \ge 0$ is true for all values of x. Because an absolute value is always nonnegative, the equation is always true.

In Example 3 we make use of the above properties to solve absolute value inequalities.

EXAMPLE 3   Solve Absolute Value Inequalities

Solve each of the following inequalities.

a. $|2 - 3x| < 7$ **b.** $|4x - 3| \ge 5$

Solution

a. $|2 - 3x| < 7$ if and only if $-7 < 2 - 3x < 7$. Solve this compound inequality.

$$-7 < 2 - 3x < 7$$
$$-9 < -3x < 5 \qquad \text{• Subtract 2 from each of the three parts of the inequality.}$$

$$3 > x > -\frac{5}{3} \qquad \text{• Multiply each part of the inequality by } -\frac{1}{3} \text{ and reverse the inequality symbols.}$$

In interval notation, the solution set is given by $\left(-\frac{5}{3}, 3\right)$. See **Figure 1.8**.

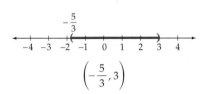

$\left(-\dfrac{5}{3}, 3\right)$

FIGURE 1.8

b. $|4x - 3| \ge 5$ implies $4x - 3 \le -5$ or $4x - 3 \ge 5$. Solving each of these inequalities produces

$$4x - 3 \le -5 \qquad \text{or} \qquad 4x - 3 \ge 5$$
$$4x \le -2 \qquad\qquad\qquad 4x \ge 8$$
$$x \le -\frac{1}{2} \qquad\qquad\qquad x \ge 2$$

The solution set is $\left(-\infty, -\dfrac{1}{2}\right] \cup [2, \infty)$. See **Figure 1.9**.

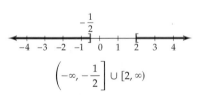

$\left(-\infty, -\dfrac{1}{2}\right] \cup [2, \infty)$

FIGURE 1.9

▶ **TRY EXERCISE 18, PAGE 140**

• THE CRITICAL VALUE METHOD

Any value of x that causes a polynomial in x to equal zero is called a **zero of the polynomial.** For example, -4 and 1 are both zeros of the polynomial $x^2 + 3x - 4$, because $(-4)^2 + 3(-4) - 4 = 0$ and $1^2 + 3 \cdot 1 - 4 = 0$.

A Sign Property of Polynomials

Polynomials in x have the property that for all values of x between two consecutive real zeros, all values of the polynomial are positive or all values of the polynomial are negative.

In our work with inequalities that involve polynomials, the real zeros of the polynomial are also referred to as **critical values of the inequality.** On a number line the critical values of an inequality separate the real numbers that make the inequality true from those that make it false. In Example 4, we use critical values and the sign property of polynomials to solve an inequality.

EXAMPLE 4   Solve a Quadratic Inequality

Solve: $x^2 + 3x - 4 < 0$

Solution

Factoring the polynomial $x^2 + 3x - 4$ produces the equivalent inequality

$$(x + 4)(x - 1) < 0$$

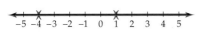

FIGURE 1.10

The zeros of the polynomial $x^2 + 3x - 4$ are -4 and 1. They are the critical values of the inequality $x^2 + 3x - 4 < 0$. They separate the real number line into the three intervals shown in **Figure 1.10.**

To determine the intervals on which $x^2 + 3x - 4 < 0$, pick a number called a **test value** from each of the three intervals and then determine whether $x^2 + 3x - 4 < 0$ for each of these test values. For example, in the interval $(-\infty, -4)$, pick a test value of, say, -5. Then

$$x^2 + 3x - 4 = (-5)^2 + 3(-5) - 4 = 6$$

Because 6 is not less than 0, by the sign property of polynomials, no number in the interval $(-\infty, -4)$ makes $x^2 + 3x - 4 < 0$.

Now pick a test value from the interval $(-4, 1)$, say, 0. When $x = 0$,

$$x^2 + 3x - 4 = 0^2 + 3(0) - 4 = -4$$

Because -4 is less than 0, by the sign property of polynomials, all numbers in the interval $(-4, 1)$ make $x^2 + 3x - 4 < 0$.

If we pick a test value of 2 from the interval $(1, \infty)$, then

$$x^2 + 3x - 4 = (2)^2 + 3(2) - 4 = 6$$

Because 6 is not less than 0, by the sign property of polynomials, no number in the interval $(1, \infty)$ makes $x^2 + 3x - 4 < 0$.

Continued ▶

The following table is a summary of our work.

Interval	Test Value x	$x^2 + 3x - 4 \overset{?}{<} 0$
$(-\infty, -4)$	-5	$(-5)^2 + 3(-5) - 4 < 0$ $6 < 0$ False
$(-4, 1)$	0	$(0)^2 + 3(0) - 4 < 0$ $-4 < 0$ True
$(1, \infty)$	2	$(2)^2 + 3(2) - 4 < 0$ $6 < 0$ False

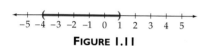

FIGURE 1.11

In interval notation, the solution set of $x^2 + 3x - 4 < 0$ is $(-4, 1)$. The solution set is graphed in **Figure 1.11.** Note that in this case the critical values -4 and 1 are not included in the solution set because they do not make $x^2 + 3x - 4$ less than 0.

▶ **TRY EXERCISE 34, PAGE 140**

To avoid the arithmetic in Example 4, we often use a *sign diagram*. For example, note that the factor $(x + 4)$ is negative for all $x < -4$ and positive for all $x > -4$. The factor $(x - 1)$ is negative for all $x < 1$ and positive for all $x > 1$. These results are shown in **Figure 1.12.**

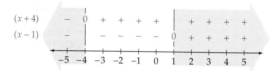

A sign diagram for $(x + 4)$ and $(x - 1)$.

FIGURE 1.12

To determine on which intervals the product $(x + 4)(x - 1)$ is negative, we examine the sign diagram to see where the factors have opposite signs. This occurs only on the interval $(-4, 1)$, where $(x + 4)$ is positive and $(x - 1)$ is negative, so the original equality is true only on the interval $(-4, 1)$.

Following is a summary of the steps used to solve polynomial inequalities by the critical value method.

Solving a Polynomial Inequality by the Critical Value Method

1. Write the inequality so that one side of the inequality is a nonzero polynomial and the other side is 0.

2. Find the real zeros of the polynomial.[3] They are the critical values of the original inequality.

3. Use test values to determine which of the consecutive intervals formed by the critical values are to be included in the solution set.

---

[3]In Chapter 3, additional ways to find the zeros of a polynomial are developed. For the present, however, we will find the zeros by factoring or by using the quadratic formula.

• RATIONAL INEQUALITIES

A rational expression is the quotient of two polynomials. **Rational inequalities** involve rational expressions, and they can be solved by an extension of the critical value method.

Critical Values of a Rational Expression

The **critical values of a rational expression** are the numbers that cause the numerator of the rational expression to equal zero or the denominator of the rational expression to equal zero.

Rational expressions also have the property that they remain either positive for all values of the variable between consecutive critical values or negative for all values of the variable between consecutive critical values.

Following is a summary of the steps used to solve rational inequalities by the critical value method.

Solving a Rational Inequality by the Critical Value Method

1. Write the inequality so that one side of the inequality is a rational expression and the other side is 0.

2. Find the real zeros of the numerator of the rational expression and the real zeros of its denominator. They are the critical values of the inequality.

3. Use test values to determine which of the consecutive intervals formed by the critical values are to be included in the solution set.

EXAMPLE 5   Solve a Rational Inequality

Solve: $\dfrac{3x + 4}{x + 1} \le 2$

Solution

Write the inequality so that 0 appears on the right side of the inequality.

$$\frac{3x + 4}{x + 1} \le 2$$

$$\frac{3x + 4}{x + 1} - 2 \le 0$$

Continued ▶

Write the left side as a rational expression.

$$\frac{3x+4}{x+1} - \frac{2(x+1)}{x+1} \le 0 \qquad \text{• The LCD is } x+1.$$

$$\frac{3x+4-2x-2}{x+1} \le 0 \qquad \text{• Simplify.}$$

$$\frac{x+2}{x+1} \le 0$$

The critical values of this inequality are -2 and -1 because the numerator $x+2$ is equal to zero when $x = -2$, and the denominator $x+1$ is equal to zero when $x = -1$. The critical values -2 and -1 separate the real number line into the three intervals $(-\infty, -2)$, $(-2, -1)$, and $(-1, \infty)$.

All values of x on the interval $(-2, -1)$ make $\dfrac{x+2}{x+1}$ negative, as desired. On the other intervals, the quotient $\dfrac{x+2}{x+1}$ is positive. See the sign diagram in **Figure 1.13**.

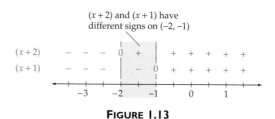

FIGURE 1.13

The solution set is $[-2, -1)$. The graph of the solution set is shown in **Figure 1.14**. Note that -2 is included in the solution set because $\dfrac{x+2}{x+1} = 0$ when $x = -2$. However, -1 is not included in the solution set because the denominator $(x+1)$ is zero when $x = -1$.

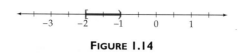

FIGURE 1.14

▶ **TRY EXERCISE 46, PAGE 140**

● APPLICATIONS

Many applied problems can be solved by using inequalities.

EXAMPLE 6   **Solve an Application Concerning Leases**

A real estate company needs a new copy machine. The company has decided to lease either the model ABC machine for $75 a month plus 5 cents per copy or the model XYZ machine for $210 a month and 2 cents per copy. Under what conditions is it less expensive to lease the XYZ machine?

Solution

Let x represent the number of copies the company produces per month. The dollar costs per month are $75 + 0.05x$ for model ABC and $210 + 0.02x$ for model XYZ. It will be less expensive to lease model XYZ provided

$$210 + 0.02x < 75 + 0.05x$$
$$210 - 0.03x < 75 \qquad \text{• Subtract 0.05x from each side.}$$
$$-0.03x < -135 \qquad \text{• Subtract 210 from each side.}$$
$$x > 4500 \qquad \text{• Divide each side by } -0.03. \text{ Reverse the inequality symbol.}$$

The company will find it less expensive to lease model XYZ if it produces over 4500 copies per month.

▶ **TRY EXERCISE 52, PAGE 141**

EXAMPLE 7   **Solve an Application Concerning Test Scores**

Tyra has test scores of 70 and 81 in her biology class. To receive a C grade, she must obtain an average greater than or equal to 72 but less than 82. What range of test scores on the one remaining test will enable Tyra to get a C for the course?

Solution

The average of three test scores is the sum of the scores divided by 3. Let x represent Tyra's next test score. The requirements for a C grade produce the following inequality:

$$72 \leq \frac{70 + 81 + x}{3} < 82$$
$$216 \leq 70 + 81 + x < 246 \qquad \text{• Multiply each part of the inequality by 3.}$$
$$216 \leq 151 + x < 246 \qquad \text{• Simplify.}$$
$$65 \leq x < 95 \qquad \text{• Solve for } x \text{ by subtracting 151 from each part of the inequality.}$$

To get a C in the course, Tyra's remaining test score must be in the interval $[65, 95)$.

▶ **TRY EXERCISE 58, PAGE 141**

In many business applications a company is interested in the cost C of manufacturing x items, the revenue R generated by selling all the items, and the profit P made by selling the items.

In the next example the cost of manufacturing x tennis racquets is given by $C = 32x + 120,000$ dollars. The 120,000 represents the fixed cost because it remains constant regardless of how many racquets are manufactured. The $32x$ represents the variable cost because this term varies depending on how many racquets are manufactured. Each additional racquet costs the company an additional $32.

The revenue received from the sale of x tennis racquets is given by $R = x(200 - 0.01x)$ dollars. The quantity $(200 - 0.01x)$ is the price the company charges for each tennis racquet. The price varies depending on the number of racquets that are manufactured. For instance, if the number of racquets x that the company manufactures is small, the company will be able to demand almost $200 for each racquet. As the number of racquets that the company manufactures increases (approaches 20,000), the company will only be able to sell *all* the racquets if it decreases the price of each racquet.

The following profit formula shows the relationship between profit P, revenue R, and cost C:

$$P = R - C$$

EXAMPLE 8   Solve a Business Application

A company determines that the cost C, in dollars, of producing x tennis racquets is $C = 32x + 120,000$. The revenue R, in dollars, from selling all of the tennis racquets is $R = x(200 - 0.01x)$.

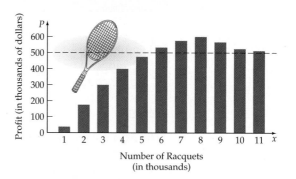

Number of Racquets
(in thousands)

How many racquets should the company manufacture and sell if the company wishes to earn a profit of at least $500,000?

Solution

The profit is given by

$$\begin{aligned}
P &= R - C \\
&= x(200 - 0.01x) - (32x + 120,000) \\
&= 200x - 0.01x^2 - 32x - 120,000 \\
&= -0.01x^2 + 168x - 120,000
\end{aligned}$$

The profit will be at least $500,000 provided

$$-0.01x^2 + 168x - 120,000 \geq 500,000$$
$$-0.01x^2 + 168x - 620,000 \geq 0$$

Using the quadratic formula, we find that the approximate critical values of this last inequality are 5474.3 and 11,325.7. Test values show that the inequality is positive only on the interval (5474.3, 11,325.7). The company should manufacture at least 5475 tennis racquets but not more than 11,325 tennis racquets to produce the desired profit.

▶ **TRY EXERCISE 54, PAGE 141**

EXAMPLE 9   **Solve an Application Involving Batting Averages**

 During a recent season, Sammy Sosa had 53 hits out of 163 at-bats. At that time his batting average was approximately 0.325. If Sosa goes into a batting slump in which he gets no hits, how many more at-bats will it take for his batting average to fall below 0.300?

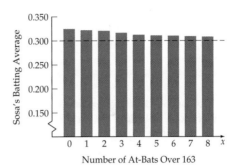

Number of At-Bats Over 163

Solution

A baseball player's batting average is determined by dividing the player's number of hits by the number of times the player has been at bat. Let x be the number of additional at-bats that Sosa takes over 163. During this period, his batting average will be $\dfrac{53}{163 + x}$, and we wish to solve

$$\frac{53}{163 + x} < 0.300$$

This rational inequality can be solved by using the critical value method, but there is an easier method. In this application we know that $163 + x$ is positive. Thus, if we multiply each side of the preceding inequality by $163 + x$, we will obtain the linear inequality $53 < 48.9 + 0.300x$, with the condition that x is a positive integer. Solving this inequality produces

$$53 < 48.9 + 0.300x$$
$$4.1 < 0.300x$$
$$x > 13.\overline{6}$$

Because x must be a positive integer, Sosa's average will fall below 0.300 if he goes hitless for 14 or more at-bats.

▶ **TRY EXERCISE 66, PAGE 142**

TOPICS FOR DISCUSSION

1. If $x < y$, then $y > x$. Do you agree?

2. Can the solution set of the compound inequality

$$x < -3 \text{ or } x > 5$$

be expressed as $-3 > x > 5$? Explain.

3. If $-a < b$, then it must be true that $a > -b$. Do you agree? Explain.

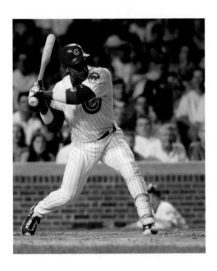

4. Do the inequalities $x < 4$ and $x^2 < 4^2$ both have the same solution set? Explain.

5. True or False: If $k < 0$, then $|k| = -k$.

EXERCISE SET 1.5

In Exercises 1 to 8, use the properties of inequalities to solve each inequality. Write the solution set using set-builder notation, and graph the solution set.

1. $2x + 3 < 11$

2. $3x - 5 > 16$

3. $x + 4 > 3x + 16$

4. $5x + 6 < 2x + 1$

5. $-3(x + 2) \le 5x + 7$

▶ 6. $-4(x - 5) \ge 2x + 15$

7. $-4(3x - 5) > 2(x - 4)$

8. $3(x + 7) \le 5(2x - 8)$

In Exercises 9 to 16, solve each compound inequality. Write the solution set using set-builder notation, and graph the solution set.

9. $4x + 1 > -2$ and $4x + 1 \le 17$

▶ 10. $2x + 5 > -16$ and $2x + 5 < 9$

11. $10 \ge 3x - 1 \ge 0$

12. $0 \le 2x + 6 \le 54$

13. $x + 2 < -1$ or $x + 3 \ge 2$

14. $x + 1 > 4$ or $x + 2 \le 3$

15. $-4x + 5 > 9$ or $4x + 1 < 5$

16. $2x - 7 \le 15$ or $3x - 1 \le 5$

In Exercises 17 to 28, use interval notation to express the solution set of each inequality.

17. $|2x - 1| > 4$

▶ 18. $|2x - 9| < 7$

19. $|x + 3| \ge 5$

20. $|x - 10| \ge 2$

21. $|3x - 10| \le 14$

22. $|2x - 5| \ge 1$

23. $|4 - 5x| \ge 24$

24. $|3 - 2x| \le 5$

25. $|x - 5| \ge 0$

26. $|x - 7| \ge 0$

27. $|x - 4| \le 0$

28. $|2x + 7| \le 0$

In Exercises 29 to 36, use the critical value method to solve each polynomial inequality. Use interval notation to write each solution set.

29. $x^2 + 7x > 0$

30. $x^2 - 5x \le 0$

31. $x^2 - 16 \le 0$

32. $x^2 - 49 > 0$

33. $x^2 + 7x + 10 < 0$

▶ 34. $x^2 + 5x + 6 < 0$

35. $x^2 - 3x \ge 28$

36. $x^2 < -x + 30$

In Exercises 37 to 50, use the critical value method to solve each rational inequality. Write each solution set in interval notation.

37. $\dfrac{x + 4}{x - 1} < 0$

38. $\dfrac{x - 2}{x + 3} > 0$

39. $\dfrac{x - 5}{x + 8} \ge 3$

40. $\dfrac{x - 4}{x + 6} \le 1$

41. $\dfrac{x}{2x + 7} \ge 4$

42. $\dfrac{x}{3x - 5} \le -5$

43. $\dfrac{(x + 1)(x - 4)}{x - 2} < 0$

44. $\dfrac{x(x - 4)}{x + 5} > 0$

45. $\dfrac{x + 2}{x - 5} \le 2$

▶ 46. $\dfrac{3x + 1}{x - 2} \ge 4$

47. $\dfrac{6x^2 - 11x - 10}{x} > 0$

48. $\dfrac{3x^2 - 2x - 8}{x - 1} \ge 0$

49. $\dfrac{x^2 - 6x + 9}{x - 5} \le 0$

50. $\dfrac{x^2 + 10x + 25}{x + 1} \ge 0$

51. PERSONAL FINANCE A bank offers two checking account plans. The monthly fee and charge per check for each plan are shown below. Under what conditions is it less expensive to use the LowCharge plan?

Account Plan	Monthly Fee	Charge per Check
LowCharge	$5.00	$.01
FeeSaver	$1.00	$.08

▶ **52. PERSONAL FINANCE** You can rent a car for the day from Company A for $29.00 plus $0.12 a mile. Company B charges $22.00 plus $0.21 a mile. Find the number of miles *m* (to the nearest mile) per day for which it is cheaper to rent from Company A.

53. SHIPPING REQUIREMENTS United Parcel Service (UPS) will only ship packages for which the length is less than or equal to 108 inches and the length plus the girth is less than or equal to 130 inches. The length of a package is defined as the length of the longest side. The girth is defined as twice the width plus twice the height of the package. If a box has a length of 34 inches and a width of 22 inches, determine the possible range of heights *h* for this package if you wish to ship it by UPS. (*Source:* http://www.iship.com.)

▶ **54. MOVIE TICKET PRICES** The average U.S. movie ticket price *P*, in dollars, can be modeled by

$$P = 0.218t + 4.02, \quad t \geq 0$$

where *t* = 0 represents the year 1994. According to this model, in what year will the average price of a movie ticket first exceed $6.50? (*Source:* National Association of Theatre Owners, http://www.natoonline.org/satistics-tickets.htm.)

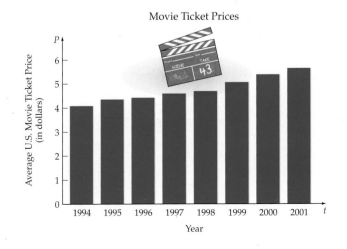

Movie Ticket Prices

55. PERSONAL FINANCE A sales clerk has a choice between two payment plans. Plan A pays $100.00 a week plus $8.00 a sale. Plan B pays $250.00 a week plus $3.50 a sale. How many sales per week must be made for plan A to yield the greater paycheck?

56. PERSONAL FINANCE A video store offers two rental plans. The yearly membership fee and the daily charge per video for each plan are shown below. How many one-night rentals can be made per year if the No-fee plan is to be the less expensive of the plans?

THE VIDEO STORE

Rental Plan	Yearly Fee	Daily Charge per Video
Low-rate	$15.00	$1.49
No-fee	None	$1.99

57. AVERAGE TEMPERATURES The average daily minimum-to-maximum temperature range for the city of Palm Springs during the month of September is 68 to 104 degrees Fahrenheit. What is the corresponding temperature range measured on the Celsius temperature scale? (*Hint:* Let *F* be the average daily temperature. Then $68 \leq F \leq 104$. Now substitute $\frac{9}{5}C + 32$ for *F* and solve the resulting inequality for *C*.)

▶ **58. AVERAGE TEMPERATURES** The average daily minimum-to-maximum temperature range for the city of Palm Springs during the month of January is 41 to 68 degrees Fahrenheit. What is the corresponding temperature range measured on the Celsius temperature scale? (*Hint:* See Exercise 57.)

59. CONSECUTIVE EVEN INTEGERS The sum of three consecutive even integers is between 36 and 54. Find all possible sets of integers that satisfy these conditions.

60. CONSECUTIVE ODD INTEGERS The sum of three consecutive odd integers is between 63 and 81. Find all possible sets of integers that satisfy these conditions.

61. FORENSIC SCIENCE Forensic specialists can estimate the height of a deceased person from the lengths of the person's bones. These lengths are substituted into mathematical inequalities. For instance, an inequality that relates the height *h*, in centimeters, of an adult female and the length *f*, in centimeters, of her femur is

$$|h - (2.47f + 54.10)| \leq 3.72$$

Use this inequality to estimate the possible range of heights, rounded to the nearest 0.1 centimeter, for an adult female whose femur measures 32.24 centimeters.

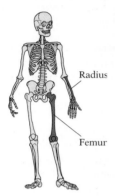

Radius

Femur

62. **FORENSIC SCIENCE** An inequality that is used to calculate the height h of an adult male from the length r of his radius is

$$|h - (3.32r + 85.43)| \leq 4.57$$

where h and r are both in centimeters. Use this inequality to estimate the possible range of heights for an adult male whose radius measures 26.36 centimeters.

63. **REVENUE** The monthly revenue R for a product is given by $R = 420x - 2x^2$, where x is the price in dollars of each unit produced. Find the interval, in terms of x, for which the monthly revenue is greater than zero.

64. **REVENUE** A shoe manufacturer finds that the monthly revenue R from a particular style of aerobics shoe is given by $R = 312x - 3x^2$, where x is the price in dollars of each pair of shoes sold. Find the interval, in terms of x, for which the monthly revenue is greater than or equal to $5925.

65. **PUBLISHING** A publisher has determined that if x books are published, the average cost per book is given by

$$\overline{C} = \frac{14.25x + 350,000}{x}$$

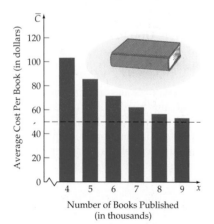

Number of Books Published
(in thousands)

How many books should be published if the company wants to bring the average cost per book below $50?

▶ **66.** **MANUFACTURING** A company manufactures running shoes. The company has determined that if it manufactures x pairs of shoes, the average cost, in dollars, per pair is

$$\overline{C} = \frac{0.00014x^2 + 12x + 400,000}{x}$$

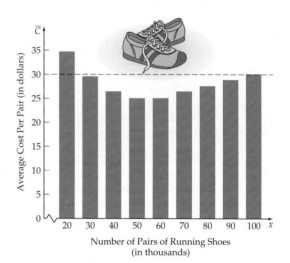

Number of Pairs of Running Shoes
(in thousands)

How many pairs of running shoes should the company manufacture if it wishes to bring the average cost below $30 per pair?

CONNECTING CONCEPTS

67. TOLERANCE A machinist is producing a circular cylinder on a lathe. The circumference of the cylinder must be 28 inches, with a tolerance of 0.15 inch. What maximum and minimum radii (to the nearest 0.001 inch) must the machinist stay between to produce an acceptable cylinder?

68. TOLERANCE A tall, narrow beaker has an inner radius of 2 centimeters. How high h (to the nearest 0.1 centimeter) should we fill the beaker if we need to measure $\dfrac{3}{4}$ liter (750 cubic centimeters) of a solution with an error of 15 cubic centimeters or less?

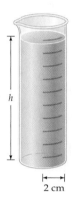

h

2 cm

In Exercises 69 to 72, use the critical value method to solve each inequality. Use interval notation to write each solution set.

69. $\dfrac{(x-3)^2}{(x-6)^2} > 0$

70. $\dfrac{(x-1)^2}{(x-4)^4} \geq 0$

71. $\dfrac{(x-4)^2}{(x+3)^3} \geq 0$

72. $\dfrac{(2x-7)}{(x-1)^2(x+2)^2} \geq 0$

In Exercises 73 to 78, use interval notation to express the solution set of each inequality.

73. $1 < |x| < 5$

74. $2 < |x| < 3$

75. $3 \leq |x| < 7$

76. $0 < |x| \leq 3$

77. $0 < |x - a| < \delta \quad (\delta > 0)$

78. $0 < |x - 5| < 2$

79. HEIGHT OF A PROJECTILE The equation

$$s = -16t^2 + v_0 t + s_0$$

gives the height s, in feet above ground level, at the time t seconds, of an object thrown directly upward from a height s_0 feet above the ground and with an initial velocity of v_0 feet per second. A ball is thrown directly upward from ground level with an initial velocity of 64 feet per second. Find the time interval during which the ball has a height of more than 48 feet.

80. HEIGHT OF A PROJECTILE A ball is thrown directly upward from a height of 32 feet above the ground with an initial velocity of 80 feet per second. Find the time interval during which the ball will be more than 96 feet above the ground. (*Hint:* See Exercise 79.)

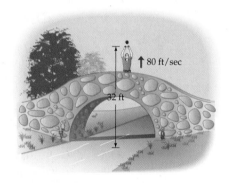

80 ft/sec

32 ft

PREPARE FOR SECTION 1.6

81. Solve $1820 = k(28)$ for k. [1.1]

82. Solve $20 = \dfrac{k}{1.5^2}$ for k. [1.1]

83. Evaluate $k\dfrac{3}{5^2}$ given that $k = 225$. [P.1]

84. Evaluate $k\dfrac{4.5 \cdot 32}{8^2}$ given that $k = 12.5$. [P.1]

85. If the length of each side of a square is doubled, what effect does this have on its area? [P.1/P.2]

86. If the radius of a cylinder is tripled, does this triple the volume of the cylinder? [P.1/P.2]

PROJECTS

1. **TRIANGLES** In any triangle, the sum of the lengths of the two shorter sides must be greater than the length of the longest side. Find all possible values of x if a triangle has sides of lengths

  a. $x, x + 5$, and $x + 9$ **b.** $x, x^2 + x$, and $2x^2 + x$

  c. $\dfrac{1}{x + 2}, \dfrac{1}{x + 1}$, and $\dfrac{1}{x}$

2. **FAIR COINS** A coin is considered a **fair** coin if it has an equal chance of landing heads up or tails up. To decide whether a coin is a fair coin, a statistician tosses it 1000 times and records the number of tails t. The statistician is prepared to state that the coin is a fair coin if

$$\left| \frac{t - 500}{15.81} \right| \le 2.33$$

  a. Determine what values of t will cause the statistician to state that the coin is a fair coin.

  b. Pick a coin and test it according to the criteria above to see whether it is a fair coin.

SECTION 1.6

VARIATION AND APPLICATIONS

- **DIRECT VARIATION**
- **INVERSE VARIATION**
- **JOINT VARIATION AND COMBINED VARIATION**

• DIRECT VARIATION

Many real-life situations involve variables that are related by a type of equation called a **variation.** For example, a stone thrown into a pond generates circular ripples whose circumference and diameter are increasing. The equation $C = \pi d$ expresses the relationship between the circumference C of a circle and its diameter d. If d increases, then C increases. The circumference C is said to *vary directly* as the diameter d.

Definition of Direct Variation

The variable y **varies directly** as the variable x, or y is **directly proportional** to x, if and only if

$$y = kx$$

where k is a constant called the **constant of proportionality** or the **variation constant.**

Direct variations occur in many daily applications. For example, suppose the cost of a newspaper is 50 cents. The cost C to purchase n newspapers is directly proportional to the number n. That is, $C = 50n$. In this example the variation constant is 50.

To solve a problem that involves a variation, we typically write a general equation that relates the variables and then use given information to solve for the variation constant.

EXAMPLE 1   **Solve a Direct Variation**

The distance sound travels varies directly as the time it travels. If sound travels 1340 meters in 4 seconds, find the distance sound will travel in 5 seconds.

Solution

Write an equation that relates the distance d to the time t. Because d varies directly as t, our equation is $d = kt$. Because $d = 1340$ when $t = 4$, we obtain

$$1340 = k \cdot 4 \quad \text{which implies} \quad k = \frac{1340}{4} = 335$$

Therefore, the specific equation that relates the distance d sound travels in t seconds is $d = 335t$. To find the distance sound travels in 5 seconds, replace t with 5 to produce

$$d = 335(5) = 1675$$

Under the same conditions, sound will travel 1675 meters in 5 seconds. See **Figure 1.15.**

▶ **TRY EXERCISE 22, PAGE 151**

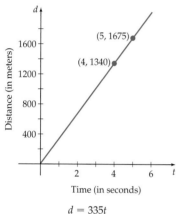

$d = 335t$

FIGURE 1.15

Direct Variation as the *n*th Power

If *y* **varies directly as the *n*th power** of *x*, then

$$y = kx^n$$

where k is a constant.

EXAMPLE 2   **Solve a Variation of the Form $y = kx^2$**

The distance s that an object falls from rest (neglecting air resistance) varies directly as the square of the time t that it has been falling. If an object falls 64 feet in 2 seconds, how far will it fall in 10 seconds?

Solution

Because s varies directly as the square of t, $s = kt^2$. The variable s is 64 when t is 2, so

$$64 = k \cdot 2^2 \quad \text{which implies} \quad k = \frac{64}{4} = 16$$

The specific equation that relates the distance s an object falls in t seconds is $s = 16t^2$. Letting $t = 10$ yields

$$s = 16(10^2) = 16(100) = 1600$$

Under the same conditions, the object will fall 1600 feet in 10 seconds. See **Figure 1.16.**

▶ **TRY EXERCISE 26, PAGE 151**

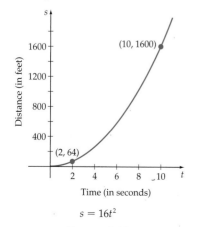

$s = 16t^2$

FIGURE 1.16

● **INVERSE VARIATION**

Two variables also can vary *inversely*.

> **Definition of Inverse Variation**
>
> The variable y **varies inversely** as the variable x, or y is **inversely proportional** to x, if and only if
>
> $$y = \frac{k}{x}$$
>
> where k is the variation constant.

In 1661, Robert Boyle made a study of the *compressibility* of gases. **Figure 1.17** shows that he used a J-shaped tube to demonstrate the inverse relationship between the volume of a gas at a given temperature and the applied pressure. The J-shaped tube on the left shows that the volume of a gas at normal atmospheric pressure is 60 milliliters. If the pressure is doubled by adding mercury (Hg), as shown in the middle tube, the volume of the gas is halved to 30 milliliters. Tripling the pressure decreases the volume of the gas to 20 milliliters, as shown in the tube at the right.

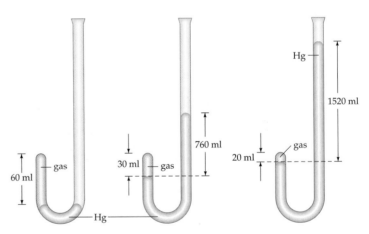

FIGURE 1.17

EXAMPLE 3   **Solve an Inverse Variation**

Boyle's Law states that the volume V of a sample of gas (at a constant temperature) varies inversely as the pressure P. The volume of a gas in a J-shaped tube is 75 milliliters when the pressure is 1.5 atmospheres. Find the volume of the gas when the pressure is increased to 2.5 atmospheres.

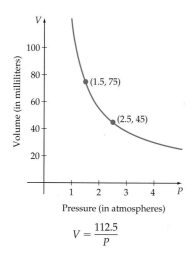

$$V = \frac{112.5}{P}$$

FIGURE 1.18

---

take note

Because the volume V varies inversely as the pressure P, the function $V = \frac{112.5}{P}$ is a decreasing function, as shown in Figure 1.18.

Solution

The volume V varies inversely as the pressure P, so $V = \frac{k}{P}$. The volume V is 75 milliliters when the pressure is 1.5 atmospheres, so

$$75 = \frac{k}{1.5} \quad \text{and} \quad k = (75)(1.5) = 112.5$$

Thus $V = \frac{112.5}{P}$. When the pressure is 2.5 atmospheres, we have

$$V = \frac{112.5}{2.5} = 45 \text{ milliliters}$$

See **Figure 1.18.**

▶ **TRY EXERCISE 30, PAGE 151**

Many real-world situations can be modeled by inverse variations that involve a power.

Inverse Variation as the *n*th Power

If y varies inversely as the *n*th power of x, then

$$y = \frac{k}{x^n}$$

where k is a constant and $n > 0$.

❓ **QUESTION** Consider the variation $y = \frac{k}{x}$, with $k > 0$. What happens to y as x increases?

EXAMPLE 4   **Solve an Inverse Variation Involving a Power**

The intensity of music, measured in decibels (dB), is inversely proportional to the square of the distance from the music source to a listener. During a party, the decibel level of the music measured 100 decibels at a distance of 12 feet from a loud speaker. What was the decibel level of the music for a dancer who was only 10 feet from the speaker?

Continued ▶

❓ **ANSWER**   As x increases, y decreases.

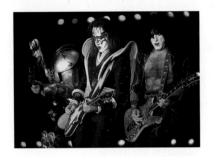

Solution

The decibel level L is inversely proportional to the square of the distance d between the loud speaker and a listener. The general variation is given by

$$L = \frac{k}{d^2}$$

We are given that $L = 100$ decibels when $d = 12$ feet. Substituting these values into the preceding variation allows us to solve for the variation constant k.

$$100 = \frac{k}{12^2}$$
$$k = 12^2(100)$$
$$= 14,400$$

The specific variation formula is

$$L = \frac{14,400}{d^2}$$

When $d = 10$ feet, the decibel level of the music is

$$L = \frac{14,400}{10^2}$$
$$= 144$$

At a distance of 10 feet from the speaker, the decibel level of the music is 144 decibels.

▶ **TRY EXERCISE 32, PAGE 152**

● JOINT VARIATION AND COMBINED VARIATION

Some variations involve more than two variables.

Definition of Joint Variation

The variable z **varies jointly** as the variables x and y if and only if

$$z = kxy$$

where k is a constant.

EXAMPLE 5   Solve a Joint Variation

The cost of insulating the ceiling of a house varies jointly as the thickness of the insulation and the area of the ceiling. It costs $175 to insulate a 2100-square-foot ceiling with insulation that is 4 inches thick. Find the cost of insulating a 2400-square-foot ceiling with insulation that is 6 inches thick.

Solution

Because the cost C varies jointly as the area A of the ceiling and the thickness T of the insulation, we know $C = kAT$. Using the fact that $C = 175$ when $A = 2100$ and $T = 4$ gives us

$$175 = k(2100)(4) \quad \text{which implies} \quad k = \frac{175}{(2100)(4)} = \frac{1}{48}$$

Consequently, the specific formula for C is $C = \dfrac{1}{48}AT$. Now, when $A = 2400$ and $T = 6$, we have

$$C = \frac{1}{48}(2400)(6) = 300$$

The cost of insulating the 2400-square-foot ceiling with 6-inch insulation is $300.

▶ **TRY EXERCISE 34, PAGE 152**

Combined variations involve more than one type of variation.

EXAMPLE 6   **Solve a Combined Variation**

The weight that a horizontal beam with a rectangular cross section can safely support varies jointly as the width and square of the depth of the cross section and inversely as the length of the beam. See **Figure 1.19.** If a 4-inch by 4-inch beam 10 feet long safely supports a load of 256 pounds, what load L can be safely supported by a beam made of the same material and with a width w of 4 inches, a depth d of 6 inches, and a length l of 16 feet?

Solution

The general variation equation is $L = k\dfrac{wd^2}{l}$. Using the given data yields

$$256 = k\frac{4(4^2)}{10}$$

Solving for k produces $k = 40$, so the specific formula for L is

$$L = 40\frac{wd^2}{l}$$

Substituting 4 for w, 6 for d, and 16 for l gives

$$L = 40\frac{4(6^2)}{16} = 360 \text{ pounds}$$

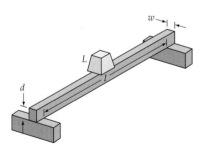

FIGURE 1.19

▶ **TRY EXERCISE 38, PAGE 152**

TOPICS FOR DISCUSSION

1. The area A of a trapezoid varies jointly as the product of its height h and the sum of its bases b and B. State an equation that represents this variation. Given that $A = 15$ square inches when $h = 6$ inches, $b = 2$ inches, and $B = 3$ inches, explain how you would determine the value of the variation constant.

2. Given that the variation constant $k > 0$ and that A varies directly as b, then A _____ when b increases, and A _____ when b decreases.

3. Given that the variation constant $k > 0$ and that S varies inversely as d, then S _____ when d increases, and S _____ when d decreases.

4. The volume V of a right circular cylinder varies jointly as the square of the radius r and the height h. Tell what happens to V when

   a. h is tripled   b. r is tripled

   c. r is doubled and h is decreased to $\frac{1}{2}h$

5. Give some examples of real situations where one quantity varies inversely as a second quantity.

EXERCISE SET 1.6

In Exercises 1 to 12, write an equation that represents the relationship between the given variables. Use k as the variation constant.

1. d varies directly as t.

2. r varies directly as the square of s.

3. y varies inversely as x.

4. p is inversely proportional to q.

5. m varies jointly as n and p.

6. t varies jointly as r and the cube of s.

7. V varies jointly as l, w, and h.

8. u varies directly as v and inversely as the square of w.

9. A is directly proportional to the square of s.

10. A varies jointly as h and the square of r.

11. F varies jointly as m_1 and m_2 and inversely as the square of d.

12. T varies jointly as t and r and the square of a.

In Exercises 13 to 20, write the equation that expresses the relationship between the variables, and then use the given data to solve for the variation constant.

13. y varies directly as x, and $y = 64$ when $x = 48$.

14. m is directly proportional to n, and $m = 92$ when $n = 23$.

15. r is directly proportional to the square of t, and $r = 144$ when $t = 108$.

16. C varies directly as r, and $C = 94.2$ when $r = 15$.

17. T varies jointly as r and the square of s, and $T = 210$ when $r = 30$ and $s = 5$.

18. u varies directly as v and inversely as the square root of w, and $u = 0.04$ when $v = 8$ and $w = 0.04$.

19. V varies jointly as l, w, and h, and $V = 240$ when $l = 8$, $w = 6$, and $h = 5$.

20. t varies directly as the cube of r and inversely as the square root of s, and $t = 10$ when $r = 5$ and $s = 0.09$.

21. CHARLES'S LAW *Charles's Law* states that the volume V occupied by a gas (at a constant pressure) is directly proportional to its absolute temperature T. An experiment with a balloon shows that the volume of the balloon is 0.85 liter at 270 K (absolute temperature).[4] What will the volume of the balloon be when its temperature is 324 K?

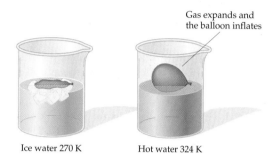

Gas expands and the balloon inflates

Ice water 270 K   Hot water 324 K

▶ **22. HOOKE'S LAW** *Hooke's Law* states that the distance a spring stretches varies directly as the weight on the spring. A weight of 80 pounds stretches a spring 6 inches. How far will a weight of 100 pounds stretch the spring?

$y = kx$

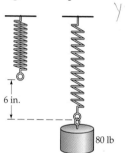

6 in.

80 lb

23. SEMESTER HOURS VS. QUARTER HOURS A student plans to transfer from a college that uses the quarter system to a college that uses the semester system. The number of semester hours a student receives credit for is directly proportional to the number of quarter hours the student has earned. A student with 51 quarter hours is given credit for 34 semester hours. How many semester hours credit should a student receive if the student has completed 93 quarter hours?

24. PRESSURE AND DEPTH The pressure a liquid exerts at a given point on a submarine is directly proportional to the depth of the point below the surface of the liquid. If the pressure at a depth of 3 feet is 187.5 pounds per square foot, find the pressure at a depth of 7 feet.

[4]Absolute temperature is measured on the Kelvin scale. A unit (called a kelvin) on the Kelvin scale is the same measure as a degree on the Celsius scale; however, 0 on the Kelvin scale corresponds to $-273°C$ on the Celsius scale.

25. AMOUNT OF JUICE CONTAINED IN A GRAPEFRUIT The amount of juice in a grapefruit is directly proportional to the cube of its diameter. A grapefruit with a 4-inch diameter contains 6 fluid ounces of juice. How much juice is contained in a grapefruit with a 5-inch diameter? Round to the nearest tenth of a fluid ounce.

▶ **26. MOTORCYCLE JUMP** The range of a projectile is directly proportional to the square of its velocity. If a motorcyclist can make a jump of 140 feet by coming off a ramp at 60 mph, find the distance the motorcyclist could expect to jump if the speed coming off the ramp were increased to 65 mph. Round to the nearest tenth of a foot.

27. PERIOD OF A PENDULUM The period T of a pendulum (the time it takes the pendulum to make one complete oscillation) varies directly as the square root of its length L. A pendulum 3 feet long has a period of 1.8 seconds.

  a. Find the period of a pendulum that is 10 feet long. Round to the nearest tenth of a second.

  b. What is the length of a pendulum that *beats seconds* (that is, has a 2-second period)? Round to the nearest tenth of a foot.

28. AREA OF A PROJECTED PICTURE The area of a projected picture on a movie screen varies directly as the square of the distance from the projector to the screen. If a distance of 20 feet produces a picture with an area of 64 square feet, what distance produces an area of 100 square feet?

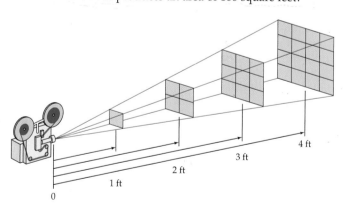

1 ft   2 ft   3 ft   4 ft

0

29. SPEED OF A BICYCLE GEAR The speed of a bicycle gear, in revolutions per minute, is inversely proportional to the number of teeth on the gear. If a gear with 64 teeth has a speed of 30 revolutions per minute, what will be the speed of a gear with 48 teeth?

▶ **30. VIBRATION OF A GUITAR STRING** The frequency of vibration of a guitar string under constant tension varies inversely as the length of the string. A guitar string with a length of 20 inches has a frequency of 144 vibrations per second. Find the frequency of a guitar string with a length of 18 inches. Assume the tension is the same for both strings.

31. DECIBELS The loudness, measured in decibels, of a stereo speaker is inversely proportional to the square of the distance of the listener from the speaker. The loudness is 28 decibels at a distance of 8 feet. What is the loudness when the listener is 4 feet from the speaker?

▶ **32. ILLUMINATION** The illumination a source of light provides is inversely proportional to the square of the distance from the source. If the illumination at a distance of 10 feet from the source is 50 footcandles, what is the illumination at a distance of 15 feet from the source? Round to the nearest tenth of a footcandle.

33. VOLUME RELATIONSHIPS The volume V of a right circular cone varies jointly as the square of the radius r and the height h. Tell what happens to V when

   a. r is tripled

   b. h is tripled

   c. both r and h are tripled

▶ **34. SAFE LOAD** The load L that a horizontal beam can safely support varies jointly as the width w and the square of the depth d. If a beam with width 2 inches and depth 6 inches safely supports up to 200 pounds, how many pounds can a beam of the same length that has width 4 inches and depth 4 inches be expected to support? Round to the nearest pound.

35. IDEAL GAS LAW The *Ideal Gas Law* states that the volume V of a gas varies jointly as the number of moles of gas n and the absolute temperature T and inversely as the

pressure P. What happens to V when n is tripled and P is reduced by a factor of one-half?

36. MAXIMUM LOAD The maximum load a cylindrical column of circular cross section can support varies directly as the fourth power of the diameter and inversely as the square of the height. If a column 2 feet in diameter and 10 feet high supports up to 6 tons, how much of a load does a column 3 feet in diameter and 14 feet high support?

37. EARNED RUN AVERAGE A pitcher's earned run average (ERA) is directly proportional to the number of earned runs the pitcher has allowed and is inversely proportional to the number of innings the pitcher has pitched. During the 2002 season, Randy Johnson of the Arizona Diamondbacks had an ERA of 2.32. He allowed 67 earned runs in 260 innings. During the same 2002 season, Tom Glavine of the Atlanta Braves allowed 74 earned runs in 224.2 innings. What was Glavine's ERA for the 2002 season? Round to the nearest hundredth. (*Source:* MLB.com)

▶ **38. SAFE LOAD** The load L a horizontal beam can safely support varies jointly as the width w and the square of the depth d and inversely as the length l. If a 12-foot beam with width 4 inches and depth 8 inches safely supports 800 pounds, how many pounds can a 16-foot beam that has width 3.5 inches and depth 6 inches be expected to support?

39. FORCE, SPEED, AND RADIUS RELATIONSHIPS The force needed to keep a car from skidding on a curve varies jointly as the weight of the car and the square of its speed and inversely as the radius of the curve. It takes 2800 pounds of force to keep an 1800-pound car from skidding on a curve with radius 425 feet at 45 mph. What force is needed to keep the same car from skidding when it takes a similar curve with radius 450 feet at 55 mph? Round to the nearest ten pounds.

CONNECTING CONCEPTS

40. STIFFNESS OF A BEAM A cylindrical log is to be cut so that it will yield a beam that has a rectangular cross section of depth d and width w. The stiffness of a beam of given length is directly proportional to the width and the cube of the depth. The diameter of the log is 18 inches.

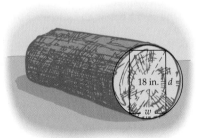

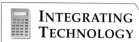
INTEGRATING TECHNOLOGY

In Example 5 it was possible to find the x-intercepts by solving a quadratic equation. In some instances, however, solving an equation to find the intercepts may be very difficult. In these cases, a graphing calculator can be used to estimate the x-intercepts.

The x-intercepts of the graph of $y = x^3 + x + 4$ can be estimated using the INTERCEPT feature of a TI-83 calculator. The keystrokes and some sample screens for this procedure are shown below.

Press Y=. Now enter X^3+X+4. Press ZOOM and select the standard viewing window.

Press 2nd CALC to access the CALCULATE menu. The y-coordinate of an x-intercept is zero. Therefore, select 2:zero. Press ENTER.

The "Left Bound?" shown on the bottom of the screen means to move the cursor until it is to the left of an x-intercept. Press ENTER.

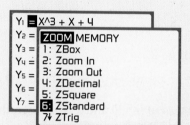

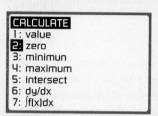

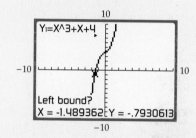

The "Right Bound?" shown on the bottom of the screen means to move the cursor until it is to the right of the desired x-intercept. Press ENTER.

"Guess?" is shown on the bottom of the screen. Move the cursor until it is approximately on the x-intercept. Press ENTER.

The "Zero" shown on the bottom of the screen means that the value of y is 0 when $x = -1.378797$. The x-intercept is about $(-1.378797, 0)$.

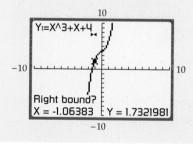

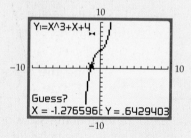

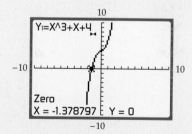

● CIRCLES, THEIR EQUATIONS, AND THEIR GRAPHS

Frequently you will sketch graphs by plotting points. However, some graphs can be sketched merely by recognizing the form of the equation. A *circle* is an example of a curve whose graph you can sketch after you have inspected its equation.

Definition of a Circle

A **circle** is the set of points in a plane that are a fixed distance from a specified point. The distance is the **radius** of the circle, and the specified point is the **center** of the circle.

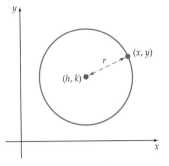

FIGURE 2.16

The standard form of the equation of a circle is derived by using this definition. To derive the standard form, we use the distance formula. **Figure 2.16** is a circle with center (h, k) and radius r. The point (x, y) is on the circle if and only if it is a distance of r units from the center (h, k). Thus (x, y) is on the circle if and only if

$$\sqrt{(x - h)^2 + (y - k)^2} = r$$
$$(x - h)^2 + (y - k)^2 = r^2 \qquad \bullet \text{ Square each side.}$$

Standard Form of the Equation of a Circle

The **standard form of the equation of a circle** with center at (h, k) and radius r is

$$(x - h)^2 + (y - k)^2 = r^2$$

For example, the equation $(x - 3)^2 + (y + 1)^2 = 4$ is the equation of a circle. The standard form of the equation is

$$(x - 3)^2 + (y - (-1))^2 = 2^2$$

from which it can be determined that $h = 3$, $k = -1$, and $r = 2$. Thus the graph is a circle centered at $(3, -1)$ with a radius of 2.

If a circle is centered at the origin $(0, 0)$ (that is, if $h = 0$ and $k = 0$), then the standard form of the equation of the circle simplifies to

$$x^2 + y^2 = r^2$$

For example, the graph of $x^2 + y^2 = 9$ is a circle with center at the origin and radius of 3.

? QUESTION What are the radius and the coordinates of the center of the circle with equation $x^2 + (y - 2)^2 = 10$?

EXAMPLE 6 **Find the Standard Form of the Equation of a Circle**

Find the standard form of the equation of the circle that has center $C(-4, -2)$ and contains the point $P(-1, 2)$.

Solution

$$\left(P_x - C_x \right)^2 + \left(P_y - C_y \right)^2$$

See the graph of the circle in **Figure 2.17**. Because the point P is on the circle, the radius r of the circle must equal the distance from C to P. Thus

$$r = \sqrt{(-1 - (-4))^2 + (2 - (-2))^2}$$
$$= \sqrt{9 + 16} = \sqrt{25} = 5$$

Using the standard form with $h = -4$, $k = -2$, and $r = 5$, we obtain

$$(x + 4)^2 + (y + 2)^2 = 5^2$$

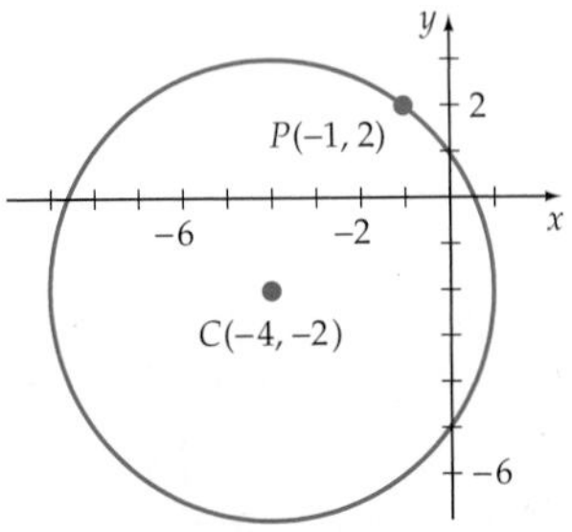

▶ **TRY EXERCISE 64, PAGE 175**

$(x + 4)^2 + (y + 2)^2 = 5^2$

FIGURE 2.17

? ANSWER The radius is $\sqrt{10}$ and the coordinates of the center are $(0, 2)$.

If we rewrite $(x + 4)^2 + (y + 2)^2 = 5^2$ by squaring and combining like terms, we produce

$$x^2 + 8x + 16 + y^2 + 4y + 4 = 25$$
$$x^2 + y^2 + 8x + 4y - 5 = 0$$

This form of the equation is known as the **general form of the equation of a circle.** By completing the square, it is always possible to write the general form $x^2 + y^2 + Ax + By + C = 0$ in the standard form

$$(x - h)^2 + (y - k)^2 = s$$

for some number s. If $s > 0$, the graph is a circle with radius $r = \sqrt{s}$. If $s = 0$, the graph is the point (h, k), and if $s < 0$, the equation has no real solutions and there is no graph.

EXAMPLE 7   **Find the Center and Radius of a Circle by Completing the Square**

Find the center and the radius of the circle that is given by

$$x^2 + y^2 - 6x + 4y - 3 = 0$$

Solution

To review **COMPLETING THE SQUARE,** *see p. 105.*

First rearrange and group the terms as shown.

$$(x^2 - 6x) + (y^2 + 4y) = 3$$

Now complete the square of $(x^2 - 6x)$ and $(y^2 + 4y)$.

$$(x^2 - 6x + 9) + (y^2 + 4y + 4) = 3 + 9 + 4 \qquad \bullet \text{ Add 9 and 4 to each side of the equation.}$$

$$(x - 3)^2 + (y + 2)^2 = 16$$
$$(x - 3)^2 + (y - (-2))^2 = 4^2$$

This equation is the standard form of the equation of a circle and indicates that the graph of the original equation is a circle centered at $(3, -2)$ with radius 4. See **Figure 2.18**.

▶ **TRY EXERCISE 66, PAGE 175**

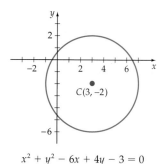

$x^2 + y^2 - 6x + 4y - 3 = 0$

FIGURE 2.18

TOPICS FOR DISCUSSION

1. The distance formula states that the distance d between the points $P_1(x_1, y_1)$ and $P_2(x_2, y_2)$ is $d = \sqrt{(x_2 - x_1)^2 + (y_2 - y_1)^2}$. Can the distance formula also be written as follows? Explain.

$$d = \sqrt{(x_1 - x_2)^2 + (y_1 - y_2)^2}$$

2. Does the equation $(x - 3)^2 + (y + 4)^2 = -6$ have a graph that is a circle? Explain.

3. Explain why the graph of $|x| + |y| = 1$ does not contain any points that have

   a. a y-coordinate that is greater than 1 or less than -1

   b. an x-coordinate that is greater than 1 or less than -1

4. Discuss the graph of $xy = 0$.

5. Explain how to determine the x- and y-intercepts of a graph defined by an equation (without using the graph).

EXERCISE SET 2.1

In Exercises 1 and 2, plot the points whose coordinates are given on a Cartesian coordinate system.

1. $(2, 4), (0, -3), (-2, 1), (-5, -3)$

2. $(-3, -5), (-4, 3), (0, 2), (-2, 0)$

3. 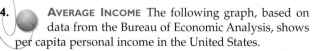 **HEALTH** A study at the Ohio State University measured the changes in heart rates of students doing stepping exercises. Students stepped onto a platform that was approximately 11 inches high at a rate of 14 steps per minute. The heart rate, in beats per minute, before and after the exercise is given in the table below.

Before	After	Before	After
63	84	96	141
72	99	69	93
87	111	81	96
90	129	75	90
90	108	84	90

a. Draw a scatter diagram for these data.

b. For these students, what is the average increase in heart rate?

4. 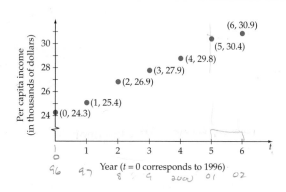 **AVERAGE INCOME** The following graph, based on data from the Bureau of Economic Analysis, shows per capita personal income in the United States.

a. From the data, does it appear that per capita personal income is increasing, decreasing, or remaining the same?

b. If per capita personal income continues to increase by the same percent as the percent increase between 2001 and 2002, what will be the per capita income in 2004?

In Exercises 5 to 16, find the distance between the points whose coordinates are given.

5. $(6, 4), (-8, 11)$

▶ **6.** $(-5, 8), (-10, 14)$

7. $(-4, -20), (-10, 15)$

8. $(40, 32), (36, 20)$

9. $(5, -8), (0, 0)$

10. $(0, 0), (5, 13)$

11. $\left(\sqrt{3}, \sqrt{8}\right), \left(\sqrt{12}, \sqrt{27}\right)$

12. $\left(\sqrt{125}, \sqrt{20}\right), \left(6, 2\sqrt{5}\right)$

13. $(a, b), (-a, -b)$

14. $(a - b, b), (a, a + b)$

15. $(x, 4x), (-2x, 3x)$ given that $x < 0$

16. $(x, 4x), (-2x, 3x)$ given that $x > 0$

17. Find all points on the x-axis that are 10 units from $(4, 6)$. (*Hint:* First write the distance formula with $(4, 6)$ as one of the points and $(x, 0)$ as the other point.)

18. Find all points on the y-axis that are 12 units from $(5, -3)$.

In Exercises 19 to 24, find the midpoint of the line segment with the following endpoints.

19. $(1, -1), (5, 5)$

20. $(-5, -2), (6, 10)$

21. $(6, -3), (6, 11)$

22. $(4, 7), (-10, 7)$

23. $(1.75, 2.25), (-3.5, 5.57)$

24. $(-8.2, 10.1), (-2.4, -5.7)$

In Exercises 25 to 38, graph each equation by plotting points that satisfy the equation.

25. $x - y = 4$

▶ **26.** $2x + y = -1$

27. $y = 0.25x^2$

28. $3x^2 + 2y = -4$

29. $y = -2|x - 3|$

▶ **30.** $y = |x + 3| - 2$

31. $y = x^2 - 3$

▶ **32.** $y = x^2 + 1$

33. $y = \dfrac{1}{2}(x - 1)^2$

34. $y = 2(x + 2)^2$

35. $y = x^2 + 2x - 8$

36. $y = x^2 - 2x - 8$

37. $y = -x^2 + 2$

38. $y = -x^2 - 1$

In Exercises 39 to 48, find the x- and y-intercepts of the graph of each equation. Use the intercepts and additional points as needed to draw the graph of the equation.

39. $2x + 5y = 12$

▶ **40.** $3x - 4y = 15$

41. $x = -y^2 + 5$

42. $x = y^2 - 6$

43. $x = |y| - 4$

44. $x = y^3 - 2$

45. $x^2 + y^2 = 4$

46. $x^2 = y^2$

47. $|x| + |y| = 4$

48. $|x - 4y| = 8$

In Exercises 49 to 56, determine the center and radius of the circle with the given equation.

49. $x^2 + y^2 = 36$

50. $x^2 + y^2 = 49$

51. $(x - 1)^2 + (y - 3)^2 = 49$

52. $(x - 2)^2 + (y - 4)^2 = 25$

53. $(x + 2)^2 + (y + 5)^2 = 25$

54. $(x + 3)^2 + (y + 5)^2 = 121$

55. $(x - 8)^2 + y^2 = \dfrac{1}{4}$

56. $x^2 + (y - 12)^2 = 1$

In Exercises 57 to 64, find an equation of a circle that satisfies the given conditions. Write your answer in standard form.

57. Center $(4, 1)$, radius $r = 2$

58. Center $(5, -3)$, radius $r = 4$

59. Center $\left(\dfrac{1}{2}, \dfrac{1}{4}\right)$, radius $r = \sqrt{5}$

60. Center $\left(0, \dfrac{2}{3}\right)$, radius $r = \sqrt{11}$

61. Center $(0, 0)$, passing through $(-3, 4)$

62. Center $(0, 0)$, passing through $(5, 12)$

63. Center $(1, 3)$, passing through $(4, -1)$

▶ **64.** Center $(-2, 5)$, passing through $(1, 7)$

Practice

In Exercises 65 to 72, find the center and the radius of the graph of the circle. The equations of the circles are written in the general form.

Complete the ☐

65. $x^2 + y^2 - 6x + 5 = 0$

▶ **66.** $x^2 + y^2 - 6x - 4y + 12 = 0$

67. $x^2 + y^2 - 14x + 8y + 56 = 0$

68. $x^2 + y^2 - 10x + 2y + 25 = 0$

69. $4x^2 + 4y^2 + 4x - 63 = 0$

70. $9x^2 + 9y^2 - 6y - 17 = 0$

71. $x^2 + y^2 - x + 3y - \dfrac{15}{4} = 0$

72. $x^2 + y^2 + 3x - 5y + \dfrac{25}{4} = 0$

73. Find an equation of a circle that has a diameter with endpoints $(2, 3)$ and $(-4, 11)$. Write your answer in standard form.

74. Find an equation of a circle that has a diameter with endpoints $(7, -2)$ and $(-3, 5)$. Write your answer in standard form.

75. Find an equation of a circle that has its center at $(7, 11)$ and is tangent to the x-axis. Write your answer in standard form.

76. Find an equation of a circle that has its center at $(-2, 3)$ and is tangent to the y-axis. Write your answer in standard form.

─── *CONNECTING CONCEPTS* ───

In Exercises 77 to 86, graph the set of all points whose x- and y-coordinates satisfy the given conditions.

77. $x = 1, y \geq 1$ line

78. $y = -3, x \geq -2$

79. $y \leq 3$

80. $x \geq 2$

81. $xy \geq 0$

82. $|y| \geq 1, \dfrac{x}{y} \leq 0$

83. $|x| = 2, |y| = 3$

84. $|x| = 4, |y| = 1$

85. $|x| \leq 2, y \geq 2$

86. $x \geq 1, |y| \leq 3$

In Exercises 87 to 90, find the other endpoint of the line segment that has the given endpoint and midpoint.

87. Endpoint $(5, 1)$, midpoint $(9, 3)$

88. Endpoint $(4, -6)$, midpoint $(-2, 11)$

89. Endpoint $(-3, -8)$, midpoint $(2, -7)$

90. Endpoint $(5, -4)$, midpoint $(0, 0)$

91. Find a formula for the set of all points (x, y) for which the distance from (x, y) to $(3, 4)$ is 5.

92. Find a formula for the set of all points (x, y) for which the distance from (x, y) to $(-5, 12)$ is 13.

93. Find a formula for the set of all points (x, y) for which the sum of the distances from (x, y) to $(4, 0)$ and from (x, y) to $(-4, 0)$ is 10.

94. Find a formula for the set of all points for which the absolute value of the difference of the distances from (x, y) to $(0, 4)$ and from (x, y) to $(0, -4)$ is 6.

95. Find an equation of a circle that is tangent to both axes, has its center in the second quadrant, and has a radius of 3.

96. Find an equation of a circle that is tangent to both axes, has its center in the third quadrant, and has a diameter of $\sqrt{5}$.

─── *PREPARE FOR SECTION 2.2* ───

97. Evaluate $x^2 + 3x - 4$ when $x = -3$. [P.3]

98. From the set of ordered pairs $A = \{(-3, 2), (-2, 4), (-1, 1), (0, 4), (2, 5)\}$, create two new sets, D and R, where D is the set of the first coordinates of the ordered pairs of A and R is the set of the second coordinates of the ordered pairs of A. [P.1]

99. Find the length of the line segment connecting $P_1(-4, 1)$ and $P_2(3, -2)$. [2.1]

100. For what values of x is $\sqrt{2x - 6}$ a real number? [P.6/1.5]

101. For what values of x is $\dfrac{x + 3}{x^2 - x - 6}$ not a real number? [P.5]

102. If $a = 3x + 4$ and $a = 6x - 5$, find the value of a. [1.1]

─── *PROJECTS* ───

1. VERIFY A GEOMETRIC THEOREM Use the midpoint formula and the distance formula to prove that the midpoint M of the hypotenuse of a right triangle is equidistant from each of the vertices of the triangle. (*Hint:* Label the vertices of the triangle as shown in the figure at the right.)

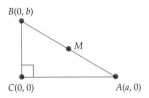

2. **Solve a Quadratic Equation Geometrically** In the 17th century, Descartes (and others) solved equations by using both algebra and geometry. This project outlines the method Descartes used to solve certain quadratic equations.

a. Consider the equation $x^2 = 2ax + b^2$. Construct a right triangle ABC with $d(A, C) = a$ and $d(C, B) = b$. Now draw a circle with center at A and radius a. Let P be the point at which the circle intersects the hypotenuse of the right triangle and Q the point at which an extension of the hypotenuse intersects the circle. Your drawing should be similar to the one at the right.

b. Show that a solution of the equation $x^2 = 2ax + b^2$ is $d(Q, B)$.

c. Show that $d(P, B)$ is a solution of the equation $x^2 = -2ax + b^2$.

d. Construct a line parallel to AC and passing through B. Let S and T be the points at which the line intersects the circle. Show that $d(S, B)$ and $d(T, B)$ are solutions of the equation $x^2 = 2ax - b^2$.

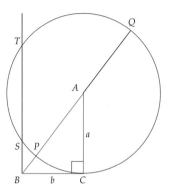

SECTION 2.2

INTRODUCTION TO FUNCTIONS

● RELATIONS

In many situations in science, business, and mathematics, a correspondence exists between two sets. The correspondence is often defined by a *table*, an *equation*, or a *graph*, each of which can be viewed from a mathematical perspective as a set of ordered pairs. In mathematics, any set of ordered pairs is called a **relation.**

Table 2.1 defines a correspondence between a set of percent scores and a set of letter grades. For each score from 0 to 100, there corresponds only one letter grade. The score 94% corresponds to the letter grade of A. Using ordered-pair notation, we record this correspondence as (94, A).

The *equation* $d = 16t^2$ indicates that the distance d that a rock falls (neglecting air resistance) corresponds to the time t that it has been falling. For each nonnegative value t, the equation assigns only one value for the distance d. According to this equation, in 3 seconds a rock will fall 144 feet, which we record as (3, 144). Some of the other ordered pairs determined by $d = 16t^2$ are (0, 0), (1, 16), (2, 64), and (2.5, 100).

TABLE 2.1

Score	Grade
[90, 100]	A
[80, 90)	B
[70, 80)	C
[60, 70)	D
[0, 60)	F

Equation: $d = 16t^2$

If $t = 3$, then $d = 16(3)^2 = 144$

The *graph* in **Figure 2.19** defines a correspondence between the length of a pendulum and the time it takes the pendulum to complete one oscillation. For each nonnegative pendulum length, the graph yields only one time. According to the graph, a pendulum length of 2 feet yields an oscillation time of 1.6 seconds, and a length of 4 feet yields an oscillation time of 2.2 seconds, where the time is measured to the nearest tenth of a second. These results can be recorded as the ordered pairs (2, 1.6) and (4, 2.2).

Graph: A pendulum's oscillation time

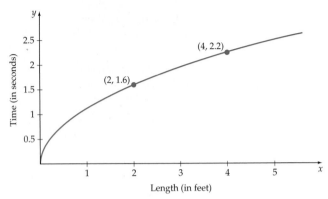

FIGURE 2.19

● FUNCTIONS

The preceding table, equation, and graph each determines a special type of relation called a *function*.

> ### Definition of a Function
>
> A **function** is a set of ordered pairs in which no two ordered pairs have the same first coordinate and different second coordinates.

Although every function is a relation, not every relation is a function. For instance, consider (94, A) from the grading correspondence. The first coordinate, 94, is paired with a second coordinate of A. It would not make sense to have 94 paired with A, (94, A), and 94 paired with B, (94, B). The same first coordinate would be paired with two different second coordinates. This would mean that two students with the same score received different grades, one student an A and the other a B!

Functions may have ordered pairs with the same second coordinate. For instance, (94, A) and (95, A) are both ordered pairs that belong to the function defined by **Table 2.1.** A function may have different first coordinates and the same second coordinate.

The equation $d = 16t^2$ represents a function because for each value of t there is only one value of d. Not every equation, however, represents a function. For instance, $y^2 = 25 - x^2$ does not represent a function. The ordered pairs $(-3, 4)$ and $(-3, -4)$ are both solutions of the equation. But these ordered pairs do not satisfy the definition of a function; there are two ordered pairs with the same first coordinate but *different* second coordinates.

❷ QUESTION Does the set $\{(0, 0), (1, 0), (2, 0), (3, 0), (4, 0)\}$ define a function?

The **domain** of a function is the set of all the first coordinates of the ordered pairs. The **range** of a function is the set of all the second coordinates. In the func-

---

❷ ANSWER Yes. There are no two ordered pairs with the same first coordinate that have different second coordinates.

tion determined by the grading correspondence in **Table 2.1,** the domain is the interval [0, 100]. The range is {A, B, C, D, F}. In a function, each domain element is paired with one and only one range element.

If a function is defined by an equation, the variable that represents elements of the domain is the **independent variable.** The variable that represents elements of the range is the **dependent variable.** In the free-fall experiment, we used the equation $d = 16t^2$. The elements of the domain represented the time the rock fell, and the elements of the range represented the distance the rock fell. Thus, in $d = 16t^2$, the independent variable is t and the dependent variable is d.

The specific letters used for the independent and the dependent variable are not important. For example, $y = 16x^2$ represents the same function as $d = 16t^2$. Traditionally, x is used for the independent variable and y for the dependent variable. Anytime we use the phrase "y is a function of x" or a similar phrase with different letters, the variable that follows "function of" is the independent variable.

● FUNCTIONAL NOTATION

Functions can be named by using a letter or a combination of letters, such as f, g, A, log, or tan. If x is an element of the domain of f, then $f(x)$, which is read "f of x" or "the value of f at x," is the element in the range of f that corresponds to the domain element x. The notation "f" and the notation "$f(x)$" mean different things. "f" is the name of the function, whereas "$f(x)$" is the value of the function at x. Finding the value of $f(x)$ is referred to as *evaluating f at x*. To evaluate $f(x)$ at $x = a$, substitute a for x, and simplify.

EXAMPLE 1　　**Evaluate Functions**

Let $f(x) = x^2 - 1$, and evaluate.

a.　$f(-5)$　　**b.**　$f(3b)$　　**c.**　$3f(b)$　　**d.**　$f(a + 3)$　　**e.**　$f(a) + f(3)$

Solution

a.　$f(-5) = (-5)^2 - 1 = 25 - 1 = 24$　　• Substitute **−5** for *x*, and simplify.

b.　$f(3b) = (3b)^2 - 1 = 9b^2 - 1$　　• Substitute **3b** for *x*, and simplify.

c.　$3f(b) = 3(b^2 - 1) = 3b^2 - 3$　　• Substitute **b** for *x*, and simplify.

d.　$f(a + 3) = (a + 3)^2 - 1$　　• Substitute **a + 3** for *x*.
　　　$= a^2 + 6a + 8$　　• Simplify.

e.　$f(a) + f(3) = (a^2 - 1) + (3^2 - 1)$　　• Substitute **a** for *x*; substitute **3** for *x*.

　　　　$= a^2 + 7$　　• Simplify.

▶ **TRY EXERCISE 2, PAGE 190**

take note

In Example 1, observe that
$$f(3b) \neq 3f(b)$$
and that
$$f(a + 3) \neq f(a) + f(3)$$

Piecewise-defined functions are functions represented by more than one expression. The function shown below is an example of a piecewise-defined function.

$$f(x) = \begin{cases} 2x, & x < -2 \\ x^2, & -2 \le x < 1 \\ 4 - x, & x \ge 1 \end{cases}$$　• This function is made up of different *pieces*, 2x, x², and 4 − x, depending on the value of x.

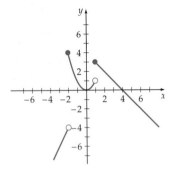

FIGURE 2.20

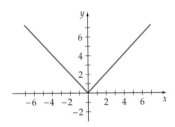

FIGURE 2.21

The expression that is used to evaluate this function depends on the value of x. For instance, to find $f(-3)$, we note that $-3 < -2$ and therefore use the expression $2x$ to evaluate the function.

$$f(-3) = 2(-3) = -6 \qquad \bullet \text{When } x < -2, \text{ use the expression } 2x.$$

Here are some additional instances of evaluating this function:

$$f(-1) = (-1)^2 = 1 \qquad \bullet \text{When } x \text{ satisfies } -2 \le x < 1,$$
$$\text{use the expression } x^2.$$

$$f(4) = 4 - 4 = 0 \qquad \bullet \text{When } x \ge 1, \text{ use the expression } 4 - x.$$

The graph of this function is shown in **Figure 2.20.** Note the use of the open and closed circles at the endpoints of the intervals. These circles are used to show the evaluation of the function at the endpoints of each interval. For instance, because -2 is in the interval $-2 \le x < 1$, the value of the function at -2 is 4 $[f(-2) = (-2)^2 = 4]$. Therefore a closed dot is placed at $(-2, 4)$. Similarly, when $x = 1$, because 1 is in the interval $x \ge 1$, the value of the function at 1 is 3 $(f(1) = 4 - 1 = 3)$.

 QUESTION Evaluate the function f defined at the bottom of page 179 when $x = 0.5$.

The absolute value function is another example of a piecewise-defined function. Below is the definition of this function, which is sometimes abbreviated abs(x). Its graph **(Figure 2.21)** is shown at the left.

$$\text{abs}(x) = \begin{cases} -x, & x < 0 \\ x, & x \ge 0 \end{cases}$$

EXAMPLE 2   **Evaluate a Piecewise-Defined Function**

The number of monthly spam email attacks is shown in **Figure 2.22.**

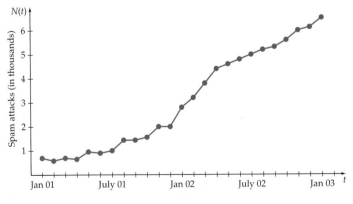

FIGURE 2.22

Source: www.brightmail.com

 ANSWER   0.5 is in the interval $-2 \le x < 1$. Therefore, $f(0.5) = 0.5^2 = 0.25$.

The data in the graph can be approximated by

$$N(t) = \begin{cases} 24.68t^2 - 170.47t + 957.73, & 0 \le t < 17 \\ 196.9t + 1164.6, & 17 \le t \le 26 \end{cases}$$

where $N(t)$ is the number of spam attacks in thousands for month t, where $t = 0$ corresponds to January 2001. Use this function to estimate, to the nearest hundred thousand, the number of monthly spam attacks for the following months.

a.   October 2001   **b.**   December 2002

Solution

a.   The month October 2001 corresponds to $t = 9$. Because $t = 9$ is in the interval $0 \le t < 17$, evaluate $24.68t^2 - 170.47t + 957.73$ at $t = 9$.

$$24.68t^2 - 170.47t + 957.73$$
$$24.68(9)^2 - 170.47(9) + 957.73 = 1422.58$$

There were approximately 1,423,000 spam attacks in October 2001.

b.   The month December 2002 corresponds to $t = 23$. Because $t = 23$ is in the interval $17 \le t \le 26$, evaluate $196.9t + 1164.6$ at 23.

$$196.9t + 1164.6$$
$$196.9(23) + 1164.6 = 5693.3$$

There were approximately 5,693,000 spam attacks in December 2002.

▶ **TRY EXERCISE 10, PAGE 191**

● IDENTIFYING FUNCTIONS

Recall that although every function is a relation, not every relation is a function. In the next example we examine four relations to determine which are functions.

EXAMPLE 3   **Identify Functions**

Which relations define y as a function of x?

a. $\{(2, 3), (4, 1), (4, 5)\}$ **b.** $3x + y = 1$ **c.** $-4x^2 + y^2 = 9$

d.   The correspondence between the x values and the y values in **Figure 2.23.**

Solution

a.   There are two ordered pairs, $(4, 1)$ and $(4, 5)$, with the same first coordinate and different second coordinates. This set does not define y as a function of x.

b.   Solving $3x + y = 1$ for y yields $y = -3x + 1$. Because $-3x + 1$ is a unique real number for each x, this equation defines y as a function of x.

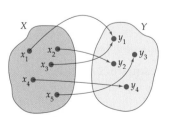

FIGURE 2.23

Continued ▶

c. Solving $-4x^2 + y^2 = 9$ for y yields $y = \pm\sqrt{4x^2 + 9}$. The right side $\pm\sqrt{4x^2 + 9}$ produces two values of y for each value of x. For example, when $x = 0$, $y = 3$ or $y = -3$. Thus $-4x^2 + y^2 = 9$ does not define y as a function of x.

d. Each x is paired with one and only one y. The correspondence in **Figure 2.23** defines y as a function of x.

▶ **TRY EXERCISE 14, PAGE 191**

take note

You may indicate the domain of a function using set notation or interval notation. For instance, the domain of $f(x) = \sqrt{x - 3}$ may be given in each of the following ways:

Set notation: $\{x \mid x \geq 3\}$

Interval notation: $[3, \infty)$

Sometimes the domain of a function is stated explicitly. For example, each of f, g, and h below is given by an equation, followed by a statement that indicates the domain of the function.

$$f(x) = x^2, x > 0 \qquad g(t) = \frac{1}{t^2 + 4}, 0 \leq t \leq 5 \qquad h(x) = x^2, x = 1, 2, 3$$

Although f and h have the same equation, they are different functions because they have different domains. If the domain of a function is not explicitly stated, then its domain is determined by the following convention.

Domain of a Function

Unless otherwise stated, the domain of a function is the set of all real numbers for which the function makes sense and yields real numbers.

EXAMPLE 4 — Determine the Domain of a Function

Determine the domain of each function.

a. $G(t) = \dfrac{1}{t - 4}$ b. $f(x) = \sqrt{x + 1}$

c. $A(s) = s^2$, where $A(s)$ is the area of a square whose sides are s units.

Solution

a. The number 4 is not an element of the domain because G is undefined when the denominator $t - 4$ equals 0. The domain of G is all real numbers except 4. In interval notation the domain is $(-\infty, 4) \cup (4, \infty)$.

b. The radical $\sqrt{x + 1}$ is a real number only when $x + 1 \geq 0$ or when $x \geq -1$. Thus, in set notation, the domain of f is $\{x \mid x \geq -1\}$.

c. Because s represents the length of the side of a square, s must be positive. In interval notation the domain of A is $(0, \infty)$.

▶ **TRY EXERCISE 28, PAGE 191**

INTEGRATING TECHNOLOGY

Many graphing calculators use the notation int(x) for the floor function. The screens at the left are from a TI-83 Plus graphing calculator.

A graphing calculator also can be used to graph the floor function. The graph in **Figure 2.31** was drawn in "connected" mode. This graph does not show the discontinuities that occur whenever x is an integer.

The graph in **Figure 2.32** was constructed by graphing the floor function in "dot" mode. In this case the discontinuities at the integers are apparent.

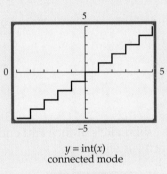

$y = \text{int}(x)$
connected mode

FIGURE 2.31

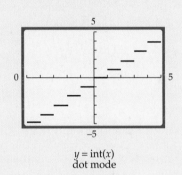

$y = \text{int}(x)$
dot mode

FIGURE 2.32

EXAMPLE 7   Use the Greatest Integer Function to Model Expenses

The cost of parking in a garage is $3 for the first hour or any part of the hour and $2 for each additional hour or any part of the hour thereafter. If x is the time in hours that you park your car, then the cost is given by

$$C(x) = 3 - 2\,\text{int}(1 - x), \quad x > 0$$

a. Evaluate $C(2)$ and $C(2.5)$.   **b.** Graph $y = C(x)$ for $0 < x \le 5$.

Solution

a.
$$
\begin{aligned}
C(2) &= 3 - 2\,\text{int}(1 - 2) \\
&= 3 - 2\,\text{int}(-1) \\
&= 3 - 2(-1) \\
&= \$5
\end{aligned}
\qquad
\begin{aligned}
C(2.5) &= 3 - 2\,\text{int}(1 - 2.5) \\
&= 3 - 2\,\text{int}(-1.5) \\
&= 3 - 2(-2) \\
&= \$7
\end{aligned}
$$

b. To graph $C(x)$ for $0 < x \le 5$, consider the value of int($1 - x$) for each of the intervals $0 < x \le 1, 1 < x \le 2, 2 < x \le 3, 3 < x \le 4$, and $4 < x \le 5$. For instance, when $0 < x \le 1, 0 \le 1 - x < 1$. Thus int($1 - x$) = 0 when $0 < x \le 1$. Now consider $1 < x \le 2$. When $1 < x \le 2, 1 \le 1 - x < 2$. Thus int($1 - x$) = 1 when $1 < x \le 2$. Applying the same reasoning to

Continued ▶

each of the other intervals gives the following table of values and the corresponding graph of C.

x	C(x) = 3 − 2 int(1 − x)
$0 < x \leq 1$	3
$1 < x \leq 2$	5
$2 < x \leq 3$	7
$3 < x \leq 4$	9
$4 < x \leq 5$	11

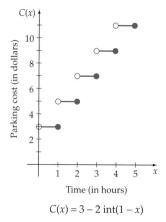

$C(x) = 3 - 2 \, \text{int}(1 - x)$

FIGURE 2.33

Because $C(1) = 3$, $C(2) = 5$, $C(3) = 7$, $C(4) = 9$, and $C(5) = 11$, we can use a solid circle at the right endpoint of each "step" and an open circle at each left endpoint.

▶ **TRY EXERCISE 48, PAGE 192**

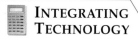 **INTEGRATING TECHNOLOGY**

Example 7 illustrates that a graphing calculator may not produce a graph that is a good representation of a function. You may be required to *make adjustments* in the MODE, SET UP, or WINDOW of the graphing calculator so that it will produce a better representation of the function. Some graphs may also require some *fine tuning*, such as open or solid circles at particular points, to accurately represent the function.

● APPLICATIONS

EXAMPLE 8   Solve an Application

A car was purchased for $16,500. Assuming the car depreciates at a constant rate of $2200 per year (*straight-line depreciation*) for the first 7 years, write the value v of the car as a function of time, and calculate the value of the car 3 years after purchase.

Solution

Let t represent the number of years that have passed since the car was purchased. Then $2200t$ is the amount that the car has depreciated after t years. The value of the car at time t is given by

$$v(t) = 16{,}500 - 2200t, \quad 0 \leq t \leq 7$$

When $t = 3$, the value of the car is

$$v(3) = 16{,}500 - 2200(3) = 16{,}500 - 6600 = \$9900$$

▶ **TRY EXERCISE 66, PAGE 193**

Often in applied mathematics, formulas are used to determine the functional relationship that exists between two variables.

EXAMPLE 9 Solve an Application

A lighthouse is 2 miles south of a port. A ship leaves port and sails east at a rate of 7 mph. Express the distance d between the ship and the lighthouse as a function of time, given that the ship has been sailing for t hours.

Solution

Draw a diagram and label it as shown in **Figure 2.34.** Note that because distance = (rate)(time) and the rate is 7, in t hours the ship has sailed a distance of $7t$.

$$[d(t)]^2 = (7t)^2 + 2^2 \qquad \bullet \text{ The Pythagorean Theorem}$$
$$[d(t)]^2 = 49t^2 + 4$$
$$d(t) = \sqrt{49t^2 + 4} \qquad \bullet \text{ The } \pm \text{ sign is not used because } d \text{ must be nonnegative.}$$

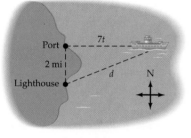

FIGURE 2.34

▶ **TRY EXERCISE 72, PAGE 194**

EXAMPLE 10 Solve an Application

An open box is to be made from a square piece of cardboard that measures 40 inches on each side. To construct the box, squares that measure x inches on each side are cut from each corner of the cardboard as shown in **Figure 2.35.**

a. Express the volume V of the box as a function of x.

b. Determine the domain of V.

Solution

a. The length l of the box is $40 - 2x$. The width w is also $40 - 2x$. The height of the box is x. The volume V of a box is the product of its length, its width, and its height. Thus

$$V = (40 - 2x)^2 x$$

b. The squares that are cut from each corner require x to be larger than 0 inches but less than 20 inches. Thus the domain is $\{x \mid 0 < x < 20\}$.

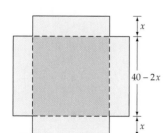

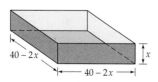

FIGURE 2.35

▶ **TRY EXERCISE 68, PAGE 193**

TOPICS FOR DISCUSSION

1. Discuss the definition of *function*. Give some examples of relationships that are functions and some that are not functions.

2. What is the difference between the domain and range of a function?

3. How many *y*-intercepts can a function have? How many *x*-intercepts can a function have?

4. Discuss how the vertical line test is used to determine whether or not a graph is the graph of a function. Explain why the vertical line test works.

5. What is the domain of $f(x) = \dfrac{\sqrt{1-x}}{x^2 - 9}$? Explain.

6. Is 2 in the range of $g(x) = \dfrac{6x - 5}{3x + 1}$? Explain the process you used to make your decision.

7. Suppose that f is a function and that $f(a) = f(b)$. Does this imply that $a = b$? Explain your answer.

EXERCISE SET 2.2

In Exercises 1 to 8, evaluate each function.

1. Given $f(x) = 3x - 1$, find

    a. $f(2)$ **b.** $f(-1)$ **c.** $f(0)$

    d. $f\left(\dfrac{2}{3}\right)$ **e.** $f(k)$ **f.** $f(k + 2)$

▶ **2.** Given $g(x) = 2x^2 + 3$, find

    a. $g(3)$ **b.** $g(-1)$ **c.** $g(0)$

    d. $g\left(\dfrac{1}{2}\right)$ **e.** $g(c)$ **f.** $g(c + 5)$

3. Given $A(w) = \sqrt{w^2 + 5}$, find

    a. $A(0)$ **b.** $A(2)$ **c.** $A(-2)$

    d. $A(4)$ **e.** $A(r + 1)$ **f.** $A(-c)$

4. Given $J(t) = 3t^2 - t$, find

    a. $J(-4)$ **b.** $J(0)$ **c.** $J\left(\dfrac{1}{3}\right)$

    d. $J(-c)$ **e.** $J(x + 1)$ **f.** $J(x + h)$

5. Given $f(x) = \dfrac{1}{|x|}$, find

    a. $f(2)$ **b.** $f(-2)$ **c.** $f\left(\dfrac{-3}{5}\right)$

    d. $f(2) + f(-2)$ **e.** $f(c^2 + 4)$ **f.** $f(2 + h)$

6. Given $T(x) = 5$, find

    a. $T(-3)$ **b.** $T(0)$ **c.** $T\left(\dfrac{2}{7}\right)$

    d. $T(3) + T(1)$ **e.** $T(x + h)$ **f.** $T(3k + 5)$

7. Given $s(x) = \dfrac{x}{|x|}$, find

    a. $s(4)$ **b.** $s(5)$ **c.** $s(-2)$

    d. $s(-3)$ **e.** $s(t), t > 0$ **f.** $s(t), t < 0$

8. Given $r(x) = \dfrac{x}{x + 4}$, find

    a. $r(0)$ **b.** $r(-1)$ **c.** $r(-3)$

d. $r\left(\dfrac{1}{2}\right)$ **e.** $r(0.1)$ **f.** $r(10,000)$

In Exercises 9 and 10, evaluate each piecewise-defined function for the indicated values.

9. $P(x) = \begin{cases} 3x + 1, & \text{if } x < 2 \\ -x^2 + 11, & \text{if } x \geq 2 \end{cases}$

     a. $P(-4)$ **b.** $P(\sqrt{5})$

     c. $P(c), \quad c < 2$ **d.** $P(k + 1), \quad k \geq 1$

▶ **10.** $Q(t) = \begin{cases} 4, & \text{if } 0 \leq t \leq 5 \\ -t + 9, & \text{if } 5 < t \leq 8 \\ \sqrt{t - 7}, & \text{if } 8 < t \leq 11 \end{cases}$

     a. $Q(0)$ **b.** $Q(e), \quad 6 < e < 7$

     c. $Q(n), \quad 1 < n < 2$ **d.** $Q(m^2 + 7), \quad 1 < m \leq 2$

In Exercises 11 to 20, identify the equations that define y as a function of x.

11. $2x + 3y = 7$ **12.** $5x + y = 8$

13. $-x + y^2 = 2$ ▶ **14.** $x^2 - 2y = 2$

15. $y = 4 \pm \sqrt{x}$ **16.** $x^2 + y^2 = 9$

17. $y = \sqrt[3]{x}$ **18.** $y = |x| + 5$

19. $y^2 = x^2$ **20.** $y^3 = x^3$

In Exercises 21 to 26, identify the sets of ordered pairs (x, y) that define y as a function of x.

21. $\{(2, 3), (5, 1), (-4, 3), (7, 11)\}$

22. $\{(5, 10), (3, -2), (4, 7), (5, 8)\}$

23. $\{(4, 4), (6, 1), (5, -3)\}$

24. $\{(2, 2), (3, 3), (7, 7)\}$

25. $\{(1, 0), (2, 0), (3, 0)\}$

26. $\left\{ \left(-\dfrac{1}{3}, \dfrac{1}{4}\right), \left(-\dfrac{1}{4}, \dfrac{1}{3}\right), \left(\dfrac{1}{4}, \dfrac{2}{3}\right) \right\}$

In Exercises 27 to 38, determine the domain of the function represented by the given equation.

27. $f(x) = 3x - 4$ ▶ **28.** $f(x) = -2x + 1$

29. $f(x) = x^2 + 2$ **30.** $f(x) = 3x^2 + 1$

31. $f(x) = \dfrac{4}{x + 2}$ **32.** $f(x) = \dfrac{6}{x - 5}$

33. $f(x) = \sqrt{7 + x}$ **34.** $f(x) = \sqrt{4 - x}$

35. $f(x) = \sqrt{4 - x^2}$ **36.** $f(x) = \sqrt{12 - x^2}$

37. $f(x) = \dfrac{1}{\sqrt{x + 4}}$ **38.** $f(x) = \dfrac{1}{\sqrt{5 - x}}$

In Exercises 39 to 46, graph each function. Insert <u>solid circles or hollow circles</u> where necessary to indicate the true nature of the function.

39. $f(x) = \begin{cases} |x|, & \text{if } x \leq 1 \\ 2, & \text{if } x > 1 \end{cases}$

▶ **40.** $g(x) = \begin{cases} -4, & \text{if } x \leq 0 \\ x^2 - 4, & \text{if } 0 < x \leq 1 \\ -x, & \text{if } x > 1 \end{cases}$

41. $J(x) = \begin{cases} 4, & \text{if } x \leq -1 \\ x^2, & \text{if } -1 < x < 1 \\ -x + 5, & \text{if } x \geq 1 \end{cases}$

42. $K(x) = \begin{cases} 1, & \text{if } x \leq -2 \\ x^2 - 4, & \text{if } -2 < x < 2 \\ \dfrac{1}{2}x, & \text{if } x \geq 2 \end{cases}$

43. $L(x) = \left[\!\left[\dfrac{1}{3}x \right]\!\right] \quad$ for $-6 \leq x \leq 6$

44. $M(x) = [\![x]\!] + 2 \quad$ for $0 \leq x \leq 4$

45. $N(x) = \text{int}(-x) \quad$ for $-3 \leq x \leq 3$

46. $P(x) = \text{int}(x) + x \quad$ for $0 \leq x \leq 4$

47. FIRST-CLASS MAIL In 2003, the cost to mail a first-class letter is given by

$$C(w) = 0.37 - 0.34 \, \text{int}(1 - w), \quad w > 0$$

where C is in dollars and w is the weight of the letter in ounces.

a. What is the cost to mail a letter that weighs 2.8 ounces?

b. Graph C for $0 < w \leq 5$.

▶ **48.** **INCOME TAX** The amount of federal income tax $T(x)$ a person owed in 2003 is given by

$$T(x) = \begin{cases} 0.10x, & 0 \le x < 6000 \\ 0.15(x - 6000) + 600, & 6000 \le x < 27{,}950 \\ 0.27(x - 27{,}950) + 3892.50, & 27{,}950 \le x < 67{,}700 \\ 0.30(x - 67{,}700) + 14{,}625, & 67{,}700 \le x < 141{,}250 \\ 0.35(x - 141{,}250) + 36{,}690, & 141{,}250 \le x < 307{,}050 \\ 0.386(x - 307{,}050) + 94{,}720, & x \ge 307{,}050 \end{cases}$$

where x is the adjusted gross income tax of the taxpayer.

a. What is the domain of this function?

b. Find the income tax owed by a taxpayer whose adjusted gross income was $31,250.

c. Find the income tax owed by a taxpayer whose adjusted gross income was $72,000.

49. Use the vertical line test to determine which of the following graphs are graphs of functions.

a.

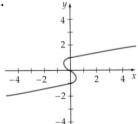

b.

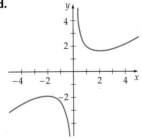

c.

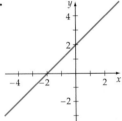

d.

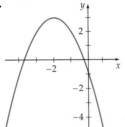

▶ **50.** Use the vertical line test to determine which of the following graphs are graphs of functions.

a.

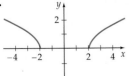

b.

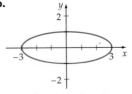

c.

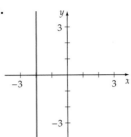

d.

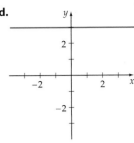

In Exercises 51 to 60, use the indicated graph to identify the intervals over which the function is increasing, constant, or decreasing.

51.

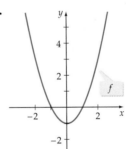

52.

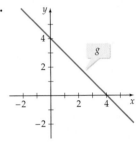

53.

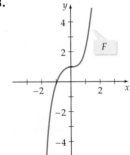

54.

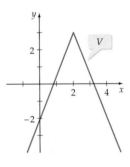

55.

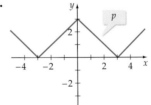

56.

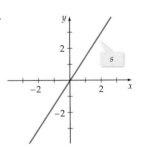

57.

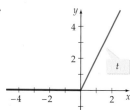

58.

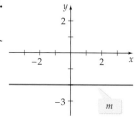

59.

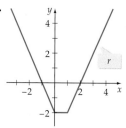

60.

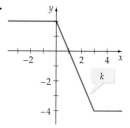

61. Use the horizontal line test to determine which of the following functions are one-to-one.

   f as shown in Exercise 51
   g as shown in Exercise 52
   F as shown in Exercise 53
   V as shown in Exercise 54
   p as shown in Exercise 55

62. Use the horizontal line test to determine which of the following functions are one-to-one.

   s as shown in Exercise 56
   t as shown in Exercise 57
   m as shown in Exercise 58
   r as shown in Exercise 59
   k as shown in Exercise 60

63. A rectangle has a length of l feet and a perimeter of 50 feet.

   a. Write the width w of the rectangle as a function of its length.

   b. Write the area A of the rectangle as a function of its length.

64. The sum of two numbers is 20. Let x represent one of the numbers.

   a. Write the second number y as a function of x.

   b. Write the product P of the two numbers as a function of x.

65. DEPRECIATION A bus was purchased for $80,000. Assuming the bus depreciates at a rate of $6500 per year (*straight-line depreciation*) for the first 10 years, write the value v of the bus as a function of the time t (measured in years) for $0 \le t \le 10$.

▶ 66. DEPRECIATION A boat was purchased for $44,000. Assuming the boat depreciates at a rate of $4200 per year (*straight-line depreciation*) for the first 8 years, write the value v of the boat as a function of the time t (measured in years) for $0 \le t \le 8$.

67. COST, REVENUE, AND PROFIT A manufacturer produces a product at a cost of $22.80 per unit. The manufacturer has a fixed cost of $400.00 per day. Each unit retails for $37.00. Let x represent the number of units produced in a 5-day period.

   a. Write the total cost C as a function of x.

   b. Write the revenue R as a function of x.

   c. Write the profit P as a function of x. (*Hint:* The profit function is given by $P(x) = R(x) - C(x)$.)

▶ 68. VOLUME OF A BOX An open box is to be made from a square piece of cardboard having dimensions 30 inches by 30 inches by cutting out squares of area x^2 from each corner, as shown in the figure.

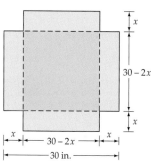

   a. Express the volume V of the box as a function of x.

   b. State the domain of V.

69. HEIGHT OF AN INSCRIBED CYLINDER A cone has an altitude of 15 centimeters and a radius of 3 centimeters. A right circular cylinder of radius r and height h is inscribed in the cone as shown in the figure. Use similar triangles to write h as a function of r.

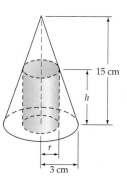

70. VOLUME OF WATER Water is flowing into a conical drinking cup that has an altitude of 4 inches and a radius of 2 inches, as shown in the figure.

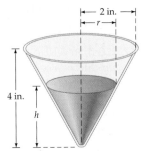

    a. Write the radius r of the water as a function of its depth h.

    b. Write the volume V of the water as a function of its depth h.

71. DISTANCE FROM A BALLOON For the first minute of flight, a hot air balloon rises vertically at a rate of 3 meters per second. If t is the time in seconds that the balloon has been airborne, write the distance d between the balloon and a point on the ground 50 meters from the point of lift-off as a function of t.

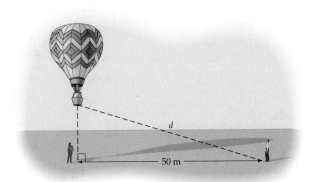

72. TIME FOR A SWIMMER An athlete swims from point A to point B at the rate of 2 mph and runs from point B to point C at a rate of 8 mph. Use the dimensions in the figure to write the time t required to reach point C as a function of x.

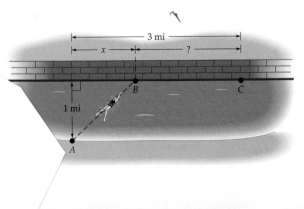

73. DISTANCE BETWEEN SHIPS At 12:00 noon Ship A is 45 miles due south of ship B and is sailing north at a rate of 8 mph. Ship B is sailing east at a rate of 6 mph. Write the distance d between the ships as a function of the time t, where $t = 0$ represents 12:00 noon.

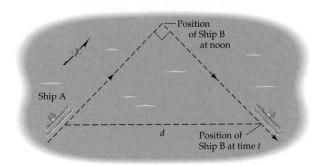

74. AREA A rectangle is bounded by the x- and y-axes and the graph of $y = -\dfrac{1}{2}x + 4$.

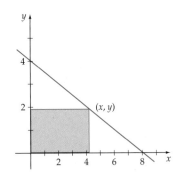

    a. Find the area of the rectangle as a function of x.

    b. Complete the table below.

x	Area
1	
2	
4	
6	
7	

    c. What is the domain of this function?

75. AREA A piece of wire 20 centimeters long is cut at a point x centimeters from the left end. The left-hand piece is formed into the shape of a circle and the right-hand piece is formed into a square.

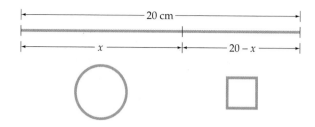

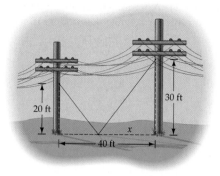

a. Find the area enclosed by the two figures as a function of x.

b. Complete the table below. Round the area to the nearest hundredth.

x	Total Area Enclosed
0	
4	
8	
12	
16	
20	

c. What is the domain of this function?

76. AREA A triangle is bounded by the x- and y-axes and must pass through $P(2, 2)$, as shown below.

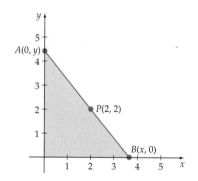

a. Find the area of the triangle as a function of x. (*Suggestion:* The slope of the line between points A and P equals the slope of the line between P and B.)

b. What is the domain of the function you found in part **a.**?

77. LENGTH Two guy wires are attached to utility poles that are 40 feet apart, as shown in the following diagram.

a. Find the total length of the two guy wires as a function of x.

b. Complete the table below. Round the length to the nearest hundredth.

x	Total Length of Wires
0	
10	
20	
30	
40	

c. What is the domain of this function?

78. SALES VS. PRICE A business finds that the number of feet f of pipe it can sell per week is a function of the price p in cents per foot as given by

$$f(p) = \frac{320{,}000}{p + 25}, \quad 40 \le p \le 90$$

Complete the following table by evaluating f (to the nearest hundred feet) for the indicated values of p.

p	40	50	60	75	90
f(p)					

79. MODEL YIELD The yield Y of apples per tree is related to the amount x of a particular type of fertilizer applied (in pounds per year) by the function

$$Y(x) = 400[1 - 5(x - 1)^{-2}], \quad 5 \le x \le 20$$

Complete the following table by evaluating Y (to the nearest apple) for the indicated applications.

x	5	10	12.5	15	20
Y(x)					

80. MODEL COST A manufacturer finds that the cost C in dollars of producing x items of a product is given by

$$C(x) = \left(225 + 1.4\sqrt{x}\right)^2, \quad 100 \le x \le 1000$$

Complete the following table by evaluating C (to the nearest dollar) for the indicated numbers of items.

x	100	200	500	750	1000
$C(x)$					

81. If $f(x) = x^2 - x - 5$ and $f(c) = 1$, find c.

82. If $g(x) = -2x^2 + 4x - 1$ and $g(c) = -4$, find c.

83. Determine whether 1 is in the range of $f(x) = \dfrac{x - 1}{x + 1}$.

84. Determine whether 0 is in the range of $g(x) = \dfrac{1}{x - 3}$.

In Exercises 85 to 90, use a graphing utility.

85. Graph $f(x) = \dfrac{[\![x]\!]}{|x|}$ for $-4.7 \le x \le 4.7$ and $x \ne 0$.

86. Graph $f(x) = \dfrac{[\![2x]\!]}{|x|}$ for $-4 \le x \le 4$ and $x \ne 0$.

87. Graph: $f(x) = x^2 - 2|x| - 3$

88. Graph: $f(x) = x^2 - |2x - 3|$

89. Graph: $f(x) = |x^2 - 1| - |x - 2|$

90. Graph: $f(x) = |x^2 - 2x| - 3$

CONNECTING CONCEPTS

The notation $f(x)|_a^b$ is used to denote the difference $f(b) - f(a)$. That is,

$$f(x)|_a^b = f(b) - f(a)$$

In Exercises 91 to 94, evaluate $f(x)|_a^b$ for the given function f and the indicated values of a and b.

91. $f(x) = x^2 - x; f(x)|_2^3$

92. $f(x) = -3x + 2; f(x)|_4^7$

93. $f(x) = 2x^3 - 3x^2 - x; f(x)|_0^2$

94. $f(x) = \sqrt{8 - x}; f(x)|_0^8$

In Exercises 95 to 98, each function has two or more independent variables.

95. Given $f(x, y) = 3x + 5y - 2$, find

    a. $f(1, 7)$ **b.** $f(0, 3)$ **c.** $f(-2, 4)$

    d. $f(4, 4)$ **e.** $f(k, 2k)$ **f.** $f(k + 2, k - 3)$

96. Given $g(x, y) = 2x^2 - |y| + 3$, find

    a. $g(3, -4)$ **b.** $g(-1, 2)$

    c. $g(0, -5)$ **d.** $g\left(\dfrac{1}{2}, -\dfrac{1}{4}\right)$

    e. $g(c, 3c), c > 0$ **f.** $g(c + 5, c - 2), c < 0$

97. AREA OF A TRIANGLE The area of a triangle with sides a, b, and c is given by the function

$$A(a, b, c) = \sqrt{s(s - a)(s - b)(s - c)}$$

where s is the semiperimeter

$$s = \frac{a + b + c}{2}$$

Find $A(5, 8, 11)$.

98. COST OF A PAINTER The cost in dollars to hire a house painter is given by the function

$$C(h, g) = 15h + 14g$$

where h is the number of hours it takes to paint the house and g is the number of gallons of paint required to paint the house. Find $C(18, 11)$.

A *fixed point* of a function is a number a such that $f(a) = a$. In Exercises 99 and 100, find all fixed points for the given function.

99. $f(x) = x^2 + 3x - 3$

100. $g(x) = \dfrac{x}{x + 5}$

In Exercises 101 and 102, sketch the graph of the piecewise-defined function.

101. $s(x) = \begin{cases} 1 & \text{if } x \text{ is an integer} \\ 2 & \text{if } x \text{ is not an integer} \end{cases}$

102. $v(x) = \begin{cases} 2x - 2 & \text{if } x \neq 3 \\ 1 & \text{if } x = 3 \end{cases}$

PREPARE FOR SECTION 2.3

103. Find the distance on a real number line between the points whose coordinates are -2 and 5. [P.1]

104. Find the product of a nonzero number and its negative reciprocal. [P.5]

105. Given the points $P_1(-3, 4)$ and $P_2(2, -4)$, evaluate $\dfrac{y_2 - y_1}{x_2 - x_1}$.
[P.1]

106. Solve $y - 3 = -2(x - 3)$ for y. [1.1]

107. Solve $3x - 5y = 15$ for y. [1.1]

108. Given $y = 3x - 2(5 - x)$, find the value of x for which $y = 0$. [1.1]

PROJECTS

1. **DAY OF THE WEEK** A formula known as Zeller's Congruence makes use of the greatest integer function $[\![x]\!]$ to determine the day of the week on which a given day fell or will fall. To use Zeller's Congruence, we first compute the integer z given by

$$z = \left[\!\left[\frac{13m - 1}{5} \right]\!\right] + \left[\!\left[\frac{y}{4} \right]\!\right] + \left[\!\left[\frac{c}{4} \right]\!\right] + d + y - 2c$$

The variables c, y, d, and m are defined as follows:

$c = $ the century
$y = $ the year of the century
$d = $ the day of the month
$m = $ the month, using 1 for March, 2 for April, ..., 10 for December. January and February are assigned the values 11 and 12 of the previous year.

For example, for the date September 12, 2001, we use $c = 20$, $y = 1$, $d = 12$, and $m = 7$. The remainder of z divided by 7 gives the day of the week. A remainder of 0 represents a Sunday, a remainder of 1 a Monday, ..., and a remainder of 6 a Saturday.

a. Verify that December 7, 1941 was a Sunday.

b. Verify that January 1, 2010 will fall on a Friday.

c. Determine on what day of the week Independence Day (July 4, 1776) fell.

d. Determine on what day of the week you were born.

LINEAR FUNCTIONS

The following function has many applications.

Definition of a Linear Function

A function of the form

$$f(x) = mx + b, \quad m \neq 0$$

where m and b are real numbers, is a **linear function** of x.

• SLOPES OF LINES

The graph of $f(x) = mx + b$, or $y = mx + b$, is a nonvertical straight line.

The graphs shown in **Figure 2.36** are the graphs of $f(x) = mx + b$ for various values of m. The graphs intersect at the point $(-2, -1)$, but they differ in *steepness*. The steepness of a line is called the *slope* of the line and is denoted by the symbol m. The slope of a line is the ratio of the change in the y values of any two points on the line to the change in the x values of the same two points. For example, the graph of the line L_1 in **Figure 2.36** passes through the points $(-2, -1)$ and $(3, 5)$. The change in the y values is determined by subtracting the two y-coordinates.

$$\text{Change in } y = 5 - (-1) = 6$$

The change in the x values is determined by subtracting the two x-coordinates.

$$\text{Change in } x = 3 - (-2) = 5$$

The slope m of L_1 is the ratio of the change in the y values of the two points to the change in the x values of the two points. That is,

$$m = \frac{\text{change in } y}{\text{change in } x} = \frac{6}{5}$$

Because the slope of a nonvertical line can be calculated by using any two arbitrary points on the line, we have the following formula.

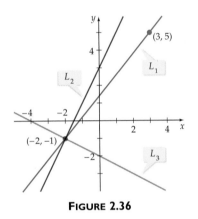

FIGURE 2.36

Slope of a Nonvertical Line

The **slope** m of the line passing through the points $P_1(x_1, y_1)$ and $P_2(x_2, y_2)$ with $x_1 \neq x_2$ is given by

$$m = \frac{y_2 - y_1}{x_2 - x_1}$$

Because the numerator $y_2 - y_1$ is the vertical **rise** and the denominator $x_2 - x_1$ is the horizontal **run** from P_1 to P_2, slope is often referred to as the *rise over the run* or the *change in y divided by the change in x*. See **Figure 2.37**. Lines that have a positive slope slant upward from left to right. Lines that have a negative slope slant downward from left to right.

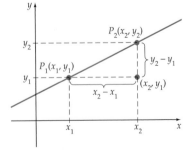

FIGURE 2.37

| EXAMPLE 1 | Find the Slope of a Line |

Find the slope of the line passing through the points whose coordinates are given.

a. $(1, 2)$ and $(3, 6)$ **b.** $(-3, 4)$ and $(1, -2)$

Solution

a. The slope of the line passing through $(1, 2)$ and $(3, 6)$ is

$$m = \frac{y_2 - y_1}{x_2 - x_1} = \frac{6 - 2}{3 - 1} = \frac{4}{2} = 2$$

Because $m > 0$, the line slants upward from left to right. See **Figure 2.38.**

b. The slope of the line passing through $(-3, 4)$ and $(1, -2)$ is

$$m = \frac{y_2 - y_1}{x_2 - x_1} = \frac{-2 - 4}{1 - (-3)} = \frac{-6}{4} = -\frac{3}{2}$$

Because $m < 0$, the line slants downward from left to right. See **Figure 2.39.**

▶ **TRY EXERCISE 2, PAGE 207**

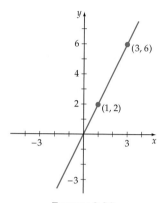

FIGURE 2.38

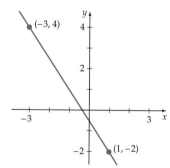

FIGURE 2.39

The definition of slope does not apply to vertical lines. Consider, for example, the points $(2, 1)$ and $(2, 3)$ on the vertical line l_1 in **Figure 2.40.** Applying the definition of slope to this line produces

$$m = \frac{3 - 1}{2 - 2}$$

which is undefined because it requires division by zero. Because division by zero is undefined, we say that the slope of any vertical line is undefined.

Every point on the vertical line through $(a, 0)$ has an x-coordinate of a. The equation of the vertical line through $(a, 0)$ is $x = a$. See **Figure 2.41.**

? QUESTION Is the graph of a vertical line the graph of a function?

All horizontal lines have 0 slope. For example, the line l_2 through $(2, 3)$ and $(4, 3)$ in **Figure 2.40** is a horizontal line. Its slope is given by

$$m = \frac{3 - 3}{4 - 2} = \frac{0}{2} = 0$$

When computing the slope of a line, it does not matter which point we label P_1 and which P_2 because

$$\frac{y_2 - y_1}{x_2 - x_1} = \frac{y_1 - y_2}{x_1 - x_2}$$

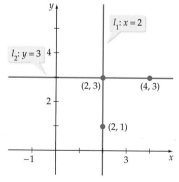

FIGURE 2.40

? ANSWER No. For example, the vertical line passing through $x = 2$ contains the ordered pairs $(2, 3)$ and $(2, -5)$. Thus there are two ordered pairs with the same first coordinate but different second coordinates.

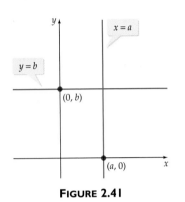

FIGURE 2.41

In functional notation, the points P_1 and P_2 can be represented by

$$(x_1, f(x_1)) \quad \text{and} \quad (x_2, f(x_2))$$

In this notation, the slope formula

$$m = \frac{y_2 - y_1}{x_2 - x_1} \quad \text{is expressed as} \quad m = \frac{f(x_2) - f(x_1)}{x_2 - x_1} \qquad (1)$$

If $m = 0$, then $f(x) = mx + b$ can be written as $f(x) = b$, or $y = b$. The graph of $y = b$ is the horizontal line through $(0, b)$. See **Figure 2.41.** Because every point on the graph of $y = b$ has a y-coordinate of b, the function $f(x) = b$ is called a **constant function.**

Horizontal Lines and Vertical Lines

The graph of $x = a$ is a vertical line through $(a, 0)$.

The graph of $y = b$ is a horizontal line through $(0, b)$.

The equation $f(x) = mx + b$ is called the **slope-intercept form** of the equation of a line because of the following theorem.

Slope-Intercept Form

The graph of $f(x) = mx + b$ is a line with slope m and y-intercept $(0, b)$.

Proof The slope of the graph of $f(x) = mx + b$ is given by Equation (1).

$$\frac{f(x_2) - f(x_1)}{x_2 - x_1} = \frac{(mx_2 + b) - (mx_1 + b)}{x_2 - x_1} = \frac{m(x_2 - x_1)}{x_2 - x_1} = m, \quad x_1 \neq x_2$$

The y-intercept of the graph of $f(x) = mx + b$ is found by letting $x = 0$.

$$f(0) = m(0) + b = b$$

Thus $(0, b)$ is the y-intercept, and m is the slope, of the graph of $f(x) = mx + b$. ◆

If a function is written in the form $f(x) = mx + b$, then its graph can be drawn by first plotting the y-intercept $(0, b)$ and then using the slope m to determine another point on the line.

EXAMPLE 2 Graph a Linear Function

Graph: $f(x) = 2x - 1$

Solution

The equation $y = 2x - 1$ is in slope-intercept form, with $b = -1$ and $m = 2$. Thus the y-intercept is $(0, -1)$, and the slope is 2. Write the slope as

$$m = \frac{2}{1} = \frac{\text{change in } y}{\text{change in } x}$$

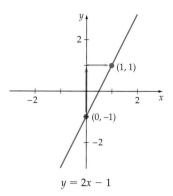

$y = 2x - 1$

FIGURE 2.42

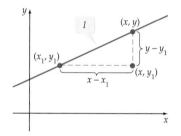

The slope of line l is $m = \dfrac{y - y_1}{x - x_1}$.

FIGURE 2.43

To graph the equation, first plot the y-intercept, and then use the slope to plot a second point. This second point is 2 units up (change in y) and 1 unit to the right (change in x) of the y-intercept. See **Figure 2.42**.

▶ **TRY EXERCISE 16, PAGE 208**

● FIND THE EQUATION OF A LINE

We can find an equation of a line provided we know its slope and at least one point on the line. **Figure 2.43** suggests that if (x_1, y_1) is a point on a line l of slope m, and (x, y) is *any other* point on the line, then

$$\frac{y - y_1}{x - x_1} = m, \quad x \neq x_1$$

Multiplying each side by $x - x_1$ produces $y - y_1 = m(x - x_1)$. This equation is called the **point-slope form** of the equation of line l.

Point-Slope Form

The graph of

$$y - y_1 = m(x - x_1)$$

is a line that has slope m and passes through (x_1, y_1).

EXAMPLE 3   Use the Point-Slope Form

Find an equation of the line with slope -3 that passes through $(-1, 4)$.

Solution

Use the point-slope form with $m = -3$, $x_1 = -1$, and $y_1 = 4$.

$$y - y_1 = m(x - x_1)$$
$$y - 4 = -3[x - (-1)] \qquad \text{• Substitute.}$$
$$y - 4 = -3x - 3 \qquad \text{• Solve for y.}$$
$$y = -3x + 1 \qquad \text{• Slope-intercept form}$$

▶ **TRY EXERCISE 28, PAGE 208**

take note

To determine an equation of a nonvertical line that passes through two points, first determine the slope of the line and then use the coordinates of either one of the points in the point-slope form.

An equation of the form $Ax + By = C$, where A, B, and C are real numbers and both A and B are not zero, is called the **general form of the equation of a line.** For example, the equation $y = -3x + 1$ in Example 3 can be written in general form as $3x + y = 1$.

<table>
<tr><td>

take note

It is not always possible to solve a linear equation in general form for y in terms of x. For instance, the linear equation $x = 4$ is in general form with $A = 1$, $B = 0$, and $C = 4$.

</td></tr>
</table>

One way to graph a linear equation that is written in general form is to first solve the equation for y in terms of x. For instance, to graph $3x - 2y = 4$, solve for y.

$$3x - 2y = 4$$
$$-2y = -3x + 4$$
$$y = \frac{3}{2}x - 2$$

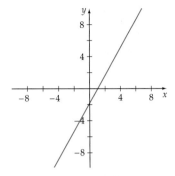

The y-intercept is $(0, -2)$ and the slope is $\frac{3}{2}$. The graph is shown in **Figure 2.44**.

FIGURE 2.44

By solving a first-degree equation, the specific relationship between an element of the domain and an element of the range of a linear function can be determined.

EXAMPLE 4   **Find the Value in the Domain of *f* for which *f*(*x*) = *b***

Find the value x in the domain of $f(x) = 3x - 4$ for which $f(x) = 5$.

Algebraic Solution

$$f(x) = 3x - 4$$
$$5 = 3x - 4 \qquad \text{• Replace } f(x) \text{ by 5 and solve for } x.$$
$$9 = 3x$$
$$3 = x$$

When $x = 3$, $f(x) = 5$. This means that 3 in the domain of f is paired with 5 in the range of f. Another way of stating this is that the ordered pair $(3, 5)$ is an element of f.

Visualize the Solution

By graphing $y = 5$ and $f(x) = 3x - 4$, we can see that $f(x) = 5$ when $x = 3$.

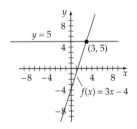

▶ **Try Exercise 42, page 208**

Although we are mainly concerned with linear functions in this section, the following theorem applies to all functions. It illustrates a powerful relationship between the real solutions of $f(x) = 0$ and the x-intercepts of the graph of $y = f(x)$.

Real Solutions and *x*-Intercepts Theorem

For every function f, the real number c is a solution of $f(x) = 0$ if and only if $(c, 0)$ is an x-intercept of the graph of $y = f(x)$.

❓ QUESTION   Is $(-2, 0)$ an x-intercept of $f(x) = x^3 - x + 6$?

The real solutions and x-intercepts theorem tells us that we can find real solutions of $f(x) = 0$ by graphing. The following example illustrates the theorem for a linear function of x.

❓ ANSWER   Yes. $f(-2) = (-2)^3 - (-2) + 6 = 0$

EXAMPLE 5   Verify the Real Solutions and x-Intercepts Theorem

Let $f(x) = -2x + 6$. Find the real solution of $f(x) = 0$ and then graph $f(x)$.
Compare the solution of $f(x) = 0$ with the x-intercept of the graph of f.

Algebraic Solution

To find the real solution of $f(x) = 0$, replace $f(x)$ by $-2x + 6$ and solve
for x.

$$f(x) = 0$$
$$-2x + 6 = 0$$
$$-2x = -6$$
$$x = 3$$

The x-coordinate of the x-intercept is 3. The real solution of $f(x) = 0$ is 3.

Visualize the Solution

Graph $f(x) = -2x + 6$ (see
Figure 2.45).

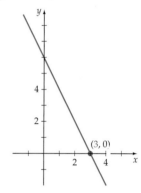

FIGURE 2.45

The x-intercept is $(3, 0)$.

▶ **TRY EXERCISE 46, PAGE 208**

EXAMPLE 6   Solve $f_1(x) = f_2(x)$

Let $f_1(x) = 2x - 1$ and $f_2(x) = -x + 11$. Find the values x for which $f_1(x) = f_2(x)$.

Algebraic Solution

$$f_1(x) = f_2(x)$$
$$2x - 1 = -x + 11$$
$$3x = 12$$
$$x = 4$$

When $x = 4$, $f_1(x) = f_2(x)$.

Visualize the Solution

The graphs of $y = f_1(x)$ and
$y = f_2(x)$ are shown on the
same coordinate axes (see
Figure 2.46). Note that the
point of intersection is $(4, 7)$.

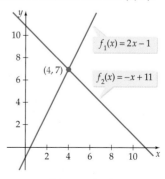

FIGURE 2.46

▶ **TRY EXERCISE 50, PAGE 208**

● **APPLICATIONS**

EXAMPLE 7   **Find a Linear Model of Data**

The bar graph in **Figure 2.47** is based on data from the Nevada Department of Motor Vehicles. The graph illustrates the distance (in feet) a car travels between the time (in seconds) a driver recognizes an emergency and the time the brakes are applied.

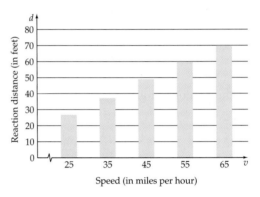

FIGURE 2.47

a. Find a linear function that models the reaction distance in terms of the speed of the car by using the ordered pairs $(25, 27)$ and $(55, 60)$.

b. What reaction distance does the model predict for a car traveling at 50 miles per hour?

Solution

a. First, calculate the slope of the line. Then use the point-slope formula to find the equation of the line.

$$m = \frac{d_2 - d_1}{v_2 - v_1} = \frac{60 - 27}{55 - 25} = \frac{33}{30} = 1.1$$ • **Find the slope.**

$$d - d_1 = m(v - v_1)$$ • **Use the point-slope formula.**

$$d - 27 = 1.1(v - 25)$$ • $d_1 = 27, v_1 = 25, m = 1.1.$

$$d = 1.1v - 0.5$$

In functional notation, the linear model is $d(v) = 1.1v - 0.5$.

b. To find the reaction distance for a car traveling at 50 miles per hour, evaluate $d(v)$ when $v = 50$.

$$d(v) = 1.1v - 0.5$$
$$d(50) = 1.1(50) - 0.5$$
$$= 54.5$$

The reaction distance is 54.5 feet.

▶ **TRY EXERCISE 56, PAGE 209**

If a manufacturer produces x units of a product that sells for p dollars per unit, then the **cost function** C, the **revenue function** R, and the **profit function** P are defined as follows:

$$C(x) = \text{cost of producing and selling } x \text{ units}$$

$$R(x) = xp = \text{revenue from the sale of } x \text{ units at } p \text{ dollars each}$$

$$P(x) = \text{profit from selling } x \text{ units}$$

Because profit equals the revenue less the cost, we have

$$P(x) = R(x) - C(x)$$

The value of x for which $R(x) = C(x)$ is called the **break-even point.** At the break-even point, $P(x) = 0$.

EXAMPLE 8 | **Find the Profit Function and the Break-even Point**

A manufacturer finds that the costs incurred in the manufacture and sale of a particular type of calculator are \$180,000 plus \$27 per calculator.

a. Determine the profit function P, given that x calculators are manufactured and sold at \$59 each.

b. Determine the break-even point.

Solution

a. The cost function is $C(x) = 27x + 180{,}000$. The revenue function is $R(x) = 59x$. Thus the profit function is

$$P(x) = R(x) - C(x)$$
$$= 59x - (27x + 180{,}000)$$
$$= 32x - 180{,}000, \quad x \geq 0 \text{ and } x \text{ is an integer}$$

b. At the break-even point, $R(x) = C(x)$.

$$59x = 27x + 180{,}000$$
$$32x = 180{,}000$$
$$x = 5625$$

The manufacturer will break even when 5625 calculators are sold.

▶ **TRY EXERCISE 66, PAGE 211**

take note

The graphs of C, R, and P are shown below. Observe that the graphs of C and R intersect at the break-even point, where $x = 5625$ and $P(5625) = 0$.

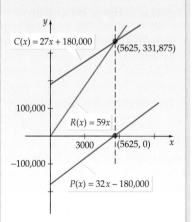

● **PARALLEL AND PERPENDICULAR LINES**

Two nonintersecting lines in a plane are **parallel.** All vertical lines are parallel to each other. All horizontal lines are parallel to each other.

Two lines are **perpendicular** if and only if they intersect and form adjacent angles each of which measures 90°. In a plane, vertical and horizontal lines are perpendicular to one another.

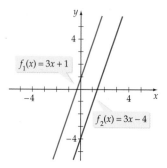

FIGURE 2.48

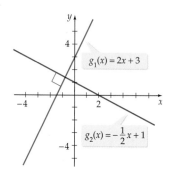

FIGURE 2.49

Parallel and Perpendicular Lines

Let l_1 be the graph of $f_1(x) = m_1 x + b$ and l_2 be the graph of $f_2(x) = m_2 x + b$. Then

- l_1 and l_2 are parallel if and only if $m_1 = m_2$.

- l_1 and l_2 are perpendicular if and only if $m_1 = -\dfrac{1}{m_2}$.

The graphs of $f_1(x) = 3x + 1$ and $f_2(x) = 3x - 4$ are shown in **Figure 2.48.** Because $m_1 = m_2 = 3$, the lines are parallel.

If $m_1 = -\dfrac{1}{m_2}$, then m_1 and m_2 are negative reciprocals of each other. The graphs of $g_1(x) = 2x + 3$ and $g_2(x) = -\dfrac{1}{2}x + 1$ are shown in **Figure 2.49.** Because 2 and $-\dfrac{1}{2}$ are negative reciprocals of each other, the lines are perpendicular. The symbol $\urcorner$ indicates an angle of 90°. In **Figure 2.49** it is used to indicate that the lines are perpendicular.

EXAMPLE 9   Determine a Point of Impact

A rock attached to a string is whirled horizontally in a circular counter-clockwise path about the origin. When the string breaks, the rock travels on a linear path perpendicular to the radius $\overline{OP}$ and hits a wall located at

$$y = x + 12 \tag{2}$$

If the string breaks when the rock is at $P(4, 3)$, determine the point at which the rock hits the wall. See **Figure 2.50.**

Solution

The slope of the radius from $(0, 0)$ to $(4, 3)$ is $\dfrac{3}{4}$. The negative reciprocal of $\dfrac{3}{4}$ is $-\dfrac{4}{3}$. Therefore, the linear path of the rock is given by

$$y - 3 = -\frac{4}{3}(x - 4)$$

$$y = -\frac{4}{3}x + \frac{25}{3} \tag{3}$$

To find the point at which the rock hits the wall, set the right side of Equation (2) equal to the right side of Equation (3) and solve for x. This is the procedure explained in Example 6.

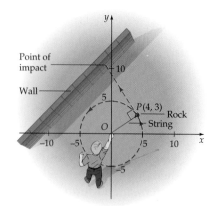

FIGURE 2.50

$$-\frac{4}{3}x + \frac{25}{3} = x + 12$$

$$-4x + 25 = 3x + 36 \qquad \bullet \text{ Multiply all terms by 3.}$$

$$-7x = 11$$

$$x = -\frac{11}{7}$$

For every point on the wall, x and y are related by $y = x + 12$. Therefore, substituting $-\frac{11}{7}$ for x in $y = x + 12$ yields $y = -\frac{11}{7} + 12 = \frac{73}{7}$, and the rock hits the wall at $\left(-\frac{11}{7}, \frac{73}{7}\right)$.

▶ **TRY EXERCISE 78, PAGE 212**

TOPICS FOR DISCUSSION

1. Can the graph of a linear function contain points in only one quadrant? only two quadrants? only four quadrants?

2. Is a "break-even point" a point or a number? Explain.

3. Some perpendicular lines do not have the property that their slopes are negative reciprocals of each other. Characterize these lines.

4. Does the real solutions and x-intercepts theorem apply only to linear functions?

5. Explain why the function $f(x) = x$ is referred to as the identity function.

EXERCISE SET 2.3

In Exercises 1 to 10, find the slope of the line that passes through the given points.

1. $(3, 4)$ and $(1, 7)$

▶ $(-2, 4)$ and $(5, 1)$

3. $(4, 0)$ and $(0, 2)$

4. $(-3, 4)$ and $(2, 4)$

5. $(0, 0)$ and $(0, 4)$

6. $(0, 0)$ and $(3, 0)$

7. $(-3, 4)$ and $(-4, -2)$

8. $(-5, -1)$ and $(-3, 4)$

9. $\left(-4, \frac{1}{2}\right)$ and $\left(\frac{7}{3}, \frac{7}{2}\right)$

10. $\left(\frac{1}{2}, 4\right)$ and $\left(\frac{7}{4}, 2\right)$

In Exercises 11 to 14, find the slope of the line that passes through the given points.

11. $(3, f(3))$ and $(3 + h, f(3 + h))$

12. $(-2, f(-2 + h))$ and $(-2 + h, f(-2 + h))$

13. $(0, f(0))$ and $(h, f(h))$

14. $(a, f(a))$ and $(a + h, f(a + h))$

In Exercises 15 to 26, graph y as a function of x by finding the slope and y-intercept of each line.

15. $y = 2x - 4$

▶ **16.** $y = -x + 1$

17. $y = -\dfrac{1}{3}x + 4$

18. $y = \dfrac{2}{3}x - 2$

19. $y = 3$

20. $y = x$

21. $y = 2x$

22. $y = -3x$

23. $2x + y = 5$

24. $x - y = 4$

25. $4x + 3y - 12 = 0$

26. $2x + 3y + 6 = 0$

In Exercises 27 to 38, find the equation of the indicated line. Write the equation in the form y = mx + b.

27. y-intercept $(0, 3)$, slope 1

▶ **28.** y-intercept $(0, 5)$, slope -2

29. y-intercept $\left(0, \dfrac{1}{2}\right)$, slope $\dfrac{3}{4}$

30. y-intercept $\left(0, \dfrac{3}{4}\right)$, slope $-\dfrac{2}{3}$

31. y-intercept $(0, 4)$, slope 0

32. y-intercept $(0, -1)$, slope $\dfrac{1}{2}$

33. Through $(-3, 2)$, slope -4

34. Through $(-5, -1)$, slope -3

35. Through $(3, 1)$ and $(-1, 4)$

36. Through $(5, -6)$ and $(2, -8)$

37. Through $(7, 11)$ and $(2, -1)$

38. Through $(-5, 6)$ and $(-3, -4)$

39. Find the value of x in the domain of $f(x) = 2x + 3$ for which $f(x) = -1$.

40. Find the value of x in the domain of $f(x) = 4 - 3x$ for which $f(x) = 7$.

41. Find the value of x in the domain of $f(x) = 1 - 4x$ for which $f(x) = 3$.

▶ **42.** Find the value of x in the domain of $f(x) = \dfrac{2x}{3} + 2$ for which $f(x) = 4$.

43. Find the value of x in the domain of $f(x) = 3 - \dfrac{x}{2}$ for which $f(x) = 5$.

44. Find the value of x in the domain of $f(x) = 4x - 3$ for which $f(x) = -2$.

In Exercises 45 to 48, find the solution f(x) = 0. Verify that the solution of f(x) = 0 is the same as the x-coordinate of the x-intercept of the graph of y = f(x).

45. $f(x) = 3x - 12$

▶ **46.** $f(x) = -2x - 4$

47. $f(x) = \dfrac{1}{4}x + 5$

48. $f(x) = -\dfrac{1}{3}x + 2$

In Exercises 49 to 52, solve $f_1(x) = f_2(x)$ by an algebraic method and by graphing.

49. $f_1(x) = 4x + 5$ \qquad $f_2(x) = x + 6$

▶ **50.** $f_1(x) = -2x - 11$ \qquad $f_2(x) = 3x + 7$

51. $f_1(x) = 2x - 4$ \qquad $f_2(x) = -x + 12$

52. $f_1(x) = \dfrac{1}{2}x + 5$ \qquad $f_2(x) = \dfrac{2}{3}x - 7$

53. OCEANOGRAPHY The graph below shows the relationship between the speed of sound in water and the temperature of the water. Find the slope of this line, and write a sentence that explains the meaning of the slope in the context of this problem.

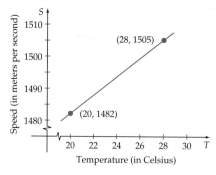

54. COMPUTER SCIENCE The graph on the following page shows the relationship between the time, in seconds, it takes to download a file and the size of the file in megabytes. Find the slope of the line between the two points shown on the graph. Write a sentence that states the meaning of the slope in the context of this problem.

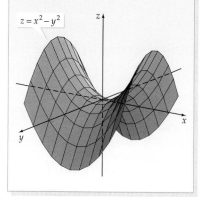
Standard Form of Quadratic Functions

Every quadratic function f given by $f(x) = ax^2 + bx + c$ can be written in the **standard form of a quadratic function:**

$$f(x) = a(x - h)^2 + k, \quad a \neq 0$$

The graph of f is a parabola with vertex (h, k). The parabola opens up if $a > 0$, and it opens down if $a < 0$. The vertical line $x = h$ is the axis of symmetry of the parabola.

The standard form is useful because it readily gives information about the vertex of the parabola and its axis of symmetry. For example, note that the graph of $f(x) = 2(x - 4)^2 - 3$ is a parabola. The coordinates of the vertex are $(4, -3)$, and the line $x = 4$ is its axis of symmetry. Because a is the positive number 2, the parabola opens upward.

EXAMPLE I   Find the Standard Form of a Quadratic Function

Use the technique of completing the square to find the standard form of $g(x) = 2x^2 - 12x + 19$. Sketch the graph.

Solution

$$\begin{aligned} g(x) &= 2x^2 - 12x + 19 \\ &= 2(x^2 - 6x) + 19 && \text{• Factor 2 from the variable terms.} \\ &= 2(x^2 - 6x + 9 - 9) + 19 && \text{• Complete the square.} \\ &= 2(x^2 - 6x + 9) - 2(9) + 19 && \text{• Regroup.} \\ &= 2(x - 3)^2 - 18 + 19 && \text{• Factor and simplify.} \\ &= 2(x - 3)^2 + 1 && \text{• Standard form} \end{aligned}$$

The vertex is $(3, 1)$. The axis of symmetry is $x = 3$. Because $a > 0$, the parabola opens up. See **Figure 2.53.**

▶ **TRY EXERCISE 10, PAGE 222**

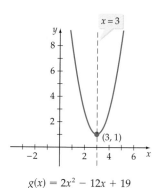

$g(x) = 2x^2 - 12x + 19$

FIGURE 2.53

● VERTEX OF A PARABOLA

We can write $f(x) = ax^2 + bx + c$ in standard form by completing the square of $ax^2 + bx + c$. This will allow us to derive a general expression for the x- and y-coordinates of the graph of $f(x) = ax^2 + bx + c$.

$$\begin{aligned} f(x) &= ax^2 + bx + c \\ &= a\left(x^2 + \frac{b}{a}x\right) + c && \text{• Factor } a \text{ from } ax^2 + bx. \\ &= a\left(x^2 + \frac{b}{a}x + \frac{b^2}{4a^2}\right) + c - \frac{b^2}{4a} && \text{• Complete the square by adding and} \\ & && \quad \text{subtracting } \left(\frac{1}{2} \cdot \frac{b}{a}\right)^2 = \frac{b^2}{4a^2}. \\ &= a\left(x + \frac{b}{2a}\right)^2 + \frac{4ac - b^2}{4a} && \text{• Factor and simplify.} \end{aligned}$$

Thus $f(x) = ax^2 + bx + c$ in standard form is $f(x) = a\left(x + \dfrac{b}{2a}\right)^2 + \dfrac{4ac - b^2}{4a}$.

Comparing this last expression with $f(x) = a(x - h)^2 + k$, we see that the coordinates of the vertex are $\left(-\dfrac{b}{2a}, \dfrac{4ac - b^2}{4a}\right)$.

Note that by evaluating $f(x) = a\left(x + \dfrac{b}{2a}\right)^2 + \dfrac{4ac - b^2}{4a}$ at $x = -\dfrac{b}{2a}$, we have

$$f(x) = a\left(x + \frac{b}{2a}\right)^2 + \frac{4ac - b^2}{4a}$$

$$f\left(-\frac{b}{2a}\right) = a\left(-\frac{b}{2a} + \frac{b}{2a}\right)^2 + \frac{4ac - b^2}{4a} = a(0) + \frac{4ac - b^2}{4a}$$

$$= \frac{4ac - b^2}{4a}$$

That is, the y-coordinate of the vertex is $f\left(-\dfrac{b}{2a}\right)$. This is summarized by the following formula.

Vertex Formula

The coordinates of the vertex of $f(x) = ax^2 + bx + c$ are $\left(-\dfrac{b}{2a}, f\left(-\dfrac{b}{2a}\right)\right)$.

The vertex formula can be used to write the standard form of the equation of a parabola. We have

$$h = -\frac{b}{2a} \quad \text{and} \quad k = f\left(-\frac{b}{2a}\right)$$

EXAMPLE 2   **Find the Vertex and Standard Form of a Quadratic Function**

Use the vertex formula to find the vertex and standard form of $f(x) = 2x^2 - 8x + 3$. See **Figure 2.54.**

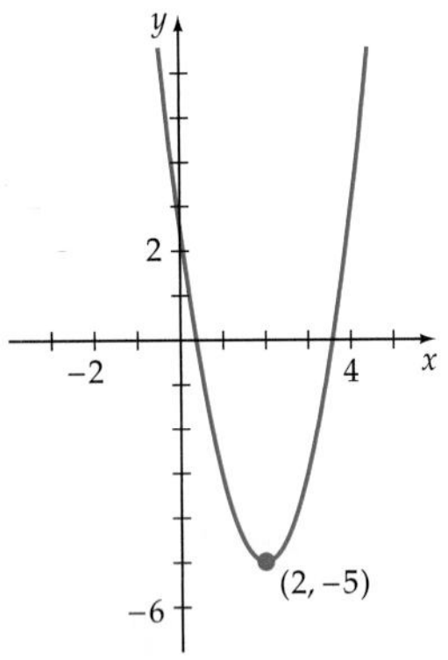

Solution

$f(x) = 2x^2 - 8x + 3$ • $a = 2, b = -8, c = 3$

$h = -\dfrac{b}{2a} = -\dfrac{-8}{2(2)} = 2$ • **x-coordinate of the vertex**

$k = f\left(-\dfrac{b}{2a}\right) = 2(2)^2 - 8(2) + 3 = -5$ • **y-coordinate of the vertex**

The vertex is $(2, -5)$. Substituting into the standard form equation $f(x) = a(x - h)^2 + k$ yields the standard form $f(x) = 2(x - 2)^2 - 5$.

$f(x) = 2x^2 - 8x + 3$

FIGURE 2.54

▶ TRY EXERCISE 20, PAGE 223

● MAXIMUM AND MINIMUM OF A QUADRATIC FUNCTION

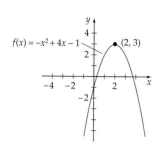

$f(x) = -x^2 + 4x - 1$

Note from Example 2 that the graph of the parabola opens up, and the vertex is the *lowest* point on the graph of the parabola. Therefore, the y-coordinate of the vertex is the *minimum* value of that function. This information can be used to determine the range of $f(x) = 2x^2 - 8x + 3$. The range is $\{y\,|\,y \geq -5\}$. Similarly, if the graph of a parabola opened down, the vertex would be the *highest* point on the graph, and the y-coordinate of the vertex would be the *maximum* value of the function. For instance, the maximum value of $f(x) = -x^2 + 4x - 1$, graphed at the left, is 3, the y-coordinate of the vertex. The range of the function is $\{y\,|\,y \leq 3\}$. For the function in Example 2 and the function whose graph is shown at the left, the domain is the set of real numbers.

EXAMPLE 3   Find the Range of $f(x) = ax^2 + bx + c$

Find the range of $f(x) = -2x^2 - 6x - 1$. Determine the values of x for which $f(x) = 3$.

Algebraic Solution

To find the range of f, determine the y-coordinate of the vertex of the graph of f.

$$f(x) = -2x^2 - 6x - 1$$

• $a = -2, b = -6, c = -1$

$$h = -\frac{b}{2a} = -\frac{-6}{2(-2)} = -\frac{3}{2}$$

• **Find the x-coordinate of the vertex.**

$$k = f\left(-\frac{3}{2}\right) = -2\left(-\frac{3}{2}\right)^2 - 6\left(-\frac{3}{2}\right) - 1 = \frac{7}{2}$$

• **Find the y-coordinate of the vertex.**

The vertex is $\left(-\dfrac{3}{2}, \dfrac{7}{2}\right)$. Because the parabola opens down, $\dfrac{7}{2}$ is the maximum value of f. Therefore, the range of f is $\left\{y\,\middle|\,y \leq \dfrac{7}{2}\right\}$.

To determine the values of x for which $f(x) = 3$, replace $f(x)$ by $-2x^2 - 6x - 1$ and solve for x.

$$f(x) = 3$$
$$-2x^2 - 6x - 1 = 3$$ • **Replace $f(x)$ by $-2x^2 - 6x - 1$.**
$$-2x^2 - 6x - 4 = 0$$ • **Solve for x.**
$$-2(x + 1)(x + 2) = 0$$ • **Factor.**
$$x + 1 = 0 \text{ or } x + 2 = 0$$ • **Use the Principle of Zero Products to solve for x.**
$$x = -1 \qquad x = -2$$

The values of x for which $f(x) = 3$ are -1 and -2.

Visualize the Solution

The graph of f is shown below. The vertex of the graph is $\left(-\dfrac{3}{2}, \dfrac{7}{2}\right)$. Note that the line $y = 3$ intersects the graph of f when $x = -2$ and when $x = -1$.

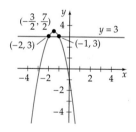

▶ **TRY EXERCISE 32, PAGE 223**

The following theorem can be used to determine the maximum value or the minimum value of a quadratic function.

Maximum or Minimum Value of a Quadratic Function

If $a > 0$, then the vertex (h, k) is the lowest point on the graph of $f(x) = a(x - h)^2 + k$, and the y-coordinate k of the vertex is the **minimum value** of the function f. See **Figure 2.55a.**

If $a < 0$, then the vertex (h, k) is the highest point on the graph of $f(x) = a(x - h)^2 + k$, and the y-coordinate k is the **maximum value** of the function f. See **Figure 2.55b.**

In either case, the maximum or minimum is achieved when $x = h$.

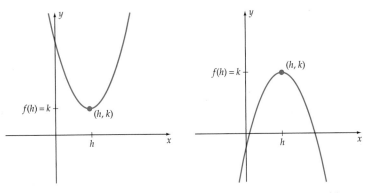

a. k is the minimum value of f.   **b.** k is the maximum value of f.

FIGURE 2.55

EXAMPLE 4   **Find the Maximum or Minimum of a Quadratic Function**

Find the maximum or minimum value of each quadratic function. State whether the value is a maximum or a minimum.

a. $F(x) = -2x^2 + 8x - 1$ **b.** $G(x) = x^2 - 3x + 1$

Solution

The maximum or minimum value of a quadratic function is the y-coordinate of the vertex of the graph of the function.

a. $\quad h = -\dfrac{b}{2a} = -\dfrac{8}{2(-2)} = 2$ **• x-coordinate of the vertex**

$\quad k = F\left(-\dfrac{b}{2a}\right) = -2(2)^2 + 8(2) - 1 = 7$ **• y-coordinate of the vertex**

Because $a < 0$, the function has a maximum value but no minimum value. The maximum value is 7. See **Figure 2.56.**

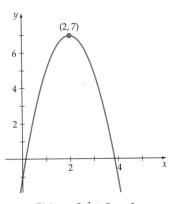

$F(x) = -2x^2 + 8x - 1$

FIGURE 2.56

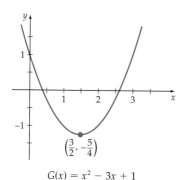

$G(x) = x^2 - 3x + 1$

FIGURE 2.57

b. $h = -\dfrac{b}{2a} = -\dfrac{-3}{2(1)} = \dfrac{3}{2}$ • **x-coordinate of the vertex**

$k = G\left(-\dfrac{b}{2a}\right) = \left(\dfrac{3}{2}\right)^2 - 3\left(\dfrac{3}{2}\right) + 1$

$= -\dfrac{5}{4}$ • **y-coordinate of the vertex**

Because $a > 0$, the function has a minimum value but no maximum value. The minimum value is $-\dfrac{5}{4}$. See **Figure 2.57.**

▶ **TRY EXERCISE 36, PAGE 223**

● **APPLICATIONS**

EXAMPLE 5 **Find the Maximum of a Quadratic Function**

A long sheet of tin 20 inches wide is to be made into a trough by bending up two sides until they are perpendicular to the bottom. How many inches should be turned up so that the trough will achieve its maximum carrying capacity?

Solution

The trough is shown in **Figure 2.58.** If x is the number of inches to be turned up on each side, then the width of the base is $20 - 2x$ inches. The maximum carrying capacity of the trough will occur when the cross-sectional area is a maximum. The cross-sectional area $A(x)$ is given by

$$A(x) = x(20 - 2x) \qquad \text{• Area} = (\text{length})(\text{width})$$
$$= -2x^2 + 20x$$

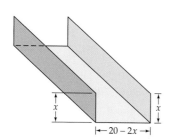

FIGURE 2.58

To find the point at which A obtains its maximum value, find the x-coordinate of the vertex of the graph of A. Using the vertex formula with $a = -2$ and $b = 20$, we have

$$x = -\dfrac{b}{2a} = -\dfrac{20}{2(-2)} = 5$$

Therefore, the maximum carrying capacity will be achieved when 5 inches are turned up. See **Figure 2.59.**

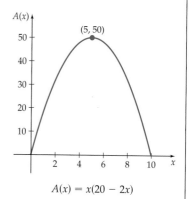

$A(x) = x(20 - 2x)$

FIGURE 2.59

▶ **TRY EXERCISE 46, PAGE 223**

EXAMPLE 6   **Solve a Business Application**

The owners of a travel agency have determined that they can sell all 160 tickets for a tour if they charge $8 (their cost) for each ticket. For each $0.25 increase in the price of a ticket, they estimate they will sell 1 ticket less. A business manager determines that their cost function is $C(x) = 8x$ and that the customer's price per ticket is

$$p(x) = 8 + 0.25(160 - x) = 48 - 0.25x$$

where x represents the number of tickets sold. Determine the maximum profit and the cost per ticket that yields the maximum profit.

Solution

The profit from selling x tickets is $P(x) = R(x) - C(x)$, where P, R, and C are the profit function, the revenue function, and the cost function as defined in Section 2.3. Thus

$$\begin{aligned} P(x) &= R(x) - C(x) \\ &= x[p(x)] - C(x) \\ &= x(48 - 0.25x) - 8x \\ &= 40x - 0.25x^2 \end{aligned}$$

The graph of the profit function is a parabola that opens down. Thus the maximum profit occurs when

$$x = -\frac{b}{2a} = -\frac{40}{2(-0.25)} = 80$$

The maximum profit is determined by evaluating $P(x)$ with $x = 80$.

$$P(80) = 40(80) - 0.25(80)^2 = 1600$$

The maximum profit is $1600.
To find the price per ticket that yields the maximum profit, we evaluate $p(x)$ with $x = 80$.

$$p(80) = 48 - 0.25(80) = 28$$

Thus the travel agency can expect a maximum profit of $1600 when 80 people take the tour at a ticket price of $28 per person. The graph of the profit function is shown in **Figure 2.60.**

❓ QUESTION In **Figure 2.60,** why have we shown only the portion of the graph that lies in quadrant I?

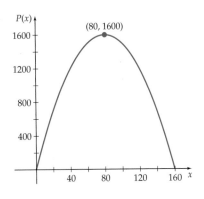

$P(x) = 40x - 0.25x^2$

FIGURE 2.60

▶ **TRY EXERCISE 68, PAGE 225**

❓ ANSWER   Since x represents the number of tickets sold, x must be greater than or equal to zero but less than or equal to 160. $P(x)$ is nonnegative for $0 \le x \le 160$.

> **EXAMPLE 7** **Solve a Projectile Application**
>
> In **Figure 2.61,** a ball is thrown vertically upward with an initial velocity of 48 feet per second. If the ball started its flight at a height of 8 feet, then its height s at time t can be determined by $s(t) = -16t^2 + 48t + 8$, where $s(t)$ is measured in feet above ground level and t is the number of seconds of flight.
>
> a. Determine the time it takes the ball to attain its maximum height.
>
> b. Determine the maximum height the ball attains.
>
> c. Determine the time it takes the ball to hit the ground.
>
> **Solution**
>
> a. The graph of $s(t) = -16t^2 + 48t + 8$ is a parabola that opens downward. See **Figure 2.62.** Therefore, s will attain its maximum value at the vertex of its graph. Using the vertex formula with $a = -16$ and $b = 48$, we get
>
> $$t = -\frac{b}{2a} = -\frac{48}{2(-16)} = \frac{3}{2}$$
>
> Therefore, the ball attains its maximum height $1\frac{1}{2}$ seconds into its flight.
>
> b. When $t = \frac{3}{2}$, the height of the ball is
>
> $$s\left(\frac{3}{2}\right) = -16\left(\frac{3}{2}\right)^2 + 48\left(\frac{3}{2}\right) + 8 = 44 \text{ feet}$$
>
> c. The ball will hit the ground when its height $s(t) = 0$. Therefore, solve $-16t^2 + 48t + 8 = 0$ for t.
>
> $$-16t^2 + 48t + 8 = 0$$
> $$-2t^2 + 6t + 1 = 0 \qquad \bullet \text{ Divide each side by 8.}$$
> $$t = \frac{-(6) \pm \sqrt{6^2 - 4(-2)(1)}}{2(-2)} \qquad \bullet \text{ Use the quadratic formula.}$$
> $$= \frac{-6 \pm \sqrt{44}}{-4} = \frac{-3 \pm \sqrt{11}}{-2}$$
>
> Using a calculator to approximate the positive root, we find that the ball will hit the ground in $t \approx 3.16$ seconds. This is also the value of the t-coordinate of the t-intercept in **Figure 2.62.**
>
> ▶ **TRY EXERCISE 70, PAGE 225**

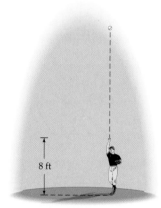

FIGURE 2.61

 To review **QUADRATIC FORMULA,** *see p. 107.*

To review **QUADRATIC FORMULA,** *see p. 107.*

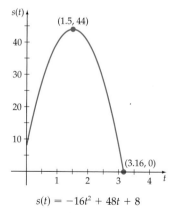

$s(t) = -16t^2 + 48t + 8$

FIGURE 2.62

TOPICS FOR DISCUSSION

1. Does the graph of every quadratic function of the form

$$f(x) = ax^2 + bx + c$$

have a y-intercept? If so, what are the coordinates of the y-intercept?

2. The graph of $f(x) = -x^2 + 6x + 11$ has a vertex of $(3, 20)$. Is this vertex point the highest point or the lowest point on the graph of f?

3. A tutor states that the graph of $f(x) = ax^2 + bx + c$ $(a \neq 0)$ is a parabola and that its axis of symmetry is $y = -\dfrac{b}{2a}$. Do you agree?

4. Every quadratic function of the form $f(x) = ax^2 + bx + c$ has a domain of all real numbers. Do you agree?

5. A classmate states that the graph of every quadratic function of the form
    $$f(x) = ax^2 + bx + c$$
    must contain points from at least two quadrants. Do you agree?

EXERCISE SET 2.4

In Exercises 1 to 8, match each graph in a. through h. with the proper quadratic function.

1. $f(x) = x^2 - 3$

2. $f(x) = x^2 + 2$

3. $f(x) = (x - 4)^2$

4. $f(x) = (x + 3)^2$

5. $f(x) = -2x^2 + 2$

6. $f(x) = -\dfrac{1}{2}x^2 + 3$

7. $f(x) = (x + 1)^2 + 3$

8. $f(x) = -2(x - 2)^2 + 2$

e.

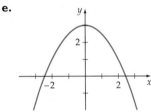

f.

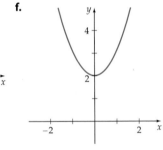

a.

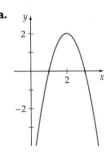

b.

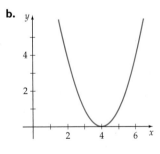

g.

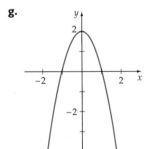

h.

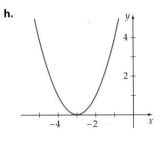

c.   **d.**

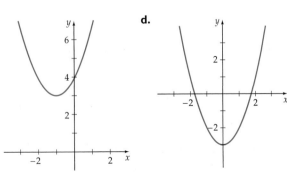

In Exercises 9 to 18, use the method of completing the square to find the standard form of the quadratic function, and then sketch its graph. Label its vertex and axis of symmetry.

9. $f(x) = x^2 + 4x + 1$

▶ **10.** $f(x) = x^2 + 6x - 1$

11. $f(x) = x^2 - 8x + 5$

12. $f(x) = x^2 - 10x + 3$

13. $f(x) = x^2 + 3x + 1$

14. $f(x) = x^2 + 7x + 2$

15. $f(x) = -x^2 + 4x + 2$

16. $f(x) = -x^2 - 2x + 5$

17. $f(x) = -3x^2 + 3x + 7$

18. $f(x) = -2x^2 - 4x + 5$

In Exercises 19 to 28, use the vertex formula to determine the vertex of the graph of the function and write the function in standard form.

19. $f(x) = x^2 - 10x$

▶ **20.** $f(x) = x^2 - 6x$

21. $f(x) = x^2 - 10$

22. $f(x) = x^2 - 4$

23. $f(x) = -x^2 + 6x + 1$

24. $f(x) = -x^2 + 4x + 1$

25. $f(x) = 2x^2 - 3x + 7$

26. $f(x) = 3x^2 - 10x + 2$

27. $f(x) = -4x^2 + x + 1$

28. $f(x) = -5x^2 - 6x + 3$

29. Find the range of $f(x) = x^2 - 2x - 1$. Determine the values of x in the domain of f for which $f(x) = 2$.

30. Find the range of $f(x) = -x^2 - 6x - 2$. Determine the values of x in the domain of f for which $f(x) = 3$.

31. Find the range of $f(x) = -2x^2 + 5x - 1$. Determine the values of x in the domain of f for which $f(x) = 2$.

▶ **32.** Find the range of $f(x) = 2x^2 + 6x - 5$. Determine the values of x in the domain of f for which $f(x) = 15$.

33. Is 3 in the range of $f(x) = x^2 + 3x + 6$? Explain your answer.

34. Is -2 in the range of $f(x) = -2x^2 - x + 1$? Explain your answer.

In Exercises 35 to 44, find the maximum or minimum value of the function. State whether this value is a maximum or a minimum.

35. $f(x) = x^2 + 8x$

▶ **36.** $f(x) = -x^2 - 6x$

37. $f(x) = -x^2 + 6x + 2$

38. $f(x) = -x^2 + 10x - 3$

39. $f(x) = 2x^2 + 3x + 1$

40. $f(x) = 3x^2 + x - 1$

41. $f(x) = 5x^2 - 11$

42. $f(x) = 3x^2 - 41$

43. $f(x) = -\dfrac{1}{2}x^2 + 6x + 17$

44. $f(x) = -\dfrac{3}{4}x^2 - \dfrac{2}{5}x + 7$

45. HEIGHT OF AN ARCH The height of an arch is given by the equation

$$h(x) = -\frac{3}{64}x^2 + 27, \quad -24 \le x \le 24$$

where $|x|$ is the horizontal distance in feet from the center of the arch.

a. What is the maximum height of the arch?

b. What is the height of the arch 10 feet to the right of center?

c. How far from the center is the arch 8 feet tall?

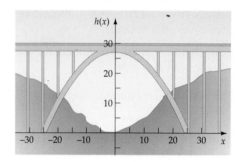

▶ **46.** The sum of the length l and the width w of a rectangular area is 240 meters.

a. Write w as a function of l.

b. Write the area A as a function of l.

c. Find the dimensions that produce the greatest area.

47. RECTANGULAR ENCLOSURE A veterinarian uses 600 feet of chain-link fencing to enclose a rectangular region and also to subdivide the region into two smaller rectangular regions by placing a fence parallel to one of the sides, as shown in the figure.

a. Write the width w as a function of the length l.

b. Write the total area A as a function of l.

c. Find the dimensions that produce the greatest enclosed area.

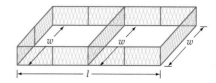

48. RECTANGULAR ENCLOSURE A farmer uses 1200 feet of fence to enclose a rectangular region and also to subdivide the region into three smaller rectangular regions by placing the fences parallel to one of the sides. Find the dimensions that produce the greatest enclosed area.

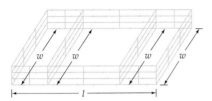

49. TEMPERATURE FLUCTUATIONS The temperature $T(t)$, in degrees Fahrenheit, during the day can be modeled by the equation $T(t) = -0.7t^2 + 9.4t + 59.3$, where t is the number of hours after 6:00 A.M.

   a. At what time is the temperature a maximum? Round to the nearest minute.

   b. What is the maximum temperature? Round to the nearest degree.

50. LARVAE SURVIVAL Soon after insect larvae are hatched, they must begin to search for food. The survival rate of the larvae depends on many factors, but the temperature of the environment is one of the most important. For a certain species of insect, a model of the number of larvae, $N(T)$, that survive this searching period is given by

$$N(T) = -0.6T^2 + 32.1T - 350$$

where T is the temperature in degrees Celsius.

   a. At what temperature will the maximum number of larvae survive? Round to the nearest degree.

   b. What is the maximum number of surviving larvae? Round to the nearest whole number.

   c. Find the x-intercepts, to the nearest whole number, for the graph of this function.

   d. Write a sentence that describes the meaning of the x-intercepts in the context of this problem.

51. REAL ESTATE The number of California homes that have sold for over $1,000,000 between 1989 and 2002 can be modeled by

$$N(t) = 1.43t^2 - 11.44t + 47.68$$

where $N(t)$ is the number (in hundreds) of homes that were sold in year t, with $t = 0$ corresponding to 1989. According to this model, in what year were the least number of million-dollar homes sold? How many million-dollar homes, to the nearest hundred, were sold that year?

52. GEOLOGY In June 2001, Mt. Etna in Sicily, Italy erupted, sending volcanic bombs (a mass of molten lava ejected from the volcano) into the air. A model of the height h, in meters, of a volcanic bomb above the crater of the volcano t seconds after the eruption is given by $h(t) = -9.8t^2 + 100t$. Find the maximum height of a volcanic bomb above the crater for this eruption. Round to the nearest meter.

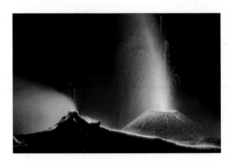

53. SPORTS For a serve to be legal in tennis, the ball must be at least 3 feet high when it is 39 feet from the server, and it must land in a spot that is less than 60 feet from the server. Does the path of a ball given by $h(x) = -0.002x^2 - 0.03x + 8$, where $h(x)$ is the height of the ball (in feet) x feet from the server, satisfy the conditions of a legal serve?

54. SPORTS A pitcher releases a baseball 6 feet above the ground at a speed of 132 feet per second (90 miles per hour) toward home plate which is 60.5 feet away. The height $h(x)$, in feet, of the ball x feet from home plate can be approximated by $h(x) = -0.0009x^2 + 6$. To be considered a strike, the ball must cross home plate and be at least 2.5 feet high and less than 5.4 feet high. Assuming the ball crosses home plate, is this particular pitch a strike? Explain.

55. AUTOMOTIVE ENGINEERING The fuel efficiency for a certain midsize car is given by

$$E(v) = -0.018v^2 + 1.476v + 3.4$$

where $E(v)$ is the fuel efficiency in miles per gallon for a car traveling v miles per hour.

   a. What speed will yield the maximum fuel efficiency? Round to the nearest mile per hour.

   b. What is the maximum fuel efficiency for this car? Round to the nearest mile per gallon.

56. SPORTS Some football fields are built in a parabolic mound shape so that water will drain off the field. A model for the shape of a certain field is given by

$$h(x) = -0.0002348x^2 + 0.0375x$$

where $h(x)$ is the height, in feet, of the field at a distance of x feet from one sideline. Find the maximum height of the field. Round to the nearest tenth of a foot.

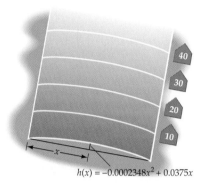

$h(x) = -0.0002348x^2 + 0.0375x$

In Exercises 57 to 60, determine the y- and x-intercepts (if any) of the quadratic function.

57. $f(x) = x^2 + 6x$

58. $f(x) = -x^2 + 4x$

59. $f(x) = -3x^2 + 5x - 6$

60. $f(x) = 2x^2 + 3x + 4$

In Exercises 61 and 62, determine the number of units x that produce a maximum revenue for the given revenue function. Also determine the maximum revenue.

61. $R(x) = 296x - 0.2x^2$

62. $R(x) = 810x - 0.6x^2$

In Exercises 63 and 64, determine the number of units x that produce a maximum profit for the given profit function. Also determine the maximum profit.

63. $P(x) = -0.01x^2 + 1.7x - 48$

64. $P(x) = -\dfrac{x^2}{14{,}000} + 1.68x - 4000$

In Exercises 65 and 66, determine the profit function for the given revenue function and cost function. Also determine the break-even point(s).

65. $R(x) = x(102.50 - 0.1x); C(x) = 52.50x + 1840$

66. $R(x) = x(210 - 0.25x); C(x) = 78x + 6399$

67. TOUR COST A charter bus company has determined that the cost of providing x people a tour is

$$C(x) = 180 + 2.50x$$

A full tour consists of 60 people. The ticket price per person is $15 plus $0.25 for each unsold ticket. Determine

a. the revenue function   **b.** the profit function

c. the company's maximum profit

d. the number of ticket sales that yields the maximum profit

▶ **68. DELIVERY COST** An air freight company has determined that the cost, in dollars, of delivering x parcels per flight is

$$C(x) = 2025 + 7x$$

The price per parcel, in dollars, the company charges to send x parcels is

$$p(x) = 22 - 0.01x$$

Determine

a. the revenue function   **b.** the profit function

c. the company's maximum profit

d. the price per parcel that yields the maximum profit

e. the minimum number of parcels the air freight company must ship to break even

69. PROJECTILE If the initial velocity of a projectile is 128 feet per second, then its height h in feet is a function of time t in seconds given by the equation $h(t) = -16t^2 + 128t$.

a. Find the time t when the projectile achieves its maximum height.

b. Find the maximum height of the projectile.

c. Find the time t when the projectile hits the ground.

▶ **70. PROJECTILE** The height in feet of a projectile with an initial velocity of 64 feet per second and an initial height of 80 feet is a function of time t in seconds given by

$$h(t) = -16t^2 + 64t + 80$$

a. Find the maximum height of the projectile.

b. Find the time t when the projectile achieves its maximum height.

c. Find the time t when the projectile has a height of 0 feet.

71. FIRE MANAGEMENT The height of a stream of water from the nozzle of a fire hose can be modeled by

$$y(x) = -0.014x^2 + 1.19x + 5$$

where $y(x)$ is the height, in feet, of the stream x feet from the firefighter. What is the maximum height that the stream of water from this nozzle can reach? Round to the nearest foot.

72. 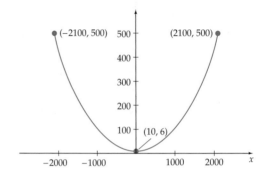 OLYMPIC SPORTS In 1988, Louise Ritter of the United States set the women's Olympic record for the high jump. A mathematical model that approximates her jump is given by

$$h(t) = -204.8t^2 + 256t$$

where $h(t)$ is her height in inches t seconds after beginning her jump. Find the maximum height of her jump.

73. NORMAN WINDOW A Norman window has the shape of a rectangle surmounted by a semicircle. The exterior perimeter of the window shown in the figure is 48 feet. Find the height h and the radius r that will allow the maximum amount of light to enter the window. (*Hint:* Write the area of the window as a quadratic function of the radius r.)

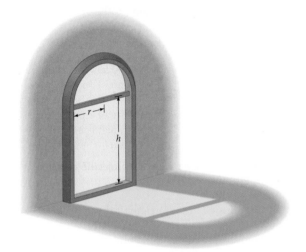

74. GOLDEN GATE BRIDGE The suspension cables of the main span of the Golden Gate Bridge are in the shape of a parabola. If a coordinate system is drawn as shown, find the quadratic function that models a suspension cable for the main span of the bridge.

CONNECTING CONCEPTS

75. Let $f(x) = x^2 - (a + b)x + ab$, where a and b are real numbers.

   a. Show that the x-intercepts are $(a, 0)$ and $(b, 0)$.

   b. Show that the minimum value of the function occurs at the x-value of the midpoint of the line segment defined by the x-intercepts.

76. Let $f(x) = ax^2 + bx + c$, where $a, b,$ and c are real numbers.

   a. What conditions must be imposed on the coefficients so that f has a maximum?

   b. What conditions must be imposed on the coefficients so that f has a minimum?

   c. What conditions must be imposed on the coefficients so that the graph of f intersects the x-axis?

77. Find the quadratic function of x whose graph has a minimum at $(2, 1)$ and passes through $(0, 4)$.

78. Find the quadratic function of x whose graph has a maximum at $(-3, 2)$ and passes through $(0, -5)$.

79. AREA OF A RECTANGLE A wire 32 inches long is bent so that it has the shape of a rectangle. The length of the rectangle is x and the width is w.

   a. Write w as a function of x.

   b. Write the area A of the rectangle as a function of x.

80. MAXIMIZE AREA Use the function A from **b.** in Exercise 79 to prove that the area A is greatest if the rectangle is a square.

81. Show that the function $f(x) = x^2 + bx - 1$ has a real zero for any value b.

82. Show that the function $g(x) = -x^2 + bx + 1$ has a real zero for any value b.

83. What effect does increasing the constant c have on the graph of $f(x) = ax^2 + bx + c$?

84. If $a > 0$, what effect does decreasing the coefficient a have on the graph of $f(x) = ax^2 + bx + c$?

85. Find two numbers whose sum is 8 and whose product is a maximum.

86. Find two numbers whose difference is 12 and whose product is a minimum.

87. Verify that the slope of the line passing through (x, x^3) and $(x + h, [x + h]^3)$ is $3x^2 + 3xh + h^2$.

88. Verify that the slope of the line passing through $(x, 4x^3 + x)$ and $(x + h, 4[x + h]^3 + [x + h])$ is given by $12x^2 + 12xh + 4h^2 + 1$.

PREPARE FOR SECTION 2.5

89. For the graph of the parabola whose equation is $f(x) = x^2 + 4x - 6$, what is the equation of the axis of symmetry? [2.4]

90. For $f(x) = \dfrac{3x^4}{x^2 + 1}$, show that $f(-3) = f(3)$. [2.2]

91. For $f(x) = 2x^3 - 5x$, show that $f(-2) = -f(2)$. [2.2]

92. Let $f(x) = x^2$ and $g(x) = x + 3$. Find $f(a) - g(a)$ for $a = -2, -1, 0, 1, 2$. [2.2]

93. What is the midpoint of the line segment between $P(-a, b)$ and $Q(a, b)$? [2.1]

94. What is the midpoint of the line segment between $P(-a, -b)$ and $Q(a, b)$? [2.1]

PROJECTS

1. THE CUBIC FORMULA Write an essay on the development of the cubic formula. An excellent source of information is the chapter "Cardano and the Solution of the Cubic" in *Journey Through Genius*, by William Dunham (New York: Wiley, 1990).

2. SIMPSON'S RULE In calculus a procedure known as *Simpson's Rule* is often used to approximate the area under a curve. The figure at the right shows the graph of a parabola that passes through $P_0(-h, y_0)$, $P_1(0, y_1)$, and $P_2(h, y_2)$. The equation of the parabola is of the form $y = Ax^2 + Bx + C$. Using calculus procedures, we can show that the area bounded by the parabola, the x-axis, and the vertical lines $x = -h$ and $x = h$ is

$$\frac{h}{3}(2Ah^2 + 6C)$$

Use algebra to show that $y_0 + 4y_1 + y_2 = 2Ah^2 + 6C$, from which we can deduce that the area of the bounded region can also be written as

$$\frac{h}{3}(y_0 + 4y_1 + y_2)$$

(*Hint:* Evaluate $Ax^2 + Bx + C$ at $x = -h$, $x = 0$, and $x = h$ to determine values of y_0, y_1, and y_2, respectively. Then compute $y_0 + 4y_1 + y_2$.)

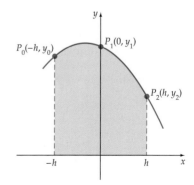

SECTION 2.5

PROPERTIES OF GRAPHS

● SYMMETRY

The graph in **Figure 2.63** is symmetric with respect to the line *l*. Note that the graph has the property that if the paper is folded along the dotted line *l*, the point A' will coincide with the point A, the point B' will coincide with the point B, and the point C' will coincide with the point C. One part of the graph is a *mirror image* of the rest of the graph across the line *l*.

A graph is **symmetric with respect to the y-axis** if, whenever the point given by (x, y) is on the graph, then $(-x, y)$ is also on the graph. The graph in **Figure 2.64** is symmetric with respect to the y-axis. A graph is **symmetric with respect to the x-axis** if, whenever the point given by (x, y) is on the graph, then $(x, -y)$ is also on the graph. The graph in **Figure 2.65** is symmetric with respect to the x-axis.

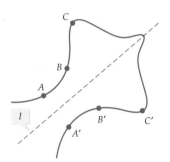

FIGURE 2.63

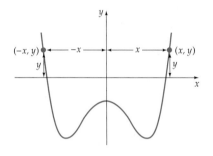

FIGURE 2.64

Symmetry with respect to the y-axis

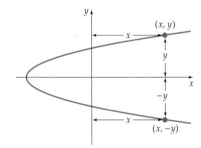

FIGURE 2.65

Symmetry with respect to the x-axis

Tests for Symmetry with Respect to a Coordinate Axis

The graph of an equation is symmetric with respect to

- the y-axis if the replacement of x with $-x$ leaves the equation unaltered.

- the x-axis if the replacement of y with $-y$ leaves the equation unaltered.

? QUESTION Which of the graphs below, I, II, or III, is **a.** symmetric with respect to the x-axis? **b.** symmetric with respect to the y-axis?

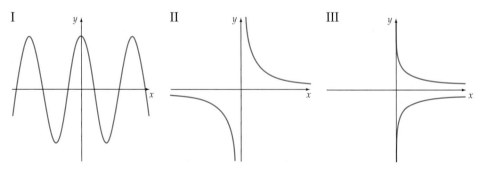

? ANSWER **a.** I is symmetric with respect to the y-axis.
b. III is symmetric with respect to the x-axis.

EXAMPLE 1 Determine Symmetries of a Graph

Determine whether the graph of the given equation has symmetry with respect to either the x- or the y-axis.

a. $y = x^2 + 2$ **b.** $x = |y| - 2$

Solution

a. The equation $y = x^2 + 2$ *is unaltered* by the replacement of x with $-x$. That is, the simplification of $y = (-x)^2 + 2$ yields the original equation $y = x^2 + 2$. Thus the graph of $y = x^2 + 2$ is symmetric with respect to the y-axis. However, the equation $y = x^2 + 2$ *is altered* by the replacement of y with $-y$. That is, the simplification of $-y = x^2 + 2$, which is $y = -x^2 - 2$, *does not* yield the original equation $y = x^2 + 2$. The graph of $y = x^2 + 2$ is not symmetric with respect to the x-axis. See **Figure 2.66.**

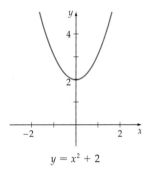

$y = x^2 + 2$

FIGURE 2.66

b. The equation $x = |y| - 2$ *is altered* by the replacement of x with $-x$. That is, the simplification of $-x = |y| - 2$, which is $x = -|y| + 2$, *does not* yield the original equation $x = |y| - 2$. This implies that the graph of $x = |y| - 2$ is not symmetric with respect to the y-axis. However, the equation $x = |y| - 2$ *is unaltered* by the replacement of y with $-y$. That is, the simplification of $x = |-y| - 2$ yields the original equation $x = |y| - 2$. The graph of $x = |y| - 2$ is symmetric with respect to the x-axis. See **Figure 2.67.**

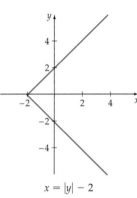

$x = |y| - 2$

FIGURE 2.67

▶ **TRY EXERCISE 14, PAGE 238**

Symmetry with Respect to a Point

A graph is **symmetric with respect to a point** Q if for each point P on the graph there is a point P' on the graph such that Q is the midpoint of the line segment PP'.

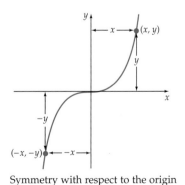

FIGURE 2.68

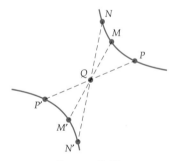

Symmetry with respect to the origin

FIGURE 2.69

The graph in **Figure 2.68** is symmetric with respect to the point Q. For any point P on the graph, there exists a point P' on the graph such that Q is the midpoint of $P'P$.

When we discuss symmetry with respect to a point, we frequently use the origin. A graph is symmetric with respect to the origin if, whenever the point given by (x, y) is on the graph, then $(-x, -y)$ is also on the graph. The graph in **Figure 2.69** is symmetric with respect to the origin.

Test for Symmetry with Respect to the Origin

The graph of an equation is symmetric with respect to the origin if the replacement of x with $-x$ and of y with $-y$ leaves the equation unaltered.

EXAMPLE 2 **Determine Symmetry with Respect to the Origin**

Determine whether the graph of each equation has symmetry with respect to the origin.

a. $xy = 4$ **b.** $y = x^3 + 1$

Solution

a. The equation $xy = 4$ is unaltered by the replacement of x with $-x$ and of y with $-y$. That is, the simplification of $(-x)(-y) = 4$ yields the original equation $xy = 4$. Thus the graph of $xy = 4$ is symmetric with respect to the origin. See **Figure 2.70.**

b. The equation $y = x^3 + 1$ *is altered* by the replacement of x with $-x$ and of y with $-y$. That is, the simplification of $-y = (-x)^3 + 1$, which is $y = x^3 - 1$, *does not* yield the original equation $y = x^3 + 1$. Thus the graph of $y = x^3 + 1$ is not symmetric with respect to the origin. See **Figure 2.71.**

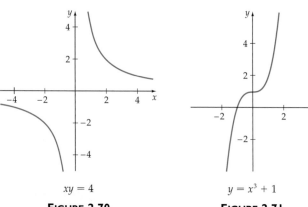

$xy = 4$ $y = x^3 + 1$

FIGURE 2.70   **FIGURE 2.71**

▶ **TRY EXERCISE 24, PAGE 238**

Some graphs have more than one symmetry. For example, the graph of $|x| + |y| = 2$ has symmetry with respect to the x-axis, the y-axis, and the origin. **Figure 2.72** is the graph of $|x| + |y| = 2$.

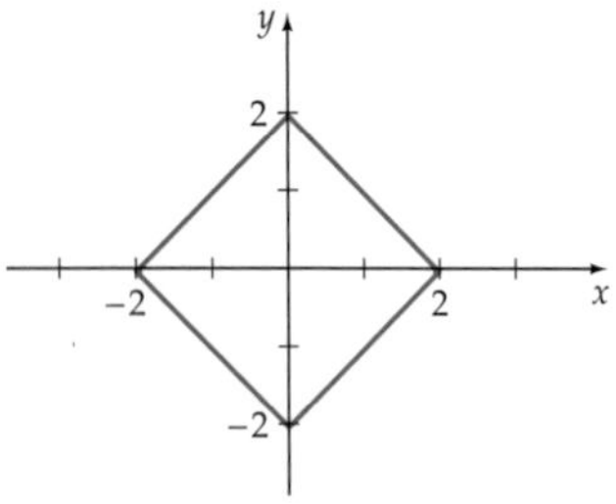

$|x| + |y| = 2$

FIGURE 2.72

● **EVEN AND ODD FUNCTIONS**

Some functions are classified as either *even* or *odd*.

Definition of Even and Odd Functions

The function f is an **even function** if

$$f(-x) = f(x) \quad \text{for all } x \text{ in the domain of } f$$

The function f is an **odd function** if

$$f(-x) = -f(x) \quad \text{for all } x \text{ in the domain of } f$$

EXAMPLE 3   **Identify Even or Odd Functions**

Determine whether each function is even, odd, or neither.

a. $f(x) = x^3$ **b.** $F(x) = |x|$ **c.** $h(x) = x^4 + 2x$

Solution

Replace x with $-x$ and simplify.

a. $f(-x) = (-x)^3 = -x^3 = -(x^3) = -f(x)$
Because $f(-x) = -f(x)$, this function is an odd function.

b. $F(-x) = |-x| = |x| = F(x)$
Because $F(-x) = F(x)$, this function is an even function.

c. $h(-x) = (-x)^4 + 2(-x) = x^4 - 2x$
This function is neither an even nor an odd function because

$$h(-x) = x^4 - 2x,$$

which is not equal to either $h(x)$ or $-h(x)$.

▶ **TRY EXERCISE 44, PAGE 238**

The following properties are a result of the tests for symmetry:

● The graph of an even function is symmetric with respect to the y-axis.

● The graph of an odd function is symmetric with respect to the origin.

The graph of f in **Figure 2.73** is symmetric with respect to the y-axis. It is the graph of an even function. The graph of g in **Figure 2.74** is symmetric with respect to the origin. It is the graph of an odd function. The graph of h in **Figure 2.75** is not symmetric with respect to the y-axis and is not symmetric with respect to the origin. It is neither an even nor an odd function.

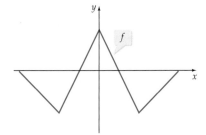

FIGURE 2.73

The graph of an even function is symmetric with respect to the y-axis.

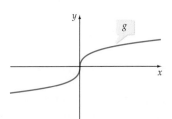

FIGURE 2.74

The graph of an odd function is symmetric with respect to the origin.

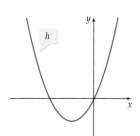

FIGURE 2.75

If the graph of a function is not symmetric to the y-axis or to the origin, then the function is neither even nor odd.

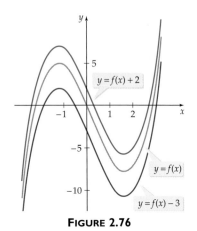

FIGURE 2.76

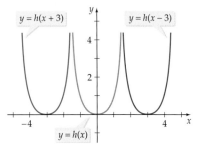

FIGURE 2.77

● TRANSLATIONS OF GRAPHS

The shape of a graph may be exactly the same as the shape of another graph; only their positions in the xy-plane may differ. For example, the graph of $y = f(x) + 2$ is the graph of $y = f(x)$ with each point moved up vertically 2 units. The graph of $y = f(x) - 3$ is the graph of $y = f(x)$ with each point moved down vertically 3 units. See **Figure 2.76.**

The graphs of $y = f(x) + 2$ and $y = f(x) - 3$ in **Figure 2.76** are called *vertical translations* of the graph of $y = f(x)$.

Vertical Translations

If f is a function and c is a positive constant, then the graph of

● $y = f(x) + c$ is the graph of $y = f(x)$ shifted up *vertically* c units.

● $y = f(x) - c$ is the graph of $y = f(x)$ shifted down *vertically* c units.

In **Figure 2.77,** the graph of $y = h(x + 3)$ is the graph of $y = h(x)$ with each point shifted to the left horizontally 3 units. Similarly, the graph of $y = h(x - 3)$ is the graph of $y = h(x)$ with each point shifted to the right horizontally 3 units.

The graphs of $y = h(x + 3)$ and $y = h(x - 3)$ in **Figure 2.77** are called *horizontal translations* of the graph of $y = h(x)$.

Horizontal Translations

If f is a function and c is a positive constant, then the graph of

● $y = f(x + c)$ is the graph of $y = f(x)$ shifted left *horizontally* c units.

● $y = f(x - c)$ is the graph of $y = f(x)$ shifted right *horizontally* c units.

INTEGRATING TECHNOLOGY

A graphing calculator can be used to draw the graphs of a *family* of functions. For instance, $f(x) = x^2 + c$ constitutes a family of functions with **parameter** c. The only feature of the graph that changes is the value of c.

A graphing calculator can be used to produce the graphs of a family of curves for specific values of the parameter. The LIST feature of the calculator can be used. For instance, to graph $f(x) = x^2 + c$ for $c = -2, 0$, and 1, we will create a list and use that list to produce the family of curves. The keystrokes for a TI-83 calculator are given below.

2nd { -2 , 0 , 1 2nd } STO 2nd L1

Now use the [Y=] key to enter

Y= X x² + 2nd L1 ZOOM 6

In Exercises 75 to 82, use a graphing utility.

75. On the same coordinate axes, graph
$$G(x) = \sqrt[3]{x} + c$$
for $c = 0, -1$, and 3.

76. On the same coordinate axes, graph
$$H(x) = \sqrt[3]{x + c}$$
for $c = 0, -1$, and 3.

77. On the same coordinate axes, graph
$$J(x) = |2(x + c) - 3| - |x + c|$$
for $c = 0, -1$, and 2.

78. On the same coordinate axes, graph
$$K(x) = |x - 1| - |x| + c$$
for $c = 0, -1$, and 2.

79. On the same coordinate axes, graph
$$L(x) = cx^2$$
for $c = 1, \frac{1}{2}$, and 2.

80. On the same coordinate axes, graph
$$M(x) = c\sqrt{x^2 - 4}$$
for $c = 1, \frac{1}{3}$, and 3.

81. On the same coordinate axes, graph
$$S(x) = c(|x - 1| - |x|)$$
for $c = 1, \frac{1}{4}$, and 4.

82. On the same coordinate axes, graph
$$T(x) = c\left(\frac{x}{|x|}\right)$$
for $c = 1, \frac{2}{3}$, and $\frac{3}{2}$.

83. Graph $V(x) = [\![cx]\!], 0 \le x \le 6$, for each value of c.

  a. $c = 1$ **b.** $c = \frac{1}{2}$ **c.** $c = 2$

84. Graph $W(x) = [\![cx]\!] - cx, 0 \le x \le 6$, for each value of c.

  a. $c = 1$ **b.** $c = \frac{1}{3}$ **c.** $c = 3$

CONNECTING CONCEPTS

85. Use the graph of $f(x) = 2/(x^2 + 1)$ to determine an equation for the graphs shown in **a.** and **b.**

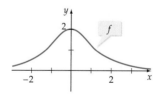

a. 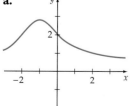   **b.**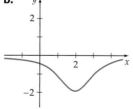

86. Use the graph of $f(x) = x\sqrt{2 + x}$ to determine an equation for the graphs shown in **a.** and **b.**

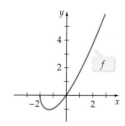

a. 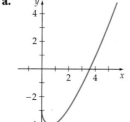   **b.**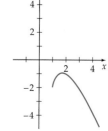

PREPARE FOR SECTION 2.6

87. Subtract: $(2x^2 + 3x - 4) - (x^2 + 3x - 5)$ [P.3]

88. Multiply: $(3x^2 - x + 2)(2x - 3)$ [P.3]

In Exercises 89 and 90, find each of the following for $f(x) = 2x^2 - 5x + 2$.

89. $f(3a)$ [2.2]

90. $f(2 + h)$ [2.2]

In Exercises 91 and 92, find the domain of each function.

91. $F(x) = \dfrac{x}{x - 1}$ [2.2]

92. $r(x) = \sqrt{2x - 8}$ [2.2]

PROJECTS

1. **DIRICHLET FUNCTION** We owe our present-day definition of a function to the German mathematician Peter Gustav Dirichlet (1805–1859). He created the following unusual function, which is now known as the *Dirichlet function.*

$$f(x) = \begin{cases} 0, & \text{if } x \text{ is a rational number} \\ 1, & \text{if } x \text{ is an irrational number} \end{cases}$$

Answer the following questions about the Dirichlet function.

a. What is its domain?   **b.** What is its range?

c. What are its *x*-intercepts?

d. What is its *y*-intercept?

e. Is it an even or an odd function?

f. Explain why a graphing calculator cannot be used to produce an accurate graph of the function.

g. Write a sentence or two that describes its graph.

2. **ISOLATED POINT** Consider the function given by

$$y = \sqrt{(x - 1)^2(x - 2)} + 1$$

Verify that the point $(1, 1)$ is a solution of the equation. Now use a graphing utility to graph the function. Does your graph include the isolated point at $(1, 1)$, as shown at the right? If the graphing utility you used failed to include the point $(1, 1)$, explain at least one reason for the omission of this iso- lated point.

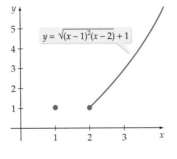

$y = \sqrt{(x-1)^2(x-2)} + 1$

3. 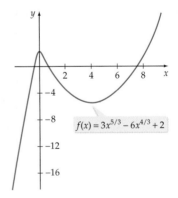 **A LINE WITH A HOLE** The function

$$f(x) = \dfrac{(x - 2)(x + 1)}{(x - 2)}$$

graphs as a line with a *y*-intercept of 1, a slope of 1, and a hole at $(2, 3)$. Use a graphing utility to graph *f*. Explain why a graphing utility might not show the hole at $(2, 3)$.

4. **FINDING A COMPLETE GRAPH** Use a graphing utility to graph the function $f(x) = 3x^{5/3} - 6x^{4/3} + 2$ for $-2 \le x \le 10$. Compare your graph with the graph below. Does your graph include the part to the left of the *y*-axis? If not, how might you enter the function in such a way that the graphing utility you used would include this part?

$f(x) = 3x^{5/3} - 6x^{4/3} + 2$

SECTION 2.6   # THE ALGEBRA OF FUNCTIONS

- ○ OPERATIONS ON FUNCTIONS
- ○ THE DIFFERENCE QUOTIENT
- ○ COMPOSITION OF FUNCTIONS

○ OPERATIONS ON FUNCTIONS

Functions can be defined in terms of other functions. For example, the function defined by $h(x) = x^2 + 8x$ is the sum of

$$f(x) = x^2 \quad \text{and} \quad g(x) = 8x$$

Thus, if we are given any two functions f and g, we can define the four new functions $f + g$, $f - g$, fg, and $\dfrac{f}{g}$ as follows.

Operations on Functions

For all values of x for which both $f(x)$ and $g(x)$ are defined, we define the following functions.

Sum	$(f + g)(x) = f(x) + g(x)$
Difference	$(f - g)(x) = f(x) - g(x)$
Product	$(fg)(x) = f(x) \cdot g(x)$
Quotient	$\left(\dfrac{f}{g}\right)(x) = \dfrac{f(x)}{g(x)}, \quad g(x) \neq 0$

Domain of $f + g$, $f - g$, fg, f/g

For the given functions f and g, the domains of $f + g$, $f - g$, and $f \cdot g$ consist of all real numbers formed by the intersection of the domains of f and g. The domain of $\dfrac{f}{g}$ is the set of all real numbers formed by the intersection of the domains of f and g, except for those real numbers x such that $g(x) = 0$.

EXAMPLE 1   Determine the Domain of a Function

If $f(x) = \sqrt{x - 1}$ and $g(x) = x^2 - 4$, find the domain of $f + g$, of $f - g$, of fg, and of $\dfrac{f}{g}$.

Solution

Note that f has the domain $\{x \mid x \geq 1\}$ and g has the domain of all real numbers. Therefore, the domain of $f + g$, $f - g$, and fg is $\{x \mid x \geq 1\}$. Because $g(x) = 0$ when $x = -2$ or $x = 2$, neither -2 nor 2 is in the domain of $\dfrac{f}{g}$. The domain of $\dfrac{f}{g}$ is $\{x \mid x \geq 1 \text{ and } x \neq 2\}$.

▶ **TRY EXERCISE 10, PAGE 251**

EXAMPLE 2 **Evaluate Functions**

Let $f(x) = x^2 - 9$ and $g(x) = 2x + 6$. Find

a. $(f + g)(5)$ b. $(fg)(-1)$ c. $\left(\dfrac{f}{g}\right)(4)$

Solution

a. $(f + g)(x) = f(x) + g(x) = (x^2 - 9) + (2x + 6) = x^2 + 2x - 3$
   Therefore, $(f + g)(5) = (5)^2 + 2(5) - 3 = 25 + 10 - 3 = 32$.

b. $(fg)(x) = f(x) \cdot g(x) = (x^2 - 9)(2x + 6) = 2x^3 + 6x^2 - 18x - 54$
   Therefore, $(fg)(-1) = 2(-1)^3 + 6(-1)^2 - 18(-1) - 54$
   $$= -2 + 6 + 18 - 54 = -32.$$

c. $\left(\dfrac{f}{g}\right)(x) = \dfrac{f(x)}{g(x)} = \dfrac{x^2 - 9}{2x + 6} = \dfrac{(x + 3)(x - 3)}{2(x + 3)} = \dfrac{x - 3}{2}, \quad x \neq -3$
   Therefore, $\left(\dfrac{f}{g}\right)(4) = \dfrac{4 - 3}{2} = \dfrac{1}{2}$.

▶ **TRY EXERCISE 14, PAGE 251**

● THE DIFFERENCE QUOTIENT

take note

The difference quotient is an important concept that plays a fundamental role in calculus.

The expression

$$\frac{f(x + h) - f(x)}{h}, \quad h \neq 0$$

is called the **difference quotient** of f. It enables us to study the manner in which a function changes in value as the independent variable changes.

EXAMPLE 3 **Determine a Difference Quotient**

Determine the difference quotient of $f(x) = x^2 + 7$.

Solution

$\dfrac{f(x + h) - f(x)}{h} = \dfrac{[(x + h)^2 + 7] - [x^2 + 7]}{h}$ • Apply the difference quotient.

$= \dfrac{[x^2 + 2xh + h^2 + 7] - [x^2 + 7]}{h}$

$= \dfrac{x^2 + 2xh + h^2 + 7 - x^2 - 7}{h}$

$= \dfrac{2xh + h^2}{h} = \dfrac{h(2x + h)}{h} = 2x + h$

▶ **TRY EXERCISE 30, PAGE 251**

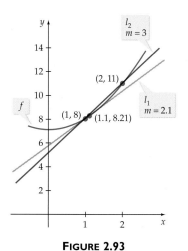

FIGURE 2.93

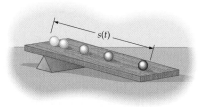

FIGURE 2.94

The difference quotient $2x + h$ of $f(x) = x^2 + 7$ from Example 3 is the slope of the secant line through the points

$$(x, f(x)) \quad \text{and} \quad (x + h, f(x + h))$$

For instance, let $x = 1$ and $h = 1$. Then the difference quotient is

$$2x + h = 2(1) + 1 = 3$$

This is the slope of the secant line l_2 through $(1, 8)$ and $(2, 11)$, as shown in **Figure 2.93**. If we let $x = 1$ and $h = 0.1$, then the difference quotient is

$$2x + h = 2(1) + 0.1 = 2.1$$

This is the slope of the secant line l_1 through $(1, 8)$ and $(1.1, 8.21)$.

The difference quotient

$$\frac{f(x + h) - f(x)}{h}$$

can be used to compute *average velocities*. In such cases it is traditional to replace f with s (for distance), the variable x with the variable a (for the time at the start of an observed interval of time), and the variable h with Δt (read as "delta t"), where Δt is the difference between the time at the end of an interval and the time at the start of the interval. For example, if an experiment is observed over the time interval from $t = 3$ seconds to $t = 5$ seconds, then the time interval is denoted as $[3, 5]$ with $a = 3$ and $\Delta t = 5 - 3 = 2$. Thus if the distance traveled by a ball that rolls down a ramp is given by $s(t)$, where t is the time in seconds after the ball is released (see **Figure 2.94**), then the **average velocity** of the ball over the interval $t = a$ to $t = a + \Delta t$ is the difference quotient

$$\frac{s(a + \Delta t) - s(a)}{\Delta t}$$

EXAMPLE 4   **Evaluate Average Velocities**

The distance traveled by a ball rolling down a ramp is given by $s(t) = 4t^2$, where t is the time in seconds after the ball is released, and $s(t)$ is measured in feet. Evaluate the average velocity of the ball for each time interval.

a. $[3, 5]$ **b.** $[3, 4]$ **c.** $[3, 3.5]$ **d.** $[3, 3.01]$

Solution

a. In this case, $a = 3$ and $\Delta t = 2$. Thus the average velocity over this interval is

$$\frac{s(a + \Delta t) - s(a)}{\Delta t} = \frac{s(3 + 2) - s(3)}{2} = \frac{s(5) - s(3)}{2} = \frac{100 - 36}{2}$$

$$= 32 \text{ feet per second}$$

b. Let $a = 3$ and $\Delta t = 4 - 3 = 1$.

$$\frac{s(a + \Delta t) - s(a)}{\Delta t} = \frac{s(3 + 1) - s(3)}{1} = \frac{s(4) - s(3)}{1} = \frac{64 - 36}{1}$$

$$= 28 \text{ feet per second}$$

Continued ▶

   c.   Let $a = 3$ and $\Delta t = 3.5 - 3 = 0.5$.

$$\frac{s(a + \Delta t) - s(a)}{\Delta t} = \frac{s(3 + 0.5) - s(3)}{0.5} = \frac{49 - 36}{0.5} = 26 \text{ feet per second}$$

   d.   Let $a = 3$ and $\Delta t = 3.01 - 3 = 0.01$.

$$\frac{s(a + \Delta t) - s(a)}{\Delta t} = \frac{s(3 + 0.01) - s(3)}{0.01} = \frac{36.2404 - 36}{0.01}$$

$$= 24.04 \text{ feet per second}$$

▶ **TRY EXERCISE 72, PAGE 253**

● COMPOSITION OF FUNCTIONS

Composition of functions is another way in which functions can be combined. This method of combining functions uses the output of one function as the input for a second function.

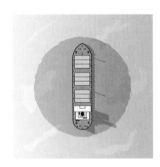

   Suppose that the spread of oil from a leak in a tanker can be approximated by a circle with the tanker at its center. The radius r (in feet) of the spill t hours after the leak begins is given by $r(t) = 150\sqrt{t}$. The area of the spill is the area of a circle and is given by the formula $A(r) = \pi r^2$. To find the area of the spill 4 hours after the leak begins, we first find the radius of the spill and then use that number to find the area of the spill.

$r(t) = 150\sqrt{t}$	$A(r) = \pi r^2$
$r(4) = 150\sqrt{4}$ • $t = 4$ **hours**	$A(300) = \pi(300^2)$ • $r = 300$ **feet**
$= 150(2)$	$= 90,000\pi$
$= 300$	$\approx 283,000$

The area of the spill after 4 hours is approximately 283,000 square feet.

   There is an alternative way to solve this problem. Because the area of the spill depends on the radius and the radius depends on the time, there is a relationship between area and time. We can determine this relationship by evaluating the formula for the area of a circle using $r(t) = 150\sqrt{t}$. This will give the area of the spill as a function of time.

$$A(r) = \pi r^2$$
$$A[r(t)] = \pi[r(t)]^2 \qquad \text{• Replace } r \text{ by } r(t).$$
$$= \pi\left[150\sqrt{t}\right]^2 \qquad \text{• } r(t) = 150\sqrt{t}$$
$$A(t) = 22,500\pi t \qquad \text{• Simplify.}$$

The area of the spill as a function of time is $A(t) = 22,500\pi t$. To find the area of the oil spill after 4 hours, evaluate this function at $t = 4$.

$$A(t) = 22,500\pi t$$
$$A(4) = 22,500\pi(4) \qquad \text{• } t = 4 \text{ hours}$$
$$= 90,000\pi$$
$$\approx 283,000$$

This is the same result we calculated earlier.

👥 📝 TOPICS FOR DISCUSSION

1. The domain of $f + g$ consists of all real numbers formed by the *union* of the domain of f and the domain of g. Do you agree?

2. Given $f(x) = 3x - 2$ and $g(x) = \dfrac{1}{3}x + \dfrac{2}{3}$, determine $f \circ g$ and $g \circ f$. Does this show that composition of functions is a commutative operation?

3. A tutor states that the difference quotient of $f(x) = x^2$ and the difference quotient of $g(x) = x^2 + 4$ are the same. Do you agree?

4. A classmate states that the difference quotient of any linear function $f(x) = mx + b$ is always m. Do you agree?

5. When we use a difference quotient to determine an average velocity, we generally replace the variable h with the variable Δt. What does Δt represent?

EXERCISE SET 2.6

In Exercises 1 to 12, use the given functions f and g to find $f + g$, $f - g$, fg, and $\dfrac{f}{g}$. State the domain of each.

1. $f(x) = x^2 - 2x - 15$, $g(x) = x + 3$

2. $f(x) = x^2 - 25$, $g(x) = x - 5$

3. $f(x) = 2x + 8$, $g(x) = x + 4$

4. $f(x) = 5x - 15$, $g(x) = x - 3$

5. $f(x) = x^3 - 2x^2 + 7x$, $g(x) = x$

6. $f(x) = x^2 - 5x - 8$, $g(x) = -x$

7. $f(x) = 2x^2 + 4x - 7$, $g(x) = 2x^2 + 3x - 5$

8. $f(x) = 6x^2 + 10$, $g(x) = 3x^2 + x - 10$

9. $f(x) = \sqrt{x - 3}$, $g(x) = x$

▶ 10. $f(x) = \sqrt{x - 4}$, $g(x) = -x$

11. $f(x) = \sqrt{4 - x^2}$, $g(x) = 2 + x$

12. $f(x) = \sqrt{x^2 - 9}$, $g(x) = x - 3$

In Exercises 13 to 28, evaluate the indicated function, where $f(x) = x^2 - 3x + 2$ and $g(x) = 2x - 4$.

13. $(f + g)(5)$

▶ **14.** $(f + g)(-7)$

15. $(f + g)\left(\dfrac{1}{2}\right)$

16. $(f + g)\left(\dfrac{2}{3}\right)$

17. $(f - g)(-3)$

18. $(f - g)(24)$

19. $(f - g)(-1)$

20. $(f - g)(0)$

21. $(fg)(7)$

22. $(fg)(-3)$

23. $(fg)\left(\dfrac{2}{5}\right)$

24. $(fg)(-100)$

25. $\left(\dfrac{f}{g}\right)(-4)$

26. $\left(\dfrac{f}{g}\right)(11)$

27. $\left(\dfrac{f}{g}\right)\left(\dfrac{1}{2}\right)$

28. $\left(\dfrac{f}{g}\right)\left(\dfrac{1}{4}\right)$

In Exercises 29 to 36, find the difference quotient of the given function.

29. $f(x) = 2x + 4$

▶ **30.** $f(x) = 4x - 5$

31. $f(x) = x^2 - 6$

32. $f(x) = x^2 + 11$

33. $f(x) = 2x^2 + 4x - 3$ **34.** $f(x) = 2x^2 - 5x + 7$

35. $f(x) = -4x^2 + 6$ **36.** $f(x) = -5x^2 - 4x$

In Exercises 37 to 48, find $g \circ f$ and $f \circ g$ for the given functions f and g.

37. $f(x) = 3x + 5, \quad g(x) = 2x - 7$

▶ **38.** $f(x) = 2x - 7, \quad g(x) = 3x + 2$

39. $f(x) = x^2 + 4x - 1, \quad g(x) = x + 2$

40. $f(x) = x^2 - 11x, \quad g(x) = 2x + 3$

41. $f(x) = x^3 + 2x, \quad g(x) = -5x$

42. $f(x) = -x^3 - 7, \quad g(x) = x + 1$

43. $f(x) = \dfrac{2}{x + 1}, \quad g(x) = 3x - 5$

44. $f(x) = \sqrt{x + 4}, \quad g(x) = \dfrac{1}{x}$

45. $f(x) = \dfrac{1}{x^2}, \quad g(x) = \sqrt{x - 1}$

46. $f(x) = \dfrac{6}{x - 2}, \quad g(x) = \dfrac{3}{5x}$

47. $f(x) = \dfrac{3}{|5 - x|}, \quad g(x) = -\dfrac{2}{x}$

48. $f(x) = |2x + 1|, \quad g(x) = 3x^2 - 1$

In Exercises 49 to 64, evaluate each composite function, where $f(x) = 2x + 3, g(x) = x^2 - 5x,$ and $h(x) = 4 - 3x^2$.

49. $(g \circ f)(4)$ ▶ **50.** $(f \circ g)(4)$

51. $(f \circ g)(-3)$ **52.** $(g \circ f)(-1)$

53. $(g \circ h)(0)$ **54.** $(h \circ g)(0)$

55. $(f \circ f)(8)$ **56.** $(f \circ f)(-8)$

57. $(h \circ g)\left(\dfrac{2}{5}\right)$ **58.** $(g \circ h)\left(-\dfrac{1}{3}\right)$

59. $(g \circ f)\left(\sqrt{3}\right)$ **60.** $(f \circ g)\left(\sqrt{2}\right)$

61. $(g \circ f)(2c)$ **62.** $(f \circ g)(3k)$

63. $(g \circ h)(k + 1)$ **64.** $(h \circ g)(k - 1)$

65. **WATER TANK** A water tank has the shape of a right circular cone, with height 16 feet and radius 8 feet. Water is running into the tank so that the radius r (in feet) of the surface of the water is given by $r = 1.5t$, where t is the time (in minutes) that the water has been running.

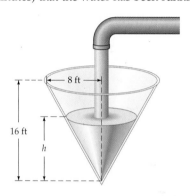

a. The area A of the surface of the water is $A = \pi r^2$. Find $A(t)$ and use it to determine the area of the surface of the water when $t = 2$ minutes.

b. The volume V of the water is given by $V = \dfrac{1}{3}\pi r^2 h$.

Find $V(t)$ and use it to determine the volume of the water when $t = 3$ minutes. (*Hint:* The height of the water in the cone is always twice the radius of the water.)

▶ **66.** [calculator icon] **SCALING A RECTANGLE** Work Example 7 of this section with the scaling as follows. The upper right corner of the original rectangle is pulled to the *left* at 0.5 inch per second and downward at 0.2 inch per second.

67. **TOWING A BOAT** A boat is towed by a rope that runs through a pulley that is 4 feet above the point where the rope is tied to the boat. The length (in feet) of the rope from the boat to the pulley is given by $s = 48 - t$, where t is the time in seconds that the boat has been in tow. The horizontal distance from the pulley to the boat is d.

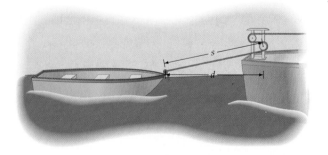

a. Find $d(t)$.   **b.** Evaluate $s(35)$ and $d(35)$.

68. **PERIMETER OF A SCALED RECTANGLE** Show by a graph that the perimeter

$$P = 2(3 + 0.5t) + 2|2 - 0.2t|$$

of the scaled rectangle in Example 7 of this section is an increasing function over $0 \le t \le 14$.

69. CONVERSION FUNCTIONS The function $F(x) = \dfrac{x}{12}$ converts x inches to feet. The function $Y(x) = \dfrac{x}{3}$ converts x feet to yards. Explain the meaning of $(Y \circ F)(x)$.

70. CONVERSION FUNCTIONS The function $F(x) = 3x$ converts x yards to feet. The function $I(x) = 12x$ converts x feet to inches. Explain the meaning of $(I \circ F)(x)$.

71. **CONCENTRATION OF A MEDICATION** The concentration $C(t)$ (in milligrams per liter) of a medication in a patient's blood is given by the data in the following table.

Concentration of Medication in Patient's Blood

t hours	$C(t)$ mg/1
0	0
0.25	47.3
0.50	78.1
0.75	94.9
1.00	99.8
1.25	95.7
1.50	84.4
1.75	68.4
2.00	50.1
2.25	31.6
2.50	15.6
2.75	4.3

The **average rate of change** of the concentration over the time interval from $t = a$ to $t = a + \Delta t$ is

$$\frac{C(a + \Delta t) - C(a)}{\Delta t}$$

Use the data in the table to evaluate the average rate of change for each of the following time intervals.

a. $[0, 1]$ (*Hint:* In this case, $a = 0$ and $\Delta t = 1$.) Compare this result to the slope of the line through $(0, C(0))$ and $(1, C(1))$.

b. $[0, 0.5]$ **c.** $[1, 2]$ **d.** $[1, 1.5]$ **e.** $[1, 1.25]$

f. The data in the table can be modeled by the function $Con(t) = 25t^3 - 150t^2 + 225t$. Use $Con(t)$ to verify that the average rate of change over $[1, 1 + \Delta t]$ is $-75(\Delta t) + 25(\Delta t)^2$. What does the average rate of change over $[1, 1 + \Delta t]$ seem to approach as Δt approaches 0?

▶ **72. BALL ROLLING ON A RAMP** The distance traveled by a ball rolling down a ramp is given by $s(t) = 6t^2$, where t is the time in seconds after the ball is released, and $s(t)$ is measured in feet. The ball travels 6 feet in 1 second and it travels 24 feet in 2 seconds. Use the difference quotient for average velocity given on page 245 to evaluate the average velocity for each of the following time intervals.

a. $[2, 3]$ (*Hint:* In this case, $a = 2$ and $\Delta t = 1$.) Compare this result to the slope of the line through $(2, s(2))$ and $(3, s(3))$.

b. $[2, 2.5]$ **c.** $[2, 2.1]$ **d.** $[2, 2.01]$ **e.** $[2, 2.001]$

f. Verify that the average velocity over $[2, 2 + \Delta t]$ is $24 + 6(\Delta t)$. What does the average velocity seem to approach as Δt approaches 0?

CONNECTING CONCEPTS

In Exercises 73 to 76, show that $(f \circ g)(x) = (g \circ f)(x)$.

73. $f(x) = 2x + 3;\ \ g(x) = 5x + 12$

74. $f(x) = 4x - 2;\ \ g(x) = 7x - 4$

75. $f(x) = \dfrac{6x}{x - 1};\ \ g(x) = \dfrac{5x}{x - 2}$

76. $f(x) = \dfrac{5x}{x + 3};\ \ g(x) = -\dfrac{2x}{x - 4}$

In Exercises 77 to 82, show that

$$(g \circ f)(x) = x \quad \text{and} \quad (f \circ g)(x) = x$$

77. $f(x) = 2x + 3, \quad g(x) = \dfrac{x - 3}{2}$

78. $f(x) = 4x - 5, \quad g(x) = \dfrac{x + 5}{4}$

79. $f(x) = \dfrac{4}{x + 1}, \quad g(x) = \dfrac{4 - x}{x}$

80. $f(x) = \dfrac{2}{1 - x}, \quad g(x) = \dfrac{x - 2}{x}$

81. $f(x) = x^3 - 1, \quad g(x) = \sqrt[3]{x + 1}$

82. $f(x) = -x^3 + 2, \quad g(x) = \sqrt[3]{2 - x}$

PREPARE FOR SECTION 2.7

In Exercises 83 and 84, find the slope and y-intercept of the graph of the equation.

83. $y = -\dfrac{x}{3} + 4$ [2.3]

84. $3x - 4y = 12$ [2.3]

85. Find the equation of the line that has a slope of -0.45 and a y-intercept of $(0, 2.3)$. [2.3]

86. Find the equation of the line that passes through $P(3, -4)$ and has a slope of $-\dfrac{2}{3}$. [2.3]

87. If $f(x) = 3x^2 + 4x - 1$, find $f(2)$. [2.2]

88. You are given $P_1(2, -1)$ and $P_2(4, 14)$. If $f(x) = x^2 - 3$, find $|f(x_1) - y_1| + |f(x_2) - y_2|$. [2.2]

PROJECTS

1. **A GRAPHING UTILITY PROJECT** For any two different real numbers x and y, the larger of the two numbers is given by

$$\text{Maximum}(x, y) = \frac{x + y}{2} + \frac{|x - y|}{2} \qquad (1)$$

a. Verify Equation (1) for $x = 5$ and $y = 9$.

b. Verify Equation (1) for $x = 201$ and $y = 80$.

For any two different functional values $f(x)$ and $g(x)$, the larger of the two is given by

$$\text{Maximum}(f(x), g(x)) = \frac{f(x) + g(x)}{2} + \frac{|f(x) - g(x)|}{2} \qquad (2)$$

To illustrate how we might make use of Equation (2), consider the functions $y_1 = x^2$ and $y_2 = \sqrt{x}$ on the interval from Xmin $= -1$ to Xmax $= 6$. The graphs of y_1 and y_2 are shown at the right.

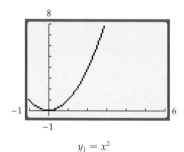

$y_1 = x^2$

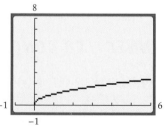

$y_2 = \sqrt{x}$

⚆ QUESTION What is the coefficient of determination for the odometer reading/trade-in value data (see page 259), and what is its significance?

⦿ QUADRATIC REGRESSION MODELS

To this point our focus has been *linear* regression equations. However, there may be a nonlinear relationship between two quantities. The accompanying scatter diagram to the left suggests that a quadratic function might be a better model of the data than a linear model. As we proceed through this text, various functional models will be discussed.

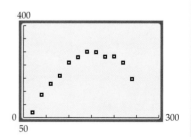

Nonlinear correlation between x and y.

EXAMPLE 2   Find a Quadratic Regression Model

The data in the table below were collected on five successive Saturdays. They show the average number of cars entering a shopping center parking lot. The value of t is the number of minutes after 9:00 A.M. The value of N is the number of cars that entered the parking lot in the 10 minutes prior to the value of t. Find a regression model for this data.

**Average Number of Cars Entering a
Shopping Center Parking Lot**

t	N	t	N
20	70	140	301
40	135	160	298
60	178	180	284
80	210	200	286
100	260	220	260
120	280	240	195

Solution

1. **Construct a scatter diagram for these data.** Enter the data into your calculator as explained on page 258.

From the scatter diagram, it appears that there is a nonlinear relationship between the variables.

Continued ▶

⚆ ANSWER $r^2 \approx 0.991$. This means that 99.1% of the total variation in trade-in value can be attributed to the odometer reading.

2. **Find the regression equation.** Try a quadratic regression model. For a TI-83 calculator, press $\boxed{\text{STAT}}$ ▶ $\boxed{\text{2ND}}$ CALC 5 $\boxed{\text{ENTER}}$.

```
QuadReg
y=ax²+bx+c
a=-.0124881369
b=3.904433067
c=-7.25
R²=.9840995401
```

take note

In the case of nonlinear regression calculations, the value of r is not shown on a TI-83 graphing calculator. In these cases, the coefficient of determination is used to determine how well the data fit the model.

3. **Examine the coefficient of determination.** The coefficient of determination is approximately 0.984. Because this number is fairly close to 1, the regression equation $y = -0.0124881369x^2 + 3.904433067x - 7.25$ provides a good model of the data.

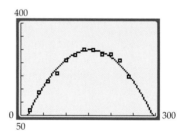

▶ **TRY EXERCISE 32, PAGE 267**

```
LinReg
y=ax+b
a=.6575174825
b=144.2727273
r²=.4193509866
r=.6475731515
```

For Example 2, we could have calculated the *linear* regression line for the data. The results are shown at the left. Note that the coefficient of determination for this calculation is approximately 0.419. Because this number is less than the coefficient of determination for the quadratic model, we choose a quadratic model of the data rather than a linear model.

Now for a final note: The regression line equation does not *prove* that the changes in the dependent variable are *caused* by the independent variable. For instance, suppose various cities throughout the United States were randomly selected and the numbers of gas stations (independent variable) and restaurants (dependent variable) were recorded in a table. If we calculated the regression equation for these data, we would find that r would be close to 1. However, this does not mean that gas stations *cause* restaurants to be built. The primary cause is that there are fewer gas stations and restaurants in cities with small populations and greater numbers of gas stations and restaurants in cities with large populations.

 TOPICS FOR DISCUSSION

1. What is the purpose of calculating the equation of a regression line?

2. Discuss the implications of the following correlation coefficients: $r = -1$, $r = 0$, and $r = 1$.

3. Discuss the coefficient of determination and what its value says about a data set.

4. What are the implications of $r^2 = 1$ for a nonlinear regression equation?

a. Find a quadratic model for these data.

b. Use the model to predict the fuel efficiency of this car when it is traveling at a speed of 50 mph.

▶ **32.** **BIOLOGY** The data in the table at the right show the oxygen consumption, in milliliters per minute, of a bird flying level at various speeds in kilometers per hour.

a. Find a quadratic model for these data.

b. Use the model to determine the speed at which the bird has minimum oxygen consumption.

Oxygen Consumption

Speed	Consumption
20	32
25	27
28	22
35	21
42	26
50	34

CONNECTING CONCEPTS

33. **PHYSICS** Galileo (1564–1642) studied the acceleration due to gravity by allowing balls of various weights to roll down an incline. This allowed him to time the descent of a ball more accurately than by just dropping the ball. The data in the table show some possible results of such an experiment using balls of different masses. Time, t, is measured in seconds; distance, s, is measured in centimeters.

Distance Traveled for Balls of Various Weights

5-Pound Ball		10-Pound Ball		15-Pound Ball	
t	s	t	s	t	s
2	2	3	5	3	5
4	10	6	22	5	15
6	22	9	49	7	30
8	39	12	87	9	49
10	61	15	137	11	75
12	86	18	197	13	103
14	120			15	137
16	156				

a. Find a quadratic model for each of the balls.

b. On the basis of a similar experiment, Galileo concluded that if air resistance is excluded, all falling objects fall with the same acceleration. Explain how one could make such a conclusion from the regression equations.

34. **ASTRONOMY** In 1929, Edwin Hubble published a paper that revolutionized astronomy ("A Relationship Between Distance and Radial Velocity Among Extra-Galactic Nebulae," *Proceedings of the National Academy of Science*, 168). His paper dealt with the distance an extragalactic nebula was from the Milky Way galaxy and the nebula's velocity with respect to the Milky Way. The data are given in the table below. Distance is measured in megaparsecs (1 megaparsec equals 1.918×10^{19} miles), and velocity (called the *recession velocity*) is measured in kilometers per second. A negative velocity means the nebula is moving toward the Milky Way; a positive velocity means the nebula is moving away from the Milky Way.

Recession Velocities

Distance	Velocity	Distance	Velocity
0.032	170	0.9	650
0.034	290	0.9	150
0.214	−130	0.9	500
0.263	−70	1.0	920
0.275	−185	1.1	450
0.275	−220	1.1	500
0.45	200	1.4	500
0.5	290	1.7	960
0.5	270	2.0	500
0.63	200	2.0	850
0.8	300	2.0	800
0.9	−30	2.0	1090

a. Find the linear regression model for these data.

b. On the basis of this model, what is the recession velocity of a nebula that is 1.5 megaparsecs from the Milky Way?

35. 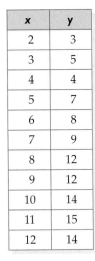 The data in the table at the right were collected on five successive Saturdays. They show the average number of cars entering a shopping center parking lot. The value of t is the number of minutes after 9:00 A.M. The value of N is the number of cars that entered the parking lot in the 10 minutes prior to the value of t. Does a linear model or a quadratic regression model better fit these data? Explain.

Average Number of Cars Entering a Parking Lot

t	N	t	N
20	70	140	301
40	135	160	298
60	178	180	284
80	210	200	286
100	260	220	260
120	280	240	195

PROJECTS

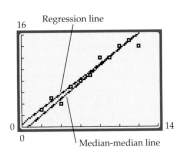

MEDIAN–MEDIAN LINE Another linear model of data is called the **median–median line**. This line employs *summary points* calculated using the medians of subsets of the independent and dependent variables. The **median** of a data set is the middle number or the average of the two middle numbers for a data set arranged in numerical order. For instance, to find the median of $\{8, 12, 6, 7, 9\}$, first arrange the data in numerical order.

$$6, 7, 8, 9, 12$$

The median is 8, the number in the middle. To find the median of $\{15, 12, 20, 9, 13, 10\}$, arrange the numbers in numerical order.

$$9, 10, 12, 13, 15, 20$$

The median is 12.5, the average of the two middle numbers.

$$\text{Median} = \frac{12 + 13}{2} = 12.5$$

The median–median line is determined by dividing a data set into three equal groups. (If the set cannot be divided into three equal groups, the first and third groups should be equal. For instance, if there are 11 data points, divide the set into groups of 4, 3, and 4.) The slope of the median–median line is the slope of the line through the x-medians and y-medians of the first and third sets of points. The median–median line passes through the average of the x- and y-medians of all three sets.

A graphing calculator can be used to find the median–median line. This line, along with the linear regression line, is shown in the next column for the data in the accompanying table.

1. Find the median–median line for the data in Exercise 17 on page 263.

2. Find the median–median line for the data in Exercise 18 on page 264.

x	y
2	3
3	5
4	4
5	7
6	8
7	9
8	12
9	12
10	14
11	15
12	14

3. Consider the data set $\{(1, 3), (2, 5), (3, 7), (4, 9), (5, 11), (6, 13), (7, 15), (8, 17)\}$.

a. Find the linear regression line for these data.

b. Find the median–median line for these data.

c. 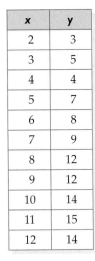 What conclusion might you draw from the answers to parts **a.** and **b.**?

4. For this exercise, use the data in the table in Project 1.

a. Calculate the median–median line and the linear regression line.

b. Change the entry (12, 14) to (12, 1) and then recalculate the median–median line and the linear regression line.

c. 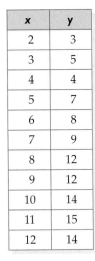 Explain why there is more change in the linear regression line than in the median–median line.

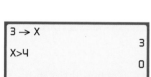

EXPLORING CONCEPTS WITH TECHNOLOGY

Graphing Piecewise Functions with a Graphing Calculator

A graphing calculator can be used to graph piecewise functions by including as part of the function the interval on which each piece of the function is defined. The method is based on the fact that a graphing calculator "evaluates" inequalities. For purposes of this Exploration, we will use keystrokes for a TI-83 calculator.

For instance, store 3 in **X** by pressing 3 | STO▶ | | X,T,Θ,*n* | | ENTER |. Now enter the inequality $x > 4$ by pressing | X,T,Θ,*n* | | 2ND | TEST 3 4 | ENTER |. Your screen should look like the one at the left. Note that the value of the inequality is 0. This occurs because the calculator replaced **X** by 3 and then determined whether the inequality $3 > 4$ was true or false. The calculator expresses the fact that the inequality is false by placing a zero on the screen. If we repeat the sequence of steps above, except that we store 5 in **X** instead of 3, the calculator will determine that the inequality is true and place a 1 on the screen.

This property of calculators is used to graph piecewise functions. Graphs of these functions work best when Dot mode rather than Connected mode is used. To switch to Dot mode, select | MODE |, use the arrow keys to highlight | DOT |, and then press | ENTER |.

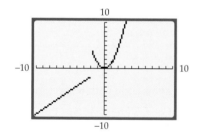

Now we will graph the piecewise function defined by $f(x) = \begin{cases} x, & x \le -2 \\ x^2, & x > -2 \end{cases}$.

Enter the function[4] as Y₁=X*(X≤-2)+X²*(X>-2) and graph this in the standard viewing window. Note that you are multiplying each piece of the function by its domain. The graph will appear as shown at the left.

To understand how the graph is drawn, we will consider two values of x, -8 and 2, and evaluate Y₁ for each of these values.

Y₁=X*(X≤-2)+X²*(X>-2)

$= -8(-8 \le -2) + (-8)^2(-8 > -2)$

$= -8(1) + 64(0) = -8$ • When $x = -8$, the value assigned to $-8 \le -2$ is 1; the value assigned to $-8 > -2$ is 0.

Y₁=X*(X≤-2)+X²*(X>-2)

$= 2(2 \le -2) + 2^2(2 > -2)$

$= 2(0) + 4(1) = 4$ • When $x = 2$, the value assigned to $2 \le -2$ is 0; the value assigned to $2 > -2$ is 1.

In a similar manner, for any value of x for which $x \le -2$, the value assigned to (X≤-2) is 1 and the value assigned to (X>-2) is 0. Thus Y₁=X*1+X²*0=X on that interval. This means that only the $f(x) = x$ piece of the function is graphed. When $x > -2$, the value assigned to (X≤-2) is 0 and the value assigned to (X>-2) is 1. Thus Y₁=X*0+X²*1=X² on that interval. This means that only the $f(x) = x^2$ piece of the function is graphed on that interval.

---

[4]Note that pressing | 2ND | TEST will display the inequality menu.

1. Graph: $f(x) = \begin{cases} x^2, & x < 2 \\ -x, & x \geq 2 \end{cases}$

2. Graph: $f(x) = \begin{cases} x^2 - x, & x < 2 \\ -x + 4, & x \geq 2 \end{cases}$

3. Graph: $f(x) = \begin{cases} -x^2 + 1, & x < 0 \\ x^2 - 1, & x \geq 0 \end{cases}$

4. Graph: $f(x) = \begin{cases} x^3 - 4x, & x < 1 \\ x^2 - x + 2, & x \geq 1 \end{cases}$

CHAPTER 2 SUMMARY

2.1 A Two-Dimensional Coordinate System and Graphs

- *The Distance Formula* The distance d between the points represented by (x_1, y_1) and (x_2, y_2) is

$$d = \sqrt{(x_2 - x_1)^2 + (y_2 - y_1)^2}$$

- The midpoint of the line segment from $P_1(x_1, y_1)$ to $P_2(x_2, y_2)$ is

$$\left(\frac{x_1 + x_2}{2}, \frac{y_1 + y_2}{2} \right)$$

- The standard form of the equation of a circle with center at (h, k) and radius r is $(x - h)^2 + (y - k)^2 = r^2$.

2.2 Introduction to Functions

- *Definition of a Function* A function is a set of ordered pairs in which no two ordered pairs that have the same first coordinate have different second coordinates.

- A graph is the graph of a function if and only if no vertical line intersects the graph at more than one point. If every horizontal line intersects the graph of a function at most once, then the graph is the graph of a one-to-one function.

2.3 Linear Functions

- A function is a linear function of x if it can be written in the form $f(x) = mx + b$, where m and b are real numbers and $m \neq 0$.

- The slope m of the line passing through the points $P_1(x_1, y_1)$ and $P_2(x_2, y_2)$ with $x_1 \neq x_2$ is given by

$$m = \frac{y_2 - y_1}{x_2 - x_1}$$

- The graph of the equation $f(x) = mx + b$ has slope m and y-intercept $(0, b)$.

- Two nonvertical lines are parallel if and only if their slopes are equal. Two lines with slopes m_1 and m_2 are perpendicular if and only if $m_1 = -\dfrac{1}{m_2}$.

2.4 Quadratic Functions

- A quadratic function of x is a function that can be represented by an equation of the form $f(x) = ax^2 + bx + c$, where a, b, and c are real numbers and $a \neq 0$.

- The vertex of the graph of $f(x) = ax^2 + bx + c$ is

$$\left(-\frac{b}{2a}, f\left(-\frac{b}{2a} \right) \right)$$

- Every quadratic function $f(x) = ax^2 + bx + c$ can be written in the standard form $f(x) = a(x - h)^2 + k$, $a \neq 0$. The graph of f is a parabola with vertex (h, k). The parabola is symmetric with respect to the vertical line $x = h$, which is called the axis of symmetry of the parabola. The parabola opens up if $a > 0$; it opens down if $a < 0$.

2.5 Properties of Graphs

- The graph of an equation is symmetric with respect to

  the y-axis if the replacement of x with $-x$ leaves the equation unaltered.

  the x-axis if the replacement of y with $-y$ leaves the equation unaltered.

  the origin if the replacement of x with $-x$ and y with $-y$ leaves the equation unaltered.

- If f is a function and c is a positive constant, then

  $y = f(x) + c$ is the graph of $y = f(x)$ shifted up *vertically* c units

  $y = f(x) - c$ is the graph of $y = f(x)$ shifted down *vertically* c units

  $y = f(x + c)$ is the graph of $y = f(x)$ shifted left *horizontally* c units

  $y = f(x - c)$ is the graph of $y = f(x)$ shifted right *horizontally* c units

- The graph of

  $y = -f(x)$ is the graph of $y = f(x)$ reflected across the x-axis.

$y = f(-x)$ is the graph of $y = f(x)$ reflected across the y-axis.

- If $a > 1$, then the graph of $y = f(ax)$ is a horizontal compressing of $y = f(x)$.

- If $0 < a < 1$, then the graph of $y = f(ax)$ is a horizontal stretching of the graph of $y = f(x)$.

2.6 The Algebra of Functions

- For all values of x for which both $f(x)$ and $g(x)$ are defined, we define the following functions.

Sum $\quad (f + g)(x) = f(x) + g(x)$

Difference $\quad (f - g)(x) = f(x) - g(x)$

Product $\quad (fg)(x) = f(x) \cdot g(x)$

Quotient $\quad \left(\dfrac{f}{g}\right)(x) = \dfrac{f(x)}{g(x)}, \quad g(x) \neq 0$

- The expression

$$\frac{f(x + h) - f(x)}{h}, \quad h \neq 0$$

is called the difference quotient of f. The difference quotient is an important function because it can be used to compute the *average rate of change* of f over the time interval $[x, x + h]$.

- For the functions f and g, the composite function, or composition, of f by g is given by $(g \circ f)(x) = g[f(x)]$ for all x in the domain of f such that $f(x)$ is in the domain of g.

2.7 Modeling Data Using Regression

- Regression analysis is used to find a mathematical model of collected data.

- The least-squares regression line is the line that minimizes the sum of the squares of the vertical deviations of all data points from the line.

- The linear correlation coefficient r is a measure of how closely the points of a data set can be modeled by a straight line. If $r = -1$, then the points of the data set can be modeled *exactly* by a straight line with negative slope. If $r = 1$, then the data set can be modeled *exactly* by a straight line with positive slope. For all data sets, $-1 \leq r \leq 1$.

- The coefficient of determination is r^2. It measures the percent of the total variation in the dependent variable that is explained by the regression line.

- It is possible to find both linear and nonlinear mathematical models of data.

CHAPTER 2 TRUE/FALSE EXERCISES

In Exercises 1 to 14, answer true or false. If the statement is false, give an example or a reason to show that the statement is false.

1. Let f be any function. Then $f(a) = f(b)$ implies that $a = b$.

2. If f and g are two functions, then $(f \circ g)(x) = (g \circ f)(x)$.

3. If f is not a one-to-one function, then there are at least two numbers u and v in the domain of f for which $f(u) = f(v)$.

4. Let f be a function such that $f(x) = f(x + 4)$ for all real numbers x. If $f(2) = 3$, then $f(18) = 3$.

5. For all functions f, $[f(x)]^2 = f[f(x)]$.

6. Let f be any function. Then for all a and b in the domain of f such that $f(b) \neq 0$ and $b \neq 0$,

$$\frac{f(a)}{f(b)} = \frac{a}{b}$$

7. The **identity function** $f(x) = x$ is its own inverse.

8. If f is a function, then $f(a + b) = f(a) + f(b)$ for all real numbers a and b in the domain of f.

9. If f is defined by $f(x) = |x|$, then $f(ab) = f(a)f(b)$ for all real numbers a and b.

10. If f is a one-to-one function and a and b are real numbers in the domain of f with $a < b$, then $f(a) \neq f(b)$.

11. The coordinates of a point on the graph of $y = f(x)$ are (a, b). If k is a positive constant, then (a, kb) are the coordinates of a point on the graph of $y = kf(x)$.

12. For every function f, the real number c is a solution of $f(x) = 0$ if and only if $(c, 0)$ is an x-intercept of the graph of $y = f(x)$.

13. The domain of every polynomial function is the set of real numbers.

14. If the linear coefficient of determination is 0.8, then the slope of the regression line is positive.

CHAPTER 2 REVIEW EXERCISES

In Exercises 1 and 2, find the distance between the points whose coordinates are given.

1. $(-3, 2)$ $(7, 11)$

2. $(5, -4)$ $(-3, -8)$

In Exercises 3 and 4, find the midpoint of the line segment with the given endpoints.

3. $(2, 8)$ $(-3, 12)$

4. $(-4, 7)$ $(8, -11)$

In Exercises 5 and 6, determine the center and radius of the circle with the given equation.

5. $(x - 3)^2 + (y + 4)^2 = 81$

6. $x^2 + y^2 + 10x + 4y + 20 = 0$

In Exercises 7 and 8, find the equation in standard form of the circle that satisfies the given conditions.

7. Center $C = (2, -3)$, radius $r = 5$

8. Center $C = (-5, 1)$, passing through $(3, 1)$

9. If $f(x) = 3x^2 + 4x - 5$, find

   a. $f(1)$ **b.** $f(-3)$ **c.** $f(t)$

   d. $f(x + h)$ **e.** $3f(t)$ **f.** $f(3t)$

10. If $g(x) = \sqrt{64 - x^2}$, find

   a. $g(3)$ **b.** $g(-5)$ **c.** $g(8)$

   d. $g(-x)$ **e.** $2g(t)$ **f.** $g(2t)$

11. If $f(x) = x^2 + 4x$ and $g(x) = x - 8$, find

   a. $(f \circ g)(3)$ **b.** $(g \circ f)(-3)$

   c. $(f \circ g)(x)$ **d.** $(g \circ f)(x)$

12. If $f(x) = 2x^2 + 7$ and $g(x) = |x - 1|$, find

   a. $(f \circ g)(-5)$ **b.** $(g \circ f)(-5)$

   c. $(f \circ g)(x)$ **d.** $(g \circ f)(x)$

13. If $f(x) = 4x^2 - 3x - 1$, find the difference quotient

$$\frac{f(x + h) - f(x)}{h}$$

14. If $g(x) = x^3 - x$, find the difference quotient

$$\frac{g(x + h) - g(x)}{h}$$

In Exercises 15 to 20, sketch the graph of f. Find the interval(s) in which f is a. increasing, b. constant, c. decreasing.

15. $f(x) = |x - 3| - 2$

16. $f(x) = x^2 - 5$

17. $f(x) = |x + 2| - |x - 2|$

18. $f(x) = [\![x + 3]\!]$

19. $f(x) = \dfrac{1}{2}x - 3$

20. $f(x) = \sqrt[3]{x}$

In Exercises 21 to 24, determine the domain of the function represented by the given equation.

21. $f(x) = -2x^2 + 3$

22. $f(x) = \sqrt{6 - x}$

23. $f(x) = \sqrt{25 - x^2}$

24. $f(x) = \dfrac{3}{x^2 - 2x - 15}$

In Exercises 25 and 26, find the slope-intercept form of the equation of the line through the two points.

25. $(-1, 3)$ $(4, -7)$

26. $(0, 0)$ $(7, 11)$

27. Find the slope-intercept form of the equation of the line that is parallel to the graph of $3x - 4y = 8$ and passes through $(2, 11)$.

28. Find the slope-intercept form of the equation of the line that is perpendicular to the graph of $2x = -5y + 10$ and passes through $(-3, -7)$.

In Exercises 29 to 34, use the method of completing the square to write each quadratic equation in its standard form.

29. $f(x) = x^2 + 6x + 10$

30. $f(x) = 2x^2 + 4x + 5$

31. $f(x) = -x^2 - 8x + 3$

32. $f(x) = 4x^2 - 6x + 1$

33. $f(x) = -3x^2 + 4x - 5$

34. $f(x) = x^2 - 6x + 9$

In Exercises 35 to 38, find the vertex of the graph of the quadratic function.

35. $f(x) = 3x^2 - 6x + 11$

36. $h(x) = 4x^2 - 10$

37. $k(x) = -6x^2 + 60x + 11$

38. $m(x) = 14 - 8x - x^2$

39. Use the formula

$$d = \frac{|mx_1 + b - y_1|}{\sqrt{1 + m^2}}$$

to find the distance from the point $(1, 3)$ to the line given by $y = 2x - 3$.

40. A freight company has determined that its cost per delivery of delivering x parcels is

$$C(x) = 1050 + 0.5x$$

The price it charges to send a parcel is $13.00 per parcel. Determine

a. the revenue function

b. the profit function

c. the minimum number of parcels the company must ship to break even

In Exercises 41 and 42, sketch a graph that is symmetric to the given graph with respect to the a. x-axis, b. y-axis, c. origin.

41.

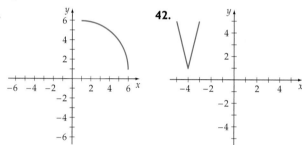

42.

In Exercises 43 to 50, determine whether the graph of each equation is symmetric with respect to the a. x-axis, b. y-axis, c. origin.

43. $y = x^2 - 7$

44. $x = y^2 + 3$

45. $y = x^3 - 4x$

46. $y^2 = x^2 + 4$

47. $\dfrac{x^2}{3^2} + \dfrac{y^2}{4^2} = 1$

48. $xy = 8$

49. $|y| = |x|$

50. $|x + y| = 4$

In Exercises 51 to 56, sketch the graph of g. a. Find the domain and the range of g. b. State whether g is even, odd, or neither even nor odd.

51. $g(x) = -x^2 + 4$

52. $g(x) = -2x - 4$

53. $g(x) = |x - 2| + |x + 2|$

54. $g(x) = \sqrt{16 - x^2}$

55. $g(x) = x^3 - x$

56. $g(x) = 2[\![x]\!]$

In Exercises 57 to 62, first write the quadratic function in standard form, and then make use of translations to graph the function.

57. $F(x) = x^2 + 4x - 7$

58. $A(x) = x^2 - 6x - 5$

59. $P(x) = 3x^2 - 4$

60. $G(x) = 2x^2 - 8x + 3$

61. $W(x) = -4x^2 - 6x + 6$

62. $T(x) = -2x^2 - 10x$

63. On the same set of coordinate axes, sketch the graph of $p(x) = \sqrt{x} + c$ for $c = 0, -1,$ and 2.

64. On the same set of coordinate axes, sketch the graph of $q(x) = \sqrt{x + c}$ for $c = 0, -1,$ and 2.

65. On the same set of coordinate axes, sketch the graph of $r(x) = c\sqrt{9 - x^2}$ for $c = 1, \dfrac{1}{2},$ and -2.

66. On the same set of coordinate axes, sketch the graph of $s(t) = [\![cx]\!]$ for $c = 1, \dfrac{1}{4},$ and 4.

In Exercises 67 and 68, graph each piecewise-defined function.

67. $f(x) = \begin{cases} x, & \text{if } x \le 0 \\ \dfrac{1}{2}x, & \text{if } x > 0 \end{cases}$

68. $g(x) = \begin{cases} -2, & \text{if } x < -3 \\ \dfrac{2}{3}x, & \text{if } -3 \le x \le 3 \\ 2, & \text{if } x > 3 \end{cases}$

In Exercises 69 and 70, use the given functions f and g to find $f + g, f - g, fg,$ and $\dfrac{f}{g}$. State the domain of each.

69. $f(x) = x^2 - 9, \quad g(x) = x + 3$

70. $f(x) = x^3 + 8, \quad g(x) = x^2 - 2x + 4$

71. Find two numbers whose sum is 50 and whose product is a maximum.

72. Find two numbers whose difference is 10 and the sum of whose squares is a minimum.

73. The distance traveled by a ball rolling down a ramp is given by $s(t) = 3t^2$, where t is the time in seconds after the ball is released and $s(t)$ is measured in feet. Evaluate the average velocity of the ball for each of the following time intervals.

a. $[2, 4]$ **b.** $[2, 3]$ **c.** $[2, 2.5]$ **d.** $[2, 2.01]$

e. What appears to be the average velocity of the ball for the time interval $[2, 2 + \Delta t]$ as Δt approaches 0?

74. The distance traveled by a ball that is pushed down a ramp is given by $s(t) = 2t^2 + t$, where t is the time in seconds after the ball is released and $s(t)$ is measured in feet.

Evaluate the average velocity of the ball for each of the following time intervals.

a. $[3, 5]$ **b.** $[3, 4]$ **c.** $[3, 3.5]$ **d.** $[3, 3.01]$

e. What appears to be the average velocity of the ball for the time interval $[3, 3 + \Delta t]$ as Δt approaches 0?

75. **COMPUTER SCIENCE** A test of an Internet service provider showed the following download times (in seconds) for files of various sizes (in kilobytes).

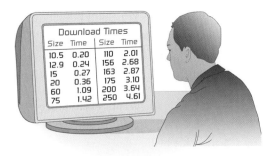

Download Times

Size	Time	Size	Time
10.5	0.20	110	2.01
12.9	0.24	156	2.68
15	0.27	163	2.87
20	0.36	175	3.10
60	1.09	200	3.64
75	1.42	250	4.61

a. Find a linear regression model for these data.

b. Judging on the basis of the value of r, is a linear model of these data a reasonable model? Explain.

c. On the basis of the model, what is the expected download time of a file that is 100 kilobytes in size? Round to the nearest tenth of a second.

76. **PHYSICS** The rate at which water will escape from the bottom of a can depends on a number of factors, including the height of the water, the size of the hole, and the diameter of the can. The table below shows the height (in millimeters) of water in a can after t seconds.

Water Escaping a Ruptured Can

Height	Time	Height	Time
0	180	60	93
10	163	70	81
20	147	80	70
30	133	90	60
40	118	100	50
50	105	110	48

a. Find the quadratic regression model for these data.

b. On the basis of this model, will the can ever empty?

c. Explain why there seems to be a contradiction between the model and reality, in that we know the can will eventually run out of water.

CHAPTER 2 TEST

1. Find the midpoint and the length of the line segment with endpoints $(-2, 3)$ and $(4, -1)$.

2. Determine the x- and y-intercepts, and then graph the equation $x = 2y^2 - 4$.

3. Graph the equation $y = |x + 2| + 1$.

4. Find the center and radius of the circle that has the general form $x^2 - 4x + y^2 + 2y - 4 = 0$.

5. Determine the domain of the function
$$f(x) = -\sqrt{x^2 - 16}$$

6. Graph $f(x) = -2|x - 2| + 1$. Identify the intervals over which the function is

a. increasing

b. constant

c. decreasing

7. An air freight company has determined that its cost per flight of delivering x parcels is
$$C(x) = 875 + 0.75x$$

The price it charges to send a parcel is $12.00 per parcel. Determine

a. the revenue function

b. the profit function

c. the minimum number of parcels the company must ship to break even

The quotient is $x^3 - 4x^2 + 12x - 41$, and the remainder is 179.

$$\frac{x^4 - 4x^2 + 7x + 15}{x + 4} = x^3 - 4x^2 + 12x - 41 + \frac{179}{x + 4}$$

▶ **TRY EXERCISE 12, PAGE 287**

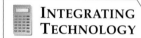
INTEGRATING TECHNOLOGY

A TI-82/83 synthetic-division program called SYDIV is available on the Internet at math.college.hmco.com. The program prompts you to enter the degree of the dividend, the coefficients of the dividend, and the constant c from the divisor $x - c$. For instance, to perform the synthetic division in Example 2, enter **4** for the degree of the dividend, followed by the coefficients **1**, **0**, **–4**, **7**, and **15**. See **Figure 3.1**. Press ENTER followed by **–4** to produce the display in **Figure 3.2**. Press ENTER to produce the display in **Figure 3.3**. Press ENTER again to produce the display in **Figure 3.4**.

```
prgmSYDIV
DEGREE? 4
DIVIDEND COEF
?1
?0
?-4
?7
?15
```

```
C? -4
```

```
COEF OF QUOTIENT
               1
              -4
              12
             -41
```

```
REMAINDER
               179
QUIT?  PRESS 1
NEW C?  PRESS 2
```

FIGURE 3.1   **FIGURE 3.2**   **FIGURE 3.3**   **FIGURE 3.4**

● THE REMAINDER THEOREM

The following theorem shows that synthetic division can be used to determine the value $P(c)$ for a given polynomial P and constant c.

The Remainder Theorem

If a polynomial $P(x)$ is divided by $x - c$, then the remainder equals $P(c)$.

The following example illustrates the Remainder Theorem by showing that the remainder of $(x^2 + 9x - 16) \div (x - 3)$ is the same as $P(x) = x^2 + 9x - 16$ evaluated at $x = 3$.

Let $x = 3$ and $P(x) = x^2 + 9x - 16$.

Then $P(3) = (3)^2 + 9(3) - 16$

$= 9 + 27 - 16$

$= 20$

$$\begin{array}{r} x + 12 \\ x - 3 \overline{)x^2 + 9x - 16} \\ \underline{x^2 - 3x} \\ 12x - 16 \\ \underline{12x - 36} \\ 20 \end{array}$$

$P(3)$ is equal to the remainder of $P(x)$ divided by $(x - 3)$.

In Example 3 we use synthetic division and the Remainder Theorem to evaluate a polynomial function.

EXAMPLE 3   **Use the Remainder Theorem to Evaluate a Polynomial Function**

Let $P(x) = 2x^3 + 3x^2 + 2x - 2$. Use the Remainder Theorem to find $P(c)$ for $c = -2$ and $c = \dfrac{1}{2}$.

Algebraic Solution

Perform synthetic division with $c = -2$ and $c = \dfrac{1}{2}$ and examine the remainders.

$$
\begin{array}{r|rrrr}
-2 & 2 & 3 & 2 & -2 \\
   &   & -4 & 2 & -8 \\
\hline
   & 2 & -1 & 4 & -10
\end{array}
$$

The remainder is -10. Therefore, $P(-2) = -10$.

$$
\begin{array}{r|rrrr}
\frac{1}{2} & 2 & 3 & 2 & -2 \\
   &   & 1 & 2 & 2 \\
\hline
   & 2 & 4 & 4 & 0
\end{array}
$$

The remainder is 0. Therefore, $P\left(\dfrac{1}{2}\right) = 0$.

Visualize the Solution

A graph of P shows that the points $(-2, -10)$ and $\left(\dfrac{1}{2}, 0\right)$ are on the graph.

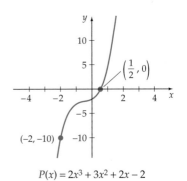

$P(x) = 2x^3 + 3x^2 + 2x - 2$

▶ **TRY EXERCISE 26, PAGE 287**

Using the Remainder Theorem to evaluate a polynomial function is often faster than evaluating the polynomial function by direct substitution. For instance, to evaluate $P(x) = x^5 - 10x^4 + 35x^3 - 50x^2 + 24x$ by substituting 7 for x, we must do the following work.

$$
\begin{aligned}
P(7) &= (7)^5 - 10(7)^4 + 35(7)^3 - 50(7)^2 + 24(7) \\
&= 16{,}807 - 10(2401) + 35(343) - 50(49) + 24(7) \\
&= 16{,}807 - 24{,}010 + 12{,}005 - 2450 + 168 \\
&= 2520
\end{aligned}
$$

> ***take note***
>
> Because $P(x)$ has a constant term of 0, we must include 0 as the last number in the first row of the synthetic division at the right.

Using the Remainder Theorem to perform the above evaluation requires only the following work.

$$
\begin{array}{r|rrrrrr}
7 & 1 & -10 & 35 & -50 & 24 & 0 \\
  &   & 7 & -21 & 98 & 336 & 2520 \\
\hline
  & 1 & -3 & 14 & 48 & 360 & 2520 \longleftarrow P(7)
\end{array}
$$

● **THE FACTOR THEOREM**

Note from Example 3 that $P\left(\dfrac{1}{2}\right) = 0$. Recall that $\dfrac{1}{2}$ is a zero of P because $P(x) = 0$ when $x = \dfrac{1}{2}$.

The following theorem is a direct result of the Remainder Theorem. It points out the important relationship between a zero of a given polynomial function and a factor of the polynomial function.

The Factor Theorem

A polynomial function $P(x)$ has a factor $(x - c)$ if and only if $P(c) = 0$. That is, $(x - c)$ is a factor of $P(x)$ if and only if c is a zero of P.

EXAMPLE 4   Apply the Factor Theorem

Use synthetic division and the Factor Theorem to determine whether $(x + 5)$ or $(x - 2)$ is a factor of $P(x) = x^4 + x^3 - 21x^2 - x + 20$.

Solution

$$
\begin{array}{r|rrrrr}
-5 & 1 & 1 & -21 & -1 & 20 \\
   &   & -5 & 20 & 5 & -20 \\
\hline
   & 1 & -4 & -1 & 4 & 0
\end{array}
$$

last # should = 0

The remainder of 0 indicates that $(x + 5)$ is a factor of $P(x)$.

$$
\begin{array}{r|rrrrr}
2 & 1 & 1 & -21 & -1 & 20 \\
  &   & 2 & 6 & -30 & -62 \\
\hline
  & 1 & 3 & -15 & -31 & -42
\end{array}
$$

The remainder of -42 indicates that $(x - 2)$ is not a factor of $P(x)$.

▶ **TRY EXERCISE 36, PAGE 287**

❓ **QUESTION** Is -5 a zero of $P(x)$ given in Example 4?

Here is a summary of the important role played by the remainder in the division of a polynomial by $(x - c)$.

The Remainder of a Polynomial Division

In the division of the polynomial function $P(x)$ by $(x - c)$, the remainder is

- equal to $P(c)$.
- 0 if and only if $(x - c)$ is a factor of P.
- 0 if and only if c is a zero of P.

Also, if c is a real number, then the remainder of $P(x) \div (x - c)$ is 0 if and only if $(c, 0)$ is an x-intercept of the graph of P.

❓ **ANSWER** Yes. Because $(x + 5)$ is a factor of $P(x)$, the Factor Theorem states that $P(-5) = 0$, and thus -5 is a zero of $P(x)$.

● REDUCED POLYNOMIALS

In Example 4 we determined that $(x + 5)$ is a factor of the polynomial function $P(x) = x^4 + x^3 - 21x^2 - x + 20$ and that the quotient of $x^4 + x^3 - 21x^2 - x + 20$ divided by $(x + 5)$ is $Q(x) = x^3 - 4x^2 - x + 4$. Thus

$$P(x) = (x + 5)(x^3 - 4x^2 - x + 4)$$

The quotient $Q(x) = x^3 - 4x^2 - x + 4$ is called a **reduced polynomial** or a **depressed polynomial** of $P(x)$ because it is a factor of $P(x)$ and its degree is 1 less than the degree of $P(x)$. Reduced polynomials will play an important role in Sections 3.3 and 3.4.

EXAMPLE 5 **Find a Reduced Polynomial**

Verify that $(x - 3)$ is a factor of $P(x) = 2x^3 - 3x^2 - 4x - 15$, and write $P(x)$ as the product of $(x - 3)$ and the reduced polynomial $Q(x)$.

Solution

$$
\begin{array}{r|rrrr}
3 & 2 & -3 & -4 & -15 \\
  &   & 6 & 9 & 15 \\
\hline
  & 2 & 3 & 5 & 0
\end{array}
$$

Coefficients of the reduced polynomial $Q(x)$

Thus $(x - 3)$ and the reduced polynomial $2x^2 + 3x + 5$ are both factors of $P(x)$. That is,

$$P(x) = 2x^3 - 3x^2 - 4x - 15 = (x - 3)(2x^2 + 3x + 5)$$

▶ **TRY EXERCISE 56, PAGE 288**

 ### TOPICS FOR DISCUSSION

1. Explain the meaning of the phrase *zero of a polynomial.*

2. If $P(x)$ is a polynomial of degree 3, what is the degree of the quotient of $\dfrac{P(x)}{x - c}$?

3. Discuss how the Remainder Theorem can be used to determine whether a number is a zero of a polynomial.

4. A zero of $P(x) = x^3 - x^2 - 14x + 24$ is -4. Discuss how this information and the Factor Theorem can be used to solve $x^3 - x^2 - 14x + 24 = 0$.

5. Discuss the advantages and disadvantages of using synthetic division rather than substitution to evaluate a polynomial function at $x = c$.

EXAMPLE 4 **Apply the Zero Location Theorem**

Use the Zero Location Theorem to verify that $S(x) = x^3 - x - 2$ has a real zero between 1 and 2.

Algebraic Solution

Use synthetic division to evaluate S for $x = 1$ and $x = 2$. If S changes sign between these two values, then S has a real zero between 1 and 2.

$$
\begin{array}{r|rrrr}
1 & 1 & 0 & -1 & -2 \\
  &   & 1 & 1 & 0 \\
\hline
  & 1 & 1 & 0 & -2
\end{array}
$$ • $S(1)$ is negative.

$$
\begin{array}{r|rrrr}
2 & 1 & 0 & -1 & -2 \\
  &   & 2 & 4 & 6 \\
\hline
  & 1 & 2 & 3 & 4
\end{array}
$$ • $S(2)$ is positive.

The graph of S is continuous because S is a polynomial function. Also, $S(1)$ is negative and $S(2)$ is positive. Thus the Zero Location Theorem indicates that there is a real zero between 1 and 2.

Visualize the Solution

The graph of S crosses the x-axis between $x = 1$ and $x = 2$. Thus S has a real zero between 1 and 2.

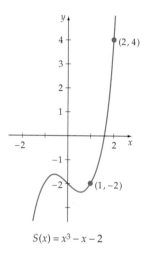

$S(x) = x^3 - x - 2$

▶ **TRY EXERCISE 28, PAGE 302**

The following theorem summarizes important relationships among the real zeros of a polynomial function, the x-intercepts of its graph, and its factors that can be written in the form $(x - c)$, where c is a real number.

Polynomial Functions, Real Zeros, Graphs, and Factors $(x - c)$

If P is a polynomial function and c is a real number, then all the following statements are equivalent in the sense that if any one statement is true, then they are all true, and if any one statement is false, then they are all false.

● $(x - c)$ is a factor of P.

● $x = c$ is a real solution of $P(x) = 0$.

● $x = c$ is a real zero of P.

● $(c, 0)$ is an x-intercept of the graph of $y = P(x)$.

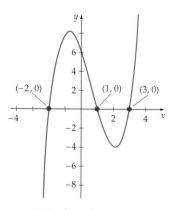

$S(x) = x^3 - 2x^2 - 5x + 6$

FIGURE 3.15

Sometimes it is possible to make use of the preceding theorem and a graph of a polynomial function to find factors of a function. For example, the graph of

$$S(x) = x^3 - 2x^2 - 5x + 6$$

is shown in **Figure 3.15.** The x-intercepts are $(-2, 0)$, $(1, 0)$, and $(3, 0)$. Hence -2, 1, and 3 are zeros of S, and $[x - (-2)]$, $(x - 1)$, and $(x - 3)$ are all factors of S.

• EVEN AND ODD POWERS OF (x − c) THEOREM

Use a graphing utility to graph $P(x) = (x + 3)(x - 4)^2$. Compare your graph with **Figure 3.16.** Examine the graph near the x-intercepts $(-3, 0)$ and $(4, 0)$. Observe that the graph of P

- crosses the x-axis at $(-3, 0)$.

- intersects the x-axis but does not cross the x-axis at $(4, 0)$.

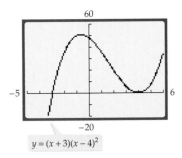

$$y = (x + 3)(x - 4)^2$$

FIGURE 3.16

The following theorem can be used to determine at which x-intercepts the graph of a polynomial function will cross the x-axis and at which x-intercepts the graph will intersect but not cross the x-axis.

Even and Odd Powers of (x − c) Theorem

If c is a real number and the polynomial function $P(x)$ has $(x - c)$ as a factor exactly k times, then the graph of P will

- intersect but not cross the x-axis at $(c, 0)$, provided k is an even positive integer.

- cross the x-axis at $(c, 0)$, provided k is an odd positive integer.

EXAMPLE 5   **Apply the Even and Odd Powers of (x − c) Theorem**

Determine where the graph of $P(x) = (x + 3)(x - 2)^2(x - 4)^3$ crosses the x-axis and where the graph intersects but does not cross the x-axis.

Solution

The exponents of the factors $(x + 3)$ and $(x - 4)$ are odd integers. Therefore, the graph of P will cross the x-axis at the x-intercepts $(-3, 0)$ and $(4, 0)$.

The exponent of the factor $(x - 2)$ is an even integer. Therefore, the graph of P will intersect but not cross the x-axis at $(2, 0)$.

Use a graphing utility to check these results.

▶ **TRY EXERCISE 34, PAGE 302**

● A PROCEDURE FOR GRAPHING POLYNOMIAL FUNCTIONS

You may find that you can sketch the graph of a polynomial function just by plotting several points; however, the following procedure will help you sketch the graph of many polynomial functions in an efficient manner.

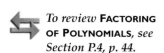

To review FACTORING OF POLYNOMIALS, *see Section P.4, p. 44.*

A Procedure for Graphing Polynomial Functions

$$P(x) = a_n x^n + a_{n-1} x^{n-1} + \cdots + a_1 x + a_0, \quad a_n \neq 0$$

To graph P:

1. ***Determine the far-left and the far-right behavior.*** Examine the leading coefficient $a_n x^n$ to determine the far-left and the far-right behavior of the graph.

2. ***Find the y-intercept.*** Determine the y-intercept by evaluating $P(0)$.

3. ***Find the x-intercept(s) and determine the behavior of the graph near the x-intercept(s).*** If possible, find the x-intercepts by factoring. If $(x - c)$, where c is a real number, is a factor of P, then $(c, 0)$ is an x-intercept of the graph. Use the Even and Odd Powers of $(x - c)$ Theorem to determine where the graph crosses the x-axis and where the graph intersects but does not cross the x-axis.

4. ***Find additional points on the graph.*** Find a few additional points (in addition to the intercepts).

5. ***Check for symmetry.***

   a.   The graph of an even function is symmetric with respect to the y-axis.

   b.   The graph of an odd function is symmetric with respect to the origin.

6. ***Sketch the graph.*** Use all the information obtained above to sketch the graph of the polynomial function. The graph should be a smooth continuous curve that passes through the points determined in steps 2 to 4. The graph should have a maximum of $n - 1$ turning points.

EXAMPLE 6   Graph a Polynomial Function

Sketch the graph of $P(x) = x^3 - 4x^2 + 4x$.

Solution

Step 1   ***Determine the far-left and the far-right behavior.*** The leading term is $1x^3$. Because the leading coefficient 1 is positive and the degree of the polynomial 3 is odd, the graph of P goes down to its far left and up to its far right.

Step 2   ***Find the y-intercept.*** $P(0) = 0^3 - 4(0)^2 + 4(0) = 0$. The y-intercept is $(0, 0)$.

Continued ▶

Step 3 ***Find the x-intercept(s) and determine the behavior of the graph near the x-intercept(s).*** Try to factor $x^3 - 4x^2 + 4x$.

$$x^3 - 4x^2 + 4x = x(x^2 - 4x + 4)$$
$$= x(x - 2)(x - 2)$$
$$= x(x - 2)^2$$

Because $(x - 2)$ is a factor of P, the point $(2, 0)$ is an x-intercept of the graph of P. Because x is a factor of P (think of x as $x - 0$), the point $(0, 0)$ is an x-intercept of the graph of P. Applying the Even and Odd Powers of $(x - c)$ Theorem allows us to determine that the graph of P crosses the x-axis at $(0, 0)$ and intersects but does not cross the x-axis at $(2, 0)$.

Step 4 ***Find additional points on the graph.***

x	P(x)
−1	−9
0.5	1.125
1	1
3	3

Step 5 ***Check for symmetry.*** The function P is not an even or an odd function, so the graph of P is *not* symmetric to either the y-axis or the origin.

Step 6 ***Sketch the graph.***

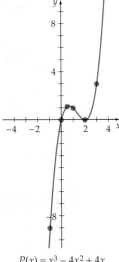

$P(x) = x^3 - 4x^2 + 4x$

 TRY EXERCISE 42, PAGE 302

TOPICS FOR DISCUSSION

1. Give an example of a polynomial function and of a function that is not a polynomial function.

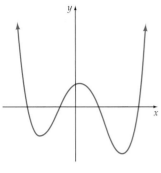

FIGURE 3.17

2. Is it possible for the graph of the polynomial function shown in **Figure 3.17** to be the graph of a polynomial function of odd degree? If so, explain how. If not, explain why not.

3. Explain the difference between a relative minimum and an absolute minimum.

4. Discuss how the Zero Location Theorem can be used to find a real zero of a polynomial function.

5. Let $P(x)$ be a polynomial function with real coefficients. Explain the relationships among the real zeros of the polynomial function, the x-coordinates of the x-intercepts of the graph of the polynomial function, and the solutions of the equation $P(x) = 0$.

EXERCISE SET 3.2

In Exercises 1 to 8, examine the leading term and determine the far-left and far-right behavior of the graph of the polynomial function.

1. $P(x) = 3x^4 - 2x^2 - 7x + 1$

2. $P(x) = -2x^3 - 6x^2 + 5x - 1$

3. $P(x) = 5x^5 - 4x^3 - 17x^2 + 2$

4. $P(x) = -6x^4 - 3x^3 + 5x^2 - 2x + 5$

5. $P(x) = 2 - 3x - 4x^2$

6. $P(x) = -16 + x^4$

7. $P(x) = \dfrac{1}{2}(x^3 + 5x^2 - 2)$

8. $P(x) = -\dfrac{1}{4}(x^4 + 3x^2 - 2x + 6)$

9. The following graph is the graph of a third-degree (cubic) polynomial function. What does the far-left and far-right behavior of the graph say about the leading coefficient a?

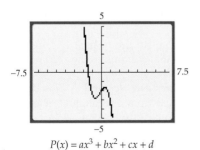

$P(x) = ax^3 + bx^2 + cx + d$

10. The following graph is the graph of a fourth-degree (quartic) polynomial function. What does the far-left and far-right behavior of the graph say about the leading coefficient a?

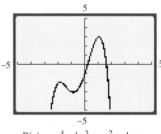

$P(x) = ax^4 + bx^3 + cx^2 + dx + e$

In Exercises 11 to 14, state the vertex of the graph of the function and use your knowledge of the vertex of a parabola to find the maximum or minimum of each function.

11. $P(x) = x^2 + 4x - 1$

12. $P(x) = x^2 + 6x + 1$

13. $P(x) = -x^2 - 8x + 1$

14. $P(x) = -2x^2 + 8x - 1$

In Exercises 15 to 20, use a graphing utility to graph each polynomial. Use the maximum and minimum features of the graphing utility to estimate, to the nearest tenth, the coordinates of the points where $P(x)$ has a relative maximum or a relative minimum. For each point, indicate whether the y value is a relative maximum or a relative minimum. The number in parentheses to the

right of the polynomial is the total number of relative maxima and minima.

15. $P(x) = x^3 + x^2 - 9x - 9$ (2)

16. $P(x) = x^3 + 4x^2 - 4x - 16$ (2)

17. $P(x) = x^3 - 3x^2 - 24x + 3$ (2)

18. $P(x) = -2x^3 - 3x^2 + 12x + 1$ (2)

19. $P(x) = x^4 - 4x^3 - 2x^2 + 12x - 5$ (3)

20. $P(x) = x^4 - 10x^2 + 9$ (3)

In Exercises 21 to 26, find the real zeros of each polynomial function by factoring. The number in parentheses to the right of each polynomial indicates the number of real zeros of the given polynomial function.

21. $P(x) = x^3 - 2x^2 - 15x$ (3)

▶ **22.** $P(x) = x^3 - 6x^2 + 8x$ (3)

23. $P(x) = x^4 - 13x^2 + 36$ (4)

24. $P(x) = 4x^4 - 37x^2 + 9$ (4)

25. $P(x) = x^5 - 5x^3 + 4x$ (5)

26. $P(x) = x^5 - 25x^3 + 144x$ (5)

In Exercises 27 to 32, use the Zero Location Theorem to verify that P has a zero between a and b.

27. $P(x) = 2x^3 + 3x^2 - 23x - 42$; $a = 3, b = 4$

▶ **28.** $P(x) = 4x^3 - x^2 - 6x + 1$; $a = 0, b = 1$

29. $P(x) = 3x^3 + 7x^2 + 3x + 7$; $a = -3, b = -2$

30. $P(x) = 2x^3 - 21x^2 - 2x + 25$; $a = 1, b = 2$

31. $P(x) = 4x^4 + 7x^3 - 11x^2 + 7x - 15$; $a = 1, b = 1\frac{1}{2}$

32. $P(x) = 5x^3 - 16x^2 - 20x + 64$; $a = 3, b = 3\frac{1}{2}$

In Exercises 33 to 40, determine the x-intercepts of the graph of P. For each x-intercept, use the Even and Odd Powers of $(x - c)$ Theorem to determine whether the graph of P crosses the x-axis or intersects but does not cross the x-axis.

33. $P(x) = (x - 1)(x + 1)(x - 3)$

▶ **34.** $P(x) = (x + 2)(x - 6)^2$

35. $P(x) = -(x - 3)^2(x - 7)^5$

36. $P(x) = (x + 2)^3(x - 6)^{10}$

37. $P(x) = (2x - 3)^4(x - 1)^{15}$

38. $P(x) = (5x + 10)^6(x - 2.7)^5$

39. $P(x) = x^3 - 6x^2 + 9x$

40. $P(x) = x^4 + 3x^3 + 4x^2$

In Exercises 41 to 46, sketch the graph of the polynomial function.

41. $P(x) = x^3 - x^2 - 2x$

▶ **42.** $P(x) = x^3 + 2x^2 - 3x$

43. $P(x) = -x^3 - 2x^2 + 5x + 6$ (*Hint:* In factored form $P(x) = (x + 3)(x + 1)(x - 2)$.)

44. $P(x) = -x^3 - 3x^2 + x + 3$ (*Hint:* In factored form $P(x) = (x + 3)(x + 1)(x - 1)$.)

45. $P(x) = x^4 - 4x^3 + 2x^2 + 4x - 3$ (*Hint:* In factored form $P(x) = (x + 1)(x - 1)^2(x - 3)$.)

46. $P(x) = x^4 - 6x^3 + 8x^2$

47. **CONSTRUCTION OF A BOX** A company constructs boxes from rectangular pieces of cardboard that measure 10 inches by 15 inches. An open box is formed by cutting squares that measure x inches by x inches from each corner of the cardboard and folding up the sides, as shown in the following figure.

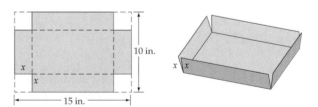

a. Express the volume V of the box as a function of x.

b. Determine (to the nearest hundredth of an inch) the x value that maximizes the volume of the box.

▶ **48.** **MAXIMIZING VOLUME** A closed box is to be constructed from a rectangular sheet of cardboard that measures 18 inches by 42 inches. The box is made by cutting rectangles that measure x inches by $2x$ inches from two of the corners and by cutting two squares that measure x inches by x inches from the top and from the bottom of the rectangle, as shown in the following figure. What value of x (to the nearest thousandth of an inch) will produce a box with maximum volume?

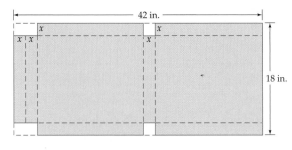

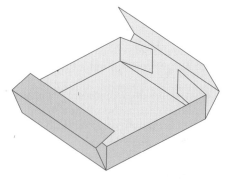

49. MAXIMIZING VOLUME An open box is to be constructed from a rectangular sheet of cardboard that measures 16 inches by 22 inches. To assemble the box, make the four cuts shown in the figure below and then fold on the dashed lines. What value of x (to the nearest thousandth of an inch) will produce a box with maximum volume?

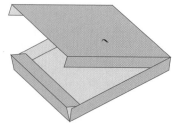

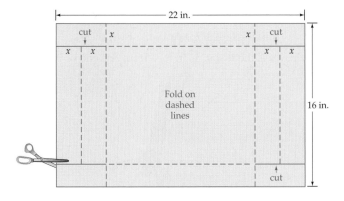

50. PROFIT A software company produces a computer game. The company has determined that its profit P, in dollars, from the manufacture and sale of x games is given by

$$P(x) = -0.000001x^3 + 96x - 98,000$$

where $0 < x \leq 9000$.

a. What is the maximum profit, to the nearest thousand dollars, the company can expect from the sale of its games?

b. How many games, to the nearest unit, does the company need to produce and sell to obtain the maximum profit?

51. ADVERTISING EXPENSES A company manufactures digital cameras. The company estimates that the profit from camera sales is

$$P(x) = -0.02x^3 + 0.01x^2 + 1.2x - 1.1$$

where P is the profit in millions of dollars and x is the amount, in hundred-thousands of dollars, spent on advertising.

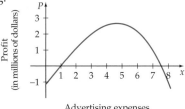

Advertising expenses
(in hundred-thousands of dollars)

Determine the amount, rounded to the nearest thousand dollars, the company needs to spend on advertising if it is to generate the maximum profit.

52. DIVORCE RATE The divorce rate for a given year is defined as the number of divorces per thousand population. The function

$$D(t) = 0.00001807t^4 - 0.001406t^3 + 0.02884t^2 \\ - 0.003466t + 2.1148$$

approximates the U.S. divorce rate for the years 1960 ($t = 0$) to 1999 ($t = 39$). Use $D(t)$ and a graphing utility to estimate

a. the year during which the U.S. divorce rate reached its absolute maximum for the period from 1960 to 1999.

b. the absolute minimum divorce rate, rounded to the nearest 0.1, during the period from 1960 to 1999.

53. MARRIAGE RATE The marriage rate for a given year is defined as the number of marriages per thousand population. The function

$$M(t) = -0.00000115t^4 + 0.000252t^3 \\ - 0.01827t^2 + 0.4438t + 9.1829$$

approximates the U.S. marriage rate for the years 1900 ($t = 0$) to 1999 ($t = 99$).

U.S. Marriage Rate, 1900–1999

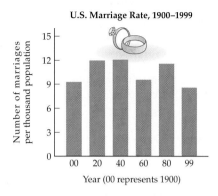

Year (00 represents 1900)

Use $M(t)$ and a graphing utility to estimate

a. during what year the U.S. marriage rate reached its maximum for the period from 1900 to 1999.

b. the relative minimum marriage rate, rounded to the nearest 0.1, during the period from 1950 to 1970.

54. **GAZELLE POPULATION** A herd of 204 African gazelles is introduced into a wild animal park. The population of the gazelles, $P(t)$, after t years is given by $P(t) = -0.7t^3 + 18.7t^2 - 69.5t + 204$, where $0 < t \le 18$.

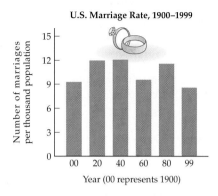

a. Use a graph of P to determine the absolute minimum gazelle population (rounded to the nearest single gazelle) that is attained during this time period.

b. Use a graph of P to determine the absolute maximum gazelle population (rounded to the nearest single gazelle) that is attained during this time period.

55. 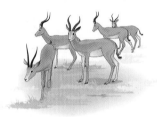 **MEDICATION LEVEL** Pseudoephedrine hydrochloride is an allergy medication. The function

$$L(t) = 0.03t^4 + 0.4t^3 - 7.3t^2 + 23.1t$$

where $0 \le t \le 5$, models the level of pseudoephedrine hydrochloride, in milligrams, in the bloodstream of a patient t hours after 30 milligrams of the medication have been taken.

a. Use a graphing utility and the function $L(t)$ to determine the maximum level of pseudoephedrine hydrochloride in the patient's bloodstream. Round your result to the nearest 0.01 milligram.

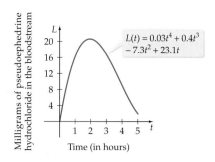

Time (in hours)

b. At what time t, to the nearest minute, is this maximum level of pseudoephedrine hydrochloride reached?

56. **SQUIRREL POPULATION** The population P of squirrels in a wilderness area is given by

$$P(t) = 0.6t^4 - 13.7t^3 + 104.5t^2 - 243.8t + 360,$$

where $0 \le t \le 12$ years.

a. What is the absolute minimum number of squirrels (rounded to the nearest single squirrel) attained on the interval $0 \le t \le 12$?

b. The absolute maximum of P is attained at the endpoint, where $t = 12$. What is this absolute maximum (rounded to the nearest single squirrel)?

57. 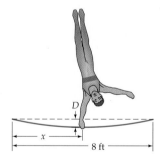 **BEAM DEFLECTION** The deflection D, in feet, of an 8-foot beam that is center loaded is given by

$$D(x) = (-0.0025)(4x^3 - 3 \cdot 8x^2), \quad 0 < x \le 4$$

where x is the distance, in feet, from one end of the beam.

a. Determine the deflection of the beam when $x = 3$ feet. Round to the nearest hundredth of an inch.

b. At what point does the beam achieve its maximum deflection? What is the maximum deflection? Round to the nearest hundredth of an inch.

c. What is the deflection at $x = 5$ feet?

58. **ENGINEERING** A cylindrical log with a diameter of 22 inches is to be cut so that it will yield a beam that has a rectangular cross section of depth d and width w.

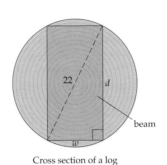

Cross section of a log

An engineer has determined that the stiffness S of the resulting beam is given by $S = 1.15wd^2$, where $0 < w < 22$ inches. Find the width and the depth that will maximize the stiffness of the beam. Round each result to the nearest hundredth of an inch. (*Hint:* Use the Pythagorean Theorem to solve for d^2 in terms of w^2.)

CONNECTING CONCEPTS

59. Use a graph of $P(x) = x^3 - x - 25$ to determine between which two consecutive integers P has a real zero.

60. Use a graph of the polynomial function $P(x) = 4x^4 - 12x^3 + 13x^2 - 12x + 9$ to determine between which two consecutive integers P has a real zero.

61. The point $(2, 0)$ is on the graph of $P(x)$. What point must be on the graph of $P(x - 3)$?

62. The point $(3, 5)$ is on the graph of $P(x)$. What point must be on the graph of $P(x + 1) - 2$?

63. Explain how to use the graph of $y = x^3$ to produce the graph of $P(x) = (x - 2)^3 + 1$.

64. Consider the following conjecture. Let $P(x)$ be a polynomial function. If a and b are real numbers such that $a < b$, $P(a) > 0$, and $P(b) > 0$, then $P(x)$ does not have a real zero between a and b. Is this conjecture true or false? Support your answer.

PREPARE FOR SECTION 3.3

65. Find the zeros of $P(x) = 6x^2 - 25x + 14$. [1.3/2.4]

66. Use synthetic division to divide $2x^3 + 3x^2 + 4x - 7$ by $x + 2$. [3.1]

67. Use synthetic division to divide $3x^4 - 21x^2 - 3x - 5$ by $x - 3$. [3.1]

68. List all natural numbers that are factors of 12. [P.1]

69. List all integers that are factors of 27. [P.1]

70. Given $P(x) = 4x^3 - 3x^2 - 2x + 5$, find $P(-x)$. [2.5]

PROJECTS

1. A student thinks that $P(n) = n^3 - n$ is always a multiple of 6 for all natural numbers n. What do you think? Provide a mathematical argument to show that the student is correct or a counterexample to show that the student is wrong.

ZEROS OF POLYNOMIAL FUNCTIONS

• MULTIPLE ZEROS OF A POLYNOMIAL FUNCTION

Recall that if $P(x)$ is a polynomial function, then the values of x for which $P(x)$ is equal to 0 are called the *zeros* of $P(x)$ or the **roots** of the equation $P(x) = 0$. A zero of a polynomial function may be a **multiple zero.** For example, $P(x) = x^2 + 6x + 9$ can be expressed in factored form as $(x + 3)(x + 3)$. Setting each factor equal to zero yields $x = -3$ in both cases. Thus $P(x) = x^2 + 6x + 9$ has a zero of -3 that occurs twice. The following definition will be most useful when we are discussing multiple zeros.

Definition of Multiple Zeros of a Polynomial Function

If a polynomial function $P(x)$ has $(x - r)$ as a factor exactly k times, then r is a **zero of multiplicity k** of the polynomial function $P(x)$.

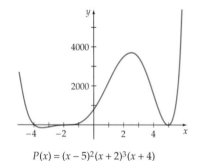

$P(x) = (x - 5)^2(x + 2)^3(x + 4)$

FIGURE 3.18

The graph of the polynomial function

$$P(x) = (x - 5)^2(x + 2)^3(x + 4)$$

is shown in **Figure 3.18.** This polynomial function has

- 5 as a zero of multiplicity 2.
- -2 as a zero of multiplicity 3.
- -4 as a zero of multiplicity 1.

A zero of multiplicity 1 is generally referred to as a **simple zero.**

When searching for the zeros of a polynomial function, it is important that we know how many zeros to expect. This question is answered completely in Section 3.4. For the work in this section, the following result is valuable.

Number of Zeros of a Polynomial Function

A polynomial function P of degree n has at most n zeros, where each zero of multiplicity k is counted k times.

• THE RATIONAL ZERO THEOREM

The rational zeros of polynomial functions with integer coefficients can be found with the aid of the following theorem.

The Rational Zero Theorem

If $P(x) = a_n x^n + a_{n-1} x^{n-1} + \cdots + a_1 x + a_0$ has *integer* coefficients ($a_n \neq 0$) and $\dfrac{p}{q}$ is a rational zero (in lowest terms) of P, then

- p is a factor of the constant term a_0 and
- q is a factor of the leading coefficient a_n.

The Rational Zero Theorem often is used to make a list of all possible rational zeros of a polynomial function. The list consists of all rational numbers of the form $\dfrac{p}{q}$, where p is an integer factor of the constant term a_0 and q is an integer factor of the leading coefficient a_n.

EXAMPLE 1   Apply the Rational Zero Theorem

Use the Rational Zero Theorem to list all possible rational zeros of

$$P(x) = 4x^4 + x^3 - 40x^2 + 38x + 12$$

Solution

List all integers p that are factors of 12 and all integers q that are factors of 4.

$$p: \quad \pm 1, \pm 2, \pm 3, \pm 4, \pm 6, \pm 12$$
$$q: \quad \pm 1, \pm 2, \pm 4$$

Form all possible rational numbers using $\pm 1, \pm 2, \pm 3, \pm 4, \pm 6,$ and ± 12 for the numerator and $\pm 1, \pm 2,$ and ± 4 for the denominator. By the Rational Zero Theorem, the possible rational zeros are

$$\pm 1, \pm \frac{1}{2}, \pm \frac{1}{4}, \pm 2, \pm 3, \pm \frac{3}{2}, \pm \frac{3}{4}, \pm 4, \pm 6, \pm 12$$

It is not necessary to list a factor that is already listed in reduced form. For example, $\pm \dfrac{6}{4}$ is not listed because it is equal to $\pm \dfrac{3}{2}$.

▶ **TRY EXERCISE 10, PAGE 316**

❓ **QUESTION** If $P(x) = a_n x^n + a_{n-1} x^{n-1} + \cdots + a_1 x + a_0$ has integer coefficients and a leading coefficient of $a_n = 1$, must all the rational zeros of P be integers?

❓ **ANSWER** Yes. By the Rational Zero Theorem, the rational zeros of P are of the form $\dfrac{p}{q}$, where p is an integer factor of a_0 and q is an integer factor of a_n. Thus $q = \pm 1$ and $\dfrac{p}{q} = \dfrac{p}{\pm 1} = \pm p$.

• UPPER AND LOWER BOUNDS FOR REAL ZEROS

A real number b is called an **upper bound** of the zeros of the polynomial function P if no zero is greater than b. A real number b is called a **lower bound** of the zeros of P if no zero is less than b. The following theorem is often used to find positive upper bounds and negative lower bounds for the real zeros of a polynomial function.

Upper- and Lower-Bound Theorem

Let $P(x)$ be a polynomial function with real coefficients. Use synthetic division to divide $P(x)$ by $x - b$, where b is a nonzero real number.

Upper bound **a.** If $b > 0$ and the leading coefficient of P is positive, then b is an upper bound for the real zeros of P provided none of the numbers in the bottom row of the synthetic division are negative.

b. If $b > 0$ and the leading coefficient of P is negative, then b is an upper bound for the real zeros of P provided none of the numbers in the bottom row of the synthetic division are positive.

Lower bound If $b < 0$ and the numbers in the bottom row of the synthetic division alternate in sign (the number zero can be considered positive or negative as needed to produce an alternating sign pattern), then b is a lower bound for the real zeros of P.

Upper and lower bounds are not unique. For example, if b is an upper bound for the real zeros of P, then any number greater than b is also an upper bound. Likewise, if a is a lower bound for the real zeros of P, then any number less than a is also a lower bound.

EXAMPLE 2   Find Upper and Lower Bounds

According to the Upper- and Lower-Bound Theorem, what is the smallest positive integer that is an upper bound and the largest negative integer that is a lower bound of the real zeros of $P(x) = 2x^3 + 7x^2 - 4x - 14$?

Solution

To find the smallest positive-integer upper bound, use synthetic division with $1, 2, \ldots,$ as test values.

$$
\begin{array}{r|rrrr}
1 & 2 & 7 & -4 & -14 \\
  &   & 2 & 9 & 5 \\
\hline
  & 2 & 9 & 5 & -9
\end{array}
\qquad
\begin{array}{r|rrrr}
2 & 2 & 7 & -4 & -14 \\
  &   & 4 & 22 & 36 \\
\hline
  & 2 & 11 & 18 & 22
\end{array}
$$

• **No negative numbers**

Thus 2 is the smallest positive-integer upper bound.

Now find the largest negative-integer lower bound.

$$
\begin{array}{r|rrrr}
-1 & 2 & 7 & -4 & -14 \\
   &   & -2 & -5 & 9 \\
\hline
   & 2 & 5 & -9 & -5
\end{array}
\qquad
\begin{array}{r|rrrr}
-2 & 2 & 7 & -4 & -14 \\
   &   & -4 & -6 & 20 \\
\hline
   & 2 & 3 & -10 & 6
\end{array}
$$

$$
\begin{array}{r|rrrr}
-3 & 2 & 7 & -4 & -14 \\
   &   & -6 & -3 & 21 \\
\hline
   & 2 & 1 & -7 & 7
\end{array}
\qquad
\begin{array}{r|rrrr}
-4 & 2 & 7 & -4 & -14 \\
   &   & -8 & 4 & 0 \\
\hline
   & 2 & -1 & 0 & -14
\end{array}
$$

• **Alternating signs**

Thus -4 is the largest negative-integer lower bound.

▶ **TRY EXERCISE 18, PAGE 316**

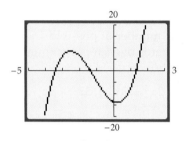

$P(x) = 2x^3 + 7x^2 - 4x - 14$

FIGURE 3.19

▦ INTEGRATING TECHNOLOGY

You can use the Upper- and Lower-Bound Theorem to determine Xmin (the lower bound) and Xmax (the upper bound) for the viewing window of a graphing utility. This will ensure that all the real zeros, which are the x-coordinates of the x-intercepts of the polynomial function, will be shown. Note in **Figure 3.19** that the zeros of $P(x) = 2x^3 + 7x^2 - 4x - 14$ are between -4 (a lower bound) and 2 (an upper bound).

● DESCARTES' RULE OF SIGNS

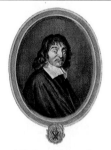
Descartes' Rule of Signs is another theorem that is often used to obtain information about the zeros of a polynomial function. In Descartes' Rule of Signs, the number of **variations in sign** of the coefficients of $P(x)$ or $P(-x)$ refers to sign changes of the coefficients from positive to negative or from negative to positive that we find when we examine successive terms of the function. The terms are assumed to appear in order of descending powers of x. For example, the polynomial function

$$P(x) = +3x^4 - 5x^3 - 7x^2 + x - 7$$

has three variations in sign. The polynomial function

$$
\begin{aligned}
P(-x) &= +3(-x)^4 - 5(-x)^3 - 7(-x)^2 + (-x) - 7 \\
      &= +\,3x^4 + 5x^3 - 7x^2 - x - 7
\end{aligned}
$$

has one variation in sign.

Terms that have a coefficient of 0 are not counted as variations in sign and may be ignored. For example,

$$P(x) = -x^5 + 4x^2 + 1$$

has one variation in sign.

Descartes' Rule of Signs

Let $P(x)$ be a polynomial function with real coefficients and with the terms arranged in order of decreasing powers of x.

1. The number of positive real zeros of $P(x)$ is equal to the number of variations in sign of $P(x)$, or to that number decreased by an even integer.

2. The number of negative real zeros of $P(x)$ is equal to the number of variations in sign of $P(-x)$, or to that number decreased by an even integer.

EXAMPLE 3 **Apply Descartes' Rule of Signs**

Use Descartes' Rule of Signs to determine both the number of possible positive and the number of possible negative real zeros of each polynomial function.

a. $P(x) = x^4 - 5x^3 + 5x^2 + 5x - 6$ b. $P(x) = 2x^5 + 3x^3 + 5x^2 + 8x + 7$

Solution

a.
$$P(x) = +x^4 - 5x^3 + 5x^2 + 5x - 6$$

There are three variations in sign. By Descartes' Rule of Signs, there are either three or one positive real zeros. Now examine the variations in sign of $P(-x)$.

$$P(-x) = x^4 + 5x^3 + 5x^2 - 5x - 6$$

There is one variation in sign of $P(-x)$. By Descartes' Rule of Signs, there is one negative real zero.

b. $P(x) = 2x^5 + 3x^3 + 5x^2 + 8x + 7$ has no variation in sign, so there are no positive real zeros.

$$P(-x) = -2x^5 - 3x^3 + 5x^2 - 8x + 7$$

$P(-x)$ has three variations in sign, so there are either three or one negative real zeros.

▶ **TRY EXERCISE 28, PAGE 316**

❓ QUESTION If $P(x) = ax^2 + bx + c$ has two variations in sign, must $P(x)$ have two positive real zeros?

❓ ANSWER No. According to Descartes' Rule of Signs, $P(x)$ will have either two positive real zeros or no positive real zeros.

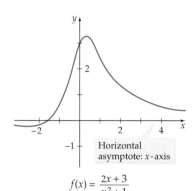

$$f(x) = \frac{2x+3}{x^2+1}$$

FIGURE 3.29

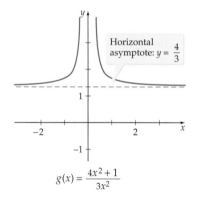

$$g(x) = \frac{4x^2+1}{3x^2}$$

FIGURE 3.30

EXAMPLE 2   Find the Horizontal Asymptote of a Rational Function

Find the horizontal asymptote of each rational function.

a. $f(x) = \dfrac{2x+3}{x^2+1}$ **b.** $g(x) = \dfrac{4x^2+1}{3x^2}$ **c.** $h(x) = \dfrac{x^3+1}{x-2}$

Solution

a. The degree of the numerator $2x + 3$ is less than the degree of the denominator $x^2 + 1$. By the Theorem on Horizontal Asymptotes, the x-axis is the horizontal asymptote of f. See the graph of f in **Figure 3.29**.

b. The numerator $4x^2 + 1$ and the denominator $3x^2$ of g are both of degree 2. By the Theorem on Horizontal Asymptotes, the line $y = \dfrac{4}{3}$ is the horizontal asymptote of g. See the graph of g in **Figure 3.30**.

c. The degree of the numerator $x^3 + 1$ is larger than the degree of the denominator $x - 2$, so by the Theorem on Horizontal Asymptotes, the graph of h has no horizontal asymptotes.

▶ **TRY EXERCISE 6, PAGE 341**

The proof of the Theorem on Horizontal Asymptotes makes use of the technique employed in the following verification. To verify that

$$y = \frac{5x^2+4}{3x^2+8x+7}$$

has a horizontal asymptote of $y = \dfrac{5}{3}$, divide the numerator and the denominator by the largest power of the variable x (x^2 in this case).

$$y = \frac{\dfrac{5x^2+4}{x^2}}{\dfrac{3x^2+8x+7}{x^2}} = \frac{5+\dfrac{4}{x^2}}{3+\dfrac{8}{x}+\dfrac{7}{x^2}}, \quad x \neq 0$$

As x increases without bound or decreases without bound, the fractions $\dfrac{4}{x^2}, \dfrac{8}{x},$ and $\dfrac{7}{x^2}$ approach zero. Thus

$$y \to \frac{5+0}{3+0+0} = \frac{5}{3} \quad \text{as} \quad x \to \pm\infty$$

and hence the line $y = \dfrac{5}{3}$ is a horizontal asymptote of the graph.

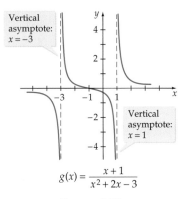

$$g(x) = \frac{x+1}{x^2+2x-3}$$

FIGURE 3.31

• A SIGN PROPERTY OF RATIONAL FUNCTIONS

The zeros and vertical asymptotes of a rational function F divide the x-axis into intervals. In each interval, $F(x)$ is positive for all x in the interval or $F(x)$ is negative for all x in the interval. For example, consider the rational function

$$g(x) = \frac{x+1}{x^2+2x-3}$$

which has vertical asymptotes of $x = -3$ and $x = 1$ and a zero of -1. These three numbers divide the x-axis into the four intervals $(-\infty, -3)$, $(-3, -1)$, $(-1, 1)$, and $(1, \infty)$. Note in **Figure 3.31** that the graph of g is negative for all x such that $x < -3$, positive for all x such that $-3 < x < -1$, negative for all x such that $-1 < x < 1$, and positive for all x such that $x > 1$.

• A GENERAL GRAPHING PROCEDURE

If $F(x) = P(x)/Q(x)$, where $P(x)$ and $Q(x)$ are polynomials that have no common factors, then the following general procedure offers useful guidelines for graphing F.

General Procedure for Graphing Rational Functions That Have No Common Factors

1. *Asymptotes*   Find the real zeros of the denominator $Q(x)$. For each zero a, draw the dashed line $x = a$. Each line is a vertical asymptote of the graph of F. Also graph any horizontal asymptotes.

2. *Intercepts*   Find the real zeros of the numerator $P(x)$. For each real zero c, plot the point $(c, 0)$. Each such point is an x-intercept of the graph of F. For each x-intercept use the even and odd powers of $(x - c)$ to determine if the graph crosses the x-axis at the intercept or if the graph intersects but does not cross the x-axis. Also evaluate $F(0)$. Plot $(0, F(0))$, the y-intercept of the graph of F.

3. *Symmetry*   Use the tests for symmetry to determine whether the graph of the function has symmetry with respect to the y-axis or symmetry with respect to the origin.

4. *Additional points*   Plot some points that lie in the intervals between and beyond the vertical asymptotes and the x-intercepts.

5. *Behavior near asymptotes*   If $x = a$ is a vertical asymptote, determine whether $F(x) \to \infty$ or $F(x) \to -\infty$ as $x \to a^-$ and also as $x \to a^+$.

6. *Complete the sketch*   Use all the information obtained above to sketch the graph of F.

EXAMPLE 3   **Graph a Rational Function**

Sketch a graph of $f(x) = \dfrac{2x^2 - 18}{x^2 + 3}$.

Solution

Asymptotes   The denominator $x^2 + 3$ has no real zeros, so the graph of f has no vertical asymptotes. The numerator and denominator both are of degree 2. The leading coefficients are 2 and 1, respectively. By the Theorem on Horizontal Asymptotes, the graph of f has a horizontal asymptote of $y = \dfrac{2}{1} = 2$.

Intercepts   The zeros of the numerator occur when $2x^2 - 18 = 0$ or, solving for x, when $x = -3$ and $x = 3$. Therefore, the x-intercepts are $(-3, 0)$ and $(3, 0)$. The factored numerator is $2(x + 3)(x - 3)$. Each linear factor has an exponent of 1, an odd number. Thus the graph crosses the x-axis at its x-intercepts. To find the y-intercept, evaluate f when $x = 0$. This gives $y = -6$. Therefore, the y-intercept is $(0, -6)$.

Symmetry   Below we show that $f(-x) = f(x)$, which means that f is an even function and therefore its graph is symmetric with respect to the y-axis.

$$f(-x) = \frac{2(-x)^2 - 18}{(-x)^2 + 3} = \frac{2x^2 - 18}{x^2 + 3} = f(x)$$

Additional Points   The intervals determined by the x-intercepts are $x < -3$, $-3 < x < 3$, and $x > 3$. Generally, it is necessary to determine points in all intervals. However, because f is an even function, its graph is symmetric with respect to the y-axis. The following table lists a few points for $x > 0$. Symmetry can be used to locate corresponding points for $x < 0$.

x	1	2	6
$f(x)$	-4	$-\dfrac{10}{7} \approx -1.43$	$\dfrac{18}{13} \approx 1.38$

Behavior Near Asymptotes   As x increases or decreases without bound, $f(x)$ approaches the horizontal asymptote $y = 2$.

To determine whether the graph of f intersects the horizontal asymptote at any point, solve the equation $f(x) = 2$.

There are no solutions of $f(x) = 2$ because

$$\frac{2x^2 - 18}{x^2 + 3} = 2 \quad \text{implies} \quad 2x^2 - 18 = 2x^2 + 6 \quad \text{implies} \quad -18 = 6$$

This is not possible. Thus the graph of f does not intersect the horizontal asymptote but approaches it from below as x increases or decreases without bound.

Continued ▶

Complete the Sketch   Use the summary in **Table 3.3,** to the left, to finish the sketch. The completed graph is shown in **Figure 3.32.**

TABLE 3.3

Vertical Asymptote	None
Horizontal Asymptote	$y = 2$
x-Intercepts	crosses at $(-3, 0)$, crosses at $(3, 0)$
y-Intercept	$(0, -6)$
Additional Points	$(1, -4)$, $(2, -1.43)$, $(6, 1.38)$

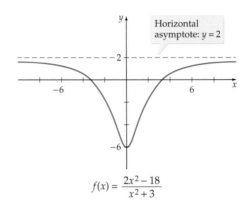

$$f(x) = \frac{2x^2 - 18}{x^2 + 3}$$

FIGURE 3.32

▶ **TRY EXERCISE 10, PAGE 341**

EXAMPLE 4   **Graph a Rational Function**

Sketch a graph of $h(x) = \dfrac{x^2 + 1}{x^2 + x - 2}$.

Solution

Asymptotes   The denominator $x^2 + x - 2 = (x + 2)(x - 1)$ has zeros -2 and 1; because there are no common factors of the numerator and the denominator, the lines $x = -2$ and $x = 1$ are vertical asymptotes.

The numerator and denominator both are of degree 2. The leading coefficients of the numerator and denominator are both 1. Thus h has the horizontal asymptote $y = \dfrac{1}{1} = 1$.

Intercepts   The numerator $x^2 + 1$ has no real zeros, so the graph of h has no x-intercepts. Because $h(0) = -0.5$, h has the y-intercept $(0, -0.5)$.

Symmetry   By applying the tests for symmetry, we can determine that the graph of h is not symmetric with respect to the origin or to the y-axis.

Additional Points   The intervals determined by the vertical asymptotes are $(-\infty, -2)$, $(-2, 1)$, and $(1, \infty)$. Plot a few points from each interval.

x	−5	−3	−1	0.5	2	3	4
h(x)	$\dfrac{13}{9}$	$\dfrac{5}{2}$	-1	-1	$\dfrac{5}{4}$	1	$\dfrac{17}{18}$

The graph of h will intersect the horizontal asymptote $y = 1$ exactly once. This can be determined by solving the equation $h(x) = 1$.

$$\frac{x^2 + 1}{x^2 + x - 2} = 1$$

$$x^2 + 1 = x^2 + x - 2 \qquad \bullet \text{ Multiply both sides by } x^2 + x - 2.$$

$$1 = x - 2$$

$$3 = x$$

The only solution is $x = 3$. Therefore, the graph of h intersects the horizontal asymptote at $(3, 1)$.

Behavior Near Asymptotes   As x approaches -2 from the left, the denominator $(x + 2)(x - 1)$ approaches 0 but remains positive. The numerator $x^2 + 1$ approaches 5, which is positive, so the quotient $h(x)$ increases without bound. Stated in mathematical notation,

$$h(x) \to \infty \quad \text{as} \quad x \to -2^-$$

Similarly, it can be determined that

$$h(x) \to -\infty \quad \text{as} \quad x \to -2^+$$
$$h(x) \to -\infty \quad \text{as} \quad x \to 1^-$$
$$h(x) \to \infty \quad \text{as} \quad x \to 1^+$$

Complete the Sketch   Use the summary in **Table 3.4** to obtain the graph sketched in **Figure 3.33**.

TABLE 3.4

Vertical Asymptote	$x = -2, x = 1$
Horizontal Asymptote	$y = 1$
x-Intercepts	None
y-Intercept	$(0, -0.5)$
Additional Points	$(-5, 1.\overline{4}), (-3, 2.5),$ $(-1, -1), (0.5, -1),$ $(2, 1.25), (3, 1),$ $(4, 0.9\overline{4})$

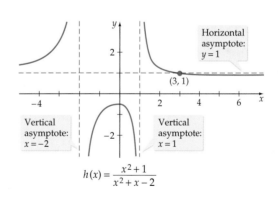

FIGURE 3.33

▶ **TRY EXERCISE 26, PAGE 341**

● **SLANT ASYMPTOTES**

Some rational functions have an asymptote that is neither vertical nor horizontal, but slanted.

Theorem on Slant Asymptotes

The rational function given by $F(x) = P(x)/Q(x)$, where $P(x)$ and $Q(x)$ have no common factors, has a **slant asymptote** if the degree of the polynomial $P(x)$ in the numerator is one greater than the degree of the polynomial $Q(x)$ in the denominator.

To find the slant asymptote, divide $P(x)$ by $Q(x)$ and write $F(x)$ in the form

$$F(x) = \frac{P(x)}{Q(x)} = (mx + b) + \frac{r(x)}{Q(x)}$$

where the degree of $r(x)$ is less than the degree of $Q(x)$. Because

$$\frac{r(x)}{Q(x)} \to 0 \quad \text{as} \quad x \to \pm\infty$$

we know that $F(x) \to mx + b$ as $x \to \pm\infty$.

The line represented by $y = mx + b$ is the slant asymptote of the graph of F.

EXAMPLE 5   Find the Slant Asymptote of a Rational Function

Find the slant asymptote of $f(x) = \dfrac{2x^3 + 5x^2 + 1}{x^2 + x + 3}$.

Solution

Because the degree of the numerator $2x^3 + 5x^2 + 1$ is exactly one larger than the degree of the denominator $x^2 + x + 3$ and f is in simplest form, f has a slant asymptote. To find the asymptote, divide $2x^3 + 5x^2 + 1$ by $x^2 + x + 3$.

$$
\begin{array}{r}
2x + 3 \\
x^2 + x + 3 \overline{)2x^3 + 5x^2 + 0x + 1} \\
\underline{2x^3 + 2x^2 + 6x} \\
3x^2 - 6x + 1 \\
\underline{3x^2 + 3x + 9} \\
-9x - 8
\end{array}
$$

Therefore,

$$f(x) = \frac{2x^3 + 5x^2 + 1}{x^2 + x + 3} = 2x + 3 + \frac{-9x - 8}{x^2 + x + 3}$$

and the line given by $y = 2x + 3$ is the slant asymptote for the graph of f. **Figure 3.34** shows the graph of f and its slant asymptote.

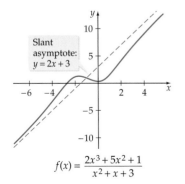

Slant asymptote: $y = 2x + 3$

$f(x) = \dfrac{2x^3 + 5x^2 + 1}{x^2 + x + 3}$

FIGURE 3.34

▶ **TRY EXERCISE 32, PAGE 341**

INTEGRATING TECHNOLOGY

In the standard viewing window of a calculator, the distance between two tic marks on the x-axis is not equal to the distance between two tic marks on the y-axis. As a result, the graph of $y = x$ does not appear to bisect the first and third quadrants. See **Figure 4.6.** This anomaly is important if a graphing calculator is being used to check whether two functions are inverses of one another. Because the graph of $y = x$ does not appear to bisect the first and third quadrants, the graphs of f and f^{-1} will not appear to be symmetric about the graph of $y = x$. The graphs of $f(x) = \dfrac{1}{3}x + 2$ and $f^{-1}(x) = 3x - 6$ from Example 2 are shown in **Figure 4.7.** Notice that the graphs do not appear to be quite symmetric about the graph of $y = x$.

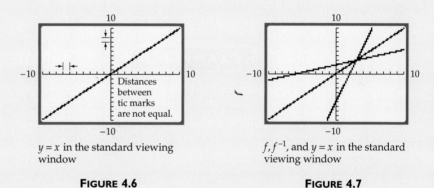

$y = x$ in the standard viewing window

FIGURE 4.6

f, f^{-1}, and $y = x$ in the standard viewing window

FIGURE 4.7

To get a better view of a function and its inverse, it is necessary to use the SQUARE viewing window, as in **Figure 4.8.** In this window, the distance between two tic marks on the x-axis is equal to the distance between two tic marks on the y-axis.

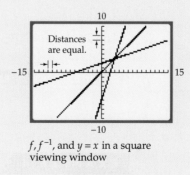

f, f^{-1}, and $y = x$ in a square viewing window

FIGURE 4.8

• FIND AN INVERSE FUNCTION

If a one-to-one function f is defined by an equation, then we can use the following method to find the equation for f^{-1}.

Steps for Finding the Inverse of a Function

To find the equation of the inverse f^{-1} of the one-to-one function f:

1. Substitute y for $f(x)$.

2. Interchange x and y.

3. Solve, if possible, for y in terms of x.

4. Substitute $f^{-1}(x)$ for y.

EXAMPLE 3 **Find the Inverse of a Function**

Find the inverse of $f(x) = 3x + 8$.

Solution

$$f(x) = 3x + 8$$
$$y = 3x + 8 \qquad \text{• Replace } f(x) \text{ by } y.$$
$$x = 3y + 8 \qquad \text{• Interchange } x \text{ and } y.$$
$$x - 8 = 3y \qquad \text{• Solve for } y.$$
$$\frac{x - 8}{3} = y$$
$$\frac{1}{3}x - \frac{8}{3} = f^{-1}(x) \qquad \text{• Replace } y \text{ by } f^{-1}(x).$$

The inverse function is given by $f^{-1}(x) = \dfrac{1}{3}x - \dfrac{8}{3}$.

▶ **TRY EXERCISE 28, PAGE 365**

EXAMPLE 4 **Find the Inverse of a Function**

Find the inverse of $f(x) = \dfrac{2x + 1}{x}$, $x \neq 0$.

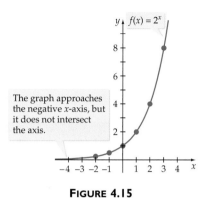

FIGURE 4.15

The graph approaches the negative x-axis, but it does not intersect the axis.

TABLE 4.2

x	$y = f(x) = 2^x$	(x, y)
−2	$f(-2) = 2^{-2} = \dfrac{1}{4}$	$\left(-2, \dfrac{1}{4}\right)$
−1	$f(-1) = 2^{-1} = \dfrac{1}{2}$	$\left(-1, \dfrac{1}{2}\right)$
0	$f(0) = 2^0 = 1$	$(0, 1)$
1	$f(1) = 2^1 = 2$	$(1, 2)$
2	$f(2) = 2^2 = 4$	$(2, 4)$
3	$f(3) = 2^3 = 8$	$(3, 8)$

Note the following properties of the graph of the exponential function $f(x) = 2^x$.

● The y-intercept is $(0, 1)$.

● The graph passes through $(1, 2)$.

● As x decreases without bound (that is, as $x \to -\infty$), $f(x) \to 0$.

● The graph is a smooth continuous increasing curve.

Now consider the graph of an exponential function for which the base is between 0 and 1. The graph of $f(x) = \left(\dfrac{1}{2}\right)^x$ is shown in **Figure 4.16**. The coordinates of some of the points on the curve are given in **Table 4.3**.

TABLE 4.3

x	$y = f(x) = \left(\dfrac{1}{2}\right)^x$	(x, y)
−3	$f(-3) = \left(\dfrac{1}{2}\right)^{-3} = 8$	$(-3, 8)$
−2	$f(-2) = \left(\dfrac{1}{2}\right)^{-2} = 4$	$(-2, 4)$
−1	$f(-1) = \left(\dfrac{1}{2}\right)^{-1} = 2$	$(-1, 2)$
0	$f(0) = \left(\dfrac{1}{2}\right)^0 = 1$	$(0, 1)$
1	$f(1) = \left(\dfrac{1}{2}\right)^1 = \dfrac{1}{2}$	$\left(1, \dfrac{1}{2}\right)$
2	$f(2) = \left(\dfrac{1}{2}\right)^2 = \dfrac{1}{4}$	$\left(2, \dfrac{1}{4}\right)$

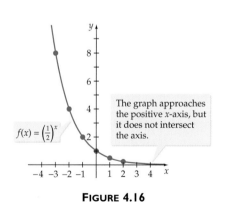

$f(x) = \left(\frac{1}{2}\right)^x$

The graph approaches the positive x-axis, but it does not intersect the axis.

FIGURE 4.16

Note the following properties of the graph of $f(x) = \left(\dfrac{1}{2}\right)^x$ in **Figure 4.16**.

● The y-intercept is $(0, 1)$.

● The graph passes through $\left(1, \dfrac{1}{2}\right)$.

- As x increases without bound, the y-values decrease toward 0. That is, as $x \to \infty$, $f(x) \to 0$.

- The graph is a smooth continuous decreasing curve.

The basic properties of exponential functions are provided in the following summary.

Properties of $f(x) = b^x$

For positive real numbers b, $b \neq 1$, the exponential function defined by $f(x) = b^x$ has the following properties:

1. The function f is a one-to-one function. It has the set of real numbers as its domain and the set of positive real numbers as its range.

2. The graph of f is a smooth continuous curve with a y-intercept of $(0, 1)$, and the graph passes through $(1, b)$.

3. If $b > 1$, f is an increasing function and the graph of f is asymptotic to the negative x-axis. [As $x \to \infty$, $f(x) \to \infty$, and as $x \to -\infty$, $f(x) \to 0$.] See **Figure 4.17a.**

4. If $0 < b < 1$, f is a decreasing function and the graph of f is asymptotic to the positive x-axis. [As $x \to -\infty$, $f(x) \to \infty$, and as $x \to \infty$, $f(x) \to 0$.] See **Figure 4.17b.**

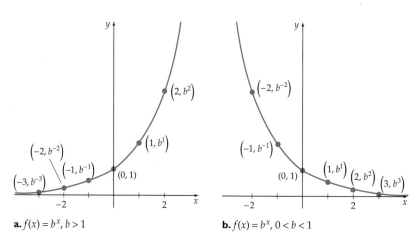

a. $f(x) = b^x$, $b > 1$ **b.** $f(x) = b^x$, $0 < b < 1$

FIGURE 4.17

❓ QUESTION What is the x-intercept of the graph of $f(x) = \left(\dfrac{1}{3}\right)^x$?

❓ ANSWER The graph does not have an x-intercept. As x increases, the graph approaches the x-axis, but it does not intersect the x-axis.

EXAMPLE 2   **Graph an Exponential Function**

Graph $g(x) = \left(\dfrac{3}{4}\right)^x$.

Solution

Because the base $\dfrac{3}{4}$ is less than 1, we know that the graph of g is a decreasing function that is asymptotic to the positive x-axis. The y-intercept of the graph is the point $(0, 1)$, and the graph also passes through $\left(1, \dfrac{3}{4}\right)$.

Plot a few additional points (see **Table 4.4**), and then draw a smooth curve through the points as in **Figure 4.18.**

TABLE 4.4

x	$y = g(x) = \left(\dfrac{3}{4}\right)^x$	(x, y)
-3	$\left(\dfrac{3}{4}\right)^{-3} = \dfrac{64}{27}$	$\left(-3, \dfrac{64}{27}\right)$
-2	$\left(\dfrac{3}{4}\right)^{-2} = \dfrac{16}{9}$	$\left(-2, \dfrac{16}{9}\right)$
-1	$\left(\dfrac{3}{4}\right)^{-1} = \dfrac{4}{3}$	$\left(-1, \dfrac{4}{3}\right)$
2	$\left(\dfrac{3}{4}\right)^{2} = \dfrac{9}{16}$	$\left(2, \dfrac{9}{16}\right)$
3	$\left(\dfrac{3}{4}\right)^{3} = \dfrac{27}{64}$	$\left(3, \dfrac{27}{64}\right)$

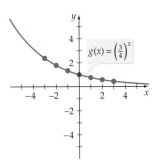

FIGURE 4.18

▶ TRY EXERCISE 22, PAGE 377

Consider the functions $F(x) = 2^x - 3$ and $G(x) = 2^{x-3}$. You can construct the graphs of these functions by plotting points; however, it is easier to construct their graphs by using translations of the graph of $f(x) = 2^x$, as shown in Example 3.

EXAMPLE 3   **Use a Translation to Produce a Graph**

a. Explain how to use the graph of $f(x) = 2^x$ to produce the graph of $F(x) = 2^x - 3$.

b. Explain how to use the graph of $f(x) = 2^x$ to produce the graph of $G(x) = 2^{x-3}$.

Solution

a. $F(x) = 2^x - 3 = f(x) - 3$. The graph of F is a vertical translation of f down 3 units, as shown in **Figure 4.19.**

Continued ▶

b. $G(x) = 2^{x-3} = f(x - 3)$. The graph of G is a horizontal translation of f to the right 3 units, as shown in **Figure 4.20.**

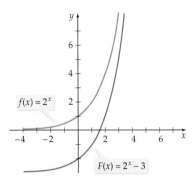

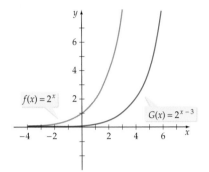

FIGURE 4.19   FIGURE 4.20

▶ **TRY EXERCISE 28, PAGE 377**

The graphs of some functions can be constructed by stretching, compressing, or reflecting the graph of an exponential function.

EXAMPLE 4   **Use Stretching or Reflecting Procedures to Produce a Graph**

a. Explain how to use the graph of $f(x) = 2^x$ to produce the graph of $M(x) = 2(2^x)$.

b. Explain how to use the graph of $f(x) = 2^x$ to produce the graph of $N(x) = 2^{-x}$.

Solution

a. $M(x) = 2(2^x) = 2f(x)$. The graph of M is a vertical stretching of f, as shown in **Figure 4.21.** If (x, y) is a point on the graph of $f(x) = 2^x$, then $(x, 2y)$ is a point on the graph of M.

b. $N(x) = 2^{-x} = f(-x)$. The graph of N is the graph of f reflected across the y-axis, as shown in **Figure 4.22.** If (x, y) is a point on the graph of $f(x) = 2^x$, then $(-x, y)$ is a point on the graph of N.

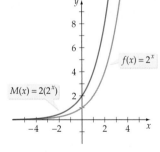

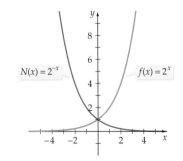

FIGURE 4.21   FIGURE 4.22

▶ **TRY EXERCISE 30, PAGE 377**

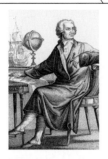

Leonhard Euler (1707–1783)

Some mathematicians consider Euler to be the greatest mathematician of all time. He certainly was the most prolific writer of mathematics of all time. He made substantial contributions in the areas of number theory, geometry, calculus, differential equations, differential geometry, topology, complex variables, and analysis, to name but a few. Euler was the first to introduce many of the mathematical notations that we use today. For instance, he introduced the symbol i for the square root of -1, the symbol π for pi, the functional notation $f(x)$, and the letter e for the base of the natural exponential function. Euler's computational skills were truly amazing. The mathematician François Arago remarked, "Euler calculated without apparent effort, as men breathe, or as eagles sustain themselves in the wind."

● THE NATURAL EXPONENTIAL FUNCTION

The irrational number π is often used in applications that involve circles. Another irrational number, denoted by the letter e, is useful in applications that involve growth or decay.

Definition of e

The **number e** is defined as the number that

$$\left(1 + \frac{1}{n}\right)^n$$

approaches as n increases without bound.

The letter e was chosen in honor of the Swiss mathematician Leonhard Euler. He was able to compute the value of e to several decimal places by evaluating $\left(1 + \frac{1}{n}\right)^n$ for large values of n, as shown in **Table 4.5.**

TABLE 4.5

Value of n	Value of $\left(1 + \dfrac{1}{n}\right)^n$
1	2
10	2.59374246
100	2.704813829
1000	2.716923932
10,000	2.718145927
100,000	2.718268237
1,000,000	2.718280469
10,000,000	2.718281693

The value of e accurate to eight decimal places is 2.71828183.

The Natural Exponential Function

For all real numbers x, the function defined by

$$f(x) = e^x$$

is called the **natural exponential function.**

A calculator can be used to evaluate e^x for specific values of x. For instance,

$$e^2 \approx 7.389056, \quad e^{3.5} \approx 33.115452, \quad \text{and} \quad e^{-1.4} \approx 0.246597$$

On a TI-83 calculator the e^x function is located above the $\boxed{\text{LN}}$ key.

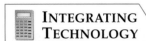

INTEGRATING TECHNOLOGY

The graph below was produced on a TI-83 graphing calculator by entering e^x in the Y= menu.

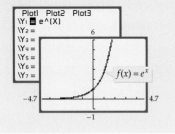

To graph $f(x) = e^x$, use a calculator to find the range values for a few domain values. The range values in **Table 4.6** have been rounded to the nearest tenth.

TABLE 4.6

x	−2	−1	0	1	2
$f(x) = e^x$	0.1	0.4	1.0	2.7	7.4

Plot the points given in **Table 4.6,** and then connect the points with a smooth curve. Because $e > 1$, we know that the graph is an increasing function. To the far left, the graph will approach the x-axis. The y-intercept is $(0, 1)$. See **Figure 4.23.** Note in **Figure 4.24** how the graph of $f(x) = e^x$ compares with the graphs of $g(x) = 2^x$ and $h(x) = 3^x$. You may have anticipated that the graph of $f(x) = e^x$ would lie between the two other graphs because e is between 2 and 3.

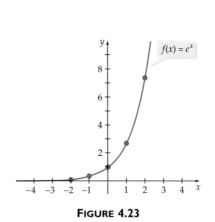

FIGURE 4.23

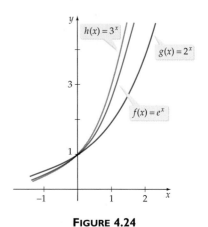

FIGURE 4.24

• APPLICATIONS OF EXPONENTIAL FUNCTIONS

Many applications can be effectively modeled by functions that involve an exponential function. For instance, in Example 5 we make use of a function that involves an exponential function to model the temperature of a cup of coffee.

EXAMPLE 5   **Use a Mathematical Model**

A cup of coffee is heated to 160°F and placed in a room that maintains a temperature of 70°F. The temperature T of the coffee, in degrees Fahrenheit, after t minutes is given by

$$T = 70 + 90e^{-0.0485t}$$

a. Find the temperature of the coffee, to the nearest degree, 20 minutes after it is placed in the room.

b. Use a graphing utility to determine when the temperature of the coffee will reach 90°F.

Solution

a. $T = 70 + 90e^{-0.0485t}$

   $\quad = 70 + 90e^{-0.0485 \cdot (20)}$ • Substitute 20 for *t*.

   $\quad \approx 70 + 34.1$

   $\quad \approx 104.1$

   After 20 minutes the temperature of the coffee is about 104°F.

b. Graph $T = 70 + 90e^{-0.0485t}$ and $T = 90$. See the following figure.

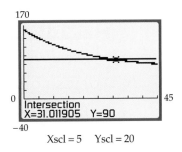

The graphs intersect at about (31.01, 90). It takes the coffee about 31 minutes to cool to 90°F.

▶ **TRY EXERCISE 44, PAGE 378**

EXAMPLE 6   **Use a Mathematical Model**

 The weekly revenue *R*, in dollars, from the sale of a product varies with time according to the function

$$R(x) = \frac{1760}{8 + 14e^{-0.03x}}$$

where *x* is the number of weeks that have passed since the product was put on the market. What will the weekly revenue approach as time goes by?

Solution

Method 1   Use a graphing utility to graph *R*(*x*), and use the TRACE feature to see what happens to the revenue as the time increases. The following graph shows that as the weeks go by, the weekly revenue will increase and approach $220.00 per week.

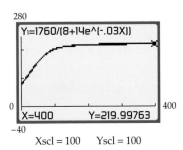

Continued ▶

Method 2   Write the revenue function in the following form.

$$R(x) = \frac{1760}{8 + \dfrac{14}{e^{0.03x}}} \quad \bullet \; 14e^{-0.03x} = \frac{14}{e^{0.03x}}$$

As x increases without bound, $e^{0.03x}$ increases without bound, and the fraction $\dfrac{14}{e^{0.03x}}$ approaches 0. Therefore, as $x \to \infty$, $R(x) \to \dfrac{1760}{8 + 0} = 220$. Both methods indicate that as the number of weeks increases, the revenue approaches $220 per week.

▶ **TRY EXERCISE 54, PAGE 379**

TOPICS FOR DISCUSSION

1. Explain how to use the graph of $f(x) = 2^x$ to produce the graph of $g(x) = 2^{(x-3)} + 4$.

2. At what point does the function $g(x) = e^{-x^2/2}$ take on its maximum value?

3. Without using a graphing utility, determine whether the revenue function $R(t) = 10 + e^{-0.05t}$ is an increasing function or a decreasing function.

4. Discuss the properties of the graph of $f(x) = b^x$ when $b > 1$.

5. What is the base of the natural exponential function? How is it calculated? What is its approximate value?

EXERCISE SET 4.2

In Exercises 1 to 8, evaluate the exponential function for the given x-values.

1. $f(x) = 3^x$; $x = 0$ and $x = 4$

▶ 2. $f(x) = 5^x$; $x = 3$ and $x = -2$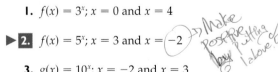

3. $g(x) = 10^x$; $x = -2$ and $x = 3$

4. $g(x) = 4^x$; $x = 0$ and $x = -1$

5. $h(x) = \left(\dfrac{3}{2}\right)^x$; $x = 2$ and $x = -3$

6. $h(x) = \left(\dfrac{2}{5}\right)^x$; $x = -1$ and $x = 3$

7. $j(x) = \left(\dfrac{1}{2}\right)^x$; $x = -2$ and $x = 4$

8. $j(x) = \left(\dfrac{1}{4}\right)^x$; $x = -1$ and $x = 5$

In Exercises 9 to 14, use a calculator to evaluate the exponential function for the given x-value. Round to the nearest hundredth.

9. $f(x) = 2^x$, $x = 3.2$

10. $f(x) = 3^x$, $x = -1.5$

11. $g(x) = e^x$, $x = 2.2$

12. $g(x) = e^x$, $x = -1.3$

13. $h(x) = 5^x$, $x = \sqrt{2}$

14. $h(x) = 0.5^x$, $x = \pi$

15. Examine the following four functions and the graphs labeled **a, b, c,** and **d.** For each graph, determine which function has been graphed.

$$f(x) = 5^x \qquad g(x) = 1 + 5^{-x}$$
$$h(x) = 5^{x+3} \qquad k(x) = 5^x + 3$$

a.
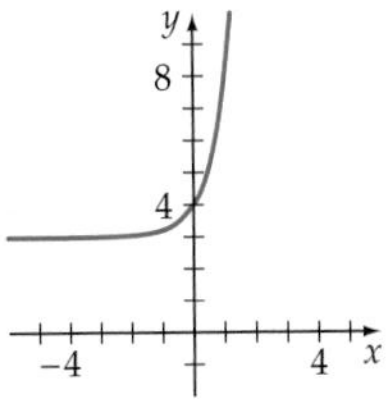

b.
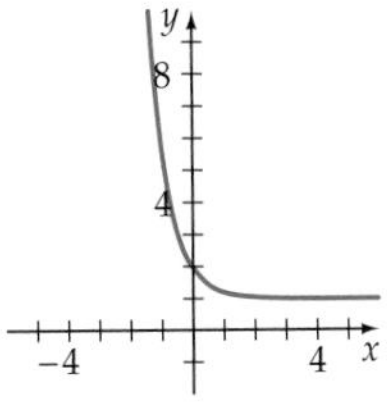

c.
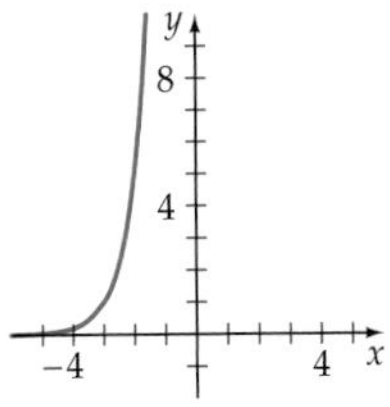

d.
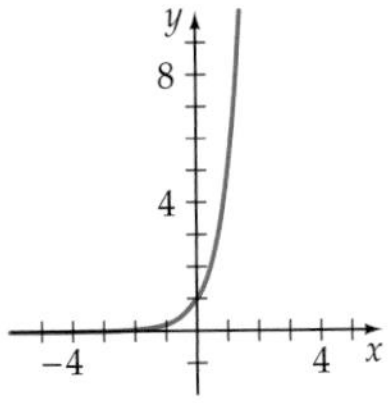

16. Examine the following four functions and the graphs labeled **a, b, c,** and **d.** For each graph, determine which function has been graphed.

$$f(x) = \left(\frac{1}{4}\right)^x \qquad g(x) = \left(\frac{1}{4}\right)^{-x}$$
$$h(x) = \left(\frac{1}{4}\right)^{x-2} \qquad k(x) = 3\left(\frac{1}{4}\right)^x$$

a.
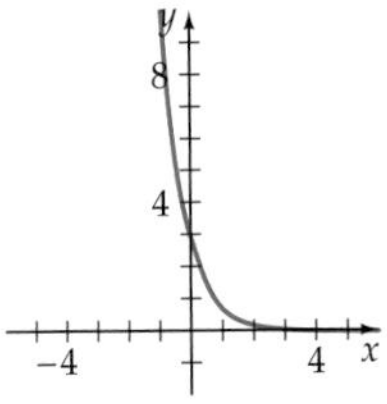

b.
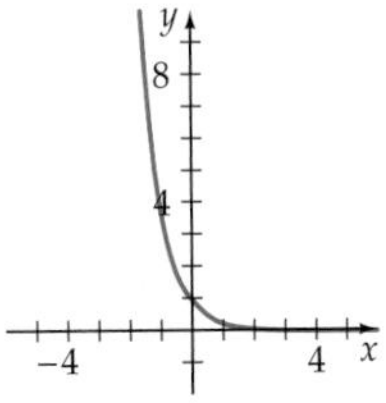

c.
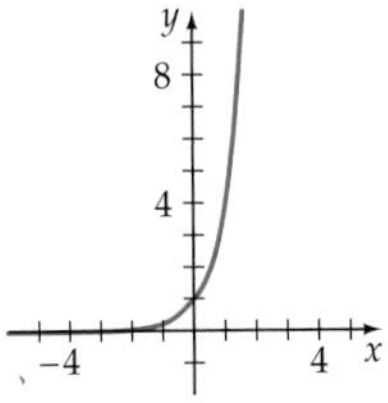

d.
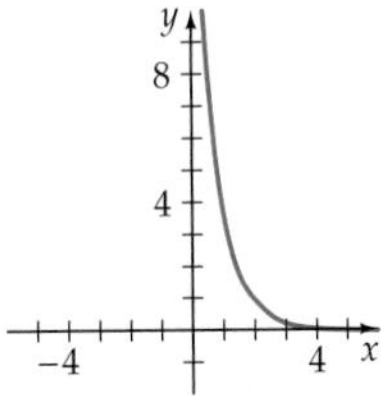

In Exercises 17 to 24, sketch the graph of each function.

17. $f(x) = 3^x$

18. $f(x) = 4^x$

19. $f(x) = 10^x$

20. $f(x) = 6^x$

21. $f(x) = \left(\frac{3}{2}\right)^x$

▶ **22.** $f(x) = \left(\frac{5}{2}\right)^x$

23. $f(x) = \left(\frac{1}{3}\right)^x$

24. $f(x) = \left(\frac{2}{3}\right)^x$

In Exercises 25 to 34, explain how to use the graph of the first function *f* to produce the graph of the second function *F*.

25. $f(x) = 3^x$, $F(x) = 3^x + 2$

26. $f(x) = 4^x$, $F(x) = 4^x - 3$

27. $f(x) = 10^x$, $F(x) = 10^{x-2}$

▶ **28.** $f(x) = 6^x$, $F(x) = 6^{x+5}$

29. $f(x) = \left(\frac{3}{2}\right)^x$, $F(x) = \left(\frac{3}{2}\right)^{-x}$

▶ **30.** $f(x) = \left(\frac{5}{2}\right)^x$, $F(x) = -\left[\left(\frac{5}{2}\right)^x\right]$

31. $f(x) = \left(\frac{1}{3}\right)^x$, $F(x) = 2\left[\left(\frac{1}{3}\right)^x\right]$

32. $f(x) = \left(\frac{2}{3}\right)^x$, $F(x) = \frac{1}{2}\left[\left(\frac{2}{3}\right)^x\right]$

33. $f(x) = e^x$, $F(x) = e^{-x} + 2$

34. $f(x) = e^x$, $F(x) = e^{x-3} + 1$

In Exercises 35 to 42, use a graphing utility to graph each function. If the function has a horizontal asymptote, state the equation of the horizontal asymptote.

35. $f(x) = \dfrac{3^x + 3^{-x}}{2}$

36. $f(x) = 4 \cdot 3^{-x^2}$

37. $f(x) = \dfrac{e^x - e^{-x}}{2}$

38. $f(x) = \dfrac{e^x + e^{-x}}{2}$

39. $f(x) = -e^{(x-4)}$

40. $f(x) = 0.5e^{-x}$

41. $f(x) = \dfrac{10}{1 + 0.4e^{-0.5x}}$, $x \geq 0$

42. $f(x) = \dfrac{10}{1 + 1.5e^{-0.5x}}$, $x \geq 0$

43. INTERNET CONNECTIONS Data from Forrester Research suggest that the number of broadband [cable and digital subscriber line (DSL)] connections to the Internet can be modeled by $f(x) = 1.353(1.9025)^x$, where x is the number of years after January 1, 1998, and $f(x)$ is the number of connections in millions.

a. How many broadband Internet connections, to the nearest million, does this model predict will exist on January 1, 2005?

b. According to the model, in what year will the number of broadband connections first reach 300 million? [*Hint:* Use the intersect feature of a graphing utility to determine the x-coordinate of the point of intersection of the graphs of $f(x)$ and $y = 300$.]

▶ **44.** MEDICATION IN BLOODSTREAM The function $A(t) = 200e^{-0.014t}$ gives the amount of medication, in milligrams, in a patient's bloodstream t minutes after the medication has been injected into the patient's bloodstream.

a. Find the amount of medication, to the nearest milligram, in the patient's bloodstream after 45 minutes.

b. Use a graphing utility to determine how long it will take, to the nearest minute, for the amount of medication in the patient's bloodstream to reach 50 milligrams.

45. DEMAND FOR A PRODUCT The demand d for a specific product, in items per month, is given by

$$d(p) = 25 + 880e^{-0.18p}$$

where p is the price, in dollars, of the product.

a. What will be the monthly demand, to the nearest unit, when the price of the product is $8 and when the price is $18?

b. What will happen to the demand as the price increases without bound?

46. SALES The monthly income I, in dollars, from a new product is given by

$$I(t) = 24,000 - 22,000e^{-0.005t}$$

where t is the time, in months, since the product was first put on the market.

a. What was the monthly income after the 10th month and after the 100th month?

b. What will the monthly income from the product approach as the time increases without bound?

47. A PROBABILITY FUNCTION The manager of a home improvement store finds that between 10 A.M. and 11 A.M., customers enter the store at the average rate of 45 customers per hour. The following function gives the probability that a customer will arrive within t minutes of 10 A.M. (*Note:* A probability of 0.6 means there is a 60% chance that a customer will arrive during a given time period.)

$$P(t) = 1 - e^{-0.75t}$$

a. Find the probability, to the nearest hundredth, that a customer will arrive within 1 minute of 10 A.M.

b. Find the probability, to the nearest hundredth, that a customer will arrive within 3 minutes of 10 A.M.

c. Use a graph of $P(t)$ to determine how many minutes, to the nearest tenth of a minute, it takes for $P(t)$ to equal 98%.

d. Write a sentence that explains the meaning of the answer in part **c.**

48. A PROBABILITY FUNCTION The owner of a sporting goods store finds that between 9 A.M. and 10 A.M., customers enter the store at the average rate of 12 customers per hour. The following function gives the probability that a customer will arrive within t minutes of 9 A.M.

$$P(t) = 1 - e^{-0.2t}$$

a. Find the probability, to the nearest hundredth, that a customer will arrive within 5 minutes of 9 A.M.

b. Find the probability, to the nearest hundredth, that a customer will arrive within 15 minutes of 9 A.M.

c. Use a graph of $P(t)$ to determine how many minutes, to the nearest 0.1 minute, it takes for $P(t)$ to equal 90%.

d. Write a sentence that explains the meaning of the answer in part **c.**

Exercises 49 and 50 involve the factorial function $x!$, which is defined for whole numbers x as

$$x! = \begin{cases} 1, & \text{if } x = 0 \\ x \cdot (x - 1) \cdot (x - 2) \cdot \cdots \cdot 3 \cdot 2 \cdot 1, & \text{if } x \geq 1 \end{cases}$$

For example, $3! = 3 \cdot 2 \cdot 1 = 6$ and $5! = 5 \cdot 4 \cdot 3 \cdot 2 \cdot 1 = 120$.

49. QUEUING THEORY During the 30-minute period before a Broadway play begins, the members of the audience arrive at the theater at the average rate of 12 people per minute. The probability that x people will arrive during a particular minute is given by $P(x) = \dfrac{12^x e^{-12}}{x!}$. Find the probability, to the nearest 0.1%, that

a. 9 people will arrive during a given minute.

b. 18 people will arrive during a given minute.

50. QUEUING THEORY During the period from 2:00 P.M. to 3:00 P.M., a bank finds that an average of seven people enter the bank every minute. The probability that x people will enter the bank during a particular minute is given by $P(x) = \dfrac{7^x e^{-7}}{x!}$. Find the probability, to the nearest 0.1%, that

a. only two people will enter the bank during a given minute.

b. 11 people will enter the bank during a given minute.

51. 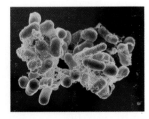 *E. COLI* INFECTION *Escherichia coli* (*E. coli*) is a bacterium that can reproduce at an exponential rate. The *E. coli* reproduce by dividing. A small number of *E. coli* bacteria in the large intestine of a human can trigger a serious infection within a few hours. Consider a particular *E. coli* infection that starts with 100 *E. coli* bacteria. Each bacterium splits into two parts every half hour. Assuming none of the bacteria die, the size of the *E. coli* population after t hours is given by $P(t) = 100 \cdot 2^{2t}$, where $0 \le t \le 16$.

a. Find $P(3)$ and $P(6)$.

b. Use a graphing utility to find the time, to the nearest tenth of an hour, it takes for the *E. coli* population to number 1 billion.

52. RADIATION Lead shielding is used to contain radiation. The percentage of a certain radiation that can penetrate x millimeters of lead shielding is given by $I(x) = 100e^{-1.5x}$.

a. What percentage of radiation, to the nearest tenth of a percent, will penetrate a lead shield that is 1 millimeter thick?

b. How many millimeters of lead shielding are required so that less than 0.05% of the radiation penetrates the shielding? Round to the nearest millimeter.

53. 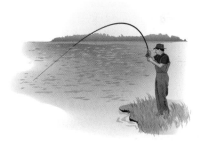 AIDS An exponential function that approximates the number of people in the United States who have been infected with AIDS is given by $N(t) = 138{,}000(1.39)^t$, where t is the number of years after January 1, 1990.

a. According to this function, how many people had been infected with AIDS as of January 1, 1994? Round to the nearest thousand.

b. Use a graph to estimate during what year the number of people in the United States who had been infected with AIDS first reached 1.5 million.

▶ **54.** FISH POPULATION The number of bass in a lake is given by

$$P(t) = \frac{3600}{1 + 7e^{-0.05t}},$$

where t is the number of months that have passed since the lake was stocked with bass.

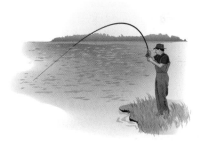

a. How many bass were in the lake immediately after it was stocked?

b. How many bass were in the lake 1 year after the lake was stocked?

c. What will happen to the bass population as t increases without bound?

55. THE PAY IT FORWARD MODEL In the movie *Pay It Forward*, Trevor McKinney, played by Haley Joel Osment, is given a school assignment to "think of an idea to change the world—and then put it into action." In response to this assignment, Trevor develops a *pay it forward* project. In this project, anyone who benefits from another person's good deed must do a good deed for

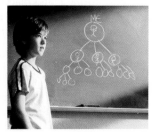

three additional people. Each of these three people is then obligated to do a good deed for another three people, and so on.

The following diagram shows the number of people who have been a beneficiary of a good deed after 1 round and after 2 rounds of this project.

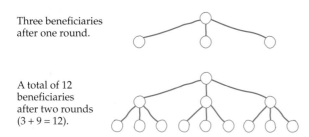

Three beneficiaries after one round.

A total of 12 beneficiaries after two rounds $(3 + 9 = 12)$.

A mathematical model for the number of pay it forward beneficiaries after n rounds is given by $B(n) = \dfrac{3^{n+1} - 3}{2}$. Use this model to determine

a. the number of beneficiaries after 5 rounds and after 10 rounds. Assume that no person is a beneficiary of more than one good deed.

b. how many rounds are required to produce at least 2 million beneficiaries.

56. **INTENSITY OF LIGHT** The percent $I(x)$ of the original intensity of light striking the surface of a lake that is available x feet below the surface of the lake is given by $I(x) = 100e^{-0.95x}$.

a. What percentage of the light, to the nearest tenth of a percent, is available 2 feet below the surface of the lake?

b. At what depth, to the nearest hundredth of a foot, is the intensity of the light one-half the intensity at the surface?

57. **A TEMPERATURE MODEL** A cup of coffee is heated to 180°F and placed in a room that maintains a temperature of 65°F. The temperature of the coffee after t minutes is given by $T(t) = 65 + 115e^{-0.042t}$.

a. Find the temperature, to the nearest degree, of the coffee 10 minutes after it is placed in the room.

b. Use a graphing utility to determine when, to the nearest tenth of a minute, the temperature of the coffee will reach 100°F.

58. **A TEMPERATURE MODEL** Soup that is at a temperature of 170°F is poured into a bowl in a room that maintains a constant temperature. The temperature of the soup decreases according to the model given by $T(t) = 75 + 95e^{-0.12t}$, where t is time in minutes after the soup is poured.

a. What is the temperature, to the nearest tenth of a degree, of the soup after 2 minutes?

b. A certain customer prefers soup at a temperature of 110°F. How many minutes, to the nearest 0.1 minute, after the soup is poured does the soup reach that temperature?

c. What is the temperature of the room?

59. **MUSICAL SCALES** Starting on the left side of a standard 88-key piano, the frequency, in vibrations per second, of the nth note is given by $f(n) = (27.5)2^{(n-1)/12}$.

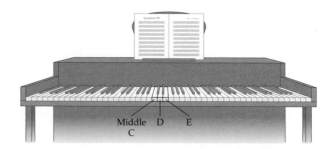

Middle C   D   E

a. Using this formula, determine the frequency, to the nearest hundredth of a vibration per second, of middle C, key number 40 on an 88-key piano.

b. Is the difference in frequency between middle C (key number 40) and D (key number 42) the same as the difference in frequency between D (key number 42) and E (key number 44)? Explain.

CONNECTING CONCEPTS

60. Verify that the hyperbolic cosine function

$\cosh(x) = \dfrac{e^x + e^{-x}}{2}$ is an even function.

61. Verify that the hyperbolic sine function $\sinh(x) = \dfrac{e^x - e^{-x}}{2}$

is an odd function.

62. Graph $g(x) = 10^x$, and then sketch the graph of g reflected across the line given by $y = x$.

63. Graph $f(x) = e^x$, and then sketch the graph of f reflected across the line given by $y = x$.

In Exercises 64 to 67, determine the domain of the given function. Write the domain using interval notation.

64. $f(x) = \dfrac{e^x - e^{-x}}{e^x + e^{-x}}$

65. $f(x) = \dfrac{e^{|x|}}{1 + e^x}$

66. $f(x) = \sqrt{1 - e^x}$

67. $f(x) = \sqrt{e^x - e^{-x}}$

PREPARE FOR SECTION 4.3

68. If $2^x = 16$, determine the value of x. [4.2]

69. If $3^{-x} = \dfrac{1}{27}$, determine the value of x. [4.2]

70. If $x^4 = 625$, determine the value of x. [4.2]

71. Find the inverse of $f(x) = \dfrac{2x}{x + 3}$. [4.1]

72. State the domain of $g(x) = \sqrt{x - 2}$. [2.2]

73. If the range of $h(x)$ is the set of all positive real numbers, then what is the domain of $h^{-1}(x)$? [4.2]

PROJECTS

1. THE SAINT LOUIS GATEWAY ARCH The Gateway Arch in Saint Louis was designed in the shape of an inverted **catenary,** as shown by the red curve in the drawing at the right. The Gateway Arch is one of the largest optical illusions ever created. As you look at the arch (and its basic shape defined by the catenary curve), it appears to be much taller than it is wide. However, this is not the case. The height of the catenary is given by

$$h(x) = 693.8597 - 68.7672\left(\dfrac{e^{0.0100333x} + e^{-0.0100333x}}{2}\right)$$

where x and $h(x)$ are measured in feet and $x = 0$ represents the position at ground level that is directly below the highest point of the catenary.

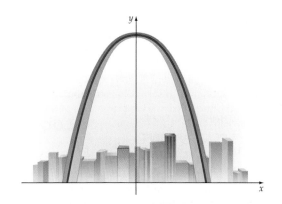

a. Use a graphing utility to graph $h(x)$.

b. Use your graph to find the height of the catenary for $x = 0$, 100, 200, and 299 feet. Round each result to the nearest tenth of a foot.

c. What is the width of the catenary at ground level and what is the maximum height of the catenary? Round each result to the nearest tenth of a foot.

d. By how much does the maximum height of the catenary exceed its width at ground level? Round to the nearest tenth of a foot.

2. **AN EXPONENTIAL REWARD** According to legend, when Sissa Ben Dahir of India invented the game of chess, King Shirham was so impressed with the game that he summoned the game's inventor and offered him the reward of his choosing. The inventor pointed to the chessboard and requested, for his reward, one grain of wheat on the first square, two grains of wheat on the second square, four grains on the third square, eight grains on the fourth square, and so on for all 64 squares on the chessboard. The King considered this a very modest reward and said he would grant the inventor's wish. The following table shows how many grains of rice are on each of the first six squares and the total number of grains of wheat needed to cover squares 1 to n for $n \leq 6$.

Square number, n	Number of grains of wheat on square n	Total number of grains of wheat on squares 1 through n
1	1	1
2	2	3
3	4	7
4	8	15
5	16	31
6	32	63

a. If all 64 squares of the chessboard are piled with wheat as requested by Sissa Ben Dahir, how many grains of wheat are on the board?

b. A grain of wheat weighs approximately 0.000008 kilogram. Find the total weight of the wheat requested by Sissa Ben Dahir.

c. In a recent year, a total of 6.5×10^8 metric tons of wheat were produced in the world. At this level, how many years, to the nearest year, of wheat production would be required to fill the request of Sissa Ben Dahir? One metric ton equals 1000 kilograms.

SECTION 4.3

LOGARITHMIC FUNCTIONS AND THEIR APPLICATIONS

- **LOGARITHMIC FUNCTIONS**
- **GRAPHS OF LOGARITHMIC FUNCTIONS**
- **DOMAINS OF LOGARITHMIC FUNCTIONS**
- **COMMON AND NATURAL LOGARITHMS**
- **APPLICATIONS OF LOGARITHMIC FUNCTIONS**

• LOGARITHMIC FUNCTIONS

Every exponential function of the form $g(x) = b^x$ is a one-to-one function and therefore has an inverse function. Sometimes we can determine the inverse of a function represented by an equation by interchanging the variables of its equation and then solving for the dependent variable. If we attempt to use this procedure for $g(x) = b^x$, we obtain

$$g(x) = b^x$$
$$y = b^x$$
$$x = b^y \qquad \bullet \text{ Interchange the variables.}$$

None of our previous methods can be used to solve the equation $x = b^y$ for the exponent y. Thus we need to develop a new procedure. One method would be to merely write

$$y = \text{the power of } b \text{ that produces } x$$

Although this would work, it is not very concise. We need a compact notation to represent "y is the power of b that produces x." This more compact notation is given in the following definition.

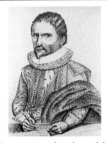
Definition of a Logarithm and a Logarithmic Function

If $x > 0$ and b is a positive constant ($b \neq 1$), then

$$y = \log_b x \quad \text{if and only if} \quad b^y = x$$

The notation $\log_b x$ is read "the **logarithm** (or log) base b of x." The function defined by $f(x) = \log_b x$ is a **logarithmic function** with base b. This function is the inverse of the exponential function $g(x) = b^x$.

It is essential to remember that $f(x) = \log_b x$ is the inverse function of $g(x) = b^x$. Because these functions are inverses and because functions that are inverses have the property that $f(g(x)) = x$ and $g(f(x)) = x$, we have the following important relationships.

Composition of Logarithmic and Exponential Functions

Let $g(x) = b^x$ and $f(x) = \log_b x$ $(x > 0, b > 0, b \neq 1)$. Then

$$g(f(x)) = b^{\log_b x} = x \quad \text{and} \quad f(g(x)) = \log_b b^x = x$$

As an example of these relationships, let $g(x) = 2^x$ and $f(x) = \log_2 x$. Then

$$2^{\log_2 x} = x \quad \text{and} \quad \log_2 2^x = x$$

The equations

$$y = \log_b x \quad \text{and} \quad b^y = x$$

are different ways of expressing the same concept.

> **take note**
>
> The notation $\log_b x$ replaces the phrase "the power of b that produces x." For instance, "3 is the power of 2 that produces 8" is abbreviated $3 = \log_2 8$. In your work with logarithms, remember that a logarithm is an *exponent*.

Exponential Form and Logarithmic Form

The **exponential form** of $y = \log_b x$ is $b^y = x$.

The **logarithmic form** of $b^y = x$ is $y = \log_b x$.

These concepts are illustrated in the next two examples.

EXAMPLE 1   **Change from Logarithmic to Exponential Form**

Write each equation in its exponential form.

a. $3 = \log_2 8$ b. $2 = \log_{10}(x + 5)$ c. $\log_e x = 4$ d. $\log_b b^3 = 3$

Solution

Use the definition $y = \log_b x$ if and only if $b^y = x$.

┌─── Logarithms are exponents. ───┐

a. $3 = \log_2 8$ if and only if $2^3 = 8$

└──────── Base ────────┘

b. $2 = \log_{10}(x + 5)$ if and only if $10^2 = x + 5$.

c. $\log_e x = 4$ if and only if $e^4 = x$.

d. $\log_b b^3 = 3$ if and only if $b^3 = b^3$.

▶ **TRY EXERCISE 4, PAGE 391**

EXAMPLE 2   **Change from Exponential to Logarithmic Form**

Write each equation in its logarithmic form.

a. $3^2 = 9$ b. $5^3 = x$ c. $a^b = c$ d. $b^{\log_b 5} = 5$

Solution

The logarithmic form of $b^y = x$ is $y = \log_b x$.

┌──── Exponent ────┐

a. $3^2 = 9$ if and only if $2 = \log_3 9$

└──────── Base ────────┘

b. $5^3 = x$ if and only if $3 = \log_5 x$.

c. $a^b = c$ if and only if $b = \log_a c$.

d. $b^{\log_b 5} = 5$ if and only if $\log_b 5 = \log_b 5$.

▶ **TRY EXERCISE 12, PAGE 391**

The definition of a logarithm and the definition of inverse functions can be used to establish many properties of logarithms. For instance:

- $\log_b b = 1$ because $b = b^1$.

- $\log_b 1 = 0$ because $1 = b^0$.

- $\log_b(b^x) = x$ because $b^x = b^x$.

- $b^{\log_b x} = x$ because $f(x) = \log_b x$ and $g(x) = b^x$ are inverse functions. Thus $g[f(x)] = x$.

▶ **50.** HYDRONIUM-ION CONCENTRATION A rainstorm in New York City produced rainwater with a pH of 5.6. Determine the hydronium-ion concentration of the rainwater.

51. DECIBEL LEVEL The range of sound intensities that the human ear can detect is so large that a special decibel scale (named after Alexander Graham Bell) is used to measure and compare sound intensities. The **decibel level** dB of a sound is given by

$$dB(I) = 10 \log \left(\frac{I}{I_0} \right)$$

where I_0 is the intensity of sound that is barely audible to the human ear. Find the decibel level for the following sounds. Round to the nearest tenth of a decibel.

Sound	Intensity
a. Automobile traffic	$I = 1.58 \times 10^8 \cdot I_0$
b. Quiet conversation	$I = 10{,}800 \cdot I_0$
c. Fender guitar	$I = 3.16 \times 10^{11} \cdot I_0$
d. Jet engine	$I = 1.58 \times 10^{15} \cdot I_0$

52. COMPARISON OF SOUND INTENSITIES A team in Arizona installed a 48,000-watt sound system in a Ford Bronco that it claims can output 175-decibel sound. The human pain threshold for sound is 125 decibels. How many times more intense is the sound from the Bronco than the human pain threshold?

53. COMPARISON OF SOUND INTENSITIES How many times more intense is a sound that measures 120 decibels than a sound that measures 110 decibels?

54. DECIBEL LEVEL If the intensity of a sound is doubled, what is the increase in the decibel level? (*Hint:* Find $dB(2I) - dB(I)$.)

55. EARTHQUAKE MAGNITUDE What is the Richter scale magnitude of an earthquake with an intensity of $I = 100{,}000 I_0$?

▶ **56.** EARTHQUAKE MAGNITUDE The Colombia earthquake of 1906 had an intensity of $I = 398{,}107{,}000 I_0$. What did it measure on the Richter scale?

57. EARTHQUAKE INTENSITY The Coalinga, California, earthquake of 1983 had a Richter scale magnitude of 6.5. Find the intensity of this earthquake.

▶ **58.** EARTHQUAKE INTENSITY The earthquake that occurred just south of Concepción, Chile, in 1960 had a Richter scale magnitude of 9.5. Find the intensity of this earthquake.

59. COMPARISON OF EARTHQUAKES Compare the intensity of an earthquake that measures 5.0 on the Richter scale to the intensity of an earthquake that measures 3.0 on the Richter scale by finding the ratio of the larger intensity to the smaller intensity.

▶ **60.** COMPARISON OF EARTHQUAKES How many times more intense was the 1960 earthquake in Chile, which measured 9.5 on the Richter scale, than the San Francisco earthquake of 1906, which measured 8.3 on the Richter scale?

61. COMPARISON OF EARTHQUAKES On March 2, 1933, an earthquake of magnitude 8.9 on the Richter scale struck Japan. In October 1989, an earthquake of magnitude 7.1 on the Richter scale struck San Francisco, California. Compare the intensity of the larger earthquake to the intensity of the smaller earthquake by finding the ratio of the larger intensity to the smaller intensity.

62. COMPARISON OF EARTHQUAKES An earthquake that occurred in China in 1978 measured 8.2 on the Richter scale. In 1988, an earthquake in California measured 6.9 on the Richter scale. Compare the intensity of the larger earthquake to the intensity of the smaller earthquake by finding the ratio of the larger intensity to the smaller intensity.

63. EARTHQUAKE MAGNITUDE Find the Richter scale magnitude of the earthquake that produced the seismogram in the following figure.

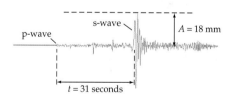

▶ **64.** EARTHQUAKE MAGNITUDE Find the Richter scale magnitude of the earthquake that produced the seismogram in the following figure.

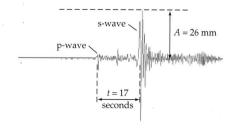

CONNECTING CONCEPTS

65. NOMOGRAMS AND LOGARITHMIC SCALES A **nomogram** is a diagram used to determine a numerical result by drawing a line across numerical scales. The following nomogram, used by Richter, determines the magnitude of an earthquake from its seismogram. To use the nomogram, mark the amplitude of a seismogram on the amplitude scale and mark the time between the s-wave and the p-wave on the S-P scale. Draw a line between these marks. The Richter scale magnitude of the earthquake that produced the seismogram is shown by the intersection of the line and the center scale. The example below shows that an earthquake with a seismogram amplitude of 23 millimeters and an S-P time of 24 seconds has a Richter scale magnitude of about 5.

The amplitude and the S-P time are shown on logarithmic scales. On the amplitude scale, the distance from 1 to 10 is the same as the distance from 10 to 100, because log 100 − log 10 = log 10 − log 1.

Use the nomogram at the left to determine the Richter scale magnitude of an earthquake with a seismogram

a. amplitude of 50 millimeters and S-P time of 40 seconds.

b. amplitude of 1 millimeter and S-P time of 30 seconds.

c. How do the results in parts **a.** and **b.** compare with the Richter scale magnitude produced by using the amplitude-time-difference formula?

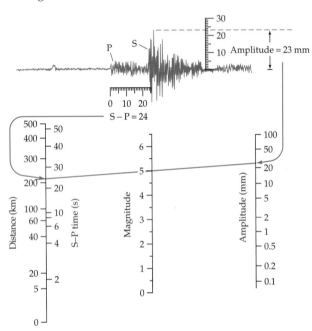

Richter's earthquake nomogram

PREPARE FOR SECTION 4.5

66. Use the definition of a logarithm to write the exponential equation $3^6 = 729$ in logarithmic form. [4.2]

67. Use the definition of a logarithm to write the logarithmic equation $\log_5 625 = 4$ in exponential form. [4.2]

68. Use the definition of a logarithm to write the exponential equation $a^{x+2} = b$ in logarithmic form. [4.2]

69. Solve for x: $4a = 7bx + 2cx$. [1.2]

70. Solve for x: $165 = \dfrac{300}{1 + 12x}$. [1.4]

71. Solve for x: $A = \dfrac{100 + x}{100 - x}$. [1.4]

PROJECTS

1. LOGARITHMIC SCALES Sometimes **logarithmic scales** are used to better view a collection of data that span a wide range of values. For instance, consider the table below, which lists the approximate masses of various marine creatures in grams. Next we have attempted to plot the masses on a number line.

Animal	Mass (g)
Rotifer	0.000000006
Dwarf goby	0.30
Lobster	15,900
Leatherback turtle	851,000
Giant squid	1,820,000
Whale shark	4,700,000
Blue whale	120,000,000

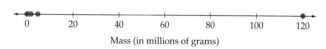

Mass (in millions of grams)

As you can see, we had to use such a large span of numbers that the data for most of the animals are bunched up at the left. Visually, this number line isn't very helpful for any comparisons.

a. Make a new number line, this time plotting the logarithm (base 10) of each of the masses.

b. Which number line is more helpful to compare the masses of the different animals?

c. If the data points for two animals on the logarithmic number line are 1 unit apart, how do the animals' masses compare? What if the points are 2 units apart?

2. LOGARITHMIC SCALES The distances of the planets in our solar system from the sun are given in the table at the top of the next column.

a. Draw a number line with an appropriate scale to plot the distances.

b. Draw a second number line, this time plotting the logarithm (base 10) of each distance.

c. Which number line do you find more helpful to compare the different distances?

d. If two distances are 3 units apart on the logarithmic number line, how do the distances of the corresponding planets compare?

Planet	Distance (million km)
Mercury	58
Venus	108
Earth	150
Mars	228
Jupiter	778
Saturn	1427
Uranus	2871
Neptune	4497
Pluto	5913

3. BIOLOGIC DIVERSITY To discuss the variety of species that live in a certain environment, a biologist needs a precise definition of *diversity*. Let $p_1, p_2, \ldots, p_n$ be the proportions of n species that live in an environment. The biologic diversity D of this system is

$$D = -(p_1 \log_2 p_1 + p_2 \log_2 p_2 + \cdots + p_n \log_2 p_n)$$

Suppose that an ecosystem has exactly five different varieties of grass: rye (R), bermuda (B), blue (L), fescue (F), and St. Augustine (A).

a. Calculate the diversity of this ecosystem if the proportions of these grasses are as shown in Table 4.1. Round to the nearest hundredth.

Table 4.1

R	B	L	F	A
$\frac{1}{5}$	$\frac{1}{5}$	$\frac{1}{5}$	$\frac{1}{5}$	$\frac{1}{5}$

b. Because bermuda and St. Augustine are virulent grasses, after a time the proportions will be as shown in Table 4.2. Calculate the diversity of this system. Does this system have more or less diversity than the system given in Table 4.1?

Table 4.2

R	B	L	F	A
$\frac{1}{8}$	$\frac{3}{8}$	$\frac{1}{16}$	$\frac{1}{8}$	$\frac{5}{16}$

c. After an even longer time period, the bermuda and St. Augustine grasses completely overrun the environment and the proportions are as shown in Table 4.3. Calculate the diversity of this system. (*Note:* Although the equation is not technically correct, for purposes of the diversity definition, we may say that $0 \log_2 0 = 0$. By using very small values of p_i, we can demonstrate that this definition makes sense.) Does this system have more or less diversity than the system given in Table 4.2?

d. Finally, the St. Augustine grasses overrun the bermuda grasses and the proportions are as shown in Table 4.4. Calculate the diversity of this system. Write a sentence that explains the meaning of the value you obtained.

Table 4.3

R	B	L	F	A
0	$\frac{1}{4}$	0	0	$\frac{3}{4}$

Table 4.4

R	B	L	F	A
0	0	0	0	1

EXPONENTIAL AND LOGARITHMIC EQUATIONS

- **SOLVE EXPONENTIAL EQUATIONS**
- **SOLVE LOGARITHMIC EQUATIONS**
- **APPLICATION**

● SOLVE EXPONENTIAL EQUATIONS

If a variable appears in an exponent of a term of an equation, such as $2^{x+1} = 32$, then the equation is called an **exponential equation**. Example 1 uses the following equality-of-exponents theorem to solve $2^{x+1} = 32$.

> **Equality of Exponents Theorem**
>
> If $b^x = b^y$, then $x = y$, provided that $b > 0$ and $b \neq 1$.

EXAMPLE 1 **Solve an Exponential Equation**

Use the Equality of Exponents Theorem to solve $2^{x+1} = 32$.

Solution

$$2^{x+1} = 32$$
$$2^{x+1} = 2^5 \qquad \text{• Write each side as a power of 2.}$$
$$x + 1 = 5 \qquad \text{• Equate the exponents.}$$
$$x = 4$$

Check: Let $x = 4$, then $2^{x+1} = 2^{4+1}$
$$= 2^5$$
$$= 32$$

▶ **TRY EXERCISE 2, PAGE 415**

A graphing utility can also be used to find solutions of an equation of the form $f(x) = g(x)$. Either of the following two methods can be employed.

> ### Using a Graphing Utility to Find the Solutions of $f(x) = g(x)$
>
> *Intersection Method*   Graph $y_1 = f(x)$ and $y_2 = g(x)$ on the same screen. The solutions of $f(x) = g(x)$ are the x-coordinates of the points of intersection of the graphs.
>
> *Intercept Method*   The solutions of $f(x) = g(x)$ are the x-coordinates of the x-intercepts of the graph of $y = f(x) - g(x)$.

Figures 4.36 and **4.37** illustrate the graphical methods for solving $2^{x+1} = 32$.

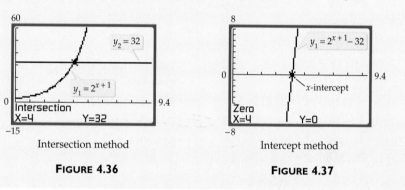

Intersection method

FIGURE 4.36

Intercept method

FIGURE 4.37

In Example 1 we were able to write both sides of the equation as a power of the same base. If you find it difficult to write both sides of an exponential equation in terms of the same base, then try the procedure of taking the logarithm of each side of the equation. This procedure is used in Example 2.

EXAMPLE 2   Solve an Exponential Equation

Solve: $5^x = 40$

Algebraic Solution

$$5^x = 40$$
$$\log(5^x) = \log 40 \qquad \text{• Take the logarithm of each side.}$$
$$x \log 5 = \log 40 \qquad \text{• Power property}$$
$$x = \frac{\log 40}{\log 5} \qquad \text{• Exact solution}$$
$$x \approx 2.3 \qquad \text{• Decimal approximation}$$

To the nearest tenth, the solution is 2.3.

Visualize the Solution

Intersection Method   The solution of $5^x = 40$ is the x-coordinate of the point of intersection of $y = 5^x$ and $y = 40$ (see **Figure 4.38**).

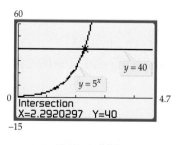

FIGURE 4.38

▶ **TRY EXERCISE 10, PAGE 415**

An alternative approach to solving the equation in Example 2 is to rewrite the exponential equation in logarithmic form: $5^x = 40$ is equivalent to the logarithmic equation $\log_5 40 = x$. Using the change-of-base formula, we find that $x = \log_5 40 = \dfrac{\log 40}{\log 5}$. In the following example, however, we must take logarithms of both sides to reach a solution.

EXAMPLE 3   **Solve an Exponential Equation**

Solve: $3^{2x-1} \doteq 5^{x+2}$

Algebraic Solution

$$3^{2x-1} = 5^{x+2}$$

$$\ln 3^{2x-1} = \ln 5^{x+2} \qquad \bullet \text{ Take the natural logarithm of each side.}$$

$$(2x - 1)\ln 3 = (x + 2)\ln 5 \qquad \bullet \text{ Power property}$$

$$2x \ln 3 - \ln 3 = x \ln 5 + 2 \ln 5 \qquad \bullet \text{ Distributive property}$$

$$2x \ln 3 - x \ln 5 = 2 \ln 5 + \ln 3 \qquad \bullet \text{ Solve for } x.$$

$$x(2 \ln 3 - \ln 5) = 2 \ln 5 + \ln 3$$

$$x = \frac{2 \ln 5 + \ln 3}{2 \ln 3 - \ln 5} \qquad \bullet \text{ Exact solution}$$

$$x \approx 7.3 \qquad \bullet \text{ Decimal approximation}$$

To the nearest tenth, the solution is 7.3.

Visualize the Solution

Intercept Method   The solution of $3^{2x-1} = 5^{x+2}$ is the x-coordinate of the x-intercept of $y = 3^{2x-1} - 5^{x+2}$ (see **Figure 4.39**).

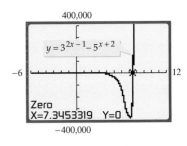

FIGURE 4.39

▶ **TRY EXERCISE 18, PAGE 415**

In Example 4 we solve an exponential equation that has two solutions.

EXAMPLE 4   **Solve an Exponential Equation Involving $b^x + b^{-x}$**

Solve: $\dfrac{2^x + 2^{-x}}{2} = 3$

Algebraic Solution

Multiplying each side by 2 produces

$$2^x + 2^{-x} = 6$$

$$2^{2x} + 2^0 = 6(2^x) \qquad \bullet \text{ Multiply each side by } 2^x \text{ to clear negative exponents.}$$

$$(2^x)^2 - 6(2^x) + 1 = 0 \qquad \bullet \text{ Write in quadratic form.}$$

$$(u)^2 - 6(u) + 1 = 0 \qquad \bullet \text{ Substitute } u \text{ for } 2^x.$$

Visualize the Solution

Intersection Method   The solutions of $\dfrac{2^x + 2^{-x}}{2} = 3$ are the x-coordinates of the points of intersection of

By the quadratic formula,

$$u = \frac{6 \pm \sqrt{36 - 4}}{2} = \frac{6 \pm 4\sqrt{2}}{2} = 3 \pm 2\sqrt{2}$$

$$2^x = 3 \pm 2\sqrt{2}$$

$$\log 2^x = \log(3 \pm 2\sqrt{2})$$

- Replace u with 2^x.
- Take the common logarithm of each side.
- Power property

$$x \log 2 = \log(3 \pm 2\sqrt{2})$$

$$x = \frac{\log(3 \pm 2\sqrt{2})}{\log 2} \approx \pm 2.54$$

The approximate solutions are -2.54 and 2.54.

▶ TRY EXERCISE 40, PAGE 415

$y = \dfrac{2^x + 2^{-x}}{2}$ and $y = 3$ (see **Figure 4.40**).

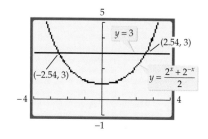

FIGURE 4.40

● SOLVE LOGARITHMIC EQUATIONS

Equations that involve logarithms are called **logarithmic equations**. The properties of logarithms, along with the definition of a logarithm, are often used to find the solutions of a logarithmic equation.

EXAMPLE 5   Solve a Logarithmic Equation

Solve: $\log(3x - 5) = 2$

Solution

$$\log(3x - 5) = 2$$

$$3x - 5 = 10^2$$ *Always*
- Definition of a logarithm

$$3x = 105$$ • Solve for x.

$$x = 35$$

Check: $\log[3(35) - 5] = \log 100 = 2$

▶ TRY EXERCISE 22, PAGE 415

❓ QUESTION Can a negative number be a solution of a logarithmic equation?

❓ ANSWER Yes. For instance, -10 is a solution of $\log(-x) = 1$.

EXAMPLE 6   **Solve a Logarithmic Equation**

Solve: $\log 2x - \log(x - 3) = 1$

Solution

$$\log 2x - \log(x - 3) = 1$$

$$\log \frac{2x}{x - 3} = 1 \qquad \bullet \textbf{ Quotient property}$$

$$\frac{2x}{x - 3} = 10^1 \qquad \bullet \textbf{ Definition of logarithm}$$

$$2x = 10x - 30 \qquad \bullet \textbf{ Solve for x.}$$

$$-8x = -30$$

$$x = \frac{15}{4}$$

Check the solution by substituting $\frac{15}{4}$ into the original equation.

▶ **TRY EXERCISE 26, PAGE 415**

In Example 7 we make use of the one-to-one property of logarithms to find the solution of a logarithmic equation. This example illustrates that the process of solving a logarithmic equation by using logarithmic properties may introduce an extraneous solution.

EXAMPLE 7   **Solve a Logarithmic Equation**

Solve: $\ln(3x + 8) = \ln(2x + 2) + \ln(x - 2)$

Algebraic Solution

$$\ln(3x + 8) = \ln(2x + 2) + \ln(x - 2)$$

$$\ln(3x + 8) = \ln[(2x + 2)(x - 2)] \qquad \bullet \textbf{ Product property}$$

$$\ln(3x + 8) = \ln(2x^2 - 2x - 4)$$

$$3x + 8 = 2x^2 - 2x - 4 \qquad \bullet \textbf{ One-to-one property of}$$
$$\textbf{logarithms}$$

$$0 = 2x^2 - 5x - 12$$

$$0 = (2x + 3)(x - 4) \qquad \bullet \textbf{ Solve for x.}$$

$$x = -\frac{3}{2} \quad \text{or} \quad x = 4$$

Thus $-\frac{3}{2}$ and 4 are possible solutions. A check will show that 4 is a solution, but $-\frac{3}{2}$ is not a solution.

▶ **TRY EXERCISE 36, PAGE 415**

Visualize the Solution

The graph of $y = \ln(3x + 8) - \ln(2x + 2) - \ln(x - 2)$ has only one x-intercept (see **Figure 4.41**). Thus there is only one real solution.

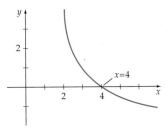

$y = \ln(3x + 8) - \ln(2x + 2) - \ln(x - 2)$

FIGURE 4.41

? QUESTION Why does $x = -\dfrac{3}{2}$ not check in Example 7?

● APPLICATION

EXAMPLE 8 | **Velocity of a Sky Diver Experiencing Air Resistance**

During the free-fall portion of a jump, the time t in seconds required for a sky diver to reach a velocity v in feet per second is given by

$$t = -\frac{175}{32} \ln\left(1 - \frac{v}{175}\right)$$

a. Determine the velocity of the diver after 5 seconds.

b. The graph of the above function has a vertical asymptote at $v = 175$. Explain the meaning of the vertical asymptote in the context of this example.

Solution

a. Substitute 5 for t and solve for v.

$$t = -\frac{175}{32} \ln\left(1 - \frac{v}{175}\right)$$

$$5 = -\frac{175}{32} \ln\left(1 - \frac{v}{175}\right) \qquad \text{• Replace } t \text{ with 5.}$$

$$\left(-\frac{32}{175}\right)5 = \ln\left(1 - \frac{v}{175}\right) \qquad \text{• Solve for } v.$$

$$-\frac{32}{35} = \ln\left(1 - \frac{v}{175}\right)$$

$$e^{-32/35} = 1 - \frac{v}{175} \qquad \text{• Write in exponential form.}$$

$$e^{-32/35} - 1 = -\frac{v}{175}$$

$$v = 175(1 - e^{-32/35})$$

$$v \approx 104.86$$

Continued ▶

take note

If air resistance is not considered, then the time in seconds required for a sky diver to reach a given velocity (in feet per second) is $t = \dfrac{v}{32}$. The function in Example 8 is a more realistic model of the time required to reach a given velocity during the free-fall of a sky diver who is experiencing air resistance.

? ANSWER If $x = -\dfrac{3}{2}$, the original equation becomes $\ln\left(\dfrac{7}{2}\right) = \ln(-1) + \ln\left(-\dfrac{7}{2}\right)$. This cannot be true, because the function $f(x) = \ln x$ is not defined for negative values of x.

After 5 seconds the velocity of the sky diver will be about 104.9 feet per second. See **Figure 4.42.**

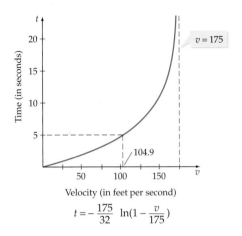

$$t = -\frac{175}{32} \ln\left(1 - \frac{v}{175}\right)$$

FIGURE 4.42

b. The vertical asymptote $v = 175$ indicates that the sky diver will not attain a velocity greater than 175 feet per second. In **Figure 4.42,** note that as $v \to 175$ from the left, $t \to \infty$.

▶ **TRY EXERCISE 68, PAGE 417**

 TOPICS FOR DISCUSSION

1. Discuss how to solve the equation $a = \log_b x$ for x.

2. What is the domain of $y = \log_4(2x - 5)$? Explain why this means that the equation $\log_4(x - 3) = \log_4(2x - 5)$ has no real number solution.

3. -8 is not a solution of the equation $\log_2 x + \log_2(x + 6) = 4$. Discuss at which step in the following solution the extraneous solution -8 was introduced.

$$\log_2 x + \log_2(x + 6) = 4$$
$$\log_2 x(x + 6) = 4$$
$$x(x + 6) = 2^4$$
$$x^2 + 6x = 16$$
$$x^2 + 6x - 16 = 0$$
$$(x + 8)(x - 2) = 0$$
$$x = -8 \quad \text{or} \quad x = 2$$

EXERCISE SET 4.5

In Exercises 1 to 46, solve for x algebraically.

1. $2^x = 64$

▶ 2. $3^x = 243$

3. $49^x = \dfrac{1}{343}$

4. $9^x = \dfrac{1}{243}$

5. $2^{5x+3} = \dfrac{1}{8}$

6. $3^{4x-7} = \dfrac{1}{9}$

7. $\left(\dfrac{2}{5}\right)^x = \dfrac{8}{125}$

8. $\left(\dfrac{2}{5}\right)^x = \dfrac{25}{4}$

9. $5^x = 70$

▶ 10. $6^x = 50$

11. $3^{-x} = 120$

12. $7^{-x} = 63$

13. $10^{2x+3} = 315$

14. $10^{6-x} = 550$

15. $e^x = 10$

16. $e^{x+1} = 20$

17. $2^{1-x} = 3^{x+1}$

▶ 18. $3^{x-2} = 4^{2x+1}$

19. $2^{2x-3} = 5^{-x-1}$

20. $5^{3x} = 3^{x+4}$

21. $\log(4x - 18) = 1$

▶ 22. $\log(x^2 + 19) = 2$

23. $\ln(x^2 - 12) = \ln x$

24. $\log(2x^2 + 3x) = \log(10x + 30)$

25. $\log_2 x + \log_2(x - 4) = 2$

▶ 26. $\log_3 x + \log_3(x + 6) = 3$

27. $\log(5x - 1) = 2 + \log(x - 2)$

28. $1 + \log(3x - 1) = \log(2x + 1)$

29. $\ln(1 - x) + \ln(3 - x) = \ln 8$

30. $\log(4 - x) = \log(x + 8) + \log(2x + 13)$

31. $\log \sqrt{x^3 - 17} = \dfrac{1}{2}$

32. $\log(x^3) = (\log x)^2$

33. $\log(\log x) = 1$

34. $\ln(\ln x) = 2$

35. $\ln(e^{3x}) = 6$

▶ 36. $\ln x = \dfrac{1}{2}\ln\left(2x + \dfrac{5}{2}\right) + \dfrac{1}{2}\ln 2$

37. $e^{\ln(x-1)} = 4$

38. $10^{\log(2x+7)} = 8$

39. $\dfrac{10^x - 10^{-x}}{2} = 20$

▶ 40. $\dfrac{10^x + 10^{-x}}{2} = 8$

41. $\dfrac{10^x + 10^{-x}}{10^x - 10^{-x}} = 5$

42. $\dfrac{10^x - 10^{-x}}{10^x + 10^{-x}} = \dfrac{1}{2}$

43. $\dfrac{e^x + e^{-x}}{2} = 15$

44. $\dfrac{e^x - e^{-x}}{2} = 15$

45. $\dfrac{1}{e^x - e^{-x}} = 4$

46. $\dfrac{e^x + e^{-x}}{e^x - e^{-x}} = 3$

In Exercises 47 to 56, use a graphing utility to approximate the solutions of the equation to the nearest hundredth.

47. $2^{-x+3} = x + 1$

48. $3^{x-2} = -2x - 1$

49. $e^{3-2x} - 2x = 1$

50. $2e^{x+2} + 3x = 2$

51. $3\log_2(x - 1) = -x + 3$

52. $2\log_3(2 - 3x) = 2x - 1$

53. $\ln(2x + 4) + \dfrac{1}{2}x = -3$

54. $2\ln(3 - x) + 3x = 4$

55. $2^{x+1} = x^2 - 1$

56. $\ln x = -x^2 + 4$

57. POPULATION GROWTH The population P of a city grows exponentially according to the function

$$P(t) = 8500(1.1)^t, \quad 0 \le t \le 8$$

where t is measured in years.

a. Find the population at time $t = 0$ and also at time $t = 2$.

b. When, to the nearest year, will the population reach 15,000?

58. PHYSICAL FITNESS After a race, a runner's pulse rate R in beats per minute decreases according to the function

$$R(t) = 145e^{-0.092t}, \quad 0 \le t \le 15$$

where t is measured in minutes.

a. Find the runner's pulse rate at the end of the race and also 1 minute after the end of the race.

b. How long, to the nearest minute, after the end of the race will the runner's pulse rate be 80 beats per minute?

59. RATE OF COOLING A can of soda at 79°F is placed in a refrigerator that maintains a constant temperature of 36°F. The temperature T of the soda t minutes after it is placed in the refrigerator is given by

$$T(t) = 36 + 43e^{-0.058t}$$

a. Find the temperature, to the nearest degree, of the soda 10 minutes after it is placed in the refrigerator.

b. When, to the nearest minute, will the temperature of the soda be 45°F?

60. MEDICINE During surgery, a patient's circulatory system requires at least 50 milligrams of an anesthetic. The amount of anesthetic present t hours after 80 milligrams of anesthetic is administered is given by

$$T(t) = 80(0.727)^t$$

a. How much, to the nearest milligram, of the anesthetic is present in the patient's circulatory system 30 minutes after the anesthetic is administered?

b. How long, to the nearest minute, can the operation last if the patient does not receive additional anesthetic?

61. PSYCHOLOGY Industrial psychologists study employee training programs to assess the effectiveness of the instruction. In one study, the percent score P on a test for a person who had completed t hours of training was given by

$$P = \frac{100}{1 + 30e^{-0.088t}}$$

a. Use a graphing utility to graph the equation for $t \geq 0$.

b. Use the graph to estimate (to the nearest hour) the number of hours of training necessary to achieve a 70% score on the test.

c. From the graph, determine the horizontal asymptote.

d. Write a sentence that explains the meaning of the horizontal asymptote.

62. PSYCHOLOGY An industrial psychologist has determined that the average percent score for an employee on a test of the employee's knowledge of the company's product is given by

$$P = \frac{100}{1 + 40e^{-0.1t}}$$

where t is the number of weeks on the job and P is the percent score.

a. Use a graphing utility to graph the equation for $t \geq 0$.

b. Use the graph to estimate (to the nearest week) the number of weeks of employment that are necessary for the average employee to earn a 70% score on the test.

c. Determine the horizontal asymptote of the graph.

d. Write a sentence that explains the meaning of the horizontal asymptote.

63. ECOLOGY A herd of bison was placed in a wildlife preserve that can support a maximum of 1000 bison. A population model for the bison is given by

$$B = \frac{1000}{1 + 30e^{-0.127t}}$$

where B is the number of bison in the preserve and t is time in years, with the year 1999 represented by $t = 0$.

a. Use a graphing utility to graph the equation for $t \geq 0$.

b. Use the graph to estimate (to the nearest year) the number of years before the bison population reaches 500.

c. Determine the horizontal asymptote of the graph.

d. Write a sentence that explains the meaning of the horizontal asymptote.

64. POPULATION GROWTH A yeast culture grows according to the equation

$$Y = \frac{50,000}{1 + 250e^{-0.305t}}$$

where Y is the number of yeast and t is time in hours.

a. Use a graphing utility to graph the equation for $t \geq 0$.

b. Use the graph to estimate (to the nearest hour) the number of hours before the yeast population reaches 35,000.

c. From the graph, estimate the horizontal asymptote.

d. Write a sentence that explains the meaning of the horizontal asymptote.

65. CONSUMPTION OF NATURAL RESOURCES A model for how long our coal resources will last is given by

$$T = \frac{\ln(300r + 1)}{\ln(r + 1)}$$

where r is the percent increase in consumption from current levels of use and T is the time (in years) before the resource is depleted.

a. Use a graphing utility to graph this equation.

b. If our consumption of coal increases by 3% per year, in how many years will we deplete our coal resources?

c. What percent increase in consumption of coal will deplete the resource in 100 years? Round to the nearest tenth of a percent.

66. CONSUMPTION OF NATURAL RESOURCES A model for how long our aluminum resources will last is given by

$$T = \frac{\ln(20{,}500r + 1)}{\ln(r + 1)}$$

where r is the percent increase in consumption from current levels of use and T is the time (in years) before the resource is depleted.

a. Use a graphing utility to graph this equation.

b. If our consumption of aluminum increases by 5% per year, in how many years (to the nearest year) will we deplete our aluminum resources?

c. What percent increase in consumption of aluminum will deplete the resource in 100 years? Round to the nearest tenth of a percent.

67. VELOCITY OF A MEDICAL CARE PACKAGE A medical care package is air lifted and dropped to a disaster area. During the free-fall portion of the drop, the time, in seconds, required for the package to obtain a velocity of v feet per second is given by the function

$$t = 2.43 \ln \frac{150 + v}{150 - v}, \quad 0 \le v < 150$$

a. Determine the velocity of the package 5 seconds after it is dropped. Round to the nearest foot per second.

b. Determine the vertical asymptote of the function.

c. Write a sentence that explains the meaning of the vertical asymptote in the context of this application.

▶ **68.** EFFECTS OF AIR RESISTANCE ON VELOCITY If we assume that air resistance is proportional to the square of the velocity, then the time t in seconds required for an object to reach a velocity v in feet per second is given by

$$t = \frac{9}{24} \ln \frac{24 + v}{24 - v}, 0 \le v < 24$$

a. Determine the velocity, to the nearest hundredth foot per second of the object after 1.5 seconds.

b. Determine the vertical asymptote for the graph of this function.

c. Write a sentence that describes the meaning of the vertical asymptote in the context of this problem.

69. TERMINAL VELOCITY WITH AIR RESISTANCE The velocity v of an object t seconds after it has been dropped from a height above the surface of the earth is given by the equation $v = 32t$ feet per second, assuming no air resistance. If we assume that air resistance is proportional to the square of the velocity, then the velocity after t seconds is given by

$$v = 100\left(\frac{e^{0.64t} - 1}{e^{0.64t} + 1}\right)$$

a. In how many seconds will the velocity be 50 feet per second?

b. Determine the horizontal asymptote for the graph of this function.

c. Write a sentence that describes the meaning of the horizontal asymptote in the context of this problem.

70. TERMINAL VELOCITY WITH AIR RESISTANCE If we assume that air resistance is proportional to the square of the velocity, then the velocity v in feet per second of an object t seconds after it has been dropped is given by

$$v = 50\left(\frac{e^{1.6t} - 1}{e^{1.6t} + 1}\right)$$

(See Exercise 69. The reason for the difference in the equations is that the proportionality constants are different.)

a. In how many seconds will the velocity be 20 feet per second?

b. Determine the horizontal asymptote for the graph of this function.

c. Write a sentence that describes the meaning of the horizontal asymptote in the context of this problem.

71. EFFECTS OF AIR RESISTANCE ON DISTANCE The distance s, in feet, that the object in Exercise 69 will fall in t seconds is given by

$$s = \frac{100^2}{32} \ln\left(\frac{e^{0.32t} + e^{-0.32t}}{2}\right)$$

a. Use a graphing utility to graph this equation for $t \ge 0$.

b. How long does it take for the object to fall 100 feet? Round to the nearest tenth of a second.

72. **EFFECTS OF AIR RESISTANCE ON DISTANCE** The distance s, in feet, that the object in Exercise 70 will fall in t seconds is given by

$$s = \frac{50^2}{40} \ln\left(\frac{e^{0.8t} + e^{-0.8t}}{2}\right)$$

a. Use a graphing utility to graph this equation for $t \geq 0$.

b. How long does it take for the object to fall 100 feet? Round to the nearest tenth of a second.

73. **RETIREMENT PLANNING** The retirement account for a graphic designer contains $250,000 on January 1, 2002, and earns interest at a rate of 0.5% per month. On February 1, 2002, the designer withdraws $2000 and plans to continue these withdrawals as retirement income each month. The value V of the account after x months is

$$V = 400,000 - 150,000(1.005)^x$$

If the designer wishes to leave $100,000 to a scholarship foundation, what is the maximum number of withdrawals (to the nearest month) the designer can make from this account and still have $100,000 to donate?

74. **HANGING CABLE** The height h, in feet, of any point P on the cable shown is given by

$$h(x) = 10(e^{x/20} + e^{-x/20}), \quad -15 \leq x \leq 15$$

where $|x|$ is the horizontal distance in feet between P and the y-axis.

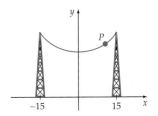

a. What is the lowest height of the cable?

b. What is the height of the cable 10 feet to the right of the y-axis? Round to the nearest tenth of a foot.

c. How far to the right of the y-axis is the cable 24 feet in height? Round to the nearest tenth of a foot.

CONNECTING CONCEPTS

75. The following argument seems to indicate that $0.125 > 0.25$. Find the first incorrect statement in the argument.

$$3 > 2$$
$$3(\log 0.5) > 2(\log 0.5)$$
$$\log 0.5^3 > \log 0.5^2$$
$$0.5^3 > 0.5^2$$
$$0.125 > 0.25$$

76. The following argument seems to indicate that $4 = 6$. Find the first incorrect statement in the argument.

$$4 = \log_2 16$$
$$4 = \log_2(8 + 8)$$
$$4 = \log_2 8 + \log_2 8$$
$$4 = 3 + 3$$
$$4 = 6$$

77. A common mistake that students make is to write $\log(x + y)$ as $\log x + \log y$. For what values of x and y does $\log(x + y) = \log x + \log y$? (*Hint:* Solve for x in terms of y.)

78. Let $f(x) = 2 \ln x$ and $g(x) = \ln x^2$. Does $f(x) = g(x)$ for all real numbers x?

79. Explain why the functions $F(x) = 1.4^x$ and $G(x) = e^{0.336x}$ represent essentially the same function.

80. Find the constant k that will make $f(t) = 2.2^t$ and $g(t) = e^{-kt}$ represent essentially the same function.

PREPARE FOR SECTION 4.6

81. Evaluate $A = 1000\left(1 + \dfrac{0.1}{12}\right)^{12t}$ for $t = 2$. Round to the nearest hundredth. [4.2]

82. Evaluate $A = 600\left(1 + \dfrac{0.04}{4}\right)^{4t}$ for $t = 8$. Round to the nearest hundredth. [4.2]

83. Solve $0.5 = e^{14k}$ for k. Round to the nearest ten-thousandth. [4.5]

84. Solve $0.85 = 0.5^{t/5730}$ for t. Round to the nearest ten. [4.5]

85. Solve $6 = \dfrac{70}{5 + 9e^{-k \cdot 12}}$ for k. Round to the nearest thousandth. [4.5]

86. Solve $2{,}000{,}000 = \dfrac{3^{n+1} - 3}{2}$ for n. Round to the nearest tenth. [4.5]

PROJECTS

1. **NAVIGATING** The pilot of a boat is trying to cross a river to a point O two miles due west of the boat's starting position by always pointing the nose of the boat toward O. Suppose the speed of the current is w miles per hour and the speed of the boat is v miles per hour. If point O is the origin and the boat's starting position is $(2, 0)$ (see the diagram at the right), then the equation of the boat's path is given by

$$y = \left(\frac{x}{2}\right)^{1-(w/v)} - \left(\frac{x}{2}\right)^{1+(w/v)}$$

a. If the speed of the current and the speed of the boat are the same, can the pilot reach point O by always having the nose of the boat pointed toward O? If not, at what point will the pilot arrive? Explain your answer.

b. If the speed of the current is greater than the speed of the boat, can the pilot reach point O by always pointing the nose of the boat toward point O? If not, where will the pilot arrive? Explain.

c. If the speed of the current is less than the speed of the boat, can the pilot reach point O by always pointing the nose of the boat toward point O? If not, where will the pilot arrive? Explain.

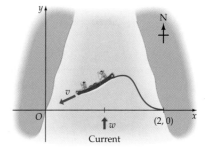

EXPONENTIAL GROWTH AND DECAY

- COMPOUND INTEREST
- EXPONENTIAL GROWTH
- EXPONENTIAL DECAY
- CARBON DATING
- THE LOGISTIC MODEL

In many applications, a quantity N grows or decays according to the function $N(t) = N_0 e^{kt}$. In this function, N is a function of time t, and N_0 is the value of N at time $t = 0$. If k is a *positive* constant, then $N(t) = N_0 e^{kt}$ is called an **exponential growth function**. If k is a *negative* constant, then $N(t) = N_0 e^{kt}$ is called an **exponential decay function**. The following examples illustrate how growth and decay functions arise naturally in the investigation of certain phenomena.

Interest is money paid for the use of money. The interest I is called **simple interest** if it is a fixed percent r, per time period t, of the amount of money invested. The amount of money invested is called the **principal** P. Simple interest is computed using the formula $I = Prt$. For example, if \$1000 is invested at 12% for 3 years, the simple interest is

$$I = Prt = \$1000(0.12)(3) = \$360$$

The balance after t years is $A = P + I = P + Prt$. In the previous example, the \$1000 invested for 3 years produced \$360 interest. Thus the balance after 3 years is \$1000 + \$360 = \$1360.

● COMPOUND INTEREST

In many financial transactions, interest is added to the principal at regular intervals so that interest is paid on interest as well as on the principal. Interest earned in this manner is called **compound interest.** For example, if \$1000 is invested at 12% annual interest compounded annually for 3 years, then the total interest after 3 years is

First-year interest	\$1000(0.12) = \$120.00
Second-year interest	\$1120(0.12) = \$134.40
Third-year interest	\$1254.40(0.12) ≈ \$150.53

$$\$404.93 \qquad \text{• Total interest}$$

This method of computing the balance can be tedious and time-consuming. A *compound interest formula* that can be used to determine the balance due after t years of compounding can be developed as follows.

Note that if P dollars is invested at an interest rate of r per year, then the balance after one year is $A_1 = P + Pr = P(1 + r)$, where Pr represents the interest earned for the year. Observe that A_1 is the product of the original principal P and $(1 + r)$. If the amount A_1 is reinvested for another year, then the balance after the second year is

$$A_2 = (A_1)(1 + r) = P(1 + r)(1 + r) = P(1 + r)^2$$

Successive reinvestments lead to the results shown in **Table 4.11**. The equation $A_t = P(1 + r)^t$ is valid if r is the annual interest rate paid during each of the t years.

TABLE 4.11

Number of Years	Balance
3	$A_3 = P(1 + r)^3$
4	$A_4 = P(1 + r)^4$
⋮	⋮
t	$A_t = P(1 + r)^t$

If r is an annual interest rate and n is the number of compounding periods per year, then the interest rate each period is r/n and the number of compounding periods after t years is nt. Thus the compound interest formula is expressed as follows:

The Compound Interest Formula

A principal P invested at an annual interest rate r, expressed as a decimal and compounded n times per year for t years, produces the balance

$$A = P\left(1 + \frac{r}{n}\right)^{nt}$$

EXAMPLE 1   Solve a Compound Interest Application

Find the balance if $1000 is invested at an annual interest rate of 10% for 2 years compounded

a.   annually   b.   monthly   c.   daily

Solution

a.   Use the compound interest formula with $P = 1000$, $r = 0.1$, $t = 2$, and $n = 1$.

$$A = \$1000\left(1 + \frac{0.1}{1}\right)^{1 \cdot 2} = \$1000(1.1)^2 = \$1210.00$$

b.   Because there are 12 months in a year, use $n = 12$.

$$A = \$1000\left(1 + \frac{0.1}{12}\right)^{12 \cdot 2} \approx \$1000(1.008333333)^{24} \approx \$1220.39$$

c.   Because there are 365 days in a year, use $n = 365$.

$$A = \$1000\left(1 + \frac{0.1}{365}\right)^{365 \cdot 2} \approx \$1000(1.000273973)^{730} \approx \$1221.37$$

▶ **TRY EXERCISE 4, PAGE 430**

To **compound continuously** means to increase the number of compounding periods without bound.

To derive a continuous compounding interest formula, substitute $\frac{1}{m}$ for $\frac{r}{n}$ in the compound interest formula

$$A = P\left(1 + \frac{r}{n}\right)^{nt} \tag{1}$$

to produce

$$A = P\left(1 + \frac{1}{m}\right)^{nt} \tag{2}$$

This substitution is motivated by the desire to express $\left(1 + \dfrac{r}{n}\right)^n$ as $\left[\left(1 + \dfrac{1}{m}\right)^m\right]^r$, which approaches e^r as m gets larger without bound.

Solving the equation $\dfrac{1}{m} = \dfrac{r}{n}$ for n yields $n = mr$, so the exponent nt can be written as mrt. Therefore Equation (2) can be expressed as

$$A = P\left(1 + \frac{1}{m}\right)^{mrt} = P\left[\left(1 + \frac{1}{m}\right)^m\right]^{rt} \qquad (3)$$

By the definition of e, we know that as m increases without bound,

$$\left(1 + \frac{1}{m}\right)^m \qquad \text{approaches} \qquad e$$

Thus, using continuous compounding, Equation (3) simplifies to $A = Pe^{rt}$.

Continuous Compounding Interest Formula

If an account with principal P and annual interest rate r is compounded continuously for t years, then the balance is $A = Pe^{rt}$.

EXAMPLE 2   Solve a Continuous Compound Interest Application

Find the balance after 4 years on $800 invested at an annual rate of 6% compounded continuously.

Algebraic Solution

Use the continuous compounding formula with $P = 800$, $r = 0.06$, and $t = 4$.

$$\begin{aligned}
A &= Pe^{rt} \\
&= 800e^{0.06(4)} \\
&= 800e^{0.24} \\
&\approx 800(1.27124915) \\
&\approx 1017.00 \qquad \bullet \text{ To the nearest cent}
\end{aligned}$$

The balance after 4 years will be $1017.00.

Visualize the Solution

Figure 4.43, a graph of $A = 800e^{0.06t}$, shows that the balance is about $1017.00 when $t = 4$.

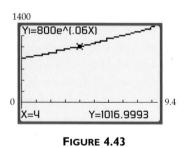

FIGURE 4.43

▶ TRY EXERCISE 6, PAGE 430

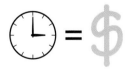

You have probably heard it said that time is money. In fact, many investors ask the question "How long will it take to double my money?" The following example answers this question for two different investments.

EXAMPLE 3   Double Your Money

Find the time required for money invested at an annual rate of 6% to double in value if the investment is compounded

a. semiannually

b. continuously

Solution

a. Use $A = P\left(1 + \dfrac{r}{n}\right)^{nt}$ with $r = 0.06$, $n = 2$, and the balance A equal to twice the principal $(A = 2P)$.

$$2P = P\left(1 + \frac{0.06}{2}\right)^{2t}$$

$$2 = \left(1 + \frac{0.06}{2}\right)^{2t} \qquad \bullet \textbf{ Divide each side by } P.$$

$$\ln 2 = \ln\left(1 + \frac{0.06}{2}\right)^{2t} \qquad \bullet \textbf{ Take the natural logarithm of each side.}$$

$$\ln 2 = 2t \ln\left(1 + \frac{0.06}{2}\right) \qquad \bullet \textbf{ Apply the power property.}$$

$$2t = \frac{\ln 2}{\ln\left(1 + \frac{0.06}{2}\right)} \qquad \bullet \textbf{ Solve for } t.$$

$$t = \frac{1}{2} \cdot \frac{\ln 2}{\ln\left(1 + \frac{0.06}{2}\right)}$$

$$t \approx 11.72$$

If the investment is compounded semiannually, it will double in value in about 11.72 years.

b. Use $A = Pe^{rt}$ with $r = 0.06$ and $A = 2P$.

$$2P = Pe^{0.06t}$$

$$2 = e^{0.06t} \qquad \bullet \textbf{ Divide each side by } P.$$

$$\ln 2 = 0.06t \qquad \bullet \textbf{ Write in logarithm form.}$$

$$t = \frac{\ln 2}{0.06} \qquad \bullet \textbf{ Solve for } t.$$

$$t \approx 11.55$$

If the investment is compounded continuously, it will double in value in about 11.55 years.

▶ **TRY EXERCISE 10, PAGE 430**

● EXPONENTIAL GROWTH

Given any two points on the graph of $N(t) = N_0e^{kt}$, you can use the given data to solve for the constants N_0 and k.

EXAMPLE 4 **Find the Exponential Growth Function That Models Given Data**

a. Find the exponential growth function for a town whose population was 16,400 in 1990 and 20,200 in 2000.

b. Use the function from part **a.** to predict, to the nearest 100, the population of the town in 2005.

Solution

a. We need to determine N_0 and k in $N(t) = N_0e^{kt}$. If we represent the year 1990 by $t = 0$, then our given data are $N(0) = 16,400$ and $N(10) = 20,200$. Because N_0 is defined to be $N(0)$, we know that $N_0 = 16,400$. To determine k, substitute $t = 10$ and $N_0 = 16,400$ into $N(t) = N_0e^{kt}$ to produce

$$N(10) = 16,400e^{k \cdot 10}$$

$$20,200 = 16,400e^{10k} \qquad \text{• Substitute 20,200 for } N(10).$$

$$\frac{20,200}{16,400} = e^{10k} \qquad \text{• Solve for } e^{10k}.$$

$$\ln\frac{20,200}{16,400} = 10k \qquad \text{• Write in logarithmic form.}$$

$$\frac{1}{10}\ln\frac{20,200}{16,400} = k \qquad \text{• Solve for } k.$$

$$0.0208 \approx k$$

The exponential growth function is $N(t) \approx 16,400e^{0.0208t}$.

b. The year 1990 was represented by $t = 0$, so we will use $t = 15$ to represent the year 2005.

$$N(t) \approx 16,400e^{0.0208t}$$

$$N(15) \approx 16,400e^{0.0208 \cdot 15}$$

$$\approx 22,400 \quad \text{(nearest 100)}$$

The exponential growth function yields 22,400 as the approximate population of the town in 2005.

> ▶ **TRY EXERCISE 18, PAGE 431**

---

take note

Because $e^{0.0208} \approx 1.021$, the growth equation can also be written as

$$N(t) \approx 16,400(1.021)^t$$

In this form we see that the population is growing by 2.1% $(1.021 - 1 = 0.021 = 2.1\%)$ per year.

● EXPONENTIAL DECAY

Many radioactive materials *decrease* in mass exponentially over time. This decrease, called radioactive decay, is measured in terms of **half-life,** which is defined as the time required for the disintegration of half the atoms in a sample of a radioactive substance. **Table 4.12** shows the half-lives of selected radioactive isotopes.

Table 4.12

Isotope	Half-Life
Carbon (^{14}C)	5730 years
Radium (^{226}Ra)	1660 years
Polonium (^{210}Po)	138 days
Phosphorus (^{32}P)	14 days
Polonium (^{214}Po)	1/10,000th of a second

EXAMPLE 5   **Find the Exponential Decay Function That Models Given Data**

Find the exponential decay function for the amount of phosphorus (^{32}P) that remains in a sample after t days.

Solution

When $t = 0$, $N(0) = N_0 e^{k(0)} = N_0$. Thus $N(0) = N_0$. Also, because the phosphorus has a half-life of 14 days (from **Table 4.12**), $N(14) = 0.5N_0$. To find k, substitute $t = 14$ into $N(t) = N_0 e^{kt}$ and solve for k.

$$N(14) = N_0 \cdot e^{k \cdot 14}$$

$$0.5N_0 = N_0 e^{14k} \qquad \text{• Substitute } 0.5N_0 \text{ for } N(14).$$

$$0.5 = e^{14k} \qquad \text{• Divide each side by } N_0.$$

$$\ln 0.5 = 14k \qquad \text{• Write in logarithmic form.}$$

$$\frac{1}{14} \ln 0.5 = k \qquad \text{• Solve for } k.$$

$$-0.0495 \approx k$$

The exponential decay function is $N(t) = N_0 e^{-0.0495t}$.

> **TRY EXERCISE 20, PAGE 431**

take note

Because $e^{-0.0495} \approx (0.5)^{1/14}$, the decay function $N(t) = N_0 e^{-0.0495t}$ can also be written as $N(t) = N_0(0.5)^{t/14}$. In this form it is easy to see that if t is increased by 14, then N will decrease by a factor of 0.5.

EXAMPLE 6 Application to Air Resistance

Assuming that air resistance is proportional to the velocity of a falling object, the velocity (in feet per second) of the object t seconds after it has been dropped is given by $v = 82(1 - e^{-0.39t})$.

a. Determine when the velocity will be 70 feet per second.

b. Write a sentence that explains the meaning of the horizontal asymptote, which is $v = 82$, in the context of this example.

Algebraic Solution

a.

$$v = 82(1 - e^{-0.39t})$$

$$70 = 82(1 - e^{-0.39t})$$ • **Replace v by 70.**

$$\frac{70}{82} = 1 - e^{-0.39t}$$ • **Divide each side by 82.**

$$e^{-0.39t} = 1 - \frac{70}{82}$$ • **Solve for $e^{-0.39t}$.**

$$-0.39t = \ln \frac{6}{41}$$ • **Write in logarithmic form.**

$$t = \frac{\ln(6/41)}{-0.39} \approx 4.9277246$$ • **Solve for t.**

The time is approximately 4.9 seconds.

b. The horizontal asymptote $v = 82$ means that as time increases, the velocity of the object will approach but never reach or exceed 82 feet per second.

▶ **TRY EXERCISE 32, PAGE 432**

Visualize the Solution

a. A graph of $y = 82(1 - e^{-0.39x})$ and $y = 70$ shows that the x-coordinate of the point of intersection is about 4.9.

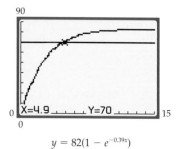

$$y = 82(1 - e^{-0.39x})$$

FIGURE 4.44

Note: The x value shown is rounded to the nearest tenth.

● CARBON DATING

The bone tissue in all living animals contains both carbon-12, which is nonradioactive, and carbon-14, which is radioactive with a half-life of approximately 5730 years. See **Figure 4.45**. As long as the animal is alive, the ratio of carbon-14 to carbon-12 remains constant. When the animal dies ($t = 0$), the carbon-14 begins to decay. Thus a bone that has a smaller ratio of carbon-14 to carbon-12 is older than a bone that has a larger ratio. The percent of carbon-14 present at time t is

$$P(t) = 0.5^{t/5730}$$

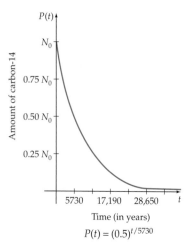

$$P(t) = (0.5)^{t/5730}$$

FIGURE 4.45

EXAMPLE 7   **Application to Archeology**

Find the age of a bone if it now has 85% of the carbon-14 it had when $t = 0$.

Solution

Let t be the time at which $P(t) = 0.85$.

$$0.85 = 0.5^{t/5730}$$

$$\ln 0.85 = \ln 0.5^{t/5730}$$ • Take the natural logarithm of each side.

$$\ln 0.85 = \frac{t}{5730} \ln 0.5$$ • Power property

$$5730\left(\frac{\ln 0.85}{\ln 0.5}\right) = t$$ • Solve for t.

$$1340 \approx t$$

The bone is about 1340 years old.

▶ **TRY EXERCISE 24, PAGE 431**

● THE LOGISTIC MODEL

The population growth function $P(t) = P_0 e^{kt}$ is called the **Malthusian growth model**. It was developed by Robert Malthus (1766–1834) in *An Essay on the Principle of Population Growth*, which was published in 1798. The Malthusian growth model is an unrestricted growth model that does not consider any limited resources that eventually will curb population growth.

The **logistic model** is a restricted growth model that takes into consideration the effects of limited resources. The logistic model was developed by Pierre Verhulst in 1836.

The Logistic Model (A Restricted Growth Model)

The magnitude of a population at time $t \geq 0$ is given by

$$P(t) = \frac{c}{1 + ae^{-bt}}$$

where c is the **carrying capacity** (the maximum population that can be supported by available resources as $t \to \infty$) and b is a positive constant called the **growth rate constant**.

The **initial population** is $P_0 = P(0)$. The constant a is related to the initial population P_0 and the carrying capacity c by the formula

$$a = \frac{c - P_0}{P_0}$$

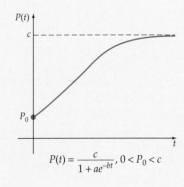

$$P(t) = \frac{c}{1 + ae^{-bt}}, 0 < P_0 < c$$

In the following example we determine a logistic growth model for a coyote population.

EXAMPLE 8   Find and Use a Logistic Model

At the beginning of 2002, the coyote population in a wilderness area was estimated at 200. By the beginning of 2004, the coyote population had increased to 250. A park ranger estimates that the carrying capacity of the wilderness area is 500 coyotes.

a. Use the given data to determine the growth rate constant for the logistic model of this coyote population.

b. Use the logistic model determined in part **a.** to predict the year in which the coyote population will first reach 400.

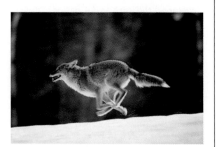

Solution

a. If we represent the beginning of the year 2002 by $t = 0$, then the beginning of the year 2004 will be represented by $t = 2$. In the logistic model, make the following substitutions: $P(2) = 250$, $c = 500$, and

$$a = \frac{c - P_0}{P_0} = \frac{500 - 200}{200} = 1.5.$$

$$P(t) = \frac{c}{1 + ae^{-bt}}$$

$$P(2) = \frac{500}{1 + 1.5e^{-b\cdot2}} \qquad \text{• Substitute the given values.}$$

$$250 = \frac{500}{1 + 1.5e^{-b\cdot2}}$$

$$250(1 + 1.5e^{-b\cdot2}) = 500 \qquad \text{• Solve for the growth rate constant } b.$$

$$1 + 1.5e^{-b\cdot2} = \frac{500}{250}$$

$$1.5e^{-b\cdot2} = 2 - 1$$

$$e^{-b\cdot2} = \frac{1}{1.5}$$

$$-2b = \ln\left(\frac{1}{1.5}\right)$$

$$b = -\frac{1}{2}\ln\left(\frac{1}{1.5}\right)$$

$$b \approx 0.20273255$$

Using $a = 1.5$, $b = 0.20273255$, and $c = 500$ gives us the following logistic model.

$$P(t) = \frac{500}{1 + 1.5e^{-0.20273255t}}$$

b. To determine during what year the logistic model predicts the coyote population will first reach 400, replace $P(t)$ with 400 and solve for t.

$$400 = \frac{500}{1 + 1.5e^{-0.20273255t}}$$

$$400(1 + 1.5e^{-0.20273255t}) = 500$$

$$1 + 1.5e^{-0.20273255t} = \frac{500}{400}$$

$$1.5e^{-0.20273255t} = 1.25 - 1$$

$$e^{-0.20273255t} = \frac{0.25}{1.5}$$

$$-0.20273255t = \ln\left(\frac{0.25}{1.5}\right) \qquad \bullet \textbf{Write in logarithmic form.}$$

$$t = \frac{1}{-0.20273255}\ln\left(\frac{0.25}{1.5}\right) \qquad \bullet \textbf{Solve for } t.$$

$$\approx 8.8$$

According to the logistic model, the coyote population will reach 400 about 8.8 years after the beginning of 2002, which is during the year 2010. The graph of the logistic model is shown in **Figure 4.46.** Note that $P(8.8) \approx 400$ and that as $t \to \infty$, $P(t) \to 500$.

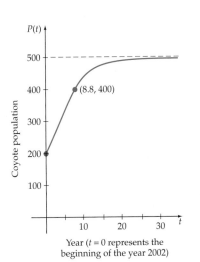

Year ($t = 0$ represents the beginning of the year 2002)

$$P(t) = \frac{500}{1 + 1.5e^{-0.20273255t}}$$

FIGURE 4.46

 TRY EXERCISE 48, PAGE 433

TOPICS FOR DISCUSSION

1. Explain the difference between compound interest and simple interest.

2. What is an exponential growth model? Give an example of an application for which the exponential growth model might be appropriate.

3. What is an exponential decay model? Give an example of an application for which the exponential decay model might be appropriate.

4. Consider the exponential model $P(t) = P_0 e^{kt}$ and the logistic model $P(t) = \dfrac{c}{1 + ae^{-bt}}$. Explain the similarities and differences between the two models.

EXERCISE SET 4.6

1. **COMPOUND INTEREST** If $8000 is invested at an annual interest rate of 5% and compounded annually, find the balance after

    a. 4 years   **b.** 7 years

2. **COMPOUND INTEREST** If $22,000 is invested at an annual interest rate of 4.5% and compounded annually, find the balance after

    a. 2 years   **b.** 10 years

3. **COMPOUND INTEREST** If $38,000 is invested at an annual interest rate of 6.5% for 4 years, find the balance if the interest is compounded

    a. annually   **b.** daily   **c.** hourly

▶ **4.** **COMPOUND INTEREST** If $12,500 is invested at an annual interest rate of 8% for 10 years, find the balance if the interest is compounded

    a. annually   **b.** daily   **c.** hourly

5. **COMPOUND INTEREST** Find the balance if $15,000 is invested at an annual rate of 10% for 5 years, compounded continuously.

▶ **6.** **COMPOUND INTEREST** Find the balance if $32,000 is invested at an annual rate of 8% for 3 years, compounded continuously.

7. **COMPOUND INTEREST** How long will it take $4000 to double if it is invested in a certificate of deposit that pays 7.84% annual interest compounded continuously? Round to the nearest tenth of a year.

8. **COMPOUND INTEREST** How long will it take $25,000 to double if it is invested in a savings account that pays 5.88% annual interest compounded continuously? Round to the nearest tenth of a year.

9. **CONTINUOUS COMPOUNDING INTEREST** Use the Continuous Compounding Interest Formula to derive an expression for the time it will take money to triple when invested at an annual interest rate of r compounded continuously.

▶ **10.** **CONTINUOUS COMPOUNDING INTEREST** How long will it take $1000 to triple if it is invested at an annual interest rate of 5.5% compounded continuously? Round to the nearest year.

11. **CONTINUOUS COMPOUNDING INTEREST** How long will it take $6000 to triple if it is invested in a savings account that pays 7.6% annual interest compounded continuously? Round to the nearest year.

12. **CONTINUOUS COMPOUNDING INTEREST** How long will it take $10,000 to triple if it is invested in a savings account that pays 5.5% annual interest compounded continuously? Round to the nearest year.

13. **POPULATION GROWTH** The number of bacteria $N(t)$ present in a culture at time t hours is given by

$$N(t) = 2200(2)^t$$

Find the number of bacteria present when

    a. $t = 0$ hours   **b.** $t = 3$ hours

14. **POPULATION GROWTH** The population of a town grows exponentially according to the function

$$f(t) = 12,400(1.14)^t$$

for $0 \le t \le 5$ years. Find, to the nearest hundred, the population of the town when t is

    a. 3 years   **b.** 4.25 years

15. **POPULATION GROWTH** A town had a population of 22,600 in 1990 and a population of 24,200 in 1995.

    a. Find the exponential growth function for the town. Use $t = 0$ to represent the year 1990.

    b. Use the growth function to predict the population of the town in 2005. Round to the nearest hundred.

16. **POPULATION GROWTH** A town had a population of 53,700 in 1996 and a population of 58,100 in 2000.

    a. Find the exponential growth function for the town. Use $t = 0$ to represent the year 1996.

    b. Use the growth function to predict the population of the town in 2008. Round to the nearest hundred.

17. **POPULATION GROWTH** The growth of the population of Los Angeles, California, for the years 1992 through 1996 can be approximated by the equation

$$P = 10,130(1.005)^t$$

where $t = 0$ corresponds to January 1, 1992 and P is in thousands.

    a. Assuming this growth rate continues, what will be the population of Los Angeles on January 1 in the year 2004?

b. In what year will the population of Los Angeles first exceed 13,000,000?

► **18.** **POPULATION GROWTH** The growth of the population of Mexico City, Mexico, for the years 1991 through 1998 can be approximated by the equation

$$P = 20,899(1.027)^t$$

where $t = 0$ corresponds to 1991 and P is in thousands.

a. Assuming this growth rate continues, what will be the population of Mexico City in the year 2003?

b. Assuming this growth rate continues, in what year will the population of Mexico City first exceed 35,000,000?

19. **MEDICINE** Sodium-24 is a radioactive isotope of sodium that is used to study circulatory dysfunction. Assuming that 4 micrograms of sodium-24 is injected into a person, the amount A in micrograms remaining in that person after t hours is given by the equation $A = 4e^{-0.046t}$.

a. Graph this equation.

b. What amount of sodium-24 remains after 5 hours?

c. What is the half-life of sodium-24?

d. In how many hours will the amount of sodium-24 be 1 microgram?

► **20.** **RADIOACTIVE DECAY** Polonium (^{210}Po) has a half-life of 138 days. Find the decay function for the amount of polonium (^{210}Po) that remains in a sample after t days.

21. **GEOLOGY** Geologists have determined that Crater Lake in Oregon was formed by a volcanic eruption. Chemical analysis of a wood chip that is assumed to be from a tree that died during the eruption has shown that it contains approximately 45% of its original carbon-14. Determine how long ago the volcanic eruption occurred. Use 5730 years as the half-life of carbon-14.

22. **RADIOACTIVE DECAY** Use $N(t) = N_0(0.5)^{t/138}$, where t is measured in days, to estimate the percentage of polonium (^{210}Po) that remains in a sample after 2 years. Round to the nearest hundredth of a percent.

23. **ARCHEOLOGY** The Rhind papyrus, named after A. Henry Rhind, contains most of what we know today of ancient Egyptian mathematics. A chemical analysis of a sample from the papyrus has shown that it contains approximately 75% of its original carbon-14. What is the age of the Rhind papyrus? Use 5730 years as the half-life of carbon-14.

► **24.** **ARCHEOLOGY** Determine the age of a bone if it now contains 65% of its original amount of carbon-14. Round to the nearest 100 years.

25. **PHYSICS** Newton's Law of Cooling states that if an object at temperature T_0 is placed into an environment at constant temperature A, then the temperature of the object, $T(t)$ (in degrees Fahrenheit), after t minutes is given by $T(t) = A + (T_0 - A)e^{-kt}$, where k is a constant that depends on the object.

a. Determine the constant k (to the nearest thousandth) for a canned soda drink that takes 5 minutes to cool from 75°F to 65°F after being placed in a refrigerator that maintains a constant temperature of 34°F.

b. What will be the temperature (to the nearest degree) of the soda drink after 30 minutes?

c. When (to the nearest minute) will the temperature of the soda drink be 36°F?

26. **PSYCHOLOGY** According to a software company, the users of its typing tutorial can expect to type $N(t)$ words per minute after t hours of practice with the product, according to the function $N(t) = 100(1.04 - 0.99^t)$.

a. How many words per minute can a student expect to type after 2 hours of practice?

b. How many words per minute can a student expect to type after 40 hours of practice?

c. According to the function N, how many hours (to the nearest hour) of practice will be required before a student can expect to type 60 words per minute?

27. **PSYCHOLOGY** In the city of Whispering Palms, which has a population of 80,000 people, the number of people $P(t)$ exposed to a rumor in t hours is given by the function $P(t) = 80,000(1 - e^{-0.0005t})$.

a. Find the number of hours until 10% of the population have heard the rumor.

b. Find the number of hours until 50% of the population have heard the rumor.

28. **LAW** A lawyer has determined that the number of people $P(t)$ in a city of 1,200,000 people who have been exposed to a news item after t days is given by the function

$$P(t) = 1,200,000(1 - e^{-0.03t})$$

a. How many days after a major crime has been reported have 40% of the population heard of the crime?

b. A defense lawyer knows it will be very difficult to pick an unbiased jury after 80% of the population have heard of the crime. After how many days will 80% of the population have heard of the crime?

29. DEPRECIATION An automobile depreciates according to the function $V(t) = V_0(1 - r)^t$, where $V(t)$ is the value in dollars after t years, V_0 is the original value, and r is the yearly depreciation rate. A car has a yearly depreciation rate of 20%. Determine, to the nearest 0.1 year, in how many years the car will depreciate to half its original value.

30. PHYSICS The current $I(t)$ (measured in amperes) of a circuit is given by the function $I(t) = 6(1 - e^{-2.5t})$, where t is the number of seconds after the switch is closed.

a. Find the current when $t = 0$.

b. Find the current when $t = 0.5$.

c. Solve the equation for t.

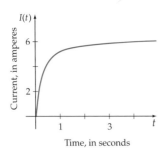

Time, in seconds

31. AIR RESISTANCE Assuming that air resistance is proportional to velocity, the velocity v, in feet per second, of a falling object after t seconds is given by $v = 32(1 - e^{-t})$.

a. Graph this equation for $t \geq 0$.

b. Determine algebraically, to the nearest 0.01 second, when the velocity is 20 feet per second.

c. Determine the horizontal asymptote of the graph of v.

d. Write a sentence that explains the meaning of the horizontal asymptote in the context of this application.

32. AIR RESISTANCE Assuming that air resistance is proportional to velocity, the velocity v, in feet per second, of a falling object after t seconds is given by $v = 64(1 - e^{-t/2})$.

a. Graph this equation for $t \geq 0$.

b. Determine algebraically, to the nearest 0.1 second, when the velocity is 50 feet per second.

c. Determine the horizontal asymptote of the graph of v.

d. Write a sentence that explains the meaning of the horizontal asymptote in the context of this application.

33. The distance s (in feet) that the object in Exercise 31 will fall in t seconds is given by $s = 32t + 32(e^{-t} - 1)$.

a. Use a graphing utility to graph this equation for $t \geq 0$.

b. Determine, to the nearest 0.1 second, the time it takes the object to fall 50 feet.

c. Calculate the slope of the secant line through $(1, s(1))$ and $(2, s(2))$.

d. Write a sentence that explains the meaning of the slope of the secant line you calculated in **c.**

34. The distance s (in feet) that the object in Exercise 32 will fall in t seconds is given by $s = 64t + 128(e^{-t/2} - 1)$.

a. Use a graphing utility to graph this equation for $t \geq 0$.

b. Determine, to the nearest 0.1 second, the time it takes the object to fall 50 feet.

c. Calculate the slope of the secant line through $(1, s(1))$ and $(2, s(2))$.

d. Write a sentence that explains the meaning of the slope of the secant line you calculated in **c.**

In Exercises 35 to 40, determine the following constants for the given logistic growth model.
a. The carrying capacity
b. The growth rate constant
c. The initial population P_0

35. $P(t) = \dfrac{1900}{1 + 8.5e^{-0.16t}}$

36. $P(t) = \dfrac{32{,}550}{1 + 0.75e^{-0.08t}}$

37. $P(t) = \dfrac{157{,}500}{1 + 2.5e^{-0.04t}}$

38. $P(t) = \dfrac{51}{1 + 1.04e^{-0.03t}}$

39. $P(t) = \dfrac{2400}{1 + 7e^{-0.12t}}$

40. $P(t) = \dfrac{320}{1 + 15e^{-0.12t}}$

In Exercises 41 to 44, use algebraic procedures to find the logistic growth model for the data.

41. $P_0 = 400$, $P(2) = 780$, and the carrying capacity is 5500.

42. $P_0 = 6200$, $P(8) = 7100$, and the carrying capacity is 9500.

43. $P_0 = 18$, $P(3) = 30$, and the carrying capacity is 100.

44. $P_0 = 3200$, $P(22) \approx 5565$, and the growth rate constant is 0.056.

45. REVENUE The annual revenue R, in dollars, of a new company can be closely modeled by the logistic growth function

$$R(t) = \frac{625{,}000}{1 + 3.1e^{-0.045t}}$$

where the *natural* number t is the time, in years, since the company was founded.

a. According to the model, what will be the company's annual revenue for its first year and its second year ($t = 1$ and $t = 2$) of operation? Round to the nearest $1000.

b. According to the model, what will the company's annual revenue approach in the long-term future?

46. NEW CAR SALES The number of cars A sold annually by an automobile dealership can be closely modeled by the logistic growth function

$$A(t) = \frac{1650}{1 + 2.4e^{-0.055t}}$$

where the *natural* number t is the time, in years, since the dealership was founded.

a. According to the model, what number of cars will the dealership sell during its first year and its second year ($t = 1$ and $t = 2$) of operation? Round to the nearest unit.

b. According to the model, what will the dealership's annual car sales approach in the long-term future?

47. POPULATION GROWTH The population of wolves in a preserve satisfies a logistic growth model in which $P_0 = 312$ in the year 2002, $c = 1600$, and $P(6) = 416$.

a. Determine the logistic growth model for this population, where t is the number of years after 2002.

b. Use the logistic growth model from part **a.** to predict the size of the wolf population in 2012.

▶ **48.** POPULATION GROWTH The population of ground-hogs on a ranch satisfies a logistic growth model in which $P_0 = 240$ in the year 2001, $c = 3400$, and $P(1) = 310$.

a. Determine the logistic growth model for this population, where t is the number of years after 2001.

b. Use the logistic growth model from part **a.** to predict the size of the groundhog population in 2008.

49. POPULATION GROWTH The population of squirrels in a nature preserve satisfies a logistic growth model in which $P_0 = 1500$ in the year 2001. The carrying capacity of the preserve is estimated at 8500 squirrels and $P(2) = 1900$.

a. Determine the logistic growth model for this population, where t is the number of years after 2001.

b. Use the logistic growth model from part **a.** to predict the year in which the squirrel population will first exceed 4000.

50. POPULATION GROWTH The population of walruses on an island satisfies a logistic growth model in which $P_0 = 800$ in the year 2000. The carrying capacity of the island is estimated at 5500 walruses, and $P(1) = 900$.

a. Determine the logistic growth model for this population, where t is the number of years after 2000.

b. Use the logistic growth model from part **a.** to predict the year in which the walrus population will first exceed 2000.

51. LEARNING THEORY The logistic model is also used in learning theory. Suppose that historical records from employee training at a company show that the percent score on a product information test is given by

$$P = \frac{100}{1 + 25e^{-0.095t}}$$

where t is the number of hours of training. What is the number of hours (to the nearest hour) of training needed before a new employee will answer 75% of the questions correctly?

52. LEARNING THEORY A company provides training in the assembly of a computer circuit to new employees. Past experience has shown that the number of correctly assembled circuits per week can be modeled by

$$N = \frac{250}{1 + 249e^{-0.503t}}$$

where t is the number of weeks of training. What is the number of weeks (to the nearest week) of training needed before a new employee will correctly make 140 circuits?

—————   *CONNECTING CONCEPTS*   —————

53. **MEDICATION LEVEL** A patient is given three dosages of aspirin. Each dosage contains 1 gram of aspirin. The second and third dosages are each taken 3 hours after the previous dosage is administered. The half-life of the aspirin is 2 hours. The amount of aspirin, A, in the patient's body t hours after the first dosage is administered is

$$A(t) = \begin{cases} 0.5^{t/2} & 0 \le t < 3 \\ 0.5^{t/2} + 0.5^{(t-3)/2} & 3 \le t < 6 \\ 0.5^{t/2} + 0.5^{(t-3)/2} + 0.5^{(t-6)/2} & t \ge 6 \end{cases}$$

Find, to the nearest hundredth of a gram, the amount of aspirin in the patient's body when

a. $t = 1$ **b.** $t = 4$ **c.** $t = 9$

54. **MEDICATION LEVEL** Use a graphing calculator and the dosage formula in Exercise 53 to determine when, to the nearest tenth of an hour, the amount of aspirin in the patient's body first reaches 0.25 gram.

Exercises 55 to 57 make use of the factorial function, which is defined as follows. For whole numbers n, the number $n!$ (which is read "n factorial") is given by

$$n! = \begin{cases} n(n-1)(n-2)\cdots 1, & \text{if } n \ge 1 \\ 1, & \text{if } n = 0 \end{cases}$$

Thus, $0! = 1$ and $4! = 4 \cdot 3 \cdot 2 \cdot 1 = 24$.

55. **QUEUEING THEORY** A study shows that the number of people who arrive at a bank teller's window averages 4.1 people every 10 minutes. The probability P that exactly x people will arrive at the teller's window in a given 10-minute period is

$$P(x) = \frac{4.1^{x}e^{-4.1}}{x!}$$

Find, to the nearest 0.1%, the probability that in a given 10-minute period, exactly

a. 0 people arrive at the window.

b. 2 people arrive at the window.

c. 3 people arrive at the window.

d. 4 people arrive at the window.

e. 9 people arrive at the window.

As $x \to \infty$, what does P approach?

56. **STIRLING'S FORMULA** *Stirling's Formula* (after James Stirling, 1692–1770),

$$n! \approx \left(\frac{n}{e}\right)^{n} \sqrt{2\pi n}$$

is often used to approximate very large factorials. Use Stirling's Formula to approximate 10!, and then compute the ratio of Stirling's approximation of 10! divided by the actual value of 10!, which is 3,628,800.

57. **RUBIK'S CUBE** The Rubik's cube shown here was invented by Erno Rubik in 1975. The small outer cubes are held together in such a way that they can be rotated around three axes. The total number of positions in which the Rubik's cube can be arranged is

$$\frac{3^{8}2^{12}8!\,12!}{2 \cdot 3 \cdot 2}$$

If you can arrange a Rubik's cube into a new arrangement every second, how many centuries would it take to place the cube into each of its arrangements? Assume that there are 365 days in a year.

58. **OIL SPILLS** Crude oil leaks from a tank at a rate that depends on the amount of oil that remains in the tank. Because $\dfrac{1}{8}$ of the oil in the tank leaks out every 2 hours, the volume of oil $V(t)$ in the tank after t hours is given by $V(t) = V_0(0.875)^{t/2}$, where $V_0 = 350{,}000$ gallons is the number of gallons in the tank at the time the tank started to leak ($t = 0$).

a. How many gallons does the tank hold after 3 hours?

b. How many gallons does the tank hold after 5 hours?

c. How long, to the nearest hour, will it take until 90% of the oil has leaked from the tank?

PREPARE FOR SECTION 4.7

59. Determine whether $N(t) = 4 - \ln t$ is an increasing or a decreasing function. [4.3]

60. Determine whether $P(t) = 1 - 2(1.05^t)$ is an increasing or a decreasing function. [4.2]

61. Evaluate $P(t) = \dfrac{108}{1 + 2e^{-0.1t}}$ for $t = 0$. [4.2]

62. Evaluate $N(t) = 840e^{1.05t}$ for $t = 0$. [4.2]

63. Solve $10 = \dfrac{20}{1 + 2.2e^{-0.05t}}$ for t. Round to the nearest tenth. [4.5]

64. Determine the horizontal asymptote of the graph of $P(t) = \dfrac{55}{1 + 3e^{-0.08t}}$. [4.2]

PROJECTS

A DECLINING LOGISTIC MODEL If $P_0 > c$ (which implies that $-1 < a < 0$), then the logistic function $P(t) = \dfrac{c}{1 + ae^{-bt}}$ decreases as t increases. Biologists often use this type of logistic function to model populations that decrease over time. See the following figure.

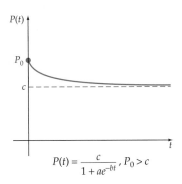

$$P(t) = \frac{c}{1 + ae^{-bt}}, \; P_0 > c$$

1. **A DECLINING FISH POPULATION** A biologist finds that the fish population in a small lake can be closely modeled by the logistic function

$$P(t) = \frac{1000}{1 + (-0.3333)e^{-0.05t}}$$

where t is the time, in years, since the lake was first stocked with fish.

a. What was the fish population when the lake was first stocked with fish?

b. According to the logistic model, what will the fish population approach in the long-term future?

2. **A DECLINING DEER POPULATION** The deer population in a reserve is given by the logistic function

$$P(t) = \frac{1800}{1 + (-0.25)e^{-0.07t}}$$

where t is the time, in years, since July 1, 2001.

a. What was the deer population on July 1, 2001? What was the deer population on July 1, 2003?

b. According to the logistic model, what will the deer population approach in the long-term future?

3. **MODELING WORLD RECORD TIMES IN THE MEN'S MILE RACE** In the early 1950s, many people speculated that no runner would ever run a mile race in under 4 minutes. During the period from 1913 to 1945, the world record in the mile event had been reduced from 4.14.4 (4 minutes, 14.4 seconds) to 4.01.4, but no one seemed capable of running a sub-four-minute mile. Then, in 1954, Roger Bannister broke through the four-minute barrier by running a mile in 3.59.6. In 1999, the current record of 3.43.13 was established. It is fun to think about future record times in the mile race. Will they ever go below 3 minutes, 30 seconds? Below 3 minutes, 20 seconds? What about a sub-three-minute mile?

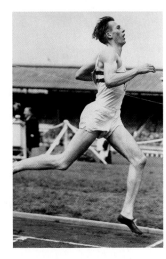

A declining logistic function that closely models the world record times *WR*, in seconds, in the men's mile run from 1913 ($t = 0$) to 1999 ($t = 86$) is given by

$$WR(t) = \frac{199.13}{1 + (-0.21726)e^{-0.0079889t}}$$

a. Use the above logistic model to predict the world record time for the men's mile run in the year 2020 and the year 2050.

b. According to the logistic function, what time will the world record in the men's mile event approach but never break through?

MODELING DATA WITH EXPONENTIAL AND LOGARITHMIC FUNCTIONS

SECTION 4.7

- ● ANALYZE SCATTER PLOTS
- ● APPLICATIONS
- ● USE REGRESSION TO FIND A LOGISTIC GROWTH MODEL

● ANALYZE SCATTER PLOTS

In Section 2.7 we used linear and quadratic functions to model several data sets. However, in some applications, data can be modeled more closely by using exponential or logarithmic functions. For instance, **Figure 4.47** illustrates some scatter plots that can be effectively modeled by exponential and logarithmic functions.

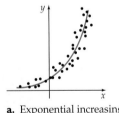

a. Exponential increasing: $y = ab^x, a > 0, b > 1$

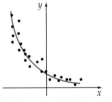

b. Exponential decreasing: $y = ab^x, a > 0, 0 < b < 1$

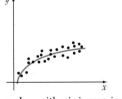

c. Logarithmic increasing: $y = a + b \ln x, b > 0$

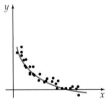

d. Logarithmic decreasing: $y = a + b \ln x, b < 0$

FIGURE 4.47

Exponential and Logarithmic Models

The terms *concave upward* and *concave downward* are often used to describe a graph. For instance, **Figures 4.48a** and **4.48b** show the graphs of two increasing functions that join the points P and Q. The graphs of f and g differ in that they bend in different directions. We can distinguish between these two types of "bending" by examining the positions of *tangent lines* to the graphs. In **Figures 4.48c** and **4.48d,** tangent lines (in red) have been drawn to the graphs of f and g. The graph of f lies above its tangent lines and the graph of g lies below its tangent lines. The function f is said to be concave upward, and g is concave downward.

a.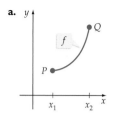

b.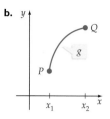

c. *f* is concave upward.   **d.** *g* is concave downward.

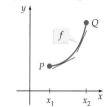

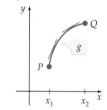

FIGURE 4.48

Definition of Concavity

If the graph of *f* lies above all of its tangents on an interval $[x_1, x_2]$, then *f* is **concave upward** on $[x_1, x_2]$.

If the graph of *f* lies below all of its tangents on an interval $[x_1, x_2]$, then *f* is **concave downward** on $[x_1, x_2]$.

An examination of the graphs in **Figure 4.47** shows that the graphs of all exponential functions of the form $y = ab^x, a > 0, b > 0, b \neq 1$ are concave upward. The graphs of increasing logarithmic functions are concave downward, and the graphs of decreasing logarithmic functions are concave upward.

In Example 1 we analyze scatter plots by determining whether the shape of the scatter plot can best be approximated by an increasing or a decreasing function, and by a function that is concave upward or concave downward.

? QUESTION Is the graph of $y = 5 - 2 \ln x$ concave upward or concave downward?

? ANSWER The equation $y = 5 - 2 \ln x$ has the form $y = a + b \ln x$, with $a > 0$ and $b < 0$. The graph of $y = 5 - 2 \ln x$ is concave upward because the *b*-value, -2, is less than zero. See **Figure 4.47d.**

EXAMPLE 1   **Analyze Scatter Plots**

For each of the following data sets, determine whether the most suitable model of the data would be an increasing exponential function or an increasing logarithmic function.

$A = \{(1, 0.6), (2, 0.7), (2.8, 0.8), (4, 1.3), (6, 1.5),$
$\qquad (6.5, 1.6), (8, 2.1), (11.2, 4.1), (12, 4.6), (15, 8.2)\}$
$B = \{(1.5, 2.8), (2, 3.5), (4.1, 5.1), (5, 5.5), (5.5, 5.7), (7, 6.1),$
$\qquad (7.2, 6.4), (8, 6.6), (9, 6.9), (11.6, 7.4), (12.3, 7.5), (14.7, 7.9)\}$

See Section 2.7 if you need to review the steps needed to create a scatter plot on a TI-83 calculator.

Solution

For each set construct a scatter plot of the data. See **Figure 4.49.**

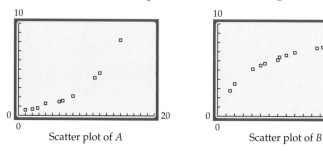

FIGURE 4.49

The scatter plot of *A* suggests that *A* is an increasing function that is concave upward. Thus *A* can be effectively modeled by an increasing exponential function.

The scatter plot of *B* suggests that *B* is an increasing function that is concave downward. Thus *B* can be effectively modeled by an increasing logarithmic function.

▶ **TRY EXERCISE 4, PAGE 444**

● **APPLICATIONS**

The methods used to model data using exponential or logarithmic functions are similar to the methods used in Section 2.7 to model data using linear or quadratic functions. Here is a summary of the modeling process.

The Modeling Process

Use a graphing utility to:

1. **Construct a** *scatter plot* **of the data** to determine which type of function will effectively model the data.

2. **Find the** *regression equation* of the modeling function and the correlation coefficient for the regression.

3. **Examine the** *correlation coefficient* and *view a graph* that displays both the modeling function and the scatter plot to determine how well your function fits the data.

In the following example we use the modeling process to find an exponential function that closely models the value of a diamond as a function of its weight.

EXAMPLE 2 | **Model an Application with an Exponential Function**

A diamond merchant has determined the values of several white diamonds that have different weights (measured in carats), but are *similar in quality.* See **Table 4.13.**

TABLE 4.13

0.50 ct	0.75 ct	1.00 ct	1.25 ct	1.50 ct	1.75 ct	2.00 ct	3.00 ct	4.00 ct
$4,600	$5,000	$5,800	$6,200	$6,700	$7,300	$7,900	$10,700	$14,500

Find a function that models the values of the diamonds as a function of their weights and use the function to predict the value of a 3.5-carat diamond of similar quality.

Solution

1. **Construct a scatter plot of the data.**

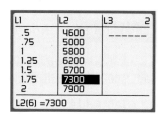

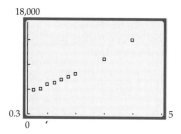

FIGURE 4.50

From the scatter plot in **Figure 4.50** it appears that the data can be closely modeled by an exponential function of the form $y = ab^x$, $a > 0$ and $b > 1$.

2. **Find the regression equation.** The calculator display in **Figure 4.51** shows that the exponential regression equation is $y \approx 4067.6(1.3816)^x$, where x is the carat weight of the diamond and y is the value of the diamond.

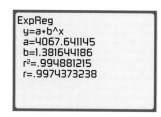

```
ExpReg
y=a*b^x
a=4067.641145
b=1.381644186
r²=.994881215
r=.9974373238
```

FIGURE 4.51

ExpReg display (DiagnosticOn)

Continued ▶

---

take note

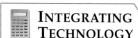

The value of a diamond is generally determined by its color, cut, clarity, and carat weight. These characteristics of a diamond are known as the four c's. In Example 2 we have assumed that the color, cut, and clarity of all the diamonds are similar. This assumption enables us to model the value of each diamond as a function of just its carat weight.

INTEGRATING TECHNOLOGY

Most graphing utilities have built-in routines that can be used to determine the exponential or logarithmic regression function that best models a set of data. On a TI-83, the ExpReg instruction is used to find the exponential regression function and the LnReg instruction is used to find the logarithmic regression function. The TI-83 does not show the value of the regression coefficient r unless the DiagnosticOn command has been entered. The DiagnosticOn command is in the CATALOG menu.

MATH MATTERS

The Hope Diamond, shown below, is the world's largest deep blue diamond. It has a weight of 45.52 carats. We should not expect the function $y \approx 4067.6 \times 1.3816^x$ in Example 2 to yield an accurate value of the Hope Diamond because the Hope Diamond is not the same type of diamond as the diamonds in **Table 4.13** and its weight is much larger than the weights of the diamonds in **Table 4.13**.

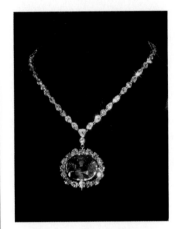

The Hope Diamond is on display at the Smithsonian Museum of Natural History in Washington, D.C.

3. **Examine the correlation coefficient.** The correlation coefficient $r \approx 0.9974$ is close to 1. This indicates that the exponential regression function $y \approx 4067.6(1.3816)^x$ provides a good fit for the data. The graph in **Figure 4.52** also shows that the exponential regression function provides a good model for the data.

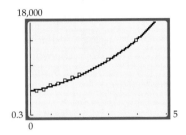

FIGURE 4.52

To estimate the value of a 3.5-carat diamond, replace x in the exponential regression function with 3.5.

$$y \approx 4067.6(1.3816)^{3.5} \approx \$12,610$$

According to the exponential regression function, the value of a 3.5-carat diamond of similar quality is about \$12,610.

▶ **TRY EXERCISE 22, PAGE 445**

In the next example we consider a data set that can be effectively modeled by more than one type of function.

EXAMPLE 3 **Choosing the Best Model**

  Table 4.14 shows the winning times in the women's Olympic 100-meter freestyle event for the years 1968 to 2000.

TABLE 4.14 Women's Olympic 100-Meter Freestyle, 1968 to 2000

Year	Time (in seconds)	Year	Time (in seconds)
1968	60.0	1988	54.93
1972	58.59	1992	54.64
1976	55.65	1996	54.50
1980	54.79	2000	53.83
1984	55.92		

Source: Time Almanac 2002.

a. Determine whether the data in **Table 4.14** can best be modeled by an exponential function or a logarithmic function.

b. Use the function you chose in part **a.** to predict the winning time in the women's Olympic 100-meter freestyle event for the year 2008.

Solution

a. Construct a scatter plot of the data. In this example we have represented the year 1968 by $x = 68$, the year 2000 by $x = 100$, and the winning time by y.

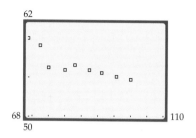

FIGURE 4.53

From the scatter plot in **Figure 4.53,** it appears that the data can be effectively modeled by a decreasing exponential function and also by a decreasing logarithmic function. Use a graphing utility to determine both an exponential regression function and a logarithmic regression function for the data. **Figure 4.54** shows the exponential regression function, and **Figure 4.55** shows the logarithmic regression function.

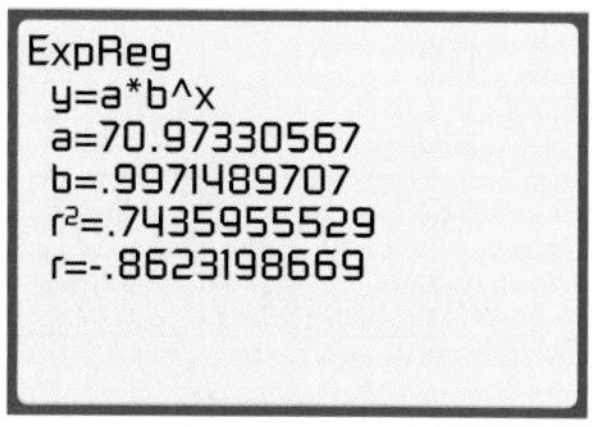

```
ExpReg
y=a*b^x
a=70.97330567
b=.9971489707
r²=.7435955529
r=-.8623198669
```

FIGURE 4.54

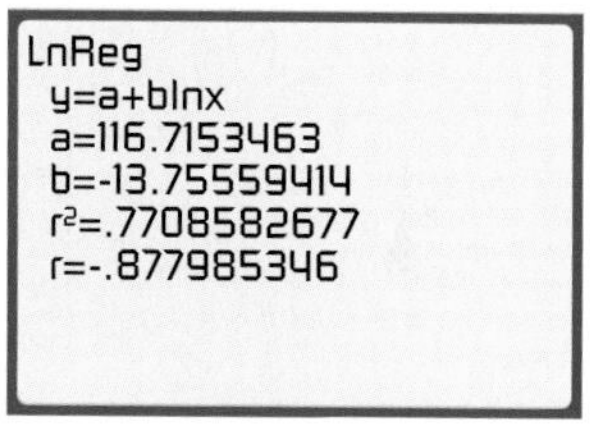

```
LnReg
y=a+blnx
a=116.7153463
b=-13.75559414
r²=.7708582677
r=-.877985346
```

FIGURE 4.55

In this example the regression coefficients are both negative. In such cases, the regression function that has a correlation coefficient closer to -1 provides the better fit for the given data. Thus the logarithmic model provides a slightly better fit for the data in this example. The logarithmic regression function is $y \approx 116.72 - 13.756 \ln x$. The graph of $y \approx 116.72 - 13.756 \ln x$, along with a scatter plot of the data, is shown in **Figure 4.56.**

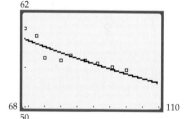

FIGURE 4.56

b. To predict the winning time for the women's Olympic 100-meter freestyle event in the year 2008, replace x in the logarithmic regression function with 108.

$$y \approx 116.72 - 13.756 \ln(108) \approx 52.31$$

According to the logarithmic regression function, the winning time for the women's Olympic 100-meter freestyle event in the year 2008 will be about 52.31 seconds.

▶ **TRY EXERCISE 24, PAGE 445**

● USE REGRESSION TO FIND A LOGISTIC GROWTH MODEL

If a scatter plot of a set of data suggests that the data can be effectively modeled by a logistic growth model, then you can use the logistic regression feature of a graphing utility to find the logistic growth model that provides the best fit for the data. This process is illustrated in Example 4.

EXAMPLE 4   Use Logistic Regression to Find a Logistic Growth Model

 Table 4.15 shows the population of deer in an animal preserve for the years 1990 to 2004.

TABLE 4.15   Deer Population at the Wild West Animal Preserve

Year	Population	Year	Population	Year	Population
1990	320	1995	1150	2000	2620
1991	410	1996	1410	2001	2940
1992	560	1997	1760	2002	3100
1993	730	1998	2040	2003	3300
1994	940	1999	2310	2004	3460

Use a graphing utility to find a logistic regression model that approximates the deer population as a function of the year. Use the model to predict the deer population in the year 2010.

Solution

1. **Construct a scatter plot of the data.** Enter the data into a graphing utility, and then use the utility to display a scatter plot of the data. In this example we represent the year 1990 by $x = 0$, the year 2004 by $x = 14$, and the deer population by y.

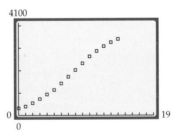

FIGURE 4.57

Figure 4.57 shows that the data can be closely approximated by a logistic growth model.

INTEGRATING TECHNOLOGY

On a TI-83 graphing calculator, the logistic growth model is given in the form

$$y = \frac{c}{1 + ae^{-bx}}$$

Think of the variable x as the time t and the variable y as $P(t)$.

SECTION 5.1

ANGLES AND ARCS

- DEGREE MEASURE
- CLASSIFICATION OF ANGLES
- CONVERSION BETWEEN UNITS
- RADIAN MEASURE
- ARCS AND ARC LENGTH
- LINEAR AND ANGULAR SPEED

A point P on a line separates the line into two parts, each of which is called a **half-line**. The union of point P and the half-line formed by P that includes point A is called a **ray**, and it is represented as $\overrightarrow{PA}$. The point P is the **endpoint** of ray $\overrightarrow{PA}$. **Figure 5.1** shows the ray $\overrightarrow{PA}$ and a second ray $\overrightarrow{QR}$.

In geometry, an *angle* is defined simply as the union of two rays that have a common endpoint. In trigonometry and many advanced mathematics courses, it is beneficial to define an angle in terms of a rotation.

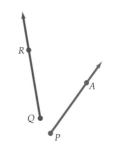

FIGURE 5.1

> ### Definition of an Angle
>
> An **angle** is formed by rotating a given ray about its endpoint to some terminal position. The original ray is the **initial side** of the angle, and the second ray is the **terminal side** of the angle. The common endpoint is the **vertex** of the angle.

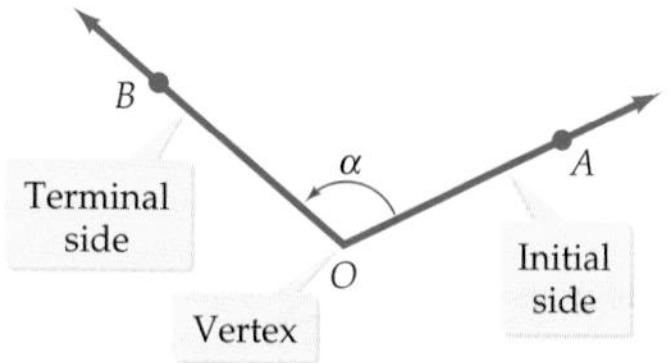

There are several methods used to name an angle. One way is to employ Greek letters. For example, the angle shown in **Figure 5.2** can be designated as α or as $\angle\alpha$. It also can be named $\angle O$, $\angle AOB$, or $\angle BOA$. If you name an angle by using three points, such as $\angle AOB$, it is traditional to list the vertex point between the other two points.

Angles formed by a counterclockwise rotation are considered **positive angles**, and angles formed by a clockwise rotation are considered **negative angles**. See **Figure 5.3**.

FIGURE 5.2

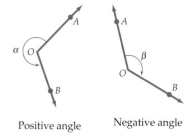

Positive angle   Negative angle

FIGURE 5.3

● DEGREE MEASURE

The **measure** of an angle is determined by the amount of rotation of the initial ray. The concept of measuring angles in *degrees* grew out of the belief of the early Sumerians and Babylonians that the seasons repeated every 360 days.

that the radius of Earth is 3960 miles. Round to the nearest hundredth of a radian per hour.

c. If the Concorde left London at 1 P.M., what time would it be expected to arrive in New York City? (*Hint:* New York City is five time zones to the west of London.)

78. RACING THE SUN A pilot is flying a supersonic jet plane from east to west along a path over the equator. How fast, in miles per hour, does the pilot need to fly to keep the sun in the same relative position to the airplane? Assume that the plane is flying at an altitude of 2 miles above Earth and that Earth has a radius of 3960 miles. Round to the nearest mile per hour.

79. ASTRONOMY At a time when the earth was 93,000,000 miles from the sun, you observed through a tinted glass that the diameter of the sun occupied an arc of 31′. Determine, to the nearest ten thousand miles, the diameter of the sun.

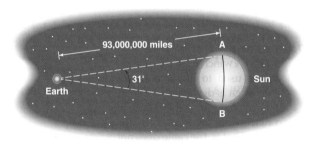

(*Hint:* Because the radius of arc *AB* is large and its central angle is small, the length of the diameter of the sun is approximately the length of the arc *AB*.)

80. ANGLE OF ROTATION AND DISTANCE The minute hand on the clock atop city hall measures 6 feet 3 inches from its tip to its axle.

a. Through what angle (in radians) does the minute hand pass between 9:12 A.M. and 9:48 A.M.?

b. What distance, to the nearest tenth of a foot, does the tip of the minute hand travel during this period?

81. VELOCITY OF THE HUBBLE SPACE TELESCOPE On April 25, 1990, the Hubble Space Telescope (HST) was deployed into a circular orbit 625 kilometers above the surface of the earth. The HST completes an earth orbit every 1.61 hours.

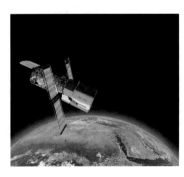

a. Find the angular velocity, with respect to the center of the earth, of the HST. Round your answer to the nearest 0.1 radian per hour.

b. Find the linear velocity of the HST. (*Hint:* The radius of the earth is about 6370 kilometers.) Round your answer to the nearest 100 kilometers per hour.

82. ESTIMATING THE RADIUS OF THE EARTH Eratosthenes, the fifth librarian of Alexandria (230 B.C.), was able to estimate the radius of the earth from the following data: The distance between the Egyptian cities of Alexandria and Syrene was 5000 stadia (520 miles). Syrene was located directly south of Alexandria. One summer, at noon, the sun was directly overhead at Syrene, whereas at the same time in Alexandria, the sun was at a 7.5° angle from the zenith.

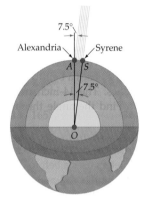

Eratosthenes reasoned that because the sun is far away, the rays of sunlight that reach the earth must be nearly parallel. From this assumption he concluded that the measure of ∠*AOS* in the accompanying figure must be 7.5°. Use this information to estimate the radius (to the nearest 10 miles) of the earth.

83. VELOCITY COMPARISONS Assume that the bicycle in the figure is moving forward at a constant rate. Point *A* is on the edge of the 30-inch rear tire, and point *B* is on the edge of the 20-inch front tire.

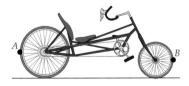

a. Which point (*A* or *B*) has the greater angular velocity? Explain.

b. Which point (*A* or *B*) has the greater linear velocity? Explain.

From the definition of the sine and cosecant functions,

$$(\sin \theta)(\csc \theta) = \frac{y}{r} \cdot \frac{r}{y} = 1 \quad \text{or} \quad (\sin \theta)(\csc \theta) = 1$$

By rewriting the last equation, we find

$$\sin \theta = \frac{1}{\csc \theta} \quad \text{and} \quad \csc \theta = \frac{1}{\sin \theta}, \text{ provided } \sin \theta \neq 0$$

The sine and cosecant functions are called **reciprocal functions.** The cosine and secant are also reciprocal functions, as are the tangent and cotangent functions. **Table 5.3** shows each trigonometric function and its reciprocal. These relationships hold for all values of θ for which both of the functions are defined.

TABLE 5.3   Trigonometric Functions and Their Reciprocals

$\sin \theta = \dfrac{1}{\csc \theta}$	$\cos \theta = \dfrac{1}{\sec \theta}$	$\tan \theta = \dfrac{1}{\cot \theta}$
$\csc \theta = \dfrac{1}{\sin \theta}$	$\sec \theta = \dfrac{1}{\cos \theta}$	$\cot \theta = \dfrac{1}{\tan \theta}$

INTEGRATING TECHNOLOGY

Some graphing calculators will allow you to display the degree symbol. For example, the TI-83 display

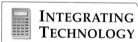

was produced by entering

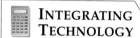

Displaying the degree symbol will cause the calculator to evaluate a trigonometric function using degree mode even if the calculator is in radian mode.

INTEGRATING TECHNOLOGY

When evaluating a trigonometric function by using a graphing calculator, be sure the calculator is in the correct mode. If the measure of an angle is written with the degree symbol, then make sure the calculator is in degree mode. If the measure of an angle is given in radians (no degree symbol is used), then make sure the calculator is in radian mode. *Many errors are made because the correct mode is not selected.*

Some graphing calculators can be used to construct a table of functional values. For instance, the TI-83 keystrokes shown below generate the table in **Figure 5.36,** in which the first column lists the domain values

$$0°, 1°, 2°, 3°, \ldots$$

and the second column lists the range values

$$\sin 0°, \sin 1°, \sin 2°, \sin 3°, \ldots$$

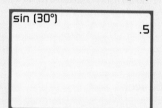

FIGURE 5.36

The graphing calculator must be in degree mode to produce this table.

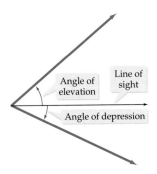

FIGURE 5.37

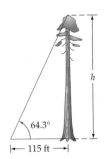

FIGURE 5.38

---

take note

The significant digits of an approximate number are

- every nonzero digit.
- the digit 0, provided it is between two nonzero digits or it is to the right of a nonzero digit in a number which includes a decimal point.

For example, the approximate number:

502 has 3 significant digits.

3700 has 2 significant digits.

47.0 has 3 significant digits.

0.0023 has 2 significant digits.

0.00840 has 3 significant digits.

• APPLICATIONS INVOLVING RIGHT TRIANGLES

One of the major reasons for the development of trigonometry was to solve application problems. In this section we will consider some applications involving right triangles. In some application problems a horizontal line of sight is used as a reference line. An angle measured above the line of sight is called an **angle of elevation,** and an angle measured below the line of sight is called an **angle of depression.** See **Figure 5.37.**

EXAMPLE 4   Solve an Angle-of-Elevation Problem

From a point 115 feet from the base of a redwood tree, the angle of elevation to the top of the tree is 64.3°. Find the height of the tree to the nearest foot.

Solution

From **Figure 5.38,** the length of the adjacent side of the angle is known (115 feet). Because we need to determine the height of the tree (length of the opposite side), we use the tangent function. Let h represent the length of the opposite side.

$$\tan 64.3° = \frac{\text{opp}}{\text{adj}} = \frac{h}{115}$$
$$h = 115 \tan 64.3° \approx 238.952 \qquad \text{• Use a calculator to evaluate } \tan 64.3°.$$

The height of the tree is approximately 239 feet.

▶ **TRY EXERCISE 64, PAGE 486**

Because the cotangent function involves the sides adjacent to and opposite an angle, we could have solved Example 4 by using the cotangent function. The solution would have been

$$\cot 64.3° = \frac{\text{adj}}{\text{opp}} = \frac{115}{h}$$
$$h = \frac{115}{\cot 64.3°} \approx 238.952 \text{ feet}$$

The accuracy of a calculator is sometimes beyond the limits of measurement. In the last example the distance from the base of the tree was given as 115 feet (three significant digits), whereas the height of the tree was shown to be 238.952 feet (six significant digits). When using approximate numbers, we will use the conventions given below for calculating with trigonometric functions.

A Rounding Convention:
Significant Digits for Trigonometric Calculations

Angle Measure to the Nearest	Significant Digits of the Lengths
Degree	Two
Tenth of a degree	Three
Hundredth of a degree	Four

EXAMPLE 5   **Solve an Angle-of-Depression Problem**

DME (Distance Measuring Equipment) is standard avionic equipment on a commercial airplane. This equipment measures the distance from a plane to a radar station. If the distance from a plane to a radar station is 160 miles and the angle of depression is 33°, find the number of ground miles from a point directly below the plane to the radar station.

Solution

From **Figure 5.39,** the length of the hypotenuse is known (160 miles). The length of the side opposite the angle of 57° is unknown. The sine function involves the hypotenuse and the opposite side, x, of the 57°angle.

$$\sin 57° = \frac{x}{160}$$

$$x = 160 \sin 57° \approx 134.1873$$

Rounded to two significant digits, the plane is 130 ground miles from the radar station.

▶ **TRY EXERCISE 66, PAGE 486**

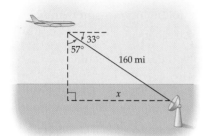

FIGURE 5.39

EXAMPLE 6   **Solve an Angle-of-Elevation Problem**

An observer notes that the angle of elevation from point A to the top of a space shuttle is 27.2°. From a point 17.5 meters further from the space shuttle, the angle of elevation is 23.9°. Find the height of the space shuttle.

Solution

From **Figure 5.40,** let x denote the distance from point A to the base of the space shuttle, and let y denote the height of the space shuttle. Then

$$(1) \quad \tan 27.2° = \frac{y}{x} \quad \text{and} \quad (2) \quad \tan 23.9° = \frac{y}{x + 17.5}$$

Solving Equation (1) for x, $x = \dfrac{y}{\tan 27.2°} = y \cot 27.2°$, and substituting into Equation (2), we have

$$\tan 23.9° = \frac{y}{y \cot 27.2° + 17.5}$$

$$y = (\tan 23.9°)(y \cot 27.2° + 17.5) \qquad \bullet \textbf{ Solve for } y.$$

$$y - y \tan 23.9° \cot 27.2° = (\tan 23.9°)(17.5)$$

$$y = \frac{(\tan 23.9°)(17.5)}{1 - \tan 23.9° \cot 27.2°} \approx 56.2993$$

To three significant digits, the height of the space shuttle is 56.3 meters.

▶ **TRY EXERCISE 74, PAGE 487**

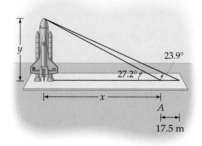

FIGURE 5.40

---

take note

The intermediate calculations in Example 6 were not rounded off. This ensures better accuracy for the final result. Using the rounding convention stated on page 482, we round off only the last result.

484 **Chapter 5** Trigonometric Functions

 TOPICS FOR DISCUSSION

1. If θ is an acute angle of a right triangle for which $\cos\theta = \dfrac{3}{8}$, then it must be the case that $\sin\theta = \dfrac{5}{8}$. Do you agree? Explain.

2. A tutor claims that $\tan 30° = \cot 60°$. Do you agree?

3. Does $\sin 2\theta = 2\sin\theta$? Explain.

4. How many significant digits are in each of the following measurements?
   a. 0.0042 inches   b. 5.03 inches   c. 62.00 inches

5. A student claims that $\sin^2 30° = (\sin 30°)^2$. Do you agree? Explain.

EXERCISE SET 5.2

In Exercises 1 to 14, find the values of the six trigonometric functions of θ for the right triangle with the given sides.

1.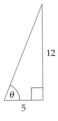

2.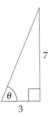

3.

4.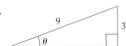

5.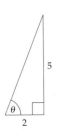

▶ **6.**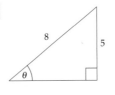

7.

8.

9.

10.

11.

12.

13.

14.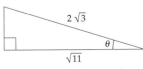

In Exercises 15 to 17, let θ be an acute angle of a right triangle for which $\sin\theta = \dfrac{3}{5}$. Find

15. $\tan\theta$ **16.** $\sec\theta$ **17.** $\cos\theta$

In Exercises 18 to 20, let θ be an acute angle of a right triangle for which $\tan\theta = \dfrac{4}{3}$. Find

18. $\sin\theta$ **19.** $\cot\theta$ ▶ **20.** $\sec\theta$

In Exercises 21 to 23, let β be an acute angle of a right triangle for which sec β = $\frac{13}{12}$. Find

21. cos β   **22.** cot β   **23.** csc β

In Exercises 24 to 26, let θ be an acute angle of a right triangle for which cos θ = $\frac{2}{3}$. Find

24. sin θ   **25.** sec θ   **26.** tan θ

In Exercises 27 to 42, find the *exact* value of each expression.

27. sin 45° + cos 45°

28. csc 45° − sec 45°

29. sin 30° cos 60° − tan 45°

30. csc 60° sec 30° + cot 45°

31. sin 30° cos 60° + tan 45°

32. sec 30° cos 30° − tan 60° cot 60°

33. 2 sin 60° − sec 45° tan 60°

34. sec 45° cot 30° + 3 tan 60°

35. $\sin \frac{\pi}{3} + \cos \frac{\pi}{6}$

36. $\csc \frac{\pi}{6} - \sec \frac{\pi}{3}$

37. $\sin \frac{\pi}{4} + \tan \frac{\pi}{6}$

▶ **38.** $\sin \frac{\pi}{3} \cos \frac{\pi}{4} - \tan \frac{\pi}{4}$

39. $\sec \frac{\pi}{3} \cos \frac{\pi}{3} - \tan \frac{\pi}{6}$

40. $\cos \frac{\pi}{4} \tan \frac{\pi}{6} + 2 \tan \frac{\pi}{3}$

41. $2 \csc \frac{\pi}{4} - \sec \frac{\pi}{3} \cos \frac{\pi}{6}$

42. $3 \tan \frac{\pi}{4} + \sec \frac{\pi}{6} \sin \frac{\pi}{3}$

In Exercises 43 to 56, use a calculator to find the value of the trigonometric function to four decimal places.

43. tan 32°   **44.** sec 88°   **45.** cos 63°20′

46. cot 55°50′   **47.** cos 34.7°   **48.** tan 81.3°

49. sec 5.9°   **50.** $\sin \frac{\pi}{5}$ **51.** $\tan \frac{\pi}{7}$

52. $\sec \frac{3\pi}{8}$ **53.** csc 1.2   **54.** sin 0.45

55. cos 1.25   **56.** $\tan \frac{3}{4}$

57. **VERTICAL HEIGHT FROM SLANT HEIGHT** A 12-foot ladder is resting against a wall and makes an angle of 52° with the ground. Find the height to which the ladder will reach on the wall.

58. **DISTANCE ACROSS A MARSH** Find the distance *AB* across the marsh shown in the accompanying figure.

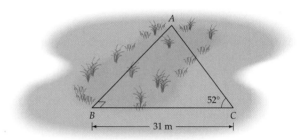

59. **SLOPE OF A LINE** Show that the slope of a line that makes an angle θ with the positive *x*-axis equals tan θ.

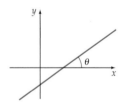

60. **WIDTH OF A SCREEN** Television screens are measured by the length of the diagonal of the screen. Find the width of a 19-inch television screen if the diagonal makes an angle of 38° with the base of the screen.

61. **CLOSEST APPROACH** A boat is 40 kilometers due east of a lighthouse and traveling in a direction that is 30° south of an east-west line as shown in the following figure. What is the closest distance the boat will come to the lighthouse?

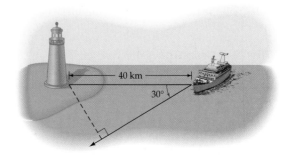

62. TIME OF CLOSEST APPROACH At 3:00 P.M., a boat is 12.5 miles due west of a radar station and traveling at 11 mph in a direction that is 57.3° south of an east-west line. At what time will the boat be closest to the radar station?

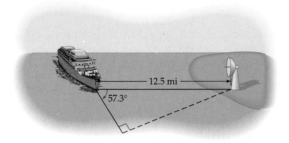

63. PLACEMENT OF A LIGHT For best illumination of a piece of art, a lighting specialist for an art gallery recommends that a ceiling-mounted light be 6 feet from the piece of art and that the angle of depression of the light be 38°. How far from a wall should the light be placed so that the recommendations of the specialist are met? Notice that the art extends outward 4 inches from the wall.

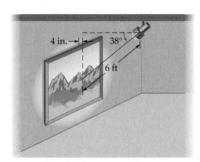

▶ **64. HEIGHT OF THE EIFFEL TOWER** The angle of elevation from a point 116 meters from the base of the Eiffel Tower to the top of the tower is 68.9°. Find the approximate height of the tower.

65. DISTANCE OF A DESCENT An airplane traveling at 240 mph is descending at an angle of depression of 6°. How many miles will the plane descend in 4 minutes?

▶ **66. TIME OF A DESCENT** A submarine traveling at 9.0 mph is descending at an angle of depression of 5°. How many minutes, to the nearest tenth, does it take the submarine to reach a depth of 80 feet?

67. HEIGHT OF AN AQUEDUCT From a point 300 feet from the base of a Roman aqueduct in southern France, the angle of elevation to the top of the aqueduct is 78°. Find the height of the aqueduct.

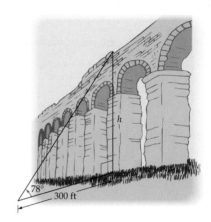

68. WIDTH OF A LAKE The angle of depression to one side of a lake, measured from a balloon 2500 feet above the lake as shown in the accompanying figure, is 43°. The angle of depression to the opposite side of the lake is 27°. Find the width of the lake.

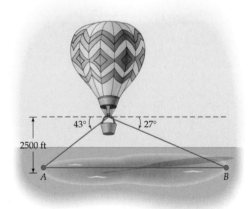

69. ASTRONOMY The moon Europa rotates in a nearly circular orbit around Jupiter. The orbital radius of Europa is approximately 670,900 kilometers. During a revolution of Europa around Jupiter, an astronomer found that the maximum value of the angle θ formed by Europa, Earth, and Jupiter was 0.056°. See the following figure on page 487. Find the distance d between Earth and Jupiter at the time the astronomer found the maximum value of θ. Round to the nearest million kilometers.

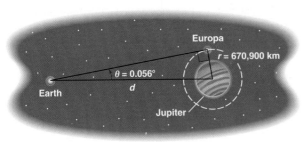

Not drawn to scale.

70. ASTRONOMY Venus rotates in a nearly circular orbit around the sun. The largest angle formed by Venus, Earth, and the sun is 46.5°. The distance from Earth to the sun is approximately 149,000,000 kilometers. What is the orbital radius *r* of Venus? Round to the nearest million kilometers.

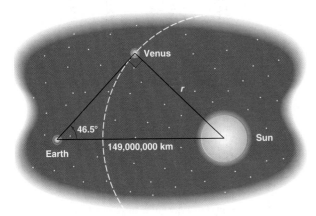

71. Consider the following *isosceles* triangle. The length of each of the two equal sides of the triangle is *a*, and each of the base angles has a measure of *θ*. Verify that the area of the triangle is $A = a^2 \sin \theta \cos \theta$.

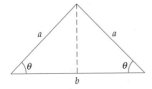

72. Find the area of the following hexagon. (*Hint:* The area consists of six isosceles triangles. Use the formula from Exercise 71 to compute the area of one of the triangles and multiply by 6.)

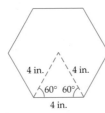

73. HEIGHT OF A PYRAMID The angle of elevation to the top of the Egyptian pyramid Cheops is 36.4°, measured from a point 350 feet from the base of the pyramid. The angle of elevation from the base of a face of the pyramid is 51.9°. Find the height of Cheops.

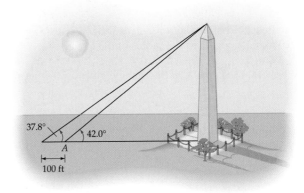

▶ **74. HEIGHT OF A BUILDING** Two buildings are 240 feet apart. The angle of elevation from the top of the shorter building to the top of the other building is 22°. If the shorter building is 80 feet high, how high is the taller building?

75. HEIGHT OF THE WASHINGTON MONUMENT From a point *A* on a line from the base of the Washington Monument, the angle of elevation to the top of the monument is 42.0°. From a point 100 feet away and on the same line, the angle to the top is 37.8°. Find the approximate height of the Washington Monument.

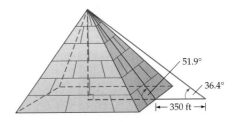

76. HEIGHT OF A BUILDING The angle of elevation to the top of a radio antenna on the top of a building is 53.4°. After moving 200 feet closer to the building, the angle of elevation is 64.3°. Find the height of the building if the height of the antenna is 180 feet.

77. THE PETRONAS TOWERS The Petronas Towers in Kuala Lumpur, Malaysia, are the world's tallest twin towers. Each tower is 1483 feet in height. The towers are connected by a skybridge at the forty-first floor. Note the information given in the accompanying figure.

a. Determine the height of the skybridge.

b. Determine the length of the skybridge.

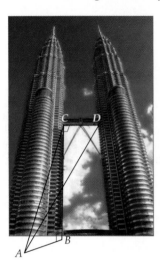

$AB = 412$ feet
$\angle CAB = 53.6°$
$\overline{AB}$ is at ground level
$\angle CAD = 15.5°$

78. AN EIFFEL TOWER REPLICA Use the information in the accompanying figure to estimate the height of the Eiffel Tower replica that stands in front of the Paris Las Vegas Hotel in Las Vegas, Nevada.

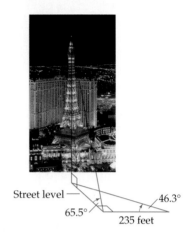

Street level   46.3°
65.5°
235 feet

CONNECTING CONCEPTS

79. A circle is inscribed in a regular hexagon with each side 6.0 meters long. Find the radius of the circle.

80. Show that the area A of the triangle given in the figure is $A = \dfrac{1}{2} ab \sin \theta$.

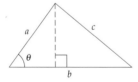

81. DETERMINE A RANGE OF HEIGHTS If an angle of 27° has been measured to the nearest degree, then the actual measure of the angle θ is such that $26.5° \le \theta < 27.5$. From a distance of exactly 100 meters from the base of a tree, the angle of elevation is measured as 27°, to the nearest degree. Find the range in which the height h of the tree must fall.

82. HEIGHT OF A TOWER Let B denote the base of a clock tower. The angle of elevation from a point A, on the ground, to the top of the clock tower is 56.3°. On a line on the

ground that is perpendicular to AB and 25 feet from A, the angle of elevation is 53.3°. Find the height of the clock tower.

83. FIND A MAXIMUM LENGTH Find the length of the longest piece of wood that can be slid around the corner of the hallway in the figure.

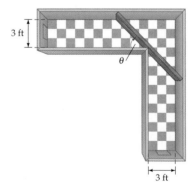

3 ft

θ

3 ft

84. FIND A MAXIMUM LENGTH In Exercise 83, suppose that the hall is 8 feet high. Find the length of the longest piece of wood that can be taken around the corner. Round to the nearest 0.1 foot.

───── *PREPARE FOR SECTION 5.3* ─────────────────────

85. Find the reciprocal of $-\dfrac{3}{4}$. [P.1]

86. Find the reciprocal of $\dfrac{2\sqrt{5}}{5}$. [P.1]

87. Find: $|120 - 180|$ [P.2]

88. Find: $2\pi - \dfrac{9\pi}{5}$ [P.1]

89. Find: $\dfrac{3}{2}\pi - \dfrac{\pi}{2}$ [P.1]

90. Find: $\sqrt{(-3)^2 + (-5)^2}$ [P.2]

───── *PROJECTS* ───────────────────────────────

1. a. PERIMETER OF A REGULAR *n*-GON Show that the perimeter P of a regular n-sided polygon (n-gon) inscribed in a circle of radius 1 is $P = 2n \sin \dfrac{180°}{n}$.

b. Let P_n denote the perimeter of a regular n-gon inscribed in a circle of radius 1. Use the result from part **a.** to complete the following table.

n	10	50	100	1000	10,000
P_n					

Write a few sentences explaining why P_n approaches 2π as n increases without bound.

2. a. AREA OF A REGULAR *n*-GON Show that the area A of a regular n-gon inscribed in a circle of radius 1 is $A = \dfrac{n}{2} \sin \dfrac{360°}{n}$.

b. Let A_n denote the area of a regular n-gon inscribed in a circle of radius 1. Use the result from part **a.** to complete the following table.

n	10	50	100	1000	10,000
A_n					

Write a few sentences explaining why A_n approaches π as n increases without bound.

SECTION 5.3   # TRIGONOMETRIC FUNCTIONS OF ANY ANGLE

- **TRIGONOMETRIC FUNCTIONS OF QUADRANTAL ANGLES**
- **SIGNS OF TRIGONOMETRIC FUNCTIONS**
- **THE REFERENCE ANGLE**

The applications of trigonometry would be quite limited if all angles had to be acute angles. Fortunately, this is not the case. In this section we extend the definition of a trigonometric function to include any angle.

Consider angle θ in **Figure 5.41** in standard position and a point $P(x, y)$ on the terminal side of the angle. We define the trigonometric functions of any angle according to the following definitions.

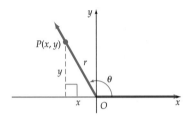

FIGURE 5.41

The Trigonometric Functions of Any Angle

Let $P(x, y)$ be any point, except the origin, on the terminal side of an angle θ in standard position. Let $r = d(O, P)$, the distance from the origin to P. The six trigonometric functions of θ are

$$\sin \theta = \frac{y}{r} \qquad \cos \theta = \frac{x}{r} \qquad \tan \theta = \frac{y}{x}, \quad x \neq 0$$

$$\csc \theta = \frac{r}{y}, \quad y \neq 0 \qquad \sec \theta = \frac{r}{x}, \quad x \neq 0 \qquad \cot \theta = \frac{x}{y}, \quad y \neq 0$$

where $r = \sqrt{x^2 + y^2}$.

take note

The measure of angle θ can be positive or negative. Note from the following figure that $\sin 120° = \sin(-240°)$ because $P(x, y)$ is on the terminal side of each angle. In a similar way, the value of any trigonometric function of 120° is equal to the value of that function of −240°.

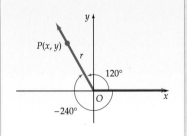

The value of a trigonometric function is independent of the point chosen on the terminal side of the angle. Consider any two points on the terminal side of an angle θ in standard position, as shown in **Figure 5.42**. The right triangles formed are similar triangles, so the ratios of the corresponding sides are equal. Thus, for example, $\dfrac{b}{a} = \dfrac{b'}{a'}$. Because $\tan \theta = \dfrac{b}{a} = \dfrac{b'}{a'}$, we have $\tan \theta = \dfrac{b'}{a'}$. Therefore, the value of the tangent function is independent of the point chosen on the terminal side of the angle. By a similar argument, we can show that the value of any trigonometric function is independent of the point chosen on the terminal side of the angle.

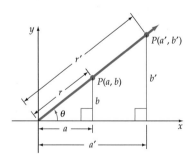

FIGURE 5.42

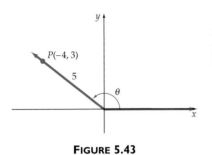

FIGURE 5.43

Any point in a rectangular coordinate system (except the origin) can determine an angle in standard position. For example, $P(-4, 3)$ in **Figure 5.43** is a point in the second quadrant and determines an angle θ in standard position with $r = \sqrt{(-4)^2 + 3^2} = 5$. The values of the trigonometric functions of θ are

$$\sin \theta = \frac{3}{5} \qquad \cos \theta = \frac{-4}{5} = -\frac{4}{5} \qquad \tan \theta = \frac{3}{-4} = -\frac{3}{4}$$

$$\csc \theta = \frac{5}{3} \qquad \sec \theta = \frac{5}{-4} = -\frac{5}{4} \qquad \cot \theta = \frac{-4}{3} = -\frac{4}{3}$$

EXAMPLE 1   **Evaluate Trigonometric Functions**

Find the value of each of the six trigonometric functions of an angle θ in standard position whose terminal side contains the point $P(-3, -2)$.

Solution

The angle is sketched in **Figure 5.44.** Find r by using the equation $r = \sqrt{x^2 + y^2}$, where $x = -3$ and $y = -2$.

$$r = \sqrt{(-3)^2 + (-2)^2} = \sqrt{9 + 4} = \sqrt{13}$$

Now use the definitions of the trigonometric functions.

$$\sin \theta = \frac{-2}{\sqrt{13}} = -\frac{2\sqrt{13}}{13} \qquad \cos \theta = \frac{-3}{\sqrt{13}} = -\frac{3\sqrt{13}}{13} \qquad \tan \theta = \frac{-2}{-3} = \frac{2}{3}$$

$$\csc \theta = \frac{\sqrt{13}}{-2} = -\frac{\sqrt{13}}{2} \qquad \sec \theta = \frac{\sqrt{13}}{-3} = -\frac{\sqrt{13}}{3} \qquad \cot \theta = \frac{-3}{-2} = \frac{3}{2}$$

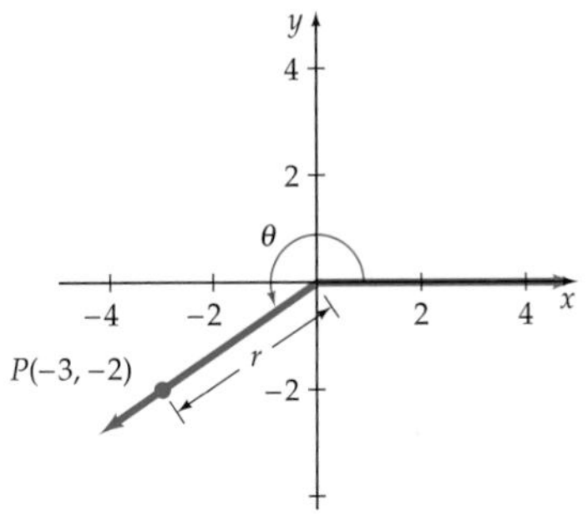

▶ **TRY EXERCISE 6, PAGE 497**

FIGURE 5.44

• TRIGONOMETRIC FUNCTIONS OF QUADRANTAL ANGLES

Recall that a quadrantal angle is an angle whose terminal side coincides with the x- or y-axis. The value of a trigonometric function of a quadrantal angle can be found by choosing any point on the terminal side of the angle and then applying the definition of that trigonometric function.

The terminal side of $0°$ coincides with the positive x-axis. Let $P(x, 0)$, $x > 0$, be any point on the x-axis, as shown in **Figure 5.45.** Then $y = 0$ and $r = x$. The values of the six trigonometric functions of $0°$ are

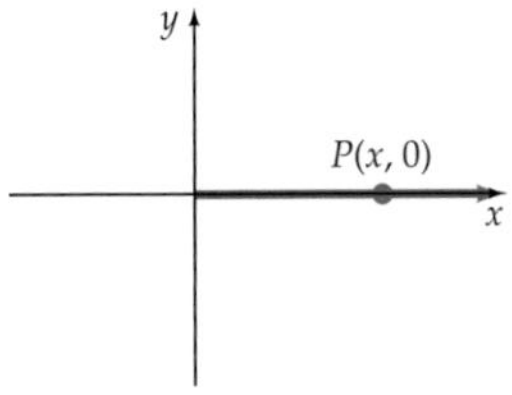

FIGURE 5.45

$$\sin 0° = \frac{0}{r} = 0 \qquad \cos 0° = \frac{x}{r} = \frac{x}{x} = 1 \qquad \tan 0° = \frac{0}{x} = 0$$

$$\csc 0° \text{ is undefined.} \qquad \sec 0° = \frac{r}{x} = \frac{x}{x} = 1 \qquad \cot 0° \text{ is undefined.}$$

? QUESTION Why are csc 0° and cot 0° undefined?

In like manner, the values of the trigonometric functions of the other quadrantal angles can be found. The results are shown in **Table 5.4**.

TABLE 5.4   Values of Trigonometric Functions for Quadrantal Angles

θ	$\sin \theta$	$\cos \theta$	$\tan \theta$	$\csc \theta$	$\sec \theta$	$\cot \theta$
0°	0	1	0	undefined	1	undefined
90°	1	0	undefined	1	undefined	0
180°	0	−1	0	undefined	−1	undefined
270°	−1	0	undefined	−1	undefined	0

● SIGNS OF TRIGONOMETRIC FUNCTIONS

The sign of a trigonometric function depends on the quadrant in which the terminal side of the angle lies. For example, if θ is an angle whose terminal side lies in Quadrant III and $P(x, y)$ is on the terminal side of θ, then both x and y are negative, and therefore, $\dfrac{y}{x}$ and $\dfrac{x}{y}$ are positive. See **Figure 5.46**. Because $\tan \theta = \dfrac{y}{x}$ and $\cot \theta = \dfrac{x}{y}$, the values of the tangent and cotangent functions are positive for any Quadrant III angle. The values of the other four trigonometric functions of any Quadrant III angle are all negative.

Table 5.5 lists the signs of the six trigonometric functions in each quadrant. **Figure 5.47** is a graphical display of the contents of **Table 5.5**.

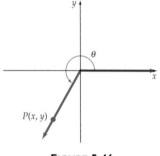

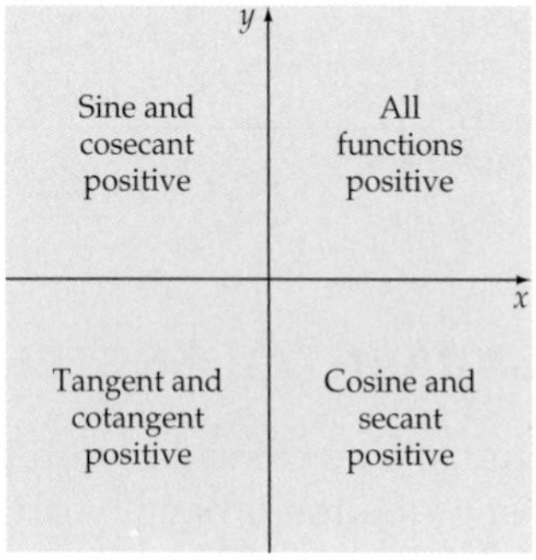

FIGURE 5.46

FIGURE 5.47

TABLE 5.5   Signs of the Trigonometric Functions

Sign of	Terminal Side of θ in Quadrant			
	I	**II**	**III**	**IV**
$\sin \theta$ and $\csc \theta$	positive	positive	negative	negative
$\cos \theta$ and $\sec \theta$	positive	negative	negative	positive
$\tan \theta$ and $\cot \theta$	positive	negative	positive	negative

In the next example we are asked to evaluate two trigonometric functions of the angle θ. A key step is to use our knowledge about trigonometric functions and their signs to determine that θ is a Quadrant IV angle.

EXAMPLE 2   Evaluate Trigonometric Functions

Given $\tan \theta = -\dfrac{7}{5}$ and $\sin \theta < 0$, find $\cos \theta$ and $\csc \theta$.

? ANSWER $P(x, 0)$ is a point on the terminal side of 0°. Thus $\csc 0° = \dfrac{r}{0}$, which is undefined. Similarly, $\cot 0° = \dfrac{x}{0}$, which is undefined.

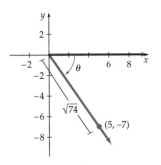

FIGURE 5.48

Solution

The terminal side of angle θ must lie in Quadrant IV; that is the only
quadrant for which $\sin \theta$ and $\tan \theta$ are both negative. Because

$$\tan \theta = -\frac{7}{5} = \frac{y}{x} \qquad (1)$$

and the terminal side of θ is in Quadrant IV, we know that y must be
negative and x must be positive. Thus Equation (1) is true for $y = -7$ and
$x = 5$. Now $r = \sqrt{5^2 + (-7)^2} = \sqrt{74}$. See **Figure 5.48**. Hence

$$\cos \theta = \frac{x}{r} = \frac{5}{\sqrt{74}} = \frac{5\sqrt{74}}{74} \qquad \text{and} \qquad \csc \theta = \frac{r}{y} = \frac{\sqrt{74}}{-7} = -\frac{\sqrt{74}}{7}$$

▶ **TRY EXERCISE 18, PAGE 497**

• THE REFERENCE ANGLE

We will often find it convenient to evaluate trigonometric functions by making use
of the concept of a *reference angle*.

Reference Angle

Given $\angle \theta$ in standard position, its **reference angle** θ' is the smallest
positive angle formed by the terminal side of $\angle \theta$ and the x-axis.

Figure 5.49 shows $\angle \theta$ and its reference angle θ' for four cases. In every case
the reference angle θ' is formed by the terminal side of $\angle \theta$ and the x-axis (never
the y-axis). The process of determining the measure of $\angle \theta'$ varies according to
what quadrant contains the terminal side of $\angle \theta$.

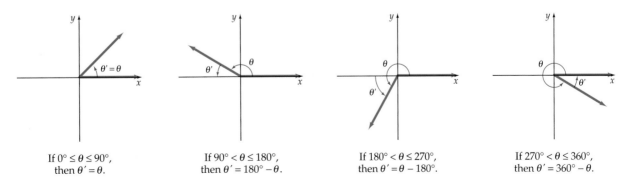

If $0° \leq \theta \leq 90°$,
then $\theta' = \theta$.

If $90° < \theta \leq 180°$,
then $\theta' = 180° - \theta$.

If $180° < \theta \leq 270°$,
then $\theta' = \theta - 180°$.

If $270° < \theta \leq 360°$,
then $\theta' = 360° - \theta$.

FIGURE 5.49

> ## EXAMPLE 3   Find the Measure of the Reference Angle

For each of the following, sketch the given angle θ (in standard position) and its reference angle θ'. Then determine the measure of θ'.

a. $\theta = 120°$ **b.** $\theta = 345°$ **c.** $\theta = 924°$

d. $\theta = \dfrac{9}{5}\pi$ **e.** $\theta = -4$ **f.** $\theta = 17$

Solution

a.

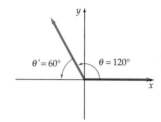

$\theta' = 180° - 120° = 60°$

b.

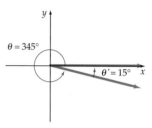

$\theta' = 360° - 345° = 15°$

c.

Because $\theta = 924° > 360°$, we first determine the coterminal angle, $\alpha = 204°$.

$\theta' = 204° - 180° = 24°$

d.

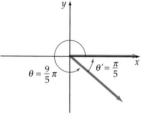

$\theta' = 2\pi - \dfrac{9}{5}\pi$

$= \dfrac{10\pi}{5} - \dfrac{9\pi}{5} = \dfrac{\pi}{5} \approx 0.628$

e.

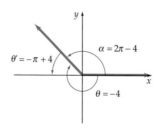

-4 radians is coterminal with $\alpha = 2\pi - 4$.

$\theta' = \pi - (2\pi - 4)$
$= -\pi + 4 \approx 0.8584$

f.

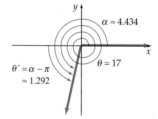

Because $17 \div (2\pi) \approx 2.70563$, a coterminal angle is $\alpha = 0.70563(2\pi) \approx 4.434$.

Because α is a Quadrant III angle, $\theta' = \alpha - \pi \approx 1.292$

▶ **TRY EXERCISE 26, PAGE 497**

The following theorem states an important relationship that exists between $\sin \theta$ and $\sin \theta'$, where θ' is the reference angle for angle θ.

Reference Angle Theorem

To evaluate $\sin \theta$, determine $\sin \theta'$. Then use either $\sin \theta'$ or its opposite as the answer, depending on which has the correct sign.

In the following example, we illustrate how to evaluate a trigonometric function of θ by first evaluating the trigonometric function of θ'.

EXAMPLE 4   Use the Reference Angle Theorem to Evaluate Trigonometric Functions

Evaluate each function.

a. $\sin 210°$ **b.** $\cos 405°$ **c.** $\tan 300°$

Solution

a. We know that $\sin 210°$ is negative (the sign chart is given in **Table 5.5**). The reference angle for $\theta = 210°$ is $\theta' = 30°$. By the Reference Angle Theorem, we know that $\sin 210°$ equals either

$$\sin 30° = \frac{1}{2} \quad \text{or} \quad -\sin 30° = -\frac{1}{2}$$

Thus $\sin 210° = -\dfrac{1}{2}$.

b. Because $\theta = 405°$ is a Quadrant I angle, we know that $\cos 405° > 0$. The reference angle for $\theta = 405°$ is $\theta' = 45°$. By the Reference Angle Theorem, $\cos 405°$ equals either

$$\cos 45° = \frac{\sqrt{2}}{2} \quad \text{or} \quad -\cos 45° = -\frac{\sqrt{2}}{2}$$

Thus $\cos 405° = \dfrac{\sqrt{2}}{2}$.

c. Because $\theta = 300°$ is a Quadrant IV angle, $\tan 300° < 0$. The reference angle for $\theta = 300°$ is $\theta' = 60°$. Hence $\tan 300°$ equals either

$$\tan 60° = \sqrt{3} \quad \text{or} \quad -\tan 60° = -\sqrt{3}$$

Thus $\tan 300° = -\sqrt{3}$.

▶ **TRY EXERCISE 38, PAGE 497**

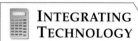
INTEGRATING
TECHNOLOGY

In many applications a calculator is used to evaluate a trigonometric function of the angle. Use *degree mode* to evaluate a trigonometric function of an angle given in degrees. Use *radian mode* to evaluate a trigonometric function of an angle given in radians. **Figure 5.50** shows the sine of 137.4 degrees, and **Figure 5.51** shows the cotangent of 3.4 radians.

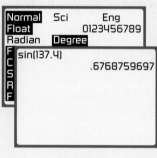

FIGURE 5.50   **FIGURE 5.51**

 TOPICS FOR DISCUSSION

1. Is every reference angle an acute angle? Explain.

2. If θ' is the reference angle for the angle θ, then $\sin \theta = \sin \theta'$. Do you agree? Explain.

3. If $\sin \theta < 0$ and $\cos \theta > 0$, then the terminal side of the angle θ lies in which quadrant?

4. A student claims that if $\theta = 160$, then $\theta' = 20$. Explain why the student is not correct.

5. Explain how to find the measure of the reference angle θ' for the angle $\theta = \dfrac{19}{5}\pi$.

EXERCISE SET 5.3

In Exercises I to 8, find the value of each of the six trigono-metric functions for the angle whose terminal side passes through the given point.

1. $P(2, 3)$ **2.** $P(3, 7)$ **3.** $P(-2, 3)$

4. $P(-3, 5)$ **5.** $P(-8, -5)$ ▶ **6.** $P(-6, -9)$

7. $P(-5, 0)$ **8.** $P(0, 2)$

In Exercises 9 to 14, let θ be an angle in standard position. State the quadrant in which the terminal side of θ lies.

9. $\sin \theta > 0$, $\cos \theta > 0$ **10.** $\tan \theta < 0$, $\sin \theta < 0$

11. $\cos \theta > 0$, $\tan \theta < 0$ **12.** $\sin \theta < 0$, $\cos \theta > 0$

13. $\sin \theta < 0$, $\cos \theta < 0$ **14.** $\tan \theta < 0$, $\cos \theta < 0$

In Exercises 15 to 24, find the value of each expression.

15. $\sin \theta = -\dfrac{1}{2}$, $180° < \theta < 270°$; find $\tan \theta$.

16. $\cot \theta = -1, 90° < \theta < 180°$; find $\cos \theta$.

17. $\csc \theta = \sqrt{2}, \dfrac{\pi}{2} < \theta < \pi$; find $\cot \theta$.

▶ **18.** $\sec \theta = \dfrac{2\sqrt{3}}{3}, \dfrac{3\pi}{2} < \theta < 2\pi$; find $\sin \theta$.

19. $\sin \theta = -\dfrac{1}{2}$ and $\cos \theta > 0$; find $\tan \theta$.

20. $\tan \theta = 1$ and $\sin \theta < 0$; find $\cos \theta$.

21. $\cos \theta = \dfrac{1}{2}$ and $\tan \theta = \sqrt{3}$; find $\csc \theta$.

22. $\tan \theta = 1$ and $\sin \theta = \dfrac{\sqrt{2}}{2}$; find $\sec \theta$.

23. $\cos \theta = -\dfrac{1}{2}$ and $\sin \theta = \dfrac{\sqrt{3}}{2}$; find $\cot \theta$.

24. $\sec \theta = \dfrac{2\sqrt{3}}{3}$ and $\sin \theta = -\dfrac{1}{2}$; find $\cot \theta$.

In Exercises 25 to 36, find the measure of the reference angle θ' for the given angle θ.

25. $\theta = 160°$ ▶ **26.** $\theta = 255°$ **27.** $\theta = 351°$

28. $\theta = 48°$ **29.** $\theta = \dfrac{11}{5}\pi$ **30.** $\theta = -6$

31. $\theta = \dfrac{8}{3}$ **32.** $\theta = \dfrac{18}{7}\pi$ **33.** $\theta = 1406°$

34. $\theta = 840°$ **35.** $\theta = -475°$ **36.** $\theta = -650°$

In Exercises 37 to 48, use the Reference Angle Theorem to find the exact value of each trigonometric function.

37. $\sin 225°$ ▶ **38.** $\cos 300°$ **39.** $\tan 405°$

40. $\sec 150°$ **41.** $\csc \dfrac{4}{3}\pi$ **42.** $\cot \dfrac{7}{6}\pi$

43. $\cos \dfrac{17\pi}{4}$ **44.** $\tan \left(-\dfrac{\pi}{3}\right)$ **45.** $\sec 765°$

46. $\csc (-510°)$ **47.** $\cot 540°$ **48.** $\cos 570°$

In Exercises 49 to 60, use a calculator to approximate the given trigonometric functions to six significant digits.

49. $\sin 127°$ **50.** $\sin (-257°)$ **51.** $\cos (-116°)$

52. $\cot 398°$ **53.** $\sec 578°$ **54.** $\sec 740°$

55. $\sin \left(-\dfrac{\pi}{5}\right)$ **56.** $\cos \dfrac{3\pi}{7}$ **57.** $\csc \dfrac{9\pi}{5}$

58. $\tan (-4.12)$ **59.** $\sec (-4.45)$ **60.** $\csc 0.34$

In Exercises 61 to 68, find (without using a calculator) the exact value of each expression.

61. $\sin 210° - \cos 330° \tan 330°$

62. $\tan 225° + \sin 240° \cos 60°$

63. $\sin^2 30° + \cos^2 30°$

64. $\cos \pi \sin \dfrac{7\pi}{4} - \tan \dfrac{11\pi}{6}$

65. $\sin\left(\dfrac{3\pi}{2}\right)\tan\left(\dfrac{\pi}{4}\right)-\cos\left(\dfrac{\pi}{3}\right)$

66. $\cos\left(\dfrac{7\pi}{4}\right)\tan\left(\dfrac{4\pi}{3}\right)+\cos\left(\dfrac{7\pi}{6}\right)$

67. $\sin^2\left(\dfrac{5\pi}{4}\right)+\cos^2\left(\dfrac{5\pi}{4}\right)$

68. $\tan^2\left(\dfrac{7\pi}{4}\right)-\sec^2\left(\dfrac{7\pi}{4}\right)$

CONNECTING CONCEPTS

In Exercises 69 to 74, find two values of θ, $0° \le \theta < 360°$, that satisfy the given trigonometric equation.

69. $\sin\theta=\dfrac{1}{2}$

70. $\tan\theta=-\sqrt{3}$

71. $\cos\theta=-\dfrac{\sqrt{3}}{2}$

72. $\tan\theta=1$

73. $\csc\theta=-\sqrt{2}$

74. $\cot\theta=-1$

In Exercises 75 to 80, find two values of θ, $0 \le \theta < 2\pi$, that satisfy the given trigonometric equation.

75. $\tan\theta=-1$

76. $\cos\theta=\dfrac{1}{2}$

77. $\tan\theta=-\dfrac{\sqrt{3}}{3}$

78. $\sec\theta=-\dfrac{2\sqrt{3}}{3}$

79. $\sin\theta=\dfrac{\sqrt{3}}{2}$

80. $\cos\theta=-\dfrac{1}{2}$

If $P(x, y)$ is a point on the terminal side of an acute angle θ in standard position and $r = \sqrt{x^2 + y^2}$, then $\sin\theta = \dfrac{y}{r}$ and $\cos\theta = \dfrac{x}{r}$. Using these definitions, we find that

$$\cos^2\theta+\sin^2\theta=\left(\dfrac{x}{r}\right)^2+\left(\dfrac{y}{r}\right)^2=\dfrac{x^2}{r^2}+\dfrac{y^2}{r^2}=\dfrac{x^2+y^2}{r^2}$$

$$=\dfrac{r^2}{r^2}=1$$

Hence $\cos^2\theta + \sin^2\theta = 1$ for all acute angles θ. This important identity is actually true for all angles θ. We will show this later. In the meantime, use the definitions of the trigonometric functions to prove the identities in Exercises 81 to 90 for the acute angle θ.

81. $1+\tan^2\theta=\sec^2\theta$

82. $\cot^2\theta+1=\csc^2\theta$

83. $\tan\theta=\dfrac{\sin\theta}{\cos\theta}$

84. $\cot\theta=\dfrac{\cos\theta}{\sin\theta}$

85. $\cos(90°-\theta)=\sin\theta$

86. $\sin(90°-\theta)=\cos\theta$

87. $\tan(90°-\theta)=\cot\theta$

88. $\cot(90°-\theta)=\tan\theta$

89. $\sin(\theta+\pi)=-\sin\theta$

90. $\cos(\theta+\pi)=-\cos\theta$

PREPARE FOR SECTION 5.4

91. Determine whether the point $(0, 1)$ is a point on the circle defined by $x^2 + y^2 = 1$. [2.1]

92. Determine whether the point $\left(\dfrac{1}{2}, \dfrac{\sqrt{3}}{2}\right)$ is a point on the circle defined by $x^2 + y^2 = 1$. [2.1]

93. Determine whether the point $\left(\dfrac{\sqrt{2}}{2}, \dfrac{\sqrt{3}}{2}\right)$ is a point on the circle defined by $x^2 + y^2 = 1$. [2.1]

94. Determine the circumference of a circle with a radius of 1. [1.2]

95. Determine whether $f(x) = x^2 - 3$ is an even function, an odd function, or a function that is neither even nor odd. [2.5]

96. Determine whether $f(x) = x^3 - x^2$ is an even function, an odd function, or a function that is neither even nor odd. [2.5]

PROJECTS

1. **FIND SUMS OR PRODUCTS** Determine the following sums or products. Do not use a calculator. (*Hint:* The Reference Angle Theorem may be helpful.) Explain to a classmate how you know you are correct.

a. $\cos 0° + \cos 1° + \cos 2° + \cdots + \cos 178° + \cos 179° + \cos 180°$

b. $\sin 0° + \sin 1° + \sin 2° + \cdots + \sin 358° + \sin 359° + \sin 360°$

c. $\cot 1° + \cot 2° + \cot 3° + \cdots + \cot 177° + \cot 178° + \cot 179°$

d. $(\cos 1°)(\cos 2°)(\cos 3°) \cdots (\cos 177°)(\cos 178°)(\cos 179°)$

e. $\cos^2 1° + \cos^2 2° + \cos^2 3° + \cdots + \cos^2 357° + \cos^2 358° + \cos^2 359°$

SECTION 5.4

TRIGONOMETRIC FUNCTIONS OF REAL NUMBERS

- **THE WRAPPING FUNCTION**
- **PROPERTIES OF TRIGONOMETRIC FUNCTIONS OF REAL NUMBERS**
- **TRIGONOMETRIC IDENTITIES**
- **AN APPLICATION INVOLVING A TRIGONOMETRIC FUNCTION OF A REAL NUMBER**

• THE WRAPPING FUNCTION

In the seventeenth century, applications of trigonometry were extended to problems in physics and engineering. These kinds of problems required trigonometric functions whose domains were sets of real numbers rather than sets of angles. During this time, the definitions of trigonometric functions were extended to real numbers by using a correspondence between an angle and a real number.

Consider a circle given by the equation $x^2 + y^2 = 1$, called a **unit circle,** and a vertical coordinate line l tangent to the unit circle at $(1, 0)$. We define a function W that pairs a real number t on the coordinate line with a point $P(x, y)$ on the unit circle. This function is called the *wrapping function* because it is analogous to wrapping a line around a circle.

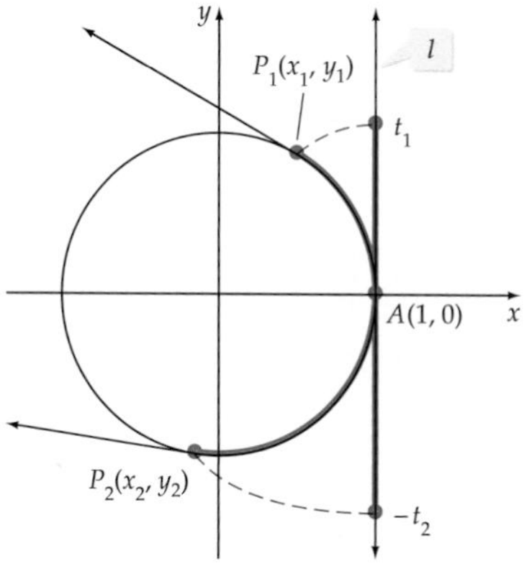

As shown in **Figure 5.52,** the positive part of the coordinate line is wrapped around the unit circle in a counterclockwise direction. The negative part of the coordinate line is wrapped around the circle in a clockwise direction. The wrapping function is defined by the equation $W(t) = P(x, y)$, where t is a real number and $P(x, y)$ is the point on the unit circle that corresponds to t.

Through the wrapping function, each real number t defines an arc $\overset{\frown}{AP}$ that subtends a central angle with a measure of θ radians. The length of the arc $\overset{\frown}{AP}$ is t (see **Figure 5.53**). From the equation $s = r\theta$ for the arc length of a circle, we have (with $t = s$) $t = r\theta$. For a unit circle, $r = 1$, and the equation becomes $t = \theta$. Thus, on a unit circle, *the measure of a central angle and the length of its arc can be represented by the same real number t.*

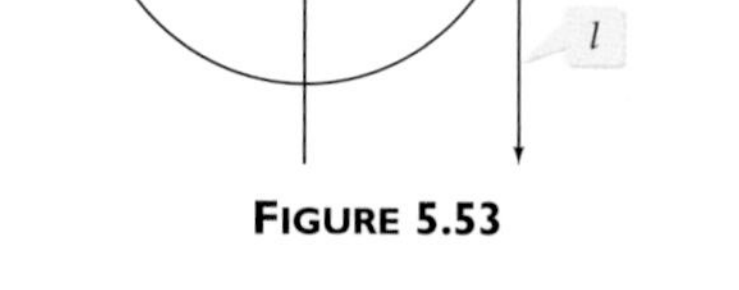

FIGURE 5.52

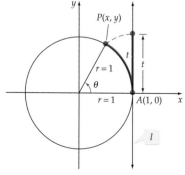

FIGURE 5.53

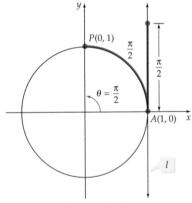

FIGURE 5.54

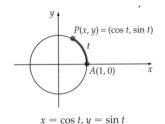

FIGURE 5.55

$P(x, y) = (\cos t, \sin t)$

$A(1, 0)$

$x = \cos t, y = \sin t$
FIGURE 5.56

| EXAMPLE I | Evaluate the Wrapping Function |

Find the values of x and y such that $W\left(\dfrac{\pi}{3}\right) = P(x, y)$.

Solution

The point $\dfrac{\pi}{3}$ on line l is shown in **Figure 5.54**. From the wrapping

function, $W\left(\dfrac{\pi}{3}\right)$ is the point P on the unit circle for which arc $\overset{\frown}{AP}$ subtends

an angle θ, the measure of which is $\dfrac{\pi}{3}$ radians. The coordinates of P can be

determined from the definitions of $\cos\theta$ and $\sin\theta$ given in Section 5.3 and
from **Table 5.2**, page 480.

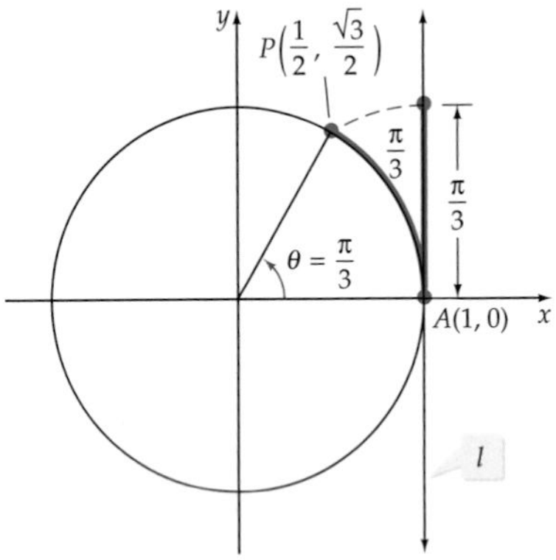

$$\cos\theta = \frac{x}{r} \qquad \sin\theta = \frac{y}{r}$$

$$\cos\frac{\pi}{3} = \frac{x}{1} = x \qquad \sin\frac{\pi}{3} = \frac{y}{1} = y \qquad \bullet\, \theta = \frac{\pi}{3}; r = 1$$

$$\frac{1}{2} = x \qquad \frac{\sqrt{3}}{2} = y \qquad \bullet\, \cos\frac{\pi}{3} = \frac{1}{2}; \sin\frac{\pi}{3} = \frac{\sqrt{3}}{2}$$

From these equations, $x = \dfrac{1}{2}$ and $y = \dfrac{\sqrt{3}}{2}$. Therefore, $W\left(\dfrac{\pi}{3}\right) = P\left(\dfrac{1}{2}, \dfrac{\sqrt{3}}{2}\right)$.

▶ **TRY EXERCISE 10, PAGE 508**

To determine $W\left(\dfrac{\pi}{2}\right)$, recall that the circumference of a unit circle is 2π. One-

fourth the circumference is $\dfrac{1}{4}(2\pi) = \dfrac{\pi}{2}$ (see **Figure 5.55**). Thus $W\left(\dfrac{\pi}{2}\right) = P(0, 1)$.

Note from the last two examples that for the given real number t, $\cos t = x$
and $\sin t = y$. That is, for a real number t and $W(t) = P(x, y)$, the value of the
cosine of t is the x-coordinate of P, and the value of the sine of t is the y-coordinate
of P. See **Figure 5.56**.

❓ **QUESTION** What is the point defined by $W\left(\dfrac{\pi}{4}\right)$?

The following definition makes use of the wrapping function $W(t)$ to define
trigonometric functions of real numbers.

Definition of the Trigonometric Functions of Real Numbers

Let W be the wrapping function, t be a real number, and $W(t) = P(x, y)$. Then

$$\sin t = y \qquad\qquad \cos t = x \qquad\qquad \tan t = \frac{y}{x}, x \neq 0$$

$$\csc t = \frac{1}{y}, \;\; y \neq 0 \qquad \sec t = \frac{1}{x}, \;\; x \neq 0 \qquad \cot t = \frac{x}{y}, \;\; y \neq 0$$

❓ **ANSWER** $\left(\dfrac{\sqrt{2}}{2}, \dfrac{\sqrt{2}}{2}\right)$

Trigonometric functions of real numbers are frequently called *circular functions* to distinguish them from trigonometric functions of angles.

The *trigonometric functions of real numbers* (or circular functions) look remarkably like the trigonometric functions defined in the last section. The difference between the two is that of domain: In one case, the domains are sets of *real numbers*; in the other case, the domains are sets of *angles*. However, there are similarities between the two functions.

Consider an angle θ (in radians) in standard position, as shown in **Figure 5.57.** Let $P(x, y)$ and $P'(x', y')$ be two points on the terminal side of θ, where $x^2 + y^2 = 1$ and $(x')^2 + (y')^2 = r^2$. Let t be the length of the arc from $A(1, 0)$ to $P(x, y)$. Then

$$\sin \theta = \frac{y'}{r} = \frac{y}{1} = \sin t$$

Thus the value of the sine function of θ, measured in radians, is equal to the value of the sine of the real number t. Similar arguments can be given to show corresponding results for the other five trigonometric functions. With this in mind, we can assert that *the value of a trigonometric function at the real number t is its value at an angle of t radians.*

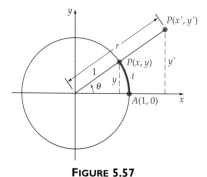

FIGURE 5.57

EXAMPLE 2   **Evaluate Trigonometric Functions of Real Numbers**

Find the exact value of each function.

a. $\cos \dfrac{\pi}{4}$ b. $\sin \left(-\dfrac{7\pi}{6} \right)$ c. $\tan \left(-\dfrac{5\pi}{4} \right)$ d. $\sec \dfrac{5\pi}{3}$

Solution

The value of a trigonometric function at the real number t is its value at an angle of t radians. Using **Table 5.2** on page 480, we have

a. $\cos \dfrac{\pi}{4} = \dfrac{\sqrt{2}}{2}$

b. $\sin \left(-\dfrac{7\pi}{6} \right) = \sin \dfrac{\pi}{6} = \dfrac{1}{2}$ • Reference angle for $-\dfrac{7\pi}{6}$ is $\dfrac{\pi}{6}$ and $\sin t > 0$ in Quadrant II.

c. $\tan \left(-\dfrac{5\pi}{4} \right) = -\tan \dfrac{\pi}{4} = -1$ • Reference angle for $-\dfrac{5\pi}{4}$ is $\dfrac{\pi}{4}$ and $\tan t < 0$ in Quadrant II.

d. $\sec \dfrac{5\pi}{3} = \sec \dfrac{\pi}{3} = 2$ • Reference angle for $\dfrac{5\pi}{3}$ is $\dfrac{\pi}{3}$ and $\sec t > 0$ in Quadrant IV.

▶ **TRY EXERCISE 16, PAGE 508**

● PROPERTIES OF TRIGONOMETRIC FUNCTIONS OF REAL NUMBERS

To review **DOMAIN AND RANGE**, *see p. 178.*

The domain and range of the trigonometric functions can be found from the definitions of these functions. If t is any real number and $P(x, y)$ is the point corresponding to $W(t)$, then by definition $\cos t = x$ and $\sin t = y$. Thus the domain of the sine and cosine functions is the set of real numbers.

Because the radius of the unit circle is 1, we have

$$-1 \le x \le 1 \qquad \text{and} \qquad -1 \le y \le 1$$

Therefore, with $x = \cos t$ and $y = \sin t$, we have

$$-1 \le \cos t \le 1 \qquad \text{and} \qquad -1 \le \sin t \le 1$$

The range of the cosine and sine functions is $[-1, 1]$.

Using the definitions of tangent and secant,

$$\tan t = \frac{y}{x} \qquad \text{and} \qquad \sec t = \frac{1}{x}$$

The domain of the tangent function is all real numbers t except those for which the x-coordinate of $W(t)$ is zero. The x-coordinate is zero when $t = \pm\frac{\pi}{2}$, $t = \pm\frac{3\pi}{2}$, and $t = \pm\frac{5\pi}{2}$ and in general when $t = \frac{(2n + 1)\pi}{2}$, where n is an integer. Thus the domain of the tangent function is the set of all real numbers t except $t = \frac{(2n + 1)\pi}{2}$, where n is an integer. The range of the tangent function is all real numbers.

Similar methods can be used to find the domain and range of the cotangent, secant, and cosecant functions. The results are summarized in **Table 5.6**.

TABLE 5.6   Domain and Range of the Trigonometric Functions (n is an integer)

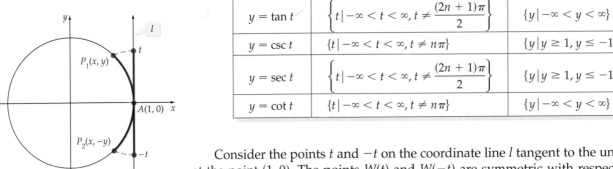

Function	Domain	Range
$y = \sin t$	$\{t \mid -\infty < t < \infty\}$	$\{y \mid -1 \le y \le 1\}$
$y = \cos t$	$\{t \mid -\infty < t < \infty\}$	$\{y \mid -1 \le y \le 1\}$
$y = \tan t$	$\left\{t \mid -\infty < t < \infty, t \ne \dfrac{(2n + 1)\pi}{2}\right\}$	$\{y \mid -\infty < y < \infty\}$
$y = \csc t$	$\{t \mid -\infty < t < \infty, t \ne n\pi\}$	$\{y \mid y \ge 1, y \le -1\}$
$y = \sec t$	$\left\{t \mid -\infty < t < \infty, t \ne \dfrac{(2n + 1)\pi}{2}\right\}$	$\{y \mid y \ge 1, y \le -1\}$
$y = \cot t$	$\{t \mid -\infty < t < \infty, t \ne n\pi\}$	$\{y \mid -\infty < y < \infty\}$

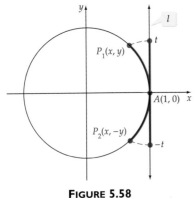

FIGURE 5.58

Consider the points t and $-t$ on the coordinate line l tangent to the unit circle at the point $(1, 0)$. The points $W(t)$ and $W(-t)$ are symmetric with respect to the x-axis. Therefore, if $P_1(x, y)$ are the coordinates of $W(t)$, then $P_2(x, -y)$ are the coordinates of $W(-t)$. See **Figure 5.58**.

From the definitions of the trigonometric functions, we have

$$\sin t = y \quad \text{and} \quad \sin(-t) = -y \qquad \text{and} \qquad \cos t = x \quad \text{and} \quad \cos(-t) = x$$

To review **ODD AND EVEN FUNCTIONS**, *see p. 230.*

Substituting $\sin t$ for y and $\cos t$ for x yields

$$\sin(-t) = -\sin t \qquad \text{and} \qquad \cos(-t) = \cos t$$

Thus the sine is an odd function, and the cosine is an even function. Because $\csc t = \dfrac{1}{\sin t}$ and $\sec t = \dfrac{1}{\cos t}$, it follows that

$$\csc(-t) = -\csc t \qquad \text{and} \qquad \sec(-t) = \sec t$$

These equations state that the cosecant is an odd function and the secant is an even function.

From the definition of the tangent function, we have $\tan t = \dfrac{y}{x}$ and $\tan(-t) = -\dfrac{y}{x}$. Substituting $\tan t$ for $\dfrac{y}{x}$ yields $\tan(-t) = -\tan t$. Because $\cot t = \dfrac{1}{\tan t}$, it follows that $\cot(-t) = -\cot t$. Thus the tangent and cotangent are odd functions.

Even and Odd Trigonometric Functions

The odd trigonometric functions are $y = \sin t$, $y = \csc t$, $y = \tan t$, and $y = \cot t$. The even trigonometric functions are $y = \cos t$ and $y = \sec t$.

Thus for all t in their domain,

$$\sin(-t) = -\sin t \qquad \cos(-t) = \cos t \qquad \tan(-t) = -\tan t$$
$$\csc(-t) = -\csc t \qquad \sec(-t) = \sec t \qquad \cot(-t) = -\cot t$$

EXAMPLE 3   Determine Whether a Function Is Even, Odd, or Neither

Is the function defined by $f(x) = x - \tan x$ an even function, an odd function, or neither?

Solution

Find $f(-x)$ and compare it to $f(x)$.

$$\begin{aligned} f(-x) &= (-x) - \tan(-x) = -x + \tan x \qquad \bullet\ \tan(-x) = -\tan x \\ &= -(x - \tan x) \\ &= -f(x) \end{aligned}$$

The function $f(x) = x - \tan x$ is an odd function.

▶ **TRY EXERCISE 44, PAGE 508**

Let W be the wrapping function, t be a point on the coordinate line tangent to the unit circle at $(1, 0)$, and $W(t) = P(x, y)$. Because the circumference of the unit circle is 2π, $W(t + 2\pi) = W(t) = P(x, y)$. Thus the value of the wrapping function repeats itself in 2π units. *The wrapping function is periodic, and the period is 2π.*

Recall the definitions of $\cos t$ and $\sin t$:

$$\cos t = x \qquad \text{and} \qquad \sin t = y$$

where $W(t) = P(x, y)$. Because $W(t + 2\pi) = W(t) = P(x, y)$ for all t,

$$\cos(t + 2\pi) = x \quad \text{and} \quad \sin(t + 2\pi) = y$$

Thus $\cos t$ and $\sin t$ have period 2π. Because

$$\sec t = \frac{1}{\cos t} = \frac{1}{\cos(t + 2\pi)} = \sec(t + 2\pi) \quad \text{and}$$

$$\csc t = \frac{1}{\sin t} = \frac{1}{\sin(t + 2\pi)} = \csc(t + 2\pi)$$

$\sec t$ and $\csc t$ have a period of 2π.

> **Period of cos t, sin t, sec t, and csc t**
>
> The period of $\cos t$, $\sin t$, $\sec t$, and $\csc t$ is 2π.

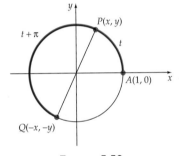

FIGURE 5.59

Although it is true that $\tan t = \tan(t + 2\pi)$, the period of $\tan t$ is not 2π. Recall that the period of a function is the *smallest* value of p for which $f(t) = f(t + p)$.

If W is the wrapping function (see **Figure 5.59**) and $W(t) = P(x, y)$, then $W(t + \pi) = P(-x, -y)$. Because

$$\tan t = \frac{y}{x} \quad \text{and} \quad \tan(t + \pi) = \frac{-y}{-x} = \frac{y}{x} = \tan t$$

we have $\tan(t + \pi) = \tan t$ for all t. A similar argument applies to $\cot t$.

> **Period of tan t and cot t**
>
> The period of $\tan t$ and $\cot t$ is π.

● TRIGONOMETRIC IDENTITIES

Recall that any equation that is true for every number in the domain of the equation is an identity. The statement

$$\csc t = \frac{1}{\sin t}, \quad \sin t \neq 0$$

is an identity because the two expressions produce the same result for all values of t for which both functions are defined.

The **ratio identities** are obtained by writing the tangent and cotangent functions in terms of the sine and cosine functions.

$$\tan t = \frac{y}{x} = \frac{\sin t}{\cos t} \quad \text{and} \quad \cot t = \frac{x}{y} = \frac{\cos t}{\sin t} \qquad \bullet \; x = \cos t \text{ and } y = \sin t$$

The **Pythagorean identities** are based on the equation of a unit circle, $x^2 + y^2 = 1$, and on the definitions of the sine and cosine functions.

$$x^2 + y^2 = 1$$
$$\cos^2 t + \sin^2 t = 1 \qquad \bullet \textbf{ Replace } x \textbf{ by } \cos t \textbf{ and } y \textbf{ by } \sin t.$$

Dividing each term of $\cos^2 t + \sin^2 t = 1$ by $\cos^2 t$, we have

$$\frac{\cos^2 t}{\cos^2 t} + \frac{\sin^2 t}{\cos^2 t} = \frac{1}{\cos^2 t} \qquad \bullet \cos t \neq 0$$

$$1 + \tan^2 t = \sec^2 t \qquad \bullet \frac{\sin t}{\cos t} = \tan t$$

Dividing each term of $\cos^2 t + \sin^2 t = 1$ by $\sin^2 t$, we have

$$\frac{\cos^2 t}{\sin^2 t} + \frac{\sin^2 t}{\sin^2 t} = \frac{1}{\sin^2 t} \qquad \bullet \sin t \neq 0$$

$$\cot^2 t + 1 = \csc^2 t \qquad \bullet \frac{\cos t}{\sin t} = \cot t$$

Here is a summary of the Fundamental Trigonometric Identities:

Fundamental Trigonometric Identities

The reciprocal identities are

$$\sin t = \frac{1}{\csc t} \qquad \cos t = \frac{1}{\sec t} \qquad \tan t = \frac{1}{\cot t}$$

The ratio identities are

$$\tan t = \frac{\sin t}{\cos t} \qquad \cot t = \frac{\cos t}{\sin t}$$

The Pythagorean identities are

$$\cos^2 t + \sin^2 t = 1 \qquad 1 + \tan^2 t = \sec^2 t \qquad 1 + \cot^2 t = \csc^2 t$$

EXAMPLE 4 Use the Unit Circle to Verify an Identity

By using the unit circle and the definitions of the trigonometric functions, show that $\sin(t + \pi) = -\sin t$.

Solution

Sketch a unit circle, and let P be the point on the unit circle such that $W(t) = P(x, y)$, as shown in **Figure 5.60.** Draw a diameter from P, and label the endpoint Q. For any line through the origin, if $P(x, y)$ is a point on the line, then $Q(-x, -y)$ is also a point on the line. Because PQ is a diameter, the length of arc PQ is π. Thus the length of arc AQ is $t + \pi$. Therefore, $W(t + \pi) = Q(-x, -y)$. From the definition of $\sin t$, we have

$$\sin t = y \qquad \text{and} \qquad \sin(t + \pi) = -y$$

Thus $\sin(t + \pi) = -\sin t$.

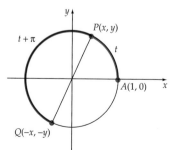

FIGURE 5.60

▶ **TRY EXERCISE 50, PAGE 509**

Using identities and basic algebra concepts, we can rewrite trigonometric expressions in different forms.

EXAMPLE 5 Simplify a Trigonometric Expression

Write the expression $\dfrac{1}{\sin^2 t} + \dfrac{1}{\cos^2 t}$ as a single term.

Solution

Express each fraction in terms of a common denominator. The common denominator is $\sin^2 t \cos^2 t$.

$$\frac{1}{\sin^2 t} + \frac{1}{\cos^2 t} = \frac{1}{\sin^2 t}\frac{\cos^2 t}{\cos^2 t} + \frac{1}{\cos^2 t}\frac{\sin^2 t}{\sin^2 t}$$

$$= \frac{\cos^2 t + \sin^2 t}{\sin^2 t \cos^2 t} = \frac{1}{\sin^2 t \cos^2 t} \qquad \bullet\ \cos^2 t + \sin^2 t = 1$$

▶ **TRY EXERCISE 68, PAGE 509**

> ### take note
>
> Because
>
> $$\frac{1}{\sin^2 t \cos^2 t} = \frac{1}{\sin^2 t} \cdot \frac{1}{\cos^2 t}$$
>
> $$= (\csc^2 t)(\sec^2 t)$$
>
> we could have written the answer to Example 5 in terms of the cosecant and secant functions.

EXAMPLE 6 Write a Trigonometric Expression in Terms of a Given Function

For $\dfrac{\pi}{2} < t < \pi$, write $\tan t$ in terms of $\sin t$.

Solution

Write $\tan t = \dfrac{\sin t}{\cos t}$. Now solve $\cos^2 t + \sin^2 t = 1$ for $\cos t$.

$$\cos^2 t + \sin^2 t = 1$$

$$\cos^2 t = 1 - \sin^2 t$$

$$\cos t = \pm\sqrt{1 - \sin^2 t}$$

Because $\dfrac{\pi}{2} < t < \pi$, $\cos t$ is negative. Therefore, $\cos t = -\sqrt{1 - \sin^2 t}$.

Thus

$$\tan t = -\frac{\sin t}{\sqrt{1 - \sin^2 t}} \qquad \bullet\ \frac{\pi}{2} < t < \pi$$

▶ **TRY EXERCISE 74, PAGE 509**

• AN APPLICATION INVOLVING A TRIGONOMETRIC FUNCTION OF A REAL NUMBER

EXAMPLE 7 Determine a Height as a Function of Time

The Millennium Wheel, in London, is the world's largest Ferris wheel. It has a diameter of 450 feet. When the Millennium Wheel is in uniform motion, it completes one revolution every 30 minutes. The height h, in feet above the Thames River, of a person riding on the Millennium Wheel can be estimated by

$$h(t) = 255 - 225 \cos\left(\frac{\pi}{15}t\right)$$

where t is the time in minutes since the person started the ride.

a. How high is the person at the start of the ride ($t = 0$)?

b. How high is the person after 18.0 minutes?

Solution

a. $h(0) = 255 - 225 \cos\left(\dfrac{\pi}{15} \cdot 0\right)$ b. $h(18.0) = 255 - 225 \cos\left(\dfrac{\pi}{15} \cdot 18.0\right)$

$\qquad\quad = 255 - 225$ $\approx 255 - (-182)$

$\qquad\quad = 30$ $= 437$

At the start of the ride, the person is 30 feet above the Thames.   After 18.0 minutes, the person is about 437 feet above the Thames.

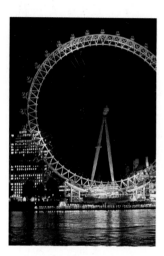

The Millennium Wheel, on the banks of the Thames River, London.

▶ **TRY EXERCISE 78, PAGE 509**

  TOPICS FOR DISCUSSION

1. Is $W(t)$ a number? Explain.

2. Explain how to find the exact value of $\cos\left(\dfrac{13\pi}{6}\right)$.

3. Explain why the equation $\cos^2 t + \sin^2 t = 1$ is called a Pythagorean identity.

4. Is $f(x) = \cos^3 x$ an even function or an odd function? Explain how you made your decision.

5. Explain how to make use of a unit circle to show that $\sin(-t) = -\sin t$.

EXERCISE SET 5.4

In Exercises 1 to 12, find W(t) for each given t.

1. $t = \dfrac{\pi}{6}$ 　　　　 **2.** $t = \dfrac{\pi}{4}$ 　　　　 **3.** $t = \dfrac{7\pi}{6}$

4. $t = \dfrac{4\pi}{3}$ 　　　　 **5.** $t = \dfrac{5\pi}{3}$ 　　　　 **6.** $t = -\dfrac{\pi}{6}$

7. $t = \dfrac{11\pi}{6}$ 　　　　 **8.** $t = 0$ 　　　　 **9.** $t = \pi$

▶ **10.** $t = -\dfrac{7\pi}{4}$ 　　 **11.** $t = -\dfrac{2\pi}{3}$ 　　 **12.** $t = -\pi$

In Exercises 13 to 22, find the exact value of each function.

13. $\tan\left(\dfrac{11\pi}{6}\right)$ 　　　　 **14.** $\cot\left(\dfrac{2\pi}{3}\right)$

15. $\cos\left(-\dfrac{2\pi}{3}\right)$ 　　 ▶ **16.** $\sec\left(-\dfrac{5\pi}{6}\right)$

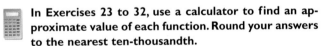

17. $\csc\left(-\dfrac{\pi}{3}\right)$ 　　　　 **18.** $\tan(12\pi)$

19. $\sin\left(\dfrac{3\pi}{2}\right)$ 　　　　 **20.** $\cos\left(\dfrac{7\pi}{3}\right)$

21. $\sec\left(-\dfrac{7\pi}{6}\right)$ 　　　 **22.** $\sin\left(-\dfrac{5\pi}{3}\right)$

In Exercises 23 to 32, use a calculator to find an approximate value of each function. Round your answers to the nearest ten-thousandth.

23. $\sin 1.22$ 　　　　 **24.** $\cos 4.22$

25. $\csc(-1.05)$ 　　　 **26.** $\sin(-0.55)$

27. $\tan\left(\dfrac{11\pi}{12}\right)$ 　　　 **28.** $\cos\left(\dfrac{2\pi}{5}\right)$

29. $\cos\left(-\dfrac{\pi}{5}\right)$ 　　　 **30.** $\csc 8.2$

31. $\sec 1.55$ 　　　　 **32.** $\cot 2.11$

In Exercises 33 to 40, use the unit circle in the next column to estimate the following values to the nearest tenth.

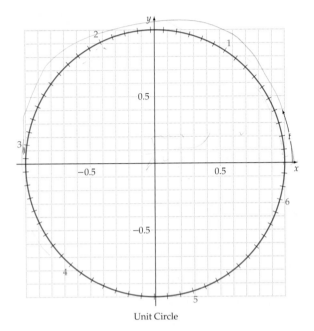

Unit Circle

33. a. $\sin 2$ 　　　　 **b.** $\cos 2$

34. a. $\sin 3$ 　　　　 **b.** $\cos 3$

35. a. $\sin 5.4$ 　　　 **b.** $\cos 5.4$

36. a. $\sin 4.1$ 　　　 **b.** $\cos 4.1$

37. All real numbers t between 0 and 2π for which $\sin t = 0.4$.

38. All real numbers t between 0 and 2π for which $\cos t = 0.8$.

39. All real numbers t between 0 and 2π for which $\sin t = -0.3$.

40. All real numbers t between 0 and 2π for which $\cos t = -0.7$.

In Exercises 41 to 48, determine whether the function defined by each equation is even, odd, or neither.

41. $f(x) = -4\sin x$ 　　　 **42.** $f(x) = -2\cos x$

43. $G(x) = \sin x + \cos x$ 　 ▶ **44.** $F(x) = \tan x + \sin x$

45. $S(x) = \dfrac{\sin x}{x},\ x \neq 0$ 　　 **46.** $C(x) = \dfrac{\cos x}{x},\ x \neq 0$

47. $v(x) = 2\sin x \cos x$ 　　 **48.** $w(x) = x \tan x$

In Exercises 49 to 56, use the unit circle to verify each identity.

49. $\cos(-t) = \cos t$ ▶ **50.** $\tan(t - \pi) = \tan t$

51. $\cos(t + \pi) = -\cos t$ **52.** $\sin(-t) = -\sin t$

53. $\sin(t - \pi) = -\sin t$ **54.** $\sec(-t) = \sec t$

55. $\csc(-t) = -\csc t$ **56.** $\tan(-t) = -\tan t$

In Exercises 57 to 72, use the trigonometric identities to write each expression in terms of a single trigonometric function or a constant. Answers may vary.

57. $\tan t \cos t$ **58.** $\cot t \sin t$

59. $\dfrac{\csc t}{\cot t}$ **60.** $\dfrac{\sec t}{\tan t}$

61. $1 - \sec^2 t$ **62.** $1 - \csc^2 t$

63. $\tan t - \dfrac{\sec^2 t}{\tan t}$ **64.** $\dfrac{\csc^2 t}{\cot t} - \cot t$

65. $\dfrac{1 - \cos^2 t}{\tan^2 t}$ **66.** $\dfrac{1 - \sin^2 t}{\cot^2 t}$

67. $\dfrac{1}{1 - \cos t} + \dfrac{1}{1 + \cos t}$ ▶ **68.** $\dfrac{1}{1 - \sin t} + \dfrac{1}{1 + \sin t}$

69. $\dfrac{\tan t + \cot t}{\tan t}$ **70.** $\dfrac{\csc t - \sin t}{\csc t}$

71. $\sin^2 t(1 + \cot^2 t)$ **72.** $\cos^2 t(1 + \tan^2 t)$

73. Write $\sin t$ in terms of $\cos t$, $0 < t < \dfrac{\pi}{2}$.

▶ **74.** Write $\tan t$ in terms of $\sec t$, $\dfrac{3\pi}{2} < t < 2\pi$.

75. Write $\csc t$ in terms of $\cot t$, $\dfrac{\pi}{2} < t < \pi$.

76. Write $\sec t$ in terms of $\tan t$, $\pi < t < \dfrac{3\pi}{2}$.

77. PATH OF A SATELLITE A satellite is launched into space from Cape Canaveral. The directed distance, in miles, that the satellite is above or below the equator is

$$d(t) = 1970 \cos\left(\frac{\pi}{64}t\right)$$

where t is the number of minutes since liftoff. A negative d value indicates that the satellite is south of the equator.

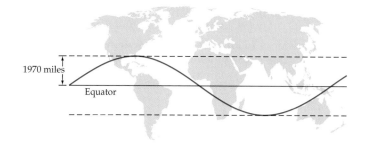

1970 miles

Equator

What distance, to the nearest 10 miles, is the satellite north of the equator 24 minutes after liftoff?

▶ **78. AVERAGE HIGH TEMPERATURE** The average high temperature T, in degrees Fahrenheit, for Fairbanks, Alaska, is given by

$$T(t) = -41 \cos\left(\frac{\pi}{6}t\right) + 36$$

where t is the number of months after January 5. Use the formula to estimate (to the nearest 0.1 degree Fahrenheit) the average high temperature in Fairbanks for March 5 and July 20.

In Exercises 79 to 90, perform the indicated operation and simplify.

79. $\cos t - \dfrac{1}{\cos t}$ **80.** $\tan t + \dfrac{1}{\tan t}$

81. $\cot t + \dfrac{1}{\cot t}$ **82.** $\sin t - \dfrac{1}{\sin t}$

83. $(1 - \sin t)^2$ **84.** $(1 - \cos t)^2$

85. $(\sin t - \cos t)^2$ **86.** $(\sin t + \cos t)^2$

87. $(1 - \sin t)(1 + \sin t)$ **88.** $(1 - \cos t)(1 + \cos t)$

89. $\dfrac{\sin t}{1 + \cos t} + \dfrac{1 + \cos t}{\sin t}$ **90.** $\dfrac{1 - \sin t}{\cos t} - \dfrac{1}{\tan t + \sec t}$

In Exercises 91 to 98, factor the expression.

91. $\cos^2 t - \sin^2 t$ **92.** $\sec^2 t - \csc^2 t$

93. $\tan^2 t - \tan t - 6$ **94.** $\cos^2 t + 3\cos t - 4$

95. $2\sin^2 t - \sin t - 1$ **96.** $4\cos^2 t + 4\cos t + 1$

97. $\cos^4 t - \sin^4 t$ **98.** $\sec^4 t - \csc^4 t$

CONNECTING CONCEPTS

In Exercises 99 to 102, use the trigonometric identities to find the value of the function.

99. Given $\csc t = \sqrt{2}, 0 < t < \dfrac{\pi}{2}$, find $\cos t$.

100. Given $\cos t = \dfrac{1}{2}, \dfrac{3\pi}{2} < t < 2\pi$, find $\sin t$.

101. Given $\sin t = \dfrac{1}{2}, \dfrac{\pi}{2} < t < \pi$, find $\tan t$.

102. Given $\cot t = \dfrac{\sqrt{3}}{3}, \pi < t < \dfrac{3\pi}{2}$, find $\cos t$.

In Exercises 103 to 106, simplify the first expression to the second expression.

103. $\dfrac{\sin^2 t + \cos^2 t}{\sin^2 t}; \csc^2 t$

104. $\dfrac{\sin^2 t + \cos^2 t}{\cos^2 t}; \sec^2 t$

105. $(\cos t - 1)(\cos t + 1); -\sin^2 t$

106. $(\sec t - 1)(\sec t + 1); \tan^2 t$

PREPARE FOR SECTION 5.5

107. Estimate, to the nearest tenth, $\sin \dfrac{3\pi}{4}$. [5.4]

108. Estimate, to the nearest tenth, $\cos \dfrac{5\pi}{4}$. [5.4]

109. Explain how to use the graph of $y = f(x)$ to produce the graph of $y = -f(x)$. [2.5]

110. Explain how to use the graph of $y = f(x)$ to produce the graph of $y = f(2x)$. [2.5]

111. Simplify: $\dfrac{2\pi}{1/3}$ [P.5]

112. Simplify: $\dfrac{2\pi}{2/5}$ [P.5]

PROJECTS

1. **VISUAL INSIGHT**

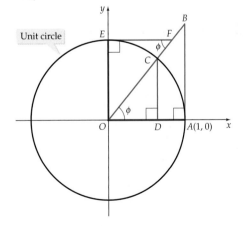

$OD = \cos \phi \quad DC = \sin \phi$

Make use of the circle and similar triangles at the left to explain why the length of line segment

a. AB is equal to $\tan \phi$ **b.** EF is equal to $\cot \phi$

c. OB is equal to $\sec \phi$ **d.** OF is equal to $\csc \phi$

2. **PERIODIC FUNCTIONS**

a. A function f is periodic with a period of 3. If $f(2) = -1$, determine $f(14)$.

b. If g is a periodic function with period 2 and h is a periodic function with period 3, determine the period of $f + g$.

c. Find two periodic functions f and g such that the function $f + g$ is not a periodic function.

3. Consider a square as shown. Start at the point $(1, 0)$ and travel counterclockwise around the square for a distance t $(t \geq 0)$.

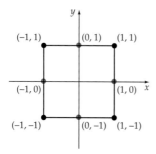

Let $WSQ(t) = P(x, y)$ be the point on the square determined by traveling counterclockwise a distance of t units from $(1, 0)$. For instance,

$$WSQ(0.5) = (1, 0.5)$$
$$WSQ(1.75) = (0.25, 1)$$

Find $WSQ(4.2)$ and $WSQ(6.4)$. We define the square sine of t, denoted by ssin t, to be the y-value of point P. The square cosine of t, denoted by scos t, is defined to be the x-value of point P. For example,

$$\text{ssin } 0.4 = 0.4 \qquad \text{scos } 0.4 = 1$$
$$\text{scos } 1.2 = 0.8 \qquad \text{scos } 5.3 = -0.7$$

The square tangent of t, denoted by stan t, is defined as

$$\text{stan } t = \frac{\text{ssin } t}{\text{scos } t} \qquad \text{scos } t \neq 0$$

Find each of the following.

a. ssin 3.2   **b.** scos 4.4   **c.** stan 5.5

d. ssin 11.2   **e.** scos −5.2   **f.** stan −6.5

GRAPHS OF THE SINE AND COSINE FUNCTIONS

- THE GRAPH OF THE SINE FUNCTION
- THE GRAPH OF THE COSINE FUNCTION

• THE GRAPH OF THE SINE FUNCTION

The trigonometric functions can be graphed on a rectangular coordinate system by plotting the points whose coordinates belong to the function. We begin with the graph of the sine function.

Table 5.7 on the following page lists some ordered pairs (x, y), where $y = \sin x, 0 \leq x \leq 2\pi$. In **Figure 5.61,** the points are plotted and a smooth curve is drawn through the points.

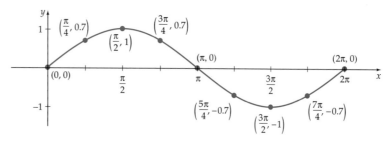

$y = \sin x, 0 \leq x \leq 2\pi$

FIGURE 5.61

TABLE 5.7

x	y = sin x
0	0
$\dfrac{\pi}{4}$	≈ 0.7
$\dfrac{\pi}{2}$	1
$\dfrac{3\pi}{4}$	≈ 0.7
π	0
$\dfrac{5\pi}{4}$	≈ -0.7
$\dfrac{3\pi}{2}$	-1
$\dfrac{7\pi}{4}$	≈ -0.7
2π	0

Because the domain of the sine function is the real numbers and the period is 2π, the graph of $y = \sin x$ is drawn by repeating the portion shown in **Figure 5.61**. Any part of the graph that corresponds to one period (2π) is one cycle of the graph of $y = \sin x$ (see **Figure 5.62**).

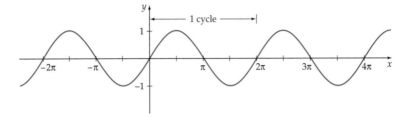

$y = \sin x$

FIGURE 5.62

The maximum value M reached by sin x is 1, and the minimum value m is -1. The amplitude of the graph of $y = \sin x$ is given by

$$\text{Amplitude} = \frac{1}{2}(M - m)$$

❷ QUESTION What is the amplitude of $y = \sin x$?

Recall that the graph of $y = a \cdot f(x)$ is obtained by *stretching* ($|a| > 1$) or *shrinking* ($0 < |a| < 1$) the graph of $y = f(x)$. **Figure 5.63** shows the graph of $y = 3 \sin x$ that was drawn by stretching the graph of $y = \sin x$. The amplitude of $y = 3 \sin x$ is 3 because

$$\text{Amplitude} = \frac{1}{2}(M - m) = \frac{1}{2}[3 - (-3)] = 3$$

Note that for $y = \sin x$ and $y = 3 \sin x$, the amplitude of the graph is the coefficient of sin x. This suggests the following theorem.

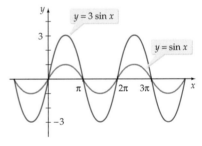

FIGURE 5.63

take note

The amplitude is defined to be half the difference between the maximum height and the minimum height. It may not be equal to the maximum height. For example, the graph of $y = 4 + \sin x$ has a maximum height of 5 and an amplitude of 1.

Amplitude of y = a sin x

The amplitude of $y = a \sin x$ is $|a|$.

---

❷ ANSWER Amplitude $= \dfrac{1}{2}(M - m) = \dfrac{1}{2}[1 - (-1)] = \dfrac{1}{2}(2) = 1$

EXAMPLE 1 Graph y = a sin x

Graph: $y = -2 \sin x$

Solution

The amplitude of $y = -2 \sin x$ is 2. The graph of $y = -f(x)$ is a *reflection* across the x-axis of $y = f(x)$. Thus the graph of $y = -2 \sin x$ is a reflection across the x-axis of $y = 2 \sin x$. See **Figure 5.64.**

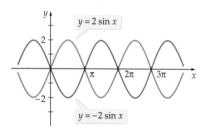

$y = 2 \sin x$

$y = -2 \sin x$

FIGURE 5.64

▶ **TRY EXERCISE 20, PAGE 518**

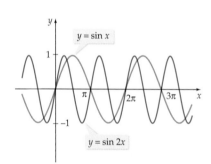

$y = \sin x$

$y = \sin 2x$

FIGURE 5.65

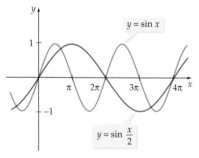

$y = \sin x$

$y = \sin \dfrac{x}{2}$

FIGURE 5.66

The graphs of $y = \sin x$ and $y = \sin 2x$ are shown in **Figure 5.65.** Because one cycle of the graph of $y = \sin 2x$ is completed in an interval of length π, the period of $y = \sin 2x$ is π.

The graphs of $y = \sin x$ and $y = \sin\left(\dfrac{x}{2}\right)$ are shown in **Figure 5.66.** Because one cycle of the graph of $y = \sin\left(\dfrac{x}{2}\right)$ is completed in an interval of length 4π, the period of $y = \sin\left(\dfrac{x}{2}\right)$ is 4π.

Generalizing the last two examples, one cycle of $y = \sin bx$, $b > 0$, is completed as bx varies from 0 to 2π. Algebraically, one cycle of $y = \sin bx$ is completed as bx varies from 0 to 2π. Therefore,

$$0 \le bx \le 2\pi$$

$$0 \le x \le \frac{2\pi}{b}$$

The length of the interval, $\dfrac{2\pi}{b}$, is the period of $y = \sin bx$. Now we consider the case when the coefficient of x is negative. If $b > 0$, then using the fact that the sine function is an odd function, we have $y = \sin(-bx) = -\sin bx$, and thus the period is still $\dfrac{2\pi}{b}$. This gives the following theorem.

Period of y = sin bx

The period of $y = \sin bx$ is $\dfrac{2\pi}{|b|}$.

Table **5.8** gives the amplitude and period of several sine functions.

TABLE 5.8

Function	$y = a \sin bx$	$y = 3\sin(-2x)$	$y = -\sin\dfrac{x}{3}$	$y = -2\sin\dfrac{3x}{4}$
Amplitude	$\lvert a\rvert$	$\lvert 3\rvert = 3$	$\lvert -1\rvert = 1$	$\lvert -2\rvert = 2$
Period	$\dfrac{2\pi}{\lvert b\rvert}$	$\dfrac{2\pi}{2} = \pi$	$\dfrac{2\pi}{1/3} = 6\pi$	$\dfrac{2\pi}{3/4} = \dfrac{8\pi}{3}$

EXAMPLE 2   **Graph $y = \sin bx$**

Graph: $y = \sin \pi x$

Solution

$$\text{Amplitude} = 1 \qquad \text{Period} = \frac{2\pi}{b} = \frac{2\pi}{\pi} = 2 \qquad \bullet\, b = \pi$$

The graph is sketched in **Figure 5.67.**

▶ **TRY EXERCISE 30, PAGE 518**

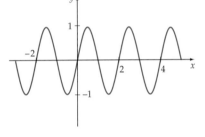

$y = \sin \pi x$

FIGURE 5.67

Figure 5.68 shows the graph of $y = a\sin bx$ for both a and b positive. Note from the graph that

- The amplitude is a.

- The period is $\dfrac{2\pi}{b}$.

- For $0 \le x \le \dfrac{2\pi}{b}$, the zeros are 0, $\dfrac{\pi}{b}$, and $\dfrac{2\pi}{b}$.

- The maximum value is a when $x = \dfrac{\pi}{2b}$, and the minimum value is $-a$ when

  $$x = \frac{3\pi}{2b}.$$

- If $a < 0$, the graph is reflected across the x-axis.

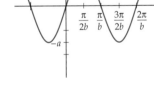

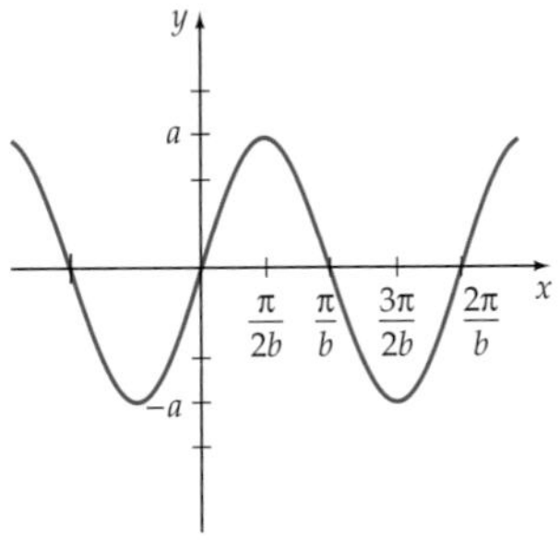

$y = a \sin bx$

FIGURE 5.68

EXAMPLE 3   **Graph $y = a \sin bx$**

Graph: $y = -\dfrac{1}{2}\sin\dfrac{x}{3}$

Solution

$$\text{Amplitude} = \left\lvert -\frac{1}{2}\right\rvert = \frac{1}{2} \qquad \text{Period} = \frac{2\pi}{1/3} = 6\pi \qquad \bullet\, b = \frac{1}{3}$$

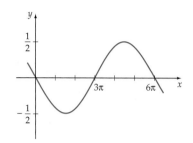

$$y = -\frac{1}{2}\sin\frac{x}{3}$$

FIGURE 5.69

The zeros in the interval $0 \le x \le 6\pi$ are 0, $\dfrac{\pi}{1/3} = 3\pi$, and $\dfrac{2\pi}{1/3} = 6\pi$.

Because $-\dfrac{1}{2} < 0$, the graph is the graph of $y = \dfrac{1}{2}\sin\dfrac{x}{3}$ reflected across the x-axis as shown in **Figure 5.69**.

▶ **TRY EXERCISE 38, PAGE 518**

● THE GRAPH OF THE COSINE FUNCTION

Table 5.9 lists some of the ordered pairs of $y = \cos x$, $0 \le x \le 2\pi$. In **Figure 5.70**, the points are plotted and a smooth curve is drawn through the points.

TABLE 5.9

x	$y = \cos x$
0	1
$\dfrac{\pi}{4}$	≈ 0.7
$\dfrac{\pi}{2}$	0
$\dfrac{3\pi}{4}$	≈ -0.7
π	-1
$\dfrac{5\pi}{4}$	≈ -0.7
$\dfrac{3\pi}{2}$	0
$\dfrac{7\pi}{4}$	≈ 0.7
2π	1

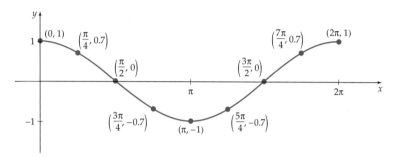

$$y = \cos x, \, 0 \le x \le 2\pi$$

FIGURE 5.70

Because the domain of $y = \cos x$ is the real numbers and the period is 2π, the graph of $y = \cos x$ is drawn by repeating the portion shown in **Figure 5.70**. Any part of the graph corresponding to one period (2π) is one cycle of $y = \cos x$ (see **Figure 5.71**).

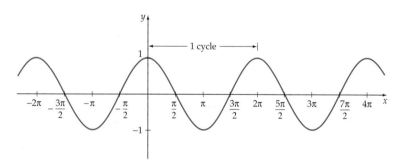

$$y = \cos x$$

FIGURE 5.71

The two theorems on the following page concerning cosine functions can be developed using methods that are analogous to those we used to determine the amplitude and period of a sine function.

Amplitude of y = a cos x

The amplitude of $y = a \cos x$ is $|a|$.

Period of y = cos bx

The period of $y = \cos bx$ is $\dfrac{2\pi}{|b|}$.

Table 5.10 gives the amplitude and period of some cosine functions.

TABLE 5.10

Function	$y = a \cos bx$	$y = 2 \cos 3x$	$y = -3 \cos \dfrac{2x}{3}$
Amplitude	$\lvert a \rvert$	$\lvert 2 \rvert = 2$	$\lvert -3 \rvert = 3$
Period	$\dfrac{2\pi}{\lvert b \rvert}$	$\dfrac{2\pi}{3}$	$\dfrac{2\pi}{2/3} = 3\pi$

EXAMPLE 4 **Graph y = cos bx**

Graph: $y = \cos \dfrac{2\pi}{3} x$

Solution

$$\text{Amplitude} = 1 \qquad \text{Period} = \dfrac{2\pi}{b} = \dfrac{2\pi}{2\pi/3} = 3 \qquad \bullet\, b = \dfrac{2\pi}{3}$$

The graph is shown in **Figure 5.72.**

▶ **TRY EXERCISE 32, PAGE 518**

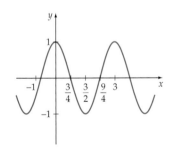

$y = \cos \dfrac{2\pi}{3} x$

FIGURE 5.72

Figure 5.73 shows the graph of $y = a \cos bx$ for both a and b positive. Note from the graph that

● The amplitude is a.

● The period is $\dfrac{2\pi}{b}$.

● For $0 \leq x \leq \dfrac{2\pi}{b}$, the zeros are $\dfrac{\pi}{2b}$ and $\dfrac{3\pi}{2b}$.

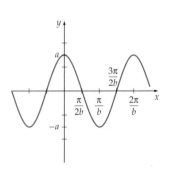

$y = a \cos bx$

FIGURE 5.73

- The maximum value is a when $x = 0$, and the minimum value is $-a$ when
$$x = \frac{\pi}{b}.$$

- If $a < 0$, then the graph is reflected across the x-axis.

EXAMPLE 5   Graph a Cosine Function

Graph: $y = -2 \cos \dfrac{\pi x}{4}$

Solution

$$\text{Amplitude} = |-2| = 2 \qquad \text{Period} = \frac{2\pi}{\pi/4} = 8 \qquad \bullet \, b = \frac{\pi}{4}$$

The zeros in the interval $0 \le x \le 8$ are $\dfrac{\pi}{2\pi/4} = 2$ and $\dfrac{3\pi}{2\pi/4} = 6$. Because

$-2 < 0$, the graph is the graph of $y = 2 \cos \dfrac{\pi x}{4}$ reflected across the x-axis

as shown in **Figure 5.74.**

▶ **TRY EXERCISE 46, PAGE 518**

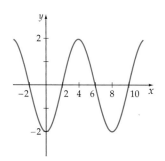

$$y = -2 \cos \frac{\pi x}{4}$$

FIGURE 5.74

EXAMPLE 6   Graph the Absolute Value of the Cosine Function

Graph $y = |\cos x|$, where $0 \le x \le 2\pi$.

Solution

Because $|\cos x| \ge 0$, the graph of $y = |\cos x|$ is drawn by reflecting the negative portion of the graph of $y = \cos x$ across the x-axis. The graph is the one shown in dark blue and light blue in **Figure 5.75.**

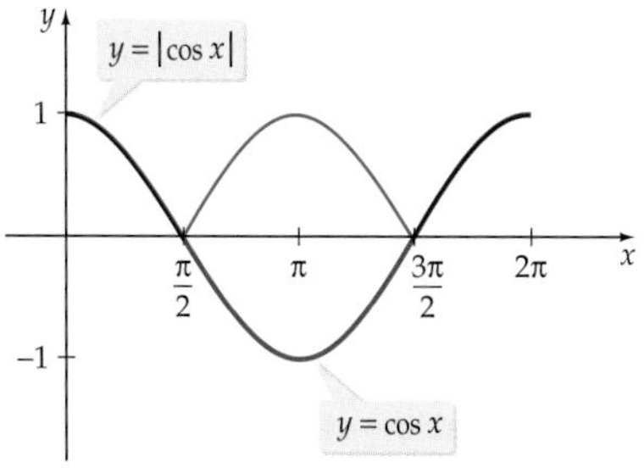

▶ **TRY EXERCISE 52, PAGE 518**

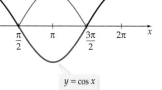

FIGURE 5.75

TOPICS FOR DISCUSSION

1. Is the graph of $f(x) = |\sin x|$ the same as the graph of $y = \sin|x|$? Explain.

2. Explain how the graph of $y = \cos 2x$ differs from the graph of $y = \cos x$.

3. Does the graph of $y = \sin(-2x)$ have the same period as the graph of $y = \sin 2x$? Explain.

4. The function $h(x) = a \sin bt$ has an amplitude of 3 and a period of 4. What are the possible values of a? What are the possible values of b?

EXERCISE SET 5.5

In Exercises 1 to 16, state the amplitude and period of the function defined by each equation.

1. $y = 2 \sin x$

2. $y = -\dfrac{1}{2} \sin x$

3. $y = \sin 2x$

4. $y = \sin \dfrac{2x}{3}$

5. $y = \dfrac{1}{2} \sin 2\pi x$

6. $y = 2 \sin \dfrac{\pi x}{3}$

7. $y = -2 \sin \dfrac{x}{2}$

8. $y = -\dfrac{1}{2} \sin \dfrac{x}{2}$

9. $y = \dfrac{1}{2} \cos x$

10. $y = -3 \cos x$

11. $y = \cos \dfrac{x}{4}$

12. $y = \cos 3x$

13. $y = 2 \cos \dfrac{\pi x}{3}$

14. $y = \dfrac{1}{2} \cos 2\pi x$

15. $y = -3 \cos \dfrac{2x}{3}$

16. $y = \dfrac{3}{4} \cos 4x$

In Exercises 17 to 54, graph at least one full period of the function defined by each equation.

17. $y = \dfrac{1}{2} \sin x$

18. $y = \dfrac{3}{2} \cos x$

19. $y = 3 \cos x$

▶ **20.** $y = -\dfrac{3}{2} \sin x$

21. $y = -\dfrac{7}{2} \cos x$

22. $y = 3 \sin x$

23. $y = -4 \sin x$

24. $y = -5 \cos x$

25. $y = \cos 3x$

26. $y = \sin 4x$

27. $y = \sin \dfrac{3x}{2}$

28. $y = \cos \pi x$

29. $y = \cos \dfrac{\pi}{2} x$

▶ **30.** $y = \sin \dfrac{3\pi}{4} x$

31. $y = \sin 2\pi x$

▶ **32.** $y = \cos 3\pi x$

33. $y = 4 \cos \dfrac{x}{2}$

34. $y = 2 \cos \dfrac{3x}{4}$

35. $y = -2 \cos \dfrac{x}{3}$

36. $y = -\dfrac{4}{3} \cos 3x$

37. $y = 2 \sin \pi x$

▶ **38.** $y = \dfrac{1}{2} \sin \dfrac{\pi x}{3}$

39. $y = \dfrac{3}{2} \cos \dfrac{\pi x}{2}$

40. $y = \cos \dfrac{\pi x}{3}$

41. $y = 4 \sin \dfrac{2\pi x}{3}$

42. $y = 3 \cos \dfrac{3\pi x}{2}$

43. $y = 2 \cos 2x$

44. $y = \dfrac{1}{2} \sin 2.5x$

45. $y = -2 \sin 1.5x$

▶ **46.** $y = -\dfrac{3}{4} \cos 5x$

47. $y = \left| 2 \sin \dfrac{x}{2} \right|$

48. $y = \left| \dfrac{1}{2} \sin 3x \right|$

49. $y = |-2 \cos 3x|$

50. $y = \left| -\dfrac{1}{2} \cos \dfrac{x}{2} \right|$

51. $y = -\left| 2 \sin \dfrac{x}{3} \right|$

▶ **52.** $y = -\left| 3 \sin \dfrac{2x}{3} \right|$

53. $y = -|3 \cos \pi x|$

54. $y = -\left| 2 \cos \dfrac{\pi x}{2} \right|$

In Exercises 55 to 60, find an equation of each graph.

55.

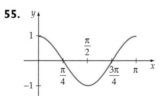

56.

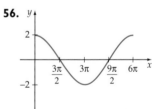

57.

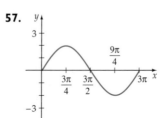

58.

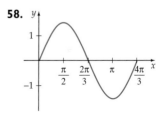

59.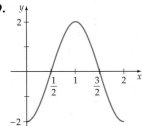

60.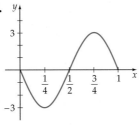

61. Sketch the graph of $y = 2 \sin \dfrac{2x}{3}$, $-3\pi \le x \le 6\pi$.

62. Sketch the graph of $y = -3 \cos \dfrac{3x}{4}$, $-2\pi \le x \le 4\pi$.

63. Sketch the graphs of

$$y_1 = 2 \cos \frac{x}{2} \quad \text{and} \quad y_2 = 2 \cos x$$

on the same set of axes for $-2\pi \le x \le 4\pi$.

64. Sketch the graphs of

$$y_1 = \sin 3\pi x \quad \text{and} \quad y_2 = \sin \frac{\pi x}{3}$$

on the same set of axes for $-2 \le x \le 4$.

In Exercises 65 to 72, use a graphing utility to graph each function.

65. $y = \cos^2 x$

66. $y = 3^{\cos^2 x} \cdot 3^{\sin^2 x}$

67. $y = \cos |x|$

68. $y = \sin |x|$

69. $y = \dfrac{1}{2} x \sin x$

70. $y = \dfrac{1}{2} x + \sin x$

71. $y = -x \cos x$

72. $y = -x + \cos x$

73. Graph $y = e^{\sin x}$. What is the maximum value of $e^{\sin x}$? What is the minimum value of $e^{\sin x}$? Is the function defined by $y = e^{\sin x}$ a periodic function? If so, what is the period?

74. Graph $y = e^{\cos x}$. What is the maximum value of $e^{\cos x}$? What is the minimum value of $e^{\cos x}$? Is the function defined by $y = e^{\cos x}$ a periodic function? If so, what is the period?

75. **EQUATION OF A WAVE** A tidal wave that is caused by an earthquake under the ocean is called a **tsunami.** A model of a tsunami is given by $f(t) = A \cos Bt$. Find the equation of a tsunami that has an amplitude of 60 feet and a period of 20 seconds.

76. **EQUATION OF HOUSEHOLD CURRENT** The electricity supplied to your home, called *alternating current*, can be modeled by $I = A \sin \omega t$, where I is the number of amperes of current at time t seconds. Write the equation of household current whose graph is given in the figure below. Calculate I when $t = 0.5$ second.

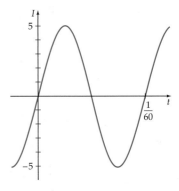

CONNECTING CONCEPTS

In Exercises 77 to 80, write an equation for a sine function using the given information.

77. Amplitude = 2; period = 3π

78. Amplitude = 5; period = $\dfrac{2\pi}{3}$

79. Amplitude = 4; period = 2

80. Amplitude = 2.5; period = 3.2

In Exercises 81 to 84, write an equation for a cosine function using the given information.

81. Amplitude = 3; period = $\dfrac{\pi}{2}$

82. Amplitude = 0.8; period = 4π

83. Amplitude = 3; period = 2.5

84. Amplitude = 4.2; period = 1

PREPARE FOR SECTION 5.6

85. Estimate, to the nearest tenth, $\tan \dfrac{\pi}{3}$. [5.4]

86. Estimate, to the nearest tenth, $\cot \dfrac{\pi}{3}$. [5.4]

87. Explain how to use the graph of $y = f(x)$ to produce the graph of $y = 2f(x)$. [2.5]

88. Explain how to use the graph of $y = f(x)$ to produce the graph of $y = f(x - 2) + 3$. [2.5]

89. Simplify: $\dfrac{\pi}{1/2}$ [P.5]

90. Simplify: $\dfrac{\pi}{\left| -\dfrac{3}{4} \right|}$ [P.1/P.5]

PROJECTS

1.　　**A TRIGONOMETRIC POWER FUNCTION**

a. Determine the domain and the range of $y = (\sin x)^{\cos x}$. Explain.

b. What is the amplitude of the function? Explain.

GRAPHS OF THE OTHER TRIGONOMETRIC FUNCTIONS

SECTION 5.6

- THE GRAPH OF THE TANGENT FUNCTION
- THE GRAPH OF THE COTANGENT FUNCTION
- THE GRAPH OF THE COSECANT FUNCTION
- THE GRAPH OF THE SECANT FUNCTION

THE GRAPH OF THE TANGENT FUNCTION

Figure 5.76 shows the graph of $y = \tan x$ for $-\dfrac{\pi}{2} < x < \dfrac{\pi}{2}$. The lines $x = \dfrac{\pi}{2}$ and $x = -\dfrac{\pi}{2}$ are vertical asymptotes for the graph of $y = \tan x$. From Section 5.4, the period of $y = \tan x$ is π. Therefore, the portion of the graph shown in **Figure 5.76** is repeated along the x-axis, as shown in **Figure 5.77**.

Because the tangent function is unbounded, there is no amplitude for the tangent function. The graph of $y = a \tan x$ is drawn by stretching ($|a| > 1$) or shrinking ($|a| < 1$) the graph of $y = \tan x$. If $a < 0$, then the graph is reflected across the x-axis. **Figure 5.78** shows the graphs of three tangent functions. Because $\tan \dfrac{\pi}{4} = 1$, the point $\left(\dfrac{\pi}{4}, a \right)$ is convenient to plot as a guide for the graph of $y = a \tan x$.

64.

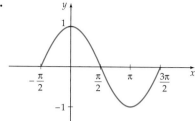

65. CARBON DIOXIDE LEVELS Because of seasonal changes in vegetation, carbon dioxide (CO_2) levels, as a product of photosynthesis, rise and fall during the year. Besides the naturally occurring CO_2 from plants, additional CO_2 is given off as a pollutant. A reasonable model of CO_2 levels in a city for the years 1982–2002 is given by $y = 2.3 \sin 2\pi t + 1.25t + 315$, where t is the number of years since 1982 and y is the concentration of CO_2 in parts per million (ppm). Find the difference in CO_2 levels between the beginning of 1982 and the beginning of 2002.

66. ENVIRONMENTAL SCIENCE Some environmentalists contend that the rate of growth of atmospheric CO_2 in parts per million is given by the equation $y = 2.54e^{0.112t} + \sin 2\pi t + 315$. See Exercise 65. Use this model to find the difference between CO_2 levels from the beginning of 1982 to the beginning of 2002.

67. HEIGHT OF A PADDLE The paddle wheel on a river boat is shown in the accompanying figure. Write an equation for the height of a paddle relative to the water at time t. The radius of the paddle wheel is 7 feet, and the distance from the center of the paddle wheel to the water is 5 feet. Assume that the paddle wheel rotates at 5 revolutions per minute and that the paddle is at its highest point at $t = 0$. Graph the equation for $0 \le t \le 0.20$ minute.

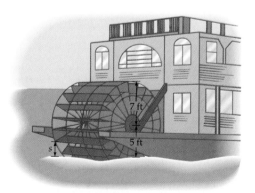

68. VOLTAGE AND AMPERAGE The graphs of the voltage and the amperage of an alternating household circuit are shown in the following figures, where t is measured in seconds. Note that there is a phase shift between the graph of the voltage and the graph of the current. The cur-

rent is said to *lag* the voltage by 0.005 second. Write an equation for the voltage and an equation for the current.

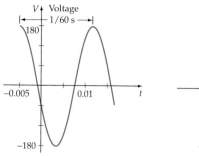

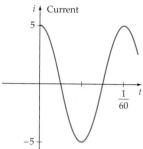

69. A LIGHTHOUSE BEACON The beacon of a lighthouse 400 meters from a straight sea wall rotates at 6 revolutions per minute. Using the accompanying figures, write an equation expressing the distance s, measured in meters, in terms of time t. Assume that when $t = 0$, the beam is perpendicular to the sea wall. Sketch a graph of the equation for $0 \le t \le 10$ seconds.

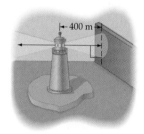

Side view

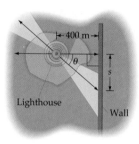

Top view

70. HOURS OF DAYLIGHT The duration of daylight for a region is dependent not only on the time of year but also on the latitude of the region. The graph gives the daylight hours for a one-year period at various latitudes. Assuming that a sine function can model these curves, write an equation for each curve.

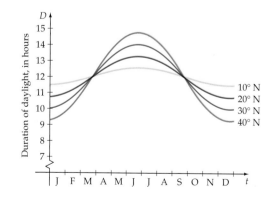

71. **TIDES** During a 24-hour day, the tides raise and lower the depth of water at a pier as shown in the figure below. Write an equation in the form $f(t) = A \cos Bt + d$, and find the depth of the water at 6 P.M.

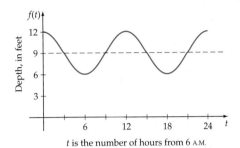

t is the number of hours from 6 A.M.

72. **TEMPERATURE** During a summer day, the ground temperature at a desert location was recorded and graphed as a function of time as shown in the following figure. The graph can be approximated by $f(t) = A \cos(bt + c) + d$. Find the equation, and approximate the temperature (to the nearest degree) at 1:00 P.M.

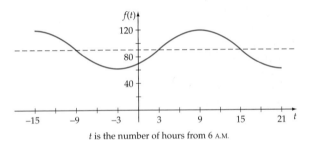

t is the number of hours from 6 A.M.

In Exercises 73 to 82, use a graphing utility to graph each function.

73. $y = \sin x - \cos \dfrac{x}{2}$

74. $y = 2 \sin 2x - \cos x$

75. $y = 2 \cos x + \sin \dfrac{x}{2}$

76. $y = -\dfrac{1}{2} \cos 2x + \sin \dfrac{x}{2}$

77. $y = \dfrac{x}{2} \sin x$

▶ **78.** $y = x \cos x$

79. $y = x \sin \dfrac{x}{2}$

80. $y = \dfrac{x}{2} \cos \dfrac{x}{2}$

81. $y = x \sin\left(x + \dfrac{\pi}{2}\right)$

82. $y = x \cos\left(x - \dfrac{\pi}{2}\right)$

BEATS When two sound waves have approximately the same frequency, the sound waves interfere with one another and produce phenomena called *beats*, which are heard as variations in the loudness of the sound. A piano tuner can use these phenomena to tune a piano. By striking a tuning fork and then tapping the corresponding key on a piano, the piano tuner listens for beats and adjusts the tension in the string until the beats disappear. Use a graphing utility to graph the functions in Exercises 83 to 86, which are based on beats.

83. $y = \sin(5\pi x) \cdot \sin\left(-\dfrac{\pi}{2}x\right)$

84. $y = \sin(9\pi x) \cdot \sin\left(-\dfrac{\pi}{2}x\right)$

85. $y = \sin(13\pi x) \cdot \sin\left(-\dfrac{\pi}{2}x\right)$

86. $y = \sin(17\pi x) \cdot \sin\left(-\dfrac{\pi}{2}x\right)$

CONNECTING CONCEPTS

87. Find an equation of the sine function with amplitude 2, period π, and phase shift $\dfrac{\pi}{3}$.

88. Find an equation of the cosine function with amplitude 3, period 3π, and phase shift $-\dfrac{\pi}{4}$.

89. Find an equation of the tangent function with period 2π and phase shift $\dfrac{\pi}{2}$.

90. Find an equation of the cotangent function with period $\dfrac{\pi}{2}$ and phase shift $-\dfrac{\pi}{4}$.

91. Find an equation of the secant function with period 4π and phase shift $\dfrac{3\pi}{4}$.

92. Find an equation of the cosecant function with period $\dfrac{3\pi}{2}$ and phase shift $\dfrac{\pi}{4}$.

93. If $g(x) = \sin^2 x$ and $h(x) = \cos^2 x$, find $g(x) + h(x)$.

94. If $g(x) = 2 \sin x - 3$ and $h(x) = 4 \cos x + 2$, find the sum $g(x) + h(x)$.

95. If $g(x) = x^2 + 2$ and $h(x) = \cos x$, find $g[h(x)]$.

96. If $g(x) = \sin x$ and $h(x) = x^2 + 2x + 1$, find $h[g(x)]$.

In Exercises 97 to 100, use a graphing utility to graph each function.

97. $y = \dfrac{\sin x}{x}$

98. $y = 2 + \sec \dfrac{x}{2}$

99. $y = |x| \sin x$

100. $y = |x| \cos x$

PREPARE FOR SECTION 5.8

101. Find the reciprocal of $\dfrac{2\pi}{3}$. [P.1]

102. Find the reciprocal of $\dfrac{2}{5}$. [P.1]

103. Evaluate $a \cos 2\pi t$ for $a = 4$ and $t = 0$. [5.4]

104. Evaluate $\sqrt{\dfrac{k}{m}}$ for $k = 18$ and $m = 2$. [P.5]

105. Evaluate $4 \cos\left(\sqrt{\dfrac{k}{m}}\, t \right)$ for $k = 16$, $m = 4$, and $t = 2\pi$. [5.4]

106. Write an equation of the form $y = a \cos bx$ whose graph has an amplitude of 4 and a period of 2. [5.5]

PROJECTS

1. **PREDATOR-PREY RELATIONSHIP** Predator-prey interactions can produce cyclic population growth for both the predator population and the prey population. Consider an animal reserve where the rabbit population r is given by

$$r(t) = 850 + 210 \sin\left(\frac{\pi}{6} t\right)$$

and the wolf population w is given by

$$w(t) = 120 + 30 \sin\left(\frac{\pi}{6} t - 2\right)$$

where t is the number of months after March 1, 2000. Graph $r(t)$ and $w(t)$ on the same coordinate system for $0 \le t \le 24$. Write a few sentences that explains a possible relationship between the two populations.

Write an equation that could be used to model the rabbit population shown by the graph at the right, where t is measured in months.

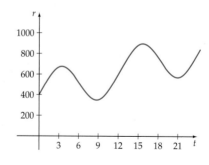

HARMONIC MOTION—AN APPLICATION OF THE SINE AND COSINE FUNCTIONS

- SIMPLE HARMONIC MOTION
- DAMPED HARMONIC MOTION

Many phenomena that occur in nature can be modeled by periodic functions, including vibrations of a swing or in a spring. These phenomena can be described by the *sinusoidal* functions, which are the sine and cosine functions or a sum of these two functions.

● SIMPLE HARMONIC MOTION

We will consider a mass on a spring to illustrate vibratory motion. Assume that we have placed a mass on a spring and allowed the spring to come to rest, as shown in **Figure 5.104.** The system is said to be in equilibrium when the mass is at rest. The point of rest is called the *origin* of the system. We consider the distance above the equilibrium point as positive and the distance below the equilibrium point as negative.

    If the mass is now lifted a distance a and released, the mass will oscillate up and down in periodic motion. If there is no friction, the motion repeats itself in a certain period of time. The distance a is called the **displacement** from the origin. The number of times the mass oscillates in 1 unit of time is called the *frequency f* of the motion, and the time one oscillation takes is the *period p* of the motion. The motion is referred to as *simple harmonic motion.* **Figure 5.105** shows the displacement y of the mass for one oscillation for $t = 0, \dfrac{p}{4}, \dfrac{p}{2}, \dfrac{3p}{4}$, and p.

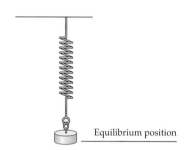

Equilibrium position

FIGURE 5.104

> *take note*
>
> The graph of the displacement y as a function of the time t is a cosine curve.

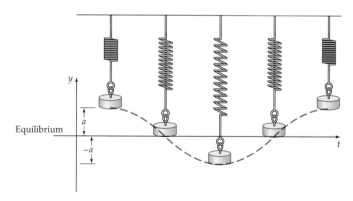

FIGURE 5.105

The frequency and the period are related by the formulas

$$f = \frac{1}{p} \qquad \text{and} \qquad p = \frac{1}{f}$$

    The maximum displacement from the equilibrium position is called the *amplitude of the motion.* Vibratory motion can be quite complicated. However, the simple harmonic motion of the mass on the spring can be described by one of the following equations.

Simple Harmonic Motion

take note

Function (1) is used if the displacement from the origin is at a maximum at time $t = 0$. Function (2) is used if the displacement at time $t = 0$ is zero.

Simple harmonic motion can be modeled by one of the following functions:

$$y = a \cos 2\pi ft \quad (1) \qquad \text{or} \qquad y = a \sin 2\pi ft \quad (2)$$

where $|a|$ is the amplitude (maximum displacement), f is the frequency, $\dfrac{1}{f}$ is the period, y is the displacement, and t is the time.

? QUESTION A simple harmonic motion has a frequency of 2 cycles per second. What is the period of the simple harmonic motion?

EXAMPLE 1   **Find the Equation of Motion of a Mass on a Spring**

A mass on a spring has been displaced 4 centimeters above the equilibrium point and released. The mass is vibrating with a frequency of $\dfrac{1}{2}$ cycle per second. Write the equation of simple harmonic motion, and graph three cycles of the displacement as a function of time.

Solution

Because the maximum displacement is 4 centimeters when $t = 0$, use $y = a \cos 2\pi ft$. See **Figure 5.106.**

$$y = a \cos 2\pi ft \qquad \bullet \text{ Equation for simple harmonic motion}$$

$$= 4 \cos 2\pi\left(\frac{1}{2}\right)t \qquad \bullet\ a = 4,\ f = \frac{1}{2}$$

$$= 4 \cos \pi t$$

▶ **TRY EXERCISE 20, PAGE 544**

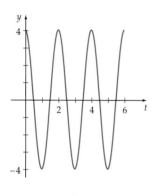

$y = 4 \cos \pi t$

FIGURE 5.106

From physical laws determined by experiment, the frequency of oscillation of a mass on a spring is given by

$$f = \frac{1}{2\pi} \sqrt{\frac{k}{m}}$$

where k is a spring constant determined by experiment and m is the mass. The simple harmonic motion of the mass on the spring (with maximum displacement

? ANSWER $\dfrac{1}{2}$ second per cycle

at $t = 0$) can then be described by

$$y = a \cos 2\,\pi ft = a \cos 2\pi \left(\frac{1}{2\pi} \sqrt{\frac{k}{m}} \right) t$$

$$= a \cos \sqrt{\frac{k}{m}}\, t \qquad\qquad (3)$$

The equation of the simple harmonic motion for zero displacement at $t = 0$ is

$$y = a \sin \sqrt{\frac{k}{m}}\, t \qquad\qquad (4)$$

EXAMPLE 2   **Find the Equation of Motion of a Mass on a Spring**

A mass of 2 units is in equilibrium suspended from a spring with a spring constant of $k = 18$. The mass is pulled down 0.5 unit and released. Find the period, frequency, and amplitude of the resulting simple harmonic motion. Write the equation of the motion, and graph two cycles of the displacement as a function of time.

Solution

At the start of the motion, the displacement is at a maximum but in the negative direction. The resulting motion is described by Equation (3), using $a = -0.5$, $k = 18$, and $m = 2$.

$$y = a \cos \sqrt{\frac{k}{m}}\, t = -0.5 \cos \sqrt{\frac{18}{2}}\, t \qquad \text{• Substitute for } a, k, \text{ and } m.$$

$$= -0.5 \cos 3t \qquad\qquad \text{• Equation of motion}$$

Period: $\dfrac{2\pi}{|b|} = \dfrac{2\pi}{3}$

Frequency: $\dfrac{1}{\text{period}} = \dfrac{3}{2\pi}$

Amplitude: $|a| = |-0.5| = 0.5$

See **Figure 5.107.**

▶ **TRY EXERCISE 28, PAGE 544**

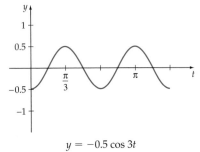

$y = -0.5 \cos 3t$

FIGURE 5.107

● DAMPED HARMONIC MOTION

The previous examples have assumed that there is no friction within the spring and no air resistance. If we consider friction and air resistance, then the motion of the mass tends to decrease as t increases. The motion is called **damped harmonic motion.**

EXAMPLE 3   **Model Damped Harmonic Motion**

 A mass on a spring has been displaced 14 inches below the equilibrium point and released. The damped harmonic motion of the mass is given by

$$f(t) = -14e^{-0.4t}\cos 2t, \quad t \geq 0$$

where $f(t)$ is measured in inches and t is the time in seconds.

a. Find the values of t for which $f(t) = 0$.

b. Use a graphing utility to determine how long it will be until the absolute value of the displacement of the mass is always less than 0.01 inch.

Solution

a. $f(t) = 0$ if and only if $\cos 2t = 0$, and $\cos 2t = 0$ when $2t = \dfrac{\pi}{2} + n\pi$.

Therefore, $f(t) = 0$ when $t = \dfrac{\pi}{4} + n\left(\dfrac{\pi}{2}\right) \approx 0.79 + n(1.57)$. The

displacement $f(t) = 0$ first occurs when $t \approx 0.79$ second. The graph of f in **Figure 5.108** confirms these results. **Figure 5.108** also shows that the graph of f lies on or between the graphs of $y = -14e^{-0.4t}$ and $y = 14e^{-0.4t}$. Recall from Section 5.7 that $-14e^{-0.4t}$ is the damping factor.

b. We need to find the smallest value of t for which the absolute value of the displacement of the mass is always less than 0.01. **Figure 5.109** shows a graph of f using a viewing window of $0 \leq t \leq 20$ and $-0.01 \leq f(t) \leq 0.01$. Use the TRACE or the INTERSECT feature to determine that $t \approx 17.59$ seconds.

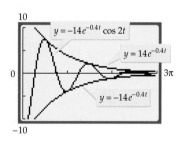

FIGURE 5.108

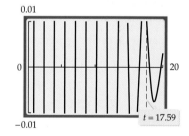

$f(t) = -14e^{-0.4t}\cos 2t$

FIGURE 5.109

▶ **TRY EXERCISE 30, PAGE 545**

---

take note

The motion in Example 3 is not periodic in a strict sense, because the motion does not repeat exactly but tends to diminish as t increases. However, because the motion cycles every π seconds, we call π the pseudoperiod of the motion. In general, the damped harmonic motion modeled by $f(t) = ae^{-kt}\cos \omega t$ has

pseudoperiod $p = \dfrac{2\pi}{\omega}$ and

frequency $f = \dfrac{\omega}{2\pi}$.

TOPICS FOR DISCUSSION

1. The period of a simple harmonic motion is the same as the frequency of the motion. Do you agree? Explain.

2. In the simple harmonic motion modeled by $y = 3\cos 2\pi t$, does the displacement y approach 0 as t increases without bound? Explain.

3. Explain how you know whether to use

$$y = a \cos 2\pi f t \qquad \text{or} \qquad y = a \sin 2\pi f t$$

to model a particular harmonic motion.

4. If the mass on a spring is increased from m to $4m$, what effect will this have on the frequency of the simple harmonic motion of the mass?

EXERCISE SET 5.8

In Exercises 1 to 8, find the amplitude, period, and frequency of the simple harmonic motion.

1. $y = 2 \sin 2t$

2. $y = \dfrac{2}{3} \cos \dfrac{t}{3}$

3. $y = 3 \cos \dfrac{2t}{3}$

4. $y = 4 \sin 3t$

5. $y = 4 \cos \pi t$

6. $y = 2 \sin \dfrac{\pi t}{3}$

7. $y = \dfrac{3}{4} \sin \dfrac{\pi t}{2}$

8. $y = 5 \cos 2\pi t$

In Exercises 9 to 12, write an equation for the simple harmonic motion that satisfies the given conditions. Assume that the maximum displacement occurs at $t = 0$. Sketch a graph of the equation.

9. Frequency $= 1.5$ cycles per second, $a = 4$ inches

10. Frequency $= 0.8$ cycle per second, $a = 4$ centimeters

11. Period $= 1.5$ seconds, $a = \dfrac{3}{2}$ feet

12. Period $= 0.6$ second, $a = 1$ meter

In Exercises 13 to 18, write an equation for the simple harmonic motion with the given conditions. Assume zero displacement at $t = 0$. Sketch a graph of the equation.

13. Amplitude 2 centimeters, period π seconds

14. Amplitude 4 inches, period $\dfrac{\pi}{2}$ seconds

15. Amplitude 1 inch, period 2 seconds

16. Amplitude 3 centimeters, period 1 second

17. Amplitude 2 centimeters, frequency 1 second

18. Amplitude 4 inches, frequency 4 seconds

In Exercises 19 to 26, write an equation for simple harmonic motion. Assume that the maximum displacement occurs when $t = 0$.

19. Amplitude $\dfrac{1}{2}$ centimeter, frequency $\dfrac{2}{\pi}$ cycle per second

▶ **20.** Amplitude 3 inches, frequency $\dfrac{1}{\pi}$ cycle per second

21. Amplitude 2.5 inches, frequency 0.5 cycle per second

22. Amplitude 5 inches, frequency $\dfrac{1}{8}$ cycle per second

23. Amplitude $\dfrac{1}{2}$ inch, period 3 seconds

24. Amplitude 5 centimeters, period 5 seconds

25. Amplitude 4 inches, period $\dfrac{\pi}{2}$ seconds

26. Amplitude 2 centimeters, period π seconds

27. SIMPLE HARMONIC MOTION   A mass of 32 units is in equilibrium suspended from a spring. The mass is pulled down 2 feet and released. Find the period, frequency, and amplitude of the resulting simple harmonic motion. Write an equation of the motion. Assume a spring constant of $k = 8$.

▶ **28.** SIMPLE HARMONIC MOTION   A mass of 27 units is in equilibrium suspended from a spring. The mass is pulled down 1.5 feet and released. Find the period, frequency, and amplitude of the resulting simple harmonic motion. Write an equation of the motion. Assume a spring constant of $k = 3$.

In Exercises 29 to 36, each of the equations models a damped harmonic motion.

a. Find the number of complete oscillations that occur during the time interval $0 \le t \le 10$ seconds.

b. Use a graph to determine how long it will be (to the nearest 0.1 second) until the absolute value of the displacement of the mass is always less than 0.01.

29. $f(t) = 4e^{-0.1t} \cos 2t$

30. $f(t) = 12e^{-0.6t} \cos t$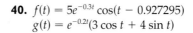

31. $f(t) = -6e^{-0.09t} \cos 2\pi t$

32. $f(t) = -11e^{-0.4t} \cos \pi t$

33. $f(t) = e^{-0.5t} \cos 2\pi t$

34. $f(t) = e^{-0.2t} \cos 3\pi t$

35. $f(t) = e^{-0.75t} \cos 2\pi t$

36. $f(t) = e^{-t} \cos 2\pi t$

 ## CONNECTING CONCEPTS

37. Assume that a mass of m pounds on the end of a spring is oscillating in simple harmonic motion. What effect will there be on the period of the motion if the mass is increased to $9m$?

38. A mass on a spring is displaced 9 inches below its equilibrium position and then released. The weight oscillates in simple harmonic motion with a frequency of 2 cycles per second. Find the period and the equation of the motion.

In Exercises 39 to 42, use a graphing utility to determine whether both of the given functions model the same damped harmonic motion.

39. $f(t) = \sqrt{2}\,e^{-0.2t} \sin\left(t + \dfrac{\pi}{4}\right)$

$g(t) = e^{-0.2t}(\cos t + \sin t)$

40. $f(t) = 5e^{-0.3t} \cos(t - 0.927295)$

$g(t) = e^{-0.2t}(3 \cos t + 4 \sin t)$

41. $f(t) = 13e^{-0.4t} \cos(t - 1.176005)$

$g(t) = e^{-0.4t}(5 \cos t + 12 \sin t)$

42. $f(t) = \sqrt{2}\,e^{-0.2t} \cos(t + 0.785398)$

$g(t) = e^{-0.2t}(\cos t - \sin t)$

PROJECTS

1.   **THREE TYPES OF DAMPED HARMONIC MOTION** In some cases the damped harmonic motion of a mass on the end of a spring does not cycle about the equilibrium point. The following three functions illustrate different types of damped harmonic motion. Use a graphing utility to graph each function, and then write a few sentences that explain the major differences among the motions.

$$f(t) = (0.5t + 1)e^{-0.5t} \qquad g(t) = -2e^{-0.4t} + 5e^{-t}$$

$$h(t) = 4e^{-0.2t} \cos 2\pi t$$

2. **LOGARITHMIC DECREMENT** If a damped harmonic motion is modeled by

$$f(t) = ae^{-\alpha t} \cos \omega t$$

then the ratio of any two consecutive relative maxima of the motion is a constant γ.

a. Use a graphing utility to determine γ for the damped harmonic motion modeled by

$$f(t) = -14e^{-0.4t} \cos 2t, \quad t \ge 0$$

b. The constant $\Delta = \dfrac{2\pi\alpha}{\omega}$ is called the **logarithmic decrement** of the motion. Compute Δ for the damped harmonic motion in the equation in part **a.** How does $\ln \gamma$ compare with Δ?

EXPLORING CONCEPTS WITH TECHNOLOGY

Sinusoidal Families

Some graphing calculators have a feature that allows you to graph a family of functions easily. For instance, entering Y₁={2,4,6}sin(X) in the Y= menu and pressing the GRAPH key on a TI-83 calculator produces a graph of the three functions $y = 2 \sin x$, $y = 4 \sin x$, and $y = 6 \sin x$, all displayed in the same window.

1. Use a graphing calculator to graph Y₁={2,4,6}sin(X). Write a sentence that indicates the similarities and the differences among the three graphs.

2. Use a graphing calculator to graph Y₁=sin({π,2π,4π}X). Write a sentence that indicates the similarities and the differences among the three graphs.

3. Use a graphing calculator to graph Y₁=sin(X+{π/4,π/6,π/12}). Write a sentence that indicates the similarities and the differences among the three graphs.

4. A student has used a graphing calculator to graph Y₁=sin(X+{π,3π,5π}) and expects to see three graphs. However, the student sees only one graph displayed on the graph window. Has the calculator displayed all three graphs? Explain.

CHAPTER 5 SUMMARY

5.1 Angles and Arcs

- An angle is in standard position when its initial side is along the positive x-axis and its vertex is at the origin of the coordinate axes.

- Angle α is an acute angle when $0° < \alpha < 90°$; it is an obtuse angle when $90° < \alpha < 180°$.

- α and β are complementary angles when $\alpha + \beta = 90°$; they are supplementary angles when $\alpha + \beta = 180°$.

- The length of the arc s that subtends the central angle θ (in radians) on a circle of radius r is given by $s = r\theta$.

- Angular speed is given by $\omega = \dfrac{\theta}{t}$.

5.2 Trigonometric Functions of Acute Angles

- Let θ be an acute angle of a right triangle. The six trigonometric functions of θ are given by

$$\sin \theta = \frac{\text{opp}}{\text{hyp}} \qquad \csc \theta = \frac{\text{hyp}}{\text{opp}}$$

$$\cos \theta = \frac{\text{adj}}{\text{hyp}} \qquad \sec \theta = \frac{\text{hyp}}{\text{adj}}$$

$$\tan \theta = \frac{\text{opp}}{\text{adj}} \qquad \cot \theta = \frac{\text{adj}}{\text{opp}}$$

5.3 Trigonometric Functions of Any Angle

- Let $P(x, y)$ be a point, except the origin, on the terminal side of an angle θ in standard position. The six trigonometric functions of θ are

$$\sin \theta = \frac{y}{r} \qquad \csc \theta = \frac{r}{y}, \quad y \neq 0$$

$$\cos \theta = \frac{x}{r} \qquad \sec \theta = \frac{r}{x}, \quad x \neq 0$$

$$\tan \theta = \frac{y}{x}, \quad x \neq 0 \qquad \cot \theta = \frac{x}{y}, \quad y \neq 0$$

5.4 Trigonometric Functions of Real Numbers

- The wrapping function pairs a real number with a point on the unit circle.

- Let W be the wrapping function, t be a real number, and $W(t) = P(x, y)$. Then the trigonometric functions of the real number t are defined as follows:

$$\sin t = y \qquad\qquad \csc t = \frac{1}{y}, \quad y \neq 0$$

$$\cos t = x \qquad\qquad \sec t = \frac{1}{x}, \quad x \neq 0$$

$$\tan t = \frac{y}{x}, \quad x \neq 0 \qquad\qquad \cot t = \frac{x}{y}, \quad y \neq 0$$

- $\sin t$, $\csc t$, $\tan t$, and $\cot t$ are odd functions.

- $\cos t$ and $\sec t$ are even functions.

- $\sin t$, $\cos t$, $\sec t$, and $\csc t$ have period 2π.

- $\tan t$ and $\cot t$ have period π.

Domain and Range of Each Trigonometric Function (n is an integer)

Function	Domain	Range
$\sin t$	$\{t \mid -\infty < t < \infty\}$	$\{y \mid -1 \leq y \leq 1\}$
$\cos t$	$\{t \mid -\infty < t < \infty\}$	$\{y \mid -1 \leq y \leq 1\}$
$\tan t$	$\left\{ t \mid -\infty < t < \infty, \; t \neq \frac{(2n+1)\pi}{2} \right\}$	$\{y \mid -\infty < y < \infty\}$
$\csc t$	$\{t \mid -\infty < t < \infty, \; t \neq n\pi\}$	$\{y \mid y \geq 1, y \leq -1\}$
$\sec t$	$\left\{ t \mid -\infty < t < \infty, \; t \neq \frac{(2n+1)\pi}{2} \right\}$	$\{y \mid y \geq 1, y \leq -1\}$
$\cot t$	$\{t \mid -\infty < t < \infty, \; t \neq n\pi\}$	$\{y \mid -\infty < y < \infty\}$

5.5 Graphs of the Sine and Cosine Functions

- The graphs of $y = a \sin bx$ and $y = a \cos bx$ both have an amplitude of $|a|$ and a period of $\frac{2\pi}{|b|}$. The graph of each for $a > 0$ and $b > 0$ is given below.

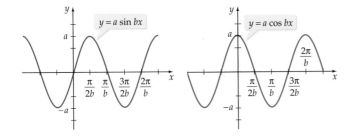

5.6 Graphs of the Other Trigonometric Functions

- The period of $y = a \tan bx$ and $y = a \cot bx$ is $\frac{\pi}{|b|}$.

- The period of $y = a \sec bx$ and $y = a \csc bx$ is $\frac{2\pi}{|b|}$.

5.7 Graphing Techniques

- Phase shift is a horizontal translation of the graph of a trigonometric function. If $y = f(bx + c)$, where f is a trigonometric function, then the phase shift is $-\frac{c}{b}$.

- The graphs of $y = a \sin(bx + c)$ and $y = a \cos(bx + c)$, $b > 0$, have amplitude $|a|$, period $\frac{2\pi}{b}$, and phase shift $-\frac{c}{b}$. One cycle of each graph is completed on the interval $-\frac{c}{b} \leq x \leq -\frac{c}{b} + \frac{2\pi}{b}$.

- Addition of ordinates is a method of graphing the sum of two functions by graphically adding the values of their y-coordinates.

- The factor $g(x)$ in $f(x) = g(x) \cos x$ is called a damping factor. The graph of f lies on or between the graphs of the equations $y = g(x)$ and $y = -g(x)$.

5.8 Harmonic Motion—An Application of the Sine and Cosine Functions

- The equations of simple harmonic motion are
$$y = a \cos 2\pi f t \qquad \text{and} \qquad y = a \sin 2\pi f t$$
where a is the amplitude and f is the frequency.

- Functions of the form $f(t) = a e^{-\alpha t} \cos \omega t$ are used to model some forms of damped harmonic motion.

CHAPTER 5 TRUE/FALSE EXERCISES

In Exercises 1 to 16, answer true or false. If the statement is false, give a reason or state an example to show that the statement is false.

1. An angle is in standard position when the vertex is at the origin of a coordinate system.

2. The angle θ in radians is in standard position with the terminal side in the second quadrant. The reference angle of θ is $\pi - \theta$.

3. In the formula $s = r\theta$, the angle θ must be measured in radians.

4. If $\tan \theta < 0$ and $\cos \theta > 0$, then the terminal side of θ is in Quadrant III.

5. $\sec^2 \theta + \tan^2 \theta = 1$ is an identity.

6. The amplitude of the graph of $y = 2 \tan x$ is 2.

7. The period of $y = \cos x$ is π.

8. The graph of $y = \sin x$ is symmetric to the origin.

9. For any acute angle θ, $\sin \theta + \cos(90° - \theta) = 1$.

10. $\sin (x + y) = \sin x + \sin y$

11. $\sin^2 x = \sin x^2$

12. The phase shift of $f(x) = 2 \sin \left(2x - \dfrac{\pi}{3} \right)$ is $\dfrac{\pi}{3}$.

13. The measure of one radian is more than 50 times the measure of one degree.

14. The measure of one radian differs depending on the radius of the circle used.

15. The graph of $y = 2^{-x} \cos x$ lies on or between the graphs of $y = 2^{-x}$ and $y = 2^x$.

16. The function $f(t) = e^{-0.1t} \cos t$, $t > 0$, models damped harmonic motion in which $|f(t)| \to 0$ as $t \to 0$.

CHAPTER 5 REVIEW EXERCISES

1. Find the complement and supplement of the angle θ whose measure is 65°.

2. Find the measure of the reference angle θ' for the angle θ whose measure is 980°.

3. Convert 2 radians to the nearest hundredth of a degree.

4. Convert 315° to radian measure.

5. Find the length (to the nearest 0.01 meter) of the arc on a circle of radius 3 meters that subtends an angle of 75°.

6. Find the radian measure of the angle subtended by an arc of length 12 centimeters on a circle whose radius is 40 centimeters.

7. A car with a 16-inch-radius wheel is moving with a speed of 50 mph. Find the angular speed (to the nearest radian per second) of the wheel in radians per second.

In Exercises 8 to 11, let θ be an acute angle of a right triangle and $\csc \theta = \dfrac{3}{2}$. Evaluate each function.

8. $\cos \theta$ 9. $\cot \theta$ 10. $\sin \theta$ 11. $\sec \theta$

12. Find the values of the six trigonometric functions of an angle in standard position with the point $P(1, -3)$ on the terminal side of the angle.

13. Find the exact value of

    a. $\sec 150°$ **b.** $\tan \left(-\dfrac{3\pi}{4} \right)$

    c. $\cot(-225°)$ **d.** $\cos \left(\dfrac{2\pi}{3} \right)$

14. Find the value of each of the following to the nearest ten-thousandth.

    a. $\cos 123°$ **b.** $\cot 4.22$

    c. $\sec 612°$ **d.** $\tan \dfrac{2\pi}{5}$

15. Given $\cos \phi = -\dfrac{\sqrt{3}}{2}$, $180° < \phi < 270°$, find the exact value of

   a. $\sin \phi$ **b.** $\tan \phi$

16. Given $\tan \phi = -\dfrac{\sqrt{3}}{3}$, $90° < \phi < 180°$, find the exact value of

   a. $\sec \phi$ **b.** $\csc \phi$

17. Given $\sin \phi = -\dfrac{\sqrt{2}}{2}$, $270° < \phi < 360°$, find the exact value of

   a. $\cos \phi$ **b.** $\cot \phi$

18. Let W be the wrapping function. Evaluate

   a. $W(\pi)$ **b.** $W\left(-\dfrac{\pi}{3}\right)$ **c.** $W\left(\dfrac{5\pi}{4}\right)$ **d.** $W(28\pi)$

19. Is the function defined by $f(x) = \sin(x)\tan(x)$ even, odd, or neither?

In Exercises 20 and 21, use the unit circle to show that each equation is an identity.

20. $\cos(\pi + t) = -\cos t$ **21.** $\tan(-t) = -\tan t$

In Exercises 22 to 27, use trigonometric identities to write each expression in terms of a single trigonometric function or as a constant.

22. $1 + \dfrac{\sin^2 \phi}{\cos^2 \phi}$ **23.** $\dfrac{\tan \phi + 1}{\cot \phi + 1}$

24. $\dfrac{\cos^2 \phi + \sin^2 \phi}{\csc \phi}$ **25.** $\sin^2 \phi(\tan^2 \phi + 1)$

26. $1 + \dfrac{1}{\tan^2 \phi}$ **27.** $\dfrac{\cos^2 \phi}{1 - \sin^2 \phi} - 1$

In Exercises 28 to 33, state the amplitude (if there is one), period, and phase shift of the graph of each function.

28. $y = 3\cos(2x - \pi)$ **29.** $y = 2\tan 3x$

30. $y = -2\sin\left(3x + \dfrac{\pi}{3}\right)$ **31.** $y = \cos\left(2x - \dfrac{2\pi}{3}\right) + 2$

32. $y = -4\sec\left(4x - \dfrac{3\pi}{2}\right)$ **33.** $y = 2\csc\left(x - \dfrac{\pi}{4}\right) - 3$

In Exercises 34 to 51, graph each function.

34. $y = 2\cos \pi x$ **35.** $y = -\sin \dfrac{2x}{3}$

36. $y = 2\sin \dfrac{3x}{2}$ **37.** $y = \cos\left(x - \dfrac{\pi}{2}\right)$

38. $y = \dfrac{1}{2}\sin\left(2x + \dfrac{\pi}{4}\right)$ **39.** $y = 3\cos 3(x - \pi)$

40. $y = -\tan \dfrac{x}{2}$ **41.** $y = 2\cot 2x$

42. $y = \tan\left(x - \dfrac{\pi}{2}\right)$ **43.** $y = -\cot\left(2x + \dfrac{\pi}{4}\right)$

44. $y = -2\csc\left(2x - \dfrac{\pi}{3}\right)$ **45.** $y = 3\sec\left(x + \dfrac{\pi}{4}\right)$

46. $y = 3\sin 2x - 3$ **47.** $y = 2\cos 3x + 3$

48. $y = -\cos\left(3x + \dfrac{\pi}{2}\right) + 2$

49. $y = 3\sin\left(4x - \dfrac{2\pi}{3}\right) - 3$

50. $y = 2 - \sin 2x$

51. $y = \sin x - \sqrt{3}\cos x$

52. A car climbs a hill that has a constant angle of $4.5°$ for a distance of 1.14 miles. What is the car's increase in altitude?

53. A tree casts a shadow of 8.55 feet when the angle of elevation of the sun is $55.3°$. Find the height of the tree.

54. Find the sine of the angle α formed by the intersection of a diagonal of a face of a cube and the diagonal of the cube originating from the same vertex.

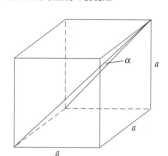

55. Find the height of a building if the angle of elevation to the top of the building changes from 18° to 37° as an observer moves a distance of 80 feet toward the building.

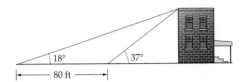

56. Find the amplitude, period, and frequency of the simple harmonic motion given by $y = 2.5 \sin 50t$.

57. A mass of 5 kilograms is in equilibrium suspended from a spring. The mass is pulled down 0.5 foot and released. Find the period, frequency, and amplitude of the motion, assuming the mass oscillates in simple harmonic motion. Write an equation of motion. Assume $k = 20$.

58. Use a graphing utility to graph the damped harmonic motion that is modeled by

$$f(t) = 3e^{-0.75t} \cos \pi t$$

where t is in seconds. Use the graph to determine how long (to the nearest 0.1 second) it will be until the absolute value of the displacement of the mass is always less than 0.01.

CHAPTER 5 TEST

1. Convert 150° to exact radian measure.

2. Find the supplement of the angle whose radian measure is $\frac{11}{12}\pi$. Express your answer in terms of π.

3. Find the length (to the nearest 0.1 centimeter) of an arc that subtends a central angle of 75° in a circle of radius 10 centimeters.

4. A wheel is rotating at 6 revolutions per second. Find the angular speed in radians per second.

5. A wheel with a diameter of 16 centimeters is rotating at 10 radians per second. Find the linear speed (in centimeters per second) of a point on the edge of the wheel.

6. If θ is an acute angle of a right triangle and $\tan \theta = \frac{3}{7}$, find $\sec \theta$.

7. Use a calculator to find the value of csc 67° to the nearest ten-thousandth.

8. Find the exact value of $\tan \frac{\pi}{6} \cos \frac{\pi}{3} - \sin \frac{\pi}{2}$.

9. Find the exact coordinates of $W\left(\frac{11\pi}{6}\right)$.

10. Express $\dfrac{\sec^2 t - 1}{\sec^2 t}$ in terms of a single trigonometric function.

11. State the period of $y = -4 \tan 3x$.

12. State the amplitude, period, and phase shift for the function $y = -3 \cos\left(2x + \frac{\pi}{2}\right)$.

13. State the period and phase shift for the function
$$y = 2 \cot\left(\frac{\pi}{3}x + \frac{\pi}{6}\right).$$

14. Graph one full period of $y = 3 \cos \frac{1}{2}x$.

15. Graph one full period of $y = -2 \sec \frac{1}{2}x$.

16. Write a sentence that explains how to obtain the graph of $y = 2 \sin\left(2x - \frac{\pi}{2}\right) - 1$ from the graph of $y = 2 \sin 2x$.

17. Graph one full period of $y = 2 - \sin \frac{x}{2}$.

18. Graph one full period of $y = \sin x - \cos 2x$.

19. The angle of elevation from point A to the top of a tree is 42.2°. At point B, 5.24 meters from A and on a line through the base of the tree and A, the angle of elevation is 37.4°. Find the height of the tree.

20. Write the equation for simple harmonic motion, given that the amplitude is 13 feet, the period is 5 seconds, and the displacement is zero when $t = 0$.

CUMULATIVE REVIEW EXERCISES

1. Factor: $x^2 - y^2$

2. Simplify: $\dfrac{\sqrt{3}}{2} \div \dfrac{1}{2}$

3. Find the area of a triangle with a base of 4 inches and a height of 6 inches.

4. Determine whether $f(x) = \dfrac{x}{x^2 + 1}$ is an even function or an odd function.

5. Find the inverse of $f(x) = \dfrac{x}{2x - 3}$.

6. Use interval notation to state the domain of $f(x) = \dfrac{2}{x - 4}$.

7. Use interval notation to state the range of $f(x) = \sqrt{4 - x^2}$.

8. Explain how to use the graph of $y = f(x)$ to produce the graph of $y = f(x - 3)$.

9. Explain how to use the graph of $y = f(x)$ to produce the graph of $y = f(-x)$.

10. Convert $300°$ to radians.

11. Convert $\dfrac{5\pi}{4}$ to degrees.

12. Evaluate $f(x) = \sin\left(x + \dfrac{\pi}{6}\right)$ for $x = \dfrac{\pi}{3}$.

13. Evaluate $f(x) = \sin x + \sin \dfrac{\pi}{6}$ for $x = \dfrac{\pi}{3}$.

14. Find the exact value of $\cos^2 45° + \sin^2 60°$.

15. Determine the sign of $\tan \theta$ given that $\dfrac{\pi}{2} < \theta < \pi$.

16. What is the measure of the reference angle for the angle $\theta = 210°$?

17. What is the measure of the reference angle for the angle $\theta = \dfrac{2\pi}{3}$?

18. Use interval notation to state the domain of $f(x) = \sin x$, where x is a real number.

19. Use interval notation to state the range of $f(x) = \cos x$, where x is a real number.

20. If θ is an acute angle of a right triangle and $\tan \theta = \dfrac{3}{4}$, find $\sin \theta$.

TRIGONOMETRIC IDENTITIES AND EQUATIONS

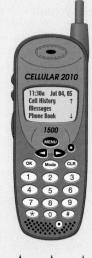

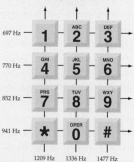

Source: Data in chart from http://www.howstuffworks.com/telephone2.htm and also found at http://hyperarchive.lcs.mit.edu/telecom-archives/tribute/touch_tone_info.html

VIDEO & DVD

SSG

Touch-Tone Phones and Trigonometry

The dial tone emitted by a telephone is produced by adding a 350-hertz (cycles per second) sound to a 440-hertz sound. An equation that models the dial tone is

$$p_1(t) = \cos(2\pi \cdot 440t) + \cos(2\pi \cdot 350t)$$

where p is the pressure on the eardrum and t is the time in seconds. Concepts from this chapter can be used to show that the dial tone can also be modeled by

$$p_2(t) = 2\cos(790\pi t)\cos(90\pi t)$$

The equation $p_1(t) = p_2(t)$ is called an *identity* because the left side of the equation equals the right side for all domain values t. Use a graphing utility to graph p_1 and p_2 on the interval $[0, 0.1]$ to see that they appear to represent the same function.

Every tone made on a touch-tone phone is produced by adding a pair of sounds. The chart to the left shows the frequencies used for each key. For instance, the sound emitted by pressing 3 on the keypad is produced by adding a 1477-hertz sound to a 697-hertz sound. An equation that models this tone is

$$p(t) = \cos(2\pi \cdot 1477t) + \cos(2\pi \cdot 697t)$$

In **Exercises 77 and 78 on page 588,** you will determine trigonometric equations that can be used to model some of the other tones that can be produced on a touch-tone phone.

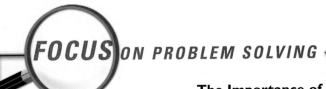

The Importance of Experimentation

In this chapter you will need to verify several trigonometric identities. The equation $\sin 2\alpha = 2 \sin \alpha \cos \alpha$ is an example of a trigonometric identity. The equation is an identity because $\sin 2\alpha$ is equal to $2 \sin \alpha \cos \alpha$ for all values of α.

Trigonometry identities are very useful. In many applications a solution can be found by making use of an identity that allows you to replace a given function with an equivalent function.

To verify an identity, you need to show that one side of the identity can be rewritten in an equivalent form that is identical to the other side. The guidelines on pages 555–556 list several procedures that may be used in a verification of a trigonometric identity. The verification process is similar to the process of solving a crossword puzzle. You examine the clues and make a guess at a possible solution. If your first guess does not work, then you make additional guesses until a proper solution is found. In verifying an identity, try one or more of the procedures listed in the guidelines. If this does not produce a verification, then experiment, by trying another approach, until a verification is found.

VERIFICATION OF TRIGONOMETRIC IDENTITIES

● FUNDAMENTAL TRIGONOMETRIC IDENTITIES

The domain of an equation consists of all values of the variable for which every term is defined. For example, the domain of

$$\frac{\sin x \cos x}{\sin x} = \cos x \tag{1}$$

includes all real numbers x except $x = n\pi$, where n is an integer, because $\sin x = 0$ for $x = n\pi$, and division by 0 is undefined. An **identity** is an equation that is true for all of its domain values. **Table 6.1** lists identities that were introduced earlier.

TABLE 6.1   Fundamental Trigonometric Identities

Reciprocal identities	$\sin x = \dfrac{1}{\csc x}$	$\cos x = \dfrac{1}{\sec x}$	$\tan x = \dfrac{1}{\cot x}$
Ratio identities	$\tan x = \dfrac{\sin x}{\cos x}$	$\cot x = \dfrac{\cos x}{\sin x}$	
Pythagorean identities	$\sin^2 x + \cos^2 x = 1$	$\tan^2 x + 1 = \sec^2 x$	$1 + \cot^2 x = \csc^2 x$
Odd-even identities	$\sin(-x) = -\sin x$ $\cos(-x) = \cos x$	$\tan(-x) = -\tan x$ $\cot(-x) = -\cot x$	$\sec(-x) = \sec x$ $\csc(-x) = -\csc x$

● VERIFICATION OF TRIGONOMETRIC IDENTITIES

To verify an identity, we show that one side of the identity can be rewritten in a form that is identical to the other side. There is no one method that can be used to verify every identity; however, the following guidelines should prove useful.

Guidelines for Verifying Trigonometric Identities

● If one side of the identity is more complex than the other, then it is generally best to try first to simplify the more complex side until it becomes identical to the other side.

● Perform indicated operations such as adding fractions or squaring a binomial. Also be aware of any factorization that may help you to achieve your goal of producing the expression on the other side.

● Make use of previously established identities that enable you to rewrite one side of the identity in an equivalent form.

● Rewrite one side of the identity so that it involves only sines and/or cosines.

Continued ▶

- Rewrite one side of the identity in terms of a single trigonometric function.

- Multiplying both the numerator and the denominator of a fraction by the same factor (such as the conjugate of the denominator or the conjugate of the numerator) may get you closer to your goal.

- Keep your goal in mind. Does it involve products, quotients, sums, radicals, or powers? Knowing exactly what your goal is may provide the insight you need to verify the identity.

EXAMPLE 1   Determine Whether an Equation is an Identity

Determine whether each equation is an identity. If the equation is an identity, then verify the identity. If the equation is not an identity, then find a value of the domain for which the left side of the equation is not equal to the right side.

a. $\sin\left(x + \dfrac{\pi}{6}\right) = \sin x + \sin \dfrac{\pi}{6}$

b. $(\sin x + \cos x)^2 = 2 \sin x \cos x + 1$

Solution

a. The graphs of $f(x) = \sin\left(x + \dfrac{\pi}{6}\right)$ and $g(x) = \sin x + \sin \dfrac{\pi}{6}$ are shown in **Figure 6.1.** Because the graphs are not identical, we know that the equation is not an identity. This can be further confirmed by letting $x = \dfrac{\pi}{3}$ and observing that

$$f\left(\frac{\pi}{3}\right) = \sin\left(\frac{\pi}{3} + \frac{\pi}{6}\right) = \sin \frac{\pi}{2} = 1$$

whereas

$$g\left(\frac{\pi}{3}\right) = \sin \frac{\pi}{3} + \sin \frac{\pi}{6} = \frac{\sqrt{3}}{2} + \frac{1}{2} \approx 1.366$$

Thus $\sin\left(x + \dfrac{\pi}{6}\right) \neq \sin x + \sin \dfrac{\pi}{6}$ for $x = \dfrac{\pi}{3}$. Therefore, the equation is not an identity.

b. The graphs of $f(x) = (\sin x + \cos x)^2$ and $g(x) = 2 \sin x \cos x + 1$ are shown in **Figure 6.2.** The graphs appear to be identical. To verify that $(\sin x + \cos x)^2 = 2 \sin x \cos x + 1$ is an identity, we expand $(\sin x + \cos x)^2$ as shown below.

$$(\sin x + \cos x)^2 = \sin^2 x + 2 \sin x \cos x + \cos^2 x$$
$$= 2 \sin x \cos x + (\sin^2 x + \cos^2 x)$$
$$= 2 \sin x \cos x + 1 \qquad \bullet \; \sin^2 x + \cos^2 x = 1$$

We have rewritten the left side in an equivalent form that is identical to the right side. Thus $(\sin x + \cos x)^2 = 2 \sin x \cos x + 1$ is an identity.

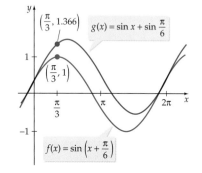

FIGURE 6.1

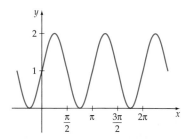

$f(x) = (\sin x + \cos x)^2$
$g(x) = 2 \sin x \cos x + 1$

FIGURE 6.2

▶ **TRY EXERCISE 2, PAGE 559**

? QUESTION Is $\cos(-x) = \cos x$ an identity?

EXAMPLE 2   Verify an Identity

Verify the identity $1 - 2\sin^2 x = 2\cos^2 x - 1$.

Solution

Rewrite the right side of the equation.

$$2\cos^2 x - 1 = 2(1 - \sin^2 x) - 1 \qquad \bullet \ \cos^2 x = 1 - \sin^2 x$$
$$= 2 - 2\sin^2 x - 1$$
$$= 1 - 2\sin^2 x$$

▶ **TRY EXERCISE 22, PAGE 560**

take note

Each of the Pythagorean identities can be written in several different forms. For instance,

$$\sin^2 x + \cos^2 x = 1$$

also can be written as

$$\sin^2 x = 1 - \cos^2 x$$

and as

$$\cos^2 x = 1 - \sin^2 x$$

Figure 6.3 shows the graph of $f(x) = 1 - 2\sin^2 x$ and the graph of $g(x) = 2\cos^2 x - 1$ on the same coordinate axes. The fact that the graphs appear to be identical on the interval $[-2\pi, 2\pi]$ supports the verification in Example 2.

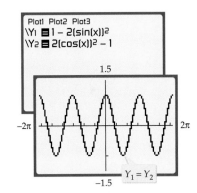

FIGURE 6.3

EXAMPLE 3   Factor to Verify an Identity

Verify the identity $\csc^2 x - \cos^2 x \csc^2 x = 1$.

Solution

Simplify the left side of the equation.

$$\csc^2 x - \cos^2 x \csc^2 x = \csc^2 x(1 - \cos^2 x) \qquad \bullet \ \textbf{Factor out } \csc^2 x.$$
$$= \csc^2 x \sin^2 x$$
$$= \frac{1}{\sin^2 x} \cdot \sin^2 x = 1 \qquad \bullet \ \csc^2 x = \frac{1}{\sin^2 x}$$

▶ **TRY EXERCISE 34, PAGE 560**

In the next example we make use of the guideline that indicates that it may be useful to multiply both the numerator and the denominator of a fraction by the same factor.

? ANSWER Yes, $\cos(-x) = \cos x$ is one of the odd-even identities shown in **Table 6.1.**

EXAMPLE 4   **Multiply by a Conjugate to Verify an Identity**

Verify the identity $\dfrac{\sin x}{1 + \cos x} = \dfrac{1 - \cos x}{\sin x}$.

Solution

Multiply the numerator and denominator of the left side of the identity by the conjugate of $1 + \cos x$, which is $1 - \cos x$.

$$\frac{\sin x}{1 + \cos x} = \frac{\sin x}{1 + \cos x} \cdot \frac{1 - \cos x}{1 - \cos x} = \frac{\sin x(1 - \cos x)}{1 - \cos^2 x}$$

$$= \frac{\sin x(1 - \cos x)}{\sin^2 x} = \frac{1 - \cos x}{\sin x}$$

> *take note*
>
> The sum $a + b$ and the difference $a - b$ are called conjugates of each other.

▶ **TRY EXERCISE 44, PAGE 560**

EXAMPLE 5   **Change to Sines and Cosines to Verify an Identity**

Verify the identity $\dfrac{\sin x + \tan x}{1 + \cos x} = \tan x$.

Solution

Rewrite the left side of the identity in terms of sines and cosines.

$$\frac{\sin x + \tan x}{1 + \cos x} = \frac{\sin x + \dfrac{\sin x}{\cos x}}{1 + \cos x}$$ • $\tan x = \dfrac{\sin x}{\cos x}$

$$= \frac{\dfrac{\sin x \cos x + \sin x}{\cos x}}{1 + \cos x}$$ • Write the terms in the numerator with a common denominator.

$$= \frac{\sin x \cos x + \sin x}{\cos x(1 + \cos x)}$$ • Simplify.

$$= \frac{\sin x(1 + \cos x)}{\cos x(1 + \cos x)}$$

$$= \tan x$$

▶ **TRY EXERCISE 54, PAGE 560**

  TOPICS FOR DISCUSSION

1. Explain why $\tan = \dfrac{\sin}{\cos}$ is not an identity.

2. Is $\cos |x| = |\cos x|$ an identity? Explain. What about $\cos |x| = \cos x$? Explain.

3. The identity $\sin^2 x + \cos^2 x = 1$ is one of the Pythagorean identities. What are the other two Pythagorean identities, and how are they derived?

The sum and difference identities can be used to simplify some trigonometric expressions.

EXAMPLE 3   **Simplify Trigonometric Expressions**

Write each expression in terms of a single trigonometric function.

a. $\sin 5x \cos 3x - \cos 5x \sin 3x$ **b.** $\dfrac{\tan 4\alpha + \tan \alpha}{1 - \tan 4\alpha \tan \alpha}$

Solution

a. $\sin 5x \cos 3x - \cos 5x \sin 3x = \sin(5x - 3x) = \sin 2x$

b. $\dfrac{\tan 4\alpha + \tan \alpha}{1 - \tan 4\alpha \tan \alpha} = \tan(4\alpha + \alpha) = \tan 5\alpha$

▶ **TRY EXERCISE 26, PAGE 570**

EXAMPLE 4   **Evaluate a Trigonometric Function**

Given $\tan \alpha = -\dfrac{4}{3}$ for α in Quadrant II and $\tan \beta = -\dfrac{5}{12}$ for β in Quadrant IV, find $\sin(\alpha + \beta)$.

Solution

See **Figure 6.5.** Because $\tan \alpha = \dfrac{y}{x} = -\dfrac{4}{3}$ and the terminal side of α is in Quadrant II, $P_1(-3, 4)$ is a point on the terminal side of α. Similarly, $P_2(12, -5)$ is a point on the terminal side of β.

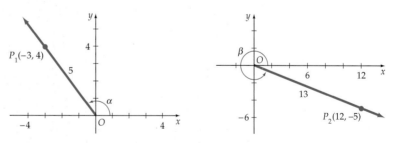

FIGURE 6.5

Using the Pythagorean Theorem, we find that the length of the line segment OP_1 is 5 and the length of OP_2 is 13.

$$\sin(\alpha + \beta) = \sin \alpha \cos \beta + \cos \alpha \sin \beta$$
$$= \frac{4}{5} \cdot \frac{12}{13} + \frac{-3}{5} \cdot \frac{-5}{13} = \frac{48}{65} + \frac{15}{65} = \frac{63}{65}$$

▶ **TRY EXERCISE 38, PAGE 570**

EXAMPLE 5   **Verify an Identity**

Verify the identity $\cos(\pi - \theta) = -\cos\theta$.

Solution

Use the identity for $\cos(\alpha - \beta)$.

$$\cos(\pi - \theta) = \cos\pi\cos\theta + \sin\pi\sin\theta = -1 \cdot \cos\theta + 0 \cdot \sin\theta = -\cos\theta$$

▶ **TRY EXERCISE 50, PAGE 571**

Figure 6.6 shows the graphs of $f(\theta) = \cos(\pi - \theta)$ and $g(\theta) = -\cos\theta$ on the same coordinate axes. The fact that the graphs appear to be identical supports the verification in Example 5.

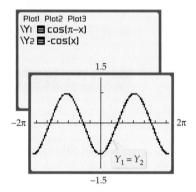

FIGURE 6.6

EXAMPLE 6   **Verify an Identity**

Verify the identity $\dfrac{\cos 4\theta}{\sin \theta} - \dfrac{\sin 4\theta}{\cos \theta} = \dfrac{\cos 5\theta}{\sin \theta \cos \theta}$.

Solution

Subtract the fractions on the left side of the equation.

$$\frac{\cos 4\theta}{\sin \theta} - \frac{\sin 4\theta}{\cos \theta} = \frac{\cos 4\theta \cos \theta - \sin 4\theta \sin \theta}{\sin \theta \cos \theta}$$

$$= \frac{\cos(4\theta + \theta)}{\sin \theta \cos \theta} = \frac{\cos 5\theta}{\sin \theta \cos \theta}$$ • Use the identity for $\cos(\alpha + \beta)$.

▶ **TRY EXERCISE 62, PAGE 571**

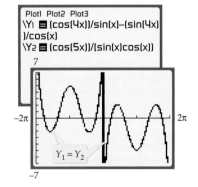

FIGURE 6.7

Figure 6.7 shows the graph of $f(\theta) = \dfrac{\cos 4\theta}{\sin \theta} - \dfrac{\sin 4\theta}{\cos \theta}$ and the graph of $g(\theta) = \dfrac{\cos 5\theta}{\sin \theta \cos \theta}$ on the same coordinate axes. The fact that the graphs appear to be identical supports the verification in Example 6.

● REDUCTION FORMULAS

The sum or difference identities can be used to write expressions such as

$$\sin(\theta + k\pi) \qquad \sin(\theta + 2k\pi) \qquad \text{and} \qquad \cos[\theta + (2k+1)\pi]$$

where k is an integer, as expressions involving only $\sin\theta$ or $\cos\theta$. The resulting formulas are called **reduction formulas.**

EXAMPLE 7   **Find Reduction Formulas**

Write as a function involving only $\sin\theta$.

$$\sin[\theta + (2k+1)\pi], \quad \text{where } k \text{ is an integer}$$

Solution

Applying the identity $\sin(\alpha + \beta) = \sin\alpha\cos\beta + \cos\alpha\sin\beta$ yields

$$\sin[\theta + (2k+1)\pi] = \sin\theta\cos[(2k+1)\pi] + \cos\theta\sin[(2k+1)\pi]$$

If k is an integer, then $2k + 1$ is an odd integer. The cosine of any odd multiple of π equals -1, and the sine of any odd multiple of π is 0. This gives us

$$\sin[\theta + (2k+1)\pi] = (\sin\theta)(-1) + (\cos\theta)(0) = -\sin\theta$$

Thus $\sin[\theta + (2k+1)\pi] = -\sin\theta$ for any integer k.

▶ **TRY EXERCISE 74, PAGE 571**

? **QUESTION**   Is $\sin(\theta + 2k\pi) = \sin\theta$ a reduction formula?

 TOPICS FOR DISCUSSION

1. Does $\sin(\alpha + \beta) = \sin\alpha + \sin\beta$ for all values of α and β? If not, find nonzero values of α and β for which $\sin(\alpha + \beta) \neq \sin\alpha + \sin\beta$.

2. If k is an integer, then $2k + 1$ is an odd integer. Do you agree? Explain.

3. What are the trigonometric cofunction identities? Explain.

4. Is $\tan(\theta + k\pi) = \tan\theta$, where k is an integer, a reduction formula? Explain.

? **ANSWER**   Yes.

EXERCISE SET 6.2

In Exercises 1 to 18, find the exact value of the expression.

1. $\sin(45° + 30°)$

2. $\sin(330° + 45°)$

3. $\cos(45° - 30°)$

▶**4.** $\cos(120° - 45°)$

5. $\tan(45° - 30°)$

6. $\tan(240° - 45°)$

7. $\sin\left(\dfrac{5\pi}{4} - \dfrac{\pi}{6}\right)$

8. $\sin\left(\dfrac{4\pi}{3} + \dfrac{\pi}{4}\right)$

9. $\cos\left(\dfrac{3\pi}{4} + \dfrac{\pi}{6}\right)$

10. $\cos\left(\dfrac{\pi}{4} - \dfrac{\pi}{3}\right)$

11. $\tan\left(\dfrac{\pi}{6} + \dfrac{\pi}{4}\right)$

12. $\tan\left(\dfrac{11\pi}{6} - \dfrac{\pi}{4}\right)$

13. $\cos 212° \cos 122° + \sin 212° \sin 122°$

14. $\sin 167° \cos 107° - \cos 167° \sin 107°$

15. $\sin\dfrac{5\pi}{12}\cos\dfrac{\pi}{4} - \cos\dfrac{5\pi}{12}\sin\dfrac{\pi}{4}$

16. $\cos\dfrac{\pi}{12}\cos\dfrac{\pi}{4} - \sin\dfrac{\pi}{12}\sin\dfrac{\pi}{4}$

17. $\dfrac{\tan\dfrac{7\pi}{12} - \tan\dfrac{\pi}{4}}{1 + \tan\dfrac{7\pi}{12}\tan\dfrac{\pi}{4}}$

18. $\dfrac{\tan\dfrac{\pi}{6} + \tan\dfrac{\pi}{3}}{1 - \tan\dfrac{\pi}{6}\tan\dfrac{\pi}{3}}$

In Exercises 19 to 24, use a cofunction identity to write an equivalent expression for the given value.

19. $\sin 42°$

▶**20.** $\cos 80°$

21. $\tan 15°$

22. $\cot 2°$

23. $\sec 25°$

24. $\csc 84°$

In Exercises 25 to 36, write each expression in terms of a single trigonometric function.

25. $\sin 7x \cos 2x - \cos 7x \sin 2x$

▶**26.** $\sin x \cos 3x + \cos x \sin 3x$

27. $\cos x \cos 2x + \sin x \sin 2x$

28. $\cos 4x \cos 2x - \sin 4x \sin 2x$

29. $\sin 7x \cos 3x - \cos 7x \sin 3x$

30. $\cos x \cos 5x - \sin x \sin 5x$

31. $\cos 4x \cos(-2x) - \sin 4x \sin(-2x)$

32. $\sin(-x) \cos 3x - \cos(-x) \sin 3x$

33. $\sin\dfrac{x}{3}\cos\dfrac{2x}{3} + \cos\dfrac{x}{3}\sin\dfrac{2x}{3}$

34. $\cos\dfrac{3x}{4}\cos\dfrac{x}{4} + \sin\dfrac{3x}{4}\sin\dfrac{x}{4}$

35. $\dfrac{\tan 3x + \tan 4x}{1 - \tan 3x \tan 4x}$

36. $\dfrac{\tan 2x - \tan 3x}{1 + \tan 2x \tan 3x}$

In Exercises 37 to 48, find the exact value of the given functions.

37. Given $\tan\alpha = -\dfrac{4}{3}$, α in Quadrant II, and $\tan\beta = \dfrac{15}{8}$, β in Quadrant III, find

   a. $\sin(\alpha - \beta)$ **b.** $\cos(\alpha + \beta)$ **c.** $\tan(\alpha - \beta)$

▶**38.** Given $\tan\alpha = \dfrac{24}{7}$, α in Quadrant I, and $\sin\beta = -\dfrac{8}{17}$, β in Quadrant III, find

   a. $\sin(\alpha + \beta)$ **b.** $\cos(\alpha + \beta)$ **c.** $\tan(\alpha - \beta)$

39. Given $\sin\alpha = \dfrac{3}{5}$, α in Quadrant I, and $\cos\beta = -\dfrac{5}{13}$, β in Quadrant II, find

   a. $\sin(\alpha - \beta)$ **b.** $\cos(\alpha + \beta)$ **c.** $\tan(\alpha - \beta)$

40. Given $\sin\alpha = \dfrac{24}{25}$, α in Quadrant II, and $\cos\beta = -\dfrac{4}{5}$, β in Quadrant III, find

   a. $\cos(\beta - \alpha)$ **b.** $\sin(\alpha + \beta)$ **c.** $\tan(\alpha + \beta)$

41. Given $\sin \alpha = -\dfrac{4}{5}$, α in Quadrant III, and $\cos \beta = -\dfrac{12}{13}$, β in Quadrant II, find

 a. $\sin(\alpha - \beta)$ **b.** $\cos(\alpha + \beta)$ **c.** $\tan(\alpha + \beta)$

42. Given $\sin \alpha = -\dfrac{7}{25}$, α in Quadrant IV, and $\cos \beta = \dfrac{8}{17}$, β in Quadrant IV, find

 a. $\sin(\alpha + \beta)$ **b.** $\cos(\alpha - \beta)$ **c.** $\tan(\alpha + \beta)$

43. Given $\cos \alpha = \dfrac{15}{17}$, α in Quadrant I, and $\sin \beta = -\dfrac{3}{5}$, β in Quadrant III, find

 a. $\sin(\alpha + \beta)$ **b.** $\cos(\alpha - \beta)$ **c.** $\tan(\alpha - \beta)$

44. Given $\cos \alpha = -\dfrac{7}{25}$, α in Quadrant II, and $\sin \beta = -\dfrac{12}{13}$, β in Quadrant IV, find

 a. $\sin(\alpha + \beta)$ **b.** $\cos(\alpha + \beta)$ **c.** $\tan(\alpha - \beta)$

45. Given $\cos \alpha = -\dfrac{3}{5}$, α in Quadrant III, and $\sin \beta = \dfrac{5}{13}$, β in Quadrant I, find

 a. $\sin(\alpha - \beta)$ **b.** $\cos(\alpha + \beta)$ **c.** $\tan(\alpha + \beta)$

46. Given $\cos \alpha = \dfrac{8}{17}$, α in Quadrant IV, and $\sin \beta = -\dfrac{24}{25}$, β in Quadrant III, find

 a. $\sin(\alpha - \beta)$ **b.** $\cos(\alpha + \beta)$ **c.** $\tan(\alpha + \beta)$

47. Given $\sin \alpha = \dfrac{3}{5}$, α in Quadrant I, and $\tan \beta = \dfrac{5}{12}$, β in Quadrant III, find

 a. $\sin(\alpha + \beta)$ **b.** $\cos(\alpha - \beta)$ **c.** $\tan(\alpha - \beta)$

48. Given $\tan \alpha = \dfrac{15}{8}$, α in Quadrant I, and $\tan \beta = -\dfrac{7}{24}$, β in Quadrant IV, find

 a. $\sin(\alpha - \beta)$ **b.** $\cos(\alpha - \beta)$ **c.** $\tan(\alpha + \beta)$

In Exercises 49 to 72, verify the identity.

49. $\cos\left(\dfrac{\pi}{2} - \theta\right) = \sin \theta$ ▶ **50.** $\cos(\theta + \pi) = -\cos \theta$

51. $\sin\left(\theta + \dfrac{\pi}{2}\right) = \cos \theta$ **52.** $\sin(\theta + \pi) = -\sin \theta$

53. $\tan\left(\theta + \dfrac{\pi}{4}\right) = \dfrac{\tan \theta + 1}{1 - \tan \theta}$

54. $\tan 2\theta = \dfrac{2 \tan \theta}{1 - \tan^2 \theta}$ **55.** $\cos\left(\dfrac{3\pi}{2} - \theta\right) = -\sin \theta$

56. $\sin\left(\dfrac{3\pi}{2} + \theta\right) = -\cos \theta$ **57.** $\cot\left(\dfrac{\pi}{2} - \theta\right) = \tan \theta$

58. $\cot(\pi + \theta) = \cot \theta$ **59.** $\csc(\pi - \theta) = \csc \theta$

60. $\sec\left(\dfrac{\pi}{2} - \theta\right) = \csc \theta$

61. $\sin 6x \cos 2x - \cos 6x \sin 2x = 2 \sin 2x \cos 2x$

▶ **62.** $\cos 5x \cos 3x + \sin 5x \sin 3x = \cos^2 x - \sin^2 x$

63. $\cos(\alpha + \beta) + \cos(\alpha - \beta) = 2 \cos \alpha \cos \beta$

64. $\cos(\alpha - \beta) - \cos(\alpha + \beta) = 2 \sin \alpha \sin \beta$

65. $\sin(\alpha + \beta) + \sin(\alpha - \beta) = 2 \sin \alpha \cos \beta$

66. $\sin(\alpha - \beta) - \sin(\alpha + \beta) = -2 \cos \alpha \sin \beta$

67. $\dfrac{\cos(\alpha - \beta)}{\sin(\alpha + \beta)} = \dfrac{\cot \alpha + \tan \beta}{1 + \cot \alpha \tan \beta}$

68. $\dfrac{\sin(\alpha + \beta)}{\sin(\alpha - \beta)} = \dfrac{1 + \cot \alpha \tan \beta}{1 - \cot \alpha \tan \beta}$

69. $\sin\left(\dfrac{\pi}{2} + \alpha - \beta\right) = \cos \alpha \cos \beta + \sin \alpha \sin \beta$

70. $\cos\left(\dfrac{\pi}{2} + \alpha + \beta\right) = -(\sin \alpha \cos \beta + \cos \alpha \sin \beta)$

71. $\sin 3x = 3 \sin x - 4 \sin^3 x$

72. $\cos 3x = 4 \cos^3 x - 3 \cos x$

In Exercises 73 to 78, write the given expression as a function that involves only sin θ, cos θ, or tan θ. (In Exercises 76, 77, and 78, assume k is an integer.)

73. $\cos(\theta + 3\pi)$ ▶ **74.** $\sin(\theta + 2\pi)$

75. $\tan(\theta + \pi)$ **76.** $\cos[\theta + (2k + 1)\pi]$

77. $\sin(\theta + 2k\pi)$ **78.** $\sin(\theta - k\pi)$

In Exercises 79 to 82, compare the graphs of each side of the equation to predict whether the equation is an identity.

79. $\sin\left(\dfrac{\pi}{2} - x\right) = \cos x$ **80.** $\cos(x + \pi) = -\cos x$

81. $\sin 7x \cos 2x - \cos 7x \sin 2x = \sin 5x$

82. $\sin 3x = 3 \sin x - 4 \sin^3 x$

───── *CONNECTING CONCEPTS* ─────

In Exercises 83 to 89, verify the identity.

83. $\sin(x - y) \cdot \sin(x + y) = \sin^2 x \cos^2 y - \cos^2 x \sin^2 y$

84. $\sin(x + y + z) = \sin x \cos y \cos z + \cos x \sin y \cos z + \cos x \cos y \sin z - \sin x \sin y \sin z$

85. $\cos(x + y + z) = \cos x \cos y \cos z - \sin x \sin y \cos z - \sin x \cos y \sin z - \cos x \sin y \sin z$

86. $\dfrac{\sin(x + y)}{\sin x \sin y} = \cot x + \cot y$

87. $\dfrac{\cos(x - y)}{\cos x \sin y} = \cot y + \tan x$

88. $\dfrac{\sin(x + h) - \sin x}{h} = \cos x \dfrac{\sin h}{h} + \sin x \dfrac{(\cos h - 1)}{h}$

89. $\dfrac{\cos(x + h) - \cos x}{h} = \cos x \dfrac{(\cos h - 1)}{h} - \sin x \dfrac{\sin h}{h}$

90. **MODEL RESISTANCE** The drag (resistance) on a fish when it is swimming is two to three times the drag when it is gliding. To compensate for this, some fish swim in a saw-tooth pattern, as shown in the accompanying figure. The ratio of the amount of energy the fish expends when swimming upward at angle β and then gliding down at angle α to the energy it expends swimming horizontally is given by

$$E_R = \dfrac{k \sin \alpha + \sin \beta}{k \sin(\alpha + \beta)}$$

where k is a value such that $2 \le k \le 3$, and k depends on the assumptions we make about the amount of drag experienced by the fish. Find E_R for $k = 2$, $\alpha = 10°$, and $\beta = 20°$.

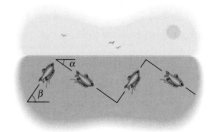

───── *PREPARE FOR SECTION 6.3* ─────

91. Use the identity for $\sin(\alpha + \beta)$ to rewrite $\sin 2\alpha$. [6.2]

92. Use the identity for $\cos(\alpha + \beta)$ to rewrite $\cos 2\alpha$. [6.2]

93. Use the identity for $\tan(\alpha + \beta)$ to rewrite $\tan 2\alpha$. [6.2]

94. Compare $\tan \dfrac{\alpha}{2}$ and $\dfrac{\sin \alpha}{1 + \cos \alpha}$ for $\alpha = 60°$, $\alpha = 90°$, and $\alpha = 120°$. [5.2]

95. Verify that $\sin 2\alpha = 2 \sin \alpha$ is *not* an identity. [5.2]

96. Verify that $\cos \dfrac{\alpha}{2} = \dfrac{1}{2} \cos \alpha$ is *not* an identity. [5.2]

PROJECTS

I. INTERSECTING LINES In the figure shown at the right, two nonvertical lines intersect in a plane. The slope of line l_1 is m_1 and the slope of line l_2 is m_2.

a. Show that the tangent of the smallest positive angle γ from l_1 to l_2 is given by

$$\tan \gamma = \frac{m_2 - m_1}{1 + m_1 m_2}$$

b. Two nonvertical lines intersect at the point $(1, 5)$. The measure of the smallest positive angle between the lines is $\gamma = 60°$. The first line is given by $y = 0.5x + 4.5$. What is the equation (in slope-intercept form) of the second line?

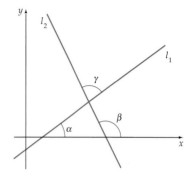

DOUBLE- AND HALF-ANGLE IDENTITIES

- **DOUBLE-ANGLE IDENTITIES**
- **HALF-ANGLE IDENTITIES**

● DOUBLE-ANGLE IDENTITIES

By using the sum identities, we can derive identities for $f(2\alpha)$, where f is a trigonometric function. These are called the *double-angle identities*. To find the sine of a double angle, substitute α for β in the identity for $\sin(\alpha + \beta)$.

$$\sin (\alpha + \beta) = \sin \alpha \cos \beta + \cos \alpha \sin \beta$$

$$\sin (\alpha + \alpha) = \sin \alpha \cos \alpha + \cos \alpha \sin \alpha \qquad \text{• Let } \beta = \alpha.$$

$$\sin 2\alpha = 2 \sin \alpha \cos \alpha$$

A double-angle identity for cosine is derived in a similar manner.

$$\cos(\alpha + \beta) = \cos \alpha \cos \beta - \sin \alpha \sin \beta$$

$$\cos(\alpha + \alpha) = \cos \alpha \cos \alpha - \sin \alpha \sin \alpha \qquad \text{• Let } \beta = \alpha.$$

$$\cos 2\alpha = \cos^2 \alpha - \sin^2 \alpha$$

There are two alternative forms of the double-angle identity for $\cos 2\alpha$. Using $\cos^2 \alpha = 1 - \sin^2 \alpha$, we can rewrite the identity for $\cos 2\alpha$ as follows:

$$\cos 2\alpha = \cos^2 \alpha - \sin^2 \alpha$$

$$\cos 2\alpha = (1 - \sin^2 \alpha) - \sin^2 \alpha \qquad \text{• } \cos^2 \alpha = 1 - \sin^2 \alpha$$

$$\cos 2\alpha = 1 - 2 \sin^2 \alpha$$

We also can rewrite $\cos 2\alpha$ as

$$\cos 2\alpha = \cos^2 \alpha - \sin^2 \alpha$$

$$\cos 2\alpha = \cos^2 \alpha - (1 - \cos^2 \alpha) \qquad \text{• } \sin^2 \alpha = 1 - \cos^2 \alpha$$

$$\cos 2\alpha = 2 \cos^2 \alpha - 1$$

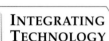

INTEGRATING TECHNOLOGY

One way of showing that $\sin 2x \neq 2 \sin x$ is by graphing $y = \sin 2x$ and $y = 2 \sin x$ and observing that the graphs are not the same.

The double-angle identity for the tangent function is derived from the identity for $\tan(\alpha + \beta)$ with $\beta = \alpha$.

$$\tan(\alpha + \beta) = \frac{\tan \alpha + \tan \beta}{1 - \tan \alpha \tan \beta}$$

$$\tan(\alpha + \alpha) = \frac{\tan \alpha + \tan \alpha}{1 - \tan \alpha \tan \alpha} \qquad \bullet \text{ Let } \beta = \alpha.$$

$$\tan 2\alpha = \frac{2 \tan \alpha}{1 - \tan^2 \alpha}$$

The double-angle identities are often used to write a trigonometric expression in terms of a single trigonometric function.

EXAMPLE 1 Simplify a Trigonometric Expression

Write $4 \sin 5\theta \cos 5\theta$ as a single trigonometric function.

Solution

$$4 \sin 5\theta \cos 5\theta = 2(2 \sin 5\theta \cos 5\theta) \qquad \bullet \text{ Use } 2 \sin \alpha \cos \alpha = \sin 2\alpha,$$
$$= 2(\sin 10\theta) = 2 \sin 10\theta \qquad \quad \text{with } \alpha = 5\theta.$$

▶ **TRY EXERCISE 2, PAGE 578**

❓ **QUESTION** Does $\sin \theta \cos \theta = \dfrac{1}{2} \sin 2\theta$?

The double-angle identities can also be used to evaluate some trigonometric expressions.

EXAMPLE 2 Evaluate a Trigonometric Function

For an angle α in Quadrant I, $\sin \alpha = \dfrac{4}{5}$. Find $\sin 2\alpha$.

Solution

Use the identity $\sin 2\alpha = 2 \sin \alpha \cos \alpha$. Find $\cos \alpha$ by substituting for $\sin \alpha$ in $\sin^2 \alpha + \cos^2 \alpha = 1$ and solving for $\cos \alpha$.

$$\cos \alpha = \sqrt{1 - \sin^2 \alpha} = \sqrt{1 - \left(\frac{4}{5}\right)^2} = \frac{3}{5} \qquad \bullet \cos \alpha > 0 \text{ if } \alpha \text{ is in Quadrant I.}$$

Substitute the values of $\sin \alpha$ and $\cos \alpha$ in the double-angle formula for $\sin 2\alpha$.

$$\sin 2\alpha = 2 \sin \alpha \cos \alpha = 2\left(\frac{4}{5}\right)\left(\frac{3}{5}\right) = \frac{24}{25}$$

▶ **TRY EXERCISE 26, PAGE 578**

❓ **ANSWER** Yes. $\sin \theta \cos \theta = \dfrac{2 \sin \theta \cos \theta}{2} = \dfrac{\sin 2\theta}{2} = \dfrac{1}{2} \sin 2\theta$.

EXAMPLE 1 Verify an Identity

Verify the identity $\cos 2x \sin 5x = \dfrac{1}{2}(\sin 7x + \sin 3x)$.

Solution

$$\cos 2x \sin 5x = \frac{1}{2}[\sin(2x + 5x) - \sin(2x - 5x)]$$

• Use the product-to-sum identity for $\cos\alpha \sin\beta$.

$$= \frac{1}{2}[\sin 7x - \sin(-3x)]$$

$$= \frac{1}{2}(\sin 7x + \sin 3x)$$

• $\sin(-3x) = -\sin 3x$

▶ **TRY EXERCISE 36, PAGE 587**

● **THE SUM-TO-PRODUCT IDENTITIES**

The *sum-to-product identities* can be derived from the product-to-sum identities. To derive the sum-to-product identity for $\sin x + \sin y$, we first let $x = \alpha + \beta$ and $y = \alpha - \beta$. Then

$$x + y = \alpha + \beta + \alpha - \beta \quad \text{and} \quad x - y = \alpha + \beta - (\alpha - \beta)$$
$$x + y = 2\alpha \qquad\qquad\qquad\qquad x - y = 2\beta$$
$$\alpha = \frac{x + y}{2} \qquad\qquad\qquad\qquad \beta = \frac{x - y}{2}$$

Substituting these expressions for α and β into the product-to-sum identity

$$\frac{1}{2}[\sin(\alpha + \beta) + \sin(\alpha - \beta)] = \sin\alpha\cos\beta$$

yields

$$\sin\left(\frac{x + y}{2} + \frac{x - y}{2}\right) + \sin\left(\frac{x + y}{2} - \frac{x - y}{2}\right) = 2\sin\frac{x + y}{2}\cos\frac{x - y}{2}$$

Simplifying the left side, we have a sum-to-product identity.

$$\sin x + \sin y = 2\sin\frac{x + y}{2}\cos\frac{x - y}{2}$$

In like manner, three other sum-to-product identities can be derived from the other product-to-sum identities. The proofs of these identities are left as exercises.

$$\sin x - \sin y = 2\cos\frac{x + y}{2}\sin\frac{x - y}{2}$$

$$\cos x + \cos y = 2\cos\frac{x + y}{2}\cos\frac{x - y}{2}$$

$$\cos x - \cos y = -2\sin\frac{x + y}{2}\sin\frac{x - y}{2}$$

> ### EXAMPLE 2 — Write the Difference of Trigonometric Expressions as a Product
>
> Write $\sin 4\theta - \sin \theta$ as the product of two functions.
>
> **Solution**
>
> $$\sin 4\theta - \sin \theta = 2 \cos \frac{4\theta + \theta}{2} \sin \frac{4\theta - \theta}{2} = 2 \cos \frac{5\theta}{2} \sin \frac{3\theta}{2}$$
>
> ▶ **TRY EXERCISE 22, PAGE 587**

? QUESTION Does $\cos 4\theta + \cos 2\theta = 2 \cos 3\theta \cos \theta$?

> ### EXAMPLE 3 — Verify a Sum-to-Product Identity
>
> Verify the identity $\dfrac{\sin 6x + \sin 2x}{\sin 6x - \sin 2x} = \tan 4x \cot 2x$.
>
> **Solution**
>
> $$\frac{\sin 6x + \sin 2x}{\sin 6x - \sin 2x} = \frac{2 \sin \dfrac{6x + 2x}{2} \cos \dfrac{6x - 2x}{2}}{2 \cos \dfrac{6x + 2x}{2} \sin \dfrac{6x - 2x}{2}} = \frac{\sin 4x \cos 2x}{\cos 4x \sin 2x}$$
>
> $$= \tan 4x \cot 2x$$
>
> ▶ **TRY EXERCISE 44, PAGE 588**

• FUNCTIONS OF THE FORM $f(x) = a \sin x + b \cos x$

The function given by $f(x) = a \sin x + b \cos x$ can be written in the form $f(x) = k \sin(x + \alpha)$. This form of the function is useful in graphing and engineering applications because the amplitude, period, and phase shift can be readily calculated.

Let $P(a, b)$ be a point on a coordinate plane, and let α represent an angle in standard position. See **Figure 6.9**. To rewrite $y = a \sin x + b \cos x$, multiply and divide the expression $a \sin x + b \cos x$ by $\sqrt{a^2 + b^2}$.

$$a \sin x + b \cos x = \frac{\sqrt{a^2 + b^2}}{\sqrt{a^2 + b^2}} (a \sin x + b \cos x)$$

$$= \sqrt{a^2 + b^2} \left(\frac{a}{\sqrt{a^2 + b^2}} \sin x + \frac{b}{\sqrt{a^2 + b^2}} \cos x \right) \qquad (1)$$

From the definition of the sine and cosine of an angle in standard position, let

$$k = \sqrt{a^2 + b^2}, \quad \cos \alpha = \frac{a}{\sqrt{a^2 + b^2}}, \quad \text{and} \quad \sin \alpha = \frac{b}{\sqrt{a^2 + b^2}}$$

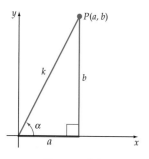

FIGURE 6.9

? ANSWER Yes. $\cos 4\theta + \cos 2\theta = 2 \cos\left(\dfrac{4\theta + 2\theta}{2}\right) \cos\left(\dfrac{4\theta - 2\theta}{2}\right) = 2 \cos 3\theta \cos \theta$.

Substituting these expressions into Equation (1) yields

$$a \sin x + b \cos x = k(\cos \alpha \sin x + \sin \alpha \cos x)$$

Now, using the identity for the sine of the sum of two angles, we have

$$a \sin x + b \cos x = k \sin(x + \alpha)$$

Thus $a \sin x + b \cos x = k \sin(x + \alpha)$, where $k = \sqrt{a^2 + b^2}$ and α is the angle for which $\sin \alpha = \dfrac{b}{\sqrt{a^2 + b^2}}$ and $\cos \alpha = \dfrac{a}{\sqrt{a^2 + b^2}}$.

EXAMPLE 4   **Rewrite $a \sin x + b \cos x$**

Rewrite $\sin x + \cos x$ in the form $k \sin(x + \alpha)$.

Solution

Comparing $\sin x + \cos x$ to $a \sin x + b \cos x$, $a = 1$ and $b = 1$. Thus $k = \sqrt{1^2 + 1^2} = \sqrt{2}$, $\sin \alpha = \dfrac{1}{\sqrt{2}}$, and $\cos \alpha = \dfrac{1}{\sqrt{2}}$. Thus $\alpha = \dfrac{\pi}{4}$.

$$\sin x + \cos x = k \sin(x + \alpha) = \sqrt{2} \sin\left(x + \frac{\pi}{4}\right)$$

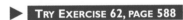

▶ **TRY EXERCISE 62, PAGE 588**

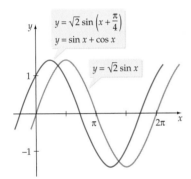

$y = \sqrt{2} \sin\left(x + \dfrac{\pi}{4}\right)$

$y = \sin x + \cos x$

$y = \sqrt{2} \sin x$

FIGURE 6.10

The graphs of $y = \sin x + \cos x$ and $y = \sqrt{2} \sin\left(x + \dfrac{\pi}{4}\right)$ are both the graph of $y = \sqrt{2} \sin x$ shifted $\dfrac{\pi}{4}$ units to the left. See **Figure 6.10.**

EXAMPLE 5   **Use an Identity to Graph a Trigonometric Function**

Graph $f(x) = -\sin x + \sqrt{3} \cos x$.

Solution

First, we write $f(x)$ as $k \sin(x + \alpha)$. Let $a = -1$ and $b = \sqrt{3}$; then $k = \sqrt{(-1)^2 + (\sqrt{3})^2} = 2$. The point $P(-1, \sqrt{3})$ is in the second quadrant (see **Figure 6.11**). Let α be an angle in standard position with P on its terminal side. Let α' be the reference angle for α. Then

$$\sin \alpha' = \frac{\sqrt{3}}{2}$$

$$\alpha' = \frac{\pi}{3}$$

$$\alpha = \pi - \alpha' = \pi - \frac{\pi}{3} = \frac{2\pi}{3}$$

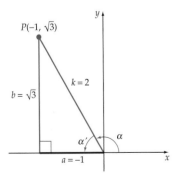

$P(-1, \sqrt{3})$

$b = \sqrt{3}$

$k = 2$

α' α

$a = -1$

FIGURE 6.11

Continued ▶

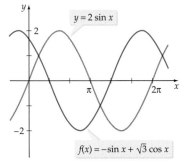

FIGURE 6.12

Substituting 2 for k and $\dfrac{2\pi}{3}$ for α in $y = k \sin(x + \alpha)$, we have

$$y = 2 \sin\left(x + \frac{2\pi}{3}\right)$$

The phase shift is $-\dfrac{c}{b} = -\dfrac{2\pi}{3}$. Thus the graph of the equation

$f(x) = -\sin x + \sqrt{3} \cos x$ is the graph of $y = 2 \sin x$ shifted $\dfrac{2\pi}{3}$ units to the left. See **Figure 6.12.**

▶ **TRY EXERCISE 70, PAGE 588**

We now list the identities that have been discussed in this section.

Product-to-Sum Identities

$$\sin \alpha \cos \beta = \frac{1}{2}[\sin(\alpha + \beta) + \sin(\alpha - \beta)]$$

$$\cos \alpha \sin \beta = \frac{1}{2}[\sin(\alpha + \beta) - \sin(\alpha - \beta)]$$

$$\cos \alpha \cos \beta = \frac{1}{2}[\cos(\alpha + \beta) + \cos(\alpha - \beta)]$$

$$\sin \alpha \sin \beta = \frac{1}{2}[\cos(\alpha - \beta) - \cos(\alpha + \beta)]$$

Sum-to-Product Identities

$$\sin x + \sin y = 2 \sin \frac{x + y}{2} \cos \frac{x - y}{2}$$

$$\cos x + \cos y = 2 \cos \frac{x + y}{2} \cos \frac{x - y}{2}$$

$$\sin x - \sin y = 2 \cos \frac{x + y}{2} \sin \frac{x - y}{2}$$

$$\cos x - \cos y = -2 \sin \frac{x + y}{2} \sin \frac{x - y}{2}$$

Sums of the Form $a \sin x + b \cos x$

$$a \sin x + b \cos x = k \sin(x + \alpha)$$

where $k = \sqrt{a^2 + b^2}$, $\sin \alpha = \dfrac{b}{\sqrt{a^2 + b^2}}$, and $\cos \alpha = \dfrac{a}{\sqrt{a^2 + b^2}}$.

TOPICS FOR DISCUSSION

1. A student claims that the *exact* value of $\sin 75° \cos 15°$ is $\dfrac{2 + \sqrt{3}}{4}$. Do you agree? Explain.

2. Do you agree with the following work? Explain.

$$\cos 195° + \cos 105° = \cos(195° + 105°) = \cos 300° = -\frac{1}{2}$$

3. The graphs of $y_1 = \sin x + \cos x$ and $y_2 = \sqrt{2} \sin\left(x + \dfrac{\pi}{4}\right)$ are identical. Do you agree? Explain.

4. Explain to a classmate how to determine the amplitude of the graph of $y = a \sin x + b \cos x$.

EXERCISE SET 6.4

In Exercises 1 to 8, write each expression as the sum or difference of two functions.

1. $2 \sin x \cos 2x$

2. $2 \sin 4x \sin 2x$

3. $\cos 6x \sin 2x$

4. $\cos 3x \cos 5x$

5. $2 \sin 5x \cos 3x$

6. $2 \sin 2x \cos 6x$

7. $\sin x \sin 5x$

8. $\cos 3x \sin x$

In Exercises 9 to 16, find the exact value of each expression. Do not use a calculator.

9. $\cos 75° \cos 15°$

10. $\sin 105° \cos 15°$

11. $\cos 157.5° \sin 22.5°$

12. $\sin 195° \cos 15°$

13. $\sin \dfrac{13\pi}{12} \cos \dfrac{\pi}{12}$

14. $\sin \dfrac{11\pi}{12} \sin \dfrac{7\pi}{12}$

15. $\sin \dfrac{\pi}{12} \cos \dfrac{7\pi}{12}$

16. $\cos \dfrac{17\pi}{12} \sin \dfrac{7\pi}{12}$

In Exercises 17 to 32, write each expression as the product of two functions.

17. $\sin 4\theta + \sin 2\theta$

18. $\cos 5\theta - \cos 3\theta$

19. $\cos 3\theta + \cos \theta$

20. $\sin 7\theta - \sin 3\theta$

21. $\cos 6\theta - \cos 2\theta$

▶ **22.** $\cos 3\theta + \cos 5\theta$

23. $\cos \theta + \cos 7\theta$

24. $\sin 3\theta + \sin 7\theta$

25. $\sin 5\theta + \sin 9\theta$

26. $\cos 5\theta - \cos \theta$

27. $\cos 2\theta - \cos \theta$

28. $\sin 2\theta + \sin 6\theta$

29. $\cos \dfrac{\theta}{2} - \cos \theta$

30. $\sin \dfrac{3\theta}{4} + \sin \dfrac{\theta}{2}$

31. $\sin \dfrac{\theta}{2} - \sin \dfrac{\theta}{3}$

32. $\cos \theta + \cos \dfrac{\theta}{2}$

In Exercises 33 to 48, verify the identity.

33. $2 \cos \alpha \cos \beta = \cos(\alpha + \beta) + \cos(\alpha - \beta)$

34. $2 \sin \alpha \sin \beta = \cos(\alpha - \beta) - \cos(\alpha + \beta)$

35. $2 \cos 3x \sin x = 2 \sin x \cos x - 8 \cos x \sin^3 x$

▶ **36.** $\sin 5x \cos 3x = \sin 4x \cos 4x + \sin x \cos x$

37. $2 \cos 5x \cos 7x = \cos^2 6x - \sin^2 6x + 2 \cos^2 x - 1$

38. $\sin 3x \cos x = \sin x \cos x(3 - 4 \sin^2 x)$

39. $\sin 3x - \sin x = 2 \sin x - 4 \sin^3 x$

40. $\cos 5x - \cos 3x = -8 \sin^2 x(2 \cos^3 x - \cos x)$

41. $\sin 2x + \sin 4x = 2 \sin x \cos x(4 \cos^2 x - 1)$

42. $\cos 3x + \cos x = 4 \cos^3 x - 2 \cos x$

43. $\dfrac{\sin 3x - \sin x}{\cos 3x - \cos x} = -\cot 2x$

▶ **44.** $\dfrac{\cos 5x - \cos 3x}{\sin 5x + \sin 3x} = -\tan x$

45. $\dfrac{\sin 5x + \sin 3x}{4 \sin x \cos^3 x - 4 \sin^3 x \cos x} = 2 \cos x$

46. $\dfrac{\cos 4x - \cos 2x}{\sin 2x - \sin 4x} = \tan 3x$

47. $\sin(x + y) \cos(x - y) = \sin x \cos x + \sin y \cos y$

48. $\sin(x + y) \sin(x - y) = \sin^2 x - \sin^2 y$

In Exercises 49 to 58, write the given equation in the form $y = k \sin (x + \alpha)$, where the measure of α is in degrees.

49. $y = -\sin x - \cos x$

50. $y = \sqrt{3} \sin x - \cos x$

51. $y = \dfrac{1}{2} \sin x - \dfrac{\sqrt{3}}{2} \cos x$

52. $y = \dfrac{\sqrt{3}}{2} \sin x - \dfrac{1}{2} \cos x$

53. $y = \dfrac{1}{2} \sin x - \dfrac{1}{2} \cos x$

54. $y = -\dfrac{\sqrt{3}}{2} \sin x - \dfrac{1}{2} \cos x$

55. $y = -3 \sin x + 3 \cos x$

56. $y = \dfrac{\sqrt{2}}{2} \sin x + \dfrac{\sqrt{2}}{2} \cos x$

57. $y = \pi \sin x - \pi \cos x$

58. $y = -0.4 \sin x + 0.4 \cos x$

In Exercises 59 to 66, write the given equation in the form $y = k \sin (x + \alpha)$, where the measure of α is in radians.

59. $y = -\sin x + \cos x$

60. $y = -\sqrt{3} \sin x - \cos x$

61. $y = \dfrac{\sqrt{3}}{2} \sin x + \dfrac{1}{2} \cos x$

▶ **62.** $y = \sin x + \sqrt{3} \cos x$

63. $y = -10 \sin x + 10 \sqrt{3} \cos x$

64. $y = 3 \sin x - 3 \sqrt{3} \cos x$

65. $y = -5 \sin x + 5 \cos x$

66. $y = 3 \sin x - 3 \cos x$

In Exercises 67 to 76, graph one cycle of each equation.

67. $y = -\sin x - \sqrt{3} \cos x$

68. $y = -\sqrt{3} \sin x + \cos x$

69. $y = 2 \sin x + 2 \cos x$

▶ **70.** $y = \sin x + \sqrt{3} \cos x$

71. $y = -\sqrt{3} \sin x - \cos x$

72. $y = -\sin x + \cos x$

73. $y = -5 \sin x + 5 \sqrt{3} \cos x$

74. $y = -\sqrt{2} \sin x + \sqrt{2} \cos x$

75. $y = 6 \sqrt{3} \sin x - 6 \cos x$

76. $y = 5 \sqrt{2} \sin x + 5 \sqrt{2} \cos x$

77. TONES ON A TOUCH-TONE PHONE

    a. Write an equation of the form $p(t) = \cos(2\pi f_1 t) + \cos(2\pi f_2 t)$ that models the tone produced by pressing the 5 key on a touch-tone phone. (*Hint:* See the Chapter Opener, page 553.)

    b. Use a sum-to-product identity to write your equation from part **a.** in the form

$$p(t) = A \cos(B\pi t) \cos(C\pi t)$$

    c. When a sound with frequency f_1 is combined with a sound with frequency f_2, the combined sound has a frequency of $\dfrac{f_1 + f_2}{2}$. What is the frequency of the tone produced when the 5 key is pressed?

78. TONES ON A TOUCH-TONE PHONE

    a. Write an equation of the form $p(t) = \cos(2\pi f_1 t) + \cos(2\pi f_2 t)$ that models the tone produced by pressing the 8 key on a touch-tone phone.

    b. Use a sum-to-product identity to write your equation from part **a.** in the form

$$p(t) = A \cos(B\pi t) \cos(C\pi t)$$

    c. What is the frequency of the tone produced when the 8 key on a touch-tone phone is pressed? (*Hint:* See part **c.** of Exercise 77.)

In Exercises 79 to 84, compare the graphs of each side of the equation to predict whether the equation is an identity.

79. $\sin 3x - \sin x = 2 \sin x - 4 \sin^3 x$

80. $\dfrac{\sin 3x - \sin x}{\cos 3x - \cos x} = -\dfrac{1}{\tan 2x}$

81. $-\sqrt{3}\sin x - \cos x = 2\sin\left(x - \dfrac{5\pi}{6}\right)$

83. $\dfrac{1}{2}\sin x - \dfrac{\sqrt{3}}{2}\cos x = \sin\left(x - \dfrac{\pi}{3}\right)$

82. $-\sqrt{3}\sin x + \cos x = 2\sin\left(x + \dfrac{5\pi}{6}\right)$

84. $\dfrac{\sqrt{3}}{2}\sin x + \dfrac{1}{2}\cos x = \sin\left(x + \dfrac{\pi}{6}\right)$

CONNECTING CONCEPTS

85. Derive the sum-to-product identity

$$\cos x + \cos y = 2\cos\dfrac{x+y}{2}\cos\dfrac{x-y}{2}$$

86. Derive the product-to-sum identity

$$\sin x \sin y = \dfrac{1}{2}[\cos(x-y) - \cos(x+y)]$$

87. If $x + y = 180°$, show that $\sin x + \sin y = 2\sin x$.

88. If $x + y = 360°$, show that $\cos x + \cos y = 2\cos x$.

In Exercises 89 to 94, verify the identity.

89. $\sin 2x + \sin 4x + \sin 6x = 4\sin 3x \cos 2x \cos x$

90. $\sin 4x - \sin 2x + \sin 6x = 4\cos 3x \sin 2x \cos x$

91. $\dfrac{\cos 10x + \cos 8x}{\sin 10x - \sin 8x} = \cot x$

92. $\dfrac{\sin 10x + \sin 2x}{\cos 10x + \cos 2x} = \dfrac{2\tan 3x}{1 - \tan^2 3x}$

93. $\dfrac{\sin 2x + \sin 4x + \sin 6x}{\cos 2x + \cos 4x + \cos 6x} = \tan 4x$

94. $\dfrac{\sin 2x + \sin 6x}{\cos 6x - \cos 2x} = -\cot 2x$

95. Verify that $\cos^2 x - \sin^2 x = \cos 2x$ by using a product-to-sum identity.

96. Verify that $2\sin x \cos x = \sin 2x$ by using a product-to-sum identity.

97. Verify that $a\sin x + b\cos x = k\cos(x - \alpha)$, where

$$k = \sqrt{a^2 + b^2} \text{ and } \tan\alpha = \dfrac{a}{b}.$$

98. Verify that $a\sin cx + b\cos cx = k\sin(cx + \alpha)$, where

$$k = \sqrt{a^2 + b^2} \text{ and } \tan\alpha = \dfrac{b}{a}.$$

PREPARE FOR SECTION 6.5

99. What is a one-to-one function? [2.2]

100. State the horizontal line test. [2.2]

101. Find $f[g(x)]$ given that $f(x) = 2x + 4$ and $g(x) = \dfrac{1}{2}x - 2$. [2.6]

102. If f and f^{-1} are inverse functions, then determine $f[f^{-1}(x)]$ for any x in the domain of f^{-1}. [4.1]

103. If f and f^{-1} are inverse functions, then explain how the graph of f^{-1} is related to the graph of f. [4.1]

104. Use the horizontal line test to determine whether the graph of $y = \sin x$, where x is any real number, is a one-to-one function. [2.2]

PROJECTS

1. **INTERFERENCE OF SOUND WAVES** The following figure on page 590 shows the waveforms of two tuning forks. The frequency of one of the tuning forks is 10 cycles per second, and the frequency of the other tuning fork is 8 cycles per second. Each of the sound waves can be modeled by an equation of the form

$$p(t) = A\cos 2\pi ft$$

where p is the pressure produced on the eardrum at time t, A is the amplitude of the sound wave, and f is the frequency of the sound.

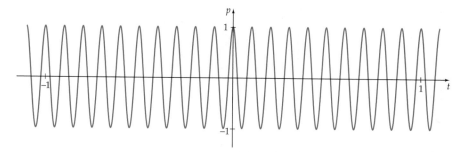

$$p_1(t) = \cos 2\pi \cdot 10t$$

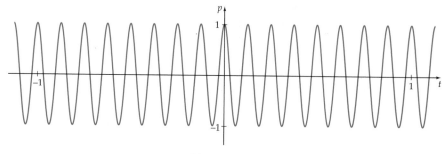

$$p_2(t) = \cos 2\pi \cdot 8t$$

If the two tuning forks are struck at the same time, with the same force, the sound that we hear fluctuates between a loud tone and silence. These regular fluctuations are called **beats.** The loud periods occur when the sound waves reinforce (interfere constructively with) one another, and the nearly silent periods occur when the waves interfere destructively with each other. The pressure p produced on the eardrum from the combined sound waves is given by

$$p(t) = p_1 + p_2 = \cos 2\pi \cdot 10t + \cos 2\pi \cdot 8t$$

The following figure shows the graph of p.

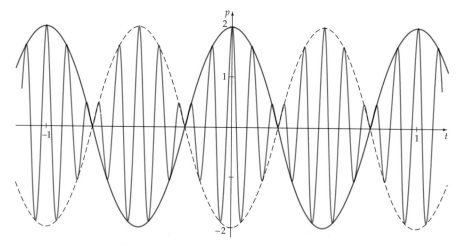

Graph of $p = p_1 + p_2 = \cos 2\pi \cdot 10t + \cos 2\pi \cdot 8t$
showing the beats in the combined sounds

a. Use the sum-to-product identity for $\cos x + \cos y$ to write p as a product.

b. Explain why the graph of $p = A \cos 2\pi f_1 t + A \cos 2\pi f_2 t$ can be thought of as a cosine curve with period $\dfrac{2}{f_1 + f_2}$ and a *variable* amplitude of

$$2A \cos\left[2\pi\left(\frac{f_1 - f_2}{2} \right)t \right]$$

c. The rate of the beats produced by two sounds, with the same intensity, is the absolute value of the difference between their frequencies. Consider two tuning forks that are struck at the same time with the same force and held on a sounding board. The tuning forks have frequencies of 564 and 568 cycles per second, respectively. How many beats will be heard each second?

d. A piano tuner strikes a tuning fork and a key on a piano that is supposed to have the same frequency as the tuning fork. The piano tuner notices that the sound produced by the piano is lower than that produced by the tuning fork. The piano tuner also notes that the combined sound of the piano and the tuning fork has 2 beats per second. How much lower is the frequency of the piano than the frequency of the tuning fork?

2. The photo shows a matched set of tuning forks. The fork on the left has an adjustable weight that can be used to vary the frequency of its tone. Check with the physics department at your school to see if a matched set of tuning forks is available. Use the tuning forks to demonstrate to your classmates the phenomenon of beats.

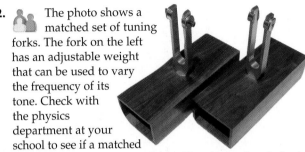

SECTION 6.5

INVERSE TRIGONOMETRIC FUNCTIONS

• INVERSE TRIGONOMETRIC FUNCTIONS

Because the graph of $y = \sin x$ fails the horizontal line test, it is not the graph of a one-to-one function. Therefore, it does not have an inverse function. **Figure 6.13** shows the graph of $y = \sin x$ on the interval $-2\pi \le x \le 2\pi$ and the graph of the inverse relation $x = \sin y$. Note that the graph of $x = \sin y$ does not satisfy the vertical line test and therefore is not the graph of a function.

If the domain of $y = \sin x$ is restricted to $-\dfrac{\pi}{2} \le x \le \dfrac{\pi}{2}$, the graph of $y = \sin x$ satisfies the horizontal line test and therefore the function has an inverse function. The graphs of $y = \sin x$ for $-\dfrac{\pi}{2} \le x \le \dfrac{\pi}{2}$ and its inverse are shown in **Figure 6.14.**

See Section 4.1 if you need to review the concept of an inverse function.

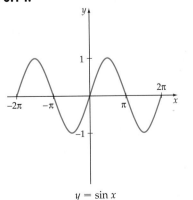

$y = \sin x$

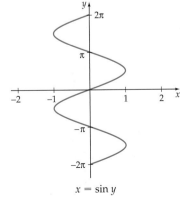

$x = \sin y$

FIGURE 6.13

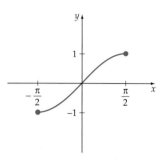

$$y = \sin x: -\frac{\pi}{2} \le x \le \frac{\pi}{2}$$

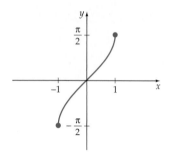

$$y = \sin^{-1} x: -1 \le x \le 1$$

FIGURE 6.14

---

take note

The -1 in $\sin^{-1} x$ is not an exponent. The -1 is used to denote the inverse function. To use -1 as an exponent for a sine function, enclose the function in parentheses.

$$(\sin x)^{-1} = \frac{1}{\sin x} = \csc x$$

$$\sin^{-1} x \ne \frac{1}{\sin x}$$

To find the inverse of the function defined by $y = \sin x$, with $-\frac{\pi}{2} \le x \le \frac{\pi}{2}$, interchange x and y. Then solve for y.

$$y = \sin x \qquad \bullet -\frac{\pi}{2} \le x \le \frac{\pi}{2}$$

$$x = \sin y \qquad \bullet \text{ Interchange } x \text{ and } y.$$

$$y = ? \qquad \bullet \text{ Solve for } y.$$

Unfortunately, there is no algebraic solution for y. Thus we establish new notation and write

$$y = \sin^{-1} x$$

which is read "y is the inverse sine of x." Some textbooks use the notation arcsin x instead of $\sin^{-1} x$.

Definition of $\sin^{-1} x$

$$y = \sin^{-1} x \quad \text{if and only if} \quad x = \sin y$$

where $-1 \le x \le 1$ and $-\frac{\pi}{2} \le y \le \frac{\pi}{2}$.

It is convenient to think of the value of an inverse trigonometric function as an angle. For instance, if $y = \sin^{-1}\left(\frac{1}{2}\right)$, then y is the angle in the interval $\left[-\frac{\pi}{2}, \frac{\pi}{2}\right]$ whose sine is $\frac{1}{2}$. Thus $y = \frac{\pi}{6}$.

Because the graph of $y = \cos x$ fails the horizontal line test, it is not the graph of a one-to-one function. Therefore, it does not have an inverse function. **Figure 6.15** shows the graph of $y = \cos x$ on the interval $-2\pi \le x \le 2\pi$ and the graph of the inverse relation $x = \cos y$. Note that the graph of $x = \cos y$ does not satisfy the vertical line test and therefore is not the graph of a function.

If the domain of $y = \cos x$ is restricted to $0 \le x \le \pi$, the graph of $y = \cos x$ satisfies the horizontal line test and therefore is the graph of a one-to-one function. The graph of $y = \cos x$ for $0 \le x \le \pi$ and that of $x = \cos y$ are shown in **Figure 6.16**.

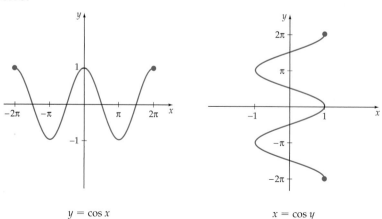

$y = \cos x$

$x = \cos y$

FIGURE 6.15

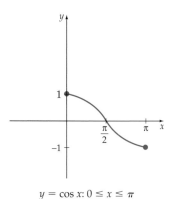

$y = \cos x: 0 \le x \le \pi$

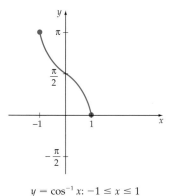

$y = \cos^{-1} x: -1 \le x \le 1$

FIGURE 6.16

To find the inverse of the function defined by $y = \cos x$, with $0 \le x \le \pi$, interchange x and y. Then solve for y.

$$y = \cos x \qquad \bullet\; 0 \le x \le \pi$$
$$x = \cos y \qquad \bullet\; \textbf{Interchange } x \textbf{ and } y.$$
$$y = ? \qquad \bullet\; \textbf{Solve for } y.$$

As in the case for the inverse sine function, there is no algebraic solution for y. Thus the notation for the inverse cosine function becomes $y = \cos^{-1} x$. We can write the following definition of the inverse cosine function.

Definition of $\cos^{-1} x$
$$y = \cos^{-1} x \quad \text{if and only if} \quad x = \cos y$$ where $-1 \le x \le 1$ and $0 \le y \le \pi$.

Because the graphs of $y = \tan x$, $y = \csc x$, $y = \sec x$, and $y = \cot x$ fail the horizontal line test, these functions are not one-to-one functions. Therefore, these functions do not have inverse functions. If the domains of all these functions are restricted in a certain way, however, the graphs satisfy the horizontal line test. Thus each of these functions has an inverse function over a restricted domain. **Table 6.2** on page 594 shows the restricted function and the inverse function for $\tan x$, $\csc x$, $\sec x$, and $\cot x$.

The choice of ranges for $y = \sec^{-1} x$ and $y = \csc^{-1} x$ is not universally accepted. For example, some calculus texts use $\left[0, \dfrac{\pi}{2} \right) \cup \left[\pi, \dfrac{3\pi}{2} \right)$ as the range of $y = \sec^{-1} x$. This definition has some advantages and some disadvantages that are explained in more advanced mathematics courses.

EXAMPLE 1 — **Evaluate Inverse Functions**

Find the exact value of each inverse function.

a. $y = \tan^{-1} \dfrac{\sqrt{3}}{3}$ **b.** $y = \cos^{-1}\left(-\dfrac{\sqrt{2}}{2}\right)$

Solution

a. Because $y = \tan^{-1} \dfrac{\sqrt{3}}{3}$, y is the angle whose measure is in the interval $\left(-\dfrac{\pi}{2}, \dfrac{\pi}{2} \right)$, and $\tan y = \dfrac{\sqrt{3}}{3}$. Therefore, $y = \dfrac{\pi}{6}$.

b. Because $y = \cos^{-1}\left(-\dfrac{\sqrt{2}}{2}\right)$, y is the angle whose measure is in the interval $[0, \pi]$, and $\cos y = -\dfrac{\sqrt{2}}{2}$. Therefore, $y = \dfrac{3}{4}\pi$.

▶ **TRY EXERCISE 2, PAGE 601**

TABLE 6.2

	$y = \tan x$	$y = \tan^{-1} x$	$y = \csc x$	$y = \csc^{-1} x$
Domain	$-\dfrac{\pi}{2} < x < \dfrac{\pi}{2}$	$-\infty < x < \infty$	$-\dfrac{\pi}{2} \leq x \leq \dfrac{\pi}{2}, x \neq 0$	$x \leq -1 \text{ or } x \geq 1$
Range	$-\infty < y < \infty$	$-\dfrac{\pi}{2} < y < \dfrac{\pi}{2}$	$y \leq -1 \text{ or } y \geq 1$	$-\dfrac{\pi}{2} \leq y \leq \dfrac{\pi}{2}, y \neq 0$
Asymptotes	$x = -\dfrac{\pi}{2}, x = \dfrac{\pi}{2}$	$y = -\dfrac{\pi}{2}, y = \dfrac{\pi}{2}$	$x = 0$	$y = 0$
Graph				

	$y = \sec x$	$y = \sec^{-1} x$	$y = \cot x$	$y = \cot^{-1} x$
Domain	$0 \leq x \leq \pi, x \neq \dfrac{\pi}{2}$	$x \leq -1 \text{ or } x \geq 1$	$0 < x < \pi$	$-\infty < x < \infty$
Range	$y \leq -1 \text{ or } y \geq 1$	$0 \leq y \leq \pi, y \neq \dfrac{\pi}{2}$	$-\infty < y < \infty$	$0 < y < \pi$
Asymptotes	$x = \dfrac{\pi}{2}$	$y = \dfrac{\pi}{2}$	$x = 0, x = \pi$	$y = 0, y = \pi$
Graph				

A calculator may not have keys for the inverse secant, cosecant, and cotangent functions. The following procedure shows an identity for the inverse cosecant function in terms of the inverse sine function. If we need to determine y, which is the angle whose cosecant is x, we can rewrite $y = \csc^{-1} x$ as follows:

$$y = \csc^{-1} x$$

- **Domain:** $x \leq -1 \text{ or } x \geq 1$

  Range: $-\dfrac{\pi}{2} \leq y \leq \dfrac{\pi}{2}, y \neq 0$

$$\csc y = x$$

- **Definition of inverse function**

$$\dfrac{1}{\sin y} = x$$

- **Substitute** $\dfrac{1}{\sin y}$ **for csc y.**

$$\sin y = \frac{1}{x} \qquad \text{• Solve for sin y.}$$

$$y = \sin^{-1}\frac{1}{x} \qquad \text{• Write using inverse notation.}$$

$$\csc^{-1}x = \sin^{-1}\frac{1}{x} \qquad \text{• Replace y with csc}^{-1} \text{ x.}$$

Thus $\csc^{-1}x$ is the same as $\sin^{-1}\dfrac{1}{x}$. There is a similar identity for $\sec^{-1}x$.

Identities for csc⁻¹ x, sec⁻¹ x, and cot⁻¹ x

If $x \le -1$ or $x \ge 1$, then

$$\csc^{-1}x = \sin^{-1}\frac{1}{x} \quad \text{and} \quad \sec^{-1}x = \cos^{-1}\frac{1}{x}$$

If x is a real number, then

$$\cot^{-1}x = \frac{\pi}{2} - \tan^{-1}x$$

● COMPOSITION OF TRIGONOMETRIC FUNCTIONS AND THEIR INVERSES

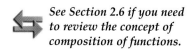

See Section 2.6 if you need to review the concept of composition of functions.

Recall that a function f and its inverse f^{-1} have the property that $f[f^{-1}(x)] = x$ for all x in the domain of f^{-1} and that $f^{-1}[f(x)] = x$ for all x in the domain of f. Applying this property to the functions $\sin x$, $\cos x$, and $\tan x$ and their inverse functions produces the following theorems.

Composition of Trigonometric Functions and Their Inverses

- If $-1 \le x \le 1$, then $\sin(\sin^{-1}x) = x$, and $\cos(\cos^{-1}x) = x$.

- If x is any real number, then $\tan(\tan^{-1}x) = x$.

- If $-\dfrac{\pi}{2} \le x \le \dfrac{\pi}{2}$, then $\sin^{-1}(\sin x) = x$.

- If $0 \le x \le \pi$, then $\cos^{-1}(\cos x) = x$.

- If $-\dfrac{\pi}{2} < x < \dfrac{\pi}{2}$, then $\tan^{-1}(\tan x) = x$.

In the next example we make use of some of the composition theorems to evaluate trigonometric expressions.

EXAMPLE 2	**Evaluate the Composition of a Function and Its Inverse**

Find the exact value of each composition of functions.

a. $\sin(\sin^{-1} 0.357)$ b. $\cos^{-1}(\cos 3)$ c. $\tan[\tan^{-1}(-11.27)]$

d. $\sin(\sin^{-1} \pi)$ e. $\cos(\cos^{-1} 0.277)$ f. $\tan^{-1}\left(\tan \dfrac{4\pi}{3}\right)$

Solution

a. Because 0.357 is in the interval $[-1, 1]$, $\sin(\sin^{-1} 0.357) = 0.357$.

b. Because 3 is in the interval $[0, \pi]$, $\cos^{-1}(\cos 3) = 3$.

c. Because -11.27 is a real number, $\tan[\tan^{-1}(-11.27)] = -11.27$.

d. Because π is not in the domain of the inverse sine function, $\sin(\sin^{-1} \pi)$ is undefined.

e. Because 0.277 is in the interval $[-1, 1]$, $\cos(\cos^{-1} 0.277) = 0.277$.

f. $\dfrac{4\pi}{3}$ is not in the interval $\left(-\dfrac{\pi}{2}, \dfrac{\pi}{2}\right)$; however, the reference angle for $\theta = \dfrac{4\pi}{3}$ is $\theta' = \dfrac{\pi}{3}$. Thus $\tan^{-1}\left(\tan \dfrac{4\pi}{3}\right) = \tan^{-1}\left(\tan \dfrac{\pi}{3}\right)$. Because $\dfrac{\pi}{3}$ is in the interval $\left(-\dfrac{\pi}{2}, \dfrac{\pi}{2}\right)$, $\tan^{-1}\left(\tan \dfrac{\pi}{3}\right) = \dfrac{\pi}{3}$. Hence $\tan^{-1}\left(\tan \dfrac{4\pi}{3}\right) = \dfrac{\pi}{3}$.

▶ **TRY EXERCISE 24, PAGE 601**

❓ QUESTION Is $\tan^{-1}(\tan x) = x$ an identity?

It is often easy to evaluate a trigonometric expression by referring to a sketch of a right triangle that satisfies given conditions. In Example 3 we make use of this technique.

EXAMPLE 3	**Evaluate a Trigonometric Expression**

Find the exact value of $\sin\left(\cos^{-1}\dfrac{2}{5}\right)$.

❓ ANSWER No. $\tan^{-1}(\tan x) = x$ only if $-\dfrac{\pi}{2} < x < \dfrac{\pi}{2}$.

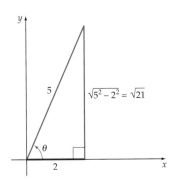

FIGURE 6.17

Solution

Let $\theta = \cos^{-1}\dfrac{2}{5}$, which implies $\cos\theta = \dfrac{2}{5}$. Because $\cos\theta$ is positive, θ is a first-quadrant angle. We draw a right triangle with base 2 and hypotenuse 5 so that we can view θ, as shown in **Figure 6.17.** The height of the triangle is $\sqrt{5^2 - 2^2} = \sqrt{21}$. Our goal is to find $\sin\theta$, which by definition is $\dfrac{\text{opp}}{\text{hyp}} = \dfrac{\sqrt{21}}{5}$. Thus

$$\sin\left(\cos^{-1}\frac{2}{5}\right) = \sin(\theta) = \frac{\sqrt{21}}{5}$$

▶ **TRY EXERCISE 46, PAGE 601**

In Example 4, we sketch two right triangles to evaluate the given expression.

EXAMPLE 4   **Evaluate a Trigonometric Expression**

Find the exact value of $\sin\left[\sin^{-1}\dfrac{3}{5} + \cos^{-1}\left(-\dfrac{5}{13}\right)\right]$.

Solution

Let $\alpha = \sin^{-1}\dfrac{3}{5}$. Thus $\sin\alpha = \dfrac{3}{5}$. Let $\beta = \cos^{-1}\left(-\dfrac{5}{13}\right)$, which implies that $\cos\beta = -\dfrac{5}{13}$. Sketch angles α and β as shown in **Figure 6.18.** We wish to evaluate

$$\sin\left[\sin^{-1}\frac{3}{5} + \cos^{-1}\left(-\frac{5}{13}\right)\right] = \sin(\alpha + \beta)$$
$$= \sin\alpha\cos\beta + \cos\alpha\sin\beta \qquad (1)$$

A close look at the triangles in **Figure 6.18** shows us that

$$\cos\alpha = \frac{4}{5} \quad\text{and}\quad \sin\beta = \frac{12}{13}$$

Substituting in Equation (1) gives us our desired result.

$$\sin\left[\sin^{-1}\frac{3}{5} + \cos^{-1}\left(-\frac{5}{13}\right)\right] = \sin\alpha\cos\beta + \cos\alpha\sin\beta$$
$$= \left(\frac{3}{5}\right)\left(-\frac{5}{13}\right) + \left(\frac{4}{5}\right)\left(\frac{12}{13}\right) = \frac{33}{65}$$

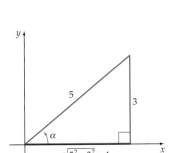

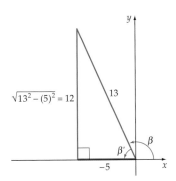

FIGURE 6.18

▶ **TRY EXERCISE 54, PAGE 602**

In Example 5 we make use of the identity $\cos(\cos^{-1}x) = x$, where $-1 \le x \le 1$, to solve an equation.

EXAMPLE 5 Solve an Inverse Trigonometric Equation

Solve $\sin^{-1}\dfrac{3}{5} + \cos^{-1}x = \pi$.

Solution

Solve for $\cos^{-1}x$, and then take the cosine of both sides of the equation.

$$\sin^{-1}\frac{3}{5} + \cos^{-1}x = \pi$$

$$\cos^{-1}x = \pi - \sin^{-1}\frac{3}{5}$$

$$\cos(\cos^{-1}x) = \cos\left(\pi - \sin^{-1}\frac{3}{5}\right)$$

$$x = \cos(\pi - \alpha)$$ • Let $\alpha = \sin^{-1}\dfrac{3}{5}$. Note that α is the angle whose sine is $\dfrac{3}{5}$. (See **Figure 6.19**.)

$$= \cos\pi\cos\alpha + \sin\pi\sin\alpha$$ • Difference identity for cosine.

$$= (-1)\cos\alpha + (0)\sin\alpha$$

$$= -\cos\alpha$$

$$= -\frac{4}{5}$$ • $\cos\alpha = \dfrac{4}{5}$ (See **Figure 6.19**.)

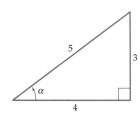

FIGURE 6.19

▶ **TRY EXERCISE 64, PAGE 602**

EXAMPLE 6 Verify a Trigonometric Identity That Involves Inverses

Verify the identity $\sin^{-1}x + \cos^{-1}x = \dfrac{\pi}{2}$.

Solution

Let $\alpha = \sin^{-1}x$ and $\beta = \cos^{-1}x$. These equations imply that $\sin\alpha = x$ and $\cos\beta = x$. From the right triangles in **Figure 6.20**,

$$\cos\alpha = \sqrt{1 - x^2} \qquad \text{and} \qquad \sin\beta = \sqrt{1 - x^2}$$

Our goal is to show $\sin^{-1}x + \cos^{-1}x$ equals $\dfrac{\pi}{2}$.

$$\sin^{-1}x + \cos^{-1}x = \alpha + \beta$$

$$= \cos^{-1}[\cos(\alpha + \beta)]$$ • Because $0 \le \alpha + \beta \le \pi$, we can apply $\alpha + \beta = \cos^{-1}[\cos(\alpha + \beta)]$.

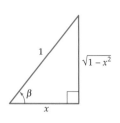

FIGURE 6.20

$$= \cos^{-1}[\cos \alpha \cos \beta - \sin \alpha \sin \beta]$$ • Addition identity for cosine

$$= \cos^{-1}\left[\left(\sqrt{1 - x^2}\right)(x) - (x)\left(\sqrt{1 - x^2}\right)\right]$$

$$= \cos^{-1} 0 = \frac{\pi}{2}$$

▶ **TRY EXERCISE 72, PAGE 602**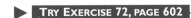

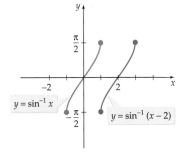

FIGURE 6.21

● GRAPHS OF INVERSE TRIGONOMETRIC FUNCTIONS

The inverse trigonometric functions can be graphed by using the procedures of stretching, shrinking, and translation that were discussed earlier in the text. For instance, the graph of $y = \sin^{-1}(x - 2)$ is a horizontal shift 2 units to the right of the graph of $y = \sin^{-1} x$, as shown in **Figure 6.21.**

EXAMPLE 7 **Graph an Inverse Function**

Graph: $y = \cos^{-1} x + 1$

Solution

Recall that the graph of $y = f(x) + c$ is a vertical translation of the graph of f. Because $c = 1$, a positive number, the graph of $y = \cos^{-1} x + 1$ is the graph of $y = \cos^{-1} x$ shifted 1 unit up. See **Figure 6.22.**

▶ **TRY EXERCISE 76, PAGE 602**

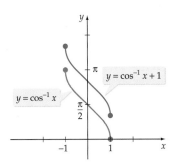

FIGURE 6.22

▦ **INTEGRATING TECHNOLOGY**

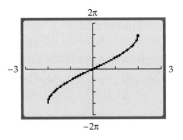

$y = 3 \sin^{-1} 0.5x$

FIGURE 6.23

When you use a graphing utility to draw the graph of an inverse trigonometric function, use the properties of these functions to verify the correctness of your graph. For instance, the graph of $y = 3 \sin^{-1} 0.5x$ is shown in **Figure 6.23.** The domain of $y = \sin^{-1} x$ is $-1 \le x \le 1$. Therefore, the domain of $y = 3 \sin^{-1} 0.5x$ is $-1 \le 0.5x \le 1$ or, multiplying the inequality by 2, $-2 \le x \le 2$. This is consistent with the graph in **Figure 6.23.**

The range of $y = \sin^{-1} x$ is $-\frac{\pi}{2} \le y \le \frac{\pi}{2}$. Thus the range of

$y = 3 \sin^{-1} 0.5x$ is $-\frac{3\pi}{2} \le y \le \frac{3\pi}{2}$. This is also consistent with the graph.

Verifying some of the properties of $y = \sin^{-1} x$ serves as a check that you have correctly entered the equation for the graph.

• AN APPLICATION INVOLVING AN INVERSE TRIGONOMETRIC FUNCTION

EXAMPLE 8 Solve an Application

 A camera is placed on a deck of a pool as shown in **Figure 6.24.** A diver is 18 feet above the camera lens. The extended length of the diver is 8 feet.

a. Show that the angle θ subtended at the lens by the diver is

$$\theta = \tan^{-1}\frac{26}{x} - \tan^{-1}\frac{18}{x}$$

b. For what values of x will $\theta = 9°$?

c. What value of x maximizes θ?

Solution

a. From **Figure 6.24** we see that $\alpha = \tan^{-1}\dfrac{26}{x}$ and $\beta = \tan^{-1}\dfrac{18}{x}$. Because $\theta = \alpha - \beta$, we have $\theta = \tan^{-1}\dfrac{26}{x} - \tan^{-1}\dfrac{18}{x}$.

b. Use a graphing utility to graph $\theta = \tan^{-1}\dfrac{26}{x} - \tan^{-1}\dfrac{18}{x}$ and

$\theta = \dfrac{\pi}{20}\left(9° = \dfrac{\pi}{20} \text{ radians}\right)$. See **Figure 6.25.** Use the "intersect" command to show that θ is 9° for $x \approx 12.22$ feet and $x \approx 38.29$ feet.

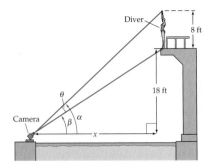

FIGURE 6.24

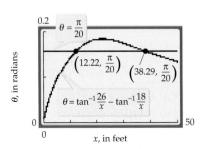

FIGURE 6.25

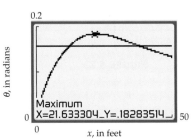

FIGURE 6.26

c. Use the "maximum" command to show that the maximum value of

$\theta = \tan^{-1}\dfrac{26}{x} - \tan^{-1}\dfrac{18}{x}$ occurs when $x \approx 21.63$ feet. See **Figure 6.26.**

▶ 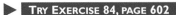 TRY EXERCISE 84, PAGE 602

TOPICS FOR DISCUSSION

1. Is the equation

$$\tan^{-1} x = \frac{1}{\tan x}$$

   true for all values of x, true for some values of x, or false for all values of x?

2. Are there real numbers x for which the following is true? Explain.

$$\sin(\sin^{-1} x) \neq \sin^{-1}(\sin x)$$

3. Explain how to find the value of $\sec^{-1} 3$ by using a scientific calculator.

4. Explain how you can determine the range of $y = (2 \cos^{-1} x) - 1$ using

   a. algebra   **b.** a graph

EXERCISE SET 6.5

In Exercises 1 to 20, find the exact radian value.

1. $\sin^{-1} 1$ ▶**2.** $\sin^{-1} \dfrac{\sqrt{2}}{2}$

3. $\cos^{-1}\left(-\dfrac{\sqrt{3}}{2}\right)$ **4.** $\cos^{-1}\left(-\dfrac{1}{2}\right)$ **5.** $\tan^{-1}(-1)$

6. $\tan^{-1} \sqrt{3}$ **7.** $\cot^{-1} \dfrac{\sqrt{3}}{3}$ **8.** $\cot^{-1} 1$

9. $\sec^{-1} 2$ **10.** $\sec^{-1} \dfrac{2\sqrt{3}}{3}$ **11.** $\csc^{-1}(-\sqrt{2})$

12. $\csc^{-1}(-2)$ **13.** $\sin^{-1}\left(-\dfrac{\sqrt{3}}{2}\right)$ **14.** $\sin^{-1} \dfrac{1}{2}$

15. $\cos^{-1}\left(-\dfrac{1}{2}\right)$ **16.** $\cos^{-1} \dfrac{\sqrt{3}}{2}$ **17.** $\tan^{-1} \dfrac{\sqrt{3}}{3}$

18. $\tan^{-1} 1$ **19.** $\cot^{-1} \sqrt{3}$ **20.** $\cot^{-1}(-1)$

In Exercises 21 to 56, find the exact value of the given expression. If an exact value cannot be given, give the value to the nearest ten-thousandth.

21. $\cos\left(\cos^{-1} \dfrac{1}{2}\right)$ **22.** $\cos(\cos^{-1} 2)$

23. $\tan(\tan^{-1} 2)$ ▶**24.** $\tan\left(\tan^{-1} \dfrac{1}{2}\right)$

25. $\sin\left(\tan^{-1} \dfrac{3}{4}\right)$ **26.** $\cos\left(\sin^{-1} \dfrac{5}{13}\right)$

27. $\tan\left(\sin^{-1} \dfrac{\sqrt{2}}{2}\right)$ **28.** $\sin\left[\cos^{-1}\left(-\dfrac{\sqrt{3}}{2}\right)\right]$

29. $\cos(\sec^{-1} 2)$ **30.** $\sin^{-1}(\sin 2)$

31. $\sin^{-1}\left(\sin \dfrac{\pi}{6}\right)$ **32.** $\sin^{-1}\left(\sin \dfrac{5\pi}{6}\right)$

33. $\cos^{-1}\left(\sin \dfrac{\pi}{4}\right)$ **34.** $\cos^{-1}\left(\cos \dfrac{5\pi}{4}\right)$

35. $\sin^{-1}\left(\tan \dfrac{\pi}{3}\right)$ **36.** $\cos^{-1}\left(\tan \dfrac{2\pi}{3}\right)$

37. $\tan^{-1}\left(\sin \dfrac{\pi}{6}\right)$ **38.** $\cot^{-1}\left(\cos \dfrac{2\pi}{3}\right)$

39. $\sin^{-1}\left[\cos\left(-\dfrac{2\pi}{3}\right)\right]$ **40.** $\cos^{-1}\left[\tan\left(-\dfrac{\pi}{3}\right)\right]$

41. $\tan\left(\sin^{-1} \dfrac{1}{2}\right)$ **42.** $\cot(\csc^{-1} 2)$

43. $\sec\left(\sin^{-1} \dfrac{1}{4}\right)$ **44.** $\csc\left(\cos^{-1} \dfrac{3}{4}\right)$

45. $\cos\left(\sin^{-1} \dfrac{7}{25}\right)$ ▶**46.** $\tan\left(\cos^{-1} \dfrac{3}{5}\right)$

47. $\sec\left(\tan^{-1}\dfrac{12}{5}\right)$

48. $\csc\left(\sin^{-1}\dfrac{12}{13}\right)$

49. $\cos\left(2\sin^{-1}\dfrac{\sqrt{2}}{2}\right)$

50. $\tan\left(2\sin^{-1}\dfrac{\sqrt{3}}{2}\right)$

51. $\sin\left(2\sin^{-1}\dfrac{4}{5}\right)$

52. $\cos(2\tan^{-1}1)$

53. $\sin\left(\sin^{-1}\dfrac{2}{3}+\cos^{-1}\dfrac{1}{2}\right)$

▶ **54.** $\cos\left(\sin^{-1}\dfrac{3}{4}+\cos^{-1}\dfrac{5}{13}\right)$

55. $\tan\left(\cos^{-1}\dfrac{1}{2}-\sin^{-1}\dfrac{3}{4}\right)$

56. $\sec\left(\cos^{-1}\dfrac{2}{3}+\sin^{-1}\dfrac{2}{3}\right)$

In Exercises 57 to 66, solve the equation for x algebraically.

57. $\sin^{-1}x=\cos^{-1}\dfrac{5}{13}$

58. $\tan^{-1}x=\sin^{-1}\dfrac{24}{25}$

59. $\sin^{-1}(x-1)=\dfrac{\pi}{2}$

60. $\cos^{-1}\left(x-\dfrac{1}{2}\right)=\dfrac{\pi}{3}$

61. $\tan^{-1}\left(x+\dfrac{\sqrt{2}}{2}\right)=\dfrac{\pi}{4}$

62. $\sin^{-1}(x-2)=-\dfrac{\pi}{6}$

63. $\sin^{-1}\dfrac{3}{5}+\cos^{-1}x=\dfrac{\pi}{4}$

▶ **64.** $\sin^{-1}x+\cos^{-1}\dfrac{4}{5}=\dfrac{\pi}{6}$

65. $\sin^{-1}\dfrac{\sqrt{2}}{2}+\cos^{-1}x=\dfrac{2\pi}{3}$

66. $\cos^{-1}x+\sin^{-1}\dfrac{\sqrt{3}}{2}=\dfrac{\pi}{2}$

In Exercises 67 to 70, evaluate each expression.

67. $\cos(\sin^{-1}x)$

68. $\tan(\cos^{-1}x)$

69. $\sin(\sec^{-1}x)$

70. $\sec(\sin^{-1}x)$

In Exercises 71 to 74, verify the identity.

71. $\sin^{-1}x+\sin^{-1}(-x)=0$

▶ **72.** $\cos^{-1}x+\cos^{-1}(-x)=\pi$

73. $\tan^{-1}x+\tan^{-1}\dfrac{1}{x}=\dfrac{\pi}{2},\; x>0$

74. $\sec^{-1}\dfrac{1}{x}+\csc^{-1}\dfrac{1}{x}=\dfrac{\pi}{2}$

In Exercises 75 to 82, use stretching, shrinking, and translation procedures to graph each equation.

75. $y=\sin^{-1}x+2$

▶ **76.** $y=\cos^{-1}(x-1)$

77. $y=\sin^{-1}(x+1)-2$

78. $y=\tan^{-1}(x-1)+2$

79. $y=2\cos^{-1}x$

80. $y=-2\tan^{-1}x$

81. $y=\tan^{-1}(x+1)-2$

82. $y=\sin^{-1}(x-2)+1$

83. **DOT-MATRIX PRINTING** In dot-matrix printing, the *blank-area factor* is the ratio of the blank area (unprinted area) to the total area of the line. If circular dots are used to print, then the blank-area factor is given by

$$\frac{A}{(S)(D)}=1-\frac{1}{2}\left[1-\left(\frac{S}{D}\right)^2+\frac{D}{S}\sin^{-1}\left(\frac{S}{D}\right)\right]$$

where $A=A_1+A_2$, A_1 and A_2 are the areas of the regions shown in the figure, S is the distance between the centers of overlapping dots, and D is the diameter of a dot.

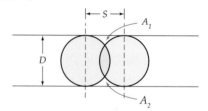

Calculate, to four decimal places, the blank-area factor where

a. $D=0.2$ millimeter and $S=0.1$ millimeter

b. $D=0.16$ millimeter and $S=0.1$ millimeter

▶ **84.** **VOLUME IN A WATER TANK** The volume V of water (measured in cubic feet) in a horizontal cylindrical tank of radius 5 feet and length 12 feet is given by

$$V(x)=12\left[25\cos^{-1}\left(\frac{5-x}{5}\right)-(5-x)\sqrt{10x-x^2}\right]$$

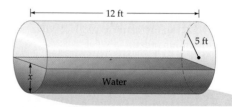

where x is the depth of the water in feet.

a. Graph V over its domain $0 \le x \le 10$.

b. Write a sentence that explains why the graph of V increases more rapidly when x increases from 4.9 feet to 5 feet than it does when x increases from 0.1 foot to 0.2 foot.

c. If $x = 4$ feet, find the volume (to the nearest 0.01 cubic foot) of the water in the tank.

d. Find the depth x (to the nearest 0.01 foot) if there are 288 cubic feet of water in the tank.

85. Graph $f(x) = \cos^{-1} x$ and $g(x) = \sin^{-1} \sqrt{1 - x^2}$ on the same coordinate axes. Does $f(x) = g(x)$ on the interval $[-1, 1]$?

86. Graph $y = \cos(\cos^{-1} x)$ on $[-1, 1]$. Graph $y = \cos^{-1}(\cos x)$ on $[-2\pi, 2\pi]$.

 In Exercises 87 to 94, use a graphing utility to graph each equation.

87. $y = \csc^{-1} 2x$

88. $y = 0.5 \sec^{-1} \dfrac{x}{2}$

89. $y = \sec^{-1}(x - 1)$

90. $y = \sec^{-1}(x + \pi)$

91. $y = 2 \tan^{-1} 2x$

92. $y = \tan^{-1}(x - 1)$

93. $y = \cot^{-1} \dfrac{x}{3}$

94. $y = 2 \cot^{-1}(x - 1)$

CONNECTING CONCEPTS

In Exercises 95 to 98, verify the identity.

95. $\cos(\sin^{-1} x) = \sqrt{1 - x^2}$

96. $\sec(\sin^{-1} x) = \dfrac{\sqrt{1 - x^2}}{1 - x^2}$

97. $\tan(\csc^{-1} x) = \dfrac{\sqrt{x^2 - 1}}{x^2 - 1}, x > 1$

98. $\sin(\cot^{-1} x) = \dfrac{\sqrt{x^2 + 1}}{x^2 + 1}$

In Exercises 99 to 102, solve for y in terms of x.

99. $5x = \tan^{-1} 3y$

100. $2x = \dfrac{1}{2} \sin^{-1} 2y$

101. $x - \dfrac{\pi}{3} = \cos^{-1}(y - 3)$

102. $x + \dfrac{\pi}{2} = \tan^{-1}(2y - 1)$

PREPARE FOR SECTION 6.6

103. Use the quadratic formula to solve $3x^2 - 5x - 4 = 0$. [1.3]

104. Use a Pythagorean identity to write $\sin^2 x$ as a function involving $\cos^2 x$. [5.4]

105. Evaluate $\dfrac{\pi}{2} + 2k\pi$ for $k = 1, 2,$ and 3. [P.1]

106. Factor by grouping: $x^2 - \dfrac{\sqrt{3}}{2} x + x - \dfrac{\sqrt{3}}{2}$. [P.4]

107. Use a graphing utility to construct a scatter plot for the data at the right. Use a viewing window with Xmin=0, Xmax=40, Ymin=0, and Ymax=100. [2.7/4.7]

x	y
3	14
7	55
11	90
15	99
19	80
23	44
27	8
31	4

108. Solve $2x^2 - 2x = 0$ by factoring. [1.3]

PROJECTS

I. VISUAL INSIGHT

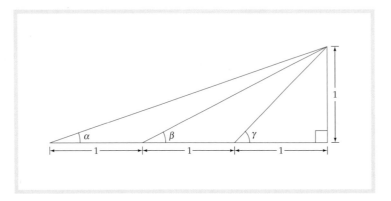

Explain how the figure above can be used to verify each identity.

a. $\tan^{-1}\dfrac{1}{3} + \tan^{-1}\dfrac{1}{2} = \dfrac{\pi}{4}$ (*Hint:* Start by using an identity to find the value of $\tan(\alpha + \beta)$.)   **b.** $\alpha + \beta = \gamma$

SECTION 6.6

TRIGONOMETRIC EQUATIONS

- SOLVE TRIGONOMETRIC EQUATIONS
- AN APPLICATION INVOLVING A TRIGONOMETRIC EQUATION
- MODEL SINUSOIDAL DATA

● SOLVE TRIGONOMETRIC EQUATIONS

Consider the equation $\sin x = \dfrac{1}{2}$. The graph of $y = \sin x$, along with the line $y = \dfrac{1}{2}$, is shown in **Figure 6.27**. The x values of the intersections of the two graphs are the solutions of $\sin x = \dfrac{1}{2}$. The solutions in the interval $0 \leq x < 2\pi$ are $x = \dfrac{\pi}{6}$ and $\dfrac{5\pi}{6}$.

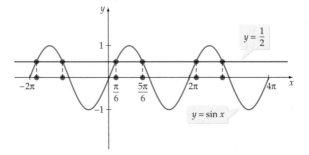

FIGURE 6.27

If we remove the restriction $0 \leq x < 2\pi$, there are many more solutions. Because the sine function is periodic with a period of 2π, other solutions are obtained by adding $2k\pi$, k an integer, to either of the previous solutions. Thus the solutions of $\sin x = \dfrac{1}{2}$ are

$$x = \frac{\pi}{6} + 2k\pi, \quad k \text{ an integer}$$

$$x = \frac{5\pi}{6} + 2k\pi, \quad k \text{ an integer}$$

❓ QUESTION How many solutions does the equation $\cos x = \dfrac{\sqrt{3}}{2}$ have on the interval $0 \leq x < 2\pi$?

Algebraic methods and trigonometric identities are used frequently to find the solutions of trigonometric equations. Algebraic methods that are often employed include solving by factoring, solving by using the quadratic formula, and squaring each side of the equation.

EXAMPLE 1 **Solve a Trigonometric Equation by Factoring**

Solve $2 \sin^2 x \cos x - \cos x = 0$, where $0 \leq x < 2\pi$.

Algebraic Solution

$2 \sin^2 x \cos x - \cos x = 0$
$\cos x (2 \sin^2 x - 1) = 0$ • **Factor cos x from each term.**

$\cos x = 0 \quad$ or $\quad 2 \sin^2 x - 1 = 0$ • **Use the Principle of Zero Products.**

$x = \dfrac{\pi}{2}, \dfrac{3\pi}{2} \qquad \sin^2 x = \dfrac{1}{2}$ • **Solve each equation for x with $0 \leq x < 2\pi$.**

$$\sin x = \pm \frac{\sqrt{2}}{2}$$

$$x = \frac{\pi}{4}, \frac{3\pi}{4}, \frac{5\pi}{4}, \frac{7\pi}{4}$$

The solutions in the interval $0 \leq x < 2\pi$ are $\dfrac{\pi}{4}, \dfrac{\pi}{2}, \dfrac{3\pi}{4}, \dfrac{5\pi}{4}, \dfrac{3\pi}{2}$, and $\dfrac{7\pi}{4}$.

Visualize the Solution

The solutions are the x-coordinates of the x-intercepts of $y = 2 \sin^2 x \cos x - \cos x$ on the interval $[0, 2\pi)$. See **Figure 6.28.**

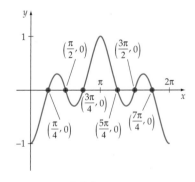

$y = 2 \sin^2 x \cos x - \cos x$

FIGURE 6.28

▶ **TRY EXERCISE 14, PAGE 614**

❓ ANSWER Two

Squaring both sides of an equation may not produce an equivalent equation. Thus, when this method is used, the proposed solutions must be checked to eliminate any extraneous solutions.

EXAMPLE 2   **Solve a Trigonometric Equation by Squaring Each Side of the Equation**

Solve $\sin x + \cos x = 1$, where $0 \le x < 2\pi$.

Algebraic Solution

$\sin x + \cos x = 1$	• **Solve for sin x.**
$\sin x = 1 - \cos x$	
$\sin^2 x = (1 - \cos x)^2$	• **Square each side.**
$\sin^2 x = 1 - 2\cos x + \cos^2 x$	
$1 - \cos^2 x = 1 - 2\cos x + \cos^2 x$	• $\sin^2 x = 1 - \cos^2 x$
$2\cos^2 x - 2\cos x = 0$	
$2\cos x(\cos x - 1) = 0$	• **Factor.**
$2\cos x = 0$ or $\cos x = 1$	
$x = \dfrac{\pi}{2}, \dfrac{3\pi}{2} \qquad x = 0$	• **Solve each equation for x with $0 \le x < 2\pi$.**

Squaring each side of an equation may introduce extraneous solutions. Therefore, we must check the solutions. A check will show that 0 and $\dfrac{\pi}{2}$ are solutions but $\dfrac{3\pi}{2}$ is not a solution.

Visualize the Solution

The solutions are the x-coordinates of the points of intersection of $y = \sin x + \cos x$ and $y = 1$ on the interval $[0, 2\pi)$. See **Figure 6.29**.

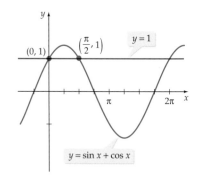

FIGURE 6.29

▶ **TRY EXERCISE 52, PAGE 614**

EXAMPLE 3   **Solve a Trigonometric Equation by Using the Quadratic Formula**

Solve $3\cos^2 x - 5\cos x - 4 = 0$, where $0 \le x < 2\pi$.

Algebraic Solution

The given equation is quadratic in form and cannot be factored easily. However, we can use the quadratic formula to solve for $\cos x$.

$$3\cos^2 x - 5\cos x - 4 = 0 \qquad • \ a = 3, b = -5, c = -4$$

$$\cos x = \frac{-(-5) \pm \sqrt{(-5)^2 - 4(3)(-4)}}{(2)(3)} = \frac{5 \pm \sqrt{73}}{6}$$

The equation $\cos x = \dfrac{5 + \sqrt{73}}{6}$ does not have a solution because $\dfrac{5 + \sqrt{73}}{6} > 2$ and for any x the maximum value of $\cos x$ is 1. Thus

Visualize the Solution

The solutions are the x-coordinates of the x-intercepts of $y = 3\cos^2 x - 5\cos x - 4$ on the interval $[0, 2\pi)$. See **Figure 6.30**.

$\cos x = \dfrac{5 - \sqrt{73}}{6}$, and because $\dfrac{5 - \sqrt{73}}{6}$ is a negative number (about

-0.59), the equation $\cos x = \dfrac{5 - \sqrt{73}}{6}$ will have two solutions on the

interval $[0, 2\pi)$. Thus

$$x = \cos^{-1}\left(\dfrac{5 - \sqrt{73}}{6}\right) \approx 2.2027 \quad \text{or}$$

$$x = 2\pi - \cos^{-1}\left(\dfrac{5 - \sqrt{73}}{6}\right) \approx 4.0805$$

To the nearest 0.0001, the solutions on the interval $[0, 2\pi)$ are 2.2027 and 4.0805.

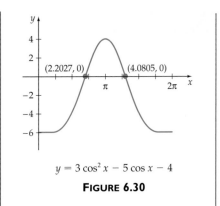

$y = 3\cos^2 x - 5\cos x - 4$

FIGURE 6.30

▶ **TRY EXERCISE 56, PAGE 614**

When solving equations that contain multiple angles, we must be sure we find all the solutions of the equation for the given interval. For example, to find all solutions of $\sin 2x = \dfrac{1}{2}$, where $0 \le x < 2\pi$, we first solve for $2x$.

$$\sin 2x = \dfrac{1}{2}$$

$$2x = \dfrac{\pi}{6} + 2k\pi \quad \text{or} \quad 2x = \dfrac{5\pi}{6} + 2k\pi \qquad \bullet \ k \text{ is an integer.}$$

Solving for x, we have $x = \dfrac{\pi}{12} + k\pi$ or $x = \dfrac{5\pi}{12} + k\pi$. Substituting integers for k, we obtain

$$k = 0: \quad x = \dfrac{\pi}{12} \quad \text{or} \quad x = \dfrac{5\pi}{12}$$

$$k = 1: \quad x = \dfrac{13\pi}{12} \quad \text{or} \quad x = \dfrac{17\pi}{12}$$

$$k = 2: \quad x = \dfrac{25\pi}{12} \quad \text{or} \quad x = \dfrac{29\pi}{12}$$

Note that for $k \ge 2$, $x \ge 2\pi$, and the solutions to $\sin 2x = \dfrac{1}{2}$ are not in the interval $0 \le x < 2\pi$. Thus, for $0 \le x < 2\pi$, the solutions are $\dfrac{\pi}{12}$, $\dfrac{5\pi}{12}$, $\dfrac{13\pi}{12}$, and $\dfrac{17\pi}{12}$.

See **Figure 6.31**.

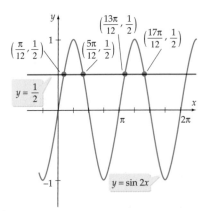

FIGURE 6.31

EXAMPLE 4 **Solve a Trigonometric Equation**

Solve: $\sin 3x = 1$

Algebraic Solution

The equation $\sin 3x = 1$ implies

$$3x = \frac{\pi}{2} + 2k\pi, \quad k \text{ an integer}$$

$$x = \frac{\pi}{6} + \frac{2k\pi}{3}, \quad k \text{ an integer} \qquad \bullet \text{ Divide each side by 3.}$$

Because x is not restricted to a finite interval, the given equation has an infinite number of solutions. All of the solutions are represented by the equation

$$x = \frac{\pi}{6} + \frac{2k\pi}{3}, \quad \text{where } k \text{ is an integer}$$

Visualize the Solution

The solutions are the x-coordinates of the points of intersection of $y = \sin 3x$ and $y = 1$. **Figure 6.32** shows eight of the points of intersection.

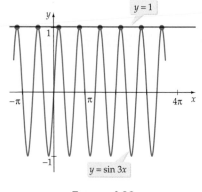

FIGURE 6.32

▶ **TRY EXERCISE 66, PAGE 615**

EXAMPLE 5 **Solve a Trigonometric Equation**

Solve $\sin^2 2x - \dfrac{\sqrt{3}}{2}\sin 2x + \sin 2x - \dfrac{\sqrt{3}}{2} = 0$, where $0° \leq x < 360°$.

Algebraic Solution

Factor the left side of the equation by grouping, and then set each factor equal to zero.

$$\sin^2 2x - \frac{\sqrt{3}}{2}\sin 2x + \sin 2x - \frac{\sqrt{3}}{2} = 0$$

$$\sin 2x\left(\sin 2x - \frac{\sqrt{3}}{2}\right) + \left(\sin 2x - \frac{\sqrt{3}}{2}\right) = 0$$

$$(\sin 2x + 1)\left(\sin 2x - \frac{\sqrt{3}}{2}\right) = 0$$

$$\sin 2x + 1 = 0 \qquad \text{or} \qquad \sin 2x - \frac{\sqrt{3}}{2} = 0$$

$$\sin 2x = -1 \qquad\qquad \sin 2x = \frac{\sqrt{3}}{2}$$

Visualize the Solution

The solutions are the x-coordinates of the x-intercepts of

$$y = \sin^2 2x - \frac{\sqrt{3}}{2}\sin 2x$$
$$+ \sin 2x - \frac{\sqrt{3}}{2}$$

on the interval $[0, 2\pi)$. See **Figure 6.33.**

The equation $\sin 2x = -1$ implies that $2x = 270° + 360° \cdot k$, k an integer. Thus $x = 135° + 180° \cdot k$. The solutions of this equation with $0° \le x < 360°$ are $135°$ and $315°$. Similarly, the equation $\sin 2x = \dfrac{\sqrt{3}}{2}$ implies

$$2x = 60° + 360° \cdot k \qquad \text{or} \qquad 2x = 120° + 360° \cdot k$$
$$x = 30° + 180° \cdot k \qquad\qquad x = 60° + 180° \cdot k$$

The solutions with $0° \le x < 360°$ are $30°, 60°, 210°$, and $240°$. Combining the solutions from each equation, we have $30°, 60°, 135°, 210°, 240°$, and $315°$ as our solutions.

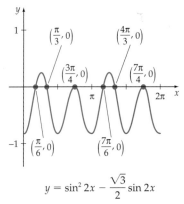

$$y = \sin^2 2x - \frac{\sqrt{3}}{2}\sin 2x$$
$$+ \sin 2x - \frac{\sqrt{3}}{2}$$

FIGURE 6.33

▶ **TRY EXERCISE 84, PAGE 615**

In Example 6, algebraic methods do not provide the solutions, so we rely on a graph.

EXAMPLE 6 Approximate Solutions Graphically

Use a graphing utility to approximate the solutions of $x + 3\cos x = 0$.

Solution

The solutions are the x-intercepts of $y = x + 3\cos x$. See **Figure 6.34.** A close-up view of the graph of $y = x + 3\cos x$ shows that, to the nearest thousandth, the solutions are

$$x_1 = -1.170, \quad x_2 = 2.663, \quad \text{and} \quad x_3 = 2.938$$

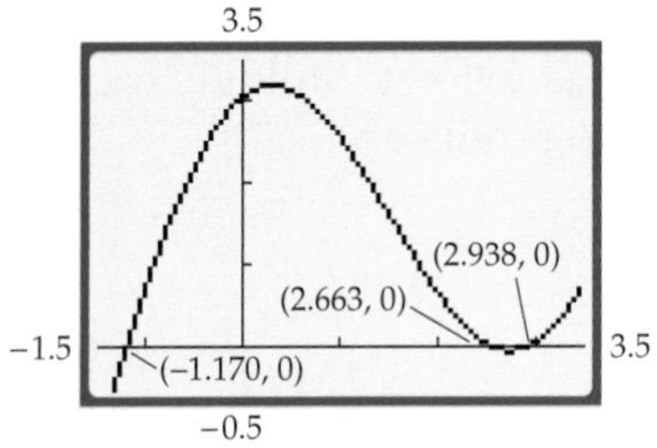

▶ **TRY EXERCISE 86, PAGE 615**

$y = x + 3\cos x$

FIGURE 6.34

● **AN APPLICATION INVOLVING A TRIGONOMETRIC EQUATION**

EXAMPLE 7 Solve a Projectile Application

A projectile is fired at an angle of inclination θ from the horizon with an initial velocity v_0. Its range d (neglecting air resistance) is given by

$$d = \frac{v_0^2}{16}\sin \theta \cos \theta$$

where v_0 is measured in feet per second and d is measured in feet. See **Figure 6.35.**

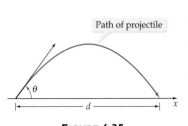

Path of projectile

FIGURE 6.35

Continued ▶

a.   If $v_0 = 325$ feet per second, find the angles θ (in degrees) for which the projectile will hit a target 2295 feet downrange.

b.   What is the maximum horizontal range for a projectile that has an initial velocity of 474 feet per second?

c.   Determine the angle of inclination that produces the maximum range.

Solution

a.   We need to solve

$$2295 = \frac{325^2}{16} \sin \theta \cos \theta \qquad (1)$$

for θ, where $0° < \theta < 90°$.

Method 1   The following solutions were obtained by using a graphing utility to graph $d = 2295$ and $d = \frac{325^2}{16} \sin \theta \cos \theta$. See **Figure 6.36.** Thus there are two angles for which the projectile will hit the target. To the nearest thousandth of a degree, they are

$$\theta = 22.025° \qquad \text{and} \qquad \theta = 67.975°$$

It should be noted that the graph in **Figure 6.36** is *not* a graph of the path of the projectile. It is a graph of the distance d as a function of the angle θ.

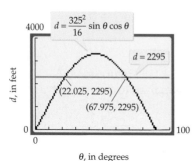

FIGURE 6.36

Method 2   To solve algebraically, we proceed as follows. Multiply each side of Equation (1) by 16 and divide by 325^2 to produce

$$\sin \theta \cos \theta = \frac{(16)(2295)}{325^2}$$

The identity $2 \sin \theta \cos \theta = \sin 2\theta$ gives us $\sin \theta \cos \theta = \dfrac{\sin 2\theta}{2}$.

Hence

$$\frac{\sin 2\theta}{2} = \frac{(16)(2295)}{325^2}$$

$$\sin 2\theta = 2\frac{(16)(2295)}{325^2} \approx 0.69529$$

There are two angles in the interval $[0°, 180°]$ whose sines are 0.69529. One is $\sin^{-1} 0.69529$, and the other one is the *reference angle* for $\sin^{-1} 0.69529$. Therefore,

$$2\theta \approx \sin^{-1} 0.69529 \qquad \text{or} \qquad 2\theta \approx 180° - \sin^{-1} 0.69529$$

$$\theta \approx \frac{1}{2} \sin^{-1} 0.69529 \qquad \text{or} \qquad \theta \approx \frac{1}{2}(180° - \sin^{-1} 0.69529)$$

$$\theta \approx 22.025° \qquad \text{or} \qquad \theta \approx 67.975°$$

These are the same angles that we obtained using Method 1.

3. **Examine the fit.** The SinReg command does not yield a correlation coefficient. However, a graph of the regression equation and the scatter plot of the data shows that the regression equation provides a good model. See **Figure 6.43**.

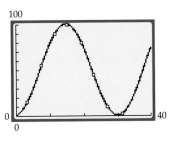

FIGURE 6.43

According to the following result, the percent of the moon illuminated at midnight (CST) on January 21, 2006 ($x = 21$) will be about 62%.

$$y \approx 49.77 \sin(0.2053(21) - 1.393) + 51.18 \approx 62$$

▶ **TRY EXERCISE 94, PAGE 616**

 TOPICS FOR DISCUSSION

1. Explain why it is not necessary for a graphing utility to have a cosine regression.

2. A student finds that $x = 0$ is a solution of $\sin x = x$. Because the function $y = \sin x$ has a period of 2π, the student reasons that $\pm 2\pi$, $\pm 4\pi$, $\pm 6\pi, \ldots$ are also solutions. Explain why the student is not correct.

3. How many solutions does $2 \sin\left(x - \dfrac{\pi}{2}\right) = 5$ have on the interval $0 \le x < 2\pi$? Explain.

4. How many solutions does $\sin\left(\dfrac{1}{x}\right) = 0$ have on the interval $0 < x < \dfrac{\pi}{2}$? Explain.

5. On the interval $0 \le x < 2\pi$, the equation $\sin x = \dfrac{1}{2}$ has solutions of $x = \dfrac{\pi}{6}$ and $x = \dfrac{5\pi}{6}$. How would you write the solutions of $\sin x = \dfrac{1}{2}$ if the real number x were not restricted to the interval $[0, 2\pi)$?

EXERCISE SET 6.6

In Exercises 1 to 22, solve each equation for exact solutions in the interval $0 \leq x < 2\pi$.

1. $\sec x - \sqrt{2} = 0$

2. $2 \sin x = \sqrt{3}$

3. $\tan x - \sqrt{3} = 0$

4. $\cos x - 1 = 0$

5. $2 \sin x \cos x = \sqrt{2} \cos x$

6. $2 \sin x \cos x = \sqrt{3} \sin x$

7. $\sin^2 x - 1 = 0$

8. $\cos^2 x - 1 = 0$

9. $4 \sin x \cos x - 2\sqrt{3} \sin x - 2\sqrt{2} \cos x + \sqrt{6} = 0$

10. $\sec^2 x + \sqrt{3} \sec x - \sqrt{2} \sec x - \sqrt{6} = 0$

11. $\csc x - \sqrt{2} = 0$

12. $3 \cot x + \sqrt{3} = 0$

13. $2 \sin^2 x + 1 = 3 \sin x$

▶ **14.** $2 \cos^2 x + 1 = -3 \cos x$

15. $4 \cos^2 x - 3 = 0$

16. $2 \sin^2 x - 1 = 0$

17. $2 \sin^3 x = \sin x$

18. $4 \cos^3 x = 3 \cos x$

19. $4 \sin^2 x + 2\sqrt{3} \sin x - \sqrt{3} = 2 \sin x$

20. $\tan^2 x + \tan x - \sqrt{3} = \sqrt{3} \tan x$

21. $\sin^4 x = \sin^2 x$

22. $\cos^4 x = \cos^2 x$

In Exercises 23 to 60, solve each equation, where $0° \leq x < 360°$. Round approximate solutions to the nearest tenth of a degree.

23. $\cos x - 0.75 = 0$

24. $\sin x + 0.432 = 0$

25. $3 \sin x - 5 = 0$

26. $4 \cos x - 1 = 0$

27. $3 \sec x - 8 = 0$

28. $4 \csc x + 9 = 0$

29. $\cos x + 3 = 0$

30. $\sin x - 4 = 0$

31. $3 - 5 \sin x = 4 \sin x + 1$

32. $4 \cos x - 5 = \cos x - 3$

33. $\dfrac{1}{2} \sin x + \dfrac{2}{3} = \dfrac{3}{4} \sin x + \dfrac{3}{5}$

34. $\dfrac{2}{5} \cos x - \dfrac{1}{2} = \dfrac{1}{3} - \dfrac{1}{2} \cos x$

35. $3 \tan^2 x - 2 \tan x = 0$

36. $4 \cot^2 x + 3 \cot x = 0$

37. $3 \cos x + \sec x = 0$

38. $5 \sin x - \csc x = 0$

39. $\tan^2 x = 3 \sec^2 x - 2$

40. $\csc^2 x - 1 = 3 \cot^2 x + 2$

41. $2 \sin^2 x = 1 - \cos x$

42. $\cos^2 x + 4 = 2 \sin x - 3$

43. $3 \cos^2 x + 5 \cos x - 2 = 0$

44. $2 \sin^2 x + 5 \sin x + 3 = 0$

45. $2 \tan^2 x - \tan x - 10 = 0$

46. $2 \cot^2 x - 7 \cot x + 3 = 0$

47. $3 \sin x \cos x - \cos x = 0$

48. $\tan x \sin x - \sin x = 0$

49. $2 \sin x \cos x - \sin x - 2 \cos x + 1 = 0$

50. $6 \cos x \sin x - 3 \cos x - 4 \sin x + 2 = 0$

51. $2 \sin x - \cos x = 1$

▶ **52.** $\sin x + 2 \cos x = 1$

53. $2 \sin x - 3 \cos x = 1$

54. $\sqrt{3} \sin x + \cos x = 1$

55. $3 \sin^2 x - \sin x - 1 = 0$

▶ **56.** $2 \cos^2 x - 5 \cos x - 5 = 0$

57. $2 \cos x - 1 + 3 \sec x = 0$

58. $3 \sin x - 5 + \csc x = 0$

59. $\cos^2 x - 3 \sin x + 2 \sin^2 x = 0$

60. $\sin^2 x = 2 \cos x + 3 \cos^2 x$

In Exercises 61 to 70, find the exact solutions, in radians, of each trigonometric equation.

61. $\tan 2x - 1 = 0$

62. $\sec 3x - \dfrac{2\sqrt{3}}{3} = 0$

63. $\sin 5x = 1$

64. $\cos 4x = -\dfrac{\sqrt{2}}{2}$

65. $\sin 2x - \sin x = 0$

▶ **66.** $\cos 2x = -\dfrac{\sqrt{3}}{2}$

67. $\sin\left(2x + \dfrac{\pi}{6}\right) = -\dfrac{1}{2}$

68. $\cos\left(2x - \dfrac{\pi}{4}\right) = -\dfrac{\sqrt{2}}{2}$

69. $\sin^2 \dfrac{x}{2} + \cos x = 1$

70. $\cos^2 \dfrac{x}{2} - \cos x = 1$

In Exercises 71 to 84, find exact solutions, where $0 \le x < 2\pi$.

71. $\cos 2x = 1 - 3\sin x$

72. $\cos 2x = 2\cos x - 1$

73. $\sin 4x - \sin 2x = 0$

74. $\sin 4x - \cos 2x = 0$

75. $\tan \dfrac{x}{2} = \sin x$

76. $\tan \dfrac{x}{2} = 1 - \cos x$

77. $\sin 2x \cos x + \cos 2x \sin x = 0$

78. $\cos 2x \cos x - \sin 2x \sin x = 0$

79. $\sin x \cos 2x - \cos x \sin 2x = \dfrac{\sqrt{3}}{2}$

80. $\cos 2x \cos x + \sin 2x \sin x = -1$

81. $\sin 3x - \sin x = 0$

82. $\cos 3x + \cos x = 0$

83. $2\sin x \cos x + 2\sin x - \cos x - 1 = 0$

▶ **84.** $2\sin x \cos x - 2\sqrt{2}\sin x - \sqrt{3}\cos x + \sqrt{6} = 0$

In Exercises 85 to 88, use a graphing utility to solve the equation. State each solution accurate to the nearest ten-thousandth.

85. $\cos x = x$, where $0 \le x < 2\pi$

▶ **86.** $2\sin x = x$, where $0 \le x < 2\pi$

87. $\sin 2x = \dfrac{1}{x}$, where $-4 \le x \le 4$

88. $\cos x = \dfrac{1}{x}$, where $0 \le x \le 5$

89. Use a graphing utility to solve $\cos x = x^3 - x$ by graphing each side and finding the x-value of all points of intersection. Round to the nearest hundredth.

90. Approximate the largest value of k for which the equation $\sin x \cos x = k$ has a solution.

PROJECTILES **Exercises 91 and 92 make use of the following. A projectile is fired at an angle of inclination θ from the horizon with an initial velocity v_0. Its range d (neglecting air resistance) is given by**

$$d = \dfrac{v_0^2}{16} \sin \theta \cos \theta$$

where v_0 is measured in feet per second and d is measured in feet.

91. If $v_0 = 288$ feet per second, use a graphing utility to find the angles θ (to the nearest hundredth of a degree) for which the projectile will hit a target 1295 feet downrange.

▶ **92.** Use a graphing utility to find the maximum horizontal range, to the nearest tenth of a foot, for a projectile that has an initial velocity of 375 feet per second. What value of θ produces this maximum horizontal range?

93. **SUNRISE TIME** The table below shows the sunrise time for Atlanta, Georgia, for selected days in 2004.

Date	Day of the Year, x	Sunrise Time
Jan. 1	1	7:42
Feb. 1	32	7:35
Mar. 1	61	7:06
April 1	92	6:25
May 1	122	5:48
June 1	153	5:28
July 1	183	5:31
Aug. 1	214	5:50
Sept. 1	245	6:12
Oct. 1	275	6:32
Nov. 1	306	6:57
Dec. 1	336	7:25

Source: The U.S. Naval Observatory. *Note:* The times do not reflect daylight savings time.

a. Use a graphing utility to find the sine regression function that models the sunrise time, in hours, as a function of the day of the year. Let $x = 1$ represent January 1, 2004. Assume that the sunrise times have a period of 365.25 days.

b. Use the regression function to estimate the sunrise time (to the nearest minute) for March 11, 2004 ($x = 71$).

▶ 94.   **SUNSET TIME** The table below shows the sunset time for Sioux City, Iowa, for selected days in 2006.

a. Use a graphing utility to find the sine regression function that models the sunset time, in hours, as a function of the day of the year. Let $x = 1$ represent January 1, 2006.

Date	Day of the Year, x	Sunset Time
Jan. 1	1	17:03
Feb. 1	32	17:39
Mar. 1	60	18:15
April 1	91	18:52
May 1	121	19:26
June 1	152	19:56
July 1	182	20:07
Aug. 1	213	19:45
Sept. 1	244	19:00
Oct. 1	274	18:07
Nov. 1	305	17:19
Dec. 1	335	16:54

Source: The U.S. Naval Observatory. *Note:* The times do not reflect daylight savings time.

Assume that the sunset times have a period of 365.25 days.

b. Use the regression function to estimate the sunset time (to the nearest minute) for May 21, 2006 ($x = 141$).

95. **PERCENT OF THE MOON ILLUMINATED** The table below shows the percent of the moon illuminated at midnight, Central Standard Time, for selected days in October and November of 2005.

Midnight of: Date in 2005	Day Number	% of Moon Illuminated
Oct. 1	1	5
Oct. 5	5	3
Oct. 9	9	33
Oct. 13	13	77
Oct. 17	17	100
Oct. 21	21	84
Oct. 25	25	48
Oct. 29	29	14
Nov. 2	33	0
Nov. 6	37	20
Nov. 10	41	63
Nov. 14	45	96
Nov. 18	49	95
Nov. 22	53	66
Nov. 26	57	29
Nov. 30	61	2

Source: The U.S. Naval Observatory.

a. Use a graphing utility to find the sine regression function that models the percent of the moon illuminated as a function of the day of the year. Let $x = 1$ represent October 1, 2005. Use 29.53 days for the period of the data.

b. Use the regression function to estimate the percent of the moon illuminated (to the nearest 1 percent) at midnight Central Standard Time on October 31, 2005.

96. 🖩 ⚫ **HOURS OF DAYLIGHT** The table below shows the hours of daylight for Houston, Texas, for selected days in 2006.

Date	Day of the Year, x	Hours of Daylight (hours:minutes)
Jan. 1	1	10:17
Feb. 1	32	10:48
Mar. 1	60	11:34
April 1	91	12:29
May 1	121	13:20
June 1	152	13:56
July 1	182	14:01
Aug. 1	213	13:33
Sept. 1	244	12:45
Oct. 1	274	11:52
Nov. 1	305	11:00
Dec. 1	335	10:23

Source: Data extracted from sunrise-sunset times given on the World Wide Web by the U.S. Naval Observatory.

a. Use a graphing utility to find the sine regression function that models the hours of daylight as a function of the day of the year. Let $x = 1$ represent January 1, 2006. Use 365.25 for the period of the data.

b. Use the regression function to estimate the hours of daylight (stated in hours and minutes, with the minutes rounded to the nearest minute) for Houston on May 12, 2006.

97. 🖩 ⚫ **ALTITUDE OF THE SUN** The table at the top of the next column shows the altitude of the sun for Detroit, Michigan, at selected times during October 19, 2007.

a. Use a graphing utility to find the sine regression function that models the altitude, in degrees, of the sun as a function of the time of day. Use 24.03 hours (the time from sunrise October 19 to sunrise October 20) for the period.

b. Use the regression function to estimate the altitude of the sun (to the nearest 0.1 degree) on October 19, 2007, at 9:25.

Time of day	Altitude (degrees)
6:00	−9.9
7:00	1.4
8:00	11.5
9:00	21.0
10:00	29.0
11:00	34.7
12:00	37.5
13:00	36.7
14:00	32.6
15:00	25.7
16:00	17.0
17:00	7.2
18:00	−3.6

Source: The U.S. Naval Observatory.

98. 🖩 ⚫ **RAINFALL TOTALS FOR SAN FRANCISCO** The table below shows some average monthly rainfall totals for San Francisco, California.

Month	Month Number	Average Rainfall (inches)
January	1	4.48
March	3	2.58
May	5	0.35
July	7	0.04
September	9	0.24
November	11	2.49

a. Use a graphing utility to find the sine regression function that models the average monthly rainfall totals as a function of the month number. Use 12 months for the period.

b. Use the regression function to estimate San Francisco's average rainfall total (to the nearest 0.01 inch) for the month of April. How does this result compare with the recorded value of 1.48 inches?

c. ✎ Use the regression function to estimate San Francisco's average rainfall total for the month of June. Explain how you know that this result is incorrect.

In Exercises 99 and 100, use a graphing utility.

99. MODEL THE DAYLIGHT HOURS For a particular day of the year t, the number of daylight hours in Mexico City can be approximated by

$$d(t) = 1.208 \sin\left(\frac{2\pi(t - 80)}{365}\right) + 12.133$$

where t is an integer and $t = 1$ corresponds to January 1. According to d, how many days per year will Mexico City have at least 12 hours of daylight?

100. MODEL THE DAYLIGHT HOURS For a particular day of the year t, the number of daylight hours in New Orleans can be approximated by

$$d(t) = 1.792 \sin\left(\frac{2\pi(t - 80)}{365}\right) + 12.145$$

where t is an integer and $t = 1$ corresponds to January 1. According to d, how many days per year will New Orleans have at least 10.75 hours of daylight?

101. CROSS-SECTIONAL AREA A rain gutter is constructed from a long sheet of aluminum that measures 9 inches in width. The aluminum is to be folded as shown by the cross section in the following diagram.

a. Verify that the area of the cross section is $A = 9 \sin \theta(\cos \theta + 1)$, where $0° < \theta \le 90°$.

b. What values of θ, to the nearest degree, produce a cross-sectional area of 10.5 square inches?

c. Determine the value of θ that produces the cross section with the maximum area.

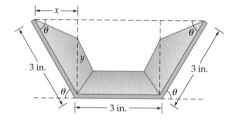

102. OBSERVATION ANGLE A person with an eye level of 5 feet 6 inches is standing in front of a painting, as shown in the following diagram. The bottom of the painting is 3 feet above floor level, and the painting is 6 feet in height. The angle θ shown in the figure is called the *observation angle* for the painting. The person is d feet from the painting.

a. Verify that $\theta = \tan^{-1}\left(\dfrac{6d}{d^2 - 8.75}\right)$.

b. Find the distance d, to the nearest tenth of a foot, for which $\theta = \dfrac{\pi}{6}$.

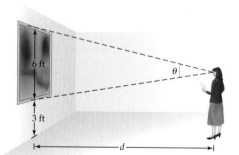

CONNECTING CONCEPTS

In Exercises 103 to 112, solve each equation for exact solutions in the interval $0 \le x < 2\pi$.

103. $\sqrt{3} \sin x + \cos x = \sqrt{3}$

104. $\sin x - \cos x = 1$

105. $-\sin x + \sqrt{3} \cos x = \sqrt{3}$

106. $-\sqrt{3} \sin x - \cos x = 1$

107. $\cos 5x - \cos 3x = 0$

108. $\cos 5x - \cos x - \sin 3x = 0$

109. $\sin 3x + \sin x = 0$

110. $\sin 3x + \sin x - \sin 2x = 0$

111. $\cos 4x + \cos 2x = 0$

112. $\cos 4x + \cos 2x - \cos 3x = 0$

113. MODEL THE MOVEMENT OF A BUS As bus A_1 makes a left turn, the back B of the bus moves to the right. If bus A_2 were waiting at a stoplight while A_1 turned left, as shown in the figure, there is a chance the two buses would scrape against one another. For a bus 28 feet long and 8 feet wide, the movement of the back of the bus to the right can be approximated by

$$x = \sqrt{(4 + 18 \cot \theta)^2 + 100} - (4 + 18 \cot \theta)$$

where θ is the angle the bus driver has turned the front of the bus. Find the value of x for $\theta = 20°$ and $\theta = 30°$. Round to the nearest hundredth of a foot.

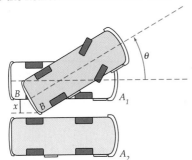

114. OPTIMAL BRANCHING OF BLOOD VESSELS It is hypothesized that the system of blood vessels in primates has evolved so that it has an optimal structure. In the case of a blood vessel splitting into two vessels, as shown in the accompanying figure, we assume that both new branches carry equal amounts of blood. A model of the angle θ is given by the equation $\cos \theta = 2^{(x-4)/(x+4)}$. The value of x is such that $1 \le x \le 2$ and depends on assumptions about the thickness of the blood vessels. Assuming this is an accurate model, find the values of the angle θ.

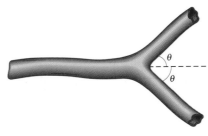

PROJECTS

1. THE MOONS OF SATURN The accompanying figure shows the east-west displacement of five moons of Saturn. The figure shows that the period of the sine curve that models the displacement of the moon Titan is about 15.95 Earth days. The period of the sine curve that models the displacement of the moon Rhea is about 4.52 Earth days.

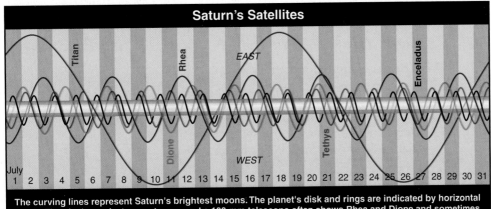

The curving lines represent Saturn's brightest moons. The planet's disk and rings are indicated by horizontal bands. Titan is the easiest moon to see, and a 100-mm telescope often shows Rhea and Dione and sometimes Tethys. Enceladus may need a 300-mm. The wavy-line chart gives only east-west displacements.

Source: "Saturn's Satellites" from *Sky & Telescope,* July 2003, p. 103. Copyright 2003 by Sky Publishing Company. Reprinted with permission.

a. Write an equation of the form

$$d(t) = A \sin (Bt + C)$$

that models the displacement of the moon Titan. Use $t = 0$ to represent the beginning of July 1, 2003, and think of "east" as a positive displacement and of "west" as a negative displacement.

b. Work part **a.** for the moon Rhea.

c. An astronomer viewed the moons Titan and Rhea at 10 P.M. on August 10, 2003. Were these moons on opposite sides of Saturn or on the same side at that time?

EXPLORING CONCEPTS WITH TECHNOLOGY

Approximate an Inverse Trigonometric Function with Polynomials

The function $y = \sin^{-1} x$ can be approximated by polynomials. For example, consider the following:

$$f_1(x) = x + \frac{x^3}{2 \cdot 3} \quad \text{where } -1 \le x \le 1$$

$$f_2(x) = x + \frac{x^3}{2 \cdot 3} + \frac{1 \cdot 3x^5}{2 \cdot 4 \cdot 5} \quad \text{where } -1 \le x \le 1$$

$$f_3(x) = x + \frac{x^3}{2 \cdot 3} + \frac{1 \cdot 3x^5}{2 \cdot 4 \cdot 5} + \frac{1 \cdot 3 \cdot 5x^7}{2 \cdot 4 \cdot 6 \cdot 7} \quad \text{where } -1 \le x \le 1$$

$$f_4(x) = x + \frac{x^3}{2 \cdot 3} + \frac{1 \cdot 3x^5}{2 \cdot 4 \cdot 5} + \frac{1 \cdot 3 \cdot 5x^7}{2 \cdot 4 \cdot 6 \cdot 7} + \frac{1 \cdot 3 \cdot 5 \cdot 7x^9}{2 \cdot 4 \cdot 6 \cdot 8 \cdot 9} \quad \text{where } -1 \le x \le 1$$

$$\vdots$$

$$f_n(x) = x + \frac{x^3}{2 \cdot 3} + \frac{1 \cdot 3x^5}{2 \cdot 4 \cdot 5} + \frac{1 \cdot 3 \cdot 5x^7}{2 \cdot 4 \cdot 6 \cdot 7} + \cdots + \frac{(2n)! \, x^{2n+1}}{(2^n n!)^2 (2n + 1)}$$

where $-1 \le x \le 1$, $n! = 1 \cdot 2 \cdot 3 \cdots (n - 1)n$

and $(2n)! = 1 \cdot 2 \cdot 3 \cdots (2n - 1)(2n)$

Use a graphing utility for the following exercises.

1. Graph $y = f_1(x)$, $y = f_2(x)$, $y = f_3(x)$, and $y = f_4(x)$ on the viewing window Xmin = −1, Xmax = 1, Ymin = −1.5708, Ymax = 1.5708.

2. Determine the values of x for which $f_3(x)$ and $\sin^{-1} x$ differ by less than 0.001. That is, determine the values of x for which

$$\left| f_3(x) - \sin^{-1} x \right| < 0.001$$

3. Determine the values of x for which

$$\left| f_4(x) - \sin^{-1} x \right| < 0.001$$

4. Write all seven terms of $f_6(x)$. Graph $y = f_6(x)$ and $y = \sin^{-1} x$ on the viewing window Xmin = −1, Xmax = 1, Ymin = $-\dfrac{\pi}{2}$, Ymax = $\dfrac{\pi}{2}$.

5. Write all seven terms of $f_6(1)$. What do you notice about the size of a term compared to that of the previous term?

6. What is the largest-degree term in $f_{10}(x)$?

CHAPTER 6 SUMMARY

6.1 Verification of Trigonometric Identities

- Trigonometric identities are verified by using algebraic methods and previously proved identities. Here are a few fundamental trigonometric identities.

$$\sin x = \frac{1}{\csc x} \qquad \cos x = \frac{1}{\sec x} \qquad \tan x = \frac{1}{\cot x}$$

$$\tan x = \frac{\sin x}{\cos x} \qquad \cot x = \frac{\cos x}{\sin x}$$

$$\sin^2 x + \cos^2 x = 1; \tan^2 x + 1 = \sec^2 x;$$
$$1 + \cot^2 x = \csc^2 x$$

6.2 Sum, Difference, and Cofunction Identities

- Sum and difference identities for the cosine function are

$$\cos(\alpha - \beta) = \cos \alpha \cos \beta + \sin \alpha \sin \beta$$
$$\cos(\alpha + \beta) = \cos \alpha \cos \beta - \sin \alpha \sin \beta$$

- Sum and difference identities for the sine function are

$$\sin(\alpha - \beta) = \sin \alpha \cos \beta - \cos \alpha \sin \beta$$
$$\sin(\alpha + \beta) = \sin \alpha \cos \beta + \cos \alpha \sin \beta$$

- Sum and difference identities for the tangent function are

$$\tan(\alpha + \beta) = \frac{\tan \alpha + \tan \beta}{1 - \tan \alpha \tan \beta}$$

$$\tan(\alpha - \beta) = \frac{\tan \alpha - \tan \beta}{1 + \tan \alpha \tan \beta}$$

- The cofunction identities are

$$\sin(90° - \theta) = \cos \theta \qquad \cos(90° - \theta) = \sin \theta$$
$$\tan(90° - \theta) = \cot \theta \qquad \cot(90° - \theta) = \tan \theta$$
$$\sec(90° - \theta) = \csc \theta \qquad \csc(90° - \theta) = \sec \theta$$

where θ is in degrees. If θ is in radian measure, replace $90°$ with $\frac{\pi}{2}$.

6.3 Double- and Half-Angle Identities

- The double-angle identities are

$$\sin 2\alpha = 2 \sin \alpha \cos \alpha$$
$$\cos 2\alpha = \cos^2 \alpha - \sin^2 \alpha$$
$$= 1 - 2 \sin^2 \alpha$$
$$= 2 \cos^2 \alpha - 1$$
$$\tan 2\alpha = \frac{2 \tan \alpha}{1 - \tan^2 \alpha}$$

- The half-angle identities are

$$\sin \frac{\alpha}{2} = \pm \sqrt{\frac{1 - \cos \alpha}{2}}$$

$$\cos \frac{\alpha}{2} = \pm \sqrt{\frac{1 + \cos \alpha}{2}}$$

$$\tan \frac{\alpha}{2} = \frac{\sin \alpha}{1 + \cos \alpha} = \frac{1 - \cos \alpha}{\sin \alpha}$$

6.4 Identities Involving the Sum of Trigonometric Functions

- The product-to-sum identities are

$$\sin \alpha \cos \beta = \frac{1}{2}[\sin(\alpha + \beta) + \sin(\alpha - \beta)]$$

$$\cos \alpha \sin \beta = \frac{1}{2}[\sin(\alpha + \beta) - \sin(\alpha - \beta)]$$

$$\cos \alpha \cos \beta = \frac{1}{2}[\cos(\alpha + \beta) + \cos(\alpha - \beta)]$$

$$\sin \alpha \sin \beta = \frac{1}{2}[\cos(\alpha - \beta) - \cos(\alpha + \beta)]$$

- The sum-to-product identities are

$$\sin x + \sin y = 2 \sin \frac{x + y}{2} \cos \frac{x - y}{2}$$

$$\cos x - \cos y = -2 \sin \frac{x + y}{2} \sin \frac{x - y}{2}$$

$$\sin x - \sin y = 2 \cos \frac{x + y}{2} \sin \frac{x - y}{2}$$

$$\cos x + \cos y = 2 \cos \frac{x + y}{2} \cos \frac{x - y}{2}$$

- For sums of the form $a \sin x + b \cos x$,

$$a \sin x + b \cos x = k \sin(x + \alpha)$$

where $k = \sqrt{a^2 + b^2}$, $\sin \alpha = \frac{b}{\sqrt{a^2 + b^2}}$, and

$$\cos \alpha = \frac{a}{\sqrt{a^2 + b^2}}.$$

6.5 Inverse Trigonometric Functions

- The inverse of $y = \sin x$ is $y = \sin^{-1} x$, with $-1 \le x \le 1$ and $-\frac{\pi}{2} \le y \le \frac{\pi}{2}$.

- The inverse of $y = \cos x$ is $y = \cos^{-1} x$, with $-1 \le x \le 1$ and $0 \le y \le \pi$.

- The inverse of $y = \tan x$ is $y = \tan^{-1} x$, with $-\infty < x < \infty$ and $-\dfrac{\pi}{2} < y < \dfrac{\pi}{2}$.

- The inverse of $y = \cot x$ is $y = \cot^{-1} x$, with $-\infty < x < \infty$ and $0 < y < \pi$.

- The inverse of $y = \csc x$ is $y = \csc^{-1} x$, with $x \leq -1$ or $x \geq 1$ and $-\dfrac{\pi}{2} \leq y \leq \dfrac{\pi}{2}$, $y \neq 0$.

- The inverse of $y = \sec x$ is $y = \sec^{-1} x$, with $x \leq -1$ or $x \geq 1$ and $0 \leq y \leq \pi$, $y \neq \dfrac{\pi}{2}$.

6.6 Trigonometric Equations

- Algebraic methods and identities are used to solve trigonometric equations. Because the trigonometric functions are periodic, there may be an infinite number of solutions. If solutions cannot be found by algebraic methods, then we often use a graphing utility to find approximate solutions.

CHAPTER 6 TRUE/FALSE EXERCISES

In Exercises 1 to 12, answer true or false. If the statement is false, give a reason or an example to show that the statement is false.

1. $\dfrac{\tan \alpha}{\tan \beta} = \dfrac{\alpha}{\beta}$

2. $\dfrac{\sin x}{\cos y} = \tan \dfrac{x}{y}$

3. $\sin^{-1} x = \csc x^{-1}$

4. $\sin 2\alpha = 2 \sin \alpha$ for all α

5. $\sin(\alpha + \beta) = \sin \alpha + \sin \beta$

6. An equation that has an infinite number of solutions is an identity.

7. If $\tan \alpha = \tan \beta$, then $\alpha = \beta$.

8. $\cos^{-1}(\cos x) = x$

9. $\cos(\cos^{-1} x) = x$

10. $\csc^{-1} \dfrac{1}{\alpha} = \dfrac{1}{\csc \alpha}$

11. If $0° \leq \theta \leq 90°$, then $\cos \theta = \sin(180° - \theta)$.

12. $\sin^2 \theta = \sin \theta^2$

CHAPTER 6 REVIEW EXERCISES

In Exercises 1 to 10, find the exact value.

1. $\cos(45° + 30°)$

2. $\tan(210° - 45°)$

3. $\sin\left(\dfrac{2\pi}{3} + \dfrac{\pi}{4}\right)$

4. $\sec\left(\dfrac{4\pi}{3} - \dfrac{\pi}{4}\right)$

5. $\sin(60° - 135°)$

6. $\cos\left(\dfrac{5\pi}{3} - \dfrac{7\pi}{4}\right)$

7. $\sin\left(22\dfrac{1}{2}\right)°$

8. $\cos 105°$

9. $\tan\left(67\dfrac{1}{2}\right)°$

10. $\sin 112.5°$

In Exercises 11 to 14, find the exact values of the given functions.

11. Given $\sin \alpha = \dfrac{1}{2}$, α in Quadrant I, and $\cos \beta = \dfrac{1}{2}$, β in Quadrant IV, find

  a. $\cos(\alpha - \beta)$ b. $\tan 2\alpha$ c. $\sin\left(\dfrac{\beta}{2}\right)$

12. Given $\sin \alpha = \dfrac{\sqrt{3}}{2}$, α in Quadrant II, and $\cos \beta = -\dfrac{1}{2}$, β in Quadrant III, find

    a. $\sin(\alpha + \beta)$ **b.** $\sec 2\beta$ **c.** $\cos\left(\dfrac{\alpha}{2}\right)$

13. Given $\sin \alpha = -\dfrac{1}{2}$, α in Quadrant IV, and $\cos \beta = -\dfrac{\sqrt{3}}{2}$, β in Quadrant III, find

    a. $\sin(\alpha - \beta)$ **b.** $\tan 2\alpha$ **c.** $\cos\left(\dfrac{\beta}{2}\right)$

14. Given $\sin \alpha = \dfrac{\sqrt{2}}{2}$, α in Quadrant I, and $\cos \beta = \dfrac{\sqrt{3}}{2}$, β in Quadrant IV, find

    a. $\cos(\alpha - \beta)$ **b.** $\tan 2\beta$ **c.** $\sin 2\alpha$

In Exercises 15 to 20, write the given expression as a single trigonometric function.

15. $2 \sin 3x \cos 3x$

16. $\dfrac{\tan 2x + \tan x}{1 - \tan 2x \tan x}$

17. $\sin 4x \cos x - \cos 4x \sin x$

18. $\cos^2 2\theta - \sin^2 2\theta$

19. $\dfrac{\sin 2\theta}{\cos 2\theta}$

20. $\dfrac{1 - \cos 2\theta}{\sin 2\theta}$

In Exercises 21 to 24, write each expression as the product of two functions.

21. $\cos 2\theta - \cos 4\theta$

22. $\sin 3\theta - \sin 5\theta$

23. $\sin 6\theta + \sin 2\theta$

24. $\sin 5\theta - \sin \theta$

In Exercises 25 to 42, verify the identity.

25. $\dfrac{1}{\sin x - 1} + \dfrac{1}{\sin x + 1} = -2 \tan x \sec x$

26. $\dfrac{\sin x}{1 - \cos x} = \csc x + \cot x, \quad 0 < x < \dfrac{\pi}{2}$

27. $\dfrac{1 + \sin x}{\cos^2 x} = \tan^2 x + 1 + \tan x \sec x$

28. $\cos^2 x - \sin^2 x - \sin 2x = \dfrac{\cos^2 2x - \sin^2 2x}{\cos 2x + \sin 2x}$

29. $\dfrac{1}{\cos x} - \cos x = \tan x \sin x$

30. $\sin(270° - \theta) - \cos(270° - \theta) = \sin \theta - \cos \theta$

31. $\sin\left(\dfrac{\pi}{4} - \alpha\right) = \dfrac{\sqrt{2}}{2}(\cos \alpha - \sin \alpha)$

32. $\sin(180° - \alpha + \beta) = \sin \alpha \cos \beta - \cos \alpha \sin \beta$

33. $\dfrac{\sin 4x - \sin 2x}{\cos 4x - \cos 2x} = -\cot 3x$

34. $2 \sin x \sin 3x = (1 - \cos 2x)(1 + 2 \cos 2x)$

35. $\sin x - \cos 2x = (2 \sin x - 1)(\sin x + 1)$

36. $\cos 4x = 1 - 8 \sin^2 x + 8 \sin^4 x$

37. $\tan 4x = \dfrac{4 \tan x - 4 \tan^3 x}{1 - 6 \tan^2 x + \tan^4 x}$

38. $\dfrac{\sin 2x - \sin x}{\cos 2x + \cos x} = \dfrac{1 - \cos x}{\sin x}$

39. $2 \cos 4x \sin 2x = 2 \sin 3x \cos 3x - 2 \sin x \cos x$

40. $2 \sin x \sin 2x = 4 \cos x \sin^2 x$

41. $\cos(x + y) \cos(x - y) = \cos^2 x + \cos^2 y - 1$

42. $\cos(x + y) \sin(x - y) = \sin x \cos x - \sin y \cos y$

In Exercises 43 to 46, evaluate each expression.

43. $\sec\left(\sin^{-1} \dfrac{12}{13}\right)$

44. $\cos\left(\sin^{-1} \dfrac{3}{5}\right)$

45. $\cos\left[\sin^{-1}\left(-\dfrac{3}{5}\right) + \cos^{-1} \dfrac{5}{13}\right]$

46. $\cos\left(2 \sin^{-1} \dfrac{3}{5}\right)$

In Exercises 47 and 48, solve each equation.

47. $2 \sin^{-1}(x - 1) = \dfrac{\pi}{3}$

48. $\sin^{-1} x + \cos^{-1} \dfrac{4}{5} = \dfrac{\pi}{2}$

In Exercises 49 and 50, solve each equation on $0° \leq x < 360°$.

49. $4 \sin^2 x + 2\sqrt{3} \sin x - 2 \sin x - \sqrt{3} = 0$

50. $2 \sin x \cos x - \sqrt{2} \cos x - 2 \sin x + \sqrt{2} = 0$

In Exercises 51 and 52, solve the trigonometric equation where x is in radians. Round approximate solutions to four decimal places.

51. $3 \cos^2 x + \sin x = 1$

52. $\tan^2 x - 2 \tan x - 3 = 0$

In Exercises 53 and 54, solve each equation on $0 \leq x < 2\pi$.

53. $\sin 3x \cos x - \cos 3x \sin x = \dfrac{1}{2}$

54. $\cos\left(2x - \dfrac{\pi}{3}\right) = -\dfrac{\sqrt{3}}{2}$

In Exercises 55 to 58, write the equation in the form $y = k \sin(x + \alpha)$, where the measure of α is in radians. Graph one period of each function.

55. $f(x) = \sqrt{3} \sin x + \cos x$

56. $f(x) = -2 \sin x - 2 \cos x$

57. $f(x) = -\sin x - \sqrt{3} \cos x$

58. $f(x) = \dfrac{\sqrt{3}}{2} \sin x - \dfrac{1}{2} \cos x$

In Exercises 59 to 62, graph each function.

59. $f(x) = 2 \cos^{-1} x$ **60.** $f(x) = \sin^{-1}(x - 1)$

61. $f(x) = \sin^{-1} \dfrac{x}{2}$ **62.** $f(x) = \sec^{-1} 2x$

63. **SUNRISE TIME** The table below shows the sunrise time for Madison, Wisconsin, for selected days in 2006.

Date	Day of the Year, x	Sunrise Time (hours:minutes)
Jan. 1	1	7:29
Feb. 1	32	7:13
Mar. 1	60	6:34
April 1	91	5:40
May 1	121	4:51
June 1	152	4:21
July 1	182	4:22
Aug. 1	213	4:48
Sept. 1	244	5:22
Oct. 1	274	5:55
Nov. 1	305	6:32
Dec. 1	335	7:09

Source: The U.S. Naval Observatory. *Note:* The times are Central Standard Times. The times do not reflect daylight savings time.

a. Use a graphing utility to find the sine regression function that models the sunrise time, in hours, as a function of the day of the year. Let $x = 1$ represent January 1, 2006. Assume that the sunrise times have a period of 365.25 days.

b. Use the regression function to estimate the sunrise time (to the nearest minute) for April 14, 2006 ($x = 104$).

CHAPTER 6 TEST

1. Verify the identity $1 + \sin^2 x \sec^2 x = \sec^2 x$.

2. Verify the identity

$$\frac{1}{\sec x - \tan x} - \frac{1}{\sec x + \tan x} = 2 \tan x$$

3. Verify the identity $\cos^3 x + \cos x \sin^2 x = \cos x$.

4. Verify the identity $\csc x - \cot x = \dfrac{1 - \cos x}{\sin x}$.

5. Find the exact value of $\sin 195°$.

6. Given $\sin \alpha = -\dfrac{3}{5}$, α in Quadrant III, and $\cos \beta = -\dfrac{\sqrt{2}}{2}$, β in Quadrant II, find $\sin(\alpha + \beta)$.

7. Verify the identity $\sin\left(\theta - \dfrac{3\pi}{2}\right) = \cos \theta$.

8. Write $\cos 6x \sin 3x + \sin 6x \cos 3x$ in terms of a single trigonometric function.

9. Find the exact value of $\cos 2\theta$ given that $\sin \theta = \dfrac{4}{5}$ and θ is in Quadrant II.

10. Verify the identity $\tan \dfrac{\theta}{2} + \dfrac{\cos \theta}{\sin \theta} = \csc \theta$.

11. Verify the identity $\sin^2 2x + 4 \cos^4 x = 4 \cos^2 x$.

12. Find the exact value of $\sin 15° \cos 75°$.

13. Write $y = -\dfrac{\sqrt{3}}{2} \sin x + \dfrac{1}{2} \cos x$ in the form $y = k \sin(x + \alpha)$, where α is measured in radians.

14. Use a calculator to approximate the radian measure of $\cos^{-1} 0.7644$ to the nearest thousandth.

15. Find the exact value of $\sin\left(\cos^{-1} \dfrac{12}{13}\right)$.

16. Graph: $y = \sin^{-1}(x + 2)$

17. Solve $3 \sin x - 2 = 0$, where $0° \le x < 360°$. (State solutions to the nearest $0.1°$.)

18. Solve $\sin x \cos x - \dfrac{\sqrt{3}}{2} \sin x = 0$, where $0 \le x < 2\pi$.

19. Find the exact solutions of $\sin 2x + \sin x - 2 \cos x - 1 = 0$, where $0 \le x < 2\pi$.

20. **ALTITUDE OF THE SUN** The table shows the altitude of the sun for Fort Lauderdale, Florida, at selected times during April 20, 2005.

Time of day	Altitude (degrees)
6:00	1.2
7:00	14.1
8:00	27.5
9:00	41.0
10:00	54.1
11:00	66.4
12:00	74.9
13:00	72.7
14:00	62.3
15:00	49.6
16:00	36.4
17:00	22.9
18:00	9.6
19:00	−3.7

Source: The U.S. Naval Observatory.

a. Use a graphing utility to find the sine regression function that models the altitude of the sun as a function of the time of day. Use 23.983 hours (the time from sunrise April 20 to sunrise April 21) for the period.

b. Use the regression function to estimate the altitude of the sun (to the nearest 0.1 degree) on April 20, 2005, at 10:40.

CUMULATIVE REVIEW EXERCISES

1. Factor: $x^3 - y^3$

2. Solve: $|x - 5| = 3$

3. Explain how to use the graph of $y = f(x)$ to produce the graph of $y = f(x + 1) + 2$.

4. Explain how to use the graph of $y = f(x)$ to produce the graph of $y = -f(x)$.

5. Find the vertical asymptote for the graph of $f(x) = \dfrac{x + 3}{x - 2}$.

6. Determine whether $f(x) = x - \sin x$ is an even function or an odd function.

7. Find the inverse of $f(x) = \dfrac{5x}{x - 1}$.

8. Write $x = 2^5$ in logarithmic form.

9. Evaluate: $\log_{10} 1000$

10. Convert 240° to radians.

11. Convert $\dfrac{5\pi}{3}$ to degrees.

12. Find $\tan \theta$, given θ is an acute angle and $\sin \theta = \dfrac{2}{3}$.

13. Determine the sign of $\cot \theta$ given that $\pi < \theta < \dfrac{3\pi}{2}$.

14. What is the measure of the reference angle for the angle $\theta = 310°$?

15. What is the measure of the reference angle for the angle $\theta = \dfrac{5\pi}{3}$?

16. Find the x- and y-coordinates of the point defined by $W\left(\dfrac{\pi}{3}\right)$.

17. Find the amplitude, the period, and the phase shift for the graph of $y = 0.43 \sin\left(2x - \dfrac{\pi}{6}\right)$.

18. Evaluate: $\sin^{-1} \dfrac{1}{2}$

19. Use interval notation to state the domain of $f(x) = \cos^{-1} x$.

20. Use interval notation to state the range of $f(x) = \tan^{-1} x$.

APPLICATIONS OF TRIGONOMETRY

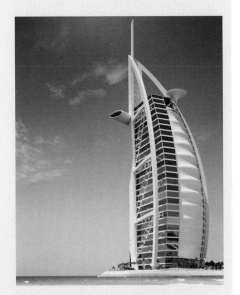

The Burji Al Arab hotel in Dubai, United Arab Emirates, is the world's tallest hotel. It is 321 meters in height and was designed in the shape of a billowing sail.

Trigonometry and Indirect Measurement

In Chapter 5 we used trigonometric functions to find the unknown length of a side of a given *right triangle.* In this chapter we present theorems that can be used to find the length of a side or the measure of an angle of a given triangle even if it is not a right triangle. The theorems in this chapter are used often in the areas of navigation, surveying, and the design of structures. Architects often use trigonometry to find the unknown distance between two points. For instance, in the following diagram, the length a of the steel brace from C to B can be determined using known values and the Law of Sines, which is a major theorem presented in this chapter.

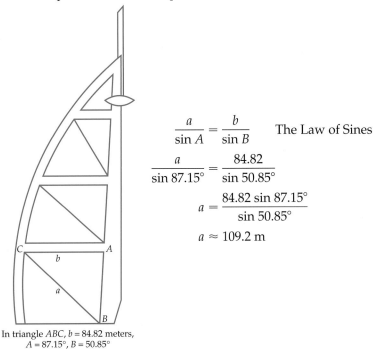

$$\frac{a}{\sin A} = \frac{b}{\sin B} \qquad \text{The Law of Sines}$$

$$\frac{a}{\sin 87.15^\circ} = \frac{84.82}{\sin 50.85^\circ}$$

$$a = \frac{84.82 \sin 87.15^\circ}{\sin 50.85^\circ}$$

$$a \approx 109.2 \text{ m}$$

In triangle ABC, $b = 84.82$ meters, $A = 87.15^\circ$, $B = 50.85^\circ$

VIDEO & DVD

SSG

See **Exercises 25 and 26, pages 634 and 635,** for additional application exercises that can be solved by applying the Law of Sines.

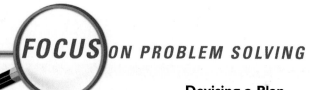

Devising a Plan

One of the most important aspects of problem solving is the process of devising a plan. In some applications there may be more than one plan (method) that can be used to obtain the solution. For instance, consider the following classic problem.

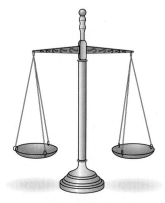

> You have eight coins. They all look identical, but one is a fake and is slightly lighter than the others. Explain how you can use a balance scale to determine which coin is the fake in exactly
> **a.** three weighings.
> **b.** two weighings.

In Chapter 7 you will often need to solve a triangle, which means to determine all unknown measures of the triangle. In most cases you will solve a triangle by using one of two theorems, which are known as the Law of Sines and the Law of Cosines. The guidelines on page 640 provide the information needed to choose between these theorems.

THE LAW OF SINES

○ THE LAW OF SINES

Solving a triangle involves finding the lengths of all sides and the measures of all angles in the triangle. In this section and the next we develop formulas for solving an **oblique triangle,** which is a triangle that does not contain a right angle. The *Law of Sines* can be used to solve oblique triangles in which either two angles and a side (AAS) or two sides and an angle opposite one of the sides (SSA) are known. In **Figure 7.1,** altitude CD is drawn from C. The length of the altitude is h. Triangles ACD and BCD are right triangles.

Using the definition of the sine of an angle of a right triangle, we have from **Figure 7.1**

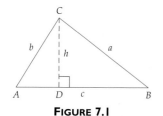

FIGURE 7.1

$$\sin B = \frac{h}{a} \qquad\qquad \sin A = \frac{h}{b}$$

$$h = a \sin B \quad (1) \qquad\qquad h = b \sin A \quad (2)$$

Equating the values of h in Equations (1) and (2), we obtain

$$a \sin B = b \sin A$$

Dividing each side of the equation by $\sin A \sin B$, we obtain

$$\frac{a}{\sin A} = \frac{b}{\sin B}$$

Similarly, when an altitude is drawn to a different side, the following formulas result:

$$\frac{c}{\sin C} = \frac{b}{\sin B} \quad \text{and} \quad \frac{c}{\sin C} = \frac{a}{\sin A}$$

take note

The Law of Sines may also be written as

$$\frac{\sin A}{a} = \frac{\sin B}{b} = \frac{\sin C}{c}$$

The Law of Sines

If A, B, and C are the measures of the angles of a triangle and a, b, and c are the lengths of the sides opposite these angles, then

$$\frac{a}{\sin A} = \frac{b}{\sin B} = \frac{c}{\sin C}$$

EXAMPLE 1 Solve a Triangle Using the Law of Sines (AAS)

Solve triangle ABC if $A = 42°$, $B = 63°$, and $c = 18$ centimeters.

Continued ▶

Solution

Find C by using the fact that the sum of the interior angles of a triangle is $180°$.

$$A + B + C = 180°$$
$$42° + 63° + C = 180°$$
$$C = 75°$$

take note

We have used the rounding conventions stated on page 482 to determine the number of significant digits to be used for a and b.

Use the Law of Sines to find a.

$$\frac{a}{\sin A} = \frac{c}{\sin C}$$

$$\frac{a}{\sin 42°} = \frac{18}{\sin 75°} \qquad \bullet \, A = 42°, c = 18, C = 75°$$

$$a = \frac{18 \sin 42°}{\sin 75°} \approx 12 \text{ centimeters}$$

Use the Law of Sines again, this time to find b.

$$\frac{b}{\sin B} = \frac{c}{\sin C}$$

$$\frac{b}{\sin 63°} = \frac{18}{\sin 75°} \qquad \bullet \, B = 63°, c = 18, C = 75°$$

$$b = \frac{18 \sin 63°}{\sin 75°} \approx 17 \text{ centimeters}$$

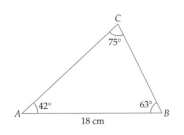

FIGURE 7.2

The solution is $C = 75°$, $a \approx 12$ centimeters, and $b \approx 17$ centimeters. A scale drawing can be used to see if these results are reasonable. See **Figure 7.2**.

▶ **TRY EXERCISE 4, PAGE 634**

● THE AMBIGUOUS CASE (SSA)

When you are given two sides of a triangle and an angle opposite one of them, you may find that the triangle is not unique. Some information may result in two triangles, and some may result in no triangle at all. It is because of this that the case of knowing two sides and an angle opposite one of them (SSA) is called the *ambiguous case* of the Law of Sines.

Suppose that sides a and c and the nonincluded angle A of a triangle are known and we are then asked to solve triangle ABC. The relationships among h, the height of the triangle, a (the side opposite $\angle A$), and c determine whether there are no, one, or two triangles.

Case I First consider the case in which $\angle A$ is an acute angle (see **Figure 7.3**). There are four possible situations.

1. $a < h$; there is no possible triangle.

2. $a = h$; there is one triangle, a right triangle.

3. $h < a < c$; there are two possible triangles. One has all acute angles, and the second has one obtuse angle.

4. $a \geq c$; there is one triangle, which is not a right triangle.

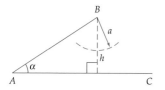

1. $a < h$; no triangle

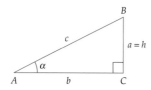

2. $a = h$; one triangle

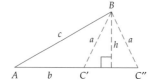

3. $h < a < c$; two triangles

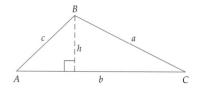

4. $a \geq c$; one triangle

FIGURE 7.3

Case 1: A is an acute angle.

Case 2 Now consider the case in which $\angle A$ is an obtuse angle (see **Figure 7.4**). Here, there are two possible situations.

1. $a \leq c$; there is no triangle.

2. $a > c$; there is one triangle.

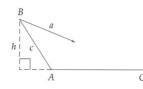

1. $a \leq c$; no triangle

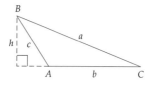

2. $a > c$; one triangle

FIGURE 7.4

Case 2: A is an obtuse angle.

EXAMPLE 2 **Solve a Triangle Using the Law of Sines (SSA)**

a. Find A, given triangle ABC with $B = 32°$, $a = 42$, and $b = 30$.

b. Find C, given triangle ABC with $A = 57°$, $a = 15$ feet, and $c = 20$ feet.

Solution

a.
$$\frac{b}{\sin B} = \frac{a}{\sin A}$$

$$\frac{30}{\sin 32°} = \frac{42}{\sin A}$$ • $B = 32°, a = 42, b = 30$

$$\sin A = \frac{42 \sin 32°}{30} \approx 0.7419$$

$$A \approx 48° \text{ or } 132°$$

• **The two angles with measure between 0° and 180° that have a sine of 0.7419 are approximately 48° and 132°.**

To check that $A \approx 132°$ is a valid result, add $132°$ to the measure of the given angle B ($32°$). Because $132° + 32° < 180°$, we know that $A \approx 132°$ is a valid result. Thus angle $A \approx 48°$ or $A \approx 132°$ ($\angle BAC$ in **Figure 7.5**).

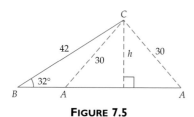

FIGURE 7.5

Continued ▶

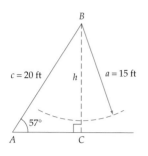

FIGURE 7.6

b.
$$\frac{a}{\sin A} = \frac{c}{\sin C}$$

$$\frac{15}{\sin 57^\circ} = \frac{20}{\sin C} \qquad \bullet \; A = 57^\circ, a = 15, c = 20$$

$$\sin C = \frac{20 \sin 57^\circ}{15} \approx 1.1182$$

Because 1.1182 is not in the range of the sine function, there is no solution of the equation. Thus there is no triangle for these values of A, a, and c. See **Figure 7.6.**

▶ **TRY EXERCISE 18, PAGE 634**

● APPLICATIONS OF THE LAW OF SINES

EXAMPLE 3   **Solve an Application Using the Law of Sines**

A radio antenna 85 feet high is located on top of an office building. At a distance AD from the base of the building, the angle of elevation to the top of the antenna is 26°, and the angle of elevation to the bottom of the antenna is 16°. Find the height of the building.

Solution

Sketch the diagram. See **Figure 7.7.** Find B and β.

$$B = 90^\circ - 26^\circ = 64^\circ$$
$$\beta = 26^\circ - 16^\circ = 10^\circ$$

Because we know the length BC and the measure of β, we can use triangle ABC and the Law of Sines to find length AC.

$$\frac{BC}{\sin \beta} = \frac{AC}{\sin B}$$

$$\frac{85}{\sin 10^\circ} = \frac{AC}{\sin 64^\circ} \qquad \bullet \; BC = 85, \beta = 10^\circ, B = 64^\circ$$

$$AC = \frac{85 \sin 64^\circ}{\sin 10^\circ}$$

Having found AC, we can now find the height of the building.

$$\sin 16^\circ = \frac{h}{AC}$$

$$h = AC \sin 16^\circ$$

$$= \frac{85 \sin 64^\circ}{\sin 10^\circ} \sin 16^\circ \approx 121 \text{ feet} \qquad \bullet \text{ Substitute for } AC.$$

The height of the building to two significant digits is 120 feet.

▶ **TRY EXERCISE 26, PAGE 635**

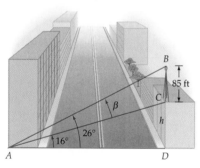

FIGURE 7.7

take note

In Example 3 we rounded the height of the building to two significant digits to comply with the rounding convention given on page 482.

In navigation and surveying problems, there are two commonly used methods for specifying direction. The angular direction in which a craft is pointed is called the **heading.** Heading is expressed in terms of an angle measured clockwise from north. **Figure 7.8** shows a heading of 65° and a heading of 285°.

The angular direction used to locate one object in relation to another object is called the **bearing.** Bearing is expressed in terms of the acute angle formed by a north–south line and the line of direction. **Figure 7.9** shows a bearing of N38°W and a bearing of S15°E.

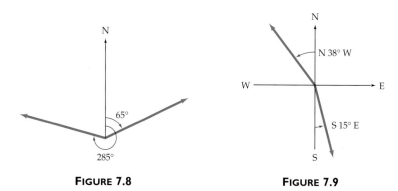

| **FIGURE 7.8** | **FIGURE 7.9** |

❓ QUESTION   Can a bearing of N50°E be written as N310°W?

EXAMPLE 4   Solve an Application

A ship with a heading of 330° first sighted a lighthouse (point *B*) at a bearing of N65°E. After traveling 8.5 miles, the ship observed the lighthouse at a bearing of S50°E. Find the distance from the ship to the lighthouse when the first sighting was made.

Solution

From **Figure 7.10** we see that the measure of $\angle CAB = 65° + 30° = 95°$, the measure of $\angle BCA = 50° - 30° = 20°$, and $B = 180° - 95° - 20° = 65°$. Use triangle *ABC* and the Law of Sines to find *c*.

$$\frac{b}{\sin B} = \frac{c}{\sin C}$$

$$\frac{8.5}{\sin 65°} = \frac{c}{\sin 20°} \qquad \bullet\ b = 8.5, B = 65°, C = 20°$$

$$c = \frac{8.5 \sin 20°}{\sin 65°} \approx 3.2$$

The lighthouse was 3.2 miles (to two significant digits) from the ship when the first sighting was made.

▶ **TRY EXERCISE 36, PAGE 636**

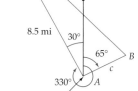

FIGURE 7.10

❓ ANSWER   No. A bearing is always expressed using an acute angle.

TOPICS FOR DISCUSSION

1. Is it possible to solve a triangle if the only given information consists of the measures of the three angles of the triangle? Explain.

2. Explain why it is not possible (in general) to use the Law of Sines to solve a triangle for which we are given only the lengths of all the sides.

3. Draw a triangle with dimensions $A = 30°$, $c = 3$ inches, and $a = 2.5$ inches. Is your answer unique? That is, can more than one triangle with the given dimensions be drawn?

4. Argue for or against the following proposition: In a scalene triangle (a triangle with no congruent sides), the largest angle is always opposite the longest side and the smallest angle is always opposite the shortest side.

EXERCISE SET 7.1

In Exercises 1 to 40, round answers according to the rounding conventions on page 482.

In Exercises 1 to 12, solve the triangles.

1. $A = 42°$, $B = 61°$, $a = 12$

2. $B = 25°$, $C = 125°$, $b = 5.0$

3. $A = 110°$, $C = 32°$, $b = 12$

▶ **4.** $B = 28°$, $C = 78°$, $c = 44$

5. $A = 132°$, $a = 22$, $b = 16$

6. $B = 82.0°$, $b = 6.0$, $c = 3.0$

7. $A = 82.0°$, $B = 65.4°$, $b = 36.5$

8. $B = 54.8°$, $C = 72.6°$, $a = 14.4$

9. $A = 33.8°$, $C = 98.5°$, $c = 102$

10. $B = 36.9°$, $C = 69.2°$, $a = 166$

11. $C = 114.2°$, $c = 87.2$, $b = 12.1$

12. $A = 54.32°$, $a = 24.42$, $c = 16.92$

In Exercises 13 to 24, solve the triangles that exist.

13. $A = 37°$, $c = 40$, $a = 28$

14. $B = 32°$, $c = 14$, $b = 9.0$

15. $C = 65°$, $b = 10$, $c = 8.0$

16. $A = 42°$, $a = 12$, $c = 18$

17. $A = 30°$, $a = 1.0$, $b = 2.4$

▶ **18.** $B = 22.6°$, $b = 5.55$, $a = 13.8$

19. $A = 14.8°$, $c = 6.35$, $a = 4.80$

20. $C = 37.9°$, $b = 3.50$, $c = 2.84$

21. $C = 47.2°$, $a = 8.25$, $c = 5.80$

22. $B = 52.7°$, $b = 12.3$, $c = 16.3$

23. $B = 117.32°$, $b = 67.25$, $a = 15.05$

24. $A = 49.22°$, $a = 16.92$, $c = 24.62$

25. **HURRICANE WATCH** A satellite weather map shows a hurricane off the coast of North Carolina. Use the information in the map to find the distance from the hurricane to Nags Head.

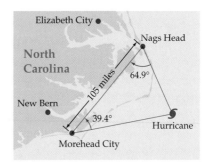

▶ **26.** **NAVAL MANEUVERS** The distance between an aircraft carrier and a Navy destroyer is 7620 feet. The angle of elevation from the destroyer to a helicopter is 77.2°, and the angle of elevation from the aircraft carrier to the helicopter is 59.0°. The helicopter is in the same vertical plane as the two ships, as shown in the following figure. Use this data to determine the distance x from the helicopter to the aircraft carrier.

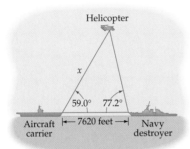

27. **CHOOSING A GOLF STRATEGY** The following diagram shows two ways to play a golf hole. One is to hit the ball down the fairway on your first shot and then hit an approach shot to the green on your second shot. A second way is to hit directly toward the pin. Due to the water hazard, this is a more risky strategy. The distance AB is 165 yards, BC is 155 yards, and angle $A = 42.0°$. Find the distance AC from the tee directly to the pin. Assume that angle B is an obtuse angle.

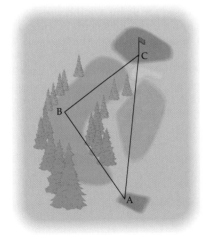

28. **DRIVING DISTANCE** A golfer drives a golf ball from the tee at point A to point B, as shown in the following diagram. The distance AC from the tee directly to the pin is 365 yards. Angle A measures 11.2°, and angle C measures 22.9°.

  a. Find the distance AB that the golfer drove the ball.

  b. Find the distance BC from the present position of the ball to the pin.

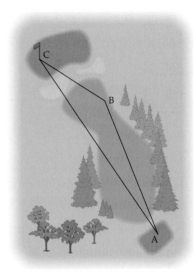

29. **DISTANCE TO A HOT AIR BALLOON** The angle of elevation to a balloon from one observer is 67°, and the angle of elevation from another observer, 220 feet away, is 31°. If the balloon is in the same vertical plane as the two observers and in between them, find the distance of the balloon from the first observer.

30. **LENGTH OF A DIAGONAL** The longer side of a parallelogram is 6.0 meters. The measure of $\angle BAD$ is 56° and α is 35°. Find the length of the longer diagonal.

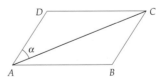

31. **DISTANCE ACROSS A CANYON** To find the distance across a canyon, a surveying team locates points A and B on one side of the canyon and point C on the other side of the canyon. The distance between A and B is 85 yards. The measure of $\angle CAB$ is 68°, and the measure of $\angle CBA$ is 75°. Find the distance across the canyon.

32. HEIGHT OF A KITE Two observers, in the same vertical plane as a kite and at a distance of 30 feet apart, observe the kite at angles 62° and 78°, as shown in the following diagram. Find the height of the kite.

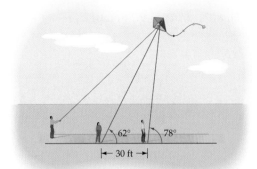

33. LENGTH OF A GUY WIRE A telephone pole 35 feet high is situated on an 11° slope from the horizontal. The measure of angle *CAB* is 21°. Find the length of the guy wire *AC*.

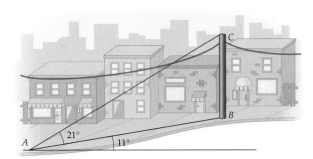

34. DIMENSIONS OF A PLOT OF LAND Three roads intersect in such a way as to form a triangular piece of land. See the accompanying figure. Find the lengths of the other two sides of the land.

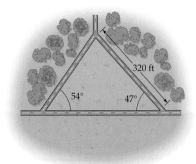

35. HEIGHT OF A HILL A surveying team determines the height of a hill by placing a 12-foot pole at the top of the hill and measuring the angles of elevation to the bottom and the top of the pole. They find the angles of elevation to be as shown in the following figure. Find the height of the hill.

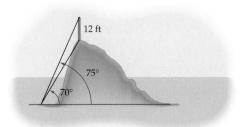

▶ **36. DISTANCE TO A FIRE** Two fire lookouts are located on mountains 20 miles apart. Lookout *B* is at a bearing of S65°E from lookout *A*. A fire was sighted at a bearing of N50°E from *A* and at a bearing of N8°E from *B*. Find the distance of the fire from lookout *A*.

37. DISTANCE TO A LIGHTHOUSE A navigator on a ship sights a lighthouse at a bearing of N36°E. After traveling 8.0 miles at a heading of 332°, the ship sights the lighthouse at a bearing of S82°E. How far is the ship from the lighthouse at the second sighting?

38. MINIMUM DISTANCE The navigator on a ship traveling due east at 8 mph sights a lighthouse at a bearing of S55°E. One hour later it is sighted at a bearing of S25°W. Find the closest the ship came to the lighthouse.

39. DISTANCE BETWEEN AIRPORTS An airplane flew 450 miles at a bearing of N65°E from airport *A* to airport *B*. The plane then flew at a bearing of S38°E to airport *C*. Find the distance from *A* to *C* if the bearing from airport *A* to airport *C* is S60°E.

40. LENGTH OF A BRACE A 12-foot solar panel is to be installed on a roof with a 15° pitch. Find the length of the vertical brace *d* if the panel must be installed to make a 40° angle with the horizontal.

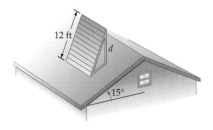

CONNECTING CONCEPTS

41. **DISTANCES BETWEEN HOUSES** House *B* is located at a bearing of N67°E from house *A*. House *C* is 300 meters from house *A* at a bearing of S68°E. House *B* is located at a bearing of N11°W from house *C*. Find the distance from house *A* to house *B*.

42. Show that for any triangle *ABC*, $\dfrac{a - b}{b} = \dfrac{\sin A - \sin B}{\sin B}$.

43. Show that for any triangle *ABC*, $\dfrac{a + b}{b} = \dfrac{\sin A + \sin B}{\sin B}$.

44. Show that for any triangle *ABC*, $\dfrac{a - b}{a + b} = \dfrac{\sin A - \sin B}{\sin A + \sin B}$.

45. **MAXIMUM LENGTH OF A ROD** The longest rod that can be carried horizontally around a corner from a hall 3 meters wide into one that is 5 meters wide is

the minimum of the length *L* of the dashed line shown in the figure below.

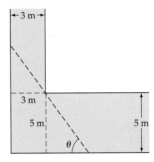

Use similar triangles to show that the length *L* is a function of the angle *θ*, given by

$$L(\theta) = \frac{5}{\sin \theta} + \frac{3}{\cos \theta}$$

Use a graphing utility to graph *L* and estimate the minimum value of *L*.

PREPARE FOR SECTION 7.2

46. Evaluate $\sqrt{a^2 + b^2 - 2ab \cos C}$ for $a = 10.0$, $b = 15.0$, and $C = 110.0°$. Round your result to the nearest tenth. [5.3]

47. Find the area of a triangle with a base of 6 inches and a height of 8.5 inches. [1.2]

48. Solve $c^2 = a^2 + b^2 - 2ab \cos C$ for *C*. [6.5]

49. The **semiperimeter** of a triangle is defined as one-half the perimeter of the triangle. Find the semiperimeter of a triangle with sides of 6 meters, 9 meters, and 10 meters. [P.1]

50. Evaluate $\sqrt{s(s - a)(s - b)(s - c)}$ for $a = 3$, $b = 4$, $c = 5$, and $s = \dfrac{a + b + c}{2}$. [P.2]

51. State a relationship between the lengths *a*, *b*, and *c* in the triangle shown at the right. [1.3]

PROJECTS

1. **FERMAT'S PRINCIPLE AND SNELL'S LAW** State Fermat's Principle and Snell's Law. The refractive index of the glass in a ring is found to be 1.82. The refractive index of a particular diamond in a ring is 2.38. Use Snell's Law to explain what this means in terms of the reflective properties of the glass and the diamond.

THE LAW OF COSINES AND AREA

• THE LAW OF COSINES

The *Law of Cosines* can be used to solve triangles in which two sides and the included angle (SAS) are known or in which three sides (SSS) are known. Consider the triangle in **Figure 7.11**. The height BD is drawn from B perpendicular to the x-axis. The triangle BDA is a right triangle, and the coordinates of B are $(a \cos C, a \sin C)$. The coordinates of A are $(b, 0)$. Using the distance formula, we can find the distance c.

$$c = \sqrt{(a \cos C - b)^2 + (a \sin C - 0)^2}$$
$$c^2 = a^2 \cos^2 C - 2ab \cos C + b^2 + a^2 \sin^2 C$$
$$c^2 = a^2(\cos^2 C + \sin^2 C) + b^2 - 2ab \cos C$$
$$c^2 = a^2 + b^2 - 2ab \cos C$$

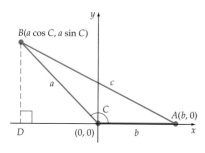

FIGURE 7.11

The Law of Cosines

If A, B, and C are the measures of the angles of a triangle and a, b, and c are the lengths of the sides opposite these angles, then

$$c^2 = a^2 + b^2 - 2ab \cos C$$
$$a^2 = b^2 + c^2 - 2bc \cos A$$
$$b^2 = a^2 + c^2 - 2ac \cos B$$

EXAMPLE I   Use the Law of Cosines (SAS)

In triangle ABC, $B = 110.0°$, $a = 10.0$ centimeters, and $c = 15.0$ centimeters. See **Figure 7.12**. Find b.

Solution

The Law of Cosines can be used because two sides and the included angle are known.

$$b^2 = a^2 + c^2 - 2ac \cos B$$
$$= 10.0^2 + 15.0^2 - 2(10.0)(15.0) \cos 110.0°$$
$$b = \sqrt{10.0^2 + 15.0^2 - 2(10.0)(15.0) \cos 110.0°}$$
$$b \approx 20.7 \text{ centimeters}$$

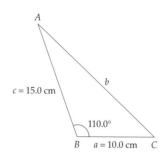

FIGURE 7.12

▶ TRY EXERCISE 12, PAGE 644

In the next example we know the length of each side, but we do not know the measure of any of the angles.

EXAMPLE 2   Use the Law of Cosines (SSS)

In triangle ABC, $a = 32$ feet, $b = 20$ feet, and $c = 40$ feet. Find B. This is the SSS case.

Solution

$$b^2 = a^2 + c^2 - 2ac \cos B$$

$$\cos B = \frac{a^2 + c^2 - b^2}{2ac} \qquad \text{• Solve for cos } B.$$

$$= \frac{32^2 + 40^2 - 20^2}{2(32)(40)} \qquad \text{• Substitute for } a, b, \text{ and } c.$$

$$B = \cos^{-1}\left(\frac{32^2 + 40^2 - 20^2}{2(32)(40)}\right) \qquad \text{• Solve for angle } B.$$

$$B \approx 30° \qquad \text{• To the nearest degree}$$

▶ **TRY EXERCISE 18, PAGE 644**

● AN APPLICATION OF THE LAW OF COSINES

EXAMPLE 3   Solve an Application Using the Law of Cosines

A car traveled 3.0 miles at a heading of 78°. The road turned, and the car traveled another 4.3 miles at a heading of 138°. Find the distance and the bearing of the car from the starting point.

Solution

Sketch a diagram (see **Figure 7.13**). First find B.

$$B = 78° + (180° - 138°) = 120°$$

Use the Law of Cosines to find b.

$$b^2 = a^2 + c^2 - 2ac \cos B$$

$$= 4.3^2 + 3.0^2 - 2(4.3)(3.0) \cos 120° \qquad \text{• Substitute for } a, c, \text{ and } B.$$

$$b = \sqrt{4.3^2 + 3.0^2 - 2(4.3)(3.0) \cos 120°}$$

$$b \approx 6.4 \text{ miles}$$

Find A.

$$\cos A = \frac{b^2 + c^2 - a^2}{2bc}$$

$$A = \cos^{-1}\left(\frac{b^2 + c^2 - a^2}{2bc}\right) \approx \cos^{-1}\left(\frac{6.4^2 + 3.0^2 - 4.3^2}{(2)(6.4)(3.0)}\right)$$

$$A \approx 35°$$

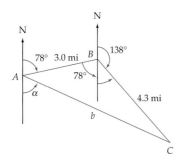

FIGURE 7.13

take note

The measure of A in Example 3 can also be determined by using the Law of Sines.

Continued ▶

The bearing of the present position of the car from the starting point A can be determined by calculating the measure of angle α in **Figure 7.13.**

$$\alpha \approx 180° - (78° + 35°) = 67°$$

The distance is approximately 6.4 miles, and the bearing (to the nearest degree) is S67°E.

▶ **TRY EXERCISE 52, PAGE 646**

There are five different cases that we may encounter when solving an oblique triangle. They are listed in the following guideline, along with the law that can be used to solve the triangle.

A Guideline for Choosing Between the Law of Sines and the Law of Cosines

Apply the Law of Sines to solve an oblique triangle for each of the following cases.

ASA The measures of two angles of the triangle and the length of the included side are known.

AAS The measures of two angles of the triangle and the length of a side opposite one of these angles are known.

SSA The lengths of two sides of the triangle and the measure of an angle opposite one of these sides are known. This case is called the ambiguous case. It may yield one solution, two solutions, or no solution.

Apply the Law of Cosines to solve an oblique triangle for each of the following cases.

SSS The lengths of all three sides of the triangle are known. After finding the measure of an angle, you can complete your solution by using the Law of Sines.

SAS The lengths of two sides of the triangle and the measure of the included angle are known. After finding the measure of the third side, you can complete your solution by using the Law of Sines.

❓ **QUESTION** In triangle ABC, $A = 40°$, $C = 60°$, and $b = 114$. Should you use the Law of Sines or the Law of Cosines to solve this triangle?

❓ **ANSWER** The Law of Sines.

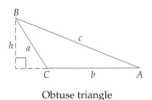

B

a h c

C b A

Acute triangle

B

h a c

C b A

Obtuse triangle

FIGURE 7.14

take note

Because each formula requires
two sides and the included angle,
it is necessary to learn only one
formula.

● AREA OF A TRIANGLE

The formula $A = \dfrac{1}{2}bh$ can be used to find the area of a triangle when the base and height are given. In this section we will find the areas of triangles when the height is not given. We will use K for the area of a triangle because the letter A is often used to represent the measure of an angle.

Consider the areas of the acute and obtuse triangles in **Figure 7.14.**

Height of each triangle: $h = c \sin A$

Area of each triangle: $K = \dfrac{1}{2}bh$

$K = \dfrac{1}{2}bc \sin A$ • **Substitute for *h*.**

Thus we have established the following theorem.

Area of a Triangle

The area K of triangle ABC is one-half the product of the lengths of any two sides and the sine of the included angle. Thus

$$K = \frac{1}{2}bc \sin A$$

$$K = \frac{1}{2}ab \sin C$$

$$K = \frac{1}{2}ac \sin B$$

EXAMPLE 4   **Find the Area of a Triangle**

Given angle $A = 62°$, $b = 12$ meters, and $c = 5.0$ meters, find the area of triangle ABC.

Solution

In **Figure 7.15,** two sides and the included angle of the triangle are given. Using the formula for area, we have

$$K = \frac{1}{2}bc \sin A = \frac{1}{2}(12)(5.0)(\sin 62°) \approx 26 \text{ square meters}$$

▶ **TRY EXERCISE 26, PAGE 645**

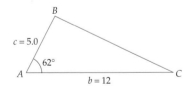

B

$c = 5.0$

$62°$

A $b = 12$ C

FIGURE 7.15

When two angles and an included side are given, the Law of Sines is used to derive a formula for the area of a triangle. First, solve for c in the Law of Sines.

$$\frac{c}{\sin C} = \frac{b}{\sin B}$$

$$c = \frac{b \sin C}{\sin B}$$

Substitute for c in the formula $K = \dfrac{1}{2}bc \sin A$.

$$K = \frac{1}{2}bc \sin A = \frac{1}{2}b\left(\frac{b \sin C}{\sin B}\right) \sin A$$

$$K = \frac{b^2 \sin C \sin A}{2 \sin B}$$

In like manner, the following two alternative formulas can be derived for the area of a triangle.

$$K = \frac{a^2 \sin B \sin C}{2 \sin A} \quad \text{and} \quad K = \frac{c^2 \sin A \sin B}{2 \sin C}$$

EXAMPLE 5 **Find the Area of a Triangle**

Given $A = 32°$, $C = 77°$, and $a = 14$ inches, find the area of triangle ABC.

Solution

To use the area formula, we need to know two angles and the included side. Therefore, we need to determine the measure of angle B.

$$B = 180° - 32° - 77° = 71°$$

Thus

$$K = \frac{a^2 \sin B \sin C}{2 \sin A} = \frac{14^2 \sin 71° \sin 77°}{2 \sin 32°} \approx 170 \text{ square inches}$$

▶ **TRY EXERCISE 28, PAGE 645**

MATH MATTERS

Recent findings indicate that Heron's formula for finding the area of a triangle was first discovered by Archimedes. However, the formula is called Heron's formula in honor of the geometer Heron of Alexandria (A.D. 50), who gave an ingenious proof of the theorem in his work *Metrica*. Because Heron of Alexandria was also known as Hero, some texts refer to Heron's formula as Hero's formula.

● HERON'S FORMULA

The Law of Cosines can be used to derive *Heron's formula* for the area of a triangle in which three sides of the triangle are given.

Heron's Formula for Finding the Area of a Triangle

If a, b, and c are the lengths of the sides of a triangle, then the area K of the triangle is

$$K = \sqrt{s(s - a)(s - b)(s - c)}, \quad \text{where } s = \frac{1}{2}(a + b + c)$$

Because s is one-half the perimeter of the triangle, it is called the **semiperimeter.**

EXAMPLE 6   **Find an Area by Heron's Formula**

Find, to two significant digits, the area of the triangle with $a = 7.0$ meters, $b = 15$ meters, and $c = 12$ meters.

Solution

Calculate the semiperimeter s.

$$s = \frac{a + b + c}{2} = \frac{7.0 + 15 + 12}{2} = 17$$

Use Heron's formula.

$$K = \sqrt{s(s - a)(s - b)(s - c)}$$
$$= \sqrt{17(17 - 7.0)(17 - 15)(17 - 12)}$$
$$= \sqrt{1700} \approx 41 \text{ square meters}$$

▶ **TRY EXERCISE 36, PAGE 645**

EXAMPLE 7   **Use Heron's Formula to Solve an Application**

The original portion of the Luxor Hotel in Las Vegas has the shape of a square pyramid. Each face of the pyramid is an isosceles triangle with a base of 646 feet and sides of length 576 feet. Assuming that the glass on the exterior of the Luxor Hotel costs \$35 per square foot, determine the cost of the glass, to the nearest \$10,000, for one of the triangular faces of the hotel.

Solution

The lengths (in feet) of the sides of a triangular face are $a = 646$, $b = 576$, and $c = 576$.

$$s = \frac{a + b + c}{2} = \frac{646 + 576 + 576}{2} = 899 \text{ feet}$$
$$K = \sqrt{s(s - a)(s - b)(s - c)}$$
$$= \sqrt{899(899 - 646)(899 - 576)(899 - 576)}$$
$$= \sqrt{23{,}729{,}318{,}063}$$
$$\approx 154{,}043 \text{ square feet}$$

The cost C of the glass is the product of the cost per square foot and the area.

$$C \approx 35 \cdot 154{,}043 = 5{,}391{,}505$$

The approximate cost of the glass for one face of the Luxor Hotel is \$5,390,000.

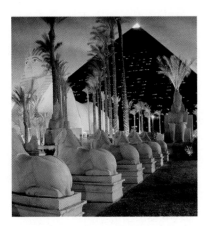

The pyramid portion of the Luxor Hotel, Las Vegas, Nevada

▶ **TRY EXERCISE 58, PAGE 646**

TOPICS FOR DISCUSSION

1. Explain why there is no triangle that has sides of lengths $a = 2$ inches, $b = 11$ inches, and $c = 3$ inches.

2. The Pythagorean Theorem is a special case of the Law of Cosines. Explain.

3. To solve a triangle in which the lengths of the three sides are given (SSS), a mathematics professor recommends the following procedure.

   (i) Use the Law of Cosines to find the measure of the largest angle.

   (ii) Use the Law of Sines to find the measure of a second angle.

   (iii) Find the measure of the third angle by using the formula $A + B + C = 180°$.

   Explain why this procedure is easier than using the Law of Cosines to find the measure of all three angles.

4. To solve a triangle in which the lengths of two sides and the measure of the included angle are given (SAS), a tutor recommends the following procedure.

   (i) Use the Law of Cosines to find the length of the third side.

   (ii) Use the Law of Sines to find the smaller of the unknown angles.

   (iii) Find the measure of the third angle by using the formula $A + B + C = 180°$.

   Explain why this procedure is easier than using the Law of Cosines three times to find each of the unknowns.

EXERCISE SET 7.2

In Exercises 1 to 52, round answers according to the rounding conventions on page 482.

In Exercises 1 to 14, find the third side of the triangle.

1. $a = 12, b = 18, C = 44°$

2. $b = 30, c = 24, A = 120°$

3. $a = 120, c = 180, B = 56°$

4. $a = 400, b = 620, C = 116°$

5. $b = 60, c = 84, A = 13°$

6. $a = 122, c = 144, B = 48°$

7. $a = 9.0, b = 7.0, C = 72°$

8. $b = 12, c = 22, A = 55°$

9. $a = 4.6, b = 7.2, C = 124°$

10. $b = 12.3, c = 14.5, A = 6.5°$

11. $a = 25.9, c = 33.4, B = 84.0°$

▶ **12.** $a = 14.2, b = 9.30, C = 9.20°$

13. $a = 122, c = 55.9, B = 44.2°$

14. $b = 444.8, c = 389.6, A = 78.44°$

In Exercises 15 to 24, given three sides of a triangle, find the specified angle.

15. $a = 25, b = 32, c = 40$; find A.

16. $a = 60, b = 88, c = 120$; find B.

17. $a = 8.0, b = 9.0, c = 12$; find C.

▶ **18.** $a = 108, b = 132, c = 160$; find A.

19. $a = 80.0, b = 92.0, c = 124$; find B.

20. $a = 166, b = 124, c = 139$; find B.

21. $a = 1025, b = 625.0, c = 1420$; find C.

22. $a = 4.7, b = 3.2, c = 5.9$; find A.

23. $a = 32.5, b = 40.1, c = 29.6$; find B.

24. $a = 112.4, b = 96.80, c = 129.2$; find C.

In Exercises 25 to 36, find the area of the given triangle. Round each area to the same number of significant digits given for each of the given sides.

25. $A = 105°, b = 12, c = 24$

▶ **26.** $B = 127°, a = 32, c = 25$

27. $A = 42°, B = 76°, c = 12$

▶ **28.** $B = 102°, C = 27°, a = 8.5$

29. $a = 16, b = 12, c = 14$

30. $a = 32, b = 24, c = 36$

31. $B = 54.3°, a = 22.4, b = 26.9$

32. $C = 18.2°, b = 13.4, a = 9.84$

33. $A = 116°, B = 34°, c = 8.5$

34. $B = 42.8°, C = 76.3°, c = 17.9$

35. $a = 3.6, b = 4.2, c = 4.8$

▶ **36.** $a = 13.3, b = 15.4, c = 10.2$

37. DISTANCE BETWEEN AIRPORTS A plane leaves airport A and travels 560 miles to airport B at a bearing of N32°E. The plane leaves airport B and travels to airport C 320 miles away at a bearing of S72°E. Find the distance from airport A to airport C.

38. LENGTH OF A STREET A developer has a triangular lot at the intersection of two streets. The streets meet at an angle of 72°, and the lot has 300 feet of frontage along one street and 416 feet of frontage along the other street. Find the length of the third side of the lot.

39. BASEBALL In a baseball game, a batter hits a ground ball 26 feet in the direction of the pitcher's mound. See the figure at the top of the next column. The pitcher runs forward and reaches for the ball. At that moment, how far is the ball from first base? (*Note*: A baseball infield is a square that measures 90 feet on each side.)

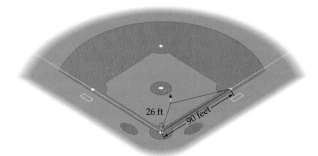

40. B-2 BOMBER The leading edge of each wing of the B-2 Stealth Bomber measures 105.6 feet in length. The angle between the wing's leading edges ($\angle ABC$) is 109.05°. What is the wing span (the distance from A to C) of the B-2 Bomber?

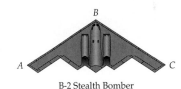

B-2 Stealth Bomber

41. ANGLE BETWEEN THE DIAGONALS OF A BOX The rectangular box in the figure measures 6.50 feet by 3.25 feet by 4.75 feet. Find the measure of the angle θ that is formed by the union of the diagonal shown on the front of the box and the diagonal shown on the right side of the box.

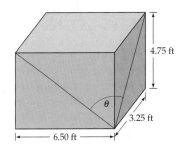

42. SUBMARINE RESCUE MISSION The surface ships shown in the figure below have determined the indicated distances. Use this data to determine the depth of the submarine below the surface of the water. Assume that the line segment between the surface ships is directly above the submarine.

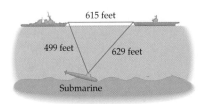

43. DISTANCE BETWEEN SHIPS Two ships left a port at the same time. One ship traveled at a speed of 18 mph at a heading of 318°. The other ship traveled at a speed of 22 mph at a heading of 198°. Find the distance between the two ships after 10 hours of travel.

44. DISTANCE ACROSS A LAKE Find the distance across a lake, using the measurements shown in the figure.

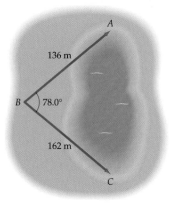

45. A regular hexagon is inscribed in a circle with a radius of 40 centimeters. Find the length of one side of the hexagon.

46. A regular pentagon is inscribed in a circle with a radius of 25 inches. Find the length of one side of the pentagon.

47. The lengths of the diagonals of a parallelogram are 20 inches and 32 inches. The diagonals intersect at an angle of 35°. Find the lengths of the sides of the parallelogram. (*Hint:* The diagonals of a parallelogram bisect one another.)

48. The sides of a parallelogram are 10 feet and 14 feet. The longer diagonal of the parallelogram is 18 feet. Find the length of the shorter diagonal of the parallelogram. (*Hint:* The diagonals of a parallelogram bisect one another.)

49. The sides of a parallelogram are 30 centimeters and 40 centimeters. The shorter diagonal of the parallelogram is 44 centimeters. Find the length of the longer diagonal of the parallelogram. (*Hint:* The diagonals of a parallelogram bisect one another.)

50. ANGLE BETWEEN BOUNDARIES OF A LOT A triangular city lot has sides of 224 feet, 182 feet, and 165 feet. Find the angle between the longer two sides of the lot.

51. DISTANCE TO A PLANE A plane traveling at 180 mph passes 400 feet directly over an observer. The plane is traveling along a straight path with an angle of elevation of 14°. Find the distance of the plane from the observer 10 seconds after the plane has passed directly overhead.

▶ **52. DISTANCE BETWEEN SHIPS** A ship leaves a port at a speed of 16 mph at a heading of 32°. One hour later another ship leaves the port at a speed of 22 mph at a heading of 254°. Find the distance between the ships 4 hours after the first ship leaves the port.

53. AREA OF A TRIANGULAR LOT Find the area of a triangular piece of land that is bounded by sides of 236 meters, 620 meters, and 814 meters.

54. Find the exact area of a parallelogram with sides of exactly 8 feet and 12 feet. The shorter diagonal is exactly 10 feet.

55. Find the exact area of a square inscribed in a circle with a radius of exactly 9 inches.

56. Find the exact area of a regular hexagon inscribed in a circle with a radius of exactly 24 centimeters.

57. COST OF A LOT A commercial piece of real estate is priced at $2.20 per square foot. Find, to the nearest $1000, the cost of a triangular lot measuring 212 feet by 185 feet by 240 feet.

▶ **58. COST OF A LOT** An industrial piece of real estate is priced at $4.15 per square foot. Find, to the nearest $1000, the cost of a triangular lot measuring 324 feet by 516 feet by 412 feet.

59. AREA OF A PASTURE Find the number of acres in a pasture whose shape is a triangle measuring 800 feet by 1020 feet by 680 feet. (An acre is 43,560 square feet.)

60. AREA OF A HOUSING TRACT Find the number of acres in a housing tract whose shape is a triangle measuring 420 yards by 540 yards by 500 yards. (An acre is 4840 square yards.)

The following identity is one of *Mollweide's formulas*. It applies to any triangle *ABC*.

$$\frac{a - b}{c} = \frac{\sin\left(\dfrac{A - B}{2}\right)}{\cos\left(\dfrac{C}{2}\right)}$$

The formula is intriguing because it contains the dimensions of all angles and all sides of $\triangle ABC$. The formula can be used to check whether a triangle has been solved correctly. Substitute the dimensions of a given triangle in the formula and compare the value of the left side of the formula with the value of the right side. If the two results are not reasonably close, then you know that at least one dimension is incorrect. The results generally will not be identical because each dimension is an approximation. In Exercises 61 and 62, you can assume that a triangle has an incorrect dimension if the value of the left side and the value of the right side of the above formula differ by more than 0.02.

61. CHECK DIMENSIONS OF TRUSSES The following diagram shows some of the steel trusses in an airplane hangar.

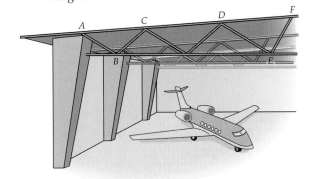

An architect has determined the following dimensions for $\triangle ABC$ and $\triangle DEF$:

$\triangle ABC$: $A = 53.5°$, $B = 86.5°$, $C = 40.0°$
　　　　　$a = 13.0$ feet, $b = 16.1$ feet, $c = 10.4$ feet

$\triangle DEF$: $D = 52.1°$, $E = 59.9°$, $F = 68.0°$
　　　　　$d = 17.2$ feet, $e = 21.3$ feet, $f = 22.8$ feet

Use Mollweide's formula to determine if either $\triangle ABC$ or $\triangle DEF$ has an incorrect dimension.

62. CHECK DIMENSIONS OF TRUSSES The following diagram shows some of the steel trusses in a railroad bridge.

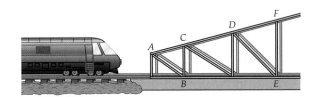

A structural engineer has determined the following dimensions for $\triangle ABC$ and $\triangle DEF$:

$\triangle ABC$: $A = 34.1°$, $B = 66.2°$, $C = 79.7°$
　　　　　$a = 9.23$ feet, $b = 15.1$ feet, $c = 16.2$ feet

$\triangle DEF$: $D = 45.0°$, $E = 56.2°$, $F = 78.8°$
　　　　　$d = 13.6$ feet, $e = 16.0$ feet, $f = 18.9$ feet

Use Mollweide's formula to determine if either $\triangle ABC$ or $\triangle DEF$ has an incorrect dimension.

CONNECTING CONCEPTS

HERON'S FORMULA FOR QUADRILATERALS The following formula is a generalization of Heron's formula. It gives the area *K* of a convex quadrilateral with sides of lengths *a*, *b*, *c*, and *d*.

$$K = \sqrt{(s - a)(s - b)(s - c)(s - d)}$$

where

$$s = \frac{a + b + c + d}{2}$$

63. Verify that Heron's formula for quadrilaterals produces the correct area for each of the following:

　a. A square with sides of length 8.

　b. A rectangle with sides of lengths 5 and 11.

64. Use Heron's formula for quadrilaterals to find the area (to the nearest 0.01 square unit) of the following convex quadrilateral.

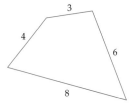

65. Find the measure of the angle formed by the sides P_1P_2 and P_1P_3 of a triangle with vertices at $P_1(-2, 4)$, $P_2(2, 1)$, and $P_3(4, -3)$.

66. A regular pentagon is inscribed in a circle with a radius of 4 inches. Find the perimeter of the pentagon.

67. An equilateral triangle is inscribed in a circle with a radius of 10 centimeters. Find the perimeter of the triangle.

68. Given a triangle ABC, prove that
$$a^2 = b^2 + c^2 - 2bc \cos A$$

69. Use the Law of Cosines to show that
$$\cos A = \frac{(b + c - a)(b + c + a)}{2bc} - 1$$

70. Prove that $K = xy \sin A$ for a parallelogram, where x and y are the lengths of adjacent sides, A is the measure of the angle between side x and side y, and K is the area of the parallelogram.

71. Show that the area of the parallelogram in the figure is $K = 2ab \sin C$.

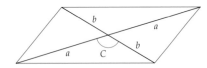

72. Given a regular hexagon inscribed in a circle with a radius of 10 inches, find the area of a segment of the circle (see the dark-shaded region in the following figure).

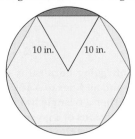

73. Find the volume of the triangular prism shown in the figure.

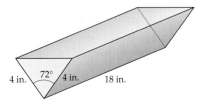

74. Show that the area of the circumscribed triangle in the figure is $K = rs$, where $s = \dfrac{a + b + c}{2}$.

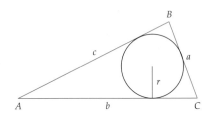

PREPARE FOR SECTION 7.3

75. Evaluate: $\sqrt{\left(\dfrac{3}{5}\right)^2 + \left(-\dfrac{4}{5}\right)^2}$ [P.2]

76. Use a calculator to evaluate $10 \cos 228°$. Round to the nearest thousandth. [5.3]

77. Solve $\tan \alpha = \left|\dfrac{-\sqrt{3}}{3}\right|$ for α, where α is an acute angle measured in degrees. [6.5]

78. Solve $\cos \alpha = \dfrac{-17}{\sqrt{338}}$ for α, where α is an obtuse angle measured in degrees. Round to the nearest tenth of a degree. [6.5]

79. Rationalize the denominator of $\dfrac{1}{\sqrt{5}}$. [P.2]

80. Rationalize the denominator of $\dfrac{28}{\sqrt{68}}$. [P.2]

PROJECTS

---

1. The following identity is known as the **Law of Tangents.** Given a triangle ABC with sides a, b, and c, then

$$\frac{a-b}{a+b} = \frac{\tan\left(\dfrac{A-B}{2}\right)}{\tan\left(\dfrac{A+B}{2}\right)}$$

The Law of Tangents can be used to solve triangles in which you are given the measures of two angles and the length of a side or you are given the lengths of two sides and the measure of the included angle (SAS).

a. Use the Law of Tangents to find the length b in triangle ABC with $A = 14.4°$, $B = 25.8°$, and $a = 123$.

b. Use the Law of Tangents to find the measures of angles A and B in triangle ABC with $C = 25.2°$, $a = 18.9$, and $b = 15.2$. (*Hint:* In triangle ABC, $A + B = 180° - C$.)

2. VISUAL INSIGHT

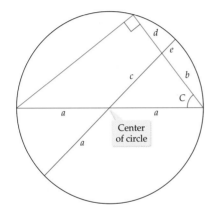

$$bd = (a+c)e$$
$$bd = (a+c)(a-c)$$
$$b(2a\cos C - b) = (a+c)(a-c)$$
$$2ab\cos C - b^2 = a^2 - c^2$$
$$c^2 = a^2 + b^2 - 2ab\cos C$$

Give the rule or reason that justifies each step in the above proof of the Law of Cosines.

SECTION 7.3

VECTORS

- **VECTORS**
- **UNIT VECTORS**
- **APPLICATIONS OF VECTORS**
- **DOT PRODUCT**
- **SCALAR PROJECTION**
- **PARALLEL AND PERPENDICULAR VECTORS**
- **WORK: AN APPLICATION OF THE DOT PRODUCT**

• VECTORS

In scientific applications, some measurements, such as area, mass, distance, speed, and time, are completely described by a real number and a unit. Examples include 30 square feet (area), 25 meters/second (speed), and 5 hours (time). These measurements are **scalar quantities,** and the number used to indicate the magnitude of the measurement is called a **scalar.** Two other examples of scalar quantities are volume and temperature.

For other quantities, besides the numerical and unit description, it is also necessary to include a *direction* to describe the quantity completely. For example, applying a force of 25 pounds at various angles to a small metal box will influence how the box moves. In **Figure 7.16,** applying the 25-pound force

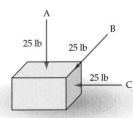

FIGURE 7.16

straight down (A) will not move the box to the left. However, applying the 25-pound force (C) parallel to the floor will move the box along the floor.

Vector quantities have a *magnitude* (numerical and unit description) and a *direction*. Force is a vector quantity. Velocity is another. Velocity includes the speed (magnitude) and a direction. A velocity of 40 mph east is different from a velocity of 40 mph north. Displacement is another vector quantity; it consists of distance (a scalar) moved in a certain direction; for example, we might speak of a displacement of 13 centimeters at an angle of 15° from the positive *x*-axis.

> ### Definition of a Vector
>
> A **vector** is a directed line segment. The length of the line segment is the magnitude of the vector, and the direction of the vector is measured by an angle.

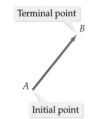

FIGURE 7.17

The point *A* for the vector in **Figure 7.17** is called the **initial point** (or tail) of the vector, and the point *B* is the **terminal point** (or head) of the vector. An arrow over the letters $(\overrightarrow{AB})$, an arrow over a single letter $(\vec{V})$, or boldface type (**AB** or **V**) is used to denote a vector. The magnitude of the vector is the length of the line segment and is denoted by $\|\overrightarrow{AB}\|$, $\|\vec{V}\|$, $\|\mathbf{AB}\|$, or $\|\mathbf{V}\|$.

Equivalent vectors have the same magnitude and the same direction. The vectors in **Figure 7.18** are equivalent. They have the same magnitude and direction.

Multiplying a vector by a positive real number (other than 1) changes the magnitude of the vector but not its direction. If **v** is any vector, then 2**v** is the vector that has the same direction as **v** but is twice the magnitude of **v**. The multiplication of 2 and **v** is called the **scalar multiplication** of the vector **v** and the scalar 2. Multiplying a vector by a negative number *a* reverses the direction of the vector and multiplies the magnitude of the vector by $|a|$. See **Figure 7.19.**

The sum of two vectors, called the **resultant vector** or the **resultant,** is the single equivalent vector that will have the same effect as the application of those two vectors. For example, a displacement of 40 meters along the positive *x*-axis and then 30 meters in the positive *y* direction is equivalent to a vector of magnitude 50 meters at an angle of approximately 37° to the positive *x*-axis. See **Figure 7.20.**

Vectors can be added graphically by using the *triangle method* or the *parallelogram method.* In the triangle method, shown in **Figure 7.21,** the tail of **V** is placed at the head of **U**. The vector connecting the tail of **U** with the head of **V** is the sum **U** + **V**.

The parallelogram method of adding two vectors graphically places the tails of the two vectors **U** and **V** together, as in **Figure 7.22.** Complete the parallelogram so that **U** and **V** are sides of the parallelogram. The diagonal beginning at the tails of the two vectors is **U** + **V**.

To find the difference between two vectors, first rewrite the expression as **V** − **U** = **V** + (−**U**). The difference is shown geometrically in **Figure 7.23.**

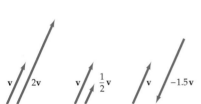

FIGURE 7.18

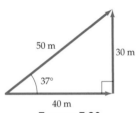

FIGURE 7.19

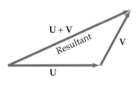

FIGURE 7.20

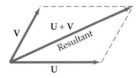

FIGURE 7.21

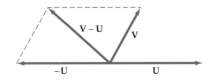

FIGURE 7.22   **FIGURE 7.23**

By introducing a coordinate plane, it is possible to develop an analytic approach to vectors. Recall from our discussion about equivalent vectors that a vector can be moved in the plane as long as *the magnitude and direction* are not changed.

With this in mind, consider **AB**, whose initial point is $A(2, -1)$ and whose terminal point is $B(-3, 4)$. If this vector is moved so that the initial point is at the origin O, the terminal point becomes $P(-5, 5)$, as shown in **Figure 7.24.** The vector **OP** is equivalent to the vector **AB**.

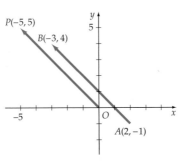

FIGURE 7.24   **FIGURE 7.25**

In **Figure 7.25,** let $P_1(x_1, y_1)$ be the initial point of a vector and $P_2(x_2, y_2)$ its terminal point. Then an equivalent vector **OP** has its initial point at the origin and its terminal point at $P(a, b)$, where $a = x_2 - x_1$ and $b = y_2 - y_1$. The vector **OP** can be denoted by $\mathbf{v} = \langle a, b \rangle$; a and b are called the **components** of the vector.

EXAMPLE 1   Find the Components of a Vector

Find the components of the vector **AB** whose tail is the point $A(2, -1)$ and whose head is the point $B(-2, 6)$. Determine a vector **v** that is equivalent to **AB** and has an initial point at the origin.

Algebraic Solution

The components of **AB** are $\langle a, b \rangle$, where

$$a = x_2 - x_1 = -2 - 2 = -4 \quad \text{and} \quad b = y_2 - y_1 = 6 - (-1) = 7$$

Thus $\mathbf{v} = \langle -4, 7 \rangle$.

Visualize the Solution

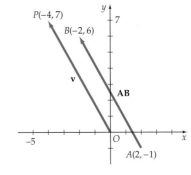

FIGURE 7.26

▶ **TRY EXERCISE 6, PAGE 661**

The magnitude and direction of a vector can be found from its components. For instance, the head of vector **v** sketched in **Figure 7.26** is the ordered pair $(-4, 7)$. Applying the Pythagorean Theorem, we find

$$\|\mathbf{v}\| = \sqrt{(-4)^2 + 7^2} = \sqrt{16 + 49} = \sqrt{65}$$

Let θ be the angle made by the positive x-axis and **v**. Let α be the reference angle for θ. Then

$$\tan \alpha = \left| \frac{b}{a} \right| = \left| \frac{7}{-4} \right| = \frac{7}{4}$$

$$\alpha = \tan^{-1} \frac{7}{4} \approx 60° \qquad \bullet \; \alpha \text{ is the reference angle.}$$

$$\theta = 180° - 60° = 120° \qquad \bullet \; \theta \text{ is the angle made by the vector} \\ \text{and the positive } x\text{-axis.}$$

The magnitude of **v** is $\sqrt{65}$, and its direction is $120°$ as measured from the positive x-axis. The angle between a vector and the positive x-axis is called the **direction angle** of the vector. Because **AB** in **Figure 7.26** is equivalent to **v**, $\| \mathbf{AB} \| = \sqrt{65}$ and the direction angle of **AB** is also $120°$.

Expressing vectors in terms of components provides a convenient method for performing operations on vectors.

Fundamental Vector Operations

If $\mathbf{v} = \langle a, b \rangle$ and $\mathbf{w} = \langle c, d \rangle$ are two vectors and k is a real number, then

1. $\| \mathbf{v} \| = \sqrt{a^2 + b^2}$

2. $\mathbf{v} + \mathbf{w} = \langle a, b \rangle + \langle c, d \rangle = \langle a + c, b + d \rangle$

3. $k\mathbf{v} = k\langle a, b \rangle = \langle ka, kb \rangle$

In terms of components, the zero vector $\mathbf{0} = \langle 0, 0 \rangle$. The additive inverse of a vector $\mathbf{v} = \langle a, b \rangle$ is given by $-\mathbf{v} = \langle -a, -b \rangle$.

EXAMPLE 2   Perform Operations on Vectors

Given $\mathbf{v} = \langle -2, 3 \rangle$ and $\mathbf{w} = \langle 4, -1 \rangle$, find

a. $\| \mathbf{w} \|$ b. $\mathbf{v} + \mathbf{w}$ c. $-3\mathbf{v}$ d. $2\mathbf{v} - 3\mathbf{w}$

Solution

a. $\| \mathbf{w} \| = \sqrt{4^2 + (-1)^2} = \sqrt{17}$ c. $-3\mathbf{v} = -3\langle -2, 3 \rangle = \langle 6, -9 \rangle$

b. $\mathbf{v} + \mathbf{w} = \langle -2, 3 \rangle + \langle 4, -1 \rangle$ d. $2\mathbf{v} - 3\mathbf{w} = 2\langle -2, 3 \rangle - 3\langle 4, -1 \rangle$

$\qquad\qquad = \langle -2 + 4, 3 + (-1) \rangle \qquad\qquad\qquad = \langle -4, 6 \rangle - \langle 12, -3 \rangle$

$\qquad\qquad = \langle 2, 2 \rangle \qquad\qquad\qquad\qquad\qquad\qquad = \langle -16, 9 \rangle$

▶ **TRY EXERCISE 20, PAGE 661**

● UNIT VECTORS

A **unit vector** is a vector whose magnitude is 1. For example, the vector $\mathbf{v} = \left\langle \dfrac{3}{5}, -\dfrac{4}{5} \right\rangle$ is a unit vector because

$$\| \mathbf{v} \| = \sqrt{ \left(\frac{3}{5} \right)^2 + \left(-\frac{4}{5} \right)^2 } = \sqrt{ \frac{9}{25} + \frac{16}{25} } = \sqrt{ \frac{25}{25} } = 1$$

Given any nonzero vector **v**, we can obtain a unit vector in the direction of **v** by dividing each component of **v** by the magnitude of **v**, $\|\mathbf{v}\|$.

EXAMPLE 3   Find a Unit Vector

Find a unit vector **u** in the direction of $\mathbf{v} = \langle -4, 2 \rangle$.

Solution

Find the magnitude of **v**.

$$\|\mathbf{v}\| = \sqrt{(-4)^2 + 2^2} = \sqrt{16 + 4} = \sqrt{20} = 2\sqrt{5}$$

Divide each component of **v** by $\|\mathbf{v}\|$.

$$\mathbf{u} = \left\langle \frac{-4}{2\sqrt{5}}, \frac{2}{2\sqrt{5}} \right\rangle = \left\langle \frac{-2}{\sqrt{5}}, \frac{1}{\sqrt{5}} \right\rangle = \left\langle -\frac{2\sqrt{5}}{5}, \frac{\sqrt{5}}{5} \right\rangle.$$

A unit vector in the direction of **v** is **u**.

▶ **TRY EXERCISE 8, PAGE 661**

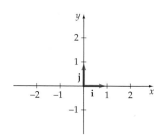

FIGURE 7.27

Two unit vectors, one parallel to the x-axis and one parallel to the y-axis, are of special importance. See **Figure 7.27**.

Definition of Unit Vectors i and j

$$\mathbf{i} = \langle 1, 0 \rangle \qquad \mathbf{j} = \langle 0, 1 \rangle$$

The vector $\mathbf{v} = \langle 3, 4 \rangle$, can be written in terms of the unit vectors **i** and **j** as shown in **Figure 7.28**.

$$\langle 3, 4 \rangle = \langle 3, 0 \rangle + \langle 0, 4 \rangle \qquad \bullet \text{Vector Addition Property}$$
$$= 3\langle 1, 0 \rangle + 4\langle 0, 1 \rangle \qquad \bullet \text{Scalar multiplication of a vector}$$
$$= 3\mathbf{i} + 4\mathbf{j} \qquad \bullet \text{Definition of i and j}$$

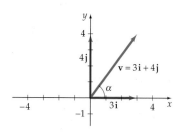

FIGURE 7.28

By means of scalar multiplication and addition of vectors, any vector can be expressed in terms of the unit vectors **i** and **j**. Let $\mathbf{v} = \langle a_1, a_2 \rangle$. Then

$$\mathbf{v} = \langle a_1, a_2 \rangle = a_1 \langle 1, 0 \rangle + a_2 \langle 0, 1 \rangle = a_1 \mathbf{i} + a_2 \mathbf{j}$$

This gives the following result.

Representation of a Vector in Terms of i and j

If **v** is a vector and $\mathbf{v} = \langle a_1, a_2 \rangle$, then $\mathbf{v} = a_1 \mathbf{i} + a_2 \mathbf{j}$.

The rules for addition and scalar multiplication of vectors can be restated in terms of **i** and **j**. If $\mathbf{v} = a_1 \mathbf{i} + a_2 \mathbf{j}$ and $\mathbf{w} = b_1 \mathbf{i} + b_2 \mathbf{j}$, then

$$\mathbf{v} + \mathbf{w} = (a_1 \mathbf{i} + a_2 \mathbf{j}) + (b_1 \mathbf{i} + b_2 \mathbf{j}) = (a_1 + b_1)\mathbf{i} + (a_2 + b_2)\mathbf{j}$$
$$k\mathbf{v} = k(a_1 \mathbf{i} + a_2 \mathbf{j}) = ka_1 \mathbf{i} + ka_2 \mathbf{j}$$

E X A M P L E 4 Operate on Vectors Written in Terms of **i** and **j**

Given $\mathbf{v} = 3\mathbf{i} - 4\mathbf{j}$ and $\mathbf{w} = 5\mathbf{i} + 3\mathbf{j}$, find $3\mathbf{v} - 2\mathbf{w}$.

Solution

$$
\begin{aligned}
3\mathbf{v} - 2\mathbf{w} &= 3(3\mathbf{i} - 4\mathbf{j}) - 2(5\mathbf{i} + 3\mathbf{j}) \\
&= (9\mathbf{i} - 12\mathbf{j}) - (10\mathbf{i} + 6\mathbf{j}) \\
&= (9 - 10)\mathbf{i} + (-12 - 6)\mathbf{j} \\
&= -\mathbf{i} - 18\mathbf{j}
\end{aligned}
$$

▶ **TRY EXERCISE 26, PAGE 661**

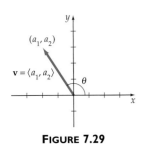

FIGURE 7.29

The components a_1 and a_2 of the vector $\mathbf{v} = \langle a_1, a_2 \rangle$ can be expressed in terms of the magnitude of $\mathbf{v}$ and the direction angle of $\mathbf{v}$ (the angle that $\mathbf{v}$ makes with the positive x-axis). Consider the vector $\mathbf{v}$ in **Figure 7.29**. Then

$$\|\mathbf{v}\| = \sqrt{(a_1)^2 + (a_2)^2}$$

From the definitions of sine and cosine, we have

$$\cos\theta = \frac{a_1}{\|\mathbf{v}\|} \quad \text{and} \quad \sin\theta = \frac{a_2}{\|\mathbf{v}\|}$$

Rewriting the last two equations, we find that the components of $\mathbf{v}$ are

$$a_1 = \|\mathbf{v}\|\cos\theta \quad \text{and} \quad a_2 = \|\mathbf{v}\|\sin\theta$$

Horizontal and Vertical Components of a Vector

Let $\mathbf{v} = \langle a_1, a_2 \rangle$, where $\mathbf{v} \neq \mathbf{0}$, the zero vector. Then

$$a_1 = \|\mathbf{v}\|\cos\theta \quad \text{and} \quad a_2 = \|\mathbf{v}\|\sin\theta$$

where θ is the angle between the positive x-axis and $\mathbf{v}$.

The **horizontal component** of $\mathbf{v}$ is $\|\mathbf{v}\|\cos\theta$. The **vertical component** of $\mathbf{v}$ is $\|\mathbf{v}\|\sin\theta$.

❓ **QUESTION** Is $\mathbf{u} = \cos\theta\,\mathbf{i} + \sin\theta\,\mathbf{j}$ a unit vector?

Any nonzero vector can be written in terms of its horizontal and vertical components. Let $\mathbf{v} = a_1\mathbf{i} + a_2\mathbf{j}$. Then

$$
\begin{aligned}
\mathbf{v} &= a_1\mathbf{i} + a_2\mathbf{j} \\
&= \big(\|\mathbf{v}\|\cos\theta\big)\mathbf{i} + \big(\|\mathbf{v}\|\sin\theta\big)\mathbf{j} \\
&= \|\mathbf{v}\|(\cos\theta\,\mathbf{i} + \sin\theta\,\mathbf{j})
\end{aligned}
$$

$\|\mathbf{v}\|$ is the magnitude of $\mathbf{v}$, and the vector $\cos\theta\,\mathbf{i} + \sin\theta\,\mathbf{j}$ is a unit vector. The last equation shows that any vector $\mathbf{v}$ can be written as the product of its magnitude and a unit vector in the direction of $\mathbf{v}$.

❓ **ANSWER** Yes, because $\|\cos\theta\,\mathbf{i} + \sin\theta\,\mathbf{j}\| = \sqrt{\cos^2\theta + \sin^2\theta} = \sqrt{1} = 1$.

| EXAMPLE 5 | **Find the Horizontal and Vertical Components of a Vector** |

Find the approximate horizontal and vertical components of a vector **v** of magnitude 10 meters with direction angle 228°. Write the vector in the form $\mathbf{v} = a_1\mathbf{i} + a_2\mathbf{j}$.

Solution

$a_1 = 10 \cos 228° \approx -6.7$

$a_2 = 10 \sin 228° \approx -7.4$

The approximate horizontal and vertical components are -6.7 and -7.4, respectively.

$$\mathbf{v} \approx -6.7\mathbf{i} - 7.4\mathbf{j}$$

▶ TRY EXERCISE 36, PAGE 661

---

take note

Ground speed is the magnitude of the resultant of the plane's velocity vector and the wind velocity vector.

● APPLICATIONS OF VECTORS

Consider an object on which two vectors are acting simultaneously. This occurs when a boat is moving in a current or an airplane is flying in a wind. The **airspeed** of a plane is the speed at which the plane would be moving if there were no wind. The actual velocity of a plane is the velocity relative to the ground. The magnitude of the actual velocity is the **ground speed.**

| EXAMPLE 6 | **Solve an Application Involving Airspeed** |

An airplane is traveling with an airspeed of 320 mph and a heading of 62°. A wind of 42 mph is blowing at a heading of 125°. Find the ground speed and the course of the airplane.

Solution

Sketch a diagram similar to **Figure 7.30** showing the relevant vectors. **AB** represents the heading and the airspeed, **AD** represents the wind velocity, and **AC** represents the course and the ground speed. By vector addition, **AC = AB + AD**. From the figure,

$$\mathbf{AB} = 320(\cos 28°\mathbf{i} + \sin 28°\mathbf{j})$$
$$\mathbf{AD} = 42[\cos(-35°)\mathbf{i} + \sin(-35°)\mathbf{j}]$$
$$\mathbf{AC} = 320(\cos 28°\mathbf{i} + \sin 28°\mathbf{j}) + 42[\cos(-35°)\mathbf{i} + \sin(-35°)\mathbf{j}]$$
$$\approx (282.5\mathbf{i} + 150.2\mathbf{j}) + (34.4\mathbf{i} - 24.1\mathbf{j})$$
$$= 316.9\mathbf{i} + 126.1\mathbf{j}$$

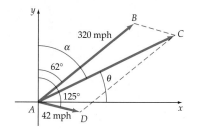

FIGURE 7.30

AC is the course of the plane. The ground speed is $\|\mathbf{AC}\|$. The heading is $\alpha = 90° - \theta$.

$$\|\mathbf{AC}\| = \sqrt{(316.9)^2 + (126.1)^2} \approx 340$$

$$\alpha = 90° - \theta = 90° - \tan^{-1}\left(\frac{126.1}{316.9}\right) \approx 68°$$

The ground speed is approximately 340 mph at a heading of 68°.

▶ TRY EXERCISE 40, PAGE 662

There are numerous problems involving force that can be solved by using vectors. One type involves objects that are resting on a ramp. For these problems, we frequently try to find the components of a force vector relative to the ramp rather than to the x-axis.

EXAMPLE 7 **Solve an Application Involving Force**

A 110-pound box is on a 24° ramp. Find the component of the force that is parallel to the ramp.

Solution

The force-of-gravity vector (**OB**) is the sum of two components, one parallel to the ramp, **AB**, and the other (called the *normal component*) perpendicular to the ramp, **OA**. (See **Figure 7.31**). **AB** is the vector that represents the force tending to move the box down the ramp. Because triangle OAB is a right triangle and $\angle AOB$ is 24°,

$$\sin 24° = \frac{\|\mathbf{AB}\|}{110}$$

$$\|\mathbf{AB}\| = 110 \sin 24° \approx 45$$

The component of the force parallel to the ramp is approximately 45 pounds.

▶ **TRY EXERCISE 42, PAGE 662**

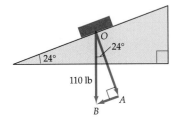

FIGURE 7.31

● DOT PRODUCT

We have considered the product of a real number (scalar) and a vector. We now turn our attention to the product of two vectors. Finding the *dot product* of two vectors is one way to multiply a vector by a vector. The dot product of two vectors is a real number and *not* a vector. The dot product is also called the *inner product* or the *scalar product*. This product is useful in engineering and physics.

Definition of Dot Product

Given $\mathbf{v} = \langle a, b \rangle$ and $\mathbf{w} = \langle c, d \rangle$, the **dot product** of $\mathbf{v}$ and $\mathbf{w}$ is given by

$$\mathbf{v} \cdot \mathbf{w} = ac + bd$$

EXAMPLE 8 **Find the Dot Product of Two Vectors**

Find the dot product of $\mathbf{v} = \langle 6, -2 \rangle$ and $\mathbf{w} = \langle -2, 4 \rangle$.

Solution

$$\mathbf{v} \cdot \mathbf{w} = 6(-2) + (-2)4 = -12 - 8 = -20$$

▶ **TRY EXERCISE 50, PAGE 662**

❷ QUESTION Is the dot product of two vectors a vector or a real number?

If the vectors in Example 8 were given in terms of the vectors **i** and **j**, then **v** = 6**i** − 2**j** and **w** = −2**i** + 4**j**. In this case,

$$\mathbf{v} \cdot \mathbf{w} = (6\mathbf{i} - 2\mathbf{j}) \cdot (-2\mathbf{i} + 4\mathbf{j}) = 6(-2) + (-2)4 = -20$$

Properties of the Dot Product

In the following properties, **u**, **v**, and **w** are vectors and a is a scalar.

1. $\mathbf{v} \cdot \mathbf{w} = \mathbf{w} \cdot \mathbf{v}$

2. $\mathbf{u} \cdot (\mathbf{v} + \mathbf{w}) = \mathbf{u} \cdot \mathbf{v} + \mathbf{u} \cdot \mathbf{w}$

3. $a(\mathbf{u} \cdot \mathbf{v}) = (a\mathbf{u}) \cdot \mathbf{v} = \mathbf{u} \cdot (a\mathbf{v})$

4. $\mathbf{v} \cdot \mathbf{v} = \|\mathbf{v}\|^2$

5. $\mathbf{0} \cdot \mathbf{v} = 0$

6. $\mathbf{i} \cdot \mathbf{i} = \mathbf{j} \cdot \mathbf{j} = 1$

7. $\mathbf{i} \cdot \mathbf{j} = \mathbf{j} \cdot \mathbf{i} = 0$

The proofs of these properties follow from the definition of dot product. Here is the proof of the fourth property. Let $\mathbf{v} = a\mathbf{i} + b\mathbf{j}$.

$$\mathbf{v} \cdot \mathbf{v} = (a\mathbf{i} + b\mathbf{j}) \cdot (a\mathbf{i} + b\mathbf{j}) = a^2 + b^2 = \|\mathbf{v}\|^2$$

Rewriting the fourth property of the dot product yields an alternative way of expressing the magnitude of a vector.

Magnitude of a Vector in Terms of the Dot Product

If $\mathbf{v} = \langle a, b \rangle$, then $\|\mathbf{v}\| = \sqrt{\mathbf{v} \cdot \mathbf{v}}$.

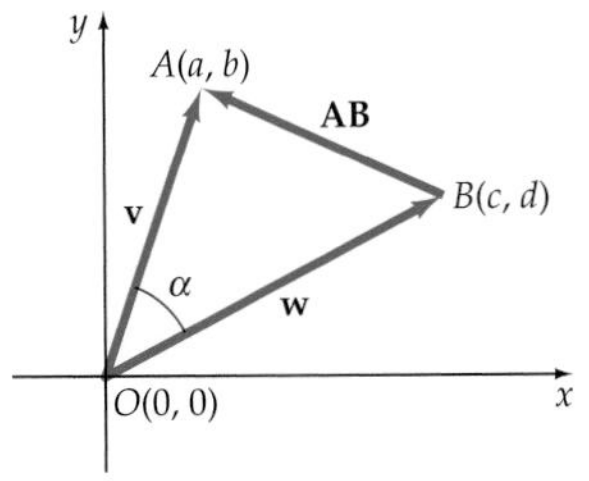

FIGURE 7.32

The Law of Cosines can be used to derive an alternative formula for the dot product. Consider the vectors $\mathbf{v} = \langle a, b \rangle$ and $\mathbf{w} = \langle c, d \rangle$ as shown in **Figure 7.32.** Using the Law of Cosines for triangle OAB, we have

$$\|\mathbf{AB}\|^2 = \|\mathbf{v}\|^2 + \|\mathbf{w}\|^2 - 2\|\mathbf{v}\|\|\mathbf{w}\| \cos \alpha$$

By the distance formula, $\|\mathbf{AB}\|^2 = (a - c)^2 + (b - d)^2$, $\|\mathbf{v}\|^2 = a^2 + b^2$, and $\|\mathbf{w}\|^2 = c^2 + d^2$. Thus

$$(a - c)^2 + (b - d)^2 = (a^2 + b^2) + (c^2 + d^2) - 2\|\mathbf{v}\|\|\mathbf{w}\|\cos \alpha$$

$$a^2 - 2ac + c^2 + b^2 - 2bd + d^2 = a^2 + b^2 + c^2 + d^2 - 2\|\mathbf{v}\|\|\mathbf{w}\| \cos \alpha$$

$$-2ac - 2bd = -2\|\mathbf{v}\|\|\mathbf{w}\| \cos \alpha$$

$$ac + bd = \|\mathbf{v}\|\|\mathbf{w}\| \cos \alpha$$

$$\mathbf{v} \cdot \mathbf{w} = \|\mathbf{v}\|\|\mathbf{w}\| \cos \alpha \qquad \bullet \mathbf{v} \cdot \mathbf{w} = ac + bd$$

❷ ANSWER A real number

> ### Alternative Formula for the Dot Product
>
> If $\mathbf{v}$ and $\mathbf{w}$ are two nonzero vectors and α is the smallest nonnegative angle between $\mathbf{v}$ and $\mathbf{w}$, then $\mathbf{v} \cdot \mathbf{w} = \|\mathbf{v}\| \|\mathbf{w}\| \cos \alpha$.

Solving the alternative formula for the dot product for $\cos \alpha$, we have a formula for the cosine of the angle between two vectors.

> ### Angle Between Two Vectors
>
> If $\mathbf{v}$ and $\mathbf{w}$ are two nonzero vectors and α is the smallest nonnegative angle between $\mathbf{v}$ and $\mathbf{w}$, then $\cos \alpha = \dfrac{\mathbf{v} \cdot \mathbf{w}}{\|\mathbf{v}\| \|\mathbf{w}\|}$ and $\alpha = \cos^{-1}\left(\dfrac{\mathbf{v} \cdot \mathbf{w}}{\|\mathbf{v}\| \|\mathbf{w}\|}\right)$.

EXAMPLE 9 Find the Angle Between Two Vectors

Find the measure of the smallest positive angle between the vectors $\mathbf{v} = 2\mathbf{i} - 3\mathbf{j}$ and $\mathbf{w} = -\mathbf{i} + 5\mathbf{j}$, as shown in **Figure 7.33.**

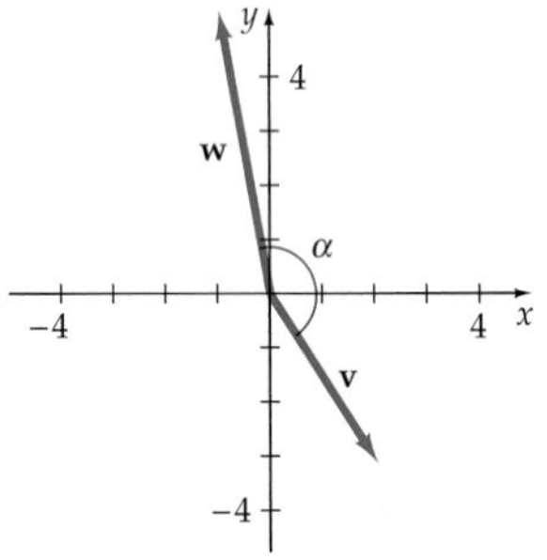

Solution

Use the equation for the angle between two vectors.

$$\cos \alpha = \frac{\mathbf{v} \cdot \mathbf{w}}{\|\mathbf{v}\|\|\mathbf{w}\|} = \frac{(2\mathbf{i} - 3\mathbf{j}) \cdot (-\mathbf{i} + 5\mathbf{j})}{(\sqrt{2^2 + (-3)^2})(\sqrt{(-1)^2 + 5^2})}$$

$$= \frac{-2 - 15}{\sqrt{13}\sqrt{26}} = \frac{-17}{\sqrt{338}}$$

$$\alpha = \cos^{-1}\left(\frac{-17}{\sqrt{338}}\right) \approx 157.6°$$

The angle between the two vectors is approximately $157.6°$.

▶ **TRY EXERCISE 60, PAGE 662**

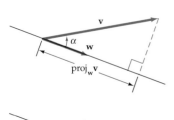

FIGURE 7.33

● SCALAR PROJECTION

Let $\mathbf{v} = \langle a_1, a_2 \rangle$ and $\mathbf{w} = \langle b_1, b_2 \rangle$ be two nonzero vectors, and let α be the angle between the vectors. Two possible configurations, one for which α is an acute angle and one for which α is an obtuse angle, are shown in **Figure 7.34.** In each case, a right triangle is formed by drawing a line segment from the head of $\mathbf{v}$ to a line through $\mathbf{w}$.

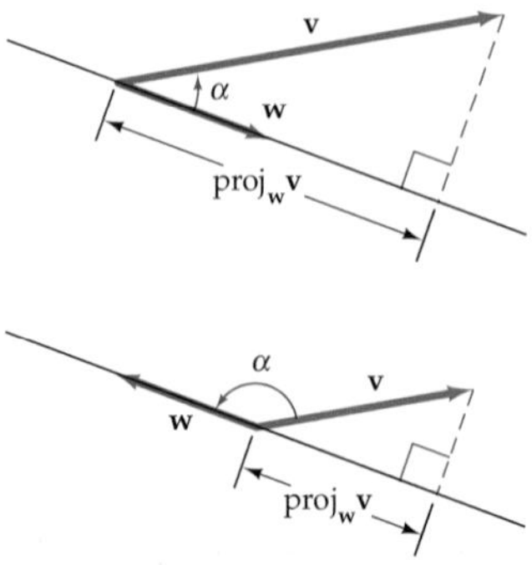

> ### Definition of the Scalar Projection of v on w
>
> If $\mathbf{v}$ and $\mathbf{w}$ are two nonzero vectors and α is the smallest positive angle between $\mathbf{v}$ and $\mathbf{w}$, then the scalar projection of $\mathbf{v}$ on $\mathbf{w}$, $\text{proj}_{\mathbf{w}}\mathbf{v}$, is given by
>
> $$\text{proj}_{\mathbf{w}}\mathbf{v} = \|\mathbf{v}\| \cos \alpha$$

FIGURE 7.34

To derive an alternate formula for $\text{proj}_w\mathbf{v}$, consider the dot product $\mathbf{v} \cdot \mathbf{w} = \|\mathbf{v}\| \|\mathbf{w}\| \cos \alpha$. Solving for $\|\mathbf{v}\| \cos \alpha$, which is $\text{proj}_w\mathbf{v}$, we have

$$\text{proj}_w\mathbf{v} = \frac{\mathbf{v} \cdot \mathbf{w}}{\|\mathbf{w}\|}$$

When the angle α between the two vectors is an acute angle, $\text{proj}_w\mathbf{v}$ is positive. When α is an obtuse angle, $\text{proj}_w\mathbf{v}$ is negative.

EXAMPLE 10   Find the Projection of v on w

Given $\mathbf{v} = 2\mathbf{i} + 4\mathbf{j}$ and $\mathbf{w} = -2\mathbf{i} + 8\mathbf{j}$ as shown in **Figure 7.35,** find $\text{proj}_w\mathbf{v}$.

Solution

Use the equation $\text{proj}_w\mathbf{v} = \dfrac{\mathbf{v} \cdot \mathbf{w}}{\|\mathbf{w}\|}$.

$$\text{proj}_w \mathbf{v} = \frac{(2\mathbf{i} + 4\mathbf{j}) \cdot (-2\mathbf{i} + 8\mathbf{j})}{\sqrt{(-2)^2 + 8^2}} = \frac{28}{\sqrt{68}} = \frac{14\sqrt{17}}{17} \approx 3.4$$

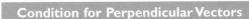 ► **TRY EXERCISE 62, PAGE 662**

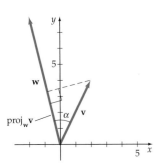

FIGURE 7.35

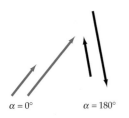

FIGURE 7.36

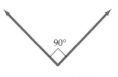

FIGURE 7.37

● **PARALLEL AND PERPENDICULAR VECTORS**

Two vectors are *parallel* when the angle α between the vectors is $0°$ or $180°$, as shown in **Figure 7.36.** When the angle α is $0°$, the vectors point in the same direction; the vectors point in opposite directions when α is $180°$.

Let $\mathbf{v} = a_1\mathbf{i} + b_1\mathbf{j}$, let c be a real number, and let $\mathbf{w} = c\mathbf{v}$. Because $\mathbf{w}$ is a constant multiple of $\mathbf{v}$, $\mathbf{w}$ and $\mathbf{v}$ are parallel vectors. When $c > 0$, the vectors point in the same direction. When $c < 0$, the vectors point in opposite directions.

Two vectors are *perpendicular* when the angle between the vectors is $90°$. See **Figure 7.37.** Perpendicular vectors are referred to as **orthogonal vectors.** If $\mathbf{v}$ and $\mathbf{w}$ are two nonzero orthogonal vectors, then from the formula for the angle between two vectors and the fact that $\cos \alpha = 0$, we have

$$0 = \frac{\mathbf{v} \cdot \mathbf{w}}{\|\mathbf{v}\| \|\mathbf{w}\|}$$

If a fraction equals zero, the numerator must be zero. Thus, for orthogonal vectors $\mathbf{v}$ and $\mathbf{w}$, $\mathbf{v} \cdot \mathbf{w} = 0$. This gives the following result.

Condition for Perpendicular Vectors

Two nonzero vectors $\mathbf{v}$ and $\mathbf{w}$ are orthogonal if and only if $\mathbf{v} \cdot \mathbf{w} = 0$.

● **WORK: AN APPLICATION OF THE DOT PRODUCT**

When a 5-pound force is used to lift a box from the ground a distance of 4 feet, *work* is done. The amount of **work** is the product of the force on the box and the distance the box is moved. In this case the work is 20 foot-pounds. When the box

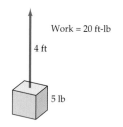

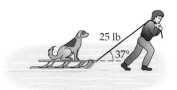

FIGURE 7.38

is lifted, the force and the displacement vector (the direction in which and the distance the box was moved) are in the same direction. (See **Figure 7.38**.)

Now consider a sled being pulled by a child along the ground by a rope attached to the sled, as shown in **Figure 7.39**. The force vector (along the rope) is *not* in the same direction as the displacement vector (parallel to the ground). In this case the dot product is used to determine the work done by the force.

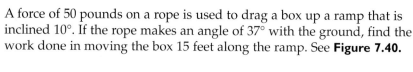

Definition of Work

The work W done by a force $\mathbf{F}$ applied along a displacement $\mathbf{s}$ is

$$W = \mathbf{F} \cdot \mathbf{s}$$

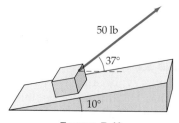

FIGURE 7.39

In the case of the child pulling the sled 7 feet, the work done is

$$
\begin{aligned}
W &= \mathbf{F} \cdot \mathbf{s} \\
&= \|\mathbf{F}\| \, \|\mathbf{s}\| \cos \alpha \qquad \text{• } \alpha \text{ is the angle between } \mathbf{F} \text{ and } \mathbf{s.} \\
&= (25)(7) \cos 37° \approx 140 \text{ foot-pounds}
\end{aligned}
$$

EXAMPLE II   Solve a Work Problem

A force of 50 pounds on a rope is used to drag a box up a ramp that is inclined 10°. If the rope makes an angle of 37° with the ground, find the work done in moving the box 15 feet along the ramp. See **Figure 7.40**.

Solution

We will provide two solutions to this example.

Method 1   From the last section, $\mathbf{u} = \cos 10°\mathbf{i} + \sin 10°\mathbf{j}$ is a unit vector parallel to the ramp. Multiplying $\mathbf{u}$ by 15 (the magnitude of the displacement vector) gives $\mathbf{s} = 15(\cos 10°\mathbf{i} + \sin 10°\mathbf{j})$. Similarly, the force vector is $\mathbf{F} = 50(\cos 37°\mathbf{i} + \sin 37°\mathbf{j})$. The work done is given by the dot product.

$$
\begin{aligned}
W &= \mathbf{F} \cdot \mathbf{s} = 50(\cos 37°\mathbf{i} + \sin 37°\mathbf{j}) \cdot 15(\cos 10°\mathbf{i} + \sin 10°\mathbf{j}) \\
&= [50 \cdot 15](\cos 37° \cos 10° + \sin 37° \sin 10°) \approx 668.3 \text{ foot-pounds}
\end{aligned}
$$

Method 2   When we write the work equation as $W = \|\mathbf{F}\|\|\mathbf{s}\| \cos \alpha$, α is the angle between the force and the displacement. Thus $\alpha = 37° - 10° = 27°$. The work done is

$$W = \|\mathbf{F}\|\|\mathbf{s}\| \cos \alpha = 50 \cdot 15 \cdot \cos 27° \approx 668.3 \text{ foot-pounds}$$

▶ **TRY EXERCISE 70, PAGE 662**

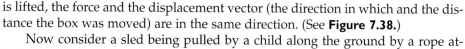

50 lb

37°

10°

FIGURE 7.40

 TOPICS FOR DISCUSSION

1.   Is the dot product of two vectors a vector or a scalar? Explain.

2.   Is the projection of $\mathbf{v}$ on $\mathbf{w}$ a vector or a scalar? Explain.

3. Is the nonzero vector $\langle a, b \rangle$ perpendicular to the vector $\langle -b, a \rangle$? Explain.

4. Explain how to determine the angle between the vector $\langle 3, 4 \rangle$ and the vector $\langle 5, -1 \rangle$.

5. Consider the nonzero vector $\mathbf{u} = \langle a, b \rangle$ and the vector
$$\mathbf{v} = \left\langle \frac{a}{\sqrt{a^2 + b^2}}, \frac{b}{\sqrt{a^2 + b^2}} \right\rangle.$$

   a. Are the vectors parallel? Explain.

   b. Which one of the vectors is a unit vector?

   c. Which vector has the larger magnitude? Explain.

EXERCISE SET 7.3

In Exercises 1 to 6, find the components of a vector with the given initial and terminal points. Write an equivalent vector in terms of its components.

1. $P_1(-3, 0); P_2(4, -1)$

2. $P_1(5, -1); P_2(3, 1)$

3. $P_1(4, 2); P_2(-3, -3)$

4. $P_1(0, -3); P_2(0, 4)$

5. $P_1(2, -5); P_2(2, 3)$

▶ **6.** $P_1(3, -2); P_2(3, 0)$

In Exercises 7 to 14, find the magnitude and direction of each vector. Find the unit vector in the direction of the given vector.

7. $\mathbf{v} = \langle -3, 4 \rangle$

▶ **8.** $\mathbf{v} = \langle 6, 10 \rangle$

9. $\mathbf{v} = \langle 20, -40 \rangle$

10. $\mathbf{v} = \langle -50, 30 \rangle$

11. $\mathbf{v} = 2\mathbf{i} - 4\mathbf{j}$

12. $\mathbf{v} = -5\mathbf{i} + 6\mathbf{j}$

13. $\mathbf{v} = 42\mathbf{i} - 18\mathbf{j}$

14. $\mathbf{v} = -22\mathbf{i} - 32\mathbf{j}$

In Exercises 15 to 23, perform the indicated operations where $\mathbf{u} = \langle -2, 4 \rangle$ and $\mathbf{v} = \langle -3, -2 \rangle$.

15. $3\mathbf{u}$

16. $-4\mathbf{v}$

17. $2\mathbf{u} - \mathbf{v}$

18. $4\mathbf{v} - 2\mathbf{u}$

19. $\frac{2}{3}\mathbf{u} + \frac{1}{6}\mathbf{v}$

▶ **20.** $\frac{3}{4}\mathbf{u} - 2\mathbf{v}$

21. $\|\mathbf{u}\|$

22. $\|\mathbf{v} + 2\mathbf{u}\|$

23. $\|3\mathbf{u} - 4\mathbf{v}\|$

In Exercises 24 to 32, perform the indicated operations where $\mathbf{u} = 3\mathbf{i} - 2\mathbf{j}$ and $\mathbf{v} = -2\mathbf{i} + 3\mathbf{j}$.

24. $-2\mathbf{u}$

25. $4\mathbf{v}$

▶ **26.** $3\mathbf{u} + 2\mathbf{v}$

27. $6\mathbf{u} + 2\mathbf{v}$

28. $\frac{1}{2}\mathbf{u} - \frac{3}{4}\mathbf{v}$

29. $\frac{2}{3}\mathbf{v} + \frac{3}{4}\mathbf{u}$

30. $\|\mathbf{v}\|$

31. $\|\mathbf{u} - 2\mathbf{v}\|$

32. $\|2\mathbf{v} + 3\mathbf{u}\|$

In Exercises 33 to 36, find the horizontal and vertical components of each vector. Write an equivalent vector in the form $\mathbf{v} = a_1\mathbf{i} + a_2\mathbf{j}$.

33. Magnitude = 5, direction angle = $27°$

34. Magnitude = 4, direction angle = $127°$

35. Magnitude = 4, direction angle = $\dfrac{\pi}{4}$

▶ **36.** Magnitude = 2, direction angle = $\dfrac{8\pi}{7}$

37. **GROUND SPEED OF A PLANE** A plane is flying at an airspeed of 340 mph at a heading of 124°. A wind of 45 mph is blowing from the west. Find the ground speed of the plane.

38. **HEADING OF A BOAT** A person who can row 2.6 mph in still water wants to row due east across a river. The river is flowing from the north at a rate of 0.8 mph. Determine the heading of the boat that will be required for it to travel due east across the river.

39. **GROUND SPEED AND COURSE OF A PLANE** A pilot is flying at a heading of 96° at 225 mph. A 50-mph wind is blowing from the southwest at a heading of 37°. Find the ground speed and course of the plane.

▶ **40.** COURSE OF A BOAT The captain of a boat is steering at a heading of 327° at 18 mph. The current is flowing at 4 mph at a heading of 60°. Find the course (to the nearest degree) of the boat.

41. MAGNITUDE OF A FORCE Find the magnitude of the force necessary to keep a 3000-pound car from sliding down a ramp inclined at an angle of 5.6°.

▶ **42.** ANGLE OF A RAMP A 120-pound force keeps an 800-pound object from sliding down an inclined ramp. Find the angle of the ramp.

43. MAGNITUDE OF THE NORMAL COMPONENT A 25-pound box is resting on a ramp that is inclined 9.0°. Find the magnitude of the normal component of force.

44. MAGNITUDE OF THE NORMAL COMPONENT Find the magnitude of the normal component of force for a 50-pound crate that is resting on a ramp that is inclined 12°.

In Exercises 45 to 52, find the dot product of the vectors.

45. $\mathbf{v} = \langle 3, -2 \rangle; \mathbf{w} = \langle 1, 3 \rangle$ **46.** $\mathbf{v} = \langle 2, 4 \rangle; \mathbf{w} = \langle 0, 2 \rangle$

47. $\mathbf{v} = \langle 4, 1 \rangle; \mathbf{w} = \langle -1, 4 \rangle$ **48.** $\mathbf{v} = \langle 2, -3 \rangle; \mathbf{w} = \langle 3, 2 \rangle$

49. $\mathbf{v} = \mathbf{i} + 2\mathbf{j}; \mathbf{w} = -\mathbf{i} + \mathbf{j}$

▶ **50.** $\mathbf{v} = 5\mathbf{i} + 3\mathbf{j}; \mathbf{w} = 4\mathbf{i} - 2\mathbf{j}$

51. $\mathbf{v} = 6\mathbf{i} - 4\mathbf{j}; \mathbf{w} = -2\mathbf{i} - 3\mathbf{j}$

52. $\mathbf{v} = -4\mathbf{i} + 2\mathbf{j}; \mathbf{w} = -2\mathbf{i} - 4\mathbf{j}$

In Exercises 53 to 60, find the angle between the two vectors. State which pairs of vectors are orthogonal.

53. $\mathbf{v} = \langle 2, -1 \rangle; \mathbf{w} = \langle 3, 4 \rangle$ **54.** $\mathbf{v} = \langle 1, -5 \rangle; \mathbf{w} = \langle -2, 3 \rangle$

55. $\mathbf{v} = \langle 0, 3 \rangle; \mathbf{w} = \langle 2, 2 \rangle$ **56.** $\mathbf{v} = \langle -1, 7 \rangle; \mathbf{w} = \langle 3, -2 \rangle$

57. $\mathbf{v} = 5\mathbf{i} - 2\mathbf{j}; \mathbf{w} = 2\mathbf{i} + 5\mathbf{j}$

58. $\mathbf{v} = 8\mathbf{i} + \mathbf{j}; \mathbf{w} = -\mathbf{i} + 8\mathbf{j}$

59. $\mathbf{v} = 5\mathbf{i} + 2\mathbf{j}; \mathbf{w} = -5\mathbf{i} - 2\mathbf{j}$

▶ **60.** $\mathbf{v} = 3\mathbf{i} - 4\mathbf{j}; \mathbf{w} = 6\mathbf{i} - 12\mathbf{j}$

In Exercises 61 to 68, find $\text{proj}_{\mathbf{w}}\mathbf{v}$.

61. $\mathbf{v} = \langle 6, 7 \rangle; \mathbf{w} = \langle 3, 4 \rangle$ ▶ **62.** $\mathbf{v} = \langle -7, 5 \rangle; \mathbf{w} = \langle -4, 1 \rangle$

63. $\mathbf{v} = \langle -3, 4 \rangle; \mathbf{w} = \langle 2, 5 \rangle$ **64.** $\mathbf{v} = \langle 2, 4 \rangle; \mathbf{w} = \langle -1, 5 \rangle$

65. $\mathbf{v} = 2\mathbf{i} + \mathbf{j}; \mathbf{w} = 6\mathbf{i} + 3\mathbf{j}$

66. $\mathbf{v} = 5\mathbf{i} + 2\mathbf{j}; \mathbf{w} = -5\mathbf{i} - 2\mathbf{j}$

67. $\mathbf{v} = 3\mathbf{i} - 4\mathbf{j}; \mathbf{w} = -6\mathbf{i} + 12\mathbf{j}$

68. $\mathbf{v} = 2\mathbf{i} + 2\mathbf{j}; \mathbf{w} = -4\mathbf{i} - 2\mathbf{j}$

69. WORK A 150-pound box is dragged 15 feet along a level floor. Find the work done if a force of 75 pounds at an angle of 32° is used.

▶ **70.** WORK A 100-pound force is pulling a sled loaded with bricks that weighs 400 pounds. The force is at an angle of 42° with the displacement. Find the work done in moving the sled 25 feet.

71. WORK A rope is being used to pull a box up a ramp that is inclined at 15°. The rope exerts a force of 75 pounds on the box, and it makes an angle of 30° with the plane of the ramp. Find the work done in moving the box 12 feet.

72. WORK A dock worker exerts a force on a box sliding down the ramp of a truck. The ramp makes an angle of 48° with the road, and the worker exerts a 50-pound force parallel to the road. Find the work done in sliding the box 6 feet.

CONNECTING CONCEPTS

73. For $\mathbf{u} = \langle -1, 1 \rangle$, $\mathbf{v} = \langle 2, 3 \rangle$, and $\mathbf{w} = \langle 5, 5 \rangle$, find the sum of the three vectors geometrically by using the triangle method of adding vectors.

74. For $\mathbf{u} = \langle 1, 2 \rangle$, $\mathbf{v} = \langle 3, -2 \rangle$, and $\mathbf{w} = \langle -1, 4 \rangle$, find $\mathbf{u} + \mathbf{v} - \mathbf{w}$ geometrically by using the triangle method of adding vectors.

75. Find a vector that has the initial point $(3, -1)$ and is equivalent to $\mathbf{v} = 2\mathbf{i} - 3\mathbf{j}$.

76. Find a vector that has the initial point $(-2, 4)$ and is equivalent to $\mathbf{v} = \langle -1, 3 \rangle$.

77. If $\mathbf{v} = 2\mathbf{i} - 5\mathbf{j}$ and $\mathbf{w} = 5\mathbf{i} + 2\mathbf{j}$ have the same initial point, is $\mathbf{v}$ perpendicular to $\mathbf{w}$? Why or why not?

78. If $\mathbf{v} = \langle 5, 6 \rangle$ and $\mathbf{w} = \langle 6, 5 \rangle$ have the same initial point, is $\mathbf{v}$ perpendicular to $\mathbf{w}$? Why or why not?

79. Let $\mathbf{v} = \langle -2, 7 \rangle$. Find a vector perpendicular to $\mathbf{v}$.

80. Let $\mathbf{w} = 4\mathbf{i} + \mathbf{j}$. Find a vector perpendicular to $\mathbf{w}$.

In Example 7 of this section, if the box were to be kept from sliding down the ramp, it would be necessary to provide a force of 45 pounds parallel to the ramp but pointed *up* the ramp. Some of this force would be provided by a frictional force between the box and the ramp. The force of friction is $\mathbf{F}_\mu = \mu \mathbf{N}$, where μ is a constant called the coefficient of friction, and N is the normal component of the force of gravity. In Exercises 81 and 82, find the frictional force.

81. FRICTIONAL FORCE A 50-pound box is resting on a ramp inclined at 12°. Find the force of friction if the coefficient of friction, μ, is 0.13.

82. FRICTIONAL FORCE A car weighing 2500 pounds is resting on a ramp inclined at 15°. Find the frictional force if the coefficient of friction, μ, is 0.21.

83. Is the dot product an associative operation? That is, given any nonzero vectors $\mathbf{u}$, $\mathbf{v}$, and $\mathbf{w}$, does

$$(\mathbf{u} \cdot \mathbf{v}) \cdot \mathbf{w} = \mathbf{u} \cdot (\mathbf{v} \cdot \mathbf{w})?$$

84. Prove that $\mathbf{v} \cdot \mathbf{w} = \mathbf{w} \cdot \mathbf{v}$.

85. Prove that $c(\mathbf{v} \cdot \mathbf{w}) = (c\mathbf{v}) \cdot \mathbf{w}$.

86. Show that the dot product of two nonzero vectors is positive if the angle between the vectors is an acute angle and negative if the angle between the two vectors is an obtuse angle.

87. COMPARISON OF WORK DONE Consider the following two situations. (1) A rope is being used to pull a box up a ramp inclined at an angle α. The rope exerts a force $\mathbf{F}$ on the box, and the rope makes an angle θ with the ramp. The box is pulled s feet. (2) A rope is being used to pull a box along a level floor. The rope exerts the same force $\mathbf{F}$ on the box. The box is pulled the same s feet. In which case is more work done?

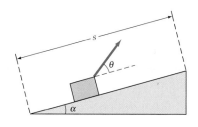

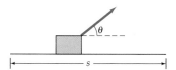

PREPARE FOR SECTION 7.4

88. Simplify: $(1 + i)(2 + i)$ [P.6]

89. Simplify: $\dfrac{2 + i}{3 - i}$ [P.6]

90. What is the conjugate of $2 + 3i$? [P.6]

91. What is the conjugate of $3 - 5i$? [P.6]

92. Use the quadratic formula to find the solutions of $x^2 + x = -1$. [1.3]

93. Solve: $x^2 + 9 = 0$ [1.3]

PROJECTS

1. SAME DIRECTION OR OPPOSITE DIRECTIONS Let $\mathbf{v} = c\mathbf{w}$, where c is a nonzero real number and $\mathbf{w}$ is a nonzero vector. Show that $\dfrac{\mathbf{v} \cdot \mathbf{w}}{\|\mathbf{v}\| \|\mathbf{w}\|} = \pm 1$ and that the result is 1 when $c > 0$ and -1 when $c < 0$.

2. THE LAW OF COSINES AND VECTORS Prove that $\|\mathbf{v} - \mathbf{w}\|^2 = \|\mathbf{v}\|^2 + \|\mathbf{w}\|^2 - 2\mathbf{v} \cdot \mathbf{w}$.

3. PROJECTION RELATIONSHIPS What is the relationship between the nonzero vectors $\mathbf{v}$ and $\mathbf{w}$ if

a. $\text{proj}_\mathbf{w} \mathbf{v} = 0$?   **b.** $\text{proj}_\mathbf{w} \mathbf{v} = \|\mathbf{v}\|$?

TRIGONOMETRIC FORM OF COMPLEX NUMBERS

● GRAPHICAL REPRESENTATION OF A COMPLEX NUMBER

Real numbers are graphed as points on a number line. Complex numbers can be graphed in a coordinate plane called the **complex plane.** The horizontal axis of the complex plane is called the **real axis;** the vertical axis is called the **imaginary axis.**

A complex number written in the form $z = a + bi$ is written in **standard form** or **rectangular form.** The graph of $a + bi$ is associated with the point $P(a, b)$ in the complex plane. **Figure 7.41** shows the graphs of several complex numbers.

● ABSOLUTE VALUE OF A COMPLEX NUMBER

See Section P.6 page 66 if you need to review complex numbers.

The length of the line segment from the origin to the point $(-3, 4)$ in the complex plane is the *absolute value* of $z = -3 + 4i$. See **Figure 7.42.** From the Pythagorean

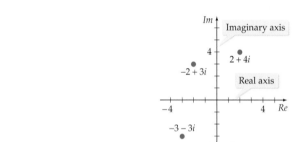

FIGURE 7.41

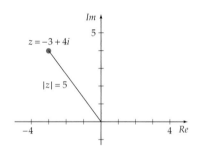

FIGURE 7.42

Theorem, the absolute value of $z = -3 + 4i$ is

$$\sqrt{(-3)^2 + 4^2} = \sqrt{25} = 5$$

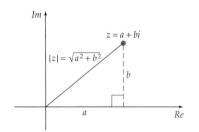

FIGURE 7.43

Definition of the Absolute Value of a Complex Number

The absolute value of the complex number $z = a + bi$, denoted by $|z|$, is

$$|z| = |a + bi| = \sqrt{a^2 + b^2}$$

Thus $|z|$ is the distance from the origin to z (see **Figure 7.43**).

❓ **QUESTION** The conjugate of $a + bi$ is $a - bi$. Does $|a + bi| = |a - bi|$?

❓ **ANSWER** Yes.

• TRIGONOMETRIC FORM OF A COMPLEX NUMBER

A complex number $z = a + bi$ can be written in terms of trigonometric functions. Consider the complex number graphed in **Figure 7.44.** We can write a and b in terms of the sine and the cosine.

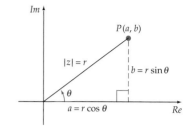

FIGURE 7.44

$$\cos \theta = \frac{a}{r} \qquad \sin \theta = \frac{b}{r}$$

$$a = r \cos \theta \qquad b = r \sin \theta$$

where $r = |z| = \sqrt{a^2 + b^2}$. Substituting for a and b in $z = a + bi$, we obtain

$$z = r \cos \theta + ir \sin \theta = r(\cos \theta + i \sin \theta)$$

The expression $z = r(\cos \theta + i \sin \theta)$ is known as the **trigonometric form** of a complex number. The trigonometric form of a complex number is also called the **polar form** of the complex number. The notation $\cos \theta + i \sin \theta$ is often abbreviated as cis θ using the c from $\cos \theta$, the imaginary unit i, and the s from $\sin \theta$.

Trigonometric Form of a Complex Number

The complex number $z = a + bi$ can be written in trigonometric form as

$$z = r(\cos \theta + i \sin \theta) = r \text{ cis } \theta$$

where $a = r \cos \theta$, $b = r \sin \theta$, $r = \sqrt{a^2 + b^2}$, and $\tan \theta = \frac{b}{a}$.

In this text we will often write the trigonometric form of a complex number in its abbreviated form $z = r$ cis θ. The value of r is called the **modulus** of the complex number z, and the angle θ is called the **argument** of the complex number z. The modulus r and the argument θ of a complex number $z = a + bi$ are given by

$$r = \sqrt{a^2 + b^2} \qquad \text{and} \qquad \cos \theta = \frac{a}{r}, \quad \sin \theta = \frac{b}{r}$$

We also can write $\alpha = \tan^{-1}\left|\frac{b}{a}\right|$, where α is the reference angle for θ. As a result of the periodic nature of the sine and cosine functions, the trigonometric form of a complex number is not unique. Because $\cos \theta = \cos(\theta + 2k\pi)$ and $\sin \theta = \sin(\theta + 2k\pi)$, where k is an integer, the following complex numbers are equal.

$$r \text{ cis } \theta = r \text{ cis}(\theta + 2k\pi) \quad \text{for } k \text{ an integer}$$

For example, $2 \text{ cis } \frac{\pi}{6} = 2 \text{ cis}\left(\frac{\pi}{6} + 2\pi\right)$.

EXAMPLE 1   **Write a Complex Number in Trigonometric Form**

Write $z = -2 - 2i$ in trigonometric form.

Solution

Find the modulus and the argument of z. Then substitute these values in the trigonometric form of z.

$$r = \sqrt{(-2)^2 + (-2)^2} = \sqrt{8} = 2\sqrt{2}$$

To determine θ, we first determine α. See **Figure 7.45**.

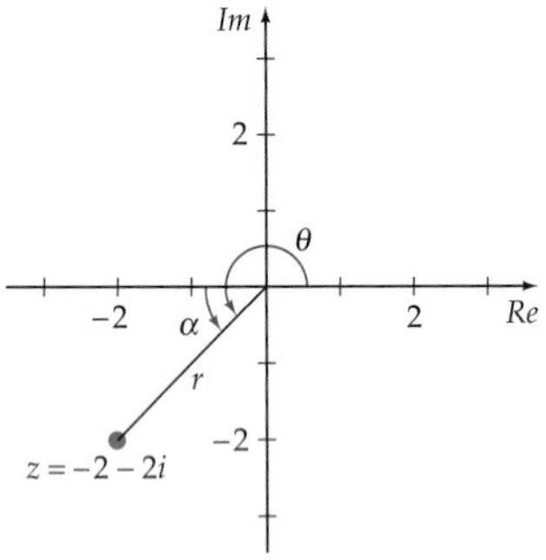

$$\alpha = \tan^{-1}\left|\frac{b}{a}\right|$$

• α is the reference angle of angle θ.

$$\alpha = \tan^{-1}\left|\frac{-2}{-2}\right| = \tan^{-1}1 = 45°$$

$$\theta = 180° + 45° = 225°$$

• Because z is in the third quadrant, $180° < \theta < 270°$.

The trigonometric form is

$$z = r \text{ cis } \theta = 2\sqrt{2} \text{ cis } 225°$$ • $r = 2\sqrt{2}, \theta = 225°$

▶ **TRY EXERCISE 12, PAGE 669**

Im ... **Re** ... $z = -2 - 2i$

FIGURE 7.45

EXAMPLE 2   **Write a Complex Number in Standard Form**

Write $z = 2 \text{ cis } 120°$ in standard form.

Solution

Write z in the form $r(\cos\theta + i\sin\theta)$ and then evaluate $\cos\theta$ and $\sin\theta$. See **Figure 7.46**.

$$z = 2 \text{ cis } 120° = 2(\cos 120° + i\sin 120°) = 2\left(-\frac{1}{2} + \frac{\sqrt{3}}{2}i\right) = -1 + i\sqrt{3}$$

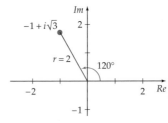

FIGURE 7.46

▶ **TRY EXERCISE 26, PAGE 670**

39. **CAPTURING THE SOUND** During televised football games, a parabolic microphone is used to capture sounds. The shield of the microphone is a paraboloid with a diameter of 18.75 inches and a depth of 3.66 inches. To pick up the sounds, a microphone is placed at the focus of the paraboloid. How far (to the nearest 0.1 of an inch) from the vertex of the paraboloid should the microphone be placed?

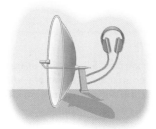

40. **THE LOVELL TELESCOPE** The Lovell Telescope is a radio telescope located at the Jodrell Bank Observatory in Cheshire, England. The dish of the telescope has the shape of a paraboloid with a diameter of 250 feet and a focal length of 75 feet.

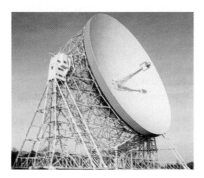

a. Find an equation of a cross section of the paraboloid that passes through the vertex of the paraboloid. Assume that the dish has its vertex at $(0, 0)$ and a vertical axis of symmetry.

b. Find the depth of the dish. Round to the nearest foot.

41. The surface area of a paraboloid with radius r and depth d is given by $S = \dfrac{\pi r}{6d^2}[(r^2 + 4d^2)^{3/2} - r^3]$.

Approximate (to the nearest 100 square feet) the surface area of:

a. The radio telescope in Exercise 38.

b. The Lovell Telescope (see Exercise 40).

42. **THE HALE TELESCOPE** The parabolic mirror in the Hale telescope at the Palomar Observatory in southern Califor-

nia has a diameter of 200 inches, and it has a concave depth of 3.75375 inches. Determine the location of its focus (to the nearest inch).

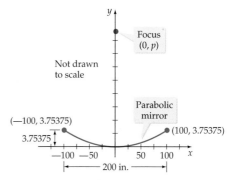

Mirror in the Hale Telescope

43. **THE LICK TELESCOPE** The parabolic mirror in the Lick telescope at the Lick Observatory on Mount Hamilton has a diameter of 120 inches, and it has a focal length of 600 inches. In the construction of the mirror, workers ground the mirror as shown in the following diagram. Determine the dimension a, which is the concave depth of the mirror.

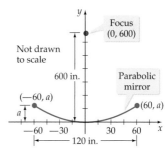

Mirror in the Lick Telescope

44. **HEADLIGHT DESIGN** A light source is to be placed on the axis of symmetry of the parabolic reflector shown in the figure below. How far to the right of the vertex point should the light source be located if the designer wishes the reflected light rays to form a beam of parallel rays?

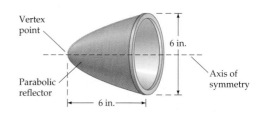

In Exercises 45 to 48, graph each equation, and find the coordinates of the points of intersection of the two graphs to the nearest ten-thousandth.

45. $y = 2x^2 - x - 1$
$y = x$

46. $y = x^2 + 2x - 4$
$y = x - 1$

47. $y = 2x^2 - 1$
$y = x^2 + x + 3$

48. $y = 2x^2 - x - 1$
$y = x^2 - 4$

CONNECTING CONCEPTS

In Exercises 49 to 51, use the following definition of latus rectum: The line segment that has endpoints on a parabola, passes through the focus of the parabola, and is perpendicular to the axis of symmetry is called the *latus rectum* of the parabola.

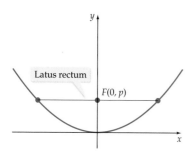

49. Find the length of the latus rectum for the parabola $x^2 = 4y$.

50. Find the length of the latus rectum for the parabola $y^2 = -8x$.

51. Find the length of the latus rectum for any parabola in terms of $|p|$, the distance from the vertex of the parabola to the focus.

The result of Exercise 51 can be stated as the following theorem: Two points on a parabola will be $2|p|$ units on each side of the axis of symmetry on the line through the focus and perpendicular to that axis.

52. Use the theorem to sketch a graph of the parabola given by the equation $(x - 3)^2 = 2(y + 1)$.

53. Use the theorem to sketch a graph of the parabola given by the equation $(y + 4)^2 = -(x - 1)$.

54. By using the definition of a parabola, find the equation in standard form of the parabola with $V(0, 0)$, $F(-c, 0)$, and directrix $x = c$.

55. Sketch a graph of $4(y - 2) = x|x| - 1$.

56. Find the equation of the directrix of the parabola with vertex at the origin and focus at the point $(1, 1)$.

57. Find the equation of the parabola with vertex at the origin and focus at the point $(1, 1)$. (*Hint:* You will need the answer to Exercise 56 and the definition of a parabola.)

PREPARE FOR SECTION 8.2

58. Find the midpoint and the length of the line segment between $P_1(5, 1)$ and $P_2(-1, 5)$. [2.1]

59. Solve: $x^2 + 6x - 16 = 0$ [1.3]

60. Solve: $x^2 - 2x = 2$ [1.3]

61. Complete the square of $x^2 - 8x$ and write the result as the square of a binomial. [1.3]

62. Solve $(x - 2)^2 + y^2 = 4$ for y. [1.3]

63. Graph: $(x - 2)^2 + (y + 3)^2 = 16$ [2.1]

─────── *PROJECTS* ───────

I. PARABOLAS AND TANGENTS Calculus procedures can be used to show that the equation of a tangent line to the parabola $4py = x^2$ at the point (x_0, y_0) is given by

$$y - y_0 = \left(\frac{1}{2p}x_0\right)(x - x_0)$$

Use this equation to verify each of the following statements.

a. If two tangent lines to a parabola intersect at right angles, then the point of intersection of the tangent lines is on the directrix of the parabola.

b. If two tangent lines to a parabola intersect at right angles, then the focus of the parabola is located on the line segment that connects the two points of tangency.

c. The tangent line to the parabola $4py = x^2$ at the point (x_0, y_0) intersects the y-axis at the point $(0, -y_0)$.

SECTION 8.2

ELLIPSES

- ● ELLIPSES WITH CENTER AT $(0, 0)$
- ● ELLIPSES WITH CENTER AT (h, k)
- ● ECCENTRICITY OF AN ELLIPSE
- ● APPLICATIONS
- ● ACOUSTIC PROPERTY OF AN ELLIPSE

take note

If the plane intersects the cone at the vertex of the cone so that the resulting figure is a point, the point is a degenerate ellipse. See the accompanying figure.

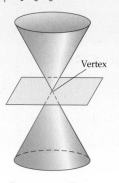

Degenerate ellipse

An ellipse is another of the conic sections formed when a plane intersects a right circular cone. If β is the angle at which the plane intersects the axis of the cone and α is the angle shown in **Figure 8.16,** an ellipse is formed when $\alpha < \beta < 90°$. If $\beta = 90°$, then a circle is formed.

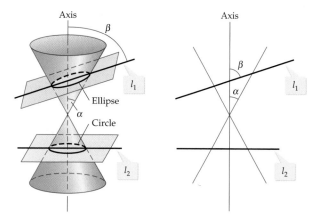

FIGURE 8.16

As was the case for a parabola, there is a definition for an ellipse in terms of a certain set of points in the plane.

Definition of an Ellipse

An **ellipse** is the set of all points in the plane, the sum of whose distances from two fixed points (**foci**) is a positive constant.

We can use this definition to draw an ellipse, equipped only with a piece of string and two tacks (see **Figure 8.17**). Tack the ends of the string to the foci, and trace a curve with a pencil held tight against the string. The resulting curve is an ellipse. The positive constant mentioned in the definition of an ellipse is the length of the string.

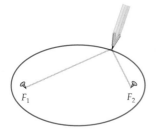

FIGURE 8.17

● ELLIPSES WITH CENTER AT $(0, 0)$

The graph of an ellipse has two axes of symmetry (see **Figure 8.18**). The longer axis is called the **major axis.** The foci of the ellipse are on the major axis. The shorter axis is called the **minor axis.** It is customary to denote the length of the major axis as $2a$ and the length of the minor axis as $2b$. The **semiaxes** are one-half the axes in length. Thus the length of the semimajor axis is denoted by a and the length of the semiminor axis by b. The **center** of the ellipse is the midpoint of the major axis. The endpoints of the major axis are the **vertices** (plural of *vertex*) of the ellipse.

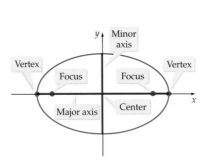

FIGURE 8.18

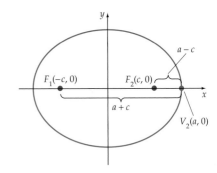

FIGURE 8.19

Consider the point $V_2(a, 0)$, which is one vertex of an ellipse, and the points $F_2(c, 0)$ and $F_1(-c, 0)$, which are the foci of the ellipse shown in **Figure 8.19**. The distance from V_2 to F_1 is $a + c$. Similarly, the distance from V_2 to F_2 is $a - c$. From the definition of an ellipse, the sum of the distances from any point on the ellipse to the foci is a positive constant. By adding the expressions $a + c$ and $a - c$, we have

$$(a + c) + (a - c) = 2a$$

Thus the positive constant referred to in the definition of an ellipse is $2a$, the length of the major axis.

Now let $P(x, y)$ be any point on the ellipse (see **Figure 8.20**). By using the definition of an ellipse, we have

$$d(P, F_1) + d(P, F_2) = 2a$$
$$\sqrt{(x + c)^2 + y^2} + \sqrt{(x - c)^2 + y^2} = 2a$$

Subtract the second radical from each side of the equation, and then square each side.

$$\left[\sqrt{(x + c)^2 + y^2}\right]^2 = \left[2a - \sqrt{(x - c)^2 + y^2}\right]^2$$
$$(x + c)^2 + y^2 = 4a^2 - 4a\sqrt{(x - c)^2 + y^2} + (x - c)^2 + y^2$$

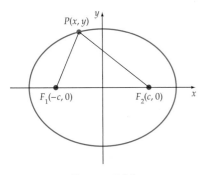

FIGURE 8.20

$$x^2 + 2cx + c^2 + y^2 = 4a^2 - 4a\sqrt{(x-c)^2 + y^2} + x^2 - 2cx + c^2 + y^2$$

$$4cx - 4a^2 = -4a\sqrt{(x-c)^2 + y^2}$$

$$[-cx + a^2]^2 = [a\sqrt{(x-c)^2 + y^2}]^2 \qquad \bullet \text{ Divide by } -4, \text{ and then} \\ \text{square each side.}$$

$$c^2x^2 - 2cxa^2 + a^4 = a^2x^2 - 2cxa^2 + a^2c^2 + a^2y^2$$

$$-a^2x^2 + c^2x^2 - a^2y^2 = -a^4 + a^2c^2 \qquad \bullet \text{ Rewrite with } x \text{ and } y \text{ terms} \\ \text{on the left side.}$$

$$-(a^2 - c^2)x^2 - a^2y^2 = -a^2(a^2 - c^2) \qquad \bullet \text{ Factor and let } b^2 = a^2 - c^2.$$

$$-b^2x^2 - a^2y^2 = -a^2b^2 \qquad \bullet \text{ Divide each side by } -a^2b^2.$$

$$\frac{x^2}{a^2} + \frac{y^2}{b^2} = 1 \qquad \bullet \text{ An equation of an ellipse} \\ \text{with center at } (0, 0)$$

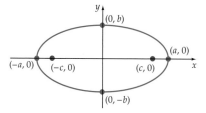

a. Major axis on x-axis

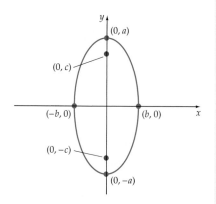

b. Major axis on y-axis

FIGURE 8.21

Standard Forms of the Equation of an Ellipse with Center at the Origin

Major Axis on the x-Axis

The standard form of the equation of an ellipse with the center at the origin and major axis on the x-axis (see **Figure 8.21a**) is given by

$$\frac{x^2}{a^2} + \frac{y^2}{b^2} = 1, \quad a > b$$

The length of the major axis is $2a$. The length of the minor axis is $2b$. The coordinates of the vertices are $(a, 0)$ and $(-a, 0)$, and the coordinates of the foci are $(c, 0)$ and $(-c, 0)$, where $c^2 = a^2 - b^2$.

Major Axis on the y-Axis

The standard form of the equation of an ellipse with the center at the origin and major axis on the y-axis (see **Figure 8.21b**) is given by

$$\frac{x^2}{b^2} + \frac{y^2}{a^2} = 1, \quad a > b$$

The length of the major axis is $2a$. The length of the minor axis is $2b$. The coordinates of the vertices are $(0, a)$ and $(0, -a)$, and the coordinates of the foci are $(0, c)$ and $(0, -c)$, where $c^2 = a^2 - b^2$.

❓ **QUESTION** For the graph of $\dfrac{x^2}{16} + \dfrac{y^2}{25} = 1$, is the major axis on the x-axis or the y-axis?

❓ **ANSWER** Because $25 > 16$, the major axis is on the y-axis.

EXAMPLE 1   **Find the Vertices and Foci of an Ellipse**

Find the vertices and foci of the ellipse given by the equation $\dfrac{x^2}{25} + \dfrac{y^2}{49} = 1.$

Sketch the graph.

Solution

Because the y^2 term has the larger denominator, the major axis is on the y-axis.

$$a^2 = 49 \qquad b^2 = 25 \qquad c^2 = a^2 - b^2$$
$$a = 7 \qquad b = 5 \qquad = 49 - 25 = 24$$
$$c = \sqrt{24} = 2\sqrt{6}$$

The vertices are $(0, 7)$ and $(0, -7)$. The foci are $\left(0, 2\sqrt{6}\right)$ and $\left(0, -2\sqrt{6}\right)$. See **Figure 8.22.**

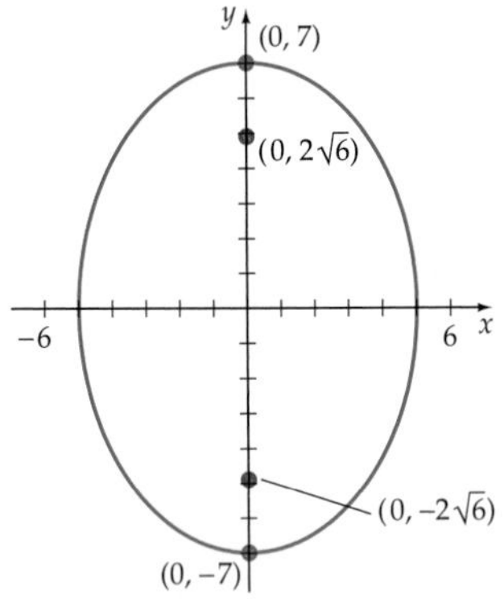

▶ **TRY EXERCISE 20, PAGE 705**

An ellipse with foci $(3, 0)$ and $(-3, 0)$ and major axis of length 10 is shown in **Figure 8.23.** To find the equation of the ellipse in standard form, we must find a^2 and b^2. Because the foci are on the major axis, the major axis is on the x-axis. The length of the major axis is $2a$. Thus $2a = 10$. Solving for a, we have $a = 5$ and $a^2 = 25$.

Because the foci are $(3, 0)$ and $(-3, 0)$ and the center of the ellipse is the midpoint between the two foci, the distance from the center of the ellipse to a focus is 3. Therefore, $c = 3$. To find b^2, use the equation

$$c^2 = a^2 - b^2$$
$$9 = 25 - b^2$$
$$b^2 = 16$$

The equation of the ellipse in standard form is $\dfrac{x^2}{25} + \dfrac{y^2}{16} = 1.$

Figure 8.22

$$\dfrac{x^2}{25} + \dfrac{y^2}{49} = 1$$

FIGURE 8.22

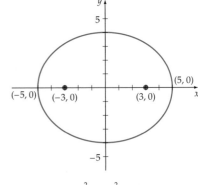

$$\dfrac{x^2}{25} + \dfrac{y^2}{16} = 1$$

FIGURE 8.23

● **ELLIPSES WITH CENTER AT (h, k)**

The equation of an ellipse with center (h, k) and with horizontal or vertical major axis can be found by using a translation of coordinates. On a coordinate system with axes labeled x' and y', the standard form of the equation of an ellipse with center at the origin of the $x'y'$-coordinate system is

$$\dfrac{(x')^2}{a^2} + \dfrac{(y')^2}{b^2} = 1$$

Now place the origin of the $x'y'$-coordinate system at (h, k) in an xy-coordinate system. See **Figure 8.24.**

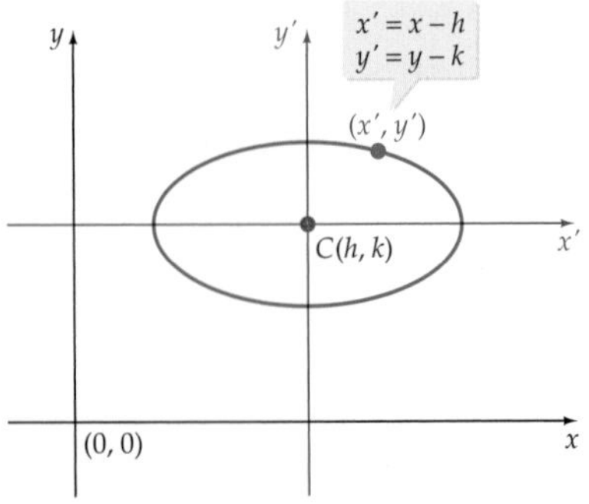

$$x' = x - h$$
$$y' = y - k$$

(x', y')

$C(h, k)$

$(0, 0)$

FIGURE 8.24

The relationship between an ordered pair in the $x'y'$-coordinate system and one in the xy-coordinate system is given by the transformation equations

$$x' = x - h$$
$$y' = y - k$$

Substitute the expressions for x' and y' into the equation of an ellipse. The equation of the ellipse with center at (h, k) is

$$\frac{(x - h)^2}{a^2} + \frac{(y - k)^2}{b^2} = 1$$

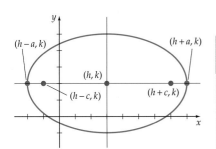

a. Major axis parallel to x-axis

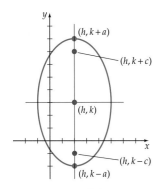

b. Major axis parallel to y-axis

FIGURE 8.25

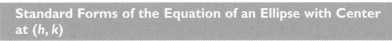
Standard Forms of the Equation of an Ellipse with Center at (h, k)

Major Axis Parallel to the x-Axis

The standard form of the equation of an ellipse with the center at (h, k) and major axis parallel to the x-axis (see **Figure 8.25a**) is given by

$$\frac{(x - h)^2}{a^2} + \frac{(y - k)^2}{b^2} = 1, \quad a > b$$

The length of the major axis is $2a$. The length of the minor axis is $2b$. The coordinates of the vertices are $(h + a, k)$ and $(h - a, k)$, and the coordinates of the foci are $(h + c, k)$ and $(h - c, k)$, where $c^2 = a^2 - b^2$.

Major Axis Parallel to the y-Axis

The standard form of the equation of an ellipse with the center at (h, k) and major axis parallel to the y-axis (see **Figure 8.25b**) is given by

$$\frac{(x - h)^2}{b^2} + \frac{(y - k)^2}{a^2} = 1, \quad a > b$$

The length of the major axis is $2a$. The length of the minor axis is $2b$. The coordinates of the vertices are $(h, k + a)$ and $(h, k - a)$, and the coordinates of the foci are $(h, k + c)$ and $(h, k - c)$, where $c^2 = a^2 - b^2$.

EXAMPLE 2   Find the Vertices and Foci of an Ellipse

Find the vertices and foci of the ellipse $4x^2 + 9y^2 - 8x + 36y + 4 = 0$. Sketch the graph.

Solution

Write the equation of the ellipse in standard form by completing the square.

$$4x^2 + 9y^2 - 8x + 36y + 4 = 0$$
$$4x^2 - 8x + 9y^2 + 36y = -4 \qquad \text{• Rearrange terms.}$$
$$4(x^2 - 2x) + 9(y^2 + 4y) = -4 \qquad \text{• Factor.}$$
$$4(x^2 - 2x + 1) + 9(y^2 + 4y + 4) = -4 + 4 + 36 \qquad \text{• Complete the square.}$$
$$4(x - 1)^2 + 9(y + 2)^2 = 36 \qquad \text{• Factor.}$$
$$\frac{(x - 1)^2}{9} + \frac{(y + 2)^2}{4} = 1 \qquad \text{• Divide by 36.}$$

Continued ▶

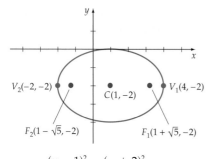

$$\frac{(x-1)^2}{9} + \frac{(y+2)^2}{4} = 1$$

FIGURE 8.26

From the equation of the ellipse in standard form, the coordinates of the center of the ellipse are $(1, -2)$. Because the larger denominator is 9, the major axis is parallel to the x-axis and $a^2 = 9$. Thus $a = 3$. The vertices are $(4, -2)$ and $(-2, -2)$.

To find the coordinates of the foci, we find c.

$$c^2 = a^2 - b^2 = 9 - 4 = 5$$
$$c = \sqrt{5}$$

The foci are $(1 + \sqrt{5}, -2)$ and $(1 - \sqrt{5}, -2)$. See **Figure 8.26**.

▶ **TRY EXERCISE 26, PAGE 705**

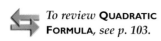 **INTEGRATING TECHNOLOGY**

A graphing utility can be used to graph an ellipse. For instance, consider the equation $4x^2 + 9y^2 - 8x + 36y + 4 = 0$ from Example 2. Rewrite the equation as

$$9y^2 + 36y + (4x^2 - 8x + 4) = 0$$

In this form, the equation is a quadratic equation in terms of the variable y with

$$A = 9, B = 36, \text{ and } C = 4x^2 - 8x + 4$$

To review **QUADRATIC FORMULA**, *see p. 103.*

Apply the quadratic formula to produce

$$y = \frac{-36 \pm \sqrt{1296 - 36(4x^2 - 8x + 4)}}{18}$$

The graph of $\text{Y1} = \dfrac{-36 + \sqrt{1296 - 36(4x^2 - 8x + 4)}}{18}$ is the part of the ellipse on or above the line $y = -2$ (see **Figure 8.27**).

The graph of $\text{Y2} = \dfrac{-36 - \sqrt{1296 - 36(4x^2 - 8x + 4)}}{18}$ is the part of the ellipse on or below the line $y = -2$, as shown in **Figure 8.27**.

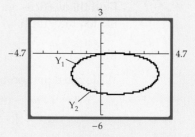

FIGURE 8.27

One advantage of this graphing procedure is that it does not require us to write the given equation in standard form. A disadvantage of the graphing procedure is that it does not indicate where the foci of the ellipse are located.

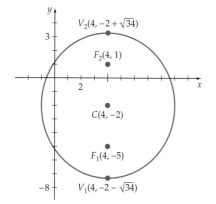

$V_2(4, -2 + \sqrt{34})$

$F_2(4, 1)$

$C(4, -2)$

$F_1(4, -5)$

$V_1(4, -2 - \sqrt{34})$

FIGURE 8.28

EXAMPLE 3   **Find the Equation of an Ellipse**

Find the standard form of the equation of the ellipse with center at $(4, -2)$, foci $F_2(4, 1)$ and $F_1(4, -5)$, and minor axis of length 10, as shown in **Figure 8.28.**

Solution

Because the foci are on the major axis, the major axis is parallel to the y-axis. The distance from the center of the ellipse to a focus is c. The distance between the center $(4, -2)$ and the focus $(4, 1)$ is 3. Therefore, $c = 3$.

The length of the minor axis is $2b$. Thus $2b = 10$ and $b = 5$.
To find a^2, use the equation $c^2 = a^2 - b^2$.

$$9 = a^2 - 25$$
$$a^2 = 34$$

Thus the equation in standard form is

$$\frac{(x - 4)^2}{25} + \frac{(y + 2)^2}{34} = 1$$

▶ **TRY EXERCISE 42, PAGE 705**

● **ECCENTRICITY OF AN ELLIPSE**

The graph of an ellipse can be very long and thin, or it can be much like a circle. The **eccentricity** of an ellipse is a measure of its "roundness."

Eccentricity (e) of an Ellipse

The eccentricity e of an ellipse is the ratio of c to a, where c is the distance from the center to a focus and a is one-half the length of the major axis. (See **Figure 8.29.**) That is,

$$e = \frac{c}{a}$$

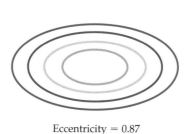

Eccentricity $= 0.87$

FIGURE 8.29

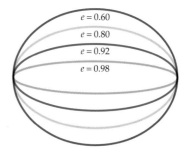

$e = 0.60$
$e = 0.80$
$e = 0.92$
$e = 0.98$

FIGURE 8.30

Because $c < a$, for an ellipse, $0 < e < 1$. When $e \approx 0$, the graph is almost a circle. When $e \approx 1$, the graph is long and thin. See **Figure 8.30.**

EXAMPLE 4 Find the Eccentricity of an Ellipse

Find the eccentricity of the ellipse given by $8x^2 + 9y^2 = 18$.

Solution

First, write the equation of the ellipse in standard form. Divide each side of the equation by 18.

$$\frac{8x^2}{18} + \frac{9y^2}{18} = 1$$

$$\frac{4x^2}{9} + \frac{y^2}{2} = 1$$

$$\frac{x^2}{9/4} + \frac{y^2}{2} = 1 \qquad \cdot \frac{4}{9} = \frac{1}{9/4}$$

The last step is necessary because the standard form of the equation has coefficients of 1 in the numerator. Thus

$$a^2 = \frac{9}{4} \qquad \text{and} \qquad a = \frac{3}{2}$$

Use the equation $c^2 = a^2 - b^2$ to find c.

$$c^2 = \frac{9}{4} - 2 = \frac{1}{4} \qquad \text{and} \qquad c = \sqrt{\frac{1}{4}} = \frac{1}{2}$$

Now find the eccentricity.

$$e = \frac{c}{a} = \frac{1/2}{3/2} = \frac{1}{3}$$

The eccentricity of the ellipse is $\frac{1}{3}$.

▶ **TRY EXERCISE 48, PAGE 705**

TABLE 8.1

Planet	Eccentricity
Mercury	0.206
Venus	0.007
Earth	0.017
Mars	0.093
Jupiter	0.049
Saturn	0.051
Uranus	0.046
Neptune	0.005
Pluto	0.250

● **APPLICATIONS**

The planets travel around the sun in elliptical orbits. The sun is located at a focus of the orbit. The eccentricities of the orbits for the planets in our solar system are given in **Table 8.1.**

❓ **QUESTION** Which planet has the most nearly circular orbit?

The terms *perihelion* and *aphelion* are used to denote the position of a planet in its orbit around the sun. The perihelion is the point nearest the sun; the aphelion is the point farthest from the sun. See **Figure 8.31.** The length of the semimajor axis of a planet's elliptical orbit is called the *mean distance* of the planet from the sun.

❓ **ANSWER** Neptune has the smallest eccentricity, so it is the planet with the most nearly circular orbit.

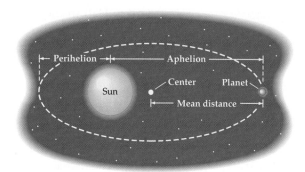

FIGURE 8.31

EXAMPLE 5 Determine an Equation for the Orbit of Earth

 Earth has a mean distance of 93 million miles and a perihelion distance of 91.5 million miles. Find an equation for Earth's orbit.

Solution

A mean distance of 93 million miles implies that the length of the semimajor axis of the orbit is $a = 93$ million miles. Earth's aphelion distance is the length of the major axis less the length of the perihelion distance. Thus

$$\text{Aphelion distance} = 2(93) - 91.5 = 94.5 \text{ million miles}$$

The distance c from the sun to the center of Earth's orbit is

$$c = \text{aphelion distance} - 93 = 94.5 - 93 = 1.5 \text{ million miles}$$

The length b of the semiminor axis of the orbit is

$$b = \sqrt{a^2 - c^2} = \sqrt{93^2 - 1.5^2} = \sqrt{8646.75}$$

An equation of Earth's orbit is

$$\frac{x^2}{93^2} + \frac{y^2}{8646.75} = 1$$

▶ **TRY EXERCISE 56, PAGE 706**

● ACOUSTIC PROPERTY OF AN ELLIPSE

Sound waves, although different from light waves, have a similar reflective property. When sound is reflected from a point P on a surface, the angle of incidence equals the angle of reflection. Applying this principle to a room with an elliptical ceiling results in what are called whispering galleries. These galleries are based on the following theorem.

The Reflective Property of an Ellipse

The lines from the foci to a point on an ellipse make equal angles with the tangent line at that point. See **Figure 8.32.**

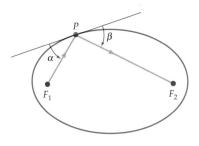

$\alpha = \beta$

FIGURE 8.32

The Statuary Hall in the Capital Building in Washington, D.C., is a whispering gallery. Two people standing at the foci of the elliptical ceiling can whisper and yet hear each other even though they are a considerable distance apart. The whisper from one person is reflected to the person standing at the other focus.

> **EXAMPLE 6**   **Locate the Foci of a Whispering Gallery**
>
> A room 88 feet long is constructed to be a whispering gallery. The room has an elliptical ceiling, as shown in **Figure 8.33.** If the maximum height of the ceiling is 22 feet, determine where the foci are located.
>
> **Solution**
>
> The length a of the semimajor axis of the elliptical ceiling is 44 feet. The height b of the semiminor axis is 22 feet. Thus
>
> $$c^2 = a^2 - b^2$$
> $$c^2 = 44^2 - 22^2$$
> $$c = \sqrt{44^2 - 22^2} \approx 38.1 \text{ feet}$$
>
> The foci are located about 38.1 feet from the center of the elliptical ceiling along its major axis.

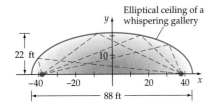

Elliptical ceiling of a whispering gallery

FIGURE 8.33

▶ **TRY EXERCISE 58, PAGE 706**

TOPICS FOR DISCUSSION

1. In every ellipse, the length of the semimajor axis a is greater than the length of the semiminor axis b and greater than the distance c from a focus to the center of the ellipse. Do you agree? Explain.

2. How many vertices does an ellipse have?

3. Every ellipse has two y-intercepts. Do you agree? Explain.

4. Explain why the eccentricity of every ellipse is a number between 0 and 1.

EXERCISE SET 8.2

In Exercises 1 to 32, find the center, vertices, and foci of the ellipse given by each equation. Sketch the graph.

1. $\dfrac{x^2}{16} + \dfrac{y^2}{25} = 1$

2. $\dfrac{x^2}{49} + \dfrac{y^2}{36} = 1$

3. $\dfrac{x^2}{9} + \dfrac{y^2}{4} = 1$

4. $\dfrac{x^2}{64} + \dfrac{y^2}{25} = 1$

5. $\dfrac{x^2}{7} + \dfrac{y^2}{9} = 1$

6. $\dfrac{x^2}{5} + \dfrac{y^2}{4} = 1$

7. $\dfrac{4x^2}{9} + \dfrac{y^2}{16} = 1$

8. $\dfrac{x^2}{9} + \dfrac{9y^2}{16} = 1$

9. $\dfrac{(x-3)^2}{25} + \dfrac{(y+2)^2}{16} = 1$

10. $\dfrac{(x+3)^2}{9} + \dfrac{(y+1)^2}{16} = 1$

11. $\dfrac{(x+2)^2}{9} + \dfrac{y^2}{25} = 1$

12. $\dfrac{x^2}{25} + \dfrac{(y-2)^2}{81} = 1$

13. $\dfrac{(x-1)^2}{21} + \dfrac{(y-3)^2}{4} = 1$ **14.** $\dfrac{(x+5)^2}{9} + \dfrac{(y-3)^2}{7} = 1$

15. $\dfrac{9(x-1)^2}{16} + \dfrac{(y+1)^2}{9} = 1$ **16.** $\dfrac{(x+6)^2}{25} + \dfrac{25y^2}{144} = 1$

17. $3x^2 + 4y^2 = 12$ **18.** $5x^2 + 4y^2 = 20$

19. $25x^2 + 16y^2 = 400$ ▶ **20.** $25x^2 + 12y^2 = 300$

21. $64x^2 + 25y^2 = 400$ **22.** $9x^2 + 64y^2 = 144$

23. $4x^2 + y^2 - 24x - 8y + 48 = 0$

24. $x^2 + 9y^2 + 6x - 36y + 36 = 0$

25. $5x^2 + 9y^2 - 20x + 54y + 56 = 0$

▶ **26.** $9x^2 + 16y^2 + 36x - 16y - 104 = 0$

27. $16x^2 + 9y^2 - 64x - 80 = 0$

28. $16x^2 + 9y^2 + 36y - 108 = 0$

29. $25x^2 + 16y^2 + 50x - 32y - 359 = 0$

30. $16x^2 + 9y^2 - 64x - 54y + 1 = 0$

31. $8x^2 + 25y^2 - 48x + 50y + 47 = 0$

32. $4x^2 + 9y^2 + 24x + 18y + 44 = 0$

In Exercises 33 to 44, find the equation in standard form of each ellipse, given the information provided.

33. Center $(0,0)$, major axis of length 10, foci at $(4,0)$ and $(-4,0)$

34. Center $(0,0)$, minor axis of length 6, foci at $(0,4)$ and $(0,-4)$

35. Vertices $(6,0)$, $(-6,0)$; ellipse passes through $(0,-4)$ and $(0,4)$

36. Vertices $(7,0)$, $(-7,0)$; ellipse passes through $(0,5)$ and $(0,-5)$

37. Major axis of length 12 on the x-axis, center at $(0,0)$; ellipse passes through $(2,-3)$

38. Major axis of length 8, center at $(0,0)$; ellipse passes through $(-2,2)$

39. Center $(-2,4)$, vertices $(-6,4)$ and $(2,4)$, foci at $(-5,4)$ and $(1,4)$

40. Center $(0,3)$, minor axis of length 4, foci at $(0,0)$ and $(0,6)$

41. Center $(2,4)$, major axis parallel to the y-axis and of length 10; ellipse passes through the point $(3,3)$

▶ **42.** Center $(-4,1)$, minor axis parallel to the y-axis and of length 8; ellipse passes through the point $(0,4)$

43. Vertices $(5,6)$ and $(5,-4)$, foci at $(5,4)$ and $(5,-2)$

44. Vertices $(-7,-1)$ and $(5,-1)$, foci at $(-5,-1)$ and $(3,-1)$

In Exercises 45 to 52, use the eccentricity of each ellipse to find its equation in standard form.

45. Eccentricity $\dfrac{2}{5}$, major axis on the x-axis and of length 10, center at $(0,0)$

46. Eccentricity $\dfrac{3}{4}$, foci at $(9,0)$ and $(-9,0)$

47. Foci at $(0,-4)$ and $(0,4)$, eccentricity $\dfrac{2}{3}$

▶ **48.** Foci at $(0,-3)$ and $(0,3)$, eccentricity $\dfrac{1}{4}$

49. Eccentricity $\dfrac{2}{5}$, foci at $(-1,3)$ and $(3,3)$

50. Eccentricity $\dfrac{1}{4}$, foci at $(-2,4)$ and $(-2,-2)$

51. Eccentricity $\dfrac{2}{3}$, major axis of length 24 on the y-axis, center at $(0,0)$

52. Eccentricity $\dfrac{3}{5}$, major axis of length 15 on the x-axis, center at $(0,0)$

53.   **MEDICINES** A *lithotripter* is an instrument used to remove a kidney stone in a patient without having to do surgery. A high-frequency sound wave is emitted from a source that is located at the focus of an ellipse. The patient is placed so that the kidney stone is located at the other focus of the ellipse. If the equation of the ellipse is

$\dfrac{(x-11)^2}{484} + \dfrac{y^2}{64} = 1$ (x and y are measured in centimeters),

where, to the nearest centimeter, should the patient's kidney stone be placed so that the reflected sound hits the kidney stone?

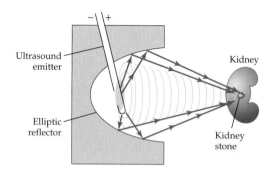

54. CONSTRUCTION A circular vent pipe is placed on a roof that has a slope of $\dfrac{4}{5}$, as shown in the figure at the right.

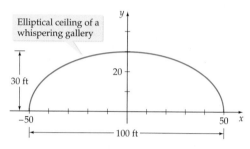

a. Use the slope to find the value of h.

b. The intersection of the vent pipe and the roof is an ellipse. To the nearest thousandth of an inch, what are the lengths of the major and minor axes?

c. Find an equation of the ellipse that should be cut from the roof so that the pipe will fit.

55. THE ORBIT OF SATURN The distance from Saturn to the sun at Saturn's aphelion is 934.34 million miles, and the distance from Saturn to the sun at its perihelion is 835.14 million miles. Find an equation for the orbit of Saturn.

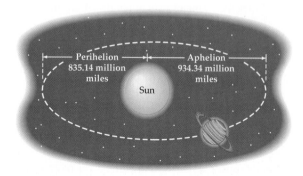

▶ 56. THE ORBIT OF VENUS Venus has a mean distance from the sun of 67.08 million miles, and the distance

from Venus to the sun at its aphelion is 67.58 million miles. Find an equation for the orbit of Venus.

57. WHISPERING GALLERY An architect wishes to design a large room that will be a whispering gallery. See Example 6. The ceiling of the room has a cross section that is an ellipse, as shown in the following figure.

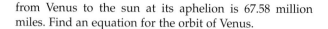

How far to the right and to the left of center are the foci located?

▶ 58. WHISPERING GALLERY An architect wishes to design a large room 100 feet long that will be a whispering gallery. The ceiling of the room has a cross section that is an ellipse, as shown in the following figure.

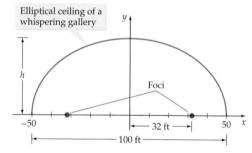

If the foci are to be located 32 feet to the right and to the left of center, find the height h of the elliptical ceiling (to the nearest 0.1 foot).

59. HALLEY'S COMET Find the equation of the path of Halley's comet in astronomical units by letting the sun (one focus) be at the origin and letting the other focus be on the positive x-axis. The length of the major axis of the orbit of Halley's comet is approximately 36 astronomical units (36 AU), and the length of the minor axis is 9 AU (1 AU = 92,960,000 miles).

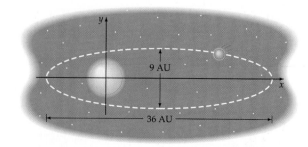

60. ELLIPTICAL RECEIVERS Some satellite receivers are made in an elliptical shape that enables the receiver to pick up signals from two satellites. The receiver shown below has a major axis of 24 inches and a minor axis of 18 inches.

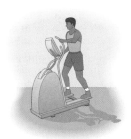

Determine, to the nearest 0.1 inch, the coordinates in the xy-plane of the foci of the ellipse. (*Note:* Because the receiver has only a slight curvature, we can estimate the location of the foci by assuming the receiver is flat.)

In Exercises 61 and 62, use the following formula for the perimeter p of an ellipse with semimajor axis a and semiminor axis b.

$$p \approx \pi \sqrt{2(a^2 + b^2)}$$

61. ELLIPTICAL EXERCISE EQUIP-MENT Many exercise clubs have installed elliptical trainers. These machines are similar to step machines except that the motion of the feet follows an elliptical path. On one elliptical trainer, the path of a person's foot is elliptical with a major axis of 16 inches and a minor axis of 10 inches. How many revolutions must the left foot make to complete a distance of 1 mile on this elliptical trainer?

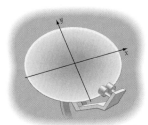

62. **ORBIT OF MARS** Mars travels around the sun in an elliptical orbit. The orbit has a major axis of 3.04 AU and a minor axis of 2.99 AU. (1 AU is 1 astronomical unit, or approximately 92,960,000 miles, the average distance of Earth from the sun.) Estimate, to the nearest million miles, the perimeter of the orbit of Mars.

63. **THE COLOSSEUM** The base of the Colosseum in Rome has an elliptical shape.

a. Find an equation in standard form for the base of the Colosseum, which has a major axis of 615 feet and a minor axis of 510 feet.

b. The area of an ellipse with a semimajor axis of length a and a semiminor axis of length b is given by $A = \pi ab$. Find, to the nearest 100 square feet, the area of the base of the Colosseum.

In Exercises 64 to 69, use the quadratic formula to solve for y in terms of x. Then use a graphing utility to graph each equation.

64. $16x^2 + 9y^2 - 64x - 80 = 0$

65. $16x^2 + 9y^2 + 36y - 108 = 0$

66. $25x^2 + 16y^2 + 50x - 32y - 359 = 0$

67. $16x^2 + 9y^2 - 64x - 54y + 1 = 0$

68. $8x^2 + 25y^2 - 48x + 50y + 47 = 0$

69. $4x^2 + 9y^2 + 24x + 18y + 44 = 0$

CONNECTING CONCEPTS

70. Explain why the graph of $4x^2 + 9y - 16x - 2 = 0$ is or is not an ellipse. Sketch the graph of this equation.

In Exercises 71 to 74, find the equation in standard form of each ellipse by using the definition of an ellipse.

71. Find the equation of the ellipse with foci at $(-3, 0)$ and $(3, 0)$ that passes through the point $\left(3, \dfrac{9}{2}\right)$.

72. Find the equation of the ellipse with foci at $(0, 4)$ and $(0, -4)$ that passes through the point $\left(\dfrac{9}{5}, 4\right)$.

73. Find the equation of the ellipse with foci at $(-1, 2)$ and $(3, 2)$ that passes through the point $(3, 5)$.

74. Find the equation of the ellipse with foci at $(-1, 1)$ and $(-1, 7)$ that passes through the point $\left(\dfrac{3}{4}, 1\right)$.

In Exercises 75 and 76, find the latus rectum of the given ellipse. A line segment with endpoints on the ellipse that is perpendicular to the major axis and passes through a focus is a *latus rectum* of the ellipse.

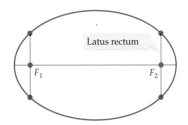

Latus rectum

75. Find the length of a latus rectum of the ellipse given by

$$\frac{(x-1)^2}{9} + \frac{(y+1)^2}{16} = 1$$

76. Find the length of a latus rectum of the ellipse given by

$$9x^2 + 16y^2 - 36x + 96y + 36 = 0$$

77. Show that for any ellipse, the length of a latus rectum is $\dfrac{2b^2}{a}$.

78. Use the definition of an ellipse to find the equation of an ellipse with center at $(0, 0)$ and foci at $(0, c)$ and $(0, -c)$.

Recall that a parabola has a directrix that is a line perpendicular to the axis of symmetry. An ellipse has two directrices, both of which are perpendicular to the major axis and outside the ellipse. For an ellipse with center at the origin and whose major axis is the *x*-axis, the equations of the directrices are $x = \dfrac{a^2}{c}$ and $x = -\dfrac{a^2}{c}$.

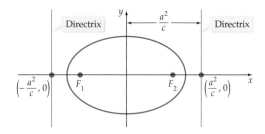

79. Find the directrices of the ellipse in Exercise 3.

80. Find the directrices of the ellipse in Exercise 4.

81. Let $P(x, y)$ be a point on the ellipse $\dfrac{x^2}{12} + \dfrac{y^2}{8} = 1$. Show that the distance from the point P to the focus $(2, 0)$ divided by the distance from the point P to the directrix $x = 6$ equals the eccentricity. (*Hint:* Solve the equation of the ellipse for y^2. Substitute this value for y^2 after applying the distance formula.)

82. Generalize the results of Exercise 81. That is, show that if $P(x, y)$ is a point on the ellipse $\dfrac{x^2}{a^2} + \dfrac{y^2}{b^2} = 1$, where $F(c, 0)$ is a focus and $x = \dfrac{a^2}{c}$ is a directrix, then the following equation is true: $e = \dfrac{d(P, F)}{d(P, D)}$. (*Hint:* Solve the equation of the ellipse for y^2. Substitute this value for y^2 after applying the distance formula.)

PREPARE FOR SECTION 8.3

83. Find the midpoint and the length of the line segment between $P_1(4, -3)$ and $P_2(-2, 1)$. [2.1]

84. Solve: $(x - 1)(x + 3) = 5$ [1.3]

85. Simplify: $\dfrac{4}{\sqrt{8}}$ [P.2]

86. Complete the square of $4x^2 + 24x$ and write the result as the square of a binomial. [1.3]

87. Solve $\dfrac{x^2}{4} - \dfrac{y^2}{9} = 1$ for y. [1.3]

88. Graph: $\dfrac{(x-2)^2}{16} + \dfrac{(y+3)^2}{9} = 1$ [8.2]

PROJECTS

1. I. M. PEI'S OVAL The poet and architect I. M. Pei suggested that the oval with the most appeal to the eye is given by the equation

$$\left(\frac{x}{a}\right)^{3/2} + \left(\frac{y}{b}\right)^{3/2} = 1$$

Use a graphing utility to graph this equation with $a = 5$ and $b = 3$. Then compare your graph with the graph of

$$\left(\frac{x}{5}\right)^{2} + \left(\frac{y}{3}\right)^{2} = 1$$

2. KEPLER'S LAWS The German astronomer Johannes Kepler (1571–1630) derived three laws that describe how the planets orbit the sun. Write an essay that includes biographical information about Kepler and a statement of Kepler's Laws. In addition, use Kepler's Laws to answer the following questions.

a. Where is a planet located in its orbit around the sun when it achieves its greatest velocity?

b. What is the period of Mars if it has a mean distance from the sun of 1.52 astronomical units? (*Hint:* Use Earth as a reference with a period of 1 year and a mean distance from the sun of 1 astronomical unit.)

3. NEPTUNE The position of the planet Neptune was discovered by using celestial mechanics and mathematics. Write an essay that tells how, when, and by whom Neptune was discovered.

4. GRAPH THE COLOSSEUM Some of the Colosseum scenes in the movie *Gladiator* (Universal Studios, 2000) were computer-generated.

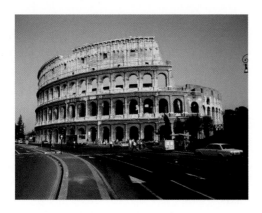

a. You can create a simple but accurate scale image of the exterior of the Colosseum by using a computer and the mathematics software program *Maple*. Open a new *Maple* worksheet and enter the following two commands.

with(plots);
plots[implicitplot3d]((x^2)/(307.5^2)+(y^2)/(255^2)= 1, x= –310..310, y= –260..260, z=0..157, scaling=CONSTRAINED, style=PATCHNOGRID, axes=FRAMED);

Execute each of the commands by placing the cursor in a command and pressing the ENTER key. After execution of the second command a three-dimensional graph will appear. Click and drag on the graph to rotate the image.

b. A graphing calculator can also be used to generate a simple "graph" of the exterior of the Colosseum. Here is a procedure for the TI-83 graphing calculator.

Enter the following in the WINDOW menu.

Xmin=–4.7 Xmax=4.7 Xscl=1 Ymin=–4
Ymax=9 Yscl=1

Enter the following formulas in the Y= menu.

Y₁=√ (9–X²) Y₂=Y₁+4 Y₃=–Y₁ Y₄=Y₃+4

Press: QUIT (2nd MODE)

Enter: Shade(Y₃,Y₄) and press ENTER.
        Note: "Shade(" is in the DRAW menu.

Explain why this "Colosseum graph" appears to be constructed with ellipses even though the functions entered in the Y= menu are the equations of semicircles.

HYPERBOLAS

The hyperbola is a conic section formed when a plane intersects a right circular cone at a certain angle. If β is the angle at which the plane intersects the axis of the cone and α is the angle shown in **Figure 8.34,** a hyperbola is formed when $0° < \beta < \alpha$ or when the plane is parallel to the axis of the cone.

As with the other conic sections, there is a definition of a hyperbola in terms of a certain set of points in the plane.

---

take note

If the plane intersects the cone along the axis of the cone, the resulting curve is two intersecting straight lines. This is the degenerate form of a hyperbola. See the accompanying figure.

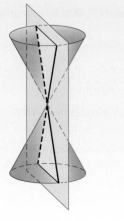

Degenerate hyperbola

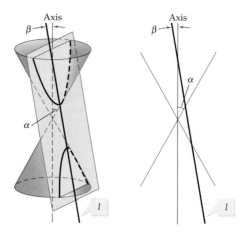

FIGURE 8.34

Definition of a Hyperbola

A **hyperbola** is the set of all points in the plane, the difference between whose distances from two fixed points (foci) is a positive constant.

This definition differs from that of an ellipse in that the ellipse was defined in terms of the *sum* of two distances, whereas the hyperbola is defined in terms of the *difference* of two distances.

● HYPERBOLAS WITH CENTER AT (0, 0)

The **transverse axis** of a hyperbola is the line segment joining the intercepts (see **Figure 8.35**). The midpoint of the transverse axis is called the **center** of the hyperbola. The **conjugate axis** passes through the center of the hyperbola and is perpendicular to the transverse axis.

The length of the transverse axis is customarily represented as $2a$, and the distance between the two foci is represented as $2c$. The length of the conjugate axis is represented as $2b$.

The **vertices** of a hyperbola are the points where the hyperbola intersects the transverse axis.

To determine the positive constant stated in the definition of a hyperbola, consider the point $V_1(a, 0)$, which is one vertex of a hyperbola, and the points $F_1(c, 0)$ and $F_2(-c, 0)$, which are the foci of the hyperbola (see **Figure 8.36**). The difference between the distance from $V_1(a, 0)$ to $F_1(c, 0)$, $c - a$, and the distance from $V_1(a, 0)$ to $F_2(-c, 0)$, $c + a$, must be a constant. By subtracting these distances, we find

$$|(c - a) - (c + a)| = |-2a| = 2a$$

Thus the constant is $2a$ and is the length of the transverse axis. The absolute value is used to ensure that the distance is a positive number.

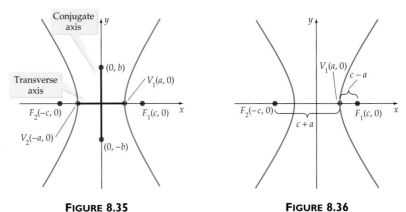

FIGURE 8.35   **FIGURE 8.36**

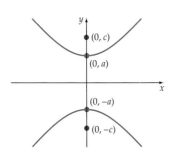

a. Transverse axis on the x-axis

b. Transverse axis on the y-axis

FIGURE 8.37

Standard Forms of the Equation of a Hyperbola with Center at the Origin

Transverse Axis on the x-Axis

The standard form of the equation of a hyperbola with the center at the origin and transverse axis on the x-axis (see **Figure 8.37a**) is given by

$$\frac{x^2}{a^2} - \frac{y^2}{b^2} = 1$$

The coordinates of the vertices are $(a, 0)$ and $(-a, 0)$, and the coordinates of the foci are $(c, 0)$ and $(-c, 0)$, where $c^2 = a^2 + b^2$.

Transverse Axis on the y-Axis

The standard form of the equation of a hyperbola with the center at the origin and transverse axis on the y-axis (see **Figure 8.37b**) is given by

$$\frac{y^2}{a^2} - \frac{x^2}{b^2} = 1$$

The coordinates of the vertices are $(0, a)$ and $(0, -a)$, and the coordinates of the foci are $(0, c)$ and $(0, -c)$, where $c^2 = a^2 + b^2$.

❓ **QUESTION** For the graph of $\dfrac{y^2}{9} - \dfrac{x^2}{4} = 1$, is the transverse axis on the x-axis or the y-axis?

❓ **ANSWER** Because the y-term is positive, the transverse axis is on the y-axis.

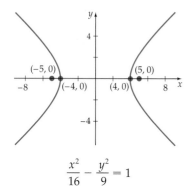

$$\frac{x^2}{16} - \frac{y^2}{9} = 1$$

FIGURE 8.38

By looking at the equations, it is possible to determine the location of the transverse axis by finding which term in the equation is positive. When the x^2 term is positive, the transverse axis is on the x-axis. When the y^2 term is positive, the transverse axis is on the y-axis.

Consider the hyperbola given by the equation $\dfrac{x^2}{16} - \dfrac{y^2}{9} = 1$. Because the x^2 term is positive, the transverse axis is on the x-axis, $a^2 = 16$, and thus $a = 4$. The vertices are $(4, 0)$ and $(-4, 0)$. To find the foci, we determine c.

$$c^2 = a^2 + b^2 = 16 + 9 = 25$$
$$c = \sqrt{25} = 5$$

The foci are $(5, 0)$ and $(-5, 0)$. The graph is shown in **Figure 8.38**.

Each hyperbola has two asymptotes that pass through the center of the hyperbola. The asymptotes of the hyperbola are a useful guide to sketching the graph of the hyperbola.

Asymptotes of a Hyperbola with Center at the Origin

The **asymptotes** of the hyperbola defined by $\dfrac{x^2}{a^2} - \dfrac{y^2}{b^2} = 1$ are given by the equations $y = \dfrac{b}{a}x$ and $y = -\dfrac{b}{a}x$ (see **Figure 8.39a**).

The asymptotes of the hyperbola defined by $\dfrac{y^2}{a^2} - \dfrac{x^2}{b^2} = 1$ are given by the equations $y = \dfrac{a}{b}x$ and $y = -\dfrac{a}{b}x$ (see **Figure 8.39b**).

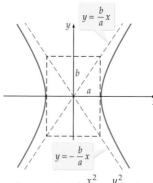

a. Asymptotes of $\dfrac{x^2}{a^2} - \dfrac{y^2}{b^2} = 1$

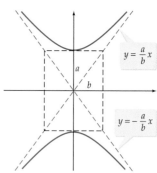

b. Asymptotes of $\dfrac{y^2}{a^2} - \dfrac{x^2}{b^2} = 1$

FIGURE 8.39

One method for remembering the equations of the asymptotes is to write the equation of a hyperbola in standard form but to replace 1 by 0 and then solve for y.

$$\frac{x^2}{a^2} - \frac{y^2}{b^2} = 0 \quad \text{so} \quad y^2 = \frac{b^2}{a^2}x^2, \text{ or } y = \pm\frac{b}{a}x$$

$$\frac{y^2}{a^2} - \frac{x^2}{b^2} = 0 \quad \text{so} \quad y^2 = \frac{a^2}{b^2}x^2, \text{ or } y = \pm\frac{a}{b}x$$

EXAMPLE 1   **Find the Vertices, Foci, and Asymptotes of a Hyperbola**

Find the vertices, foci, and asymptotes of the hyperbola given by the equation $\dfrac{y^2}{9} - \dfrac{x^2}{4} = 1$. Sketch the graph.

Solution

Because the y^2 term is positive, the transverse axis is on the y-axis. We know $a^2 = 9$; thus $a = 3$. The vertices are $V_1(0, 3)$ and $V_2(0, -3)$.

$$c^2 = a^2 + b^2 = 9 + 4$$
$$c = \sqrt{13}$$

The foci are $F_1(0, \sqrt{13})$ and $F_2(0, -\sqrt{13})$.

Because $a = 3$ and $b = 2$ ($b^2 = 4$), the equations of the asymptotes are $y = \dfrac{3}{2}x$ and $y = -\dfrac{3}{2}x$.

To sketch the graph, we draw a rectangle that has its center at the origin and has dimensions equal to the lengths of the transverse and conjugate axes. The asymptotes are extensions of the diagonals of the rectangle. See **Figure 8.40.**

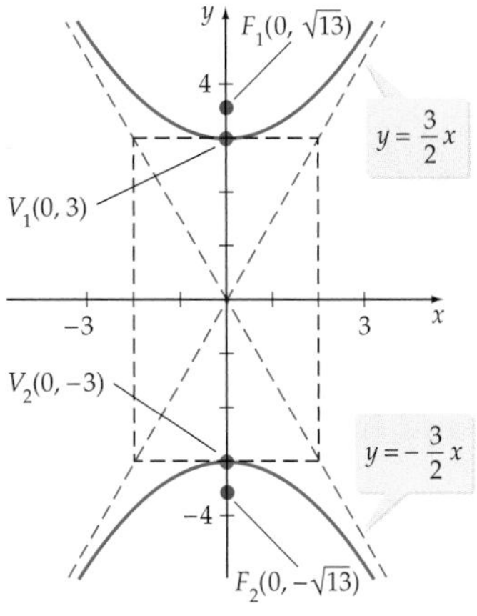

▶ **TRY EXERCISE 4, PAGE 719**

$$\frac{y^2}{9} - \frac{x^2}{4} = 1$$

FIGURE 8.40

● **HYPERBOLAS WITH CENTER AT (h, k)**

Using a translation of coordinates similar to that used for ellipses, we can write the equation of a hyperbola with its center at the point (h, k). Given coordinate axes labeled x' and y', an equation of a hyperbola with center at the origin is

$$\frac{(x')^2}{a^2} - \frac{(y')^2}{b^2} = 1 \qquad (1)$$

Now place the origin of this coordinate system at the point (h, k) of the xy-coordinate system, as shown in **Figure 8.41.** The relationship between an ordered pair in the $x'y'$-coordinate system and one in the xy-coordinate system is given by the transformation equations

$$x' = x - h$$
$$y' = y - k$$

Substitute the expressions for x' and y' into Equation (1). The equation of the hyperbola with center at (h, k) is

$$\frac{(x - h)^2}{a^2} - \frac{(y - k)^2}{b^2} = 1$$

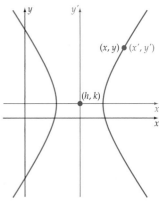

FIGURE 8.41

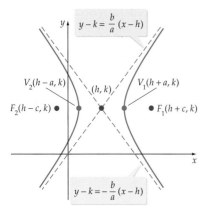

a. Transverse axis parallel to the x-axis

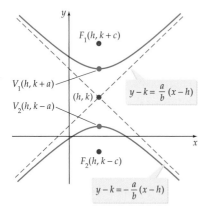

b. Transverse axis parallel to the y-axis

FIGURE 8.42

Standard Forms of the Equation of a Hyperbola with Center at (h, k)

Transverse Axis Parallel to the x-Axis

The standard form of the equation of a hyperbola with center at (h, k) and transverse axis parallel to the x-axis (see **Figure 8.42a**) is given by

$$\frac{(x - h)^2}{a^2} - \frac{(y - k)^2}{b^2} = 1$$

The coordinates of the vertices are $V_1(h + a, k)$ and $V_2(h - a, k)$. The coordinates of the foci are $F_1(h + c, k)$ and $F_2(h - c, k)$, where $c^2 = a^2 + b^2$.

The equations of the asymptotes are $y - k = \pm\dfrac{b}{a}(x - h)$.

Transverse Axis Parallel to the y-Axis

The standard form of the equation of a hyperbola with center at (h, k) and transverse axis parallel to the y-axis (see **Figure 8.42b**) is given by

$$\frac{(y - k)^2}{a^2} - \frac{(x - h)^2}{b^2} = 1$$

The coordinates of the vertices are $V_1(h, k + a)$ and $V_2(h, k - a)$. The coordinates of the foci are $F_1(h, k + c)$ and $F_2(h, k - c)$, where $c^2 = a^2 + b^2$.

The equations of the asymptotes are $y - k = \pm\dfrac{a}{b}(x - h)$.

EXAMPLE 2   **Find the Vertices, Foci, and Asymptotes of a Hyperbola**

Find the vertices, foci, and asymptotes of the hyperbola given by the equation $4x^2 - 9y^2 - 16x + 54y - 29 = 0$. Sketch the graph.

Solution

Write the equation of the hyperbola in standard form by completing the square.

$$4x^2 - 9y^2 - 16x + 54y - 29 = 0$$

$$4x^2 - 16x - 9y^2 + 54y = 29 \qquad \text{• Rearrange terms.}$$

$$4(x^2 - 4x) - 9(y^2 - 6y) = 29 \qquad \text{• Factor.}$$

$$4(x^2 - 4x + 4) - 9(y^2 - 6y + 9) = 29 + 16 - 81 \qquad \text{• Complete the square.}$$

$$4(x - 2)^2 - 9(y - 3)^2 = -36 \qquad \text{• Factor.}$$

$$\frac{(y - 3)^2}{4} - \frac{(x - 2)^2}{9} = 1 \qquad \text{• Divide by } -36.$$

The coordinates of the center are $(2, 3)$. Because the term containing $(y - 3)^2$ is positive, the transverse axis is parallel to the y-axis. We know $a^2 = 4$; thus

$a = 2$. The vertices are $(2, 5)$ and $(2, 1)$. See **Figure 8.43**.

$$c^2 = a^2 + b^2 = 4 + 9$$
$$c = \sqrt{13}$$

The foci are $\left(2, 3 + \sqrt{13}\right)$ and $\left(2, 3 - \sqrt{13}\right)$. We know $b^2 = 9$; thus $b = 3$. The equations of the asymptotes are $y - 3 = \pm\left(\dfrac{2}{3}\right)(x - 2)$, which simplifies to

$$y = \frac{2}{3}x + \frac{5}{3} \quad \text{and} \quad y = -\frac{2}{3}x + \frac{13}{3}$$

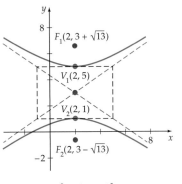

$$\frac{(y - 3)^2}{4} - \frac{(x - 2)^2}{9} = 1$$

FIGURE 8.43

▶ **TRY EXERCISE 26, PAGE 720**

INTEGRATING TECHNOLOGY

A graphing utility can be used to graph a hyperbola. For instance, consider the equation $4x^2 - 9y^2 - 16x + 54y - 29 = 0$ from Example 2. Rewrite the equation as

$$-9y^2 + 54y + (4x^2 - 16x - 29) = 0$$

In this form, the equation is a quadratic equation in terms of the variable y with

$$A = -9, B = 54, \text{ and } C = 4x^2 - 16x - 29$$

Apply the quadratic formula to produce

$$y = \frac{-54 \pm \sqrt{2916 + 36(4x^2 - 16x - 29)}}{-18}$$

The graph of $Y1 = \dfrac{-54 + \sqrt{2916 + 36(4x^2 - 16x - 29)}}{-18}$ is the upper branch of the hyperbola (see **Figure 8.44**).

The graph of $Y2 = \dfrac{-54 - \sqrt{2916 + 36(4x^2 - 16x - 29)}}{-18}$ is the lower branch of the hyperbola, as shown in **Figure 8.44**.

One advantage of this graphing procedure is that it does not require us to write the given equation in standard form. A disadvantage of the graphing procedure is that it does not indicate where the foci of the hyperbola are located.

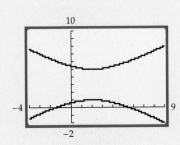

FIGURE 8.44

• ECCENTRICITY OF A HYPERBOLA

The graph of a hyperbola can be very wide or very narrow. The **eccentricity** of a hyperbola is a measure of its "wideness."

Eccentricity (e) of a Hyperbola

The eccentricity e of a hyperbola is the ratio of c to a, where c is the distance from the center to a focus and a is the length of the semitransverse axis.

$$e = \frac{c}{a}$$

For a hyperbola, $c > a$ and therefore $e > 1$. As the eccentricity of the hyperbola increases, the graph becomes wider and wider, as shown in **Figure 8.45.**

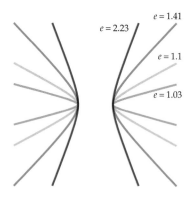

$e = 1.41$
$e = 2.23$
$e = 1.1$
$e = 1.03$

FIGURE 8.45

EXAMPLE 3   **Find the Equation of a Hyperbola Given Its Eccentricity**

Find the standard form of the equation of the hyperbola that has eccentricity $\frac{3}{2}$, center at the origin, and a focus $(6, 0)$.

Solution

Because the focus is located at $(6, 0)$ and the center is at the origin, $c = 6$. An extension of the transverse axis contains the foci, so the transverse axis is on the x-axis.

$$e = \frac{3}{2} = \frac{c}{a}$$

$$\frac{3}{2} = \frac{6}{a} \qquad \text{• Substitute 6 for } c.$$

$$a = 4 \qquad \text{• Solve for } a.$$

To find b^2, use the equation $c^2 = a^2 + b^2$ and the values for c and a.

$$c^2 = a^2 + b^2$$
$$36 = 16 + b^2$$
$$b^2 = 20$$

The equation of the hyperbola is $\dfrac{x^2}{16} - \dfrac{y^2}{20} = 1$. See **Figure 8.46.**

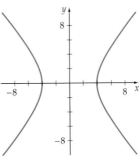

$$\dfrac{x^2}{16} - \dfrac{y^2}{20} = 1$$

FIGURE 8.46

▶ **TRY EXERCISE 48, PAGE 720**

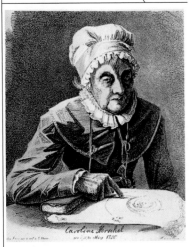
● APPLICATIONS

Orbits of Comets   In Section 8.2 we noted that the orbits of the planets are elliptical. Some comets have elliptical orbits also, the most notable being Halley's comet, whose eccentricity is 0.97.

Other comets have hyperbolic orbits with the sun at a focus. These comets pass by the sun only once. The velocity of a comet determines whether its orbit is elliptical or hyperbolic. See **Figure 8.47.**

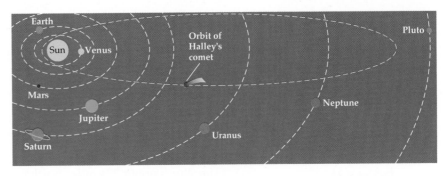

FIGURE 8.47

Hyperbolas as an Aid to Navigation   Consider two radio transmitters, T_1 and T_2, placed some distance apart. A ship with electronic equipment measures the difference between the times it takes signals from the transmitters to reach the ship.

Because the difference between the times is proportional to the difference between the distances of the ship from the transmitters, the ship must be located on the hyperbola with foci at the two transmitters.

Using a third transmitter, T_3, we can find a second hyperbola with foci T_2 and T_3. The ship lies on the intersection of the two hyperbolas, as shown in **Figure 8.48.**

EXAMPLE 4   **Determine the Position of a Ship**

Two radio transmitters are positioned along a coastline, 500 miles apart. See **Figure 8.49.** Using a LORAN (LOng RAnge Navigation) system, a ship determines that a radio signal from transmitter T_1 reaches the ship 1600 microseconds before it receives a simultaneous signal from transmitter T_2.

a. Find an equation of a hyperbola (with foci located at T_1 and T_2) on which the ship lies. See **Figure 8.49.** (Assume the radio signals travel at 0.186 mile per microsecond.)

b. If the ship is directly north of transmitter T_1, determine how far (to the nearest mile) the ship is from the transmitter.

Solution

a. The ship lies on a hyperbola at point B, with foci at T_1 and T_2. The difference of the distances $d(T_2, B)$ and $d(T_1, B)$ is given by

$$\text{Distance} = \text{rate} \times \text{time}$$
$$= 0.186 \text{ mile/microsecond} \times 1600 \text{ microseconds}$$
$$= 297.6 \text{ mile}$$

Thus the ship is located on a hyperbola with transverse axis of 297.6 miles and semitransverse axis $a = 148.8$ miles. **Figure 8.49** shows that the foci are located at $(250, 0)$ and $(-250, 0)$. Thus $c = 250$ miles, and

$$b = \sqrt{c^2 - a^2} = \sqrt{250^2 - 148.8^2} \approx 200.9 \text{ miles}$$

The ship is located on the hyperbola given by

$$\frac{x^2}{148.8^2} - \frac{y^2}{200.9^2} = 1$$

b. If the ship is directly north of T_1, then $x = 250$, and the distance from the ship to the transmitter T_1 is y, where

$$-\frac{y^2}{200.9^2} = 1 - \frac{250^2}{148.8^2}$$
$$y = \frac{200.9}{148.8} \sqrt{250^2 - 148.8^2} \approx 271 \text{ miles}$$

The ship is about 271 miles north of transmitter T_1.

▶ **TRY EXERCISE 54, PAGE 720**

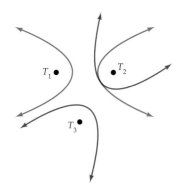

FIGURE 8.48

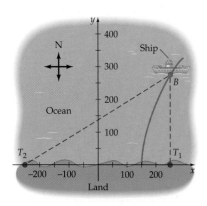

FIGURE 8.49

Hyperbolas also have a reflective property that makes them useful in many applications.

Reflective Property of a Hyperbola

A ray of light directed toward one focus of a hyperbolic mirror is reflected toward the other focus. See **Figures 8.50** and **8.51.**

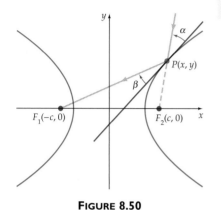

FIGURE 8.50

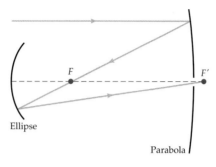 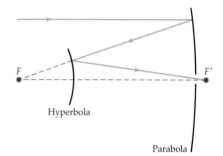

FIGURE 8.51

TOPICS FOR DISCUSSION

1. In every hyperbola, the distance c from a focus to the center of the hyperbola is greater than the length of the semitransverse axis a. Do you agree? Explain.

2. How many vertices does a hyperbola have?

3. Explain why the eccentricity of every hyperbola is a number greater than 1.

4. Is the conjugate axis of a hyperbola perpendicular to the transverse axis of the hyperbola?

EXERCISE SET 8.3

In Exercises 1 to 26, find the center, vertices, foci, and asymptotes for the hyperbola given by each equation. Graph each equation.

1. $\dfrac{x^2}{16} - \dfrac{y^2}{25} = 1$

2. $\dfrac{x^2}{16} - \dfrac{y^2}{9} = 1$

3. $\dfrac{y^2}{4} - \dfrac{x^2}{25} = 1$

▶ 4. $\dfrac{y^2}{25} - \dfrac{x^2}{36} = 1$

5. $\dfrac{x^2}{7} - \dfrac{y^2}{9} = 1$

6. $\dfrac{x^2}{5} - \dfrac{y^2}{4} = 1$

7. $\dfrac{4x^2}{9} - \dfrac{y^2}{16} = 1$

8. $\dfrac{x^2}{9} - \dfrac{9y^2}{16} = 1$

9. $\dfrac{(x-3)^2}{16} - \dfrac{(y+4)^2}{9} = 1$

10. $\dfrac{(x+3)^2}{25} - \dfrac{y^2}{4} = 1$

11. $\dfrac{(y+2)^2}{4} - \dfrac{(x-1)^2}{16} = 1$

12. $\dfrac{(y-2)^2}{36} - \dfrac{(x+1)^2}{49} = 1$

13. $\dfrac{(x+2)^2}{9} - \dfrac{y^2}{25} = 1$

14. $\dfrac{x^2}{25} - \dfrac{(y-2)^2}{81} = 1$

15. $\dfrac{9(x-1)^2}{16} - \dfrac{(y+1)^2}{9} = 1$

16. $\dfrac{(x+6)^2}{25} - \dfrac{25y^2}{144} = 1$

17. $x^2 - y^2 = 9$

18. $4x^2 - y^2 = 16$

19. $16y^2 - 9x^2 = 144$

20. $9y^2 - 25x^2 = 225$

21. $9y^2 - 36x^2 = 4$

22. $16x^2 - 25y^2 = 9$

23. $x^2 - y^2 - 6x + 8y - 3 = 0$

24. $4x^2 - 25y^2 + 16x + 50y - 109 = 0$

25. $9x^2 - 4y^2 + 36x - 8y + 68 = 0$

▶ **26.** $16x^2 - 9y^2 - 32x - 54y + 79 = 0$

In Exercises 27 to 32, use the quadratic formula to solve for y in terms of x. Then use a graphing utility to graph each equation.

27. $4x^2 - y^2 + 32x + 6y + 39 = 0$

28. $x^2 - 16y^2 + 8x - 64y + 16 = 0$

29. $9x^2 - 16y^2 - 36x - 64y + 116 = 0$

30. $2x^2 - 9y^2 + 12x - 18y + 18 = 0$

31. $4x^2 - 9y^2 + 8x - 18y - 6 = 0$

32. $2x^2 - 9y^2 - 8x + 36y - 46 = 0$

In Exercises 33 to 46, find the equation in standard form of the hyperbola that satisfies the stated conditions.

33. Vertices $(3, 0)$ and $(-3, 0)$, foci $(4, 0)$ and $(-4, 0)$

34. Vertices $(0, 2)$ and $(0, -2)$, foci $(0, 3)$ and $(0, -3)$

35. Foci $(0, 5)$ and $(0, -5)$, asymptotes $y = 2x$ and $y = -2x$

36. Foci $(4, 0)$ and $(-4, 0)$, asymptotes $y = x$ and $y = -x$

37. Vertices $(0, 3)$ and $(0, -3)$, passing through $(2, 4)$

38. Vertices $(5, 0)$ and $(-5, 0)$, passing through $(-1, 3)$

39. Asymptotes $y = \frac{1}{2}x$ and $y = -\frac{1}{2}x$, vertices $(0, 4)$ and $(0, -4)$

40. Asymptotes $y = \frac{2}{3}x$ and $y = -\frac{2}{3}x$, vertices $(6, 0)$ and $(-6, 0)$

41. Vertices $(6, 3)$ and $(2, 3)$, foci $(7, 3)$ and $(1, 3)$

42. Vertices $(-1, 5)$ and $(-1, -1)$, foci $(-1, 7)$ and $(-1, -3)$

43. Foci $(1, -2)$ and $(7, -2)$, slope of an asymptote $\frac{5}{4}$

44. Foci $(-3, -6)$ and $(-3, -2)$, slope of an asymptote 1

45. Passing through $(9, 4)$, slope of an asymptote $\frac{1}{2}$, center $(7, 2)$, transverse axis parallel to the y-axis

46. Passing through $(6, 1)$, slope of an asymptote 2, center $(3, 3)$, transverse axis parallel to the x-axis

In Exercises 47 to 52, use the eccentricity to find the equation in standard form of each hyperbola.

47. Vertices $(1, 6)$ and $(1, 8)$, eccentricity 2

▶ **48.** Vertices $(2, 3)$ and $(-2, 3)$, eccentricity $\frac{5}{2}$

49. Eccentricity 2, foci $(4, 0)$ and $(-4, 0)$

50. Eccentricity $\frac{4}{3}$, foci $(0, 6)$ and $(0, -6)$

51. Center $(4, 1)$, conjugate axis of length 4, eccentricity $\frac{4}{3}$ (*Hint:* There are two answers.)

52. Center $(-3, -3)$, conjugate axis of length 6, eccentricity 2 (*Hint:* There are two answers.)

53. **LORAN** Two radio transmitters are positioned along the coast, 250 miles apart. A signal is sent simultaneously from each transmitter. The signal from transmitter T_2 is received by a ship's LORAN 500 microseconds after it receives the signal from T_1. The radio signals travel 0.186 mile per microsecond.

a. Find an equation of a hyperbola, with foci at T_1 and T_2, on which the ship is located.

b. If the ship is 100 miles east of the y-axis, determine its distance from the coastline (to the nearest mile).

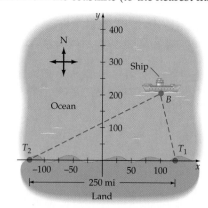

54. **LORAN** Two radio transmitters are positioned along the coast, 300 miles apart. A signal is sent simultaneously from each transmitter. The signal from transmitter T_1 is received by a ship's LORAN 800 microseconds after it receives the signal from T_2. The radio signals travel 0.186 mile per microsecond.

a. Find an equation of a hyperbola, with foci at T_1 and T_2, on which the ship is located.

b. If the ship continues to travel so that the difference of 800 microseconds is maintained, determine the point at which the ship will reach the coastline.

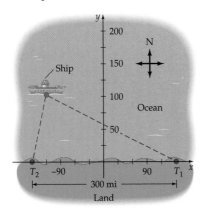

55. SONIC BOOMS When a plane exceeds the speed of sound, a sonic boom is produced by the wake of the sound waves. (See the chapter opener on page 683.) For a plane flying at 10,000 feet, the circular wave front can be given by

$$y^2 = x^2 + (z - 10{,}000)^2$$

where z is the height of the wave front above Earth. See the diagram below. Note that the xy-plane is Earth's surface, which is approximately flat over small distances.

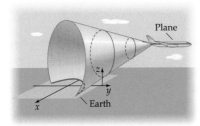

Find and name the equation formed when the wave front hits Earth.

56. WATER WAVES If two pebbles are dropped into a pond at different places $F_1(-2, 0)$ and $F_2(2, 0)$, circular waves are propagated with F_1 as the center of one set of circular waves (in green) and F_2 as the center of the other set of circular waves (in red).

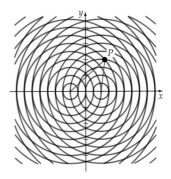

Let P be a point at which the waves intersect. In the diagram above, $|F_1P - F_2P| = 2$.

a. What curve is generated by connecting all points P for which $|F_1P - F_2P| = 2$?

b. What is the equation of the curve in part **a.**?

In Exercises 57 to 64, identify the graph of each equation as a parabola, an ellipse, or a hyperbola. Graph each equation.

57. $4x^2 + 9y^2 - 16x - 36y + 16 = 0$

58. $2x^2 + 3y - 8x + 2 = 0$

59. $5x - 4y^2 + 24y - 11 = 0$

60. $9x^2 - 25y^2 - 18x + 50y = 0$

61. $x^2 + 2y - 8x = 0$

62. $9x^2 + 16y^2 + 36x - 64y - 44 = 0$

63. $25x^2 + 9y^2 - 50x - 72y - 56 = 0$

64. $(x - 3)^2 + (y - 4)^2 = (x + 1)^2$

CONNECTING CONCEPTS

In Exercises 65 to 68, use the definition of a hyperbola to find the equation of the hyperbola in standard form.

65. Foci $(2, 0)$ and $(-2, 0)$; passes through the point $(2, 3)$

66. Foci $(0, 3)$ and $(0, -3)$; passes through the point $\left(\dfrac{5}{2}, 3\right)$

67. Foci $(0, 4)$ and $(0, -4)$; passes through the point $\left(\dfrac{7}{3}, 4\right)$

68. Foci $(5, 0)$ and $(-5, 0)$; passes through the point $\left(5, \dfrac{9}{4}\right)$

Recall that an ellipse has two directrices that are lines perpendicular to the line containing the foci. A hyperbola also has two directrices; they are perpendicular to the transverse axis and outside the hyperbola. For a hyperbola with center at the origin and transverse axis on the x-axis, the equations of the directrices are $x = \dfrac{a^2}{c}$ and $x = -\dfrac{a^2}{c}$. In Exercises 69 to 72, use this information to solve each exercise.

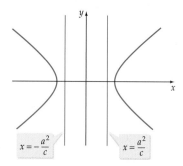

69. Find the directrices for the hyperbola in Exercise 1.

70. Find the directrices for the hyperbola in Exercise 2.

71. Let $P(x, y)$ be a point on the hyperbola $\dfrac{x^2}{9} - \dfrac{y^2}{16} = 1$. Show that the distance from the point P to the focus $(5, 0)$ divided by the distance from the point P to the directrix $x = \dfrac{9}{5}$ equals the eccentricity.

72. Generalize the results of Exercise 71. That is, show that if $P(x, y)$ is a point on the hyperbola $\dfrac{x^2}{a^2} - \dfrac{y^2}{b^2} = 1$, $F(c, 0)$ is a focus, and $x = \dfrac{a^2}{c}$ is a directrix, then the following equation is true:

$$e = \frac{d(P, F)}{d(P, D)}$$

73. Sketch a graph of $\dfrac{x|x|}{16} - \dfrac{y|y|}{9} = 1$.

74. Sketch a graph of $\dfrac{x|x|}{16} + \dfrac{y|y|}{9} = 1$.

PREPARE FOR SECTION 8.4

75. Expand $\cos(\alpha + \beta)$. [6.2]

76. Expand $\sin(\alpha + \beta)$. [6.2]

77. Solve $\cot 2\alpha = \dfrac{\sqrt{3}}{3}$ for $0 < \alpha < \dfrac{\pi}{2}$. [6.6]

78. If $\sin \alpha = \dfrac{1}{2}$ and $\cos \alpha = -\dfrac{\sqrt{3}}{2}$, find α. Write the answer in degrees. [6.6]

79. Identify the graph of $4x^2 - 6y^2 + 9x + 16y - 8 = 0$. [8.3]

80. Graph: $4x - y^2 - 2y + 3 = 0$ [8.1]

PROJECTS

1.   **A HYPERBOLIC PARABOLOID** A *hyperbolic paraboloid* is a three-dimensional figure. Some of its cross sections are parabolas and some are hyperbolas. Make a drawing of a hyperbolic paraboloid. Explain the relationship that exists between the equations of the parabolic cross sections and the relationship that exists between the equations of the hyperbolic cross sections.

2.   **A HYPERBOLOID OF ONE SHEET** Make a sketch of a *hyperboloid of one sheet*. Explain the different cross sections of the hyperboloid of one sheet. Do some research on nuclear power plants, and explain why nuclear cooling towers are designed in the shape of hyperboloids of one sheet.

ROTATION OF AXES

• THE ROTATION THEOREM FOR CONICS

The equation of a conic with axes parallel to the coordinate axes can be written in a general form.

> *take note*
>
> Some choices of the constants A, B, C, D, E, and F may result in a degenerate conic or an equation that has no solutions.

General Equation of a Conic with Axes Parallel to Coordinate Axes

The **general equation of a conic** with axes parallel to the coordinate axes and not both A and C equal to zero is

$$Ax^2 + Cy^2 + Dx + Ey + F = 0$$

The graph of the equation is a parabola when $AC = 0$, an ellipse when $AC > 0$, and a hyperbola when $AC < 0$.

The terms Dx, Ey, and F determine the translation of the conic from the origin. The general equation of a conic is a *second-degree equation* in two variables. A more general second-degree equation can be written that contains a Bxy term.

General Second-Degree Equation in Two Variables

The **general second-degree equation in two variables** is

$$Ax^2 + Bxy + Cy^2 + Dx + Ey + F = 0$$

The Bxy term ($B \neq 0$) determines a rotation of the conic so that its axes are no longer parallel to the coordinate axes.

A **rotation of axes** is a rotation of the x- and y-axes about the origin to another position denoted by x' and y'. We denote the measure of the **angle of rotation** by α.

Let P be some point in the plane, and let r represent the distance of P from the origin. The coordinates of P relative to the xy-coordinate system and the $x'y'$-coordinate system are $P(x, y)$ and $P(x', y')$, respectively.

Let $Q(x, 0)$ and $R(0, y)$ be the projections of P onto the x- and the y-axis and let $Q'(x', 0')$ and $R'(0', y')$ be the projections of P onto the x'- and the y'-axis. (See **Figure 8.52**.) The angle between the x'-axis and OP is denoted by θ. We can express the coordinates of P in each coordinate system in terms of α and θ.

$$x = r\cos(\theta + \alpha) \qquad x' = r\cos\theta$$
$$y = r\sin(\theta + \alpha) \qquad y' = r\sin\theta$$

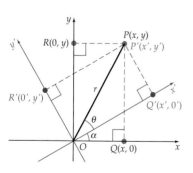

FIGURE 8.52

Applying the addition formulas for $\cos(\theta + \alpha)$ and $\sin(\theta + \alpha)$, we get

$$x = r\cos(\theta + \alpha) = r\cos\theta\cos\alpha - r\sin\theta\sin\alpha$$

$$y = r\sin(\theta + \alpha) = r\sin\theta\cos\alpha + r\cos\theta\sin\alpha$$

Now, substituting x' for $r\cos\theta$ and y' for $r\sin\theta$ into these equations yields

$$x = x'\cos\alpha - y'\sin\alpha \qquad \bullet\, x' = r\cos\theta,\, y' = r\sin\alpha$$

$$y = y'\cos\alpha + x'\sin\alpha$$

This proves the equations labeled (1) of the following theorem.

Rotation-of-Axes Formulas

Suppose that an xy-coordinate system and an $x'y'$-coordinate system have the same origin and that α is the angle between the positive x-axis and the positive x'-axis. If the coordinates of a point P are (x, y) in one system and (x', y') in the rotated system, then

$$\left.\begin{array}{l} x = x'\cos\alpha - y'\sin\alpha \\ y = y'\cos\alpha + x'\sin\alpha \end{array}\right\} \quad (1) \qquad\qquad \left.\begin{array}{l} x' = x\cos\alpha + y\sin\alpha \\ y' = y\cos\alpha - x\sin\alpha \end{array}\right\} \quad (2)$$

The derivations of the formulas for x' and y' are left as an exercise.

As we have noted, the appearance of the Bxy $(B \neq 0)$ term in the general second-degree equation indicates that the graph of the conic has been rotated. The angle through which the axes have been rotated can be determined from the following theorem.

Rotation Theorem for Conics

Let $Ax^2 + Bxy + Cy^2 + Dx + Ey + F = 0$, $B \neq 0$, be the equation of a conic in an xy-coordinate system, and let α be an angle of rotation such that

$$\cot 2\alpha = \frac{A - C}{B}, \quad 0° < 2\alpha < 180°$$

Then the equation of the conic in the rotated coordinate system will be

$$A'x'^2 + C'y'^2 + D'x' + E'y' + F' = 0$$

where $0° < 2\alpha < 180°$ and

$$A' = A\cos^2\alpha + B\cos\alpha\sin\alpha + C\sin^2\alpha \qquad (3)$$

$$C' = A\sin^2\alpha - B\cos\alpha\sin\alpha + C\cos^2\alpha \qquad (4)$$

$$D' = D\cos\alpha + E\sin\alpha \qquad (5)$$

$$E' = -D\sin\alpha + E\cos\alpha \qquad (6)$$

$$F' = F \qquad (7)$$

❓ QUESTION For $x^2 - 4xy + 9y^2 + 6x - 8y - 20 = 0$, what is the value of $\cot 2\alpha$?

EXAMPLE 1 **Use the Rotation Theorem to Sketch a Conic**

Sketch the graph of $7x^2 - 6\sqrt{3}xy + 13y^2 - 16 = 0$.

Solution

We are given

$$A = 7, \quad B = -6\sqrt{3}, \quad C = 13, \quad D = 0, \quad E = 0, \quad \text{and} \quad F = -16$$

The angle of rotation α can be determined by solving

$$\cot 2\alpha = \frac{A - C}{B} = \frac{7 - 13}{-6\sqrt{3}} = \frac{-6}{-6\sqrt{3}} = \frac{1}{\sqrt{3}} = \frac{\sqrt{3}}{3}$$

This gives us $2\alpha = 60°$, or $\alpha = 30°$. Because $\alpha = 30°$, we have

$$\sin \alpha = \frac{1}{2} \quad \text{and} \quad \cos \alpha = \frac{\sqrt{3}}{2}$$

We determine the coefficients $A', C', D', E',$ and F' by using Equations (3) to (7).

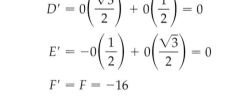

$$A' = 7\left(\frac{\sqrt{3}}{2}\right)^2 + \left(-6\sqrt{3}\right)\left(\frac{\sqrt{3}}{2}\right)\left(\frac{1}{2}\right) + 13\left(\frac{1}{2}\right)^2 = 4$$

$$C' = 7\left(\frac{1}{2}\right)^2 - \left(-6\sqrt{3}\right)\left(\frac{\sqrt{3}}{2}\right)\left(\frac{1}{2}\right) + 13\left(\frac{\sqrt{3}}{2}\right)^2 = 16$$

$$D' = 0\left(\frac{\sqrt{3}}{2}\right)^2 + 0\left(\frac{1}{2}\right) = 0$$

$$E' = -0\left(\frac{1}{2}\right) + 0\left(\frac{\sqrt{3}}{2}\right) = 0$$

$$F' = F = -16$$

The equation of the conic in the $x'y'$-plane is $4(x')^2 + 16(y')^2 - 16 = 0$ or

$$\frac{(x')^2}{2^2} + \frac{(y')^2}{1^2} = 1$$

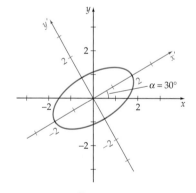

$7x^2 - 6\sqrt{3}xy + 13y^2 - 16 = 0$

FIGURE 8.53

This is the equation of an ellipse that is centered at the origin of an $x'y'$-coordinate system. The ellipse has a semimajor axis $a = 2$ and a semiminor axis $b = 1$. See **Figure 8.53**.

▶ **TRY EXERCISE 8, PAGE 730**

❓ ANSWER $\cot 2\alpha = \dfrac{1 - 9}{-4} = 2$

In Example 1, the angle of rotation α was 30°, which is a special angle. In the next example, we demonstrate a technique that is often used when the angle of rotation is not a special angle.

EXAMPLE 2   Use the Rotation Theorem to Sketch a Conic

Sketch the graph of $32x^2 - 48xy + 18y^2 - 15x - 20y = 0$.

Solution

We are given

$$A = 32, \quad B = -48, \quad C = 18, \quad D = -15, \quad E = -20, \quad \text{and} \quad F = 0$$

Therefore,

$$\cot 2\alpha = \frac{A - C}{B} = \frac{32 - 18}{-48} = -\frac{7}{24}$$

Figure 8.54 shows an angle 2α for which $\cot 2\alpha = -\dfrac{7}{24}$. From **Figure 8.54** we conclude that $\cos 2\alpha = -\dfrac{7}{25}$. The half-angle identities can be used to determine $\sin \alpha$ and $\cos \alpha$.

$$\sin \alpha = \sqrt{\frac{1 - (-7/25)}{2}} = \frac{4}{5} \quad \text{and} \quad \cos \alpha = \sqrt{\frac{1 + (-7/25)}{2}} = \frac{3}{5}$$

A calculator can be used to determine that $\alpha \approx 53.1°$.

Equations (3) to (7) give us

$$A' = 32\left(\frac{3}{5}\right)^2 + (-48)\left(\frac{3}{5}\right)\left(\frac{4}{5}\right) + 18\left(\frac{4}{5}\right)^2 = 0$$

$$C' = 32\left(\frac{4}{5}\right)^2 - (-48)\left(\frac{3}{5}\right)\left(\frac{4}{5}\right) + 18\left(\frac{3}{5}\right)^2 = 50$$

$$D' = (-15)\left(\frac{3}{5}\right) + (-20)\left(\frac{4}{5}\right) = -25$$

$$E' = -(-15)\left(\frac{4}{5}\right) + (-20)\left(\frac{3}{5}\right) = 0$$

$$F' = F = 0$$

The equation of the conic in the $x'y'$-plane is $50(y')^2 - 25x' = 0$, or

$$(y')^2 = \frac{1}{2}x'$$

This is the equation of a parabola. Because $4p = \dfrac{1}{2}$, we know $p = \dfrac{1}{8}$, and the focus of the parabola is at $\left(\dfrac{1}{8}, 0\right)$ on the x'-axis. See **Figure 8.55**.

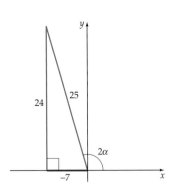

FIGURE 8.54

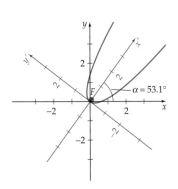

$32x^2 - 48xy + 18y^2 - 15x - 20y = 0$

FIGURE 8.55

▶ TRY EXERCISE 18, PAGE 730

EXAMPLE 6   Transform from Polar to Rectangular Coordinates

Find the rectangular coordinates of the points whose polar coordinates are: **a.** $\left(6, \dfrac{3\pi}{4}\right)$ **b.** $(-4, 30°)$

Solution

Use the equations $x = r \cos \theta$ and $y = r \sin \theta$.

a. $x = 6 \cos \dfrac{3\pi}{4} = -3\sqrt{2}$ $y = 6 \sin \dfrac{3\pi}{4} = 3\sqrt{2}$

The rectangular coordinates of $\left(6, \dfrac{3\pi}{4}\right)$ are $\left(-3\sqrt{2}, 3\sqrt{2}\right)$.

b. $x = -4 \cos 30° = -2\sqrt{3}$ $y = -4 \sin 30° = -2$

The rectangular coordinates of $(-4, 30°)$ are $\left(-2\sqrt{3}, -2\right)$.

▶ **TRY EXERCISE 44, PAGE 743**

EXAMPLE 7   Transform from Rectangular to Polar Coordinates

Find the polar coordinates of the point whose rectangular coordinates are $\left(-2, -2\sqrt{3}\right)$.

Solution

Use the equations $r = \sqrt{x^2 + y^2}$ and $\tan \theta = \dfrac{y}{x}$.

$$r = \sqrt{(-2)^2 + (-2\sqrt{3})^2} = \sqrt{4 + 12} = \sqrt{16} = 4$$

$$\tan \theta = \frac{-2\sqrt{3}}{-2} = \sqrt{3}$$

From this and the fact that $\left(-2, -2\sqrt{3}\right)$ lies in the third quadrant, $\theta = \dfrac{4\pi}{3}$.

The polar coordinates of $\left(-2, -2\sqrt{3}\right)$ are $\left(4, \dfrac{4\pi}{3}\right)$.

▶ **TRY EXERCISE 48, PAGE 743**

● WRITE POLAR COORDINATE EQUATIONS AS RECTANGULAR EQUATIONS AND RECTANGULAR COORDINATE EQUATIONS AS POLAR EQUATIONS

Using the transformation equations, it is possible to write a polar coordinate equation in rectangular form or a rectangular coordinate equation in polar form.

EXAMPLE 8	Write a Polar Coordinate Equation in Rectangular Form

Find a rectangular form of the equation $r^2 \cos 2\theta = 3$.

Solution

$$r^2 \cos 2\theta = 3$$
$$r^2(1 - 2\sin^2 \theta) = 3 \qquad \bullet \cos 2\theta = 1 - 2\sin^2 \theta$$
$$r^2 - 2r^2 \sin^2 \theta = 3$$
$$r^2 - 2(r \sin \theta)^2 = 3$$
$$x^2 + y^2 - 2y^2 = 3 \qquad \bullet \, r^2 = x^2 + y^2; \, \sin \theta = \frac{y}{r}$$
$$x^2 - y^2 = 3$$

A rectangular form of $r^2 \cos 2\theta = 3$ is $x^2 - y^2 = 3$.

▶ **TRY EXERCISE 56, PAGE 743**

EXAMPLE 9	Write a Rectangular Coordinate Equation in Polar Form

Find a polar form of the equation $x^2 + y^2 - 2x = 3$.

Solution

$$x^2 + y^2 - 2x = 3$$
$$(r \cos \theta)^2 + (r \sin \theta)^2 - 2r \cos \theta = 3 \qquad \bullet \text{ Use the transformation equations } x = r \cos \theta \text{ and } y = r \sin \theta.$$
$$r^2(\cos^2 \theta + \sin^2 \theta) - 2r \cos \theta = 3 \qquad \bullet \text{ Simplify.}$$
$$r^2 - 2r \cos \theta = 3$$

A polar form of $x^2 + y^2 - 2x = 3$ is $r^2 - 2r \cos \theta = 3$.

▶ **TRY EXERCISE 64, PAGE 743**

 TOPICS FOR DISCUSSION

1. In what quadrant is the point $(-2, 150°)$ located?

2. To explain why the graph of $\theta = \dfrac{\pi}{6}$ is a line, a tutor rewrites the equation in the form $\theta = \dfrac{\pi}{6} + 0 \cdot r$. In this form, regardless of the value of r, $\theta = \dfrac{\pi}{6}$. Use an analogous approach to explain why the graph of $r = a$ is a circle.

3. Two students use a graphing calculator to graph the polar equation $r = 2 \sin \theta$. One graph appears to be a circle, and the other graph appears to be an ellipse. Both graphs are correct. Explain.

4. Is the graph of $r^2 = 6 \cos 2\theta$ a rose curve with four petals? Explain your answer.

EXERCISE SET 8.5

In Exercises 1 to 8, plot the point on a polar coordinate system.

1. $(2, 60°)$

2. $(3, -90°)$

3. $(1, 315°)$

4. $(2, 400°)$

5. $\left(-2, \dfrac{\pi}{4}\right)$

6. $\left(4, \dfrac{7\pi}{6}\right)$

7. $\left(-3, \dfrac{5\pi}{3}\right)$

8. $(-3, \pi)$

In Exercises 9 to 24, sketch the graph of each polar equation.

9. $r = 3$

10. $r = 5$

11. $\theta = 2$

12. $\theta = -\dfrac{\pi}{3}$

13. $r = 6 \cos \theta$

▶ **14.** $r = 4 \sin \theta$

15. $r = 4 \cos 2\theta$

▶ **16.** $r = 5 \cos 3\theta$

17. $r = 2 \sin 5\theta$

18. $r = 3 \cos 5\theta$

19. $r = 2 - 3 \sin \theta$

▶ **20.** $r = 2 - 2 \cos \theta$

21. $r = 4 + 3 \sin \theta$

22. $r = 2 + 4 \sin \theta$

23. $r = 2(1 - 2 \sin \theta)$

24. $r = 4(1 - \sin \theta)$

In Exercises 25 to 40, use a graphing utility to graph each equation.

25. $r = 3 + 3 \cos \theta$

▶ **26.** $r = 4 - 4 \sin \theta$

27. $r = 4 \cos 3\theta$

28. $r = 2 \sin 4\theta$

29. $r = 3 \sec \theta$

30. $r = 4 \csc \theta$

31. $r = -5 \csc \theta$

▶ **32.** $r = -4 \sec \theta$

33. $r = 4 \sin (3.5\theta)$

34. $r = 6 \cos (2.25\theta)$

35. $r = \theta, 0 \le \theta \le 6\pi$

36. $r = -\theta, 0 \le \theta \le 6\pi$

37. $r = 2^\theta, 0 \le \theta \le 2\pi$

38. $r = \dfrac{1}{\theta}, 0 \le \theta \le 4\pi$

39. $r = \dfrac{6 \cos 7\theta + 2 \cos 3\theta}{\cos \theta}$

40. $r = \dfrac{4 \cos 3\theta + \cos 5\theta}{\cos \theta}$

In Exercises 41 to 48, transform the given coordinates to the indicated ordered pair.

41. $\left(1, -\sqrt{3}\right)$ to (r, θ)

42. $\left(-2\sqrt{3}, 2\right)$ to (r, θ)

43. $\left(-3, \dfrac{2\pi}{3}\right)$ to (x, y)

▶ **44.** $\left(2, -\dfrac{\pi}{3}\right)$ to (x, y)

45. $\left(0, -\dfrac{\pi}{2}\right)$ to (x, y)

46. $\left(3, \dfrac{5\pi}{6}\right)$ to (x, y)

47. $(3, 4)$ to (r, θ)

▶ **48.** $(12, -5)$ to (r, θ)

In Exercises 49 to 60, find a rectangular form of each of the equations.

49. $r = 3 \cos \theta$

50. $r = 2 \sin \theta$

51. $r = 3 \sec \theta$

52. $r = 4 \csc \theta$

53. $r = 4$

54. $\theta = \dfrac{\pi}{4}$

55. $r = \tan \theta$

▶ **56.** $r = \cot \theta$

57. $r = \dfrac{2}{1 + \cos \theta}$

58. $r = \dfrac{2}{1 - \sin \theta}$

59. $r(\sin \theta - 2 \cos \theta) = 6$

60. $r(2 \cos \theta + \sin \theta) = 3$

In Exercises 61 to 68, find a polar form of each of the equations.

61. $y = 2$

62. $x = -4$

63. $x^2 + y^2 = 4$

▶ **64.** $2x - 3y = 6$

65. $x^2 = 8y$

66. $y^2 = 4y$

67. $x^2 - y^2 = 25$

68. $x^2 + 4y^2 = 16$

In Exercises 69 to 76, use a graphing utility to graph each equation.

69. $r = 3 \cos \left(\theta + \dfrac{\pi}{4}\right)$

70. $r = 2 \sin \left(\theta - \dfrac{\pi}{6}\right)$

71. $r = 2 \sin \left(2\theta - \dfrac{\pi}{3}\right)$

72. $r = 3 \cos \left(2\theta + \dfrac{\pi}{4}\right)$

73. $r = 2 + 2 \sin\left(\theta - \dfrac{\pi}{6}\right)$ **74.** $r = 3 - 2 \cos\left(\theta + \dfrac{\pi}{3}\right)$ **75.** $r = 1 + 3 \cos\left(\theta + \dfrac{\pi}{3}\right)$ **76.** $r = 2 - 4 \sin\left(\theta - \dfrac{\pi}{4}\right)$

CONNECTING CONCEPTS

77. Explain why the graph of $r^2 = \cos^2 \theta$ and the graph of $r = \cos \theta$ are not the same.

78. Explain why the graph of $r = \cos 2\theta$ and the graph of $r = 2 \cos^2 \theta - 1$ are identical.

 In Exercises 79 to 86, use a graphing utility to graph each equation.

79. $r^2 = 4 \cos 2\theta$ (lemniscate)

80. $r^2 = -2 \sin 2\theta$ (lemniscate)

81. $r = 2(1 + \sec \theta)$ (conchoid)

82. $r = 2 \cos 2\theta \sec \theta$ (strophoid)

83. $r\theta = 2$ (spiral)

84. $r = 2 \sin \theta \cos^2 2\theta$ (bifolium)

85. $r = |\theta|$

86. $r = \ln \theta$

87. The graph of

$$r = 1.5^{\sin \theta} - 2.5 \cos 4\theta + \sin^7 \dfrac{\theta}{15}$$

is a *butterfly curve* similar to the one shown below.

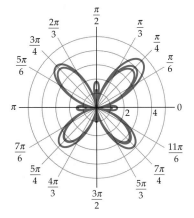

Use a graphing utility to graph the butterfly curve for

a. $0 \le \theta \le 5\pi$ **b.** $0 \le \theta \le 20\pi$

For additional information on butterfly curves, read "The Butterfly Curve" by Temple H. Fay, *The American Mathematical Monthly*, vol. 96, no. 5 (May 1989), p. 442.

PREPARE FOR SECTION 8.6

88. Find the eccentricity of the graph of $\dfrac{x^2}{25} + \dfrac{y^2}{16} = 1$. [8.2]

89. What is the equation of the directrix of the graph of $y^2 = 4x$? [8.1]

90. Solve $y = 2(1 + yx)$ for y. [1.1]

91. For the function $f(x) = \dfrac{3}{1 + \sin x}$, what is the smallest positive value of x that must be excluded from the domain of f? [5.2]

92. Let e be the eccentricity of a hyperbola. Which of the following statements is true: $e = 0$, $0 < e < 1$, $e = 1$, or $e > 1$? [8.3]

93. Write $\dfrac{4 \sec x}{2 \sec x - 1}$ in terms of $\cos x$. [6.1]

—— PROJECTS ——

1. A POLAR DISTANCE FORMULA Let $P_1(r_1, \theta_1)$ and $P_2(r_2, \theta_2)$ be two distinct points in the $r\theta$-plane.

a. Verify that the distance d between the points is
$$d = \sqrt{r_1^2 + r_2^2 - 2r_1r_2 \cos(\theta_2 - \theta_1)}$$

b. Use the above formula to find the distance (to the nearest hundredth) between $(3, 60°)$ and $(5, 170°)$.

c. 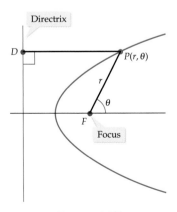 Does the formula $d = \sqrt{r_1^2 + r_2^2 - 2r_1r_2 \cos(\theta_1 - \theta_2)}$ also produce the correct distance between P_1 and P_2? Explain.

2. ANOTHER POLAR FORM FOR A CIRCLE

a. Verify that the graph of the polar equation $r = a \sin \theta + b \cos \theta$ is a circle. Assume that a and b are not both 0.

b. What are the center (in rectangular coordinates) and the radius of the circle?

SECTION 8.6

POLAR EQUATIONS OF THE CONICS

- POLAR EQUATIONS OF THE CONICS
- GRAPH A CONIC GIVEN IN POLAR FORM
- WRITE THE POLAR EQUATION OF A CONIC

• POLAR EQUATIONS OF THE CONICS

The definition of a parabola was given in terms of a point (the focus) and a line (the directrix). The definitions of both the ellipse and the hyperbola were given in terms of two points (the foci). It is possible to define each conic in terms of a point and a line.

Directrix

D

$P(r, \theta)$

r

θ

F

Focus

FIGURE 8.76

Focus-Directrix Definitions of the Conics

Let F be a fixed point and D a fixed line in a plane. Consider the set of all points P such that $\dfrac{d(P, F)}{d(P, D)} = e$, where e is a constant. The graph is a parabola for $e = 1$, an ellipse for $0 < e < 1$, and a hyperbola for $e > 1$. See **Figure 8.76**.

The fixed point is a focus of the conic, and the fixed line is a directrix. The constant e is the eccentricity of the conic. Using this definition, we can derive the polar equations of the conics.

Standard Form of the Polar Equations of the Conics

Let the pole be a focus of a conic section of eccentricity e with directrix d units from the focus. Then the equation of the conic is given by one of the following:

$$r = \frac{ed}{1 + e \cos \theta} \quad (1) \qquad r = \frac{ed}{1 - e \cos \theta} \quad (2)$$

Vertical directrix to the right of the pole

Vertical directrix to the left of the pole

$$r = \frac{ed}{1 + e \sin \theta} \quad (3) \qquad r = \frac{ed}{1 - e \sin \theta} \quad (4)$$

Horizontal directrix above the pole

Horizontal directrix below the pole

When the equation involves $\cos \theta$, the polar axis is an axis of symmetry.

When the equation involves $\sin \theta$, the line $\theta = \dfrac{\pi}{2}$ is an axis of symmetry.

Graphs of examples are shown in **Figure 8.77.**

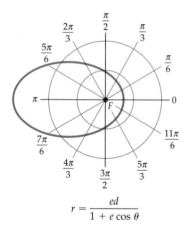

$$r = \frac{ed}{1 + e \cos \theta}$$

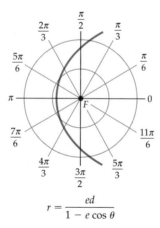

$$r = \frac{ed}{1 - e \cos \theta}$$

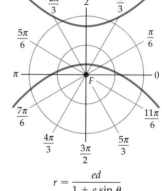

$$r = \frac{ed}{1 + e \sin \theta}$$

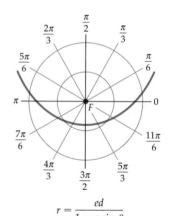

$$r = \frac{ed}{1 - e \sin \theta}$$

FIGURE 8.77

❓ **QUESTION** Is the graph of $r = \dfrac{4}{5 - 3 \sin \theta}$ a parabola, an ellipse, or a hyperbola?

We will derive Equation (2). Let $P(r, \theta)$ be any point on a conic section. Then, by definition,

$$\frac{d(P, F)}{d(P, D)} = e \quad \text{or} \quad d(P, F) = e \cdot d(P, D)$$

❓ **ANSWER** The eccentricity is $\dfrac{3}{5}$, which is less than 1. The graph is an ellipse.

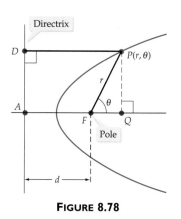

FIGURE 8.78

From **Figure 8.78,** $d(P, F) = r$ and $d(P, D) = d(A, Q)$. But note that

$$d(A, Q) = d(A, F) + d(F, Q) = d + r \cos \theta$$

Thus

$$r = e(d + r \cos \theta) \qquad \bullet \ d(P, F) = e \cdot d(P, D)$$

$$= ed + er \cos \theta$$

$$r - er \cos \theta = ed \qquad \bullet \ \text{Subtract } er \cos \theta.$$

$$r = \frac{ed}{1 - e \cos \theta} \qquad \bullet \ \text{Solve for } r.$$

The remaining equations can be derived in a similar manner.

● **GRAPH A CONIC GIVEN IN POLAR FORM**

EXAMPLE 1   **Sketch the Graph of a Hyperbola Given in Polar Form**

Describe and sketch the graph of $r = \dfrac{8}{2 - 3 \sin \theta}$.

Solution

Write the equation in standard form by dividing the numerator and denominator by 2, the constant term in the denominator.

$$r = \frac{4}{1 - \dfrac{3}{2} \sin \theta}$$

Because e is the coefficient of $\sin \theta$ and $e = \dfrac{3}{2} > 1$, the graph is a hyperbola with a focus at the pole. Because the equation contains the expression $\sin \theta$, the transverse axis is on the line $\theta = \dfrac{\pi}{2}$.

To find the vertices, choose θ equal to $\dfrac{\pi}{2}$ and $\dfrac{3\pi}{2}$. The corresponding values of r are -8 and $\dfrac{8}{5}$. The vertices are $\left(-8, \dfrac{\pi}{2} \right)$ and $\left(\dfrac{8}{5}, \dfrac{3\pi}{2} \right)$. By choosing θ equal to 0 and π, we can determine the points $(4, 0)$ and $(4, \pi)$ on the upper branch of the hyperbola. The lower branch can be determined by symmetry.

Plot some points (r, θ) for additional values of θ and corresponding values of r. See **Figure 8.79.**

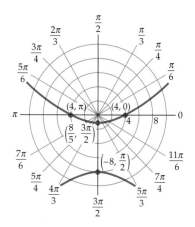

FIGURE 8.79

▶ **TRY EXERCISE 2, PAGE 750**

EXAMPLE 2   Sketch the Graph of an Ellipse Given in Polar Form

Describe and sketch the graph of $r = \dfrac{4}{2 + \cos \theta}$.

Solution

Write the equation in standard form by dividing the numerator and denominator by 2, which is the constant term in the denominator.

$$r = \frac{2}{1 + \dfrac{1}{2} \cos \theta}$$

Thus $e = \dfrac{1}{2}$ and the graph is an ellipse with a focus at the pole. Because the equation contains the expression $\cos \theta$, the major axis is on the polar axis.

To find the vertices, choose θ equal to 0 and π. The corresponding values for r are $\dfrac{4}{3}$ and 4. The vertices on the major axis are $\left(\dfrac{4}{3}, 0 \right)$ and $(4, \pi)$. Plot some points (r, θ) for additional values of θ and the corresponding values of r. Two possible points are $\left(2, \dfrac{\pi}{2} \right)$ and $\left(2, \dfrac{3\pi}{2} \right)$. See the graph of the ellipse in **Figure 8.80**.

▶ **TRY EXERCISE 4, PAGE 750**

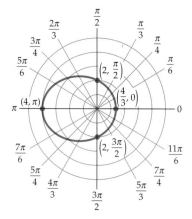

FIGURE 8.80

● **WRITE THE POLAR EQUATION OF A CONIC**

EXAMPLE 3   Find the Equation of a Conic in Polar Form

Find the equation of the parabola, shown in **Figure 8.81**, with vertex at $\left(2, \dfrac{\pi}{2} \right)$ and focus at the pole.

Solution

Because the vertex is on the line $\theta = \dfrac{\pi}{2}$ and the focus is at the pole, the axis of symmetry is the line $\theta = \dfrac{\pi}{2}$. Thus the equation of the parabola must involve $\sin \theta$. The parabola has a horizontal directrix above the pole, so the equation has the form

$$r = \frac{ed}{1 + e \sin \theta}$$

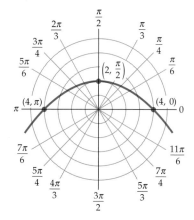

FIGURE 8.81

The distance from the vertex to the focus is 2, so the distance from the focus to the directrix is 4. Because the graph of the equation is a parabola, the eccentricity is 1. The equation is

$$r = \frac{(1)(4)}{1 + (1) \sin \theta} \qquad \bullet\, e = 1, d = 4$$

$$r = \frac{4}{1 + \sin \theta}$$

▶ **TRY EXERCISE 24, PAGE 750**

❓ QUESTION In Example 3, why is there no point on the parabola that corresponds to $\theta = \dfrac{3\pi}{2}$?

 TOPICS FOR DISCUSSION

1. Is the graph of $r = \dfrac{12}{2 + \cos \theta}$ a parabola? Explain.

2. The graph of $r = \dfrac{2 \sec \theta}{2 \sec \theta + 1}$ is an ellipse except for the fact that it has two holes. Where are the holes located?

3. Are there two different ellipses that have a focus at the pole and a vertex at $\left(1, \dfrac{\pi}{2} \right)$? Explain.

4. Does the parabola given by $r = \dfrac{6}{1 + \sin \theta}$ have a horizontal axis of symmetry or a vertical axis of symmetry? Explain.

❓ ANSWER When $\theta = \dfrac{3\pi}{2}$, $\sin \theta = -1$. Thus $1 + \sin \theta = 0$, and $r = \dfrac{4}{1 + \sin \theta}$ is undefined.

EXERCISE SET 8.6

In Exercises 1 to 14, describe and sketch the graph of each equation.

1. $r = \dfrac{12}{3 - 6 \cos \theta}$

▶ **2.** $r = \dfrac{8}{2 - 4 \cos \theta}$

3. $r = \dfrac{8}{4 + 3 \sin \theta}$

▶ **4.** $r = \dfrac{6}{3 + 2 \cos \theta}$

5. $r = \dfrac{9}{3 - 3 \sin \theta}$

6. $r = \dfrac{5}{2 - 2 \sin \theta}$

7. $r = \dfrac{10}{5 + 6 \cos \theta}$

8. $r = \dfrac{8}{2 + 4 \cos \theta}$

9. $r = \dfrac{4 \sec \theta}{2 \sec \theta - 1}$

10. $r = \dfrac{3 \sec \theta}{2 \sec \theta + 2}$

11. $r = \dfrac{12 \csc \theta}{6 \csc \theta - 2}$

12. $r = \dfrac{3 \csc \theta}{2 \csc \theta + 2}$

13. $r = \dfrac{3}{\cos \theta - 1}$

14. $r = \dfrac{2}{\sin \theta + 2}$

In Exercises 15 to 20, find a rectangular equation for the graphs in Exercises 1 to 6.

In Exercises 21 to 28, find a polar equation of the conic with focus at the pole and the given eccentricity and directrix.

21. $e = 2, r \cos \theta = -1$

22. $e = \dfrac{3}{2}, r \sin \theta = 1$

23. $e = 1, r \sin \theta = 2$

▶ **24.** $e = 1, r \cos \theta = -2$

25. $e = \dfrac{2}{3}, r \sin \theta = -4$

26. $e = \dfrac{1}{2}, r \cos \theta = 2$

27. $e = \dfrac{3}{2}, r = 2 \sec \theta$

28. $e = \dfrac{3}{4}, r = 2 \csc \theta$

29. Find the polar equation of the parabola with a focus at the pole and vertex $(2, \pi)$.

30. Find the polar equation of the ellipse with a focus at the pole, vertex at $(4, 0)$, and eccentricity $\dfrac{1}{2}$.

31. Find the polar equation of the hyperbola with a focus at the pole, vertex at $\left(1, \dfrac{3\pi}{2}\right)$, and eccentricity 2.

32. Find the polar equation of the ellipse with a focus at the pole, vertex at $\left(2, \dfrac{3\pi}{2}\right)$, and eccentricity $\dfrac{2}{3}$.

In Exercises 33 to 40, use a graphing utility to graph each equation. Write a sentence that explains how to obtain the graph from the graph of r as given in the exercise listed to the right of each equation.

33. $r = \dfrac{12}{3 - 6 \cos\left(\theta - \dfrac{\pi}{6}\right)}$ (Compare with Exercise 1.)

34. $r = \dfrac{8}{2 - 4 \cos\left(\theta - \dfrac{\pi}{2}\right)}$ (Compare with Exercise 2.)

35. $r = \dfrac{8}{4 + 3 \sin(\theta - \pi)}$ (Compare with Exercise 3.)

36. $r = \dfrac{6}{3 + 2 \cos\left(\theta - \dfrac{\pi}{3}\right)}$ (Compare with Exercise 4.)

37. $r = \dfrac{9}{3 - 3 \sin\left(\theta + \dfrac{\pi}{6}\right)}$ (Compare with Exercise 5.)

38. $r = \dfrac{5}{2 - 2 \sin\left(\theta + \dfrac{\pi}{2}\right)}$ (Compare with Exercise 6.)

39. $r = \dfrac{10}{5 + 6 \cos(\theta + \pi)}$ (Compare with Exercise 7.)

40. $r = \dfrac{8}{2 + 4 \cos\left(\theta + \dfrac{\pi}{3}\right)}$ (Compare with Exercise 8.)

CONNECTING CONCEPTS

 In Exercises 41 to 46, use a graphing utility to graph each equation.

41. $r = \dfrac{3}{3 - \sec \theta}$

42. $r = \dfrac{5}{4 - 2 \csc \theta}$

43. $r = \dfrac{3}{1 + 2 \csc \theta}$

44. $r = \dfrac{4}{1 + 3 \sec \theta}$

45. $r = 4 \sin \sqrt{2}\theta, 0 \le \theta \le 12\pi$

46. $r = 4 \cos \sqrt{3}\theta, 0 \le \theta \le 8\pi$

47. Let $P(r, \theta)$ satisfy the equation $r = \dfrac{ed}{1 - e \cos \theta}$. Show that $\dfrac{d(P, F)}{d(P, D)} = e$.

48. Show that the equation of a conic with a focus at the pole and directrix $r \sin \theta = d$ is given by $r = \dfrac{ed}{1 + e \sin \theta}$.

PREPARE FOR SECTION 8.7

49. Complete the square of $y^2 + 3y$ and write the result as the square of a binomial. [1.3]

50. If $x = 2t + 1$ and $y = x^2$, write y in terms of t. [P.1]

51. Identify the graph of $\left(\dfrac{x - 2}{3}\right)^2 + \left(\dfrac{y - 3}{2}\right)^2 = 1$. [8.2]

52. Let $x = \sin t$ and $y = \cos t$. What is the value of $x^2 + y^2$? [6.1]

53. Solve $y = \ln t$ for t. [4.5]

54. What are the domain and range of $f(t) = 3 \cos 2t$? Write the answer using interval notation. [5.5]

PROJECTS

1. POLAR EQUATION OF A LINE Verify that the polar equation of a line that is d units from the pole is given by

$$r = \frac{d}{\cos(\theta - \theta_p)}$$

where θ_p is the angle from the polar axis to a line segment that passes through the pole and is perpendicular to the line.

2. POLAR EQUATION OF A CIRCLE THAT PASSES THROUGH THE POLE Verify that the polar equation of a circle with center (a, θ_c) that passes through the pole is given by

$$r = 2a \cos(\theta - \theta_c)$$

PARAMETRIC EQUATIONS

● PARAMETRIC EQUATIONS

The graph of a function is a graph for which no vertical line can intersect the graph more than once. For a graph that is not the graph of a function (an ellipse or a hyperbola, for example), it is frequently useful to describe the graph by *parametric equations*.

Curve and Parametric Equations

Let t be a number in an interval I. A **curve** is the set of ordered pairs (x, y), where

$$x = f(t), \qquad y = g(t) \quad \text{for } t \in I$$

The variable t is called a **parameter,** and the equations $x = f(t)$ and $y = g(t)$ are **parametric equations.**

For instance,

$$x = 2t - 1, \qquad y = 4t + 1 \quad \text{for } t \in (-\infty, \infty)$$

is an example of a set of parametric equations. By choosing arbitrary values of t, ordered pairs (x, y) can be created, as shown in the table below.

t	$x = 2t - 1$	$y = 4t + 1$	(x, y)
-2	-5	-7	$(-5, -7)$
0	-1	1	$(-1, 1)$
$\dfrac{1}{2}$	0	3	$(0, 3)$
2	3	9	$(3, 9)$

By plotting the points and drawing a curve through the points, a graph of the parametric equations is produced. See **Figure 8.82.**

❓ **QUESTION** If $x = t^2 + 1$ and $y = 3 - t$, what ordered pair corresponds to $t = -3$?

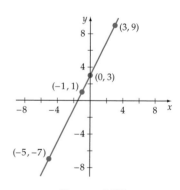

FIGURE 8.82

❓ **ANSWER** $(10, 6)$

● GRAPH A CURVE GIVEN BY PARAMETRIC EQUATIONS

EXAMPLE 1   Sketch the Graph of a Curve Given in Parametric Form

Sketch the graph of the curve given by the parametric equations

$$x = t^2 + t, \quad y = t - 1 \quad \text{for } t \in (-\infty, \infty)$$

Solution

Begin by making a table of values of t and the corresponding values of x and y. Five values of t were arbitrarily chosen for the table that follows. Many more values might be necessary to determine an accurate graph.

t	$x = t^2 + t$	$y = t - 1$	(x, y)
-2	2	-3	$(2, -3)$
-1	0	-2	$(0, -2)$
0	0	-1	$(0, -1)$
1	2	0	$(2, 0)$
2	6	1	$(6, 1)$

Graph the ordered pairs (x, y) and then draw a smooth curve through the points. See **Figure 8.83**.

▶ TRY EXERCISE 6, PAGE 758

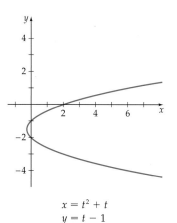

$$x = t^2 + t$$
$$y = t - 1$$

FIGURE 8.83

● ELIMINATE THE PARAMETER OF A PARAMETRIC EQUATION

It may not be clear from Example 1 and the corresponding graph that the curve is a parabola. By **eliminating the parameter**, we can write one equation in x and y that is equivalent to the two parametric equations.

To eliminate the parameter, solve $y = t - 1$ for t.

$$y = t - 1 \quad \text{or} \quad t = y + 1$$

Substitute $y + 1$ for t in $x = t^2 + t$ and then simplify.

$$x = (y + 1)^2 + (y + 1)$$
$$= y^2 + 3y + 2 \qquad \text{• The equation of a parabola}$$

Complete the square and write the equation in standard form.

$$\left(x + \frac{1}{4}\right) = \left(y + \frac{3}{2}\right)^2 \qquad \begin{array}{l}\text{• This is the equation of a parabola with} \\ \text{vertex at } \left(-\frac{1}{4}, -\frac{3}{2}\right).\end{array}$$

| EXAMPLE 2 | **Eliminate the Parameter and Sketch the Graph of a Curve** |

Eliminate the parameter and sketch the curve of the parametric equations

$$x = \sin t, \quad y = \cos t \quad \text{for } 0 \le t < 2\pi$$

Solution

The process of eliminating the parameter sometimes involves trigonometric identities. To eliminate the parameter for the equations, square each side of each equation and then add.

$$x^2 = \sin^2 t$$

$$y^2 = \cos^2 t$$

$$x^2 + y^2 = \sin^2 t + \cos^2 t$$

Thus, using the trigonometric identity $\sin^2 t + \cos^2 t = 1$, we get

$$x^2 + y^2 = 1$$

This is the equation of a circle with center $(0, 0)$ and radius equal to 1. See **Figure 8.84**.

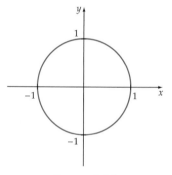

FIGURE 8.84

▶ **TRY EXERCISE 12, PAGE 758**

A parametric representation of a curve is not unique. That is, it is possible that a curve may be given by many different pairs of parametric equations. We will demonstrate this by using the equation of a line and providing two different parametric representations of the line.

Consider a line with slope m passing through the point (x_1, y_1). By the point-slope formula, the equation of the line is

$$y - y_1 = m(x - x_1)$$

Let $t = x - x_1$. Then $y - y_1 = mt$. A parametric representation is

$$x = x_1 + t, \quad y = y_1 + mt \quad \text{for } t \text{ a real number} \tag{1}$$

Let $x - x_1 = \cot t$. Then $y - y_1 = m \cot t$. A parametric representation is

$$x = x_1 + \cot t, \quad y = y_1 + m \cot t \quad \text{for } 0 < t < \pi \tag{2}$$

It can be verified that Equations (1) and (2) represent the original line.

Example 3 illustrates that the domain of the parameter t can be used to determine the domain and range of the function.

| | EXAMPLE 3 | Sketch the Graph of a Curve Given by Parametric Equations |

Eliminate the parameter and sketch the graph of the curve that is given by the parametric equations

$$x = 2 + 3\cos t, \qquad y = 3 + 2\sin t \quad \text{for } 0 \le t \le \pi$$

Solution

Rewrite each equation in terms of the trigonometric function.

$$\frac{x-2}{3} = \cos t \qquad \frac{y-3}{2} = \sin t \tag{3}$$

Using the trigonometric identity $\cos^2 t + \sin^2 t = 1$, we have

$$\cos^2 t + \sin^2 t = \left(\frac{x-2}{3}\right)^2 + \left(\frac{y-3}{2}\right)^2 = 1$$

$$\frac{(x-2)^2}{9} + \frac{(y-3)^2}{4} = 1$$

This is the equation of an ellipse with center at $(2, 3)$ and major axis parallel to the x-axis. However, because $0 \le t \le \pi$, it follows that $-1 \le \cos t \le 1$ and $0 \le \sin t \le 1$. Therefore, we have

$$-1 \le \frac{x-2}{3} \le 1 \qquad 0 \le \frac{y-3}{2} \le 1 \qquad \bullet \textbf{ Using Equations (3)}$$

Solving these inequalities for x and y yields

$$-1 \le x \le 5 \quad \text{and} \quad 3 \le y \le 5$$

Because the values of y are between 3 and 5, the graph of the parametric equations is only the top half of the ellipse. See **Figure 8.85.**

▶ | TRY EXERCISE 14, PAGE 758 |

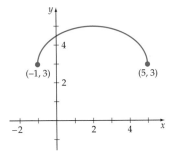

FIGURE 8.85

● THE BRACHISTOCHRONE PROBLEM

Parametric equations are useful in writing the equation of a moving point. One famous problem, involving a bead traveling down a frictionless wire, was posed in 1696 by the mathematician Johann Bernoulli. The problem was to determine the shape of a wire a bead could slide down such that the distance between two points was traveled in the shortest time. Problems that involve "shortest time" are called *brachistochrone problems.* They are very important in physics and form the basis for much of the classical theory of light propagation.

The answer to Bernoulli's problem is an arc of an inverted cycloid. See **Figure 8.86.** A **cycloid** is formed by letting a circle of radius a roll on a straight line L without slipping. See **Figure 8.87.** The curve traced by a point on the circumference of the circle is a cycloid. To find an equation for this curve, begin by placing a circle tangent to the x-axis with a point P on the circle and at the origin of a rectangular coordinate system.

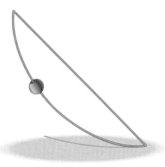

FIGURE 8.86

Roll the circle along the *x*-axis. After the radius of the circle has rotated through an angle θ, the coordinates of the point $P(x, y)$ can be given by

$$x = h - a \sin \theta, \qquad y = k - a \cos \theta \qquad (4)$$

where $C(h, k)$ is the current center of the circle.

Because the radius of the circle is a, $k = a$. See **Figure 8.87.** Because the circle rolls without slipping, the arc length subtended by θ equals h. Thus $h = a\theta$. Substituting for h and k in Equations (4), we have, after factoring,

$$x = a(\theta - \sin \theta), \qquad y = a(1 - \cos \theta) \quad \text{for } \theta \geq 0$$

See **Figure 8.88.**

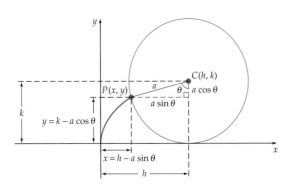

FIGURE 8.87

$$x = a(\theta - \sin \theta), y = a(1 - \cos \theta)$$

FIGURE 8.88

A cycloid

EXAMPLE 4 Graph a Cycloid

 Use a graphing utility to graph the cycloid given by

$$x = 4(\theta - \sin \theta), \qquad y = 4(1 - \cos \theta) \quad \text{for } 0 \leq \theta \leq 4\pi$$

Solution

Although θ is the parameter in the above equations, many graphing utilities, such as the TI-83, use T as the parameter for parametric equations. Thus to graph the equations for $0 \leq \theta \leq 4\pi$, we use Tmin $= 0$ and Tmax $= 4\pi$, as shown below. Use radian mode and parametric mode to produce the graph in **Figure 8.89.**

Tmin=0	Xmin=-6	Ymin=-4
Tmax=4π	Xmax=16π	Ymax=10
Tstep=0.5	Xscl=2π	Yscl=1

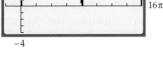

$$x = 4(\theta - \sin \theta)$$
$$y = 4(1 - \cos \theta)$$

FIGURE 8.89

▶ **TRY EXERCISE 26, PAGE 758**

● **PARAMETRIC EQUATIONS AND PROJECTILE MOTION**

The path of a projectile (assume air resistance is negligible) that is launched at an angle θ from the horizon with an initial velocity of v_0 feet per second is given by

the parametric equations

$$x = (v_0 \cos \theta)t, \qquad y = -16t^2 + (v_0 \sin \theta)t$$

where t is the time in seconds since the projectile was launched.

EXAMPLE 5 Sketch the Path of a Projectile

Use a graphing utility to sketch the path of a projectile that is launched at an angle of $\theta = 32°$ with an initial velocity of 144 feet per second. Use the graph to determine (to the nearest foot) the maximum height of the projectile and the range of the projectile. Assume the ground is level.

Solution

Use degree mode and parametric mode. Graph the parametric equations

$$x = (144 \cos 32°)t, \qquad y = -16t^2 + (144 \sin 32°)t \quad \text{for } 0 \le t \le 5$$

to produce the graph in **Figure 8.90.** Use the TRACE feature to determine that the maximum height of 91 feet is attained when $t \approx 2.38$ seconds and that the projectile strikes the ground about 581 feet downrange when $t \approx 4.76$ seconds.

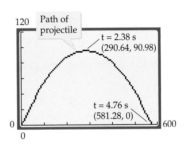

$$x = (144 \cos 32°)t$$
$$y = -16t^2 + (144 \sin 32°)t$$

FIGURE 8.90

In **Figure 8.90,** the angle of launch does not appear to be $32°$ because 1 foot on the x-axis is smaller than 1 foot on the y-axis.

▶ **TRY EXERCISE 32, PAGE 759**

 ## TOPICS FOR DISCUSSION

1. It is always possible to eliminate the parameter of a pair of parametric equations. Do you agree? Explain.

2. The line $y = 3x + 5$ has more than one parametric representation. Do you agree? Explain.

3. Parametric equations are used only to graph functions. Do you agree? Explain.

4. Every function $y = f(x)$ can be written in parametric form by letting $x = t$ and $y = f(t)$. Do you agree? Explain.

EXERCISE SET 8.7

In Exercises 1 to 10, graph the parametric equations by plotting several points.

1. $x = 2t, y = -t$, for $t \in R$

2. $x = -3t, y = 6t$, for $t \in R$

3. $x = -t, y = t^2 - 1$, for $t \in R$

4. $x = 2t, y = 2t^2 - t + 1$, for $t \in R$

5. $x = t^2, y = t^3$, for $t \in R$

▶ **6.** $x = t^2 + 1, y = t^2 - 1$, for $t \in R$

7. $x = 2 \cos t, y = 3 \sin t$, for $0 \le t < 2\pi$

8. $x = 1 - \sin t, y = 1 + \cos t$, for $0 \le t < 2\pi$

9. $x = 2^t, y = 2^{t+1}$, for $t \in R$

10. $x = t^2, y = 2 \log_2 t$, for $t \ge 1$

In Exercises 11 to 20, eliminate the parameter and graph the equation.

11. $x = \sec t, y = \tan t$, for $-\dfrac{\pi}{2} < t < \dfrac{\pi}{2}$

▶ **12.** $x = 3 + 2 \cos t, y = -1 - 3 \sin t$, for $0 \le t < 2\pi$

13. $x = 2 - t^2, y = 3 + 2t^2$, for $t \in R$

▶ **14.** $x = 1 + t^2, y = 2 - t^2$, for $t \in R$

15. $x = \cos^3 t, y = \sin^3 t$, for $0 \le t < 2\pi$

16. $x = e^{-t}, y = e^t$, for $t \in R$

17. $x = \sqrt{t + 1}, y = t$, for $t \ge -1$

18. $x = \sqrt{t}, y = 2t - 1$, for $t \ge 0$

19. $x = t^3, y = 3 \ln t$, for $t > 0$

20. $x = e^t, y = e^{2t}$, for $t \in R$

21. Eliminate the parameter for the curves

$$C_1: \quad x = 2 + t^2, \quad y = 1 - 2t^2$$

and $\qquad C_2: \quad x = 2 + t, \quad y = 1 - 2t$

and then discuss the differences between their graphs.

22. Eliminate the parameter for the curves

$$C_1: \quad x = \sec^2 t, \quad y = \tan^2 t$$

and $\qquad C_2: \quad x = 1 + t^2, \quad y = t^2$

for $0 \le t < \dfrac{\pi}{2}$, and then discuss the differences between their graphs.

23. Sketch the graph of

$$x = \sin t, \qquad y = \csc t \quad \text{for } 0 < t \le \dfrac{\pi}{2}$$

Sketch another graph for the same pair of equations but choose the domain of t to be $\pi < t \le \dfrac{3\pi}{2}$.

24. Discuss the differences between

$$C_1: \quad x = \cos t, \quad y = \cos^2 t$$

and $\qquad C_2: \quad x = \sin t, \quad y = \sin^2 t$

for $0 \le t \le \pi$.

25. Use a graphing utility to graph the cycloid $x = 2(t - \sin t), y = 2(1 - \cos t)$ for $0 \le t < 2\pi$.

▶ **26.** Use a graphing utility to graph the cycloid $x = 3(t - \sin t), y = 3(1 - \cos t)$ for $0 \le t \le 12\pi$.

Parametric equations of the form $x = a \sin \alpha t$, $y = b \cos \beta t$, for $t \ge 0$, are encountered in electrical circuit theory. The graphs of these equations are called *Lissajous figures*. Use Tstep $= 0.5$ and $0 \le$ T $\le 2\pi$.

27. Graph: $x = 5 \sin 2t, y = 5 \cos t$

28. Graph: $x = 5 \sin 3t, y = 5 \cos 2t$

29. Graph: $x = 4 \sin 2t, y = 4 \cos 3t$

30. Graph: $x = 5 \sin 10t, y = 5 \cos 9t$

In Exercises 31 to 34, graph the path of the projectile that is launched at an angle of θ with the horizon with an initial velocity of v_0. In each exercise, use the graph to determine the maximum height and the range of the projectile (to the nearest foot). Also state the time t at which the projectile reaches its maximum height and the time it hits the ground. Assume the ground is level and the only force acting on the projectile is gravity.

31. $\theta = 55°$, $v_0 = 210$ feet per second

▶ **32.** $\theta = 35°$, $v_0 = 195$ feet per second

33. $\theta = 42°$, $v_0 = 315$ feet per second

34. $\theta = 52°$, $v_0 = 315$ feet per second

CONNECTING CONCEPTS

35. Let $P_1(x_1, y_1)$ and $P_2(x_2, y_2)$ be two distinct points in the plane, and consider the line L passing through those points. Choose a point $P(x, y)$ on the line L. Show that

$$\frac{x - x_1}{x_2 - x_1} = \frac{y - y_1}{y_2 - y_1}$$

Use this result to demonstrate that $x = (x_2 - x_1)t + x_1$, $y = (y_2 - y_1)t + y_1$ is a parametric representation of the line through the two points.

36. Show that $x = h + a \sin t$, $y = k + b \cos t$, for $a > 0$, $b > 0$, and $0 \leq t < 2\pi$, are parametric equations for an ellipse with center at (h, k).

37. Suppose a string, held taut, is unwound from the circumference of a circle of radius a. The path traced by the end of the string is called the *involute* of a circle. Find parametric equations for the involute of a circle.

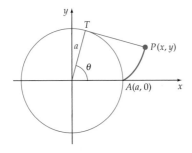

38. A circle of radius a rolls without slipping on the outside of a circle of radius $b > a$. Find the parametric equations of a point P on the smaller circle. The curve is called an *epicycloid*.

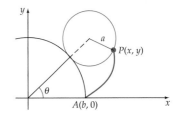

39. A circle of radius a rolls without slipping on the inside of a circle of radius $b > a$. Find the parametric equations of a point P on the smaller circle. The curve is called a *hypocycloid*.

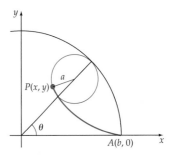

PROJECTS

1. PARAMETRIC EQUATIONS IN AN *XYZ*-COORDINATE SYSTEM

a. Graph the three-dimensional curve given by

$$x = 3 \cos t, \quad y = 3 \sin t, \quad z = 0.5t$$

b. Graph the three-dimensional curve given by

$$x = 3 \cos t, \quad y = 6 \sin t, \quad z = 0.5t$$

c. What is the main difference between these curves?

d. What name is given to curves of this type?

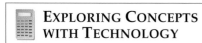

EXPLORING CONCEPTS
WITH TECHNOLOGY

Using a Graphing Calculator to Find the *n*th Roots of *z*

In Chapter 7 we used De Moivre's Theorem to find the *n*th roots of a number. The parametric feature of a graphing calculator can also be used to find and display the *n*th roots of $z = r(\cos \theta + i \sin \theta)$. Here is the procedure for a TI-83 graphing calculator. Put the calculator in parametric and degree mode. See **Figure 8.91**. To find the *n*th roots of $z = r(\cos \theta + i \sin \theta)$, enter in the ⃞Y= menu

$$X_{1T} = r^{(1/n)}\cos(\theta/n + T) \quad \text{and} \quad Y_{1T} = r^{(1/n)}\sin(\theta/n + T)$$

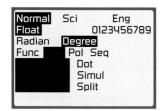

FIGURE 8.91

In the **WINDOW** menu, set **Tmin=0**, **Tmax=360**, and **Tstep=360/n**. Set **Xmin**, **Xmax**, **Ymin**, and **Ymax** to appropriate values that will allow the roots to be seen in the graph window. Press **GRAPH** to display a polygon. The *x*- and *y*-coordinates of each vertex of the polygon represent a root of *z* in the rectangular form $x + yi$. Here is a specific example that illustrates this procedure.

Example Find the fourth roots of $z = 16i$.

In trigonometric form, $z = 16(\cos 90° + i \sin 90°)$. Thus in this example, $r = 16$, $\theta = 90°$, and $n = 4$. In the ⃞Y= menu, enter

$$X_{1T} = 16^{(1/4)}\cos(90/4 + T) \quad \text{and} \quad Y_{1T} = 16^{(1/4)}\sin(90/4 + T)$$

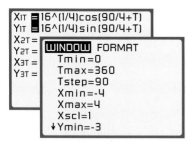

FIGURE 8.92

In the **WINDOW** menu, set

Tmin=0	Xmin=-4	Ymin=-3
Tmax=360	Xmax=4	Ymax=3
Tstep=360/4	Xscl=1	Yscl=1

See **Figure 8.92.** Press **GRAPH** to produce the quadrilateral in **Figure 8.93.** Use **TRACE** and the arrow key ⃞▷ to move to each of the vertices of the quadrilateral. **Figure 8.93** shows that one of the roots of $z = 16i$ is $1.8477591 + 0.76536686i$. Continue to press the arrow key ⃞▷ to find the other three roots, which are

$$-0.7653669 + 1.8477591i, \ -1.847759 - 0.7653669i, \text{ and}$$
$$0.76536686 - 1.847759i$$

Use a graphing calculator to estimate, in rectangular form, each of the following.

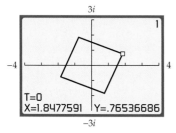

The vertices of the quadrilateral represent the fourth roots of $16i$ in the complex plane.

FIGURE 8.93

1. The cube roots of -27.

2. The fifth roots of $32i$.

3. The fourth roots of $\sqrt{8} + \sqrt{8}i$.

4. The sixth roots of $-64i$.

ordered pairs $\left(x, -\dfrac{1}{3}x + 2\right)$ are solutions of Equation (2), substitute

$\left(x, -\dfrac{1}{3}x + 2\right)$ into Equation (2) and solve for x.

$$2x + 6y = -18 \qquad \bullet \text{ Equation (2)}$$

$$2x + 6\left(-\frac{1}{3}x + 2\right) = -18 \qquad \bullet \text{ Substitute } -\frac{1}{3}x + 2 \text{ for } y.$$

$$2x - 2x + 12 = -18$$

$$12 = -18 \qquad \bullet \text{ A false statement}$$

The false statement $12 = -18$ means that no ordered pair that is a solution of Equation (1) is also a solution of Equation (2). The equations have no ordered pairs in common and thus the system of equations has no solution. This is an inconsistent system of equations.

Therefore, the graphs of the two lines are parallel and never intersect. See **Figure 9.5**.

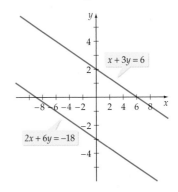

FIGURE 9.5

An inconsistent system of equations

▶ **TRY EXERCISE 18, PAGE 778**

EXAMPLE 3   **Identify a Dependent System of Equations**

Solve: $\begin{cases} 8x - 4y = 16 & (1) \\ 2x - y = 4 & (2) \end{cases}$

Algebraic Solution

Solve Equation (2) for y:

$$2x - y = 4$$

$$y = 2x - 4$$

The solutions of $y = 2x - 4$ are the ordered pairs $(x, 2x - 4)$. For the system of equations to have a solution, ordered pairs of the form $(x, 2x - 4)$ also must be solutions of $8x - 4y = 16$. To determine whether the ordered pairs $(x, 2x - 4)$ are solutions of Equation (1), substitute $(x, 2x - 4)$ into Equation (1) and solve for x.

$$8x - 4y = 16 \qquad \bullet \text{ Equation (1)}$$

$$8x - 4(2x - 4) = 16 \qquad \bullet \text{ Substitute } 2x - 4 \text{ for } y.$$

$$8x - 8x + 16 = 16$$

$$16 = 16 \qquad \bullet \text{ A true statement}$$

The true statement $16 = 16$ means that the ordered pairs $(x, 2x - 4)$ that are solutions of Equation (2) are also solutions of Equation (1). Because x can be replaced by any real number c, the solution of the system of equations is the set of ordered pairs $(c, 2c - 4)$. This is a dependent system of equations.

Visualize the Solution

Solving Equations (1) and (2) for y gives $y = 2x - 4$ and $y = 2x - 4$. Note that these two equations have the same slope, 2, and the same y-intercept, $(0, -4)$. Therefore, the graphs of the two lines are exactly the same. One graph intersects the second graph infinitely often. See **Figure 9.6**.

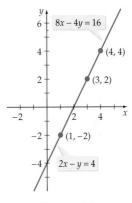

FIGURE 9.6

▶ **TRY EXERCISE 20, PAGE 778**

Some of the specific ordered-pair solutions in Example 3 can be found by choosing various values for c. The table below shows the ordered pairs that result from choosing c as 1, 3, and 4. The ordered pairs $(1, -2)$, $(3, 2)$, and $(4, 4)$ are specific solutions of the system of equations. These points are on the graphs of Equation (1) and Equation (2), as shown in **Figure 9.6.**

c	$(c, 2c - 4)$	(x, y)
1	$(1, 2(1) - 4)$	$(1, -2)$
3	$(3, 2(3) - 4)$	$(3, 2)$
4	$(4, 2(4) - 4)$	$(4, 4)$

Before leaving Example 3, note that there is more than one way to represent the ordered-pair solutions. To illustrate this point, solve Equation (2) for x.

$$2x - y = 4 \qquad \bullet \text{ Equation (2)}$$

$$x = \frac{1}{2}y + 2 \qquad \bullet \text{ Solve for } x.$$

Because y can be replaced by any real number b, there are an infinite number of ordered pairs $\left(\frac{1}{2}b + 2, b\right)$ that are solutions of the system of equations. Choosing b as -2, 2, and 4 gives the same ordered pairs: $(1, -2)$, $(3, 2)$, and $(4, 4)$. There is always more than one way to describe the ordered pairs when writing the solution of a dependent system of equations. For Example 3, either the ordered pairs $(c, 2c - 4)$ or the ordered pairs $\left(\frac{1}{2}b + 2, b\right)$ would generate all the solutions of the system of equations.

> ***take note***
>
> When a system of equations is dependent, there is more than one way to write the solutions of the solution set. The solution to Example 3 is the set of ordered pairs
>
> $(c, 2c - 4)$ or $\left(\frac{1}{2}b + 2, b\right)$
>
> However, there are infinitely more ways in which the ordered pairs could be expressed. For instance, let $b = 2w$. Then
>
> $\frac{1}{2}b + 2 = \frac{1}{2}(2w) + 2 = w + 2$
>
> The ordered-pair solutions, written in terms of w, are $(w + 2, 2w)$.

● ELIMINATION METHOD FOR SOLVING A SYSTEM OF EQUATIONS

Two systems of equations are **equivalent** if each system has exactly the same solutions. The systems

$$\begin{cases} 3x + 5y = 9 \\ 2x - 3y = -13 \end{cases} \quad \text{and} \quad \begin{cases} x = -2 \\ y = 3 \end{cases}$$

are equivalent systems of equations. Each system has the solution $(-2, 3)$, as shown in **Figure 9.7.**

A second technique for solving a system of equations is similar to the strategy for solving first-degree equations in one variable. The system of equations is replaced by a series of equivalent systems until the solution is obvious.

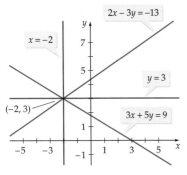

FIGURE 9.7

Operations That Produce Equivalent Systems of Equations

1. Interchange any two equations.

2. Replace an equation with a nonzero multiple of that equation.

3. Replace an equation with the sum of that equation and a nonzero constant multiple of another equation in the system.

Because the order in which the equations are written does not affect the system of equations, interchanging the equations does not affect its solution. The second operation restates the property that says that multiplying each side of an equation by the same nonzero constant does not change the solutions of the equation.

The third operation can be illustrated as follows. Consider the system of equations

$$\begin{cases} 3x + 2y = 10 & (1) \\ 2x - 3y = -2 & (2) \end{cases}$$

Multiply each side of Equation (2) by 2. (Any nonzero number would work.) Add the resulting equation to Equation (1).

$$\begin{aligned} 3x + 2y &= 10 \qquad & \bullet \text{ Equation (1)} \\ \underline{4x - 6y} &= \underline{-4} & \bullet \text{ 2 times Equation (2)} \\ 7x - 4y &= 6 \quad (3) & \bullet \text{ Add the equations.} \end{aligned}$$

Replace Equation (1) with the new Equation (3) to produce the following equivalent system of equations.

$$\begin{cases} 7x - 4y = 6 & (3) \\ 2x - 3y = -2 & (2) \end{cases}$$

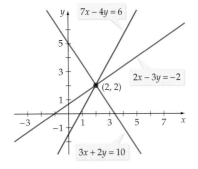

$7x - 4y = 6$

$2x - 3y = -2$

(2, 2)

$3x + 2y = 10$

FIGURE 9.8

The third property states that the resulting system of equations has the same solutions as the original system and is therefore equivalent to the original system of equations. **Figure 9.8** shows the graph of $7x - 4y = 6$. Note that the line passes through the same point at which the lines of the original system of equations intersect, the point (2, 2).

EXAMPLE 4 **Solve a System of Equations by the Elimination Method**

Solve: $\begin{cases} 3x - 4y = 10 & (1) \\ 2x + 5y = -1 & (2) \end{cases}$

Algebraic Solution

Use the operations that produce equivalent equations to eliminate a variable from one of the equations. We will eliminate x from Equation (2) by multiplying each equation by a different constant so as to have a new system of equations in which the coefficients of x are additive inverses.

$$\begin{aligned} 6x - 8y &= 20 & \bullet \text{ 2 times Equation (1)} \\ \underline{-6x - 15y} &= \underline{3} & \bullet -3 \text{ times Equation (2)} \\ -23y &= 23 & \bullet \text{ Add the equations.} \\ y &= -1 & \bullet \text{ Solve for y.} \end{aligned}$$

Solve Equation (1) for x by substituting -1 for y.

$$\begin{aligned} 3x - 4(-1) &= 10 \\ 3x &= 6 \\ x &= 2 \end{aligned}$$

The solution of the system of equations is $(2, -1)$.

Visualize the Solution

Graphing $3x - 4y = 10$ and $2x + 5y = -1$ shows that $(2, -1)$ belongs to both lines. Therefore, $(2, -1)$ is a solution of the system of equations. See **Figure 9.9**.

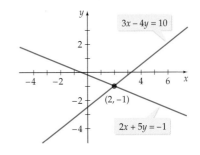

$3x - 4y = 10$

$(2, -1)$

$2x + 5y = -1$

FIGURE 9.9

▶ **TRY EXERCISE 24, PAGE 778**

The method just described is called the **elimination method** for solving a system of equations, because it involves *eliminating* a variable from one of the equations.

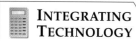

INTEGRATING TECHNOLOGY

You can use a graphing calculator to solve a system of equations in two variables. First, algebraically solve each equation for y.

Solve for y.

$$3x - 4y = 10 \quad \rightarrow \quad y = 0.75x - 2.5$$
$$2x + 5y = -1 \quad \rightarrow \quad y = -0.4x - 0.2$$

Now graph the equations. Enter **0.75X-2.5** into Y₁ and **-0.4X-0.2** into Y₂ and graph the two equations in the standard viewing window. The sequence of steps shown in **Figure 9.10** can be used to find the point of intersection with a TI-83 calculator.

Press 2nd CALC. Select 5: intersect. Press ENTER.

The "First curve?" shown on the bottom of the screen means to select the first of the two graphs that intersect. Just press ENTER.

The "Second curve?" shown on the bottom of the screen means to select the second of the two graphs that intersect. Just press ENTER.

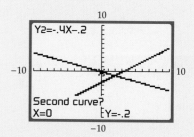

"Guess?" is shown on the bottom of the screen. Move the cursor until it is approximately on the point of intersection. Press ENTER.

The coordinates of the point of intersection $(2, -1)$ are shown at the bottom of the screen.

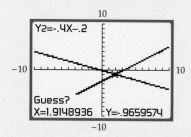

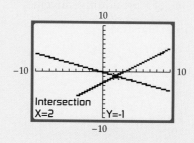

FIGURE 9.10

For the system of equations in Example 4, the intersection of the two graphs occurs at a point in the standard viewing window. If the point of intersection does not appear on the screen, you must adjust the viewing window so that the point of intersection is visible.

EXAMPLE 5   Solve a Dependent System of Equations

Solve: $\begin{cases} x - 2y = 2 & (1) \\ 3x - 6y = 6 & (2) \end{cases}$

Solution

Eliminate x by multiplying Equation (2) by $-\dfrac{1}{3}$ and then adding the result to Equation (1).

$$x - 2y = 2 \qquad \bullet \textbf{ Equation (1)}$$
$$\underline{-x + 2y = -2} \qquad \bullet -\dfrac{1}{3} \textbf{ times Equation (2)}$$
$$0 = 0 \qquad \bullet \textbf{ Add the two equations.}$$

Replace Equation (2) by $0 = 0$.

$$\begin{cases} x - 2y = 2 \\ 0 = 0 \end{cases} \qquad \bullet \textbf{This is an equivalent system of equations.}$$

Because the equation $0 = 0$ is an identity, an ordered pair that is a solution of Equation (1) is also a solution of $0 = 0$. Thus the solutions are the solutions of $x - 2y = 2$. Solving for y, we find that $y = \dfrac{1}{2}x - 1$.

Because x can be replaced by any real number c, the solutions of the system of equations are the ordered pairs $\left(c, \dfrac{1}{2}c - 1 \right)$. See **Figure 9.11.**

▶ **TRY EXERCISE 28, PAGE 778**

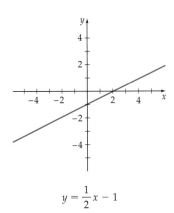

$y = \dfrac{1}{2}x - 1$

FIGURE 9.11

---

take note

Referring again to Example 5 and solving Equation (1) for x, we have $x = 2y + 2$. Because y can be any real number b, the ordered-pair solutions of the system of equations can also be written as $(2b + 2, b)$.

If one equation of the system of equations is replaced by a false equation, the system of equations has no solution. For example, the system of equations

$$\begin{cases} x + y = 4 \\ 0 = 5 \end{cases}$$

has no solution because the second equation is false for any choice of x and y.

● **APPLICATIONS OF SYSTEMS OF EQUATIONS**

Consider the situation of a Corvette car dealership. If the dealership were willing to sell a Corvette for $10, there would be many consumers willing to buy a Corvette. The problem with this plan is that the dealership would soon be out of business. On the other hand, if the dealership tried to sell each Corvette for $1 million, the dealership would not sell any cars and would still go out of business. Between $10 and $1 million, there is a price at which a dealership can sell Corvettes (and stay in business) and at which consumers are willing to pay that price. This price is referred to as the **equilibrium price.**

Economists refer to these types of problems as **supply-demand problems.** Businesses are willing to *supply* a product at a certain price and there is consumer *demand* for the product at that price. To find the equilibrium point, a system of equations is created. One equation of the system is the demand model of the business. The second equation is the supply model of the consumer.

EXAMPLE 6   Solve a Supply-Demand Problem

Suppose that the number of bushels x of apples a farmer is willing to sell is given by $x = 100p - 25$, where p is the price, in dollars, per bushel of apples. The number of bushels x of apples a grocer is willing to purchase is given by $x = -150p + 655$, where p is the price per bushel of apples. Find the equilibrium price.

Solution

Using the supply and demand equations, we have the system of equations

$$\begin{cases} x = 100p - 25 \\ x = -150p + 655 \end{cases}$$

Solve the system of equations by substitution.

$$-150p + 655 = 100p - 25$$

$-250p + 655 = -25$ • Subtract $100p$ from each side.

$-250p = -680$ • Subtract 655 from each side.

$p = 2.72$ • Divide by -250.

The equilibrium price is \$2.72 per bushel.

▶ **TRY EXERCISE 42, PAGE 778**

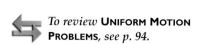

To review **UNIFORM MOTION PROBLEMS,** *see p. 94.*

As the application problems we studied earlier in the text become more complicated, a system of equations may be the method for solving these problems. The next example involves the distance-rate-time equation $d = rt$.

EXAMPLE 7   Solve an Application to River Currents

A rowing team rowing with the current traveled 18 miles in 2 hours. Against the current, the team rowed 10 miles in 2 hours. Find the rate of the boat in calm water and the rate of the current.

Solution

Let r_1 represent the rate of the boat in calm water, and let r_2 represent the rate of the current.

The rate of the boat *with the current* is $r_1 + r_2$.

The rate of the boat *against the current* is $r_1 - r_2$.

Because the rowing team traveled 18 miles in 2 hours with the current, we use the equation $d = rt$.

$$d = r \cdot t$$
$$18 = (r_1 + r_2) \cdot 2 \qquad \bullet \, d = 18, t = 2$$
$$9 = r_1 + r_2 \qquad \bullet \, \text{Divide each side by 2.}$$

Because the team rowed 10 miles in 2 hours against the current, we write

$$10 = (r_1 - r_2) \cdot 2 \qquad \bullet \, d = 10, t = 2$$
$$5 = r_1 - r_2 \qquad \bullet \, \text{Divide each side by 2.}$$

Thus we have a system of two linear equations in the variables r_1 and r_2.

$$\begin{cases} 9 = r_1 + r_2 \\ 5 = r_1 - r_2 \end{cases}$$

Solving the system by using the elimination method, we find that r_1 is 7 mph and r_2 is 2 mph. Thus the rate of the boat in calm water is 7 mph and the rate of the current is 2 mph. You should verify these solutions.

▶ **TRY EXERCISE 46, PAGE 778**

TOPICS FOR DISCUSSION

1. Explain how to use the substitution method to solve a system of equations.

2. Explain how to use the elimination method to solve a system of equations.

3. Give an example of a system of equations in two variables that is

   a. independent   b. dependent   c. inconsistent

4. If a linear system of equations in two variables has no solution, what does that mean about the graphs of the equations of the system?

5. If $A = \{(x, y) \mid x + y = 5\}$ and $B = \{(x, y) \mid x - y = 3\}$, explain the meaning of $A \cap B$.

EXERCISE SET 9.1

In Exercises 1 to 20, solve each system of equations by the substitution method.

1. $\begin{cases} 2x - 3y = 16 \\ x = 2 \end{cases}$

2. $\begin{cases} 3x - 2y = -11 \\ y = 1 \end{cases}$

3. $\begin{cases} 3x + 4y = 18 \\ y = -2x + 3 \end{cases}$

4. $\begin{cases} 5x - 4y = -22 \\ y = 5x - 2 \end{cases}$

5. $\begin{cases} -2x + 3y = 6 \\ x = 2y - 5 \end{cases}$

▶ $\begin{cases} 8x + 3y = -7 \\ x = 3y + 15 \end{cases}$

7. $\begin{cases} 6x + 5y = 1 \\ x - 3y = 4 \end{cases}$

8. $\begin{cases} -3x + 7y = 14 \\ 2x - y = -13 \end{cases}$

9. $\begin{cases} 7x + 6y = -3 \\ y = \dfrac{2}{3}x - 6 \end{cases}$

10. $\begin{cases} 9x - 4y = 3 \\ x = \dfrac{4}{3}y + 3 \end{cases}$

11. $\begin{cases} y = 4x - 3 \\ y = 3x - 1 \end{cases}$

12. $\begin{cases} y = 5x + 1 \\ y = 4x - 2 \end{cases}$

13. $\begin{cases} y = 5x + 4 \\ x = -3y - 4 \end{cases}$

14. $\begin{cases} y = -2x - 6 \\ x = -2y - 2 \end{cases}$

15. $\begin{cases} 3x - 4y = 2 \\ 4x + 3y = 14 \end{cases}$

16. $\begin{cases} 6x + 7y = -4 \\ 2x + 5y = 4 \end{cases}$

17. $\begin{cases} 3x - 3y = 5 \\ 4x - 4y = 9 \end{cases}$

▶ **18.** $\begin{cases} 3x - 4y = 8 \\ 6x - 8y = 9 \end{cases}$

19. $\begin{cases} 4x + 3y = 6 \\ y = -\dfrac{4}{3}x + 2 \end{cases}$

▶ **20.** $\begin{cases} 5x + 2y = 2 \\ y = -\dfrac{5}{2}x + 1 \end{cases}$

In Exercises 21 to 40, solve each system of equations by the elimination method.

21. $\begin{cases} 3x - y = 10 \\ 4x + 3y = -4 \end{cases}$

22. $\begin{cases} 3x + 4y = -5 \\ x - 5y = -8 \end{cases}$

23. $\begin{cases} 4x + 7y = 21 \\ 5x - 4y = -12 \end{cases}$

▶ **24.** $\begin{cases} 3x - 8y = -6 \\ -5x + 4y = 10 \end{cases}$

25. $\begin{cases} 5x - 3y = 0 \\ 10x - 6y = 0 \end{cases}$

26. $\begin{cases} 3x + 2y = 0 \\ 2x + 3y = 0 \end{cases}$

27. $\begin{cases} 6x + 6y = 1 \\ 4x + 9y = 4 \end{cases}$

▶ **28.** $\begin{cases} 4x + 5y = 2 \\ 8x - 15y = 9 \end{cases}$

29. $\begin{cases} 3x + 6y = 11 \\ 2x + 4y = 9 \end{cases}$

30. $\begin{cases} 4x - 2y = 9 \\ 2x - y = 3 \end{cases}$

31. $\begin{cases} \dfrac{5}{6}x - \dfrac{1}{3}y = -6 \\ \dfrac{1}{6}x + \dfrac{2}{3}y = 1 \end{cases}$

32. $\begin{cases} \dfrac{3}{4}x + \dfrac{2}{5}y = 1 \\ \dfrac{1}{2}x - \dfrac{3}{5}y = -1 \end{cases}$

33. $\begin{cases} \dfrac{3}{4}x + \dfrac{1}{3}y = 1 \\ \dfrac{1}{2}x + \dfrac{2}{3}y = 0 \end{cases}$

34. $\begin{cases} \dfrac{3}{5}x - \dfrac{2}{3}y = 7 \\ \dfrac{2}{5}x - \dfrac{5}{6}y = 7 \end{cases}$

35. $\begin{cases} 2\sqrt{3}x - 3y = 3 \\ 3\sqrt{3}x + 2y = 24 \end{cases}$

36. $\begin{cases} 4x - 3\sqrt{5}y = -19 \\ 3x + 4\sqrt{5}y = 17 \end{cases}$

37. $\begin{cases} 3\pi x - 4y = 6 \\ 2\pi x + 3y = 5 \end{cases}$

38. $\begin{cases} 2x - 5\pi y = 3 \\ 3x + 4\pi y = 2 \end{cases}$

39. $\begin{cases} 3\sqrt{2}x - 4\sqrt{3}y = -6 \\ 2\sqrt{2}x + 3\sqrt{3}y = 13 \end{cases}$

40. $\begin{cases} 2\sqrt{2}x + 3\sqrt{5}y = 7 \\ 3\sqrt{2}x - \sqrt{5}y = -17 \end{cases}$

In Exercises 41 to 61, solve by using a system of equations.

41. SUPPLY/DEMAND The number x of MP3 players a manufacturer is willing to sell is given by $x = 20p - 2000$, where p is the price, in dollars, per MP3 player. The number x of MP3 players a store is willing to purchase is given by $x = -4p + 1000$, where p is the price per MP3 player. Find the equilibrium price.

▶ **42. SUPPLY/DEMAND** The number x of digital cameras a manufacturer is willing to sell is given by $x = 25p - 500$, where p is the price, in dollars, per digital camera. The number x of digital cameras a store is willing to purchase is given by $x = -7p + 1100$, where p is the price per digital camera. Find the equilibrium price.

43. RATE OF WIND Flying with the wind, a plane traveled 450 miles in 3 hours. Flying against the wind, the plane traveled the same distance in 5 hours. Find the rate of the plane in calm air and the rate of the wind.

44. RATE OF WIND A plane flew 800 miles in 4 hours while flying with the wind. Against the wind, it took the plane 5 hours to travel 800 miles. Find the rate of the plane in calm air and the rate of the wind.

45. RATE OF CURRENT A motorboat traveled a distance of 120 miles in 4 hours while traveling with the current. Against the current, the same trip took 6 hours. Find the rate of the boat in calm water and the rate of the current.

▶ **46. RATE OF CURRENT** A canoeist can row 12 miles with the current in 2 hours. Rowing against the current, it takes the canoeist 4 hours to travel the same distance. Find the rate of the canoeist in calm water and the rate of the current.

47. METALLURGY A metallurgist made two purchases. The first purchase, which cost $1080, included 30 kilograms of an iron alloy and 45 kilograms of a lead alloy. The second purchase, at the same prices, cost $372 and included 15 kilograms of the iron alloy and 12 kilograms of the lead alloy. Find the cost per kilogram of the iron and lead alloys.

48. CHEMISTRY For $14.10, a chemist purchased 10 liters of hydrochloric acid and 15 liters of silver nitrate. A second purchase, at the same prices, cost $18.16 and included 12 liters of hydrochloric acid and 20 liters of silver nitrate. Find the cost per liter of each of the two chemicals.

49. CHEMISTRY A goldsmith has two gold alloys. The first alloy is 40% gold; the second alloy is 60% gold. How many grams of each should be mixed to produce 20 grams of an alloy that is 52% gold?

50. CHEMISTRY One acetic acid solution is 70% water and another is 30% water. How many liters of each solution should be mixed to produce 20 liters of a solution that is 40% water?

51. GEOMETRY A right triangle in the first quadrant is bounded by the lines $y = 0$, $y = \dfrac{1}{2}x$, and $y = -2x + 6$. Find its area.

52. GEOMETRY The lines whose equations are $2x + 3y = 1$, $3x - 4y = 10$, and $4x + ky = 5$ all intersect at the same point. What is the value of k?

53. NUMBER THEORY Adding a three-digit number 5Z7 to 256 gives XY3. If XY3 is divisible by 3, then what is the largest possible value of Z?

54. NUMBER THEORY Find the value of k if $2x + 5 = 6x + k = 4x - 7$.

55. NUMBER THEORY A *Pythagorean triple* is three positive integers a, b, and c for which $a^2 + b^2 = c^2$. Given $a = 42$, find all the values of b and c such that a, b, and c form a Pythagorean triple. (*Suggestion:* If $a = 42$, then $1764 + b^2 = c^2$ or $1764 = c^2 - b^2 = (c - b)(c + b)$. Because $(c - b)(c + b) = 1764$, $c - b$ and $b + c$ must be factors of 1764. For instance, one possibility is $2 = c - b$ and $882 = c + b$. Solving this system of equations yields one set of Pythagorean triples. Now repeat for other possible factors of 1764. Remember that answers must be positive integers.)

56. NUMBER THEORY Given $a = 30$, find all the values of b and c such that a, b, and c form a Pythagorean triple. (See the preceding exercise.)

57. MARKETING A marketing company asked 100 people whether they liked a new skin cream and lip balm. The company found that 80% of the people who liked the new skin cream also liked the new lip balm and that 50% of the people who did not like the skin cream liked the new lip balm. If 77 people liked the lip balm, how many people liked the skin cream?

58. FIRE SCIENCE An analysis of 200 scores on a firefighter qualifying exam found that 75% of those who passed the basic fire science exam also passed the exam on containing chemical fires. Of those who did not pass the basic fire science exam, 25% passed the exam on containing chemical fires. If 120 people passed the exam on containing chemical fires, how many people passed the basic fire science exam?

59. INCOME TAX The chapter opener on page 767 shows the graph of the income tax a taxpayer pays for various income levels for both a flat tax and variable tax brackets. Given that the equation of the line in red is $T = 0.25I - 3190$ for $28,400 < I \le 68,800$, find and interpret the point at which the flat tax graph crosses the variable tax bracket graph. (*Suggestion:* Use the fact that flat tax is 20% of adjusted gross income to find the equation of the line in black. Now solve a system of equations.)

60. INCOME TAX The chapter opener on page 767 shows the graph of the income tax a taxpayer pays for various income levels for both a flat tax and variable tax brackets. The equation of the line segment in green is $T = 0.10I$ for $0 \le I \le 7000$; the equation of the line segment in blue is $T = 0.15I - 350$ for $7000 < I \le 28,400$; the equation of the line segment in red is $T = 0.25I - 3190$ for $28,400 < I \le 68,800$. If the flat-tax proposal for each taxpayer is reduced to 14%, does the line representing the flat-tax proposal intersect the green, blue, or red line segment? What is the point of intersection?

61. INVESTMENT A broker invests $25,000 of a client's money in two different bond funds. The annual rate of return on one bond fund is 6%, and the annual rate of return on the second bond fund is 6.5%. The investor receives a total annual interest payment from the two bond funds of $1555. Find the amount invested in each fund.

CONNECTING CONCEPTS

In Exercises 62 to 71, solve for x and y. Use the fact that if $z_1 = a_1 + b_1 i$ and $z_2 = a_2 + b_2 i$ are two complex numbers, then $z_1 = z_2$ if and only if $a_1 = a_2$ and $b_1 = b_2$.

62. $(2 + i)x + (3 - i)y = 7$

63. $(3 + 2i)x + (4 - 3i)y = 2 - 16i$

64. $(4 - 3i)x + (5 + 2i)y = 11 + 9i$

65. $(2 + 6i)x + (4 - 5i)y = -8 - 7i$

66. $(-3 - i)x - (4 + 2i)y = 1 - i$

67. $(5 - 2i)x + (-3 - 4i)y = 12 - 35i$

68. $\begin{cases} 2x + 5y = 11 + 3i \\ 3x + y = 10 - 2i \end{cases}$

69. $\begin{cases} 4x + 3y = 11 + 6i \\ 3x - 5y = 1 + 19i \end{cases}$

70. $\begin{cases} 2x + 3y = 11 + 5i \\ 3x - 3y = 9 - 15i \end{cases}$

71. $\begin{cases} 5x - 4y = 15 - 41i \\ 3x + 5y = 9 + 5i \end{cases}$

━━━━ *PREPARE FOR SECTION 9.2* ━━━━━━━━━━━━━━━━━━━━━━

72. Solve $2x - 5y = 15$ for y. [1.1]

73. If $x = 2c + 1$, $y = -c + 3$, and $z = 2x + 5y - 4$, write z in terms of c. [P.1]

74. Solve: $\begin{cases} 5x - 2y = 10 \\ 2y = 8 \end{cases}$ [9.1]

75. Solve: $\begin{cases} 3x - y = 11 \\ 2x + 3y = -11 \end{cases}$ [9.1]

76. Solve: $\begin{cases} y = 3x - 4 \\ y = 4x - 2 \end{cases}$ [9.1]

77. Solve: $\begin{cases} 4x + y = 9 \\ -8x - 2y = -18 \end{cases}$ [9.1]

━━━━ *PROJECTS* ━━━━━━━━━━━━━━━━━━━━━━━━━━━━━━━━━━

I. **INDEPENDENT AND DEPENDENT CONDITIONS** Consider the following problem: "Maria and Michael drove from Los Angeles to New York in 60 hours. How long did Maria drive?" It is difficult to answer this question. She may have driven all 60 hours while Michael relaxed, or she may have relaxed while Michael drove all 60 hours. The difficulty is that there are two unknowns (how long each drove) and only one condition (the total driving time) relating the unknowns. If we added another condition, such as Michael drove 25 hours, then we could determine how long Maria drove, 35 hours. In most cases, an application problem will have a single answer only when there are as many *independent* conditions as there are variables. Conditions are independent if knowing one does *not* enable you to know the other.

Here is an example of conditions that are not independent. "The perimeter of a rectangle is 50 meters. The sum of the width and length is 25 meters." To see that these conditions are dependent, write the perimeter equation and divide each side by 2.

$$2w + 2l = 50$$
$$w + l = 25 \qquad \bullet \text{ Divide each side by 2.}$$

Note that the resulting equation is the second condition: the sum of the width and length is 25. Thus, knowing the first condition enables us to determine the second condition. The conditions are not independent, so there is no one solution to this problem.

For each of the following problems, determine whether the conditions are independent or dependent. For those problems that have independent conditions, find the solution (if possible). For those problems for which the conditions are dependent, find two solutions.

a. The sum of two numbers is 30. The difference between the two numbers is 10. Find the numbers.

b. The area of a square is 25 square meters. Find the length of each side.

c. The area of a rectangle is 25 square meters. Find the length of each side.

d. Emily spent $1000 for carpeting and tile. Carpeting cost $20 per square yard and tile cost $30 per square yard. How many square yards of each did she purchase?

e. The sum of two numbers is 20. Twice the smaller number is 10 minus twice the larger number. Find the two numbers.

f. Make up a word problem for which there are two independent conditions. Solve the problem.

g. Make up a word problem for which there are two dependent conditions. Find at least two solutions.

SYSTEMS OF LINEAR EQUATIONS IN MORE THAN TWO VARIABLES

● SYSTEMS OF EQUATIONS IN THREE VARIABLES

An equation of the form $Ax + By + Cz = D$, with A, B, and C not all zero, is a linear equation in three variables. A solution of an equation in three variables is an **ordered triple** (x, y, z).

The ordered triple $(2, -1, -3)$ is one of the solutions of the equation $2x - 3y + z = 4$. The ordered triple $(3, 1, 1)$ is another solution. In fact, an infinite number of ordered triples are solutions of the equation.

Graphing an equation in three variables requires a third coordinate axis perpendicular to the xy-plane. This third axis is commonly called the **z-axis.** The result is a three-dimensional coordinate system called the xyz-coordinate system (**Figure 9.12**). To help visualize a three-dimensional coordinate system, think of a corner of a room: the floor is the xy-plane, one wall is the yz-plane, and the other wall is the xz-plane.

Graphing an ordered triple requires three moves, the first along the x-axis, the second along the y-axis, and the third along the z-axis. **Figure 9.13** is the graph of the points $(-5, -4, 3)$ and $(4, 5, -2)$.

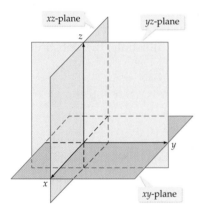

FIGURE 9.12

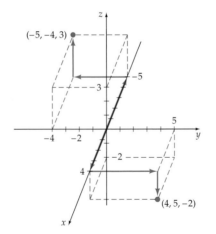

FIGURE 9.13

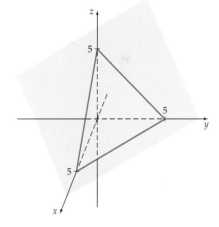

FIGURE 9.14

The graph of a linear equation in three variables is a plane. That is, if all the solutions of a linear equation in three variables were plotted in an xyz-coordinate system, the graph would look like a large, flat piece of paper with infinite extent. **Figure 9.14** is a portion of the graph of $x + y + z = 5$.

There are different ways in which three planes can be oriented in an xyz-coordinate system. **Figure 9.15** illustrates several ways.

For a linear system of equations in three variables to have a solution, the graphs of the planes must intersect at a single point, they must intersect along a common line, or all equations must have a graph that is the same plane. In **Figure 9.15,** the graphs in (a), (b), and (c) represent systems of equations that

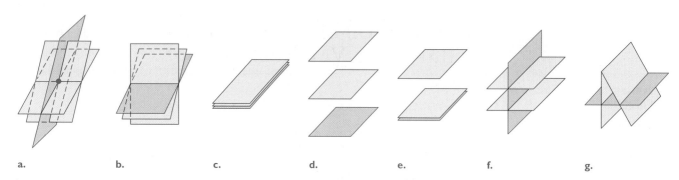

a.   b.   c.   d.   e.   f.   g.

FIGURE 9.15

have a solution. The system of equations represented in **Figure 9.15a** is a consistent system of equations. **Figures 9.15b** and **9.15c** are graphs of a dependent system of equations. The remaining graphs are examples of inconsistent systems of equations.

A system of equations in more than two variables can be solved by using the substitution method or the elimination method. To illustrate the substitution method, consider the system of equations

$$\begin{cases} x - 2y + z = 7 & (1) \\ 2x + y - z = 0 & (2) \\ 3x + 2y - 2z = -2 & (3) \end{cases}$$

Solve Equation (1) for x and substitute the result into Equations (2) and (3).

$$x = 2y - z + 7 \quad (4)$$

$2(2y - z + 7) + y - z = 0$ • **Substitute $2y - z + 7$ for x in Equation (2).**

$4y - 2z + 14 + y - z = 0$ • **Simplify.**

$$5y - 3z = -14 \quad (5)$$

$3(2y - z + 7) + 2y - 2z = -2$ • **Substitute $2y - z + 7$ for x in Equation (3).**

$6y - 3z + 21 + 2y - 2z = -2$ • **Simplify.**

$$8y - 5z = -23 \quad (6)$$

Now solve the system of equations formed from Equations (5) and (6).

$$\begin{cases} 5y - 3z = -14 & \text{multiply by 8} \rightarrow & 40y - 24z = -112 \\ 8y - 5z = -23 & \text{multiply by } -5 \rightarrow & \underline{-40y + 25z = 115} \\ & & z = 3 \end{cases}$$

Substitute 3 for z into Equation (5) and solve for y.

$$5y - 3z = -14 \qquad \text{• Equation (5)}$$
$$5y - 3(3) = -14$$
$$5y - 9 = -14$$
$$5y = -5$$
$$y = -1$$

Substitute -1 for y and 3 for z into Equation (4) and solve for x.

$$x = 2y - z + 7 = 2(-1) - (3) + 7 = 2$$

The ordered-triple solution is $(2, -1, 3)$. The graphs of the three planes intersect at a single point.

● TRIANGULAR FORM

There are many approaches one can take to determine the solution of a system of equations by the elimination method. For consistency, we will always follow a plan that produces an equivalent system of equations in **triangular form.** Three examples of systems of equations in triangular form are

$$\begin{cases} 2x - 3y + z = -4 \\ \quad\quad 2y + 3z = 9 \\ \quad\quad\quad\quad -2z = -2 \end{cases} \quad \begin{cases} w + 3x - 2y + 3z = 0 \\ \quad 2x - y + 4z = 8 \\ \quad\quad -3y - 2z = -1 \\ \quad\quad\quad\quad 3z = 9 \end{cases} \quad \begin{cases} 3x - 4y + z = 1 \\ \quad 3y + 2z = 3 \end{cases}$$

Once a system of equations is written in triangular form, the solution can be found by *back substitution*—that is, by solving the last equation of the system and substituting *back* into the previous equation. This process is continued until the value of each variable has been found.

As an example of solving a system of equations by back substitution, consider the following system of equations in triangular form.

$$\begin{cases} 2x - 4y + z = -3 \quad (1) \\ \quad\quad 3y - 2z = 9 \quad (2) \\ \quad\quad\quad\quad 3z = -9 \quad (3) \end{cases}$$

Solve Equation (3) for z. Substitute the value of z into Equation (2) and solve for y.

$3z = -9$	• Equation (3)	$3y - 2z = 9$	• Equation (2)
$z = -3$		$3y - 2(-3) = 9$	• $z = -3$
		$3y = 3$	
		$y = 1$	

Replace z by -3 and y by 1 in Equation (1) and then solve for x.

$$2x - 4y + z = -3 \quad \text{• Equation (1)}$$
$$2x - 4(1) + (-3) = -3$$
$$2x - 7 = -3$$
$$x = 2$$

The solution is the ordered triple $(2, 1, -3)$.

? QUESTION What is the solution of $\begin{cases} x + 2y + z = 2 \\ \quad\quad y - z = 3? \\ \quad\quad\quad z = 2 \end{cases}$

? ANSWER $(-10, 5, 2)$

EXAMPLE 1 **Solve an Independent System of Equations**

Solve: $\begin{cases} x + 2y - z = 1 & (1) \\ 2x - y + z = 6 & (2) \\ 2x - y - z = 0 & (3) \end{cases}$

Solution

Eliminate x from Equation (2) by multiplying Equation (1) by -2 and then adding it to Equation (2). Replace Equation (2) by the new equation.

$$\begin{array}{ll} -2x - 4y + 2z = -2 & \bullet \ -2 \textbf{ times Equation (1)} \\ \underline{2x - y + z = 6} & \bullet \textbf{ Equation (2)} \\ -5y + 3z = 4 & \bullet \textbf{ Add the equations.} \end{array}$$

$\begin{cases} x + 2y - z = 1 & (1) \\ -5y + 3z = 4 & (4) \\ 2x - y - z = 0 & (3) \end{cases}$ • **Replace Equation (2).**

Eliminate x from Equation (3) by multiplying Equation (1) by -2 and adding it to Equation (3). Replace Equation (3) by the new equation.

$$\begin{array}{ll} -2x - 4y + 2z = -2 & \bullet \ -2 \textbf{ times Equation (1)} \\ \underline{2x - y - z = 0} & \bullet \textbf{ Equation (3)} \\ -5y + z = -2 & \bullet \textbf{ Add the equations.} \end{array}$$

$\begin{cases} x + 2y - z = 1 & (1) \\ -5y + 3z = 4 & (4) \\ -5y + z = -2 & (5) \end{cases}$ • **Replace Equation (3).**

Eliminate y from Equation (5) by multiplying Equation (4) by -1 and then adding it to Equation (5). Replace Equation (5) by the new equation.

$$\begin{array}{ll} 5y - 3z = -4 & \bullet \ -1 \textbf{ times Equation (4)} \\ \underline{-5y + z = -2} & \bullet \textbf{ Equation (5)} \\ -2z = -6 & \bullet \textbf{ Add the equations.} \end{array}$$

$\begin{cases} x + 2y - z = 1 & (1) \\ -5y + 3z = 4 & (4) \\ -2z = -6 & (6) \end{cases}$ • **Replace Equation (5).**

The system of equations is now in triangular form. Solve the system of equations by back substitution.

Solve Equation (6) for z. Substitute the value into Equation (4) and then solve for y.

$$\begin{array}{ll} -2z = -6 & \bullet \textbf{ Equation (6)} \\ z = 3 & \end{array} \qquad \begin{array}{ll} -5y + 3z = 4 & \bullet \textbf{ Equation (4)} \\ -5y + 3(3) = 4 & \bullet \textbf{ Replace } z \textbf{ by 3.} \\ -5y = -5 & \bullet \textbf{ Solve for } y. \\ y = 1 & \end{array}$$

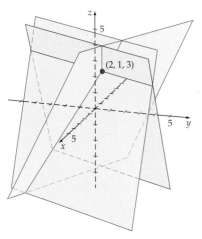

FIGURE 9.16

Replace z by 3 and y by 1 in Equation (1) and solve for x.

$$x + 2y - z = 1 \qquad \text{• Equation (1)}$$
$$x + 2(1) - 3 = 1 \qquad \text{• Replace y by 1; replace z by 3.}$$
$$x = 2$$

The system of equations is consistent. The solution is the ordered triple $(2, 1, 3)$. See **Figure 9.16**.

▶ **TRY EXERCISE 12, PAGE 792**

EXAMPLE 2 Solve a Dependent System of Equations

Solve: $\begin{cases} 2x - y - z = -1 & (1) \\ -x + 3y - z = -3 & (2) \\ -5x + 5y + z = -1 & (3) \end{cases}$

Solution

Eliminate x from Equation (2) by multiplying Equation (2) by 2 and then adding it to Equation (1). Replace Equation (2) by the new equation.

$$\begin{array}{ll} 2x - y - z = -1 & \text{• Equation (1)} \\ \underline{-2x + 6y - 2z = -6} & \text{• 2 times Equation (2)} \\ \quad\;\; 5y - 3z = -7 & \text{• Add the equations.} \;\checkmark \end{array}$$

$\begin{cases} 2x - y - z = -1 & (1) \\ \quad\;\; 5y - 3z = -7 & (4) \qquad \text{• Replace Equation (2).} \\ -5x + 5y + z = -1 & (3) \end{cases}$

Eliminate x from Equation (3) by multiplying Equation (1) by 5 and multiplying Equation (3) by 2. Then add. Replace Equation (3) by the new equation.

$$\begin{array}{ll} 10x - 5y - 5z = -5 & \text{• 5 times Equation (1)} \\ \underline{-10x + 10y + 2z = -2} & \text{• 2 times Equation (3)} \\ \quad\;\;\;\; 5y - 3z = -7 & \text{• Add the equations.} \end{array}$$

$\begin{cases} 2x - y - z = -1 & (1) \\ \quad\;\; 5y - 3z = -7 & (4) \\ \quad\;\; 5y - 3z = -7 & (5) \qquad \text{• Replace Equation (3).} \end{cases}$

Eliminate y from Equation (5) by multiplying Equation (4) by -1 and then adding it to Equation (5). Replace Equation (5) by the new equation.

$$\begin{array}{ll} -5y + 3z = 7 & \text{• -1 times Equation (4)} \\ \underline{5y - 3z = -7} & \text{• Equation (5)} \\ \; 0 = 0 & \text{• Add the equations.} \end{array}$$

$\begin{cases} 2x - y - z = -1 & (1) \\ \quad\;\; 5y - 3z = -7 & (4) \\ \; 0 = 0 & (6) \qquad \text{• Replace Equation (5).} \end{cases}$

Continued ▶

Because any ordered triple (x, y, z) is a solution of Equation (6), the solutions of the system of equations will be the ordered triples that are solutions of Equations (1) and (4).

Solve Equation (4) for y.

$$5y - 3z = -7$$
$$5y = 3z - 7$$
$$y = \frac{3}{5}z - \frac{7}{5}$$

Substitute $\frac{3}{5}z - \frac{7}{5}$ for y in Equation (1) and solve for x.

$$2x - y - z = -1 \qquad \bullet \text{ Equation (1)}$$

$$2x - \left(\frac{3}{5}z - \frac{7}{5}\right) - z = -1 \qquad \bullet \text{ Replace } y \text{ by } \frac{3}{5}z - \frac{7}{5}.$$

$$2x - \frac{8}{5}z + \frac{7}{5} = -1 \qquad \bullet \text{ Simplify and solve for } x.$$

$$2x = \frac{8}{5}z - \frac{12}{5}$$

$$x = \frac{4}{5}z - \frac{6}{5}$$

By choosing any real number c for z, we have $y = \frac{3}{5}c - \frac{7}{5}$ and $x = \frac{4}{5}c - \frac{6}{5}$. For any real number c, the ordered-triple solutions of the system of equations are $\left(\frac{4}{5}c - \frac{6}{5}, \frac{3}{5}c - \frac{7}{5}, c\right)$. The solid red line shown in **Figure 9.17** is a graph of the solutions.

▶ **TRY EXERCISE 16, PAGE 792**

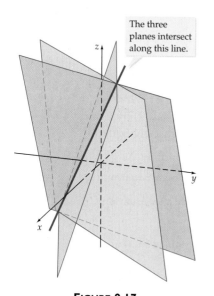

The three planes intersect along this line.

FIGURE 9.17

As in the case of a dependent system of equations in two variables, there is more than one way to represent the solutions of a dependent system of equations in three variables. For instance, from Example 2, let $a = \frac{4}{5}c - \frac{6}{5}$, the x-coordinate of the ordered triple $\left(\frac{4}{5}c - \frac{6}{5}, \frac{3}{5}c - \frac{7}{5}, c\right)$, and solve for c.

$$a = \frac{4}{5}c - \frac{6}{5} \quad \rightarrow \quad c = \frac{5}{4}a + \frac{3}{2}$$

Substitute this value of c into each component of the ordered triple.

$$\left(\frac{4}{5}\left(\frac{5}{4}a + \frac{3}{2}\right) - \frac{6}{5}, \frac{3}{5}\left(\frac{5}{4}a + \frac{3}{2}\right) - \frac{7}{5}, \frac{5}{4}a + \frac{3}{2}\right) = \left(a, \frac{3}{4}a - \frac{1}{2}, \frac{5}{4}a + \frac{3}{2}\right)$$

Thus the solutions of the system of equations can also be written as

$$\left(a, \frac{3}{4}a - \frac{1}{2}, \frac{5}{4}a + \frac{3}{2}\right)$$

EXAMPLE 3   Identify an Inconsistent System of Equations

Solve: $\begin{cases} x + 2y + 3z = 4 & (1) \\ 2x - y - z = 3 & (2) \\ 3x + y + 2z = 5 & (3) \end{cases}$

Solution

Eliminate x from Equation (2) by multiplying Equation (1) by -2 and then adding it to Equation (2). Replace Equation (2). Eliminate x from Equation (3) by multiplying Equation (1) by -3 and adding it to Equation (3). Replace Equation (3). The equivalent system is

$\begin{cases} x + 2y + 3z = 4 & (1) \\ -5y - 7z = -5 & (4) \\ -5y - 7z = -7 & (5) \end{cases}$

Eliminate y from Equation (5) by multiplying Equation (4) by -1 and adding it to Equation (5). Replace Equation (5). The equivalent system is

$\begin{cases} x + 2y + 3z = 4 & (1) \\ -5y - 7z = -5 & (4) \\ 0 = -2 & (6) \end{cases}$

This system of equations contains a false equation. The system is inconsistent and has no solution. There is no point on all three planes, as shown in **Figure 9.18**.

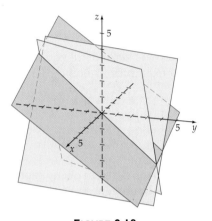

FIGURE 9.18

▶ **TRY EXERCISE 18, PAGE 792**

● NONSQUARE SYSTEMS OF EQUATIONS

The linear systems of equations that we have solved so far contain the same number of variables as equations. These are *square systems of equations.* If there are fewer equations than variables—a *nonsquare system of equations*—the system has either no solution or an infinite number of solutions.

EXAMPLE 4   Solve a Nonsquare System of Equations

Solve: $\begin{cases} x - 2y + 2z = 3 & (1) \\ 2x - y - 2z = 15 & (2) \end{cases}$

Solution

Eliminate x from Equation (2) by multiplying Equation (1) by -2 and adding it to Equation (2). Replace Equation (2).

$\begin{cases} x - 2y + 2z = 3 & (1) \\ 3y - 6z = 9 & (3) \end{cases}$

Solve Equation (3) for y.

$$3y - 6z = 9$$
$$y = 2z + 3$$

Continued ▶

Substitute $2z + 3$ for y into Equation (1) and solve for x.

$$x - 2y + 2z = 3$$
$$x - 2(2z + 3) + 2z = 3$$
$$x = 2z + 9$$

$\cdot y = 2z + 3$

For each value of z selected, there correspond values for x and y. If z is any real number c, then the solutions of the system are the ordered triples $(2c + 9, 2c + 3, c)$.

▶ **TRY EXERCISE 20, PAGE 792**

● HOMOGENEOUS SYSTEMS OF EQUATIONS

A linear system of equations for which the constant term is zero for all equations is called a **homogeneous system of equations.** Two examples of homogeneous systems of equations are

$$\begin{cases} 3x + 4y = 0 \\ 2x + 3y = 0 \end{cases} \qquad \begin{cases} 2x - 3y + 5z = 0 \\ 3x + 2y + z = 0 \\ x - 4y + 5z = 0 \end{cases}$$

The solution $(0, 0)$ is always a solution of a homogeneous system of equations in two variables, and $(0, 0, 0)$ is always a solution of a homogeneous system of equations in three variables. This solution is called the **trivial solution.**

Sometimes a homogeneous system of equations may have solutions other than the trivial solution. For example, $(1, -1, -1)$ is a solution of the homogeneous system of three equations in three variables above.

If a homogeneous system of equations has a unique solution, the graphs intersect only at the origin. Solutions of a homogeneous system of equations can be found by using the substitution method or the elimination method.

EXAMPLE 5 Solve a Homogeneous System of Equations

Solve: $\begin{cases} x + 2y - 3z = 0 & (1) \\ 2x - y + z = 0 & (2) \\ 3x + y - 2z = 0 & (3) \end{cases}$

Solution

Eliminate x from Equations (2) and (3) and replace these equations by the new equations.

$$\begin{cases} x + 2y - 3z = 0 & (1) \\ -5y + 7z = 0 & (4) \\ -5y + 7z = 0 & (5) \end{cases}$$

Eliminate y from Equation (5). Replace Equation (5).

$$\begin{cases} x + 2y - 3z = 0 & (1) \\ -5y + 7z = 0 & (4) \\ 0 = 0 & (6) \end{cases}$$

Because Equation (6) is an identity, the solutions of the system are the solutions of Equations (1) and (4).

Solve Equation (4) for y.

$$y = \frac{7}{5}z$$

Substitute the expression for y into Equation (1) and solve for x.

$$x + 2y - 3z = 0 \qquad \text{• Equation (1)}$$

$$x + 2\left(\frac{7}{5}z\right) - 3z = 0 \qquad \text{• } y = \frac{7}{5}z$$

$$x = \frac{1}{5}z$$

Letting z be any real number c, we find that the solutions of the system are $\left(\frac{1}{5}c, \frac{7}{5}c, c\right)$.

▶ **TRY EXERCISE 32, PAGE 792**

● **APPLICATIONS**

One application of a system of equations is "curve fitting." Given a set of points in the plane, try to find an equation whose graph passes through those points, or "fits" those points.

EXAMPLE 6 | **Solve an Application of a System of Equations to Curve Fitting**

Find an equation of the form $y = ax^2 + bx + c$ whose graph passes through the points whose coordinates are $(1, 4)$, $(-1, 6)$, and $(2, 9)$.

Solution

Substitute each of the given ordered pairs into the equation $y = ax^2 + bx + c$. Write the resulting system of equations.

$$\begin{cases} 4 = a(1)^2 + b(1) + c \\ 6 = a(-1)^2 + b(-1) + c \\ 9 = a(2)^2 + b(2) + c \end{cases} \quad \text{or} \quad \begin{cases} a + b + c = 4 & (1) \\ a - b + c = 6 & (2) \\ 4a + 2b + c = 9 & (3) \end{cases}$$

Solve the resulting system of equations for a, b, and c.

Eliminate a from Equation (2) by multiplying Equation (1) by -1 and then adding it to Equation (2). Now eliminate a from Equation (3) by multiplying Equation (1) by -4 and adding it to Equation (3). The result is

$$\begin{cases} a + b + c = 4 \\ -2b = 2 \\ -2b - 3c = -7 \end{cases}$$

Continued ▶

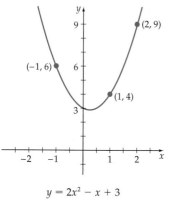

$y = 2x^2 - x + 3$

FIGURE 9.19

Although this system of equations is not in triangular form, we can solve the second equation for b and use this value to find a and c.

Solving by substitution, we obtain $a = 2$, $b = -1$, $c = 3$. The equation of the form $y = ax^2 + bx + c$ whose graph passes through $(1, 4)$, $(-1, 6)$, and $(2, 9)$ is $y = 2x^2 - x + 3$. See **Figure 9.19**.

▶ **TRY EXERCISE 36, PAGE 792**

Traffic engineers use systems of equations to study the flow of traffic. The analysis of traffic flow is based on the principle that the numbers of cars that enter and leave an intersection must be equal.

EXAMPLE 7   **Traffic Flow**

Suppose the traffic flow for some one-way streets can be modeled by the diagram below, where the numbers and the variables represent the numbers of cars entering or leaving an intersection per hour.

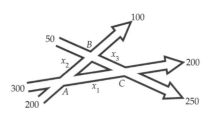

If the street connecting intersections A and C has an estimated traffic flow of between 100 and 200 cars per hour, what is the estimated traffic flow between A and B (which is x_2) and between B and C (which is x_3)?

Solution

Let x_1, x_2, and x_3 represent the numbers of cars per hour that are traveling on AC, AB, and BC, respectively. Now consider intersection A. There are $300 + 200 = 500$ cars per hour entering A and $x_1 + x_2$ cars leaving A. Therefore, $x_1 + x_2 = 500$. For intersection B, we have $50 + x_2$ cars per hour entering the intersection and $100 + x_3$ cars leaving the intersection. Thus $50 + x_2 = 100 + x_3$, or $x_2 - x_3 = 50$. Applying the same reasoning to C, we have $x_1 + x_3 = 450$. These equations result in the system of equations

$$\begin{cases} x_1 + x_2 = 500 & (1) \\ x_2 - x_3 = 50 & (2) \\ x_1 + x_3 = 450 & (3) \end{cases}$$

Subtracting Equation (2) from Equation (1) gives

$$\begin{aligned} x_1 + x_2 &= 500 & (1) \\ \underline{x_2 - x_3} &= \underline{50} & (2) \\ x_1 + x_3 &= 450 & (4) \end{aligned}$$

Subtracting Equation (4) from Equation (3) gives

$$x_1 + x_3 = 450 \quad (3)$$
$$\underline{x_1 + x_3 = 450} \quad (4)$$
$$0 = 0$$

This indicates that the system of equations is dependent. Because we are given that between 100 and 200 cars per hour flow between A and C (the value of x_1), we will solve each equation in terms of x_1. From Equation (1) we have $x_2 = -x_1 + 500$ and from Equation (3) we have $x_3 = -x_1 + 450$. Because $100 \le x_1 \le 200$, we have, by substituting for x_1, $300 \le x_2 \le 400$ and $250 \le x_3 \le 350$.

▶ **TRY EXERCISE 42, PAGE 793**

TOPICS FOR DISCUSSION

1. Can a system of equations contain more equations than variables? If not, explain why not. If so, give an example.

2. If a linear system of three equations in three variables is dependent, what does that mean about the graphs of the equations of the system?

3. If a linear system of three equations in three variables is inconsistent, what does that mean about the graphs of the equations of the system?

4. The equation of a circle centered at the origin with radius 5 is given by $x^2 + y^2 = 25$. Discuss the shape of $x^2 + y^2 + z^2 = 25$ in an xyz-coordinate system.

5. Consider the plane P given by $2x + 4y - 3z = 12$. The *trace* of the graph of P is obtained by letting one of the variables equal zero. For instance, the trace in the xy-plane is the graph of $2x + 4y = 12$ that is obtained by letting $z = 0$. Determine the traces of P in the xz- and yz-planes, and discuss how the traces can be used to visualize the graph of P.

EXERCISE SET 9.2

In Exercises 1 to 24, solve each system of equations.

1. $\begin{cases} 2x - y + z = 8 \\ 2y - 3z = -11 \\ 3y + 2z = 3 \end{cases}$

2. $\begin{cases} 3x + y + 2z = -4 \\ -3y - 2z = -5 \\ 2y + 5z = -4 \end{cases}$

3. $\begin{cases} x + 3y - 2z = 8 \\ 2x - y + z = 1 \\ 3x + 2y - 3z = 15 \end{cases}$

4. $\begin{cases} x - 2y + 3z = 5 \\ 3x - 3y + z = 9 \\ 5x + y - 3z = 3 \end{cases}$

5. $\begin{cases} 3x + 4y - z = -7 \\ x - 5y + 2z = 19 \\ 5x + y - 2z = 5 \end{cases}$

6. $\begin{cases} 2x - 3y - 2z = 12 \\ x + 4y + z = -9 \\ 4x + 2y - 3z = 6 \end{cases}$

7. $\begin{cases} 2x - 5y + 3z = -18 \\ 3x + 2y - z = -12 \\ x - 3y - 4z = -4 \end{cases}$

8. $\begin{cases} 4x - y + 2z = -1 \\ 2x + 3y - 3z = -13 \\ x + 5y + z = 7 \end{cases}$

9. $\begin{cases} x + 2y - 3z = -7 \\ 2x - y + 4z = 11 \\ 4x + 3y - 4z = -3 \end{cases}$

10. $\begin{cases} x - 3y + 2z = -11 \\ 3x + y + 4z = 4 \\ 5x - 5y + 8z = -18 \end{cases}$

11. $\begin{cases} 2x - 5y + 2z = -4 \\ 3x + 2y + 3z = 13 \\ 5x - 3y - 4z = -18 \end{cases}$

▶ **12.** $\begin{cases} 3x + 2y - 5z = 6 \\ 5x - 4y + 3z = -12 \\ 4x + 5y - 2z = 15 \end{cases}$

13. $\begin{cases} 2x + y - z = -2 \\ 3x + 2y + 3z = 21 \\ 7x + 4y + z = 17 \end{cases}$

14. $\begin{cases} 3x + y + 2z = 2 \\ 4x - 2y + z = -4 \\ 11x - 3y + 4z = -6 \end{cases}$

15. $\begin{cases} 3x - 2y + 3z = 11 \\ 2x + 3y + z = 3 \\ 5x + 14y - z = 1 \end{cases}$

▶ **16.** $\begin{cases} 2x + 3y + 2z = 14 \\ x - 3y + 4z = 4 \\ -x + 12y - 6z = 2 \end{cases}$

17. $\begin{cases} 2x - 3y + 6z = 3 \\ x + 2y - 4z = 5 \\ 3x + 4y - 8z = 7 \end{cases}$

▶ **18.** $\begin{cases} 2x + 3y - 6z = 4 \\ 3x - 2y - 9z = -7 \\ 2x + 5y - 6z = 8 \end{cases}$

19. $\begin{cases} 2x - 3y + 5z = 14 \\ x + 4y - 3z = -2 \end{cases}$

▶ **20.** $\begin{cases} x - 3y + 4z = 9 \\ 3x - 8y - 2z = 4 \end{cases}$

21. $\begin{cases} 6x - 9y + 6z = 7 \\ 4x - 6y + 4z = 9 \end{cases}$

22. $\begin{cases} 4x - 2y + 6z = 5 \\ 2x - y + 3z = 2 \end{cases}$

23. $\begin{cases} 5x + 3y + 2z = 10 \\ 3x - 4y - 4z = -5 \end{cases}$

24. $\begin{cases} 3x - 4y - 7z = -5 \\ 2x + 3y - 5z = 2 \end{cases}$

In Exercises 25 to 32, solve each homogeneous system of equations.

25. $\begin{cases} x + 3y - 4z = 0 \\ 2x + 7y + z = 0 \\ 3x - 5y - 2z = 0 \end{cases}$

26. $\begin{cases} x - 2y + 3z = 0 \\ 3x - 7y - 4z = 0 \\ 4x - 4y + z = 0 \end{cases}$

27. $\begin{cases} 2x - 3y + z = 0 \\ 2x + 4y - 3z = 0 \\ 6x - 2y - z = 0 \end{cases}$

28. $\begin{cases} 5x - 4y - 3z = 0 \\ 2x + y + 2z = 0 \\ x - 6y - 7z = 0 \end{cases}$

29. $\begin{cases} 3x - 5y + 3z = 0 \\ 2x - 3y + 4z = 0 \\ 7x - 11y + 11z = 0 \end{cases}$

30. $\begin{cases} 5x - 2y - 3z = 0 \\ 3x - y - 4z = 0 \\ 4x - y - 9z = 0 \end{cases}$

31. $\begin{cases} 4x - 7y - 2z = 0 \\ 2x + 4y + 3z = 0 \\ 3x - 2y - 5z = 0 \end{cases}$

▶ **32.** $\begin{cases} 5x + 2y + 3z = 0 \\ 3x + y - 2z = 0 \\ 4x - 7y + 5z = 0 \end{cases}$

In Exercises 33 to 44, solve each exercise by solving a system of equations.

33. CURVE FITTING Find an equation of the form $y = ax^2 + bx + c$ whose graph passes through the points $(2, 3)$, $(-2, 7)$, and $(1, -2)$.

34. CURVE FITTING Find an equation of the form $y = ax^2 + bx + c$ whose graph passes through the points $(1, -2)$, $(3, -4)$, and $(2, -2)$.

35. CURVE FITTING Find the equation of the circle whose graph passes through the points $(5, 3)$, $(-1, -5)$, and $(-2, 2)$. (*Hint:* Use the equation $x^2 + y^2 + ax + by + c = 0$.)

▶ **36. CURVE FITTING** Find the equation of the circle whose graph passes through the points $(0, 6)$, $(1, 5)$, and $(-7, -1)$. (*Hint:* See Exercise 35.)

37. CURVE FITTING Find the center and radius of the circle whose graph passes through the points $(-2, 10)$, $(-12, -14)$, and $(5, 3)$. (*Hint:* See Exercise 35.)

38. CURVE FITTING Find the center and radius of the circle whose graph passes through the points $(2, 5)$, $(-4, -3)$, and $(3, 4)$. (*Hint:* See Exercise 35.)

39. TRAFFIC FLOW Suppose that the traffic flow for some one-way streets can be modeled by the diagram below, where each number or variable represents the number of cars entering or leaving an intersection per hour.

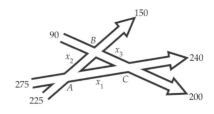

What is the minimum number of cars that can travel between A and C?

40. TRAFFIC FLOW A *roundabout* is a type of intersection that accommodates traffic flow in one direction, around a circular island. The graphic model on the following page shows the numbers of cars per hour that are entering or leaving a roundabout.

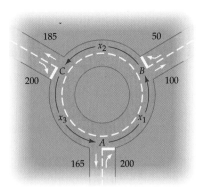

If the portion of the roundabout between A and B has an estimated traffic flow of from 60 to 80 cars per hour, what is the estimated traffic flow between C and A and between B and C?

41. TRAFFIC FLOW Suppose that the traffic flow for some one-way streets can be modeled by the diagram below, where each number or variable represents the number of cars entering or leaving an intersection per hour.

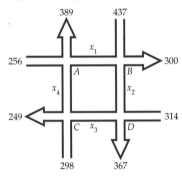

If the street connecting intersections A and B has an estimated traffic flow of from 125 to 175 cars per hour, what is the estimated traffic flow between C and A, D and C, and B and D?

▶ **42. TRAFFIC FLOW** A *roundabout* is a type of intersection that accommodates traffic flow in one direction, around a circular island. The graphic model below shows the numbers of cars per hour that are entering and leaving a roundabout.

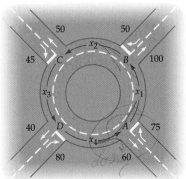

What is the minimum number of cars per hour that can travel between B and C?

43. ART A sculptor is creating a windchime consisting of three chimes that will be suspended from a rod 13 inches long. The weights, in ounces, of the chimes are shown in the diagram. For the rod to remain horizontal, the chimes must be positioned so that $w_1 d_1 + w_2 d_2 = w_3 d_3$. If the sculptor wants d_2 to be one-third of d_1, find the position of each chime so that the windchime will balance.

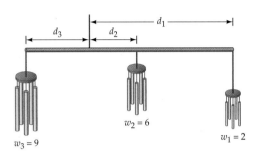

44. ART A designer wants to create a mobile of colored blocks as shown in the diagram below. The weight, in ounces, of each of the blocks is shown next to the block.

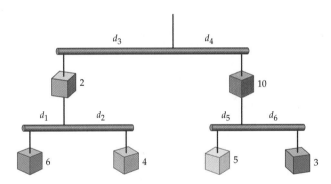

Given that $d_3 + d_4 = 20$ inches, $d_1 + d_2 = 10$ inches, and $d_5 + d_6 = 8$ inches, find the values of d_1 through d_6 so that each bar is horizontal. (A bar is horizontal when the value of weight times distance on each side of a vertical support is equal. For instance, for the above diagram, $6d_1$ must equal $4d_2$. Because there are six variables, the resulting system of equations must contain six equations.)

CONNECTING CONCEPTS

In Exercises 45 to 50, solve each system of equations.

45.
$$\begin{cases} 2x + y - 3z + 2w = -1 \\ 2y - 5z - 3w = 9 \\ 3y - 8z + w = -4 \\ 2y - 2z + 3w = -3 \end{cases}$$

46.
$$\begin{cases} 3x - y + 2z - 3w = 5 \\ 2y - 5z + 2w = -7 \\ 4y - 9z + w = -19 \\ 3y + z - 2w = -12 \end{cases}$$

47.
$$\begin{cases} x - 3y + 2z - w = 2 \\ 2x - 5y - 3z + 2w = 21 \\ 3x - 8y - 2z - 3w = 12 \\ -2x + 8y + z + 2w = -13 \end{cases}$$

48.
$$\begin{cases} x - 2y + 3z + 2w = 8 \\ 3x - 7y - 2z + 3w = 18 \\ 2x - 5y + 2z - w = 19 \\ 4x - 8y + 3z + 2w = 29 \end{cases}$$

49.
$$\begin{cases} x + 2y - 2z + 3w = 2 \\ 2x + 5y + 2z + 4w = 9 \\ 4x + 9y - 2z + 10w = 13 \\ -x - y + 8z - 5w = 3 \end{cases}$$

50.
$$\begin{cases} x - 2y + 3z - 2w = -1 \\ 3x - 7y - 2z - 3w = -19 \\ 2x - 5y + 2z - w = -11 \\ -x + 3y - 2z - w = 3 \end{cases}$$

In Exercises 51 and 52, use the system of equations
$$\begin{cases} x - 3y - 2z = A^2 \\ 2x - 5y + Az = 9 \\ 2x - 8y + z = 18 \end{cases}$$

51. Find all values of A for which the system has no solution.

52. Find all values of A for which the system has a unique solution.

In Exercises 53 to 55, use the system of equations
$$\begin{cases} x + 2y + z = A^2 \\ -2x - 3y + Az = 1 \\ 7x + 12y + A^2z = 4A^2 - 3 \end{cases}$$

53. Find all values of A for which the system has a unique solution.

54. Find all values of A for which the system has an infinite number of solutions.

55. Find all values of A for which the system has no solution.

56. Find an equation of the plane that contains the points $(2, 1, 1)$, $(-1, 2, 12)$, and $(3, 2, 0)$. (*Hint:* The equation of a plane can be written as $z = ax + by + c$.)

57. Find an equation of the plane that contains the points $(1, -1, 5)$, $(2, -2, 9)$, and $(-3, -1, -1)$. (*Hint:* The equation of a plane can be written as $z = ax + by + c$.)

PREPARE FOR SECTION 9.3

58. Solve $x^2 + 2x - 2 = 0$ for x. [1.3]

59. Solve: $\begin{cases} x + 4y = -11 \\ 3x - 2y = 9 \end{cases}$ [9.1]

60. Name the graph of $(y + 3)^2 = 8x$. [8.1]

61. Name the graph of $\dfrac{(x - 2)^2}{4} - \dfrac{(y + 3)^2}{9} = 1$. [8.3]

62. How many times do the graphs of $y = 2x - 1$ and $x^2 + y^2 = 4$ intersect? [2.1/2.2]

63. How many times do the graphs of $\dfrac{x^2}{4} + \dfrac{y^2}{9} = 1$ and $\dfrac{x^2}{9} + \dfrac{y^2}{4} = 1$ intersect? [8.2]

PROJECTS

---

1. ![pencil icon] **CONCEPT OF DIMENSION** In this chapter we graphed first-degree equations in three variables. If we were to attempt to graph an equation in four variables, we would need a fourth axis perpendicular to the three axes of an *xyz*-coordinate system. It seems impossible to imagine a fourth dimension, but incorporating it is really a quite practical matter in mathematics. In fact, there are some systems that require an infinite-dimensional coordinate system. To gain some insight into the concept of dimension, read the book *Flatland* by Edwin A. Abbott, and then write an essay explaining what this book has to do with dimension.

2. ![pencil icon] **ABILITIES OF A FOUR-DIMENSIONAL HUMAN** There have been a number of attempts to describe the abilities of a four-dimensional human in a three-dimensional world. Read some of these accounts, and then write an essay on some of the actions a four-dimensional person could perform. Answer the following question in your essay. Can a four-dimensional person remove the money from a locked safe without first opening the safe?

SECTION 9.3

NONLINEAR SYSTEMS OF EQUATIONS

- **SOLVING NONLINEAR SYSTEMS OF EQUATIONS**

• SOLVING NONLINEAR SYSTEMS OF EQUATIONS

A **nonlinear system of equations** is one in which one or more equations of the system are not linear equations. **Figure 9.20** shows examples of nonlinear systems of equations and the corresponding graphs of the equations. Each point of intersection of the graphs is a solution of the system of equations. In the third example, the graphs do not intersect; therefore, the system of equations has no real number solution.

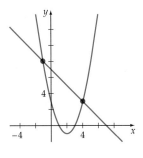

$$\begin{cases} y = x^2 - 4x + 3 \\ y = -x + 7 \end{cases}$$

2 solutions

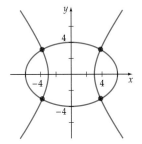

$$\begin{cases} \dfrac{x^2}{36} + \dfrac{y^2}{16} = 1 \\ \dfrac{x^2}{9} - \dfrac{y^2}{16} = 1 \end{cases}$$

4 solutions

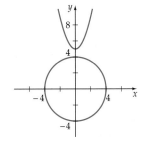

$$\begin{cases} y = x^2 + 5 \\ x^2 + y^2 = 16 \end{cases}$$

no solution

FIGURE 9.20

To solve a nonlinear system of equations, use the substitution method or the elimination method. The substitution method is usually easier for solving a nonlinear system that contains a linear equation.

❓ QUESTION What is the solution of the equation $3x - y = 5$ for y in terms of x?

EXAMPLE 1   Solve a Nonlinear System by the Substitution Method

Solve: $\begin{cases} y = x^2 - x - 1 & (1) \\ 3x - y = 4 & (2) \end{cases}$

Algebraic Solution

We will use the substitution method. Using the equation $y = x^2 - x - 1$, substitute the expression for y into $3x - y = 4$.

$$3x - y = 4$$
$$3x - (x^2 - x - 1) = 4 \qquad \bullet\, y = x^2 - x - 1$$
$$-x^2 + 4x + 1 = 4 \qquad \bullet\, \text{Simplify.}$$
$$x^2 - 4x + 3 = 0 \qquad \bullet\, \text{Write the quadratic}$$
$$\qquad\qquad\qquad\qquad\quad \text{equation in standard form.}$$
$$(x - 3)(x - 1) = 0 \qquad \bullet\, \text{Solve for } x.$$
$$x - 3 = 0 \quad \text{or} \quad x - 1 = 0$$
$$x = 3 \quad \text{or} \quad x = 1$$

Substitute these values into Equation (1) and solve for y.

$$y = 3^2 - 3 - 1 = 5 \qquad \text{or} \qquad y = 1^2 - 1 - 1 = -1$$

The solutions are $(3, 5)$ and $(1, -1)$. See **Figure 9.21.**

Visualize the Solution

Graphing $y = x^2 - x - 1$ and $3x - y = 4$ shows that $(1, -1)$ and $(3, 5)$ belong to each graph. Therefore, these ordered pairs are solutions of the system of equations.

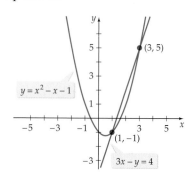

FIGURE 9.21

▶ **TRY EXERCISE 8, PAGE 800**

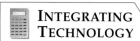

**INTEGRATING
TECHNOLOGY**

You can use a graphing calculator to solve some nonlinear systems of equations in two variables. For instance, to solve

$$\begin{cases} y = x^2 - 2x + 2 \\ y = x^3 + 2x^2 - 7x - 3 \end{cases}$$

enter X²-2X+2 into Y₁ and X^3+2X²-7X-3 into Y₂ and graph the two equations. Be sure to use a viewing window that will show all points of intersection. The sequence of steps shown in **Figure 9.22** can be used to find the points of intersection with a TI-83 calculator.

Press [2nd] CALC.
Select 5: intersect.
Press [ENTER].

```
CALCULATE
1: value
2: zero
3: minimum
4: maximum
5: intersect
6: dy/dx
7: ∫f(x)dx
```

The "First curve?" shown on the bottom of the screen means to select the first of the two graphs that intersect. Just press [ENTER].

The "Second curve?" shown on the bottom of the screen means to select the second of the two graphs that intersect. Just press [ENTER].

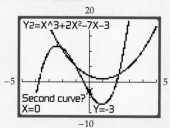

"Guess?" is shown on the bottom of the screen. Move the cursor until it is approximately at the first point of intersection. Press [ENTER].

The approximate coordinates of the point of intersection, $(-2.24, 11.47)$, are shown at the bottom of the screen.

Repeat these steps two more times to find the remaining points of intersection. The graphs are shown below.

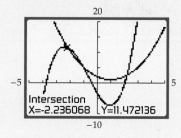

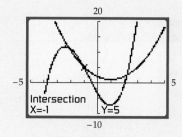

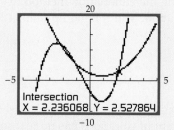

FIGURE 9.22

The coordinates of the points of intersection are $(-2.24, 11.47)$, $(-1, 5)$, and $(2.24, 2.53)$.

EXAMPLE 2   Solve a Nonlinear System by the Elimination Method

Solve: $\begin{cases} 4x^2 + 3y^2 = 48 & (1) \\ 3x^2 + 2y^2 = 35 & (2) \end{cases}$

Algebraic Solution

We will eliminate the x^2 term. Multiply Equation (1) by -3 and Equation (2) by 4. Then add the two equations.

$$-12x^2 - 9y^2 = -144$$
$$\underline{12x^2 + 8y^2 = 140}$$
$$-y^2 = -4$$
$$y^2 = 4$$
$$y = \pm 2$$

Substitute 2 for y into Equation (1) and solve for x.

$$4x^2 + 3(2)^2 = 48$$
$$4x^2 = 36$$
$$x^2 = 9$$
$$x = \pm 3$$

Because $(-2)^2 = 2^2$, replacing y by -2 yields the same values of x: $x = 3$ or $x = -3$. The solutions are $(3, 2)$, $(3, -2)$, $(-3, 2)$, and $(-3, -2)$. See **Figure 9.23.**

Visualize the Solution

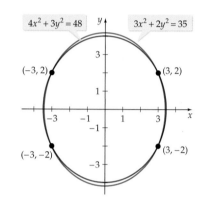

FIGURE 9.23

▶ **TRY EXERCISE 16, PAGE 800**

EXAMPLE 3   Identify an Inconsistent System of Equations

Solve: $\begin{cases} 4x^2 + 9y^2 = 36 & (1) \\ x^2 - y^2 = 25 & (2) \end{cases}$

Algebraic Solution

Using the elimination method, we will eliminate the x^2 term from each equation. Multiplying Equation (2) by -4 and then adding, we have

$$4x^2 + 9y^2 = 36$$
$$\underline{-4x^2 + 4y^2 = -100}$$
$$13y^2 = -64$$

Because the equation $13y^2 = -64$ has no real number solutions, the system of equations has no real solutions. The graphs of the equations do not intersect. See **Figure 9.24.**

Visualize the Solution

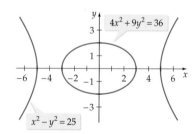

FIGURE 9.24

▶ **TRY EXERCISE 20, PAGE 800**

EXAMPLE 4   **Solve a Nonlinear System of Equations**

Solve: $\begin{cases} (x+3)^2 + (y-4)^2 = 20 \\ (x+4)^2 + (y-3)^2 = 26 \end{cases}$

Algebraic Solution

Expand the binomials in each equation. Then subtract the two equations and simplify.

$$
\begin{array}{rl}
x^2 + 6x + 9 + y^2 - 8y + 16 = 20 & \quad (1) \\
\underline{x^2 + 8x + 16 + y^2 - 6y + 9 = 26} & \quad (2) \\
-2x - 7 - 2y + 7 = -6 & \\
x + y = 3 &
\end{array}
$$

Now solve the resulting equation for y.

$$ y = -x + 3 $$

Substitute $-x + 3$ for y into Equation (1) and solve for x.

$$
\begin{aligned}
x^2 + 6x + 9 + (-x+3)^2 - 8(-x+3) + 16 &= 20 \\
2(x^2 + 4x - 5) &= 0 \\
2(x+5)(x-1) &= 0 \\
x = -5 \quad \text{or} \quad x &= 1
\end{aligned}
$$

Substitute -5 and 1 for x into the equation $y = -x + 3$ and solve for y. This yields $y = 8$ or $y = 2$. The solutions of the system of equations are $(-5, 8)$ and $(1, 2)$. See **Figure 9.25.**

Visualize the Solution

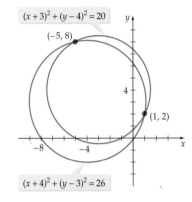

FIGURE 9.25

▶ **TRY EXERCISE 28, PAGE 800**

👥 🖊 **TOPICS FOR DISCUSSION**

1. What distinguishes a system of linear equations from a system of nonlinear equations? Give an example of both types of systems of equations.

2. Is the system of equations
$$ \begin{cases} xy = 1 \\ x + y = 1 \end{cases} $$
a nonlinear system of equations? Why or why not?

3. Can a nonlinear system of equations have no solution? If so, give an example. If not, explain why not.

4. Create a nonlinear system of equations in two variables that has at least $(2, -3)$ as a solution, contains one nonlinear equation, and contains one linear equation.

EXERCISE SET 9.3

In Exercises 1 to 32, solve the system of equations.

1. $\begin{cases} y = x^2 - x \\ y = 2x - 2 \end{cases}$

2. $\begin{cases} y = x^2 + 2x - 3 \\ y = x - 1 \end{cases}$

3. $\begin{cases} y = 2x^2 - 3x - 3 \\ y = x - 4 \end{cases}$

4. $\begin{cases} y = -x^2 + 2x - 4 \\ y = \dfrac{1}{2}x + 1 \end{cases}$

5. $\begin{cases} y = x^2 - 2x + 3 \\ y = x^2 - x - 2 \end{cases}$

6. $\begin{cases} y = 2x^2 - x + 1 \\ y = x^2 + 2x + 5 \end{cases}$

7. $\begin{cases} x + y = 10 \\ xy = 24 \end{cases}$

▶ 8. $\begin{cases} x - 2y = 3 \\ xy = -1 \end{cases}$

9. $\begin{cases} 2x - y = 1 \\ xy = 6 \end{cases}$

10. $\begin{cases} x - 3y = 7 \\ xy = -4 \end{cases}$

11. $\begin{cases} 3x^2 - 2y^2 = 1 \\ y = 4x - 3 \end{cases}$

12. $\begin{cases} x^2 + 3y^2 = 7 \\ x + 4y = 6 \end{cases}$

13. $\begin{cases} y = x^3 + 4x^2 - 3x - 5 \\ y = 2x^2 - 2x - 3 \end{cases}$

14. $\begin{cases} y = x^3 - 2x^2 + 5x + 1 \\ y = x^2 + 7x - 5 \end{cases}$

15. $\begin{cases} 2x^2 + y^2 = 9 \\ x^2 - y^2 = 3 \end{cases}$

▶ 16. $\begin{cases} 3x^2 - 2y^2 = 19 \\ x^2 - y^2 = 5 \end{cases}$

17. $\begin{cases} x^2 - 2y^2 = 8 \\ x^2 + 3y^2 = 28 \end{cases}$

18. $\begin{cases} 2x^2 + 3y^2 = 5 \\ x^2 - 3y^2 = 4 \end{cases}$

19. $\begin{cases} 2x^2 + 4y^2 = 5 \\ 3x^2 + 8y^2 = 14 \end{cases}$

▶ 20. $\begin{cases} 2x^2 + 3y^2 = 11 \\ 3x^2 + 2y^2 = 19 \end{cases}$

21. $\begin{cases} x^2 - 2x + y^2 = 1 \\ 2x + y = 5 \end{cases}$

22. $\begin{cases} x^2 + y^2 + 3y = 22 \\ 2x + y = -1 \end{cases}$

23. $\begin{cases} (x - 3)^2 + (y + 1)^2 = 5 \\ x - 3y = 7 \end{cases}$

24. $\begin{cases} (x + 2)^2 + (y - 2)^2 = 13 \\ 2x + y = 6 \end{cases}$

25. $\begin{cases} x^2 - 3x + y^2 = 4 \\ 3x + y = 11 \end{cases}$

26. $\begin{cases} x^2 + y^2 - 4y = 4 \\ 5x - 2y = 2 \end{cases}$

27. $\begin{cases} (x - 1)^2 + (y + 2)^2 = 14 \\ (x + 2)^2 + (y - 1)^2 = 2 \end{cases}$

▶ 28. $\begin{cases} (x + 2)^2 + (y - 3)^2 = 10 \\ (x - 3)^2 + (y + 1)^2 = 13 \end{cases}$

29. $\begin{cases} (x + 3)^2 + (y - 2)^2 = 20 \\ (x - 2)^2 + (y - 3)^2 = 2 \end{cases}$

30. $\begin{cases} (x - 4)^2 + (y - 5)^2 = 8 \\ (x + 1)^2 + (y + 2)^2 = 34 \end{cases}$

31. $\begin{cases} (x - 1)^2 + (y + 1)^2 = 2 \\ (x + 2)^2 + (y - 3)^2 = 3 \end{cases}$

32. $\begin{cases} (x + 1)^2 + (y - 3)^2 = 4 \\ (x - 3)^2 + (y + 2)^2 = 2 \end{cases}$

33. GEOMETRY Find the perimeter of the rectangle below.

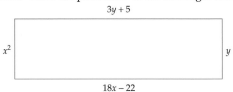

$3y + 5$

x^2

y

$18x - 22$

34. CONSTRUCTION A painter leans a ladder against a vertical wall. The top of the ladder is 7 meters above the ground. When the bottom of the ladder is moved 1 meter farther away from the wall, the top of the ladder is 5 meters above the ground. What is the length of the ladder? Round to the nearest hundredth of a meter.

35. ANALYTIC GEOMETRY For what values of the radius does the line $y = 2x + 1$ intersect (at one or more points) the circle whose equation is $x^2 + y^2 = r^2$?

36. GEOMETRY Three rectangles have exactly the same area. The dimensions of each rectangle (as length and width) are a and b; $a - 3$ and $b + 2$; and $a + 3$ and $b - 1$. Find the area of the rectangles.

37. SUPPLY/DEMAND The number x of picture cellphones a manufacturer is willing to sell at price p is given by

$$x = \frac{p^2}{5} - 20,$$

where p is the price, in dollars, per picture cellphone. The number x of picture cellphones a distributor is willing to purchase is given by $x = \dfrac{17{,}710}{p+1}$, where p is the price, in dollars, per picture cellphone. Find the equilibrium price. (See Section 9.1 for a discussion of supply-demand equations.)

38. SUPPLY/DEMAND The number x of a certain type of personal digital assistant (PDA) a manufacturer is willing to sell at price p is given by $x = \dfrac{p^2}{6} - 384$, where p is the price, in dollars, per PDA. The number x of these PDAs an office supply store is willing to purchase is given by $x = \dfrac{22{,}914}{p+1}$, where p is the price per PDA. Find the equilibrium price. (See Section 9.1 for a discussion of supply-demand equations.)

In Exercises 39 to 46, approximate the real number solutions of each system of equations to the nearest ten-thousandth.

39. $\begin{cases} y = 2^x \\ y = x + 1 \end{cases}$

40. $\begin{cases} y = \log_2 x \\ y = x - 3 \end{cases}$

41. $\begin{cases} y = e^{-x} \\ y = x^2 \end{cases}$

42. $\begin{cases} y = \ln x \\ y = -x + 4 \end{cases}$

43. $\begin{cases} y = \sqrt{x} \\ y = \dfrac{1}{x-1} \end{cases}$

44. $\begin{cases} y = \dfrac{6}{x+1} \\ y = \dfrac{x}{x-1} \end{cases}$

45. $\begin{cases} y = |x| \\ y = 2^{-x^2} \end{cases}$

46. $\begin{cases} y = \dfrac{2^x + 2^{-x}}{2} \\ y = \dfrac{2^x - 2^{-x}}{2} \end{cases}$

CONNECTING CONCEPTS

In Exercises 47 to 52, solve the system of equations for *rational number* ordered pairs.

47. $\begin{cases} y = x^2 + 4 \\ x = y^2 - 24 \end{cases}$

48. $\begin{cases} y = x^2 - 5 \\ x = y^2 - 13 \end{cases}$

49. $\begin{cases} x^2 - 3xy + y^2 = 5 \\ x^2 - xy - 2y^2 = 0 \end{cases}$

(*Hint:* Factor the second equation. Now use the principle of zero products and the substitution principle.)

50. $\begin{cases} x^2 + 2xy - y^2 = 1 \\ x^2 + 3xy + 2y^2 = 0 \end{cases}$

(*Hint:* See Exercise 49.)

51. $\begin{cases} 2x^2 - 4xy - y^2 = 6 \\ 4x^2 - 3xy - y^2 = 6 \end{cases}$

(*Hint:* Subtract the two equations.)

52. $\begin{cases} 3x^2 + 2xy - 5y^2 = 11 \\ x^2 + 3xy + y^2 = 11 \end{cases}$

(*Hint:* Subtract the two equations.)

53. Show that the line $y = mx$ intersects the hyperbola given by the equation $\dfrac{x^2}{a^2} - \dfrac{y^2}{b^2} = 1$ if and only if $|m| < \left| \dfrac{b}{a} \right|$.

PREPARE FOR SECTION 9.4

54. Factor $x^4 + 14x^2 + 49$ over the real numbers. [P.4]

55. Add: $\dfrac{5}{x-1} + \dfrac{1}{x+2}$ [P.5]

56. Simplify: $\dfrac{7}{x} - \dfrac{6}{x-1} + \dfrac{10}{(x-1)^2}$ [P.5]

57. Solve: $\begin{cases} 1 = A + B \\ 11 = -5A + 3B \end{cases}$ [9.1]

58. Solve: $\begin{cases} 0 = A + B \\ 3 = -2B + C \\ 16 = 7A - 2C \end{cases}$ [9.2]

59. Divide: $\dfrac{x^3 - 4x^2 - 19x - 35}{x^2 - 7x}$ [3.1]

■ PROJECTS

I. FINDING ZEROS OF A POLYNOMIAL One zero of $P(x) = x^3 + 2x^2 + Cx - 6$ is the sum of the other two zeros of $P(x)$. Find C and the three zeros of $P(x)$.

2. PROVING A GEOMETRY THEOREM Consider the triangle that is shown at the right inscribed in a circle of radius a with one side along the diameter of the circle. Prove that the triangle is a right triangle by completing the following steps.

a. Let $y = mx$, $m \geq 0$. Show that the graph of $y = mx$, $m \geq 0$, intersects the circle whose equation is $(x - a)^2 + y^2 = a^2, a > 0$, at $P\left(\dfrac{2a}{1 + m^2}, \dfrac{2ma}{1 + m^2}\right)$.

b. Show that the slope of the line through P and $Q(2a, 0)$ is $-\dfrac{1}{m}$.

c. What is the slope of the line between O and P?

d. Prove that line segment OP is perpendicular to line segment PQ.

e. How can you conclude from the foregoing that triangle OPQ is a right triangle?

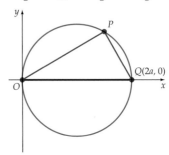

<table>
<tr><td>

SECTION 9.4

</td><td>

PARTIAL FRACTIONS

</td></tr>
</table>

● **PARTIAL FRACTION
DECOMPOSITION**

● PARTIAL FRACTION DECOMPOSITION

An algebraic application of systems of equations is a technique known as *partial fractions*. In Chapter P, we reviewed the problem of adding two rational expressions. For example,

$$\frac{5}{x - 1} + \frac{1}{x + 2} = \frac{6x + 9}{(x - 1)(x + 2)}$$

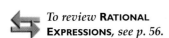
To review **RATIONAL EXPRESSIONS,** *see p. 56.*

Now we will take an opposite approach. That is, given a rational expression, we will find simpler rational expressions whose sum is the given expression. The method by which a more complicated rational expression is written as a sum of rational expressions is called **partial fraction decomposition.** This technique is based on the following theorem.

Partial Fraction Decomposition Theorem

If
$$f(x) = \frac{p(x)}{q(x)}$$

is a rational expression in which the degree of the numerator is less than the degree of the denominator, and $p(x)$ and $q(x)$ have no common factors, then $f(x)$ can be written as a partial fraction decomposition in the form

$$f(x) = f_1(x) + f_2(x) + \cdots + f_n(x)$$

where each $f_i(x)$ has one of the following forms:

$$\frac{A}{(px + q)^m} \quad \text{or} \quad \frac{Bx + C}{(ax^2 + bx + c)^m}$$

The procedure for finding a partial fraction decomposition of a rational expression depends on factorization of the denominator of the rational expression. There are four cases.

Case 1 Nonrepeated Linear Factors

The partial fraction decomposition will contain an expression of the form $\dfrac{A}{x+a}$ for each nonrepeated linear factor of the denominator. Example:

$$\frac{3x-1}{x(3x+4)(x-2)}$$

• **Each linear factor of the denominator occurs only once.**

Partial fraction decomposition:

$$\frac{3x-1}{x(3x+4)(x-2)} = \frac{A}{x} + \frac{B}{3x+4} + \frac{C}{x-2}$$

Case 2 Repeated Linear Factors

The partial fraction decomposition will contain an expression of the form

$$\frac{A_1}{(x+a)} + \frac{A_2}{(x+a)^2} + \cdots + \frac{A_m}{(x+a)^m}$$

for each repeated linear factor of multiplicity m. Example:

$$\frac{4x+5}{(x-2)^2(2x+1)}$$

• $(x-2)^2 = (x-2)(x-2)$, a repeated linear factor.

Partial fraction decomposition:

$$\frac{4x+5}{(x-2)^2(2x+1)} = \frac{A_1}{x-2} + \frac{A_2}{(x-2)^2} + \frac{B}{2x+1}$$

Case 3 Nonrepeated Quadratic Factors

The partial fraction decomposition will contain an expression of the form

$$\frac{Ax+B}{ax^2+bx+c}$$

for each quadratic factor irreducible over the real numbers. Example:

$$\frac{x-4}{(x^2+x+1)(x-4)}$$

• $x^2 + x + 1$ is irreducible over the real numbers.

Partial fraction decomposition:

$$\frac{x-4}{(x^2+x+1)(x-4)} = \frac{Ax+B}{x^2+x+1} + \frac{C}{x-4}$$

Case 4 Repeated Quadratic Factors

The partial fraction decomposition will contain an expression of the form

$$\frac{A_1x+B_1}{ax^2+bx+c} + \frac{A_2x+B_2}{(ax^2+bx+c)^2} + \cdots + \frac{A_mx+B_m}{(ax^2+bx+c)^m}$$

for each quadratic factor irreducible over the real numbers. Example:

$$\frac{2x}{(x-2)(x^2+4)^2}$$ • $(x^2+4)^2$ is a repeated quadratic factor.

Partial fraction decomposition:

$$\frac{2x}{(x-2)(x^2+4)^2} = \frac{A_1x+B_1}{x^2+4} + \frac{A_2x+B_2}{(x^2+4)^2} + \frac{C}{x-2}$$

? QUESTION Which of the four cases of a partial fraction decomposition apply
to $\dfrac{x+2}{(x-2)(x^2+4)}$?

There are various methods for finding the constants of a partial fraction decomposition. One such method is based on a property of polynomials.

Equality of Polynomials

If the two polynomials $p(x) = a_nx^n + a_{n-1}x^{n-1} + \cdots + a_1x + a_0$ and
$r(x) = b_nx^n + b_{n-1}x^{n-1} + \cdots + b_1x + b_0$ are of degree n, then $p(x) = r(x)$ if
and only if $a_0 = b_0, a_1 = b_1, a_2 = b_2, \ldots, a_n = b_n$.

EXAMPLE 1   **Find a Partial Fraction Decomposition**
Case 1: Nonrepeated Linear Factors

Find a partial fraction decomposition of $\dfrac{x+11}{x^2-2x-15}$.

Solution

First factor the denominator.

$$x^2 - 2x - 15 = (x+3)(x-5)$$

The factors are nonrepeated linear factors. Therefore, the partial fraction
decomposition will have the form

$$\frac{x+11}{(x+3)(x-5)} = \frac{A}{x+3} + \frac{B}{x-5} \tag{1}$$

To solve for A and B, multiply each side of the equation by the least
common multiple of the denominators, $(x+3)(x-5)$.

$$x+11 = A(x-5) + B(x+3)$$
$$x+11 = (A+B)x + (-5A+3B)$$ • **Combine like terms.**

? ANSWER Cases 1 and 4.

Using the Equality of Polynomials Theorem, equate coefficients of like powers. The result will be the system of equations

$$\begin{cases} 1 = A + B \\ 11 = -5A + 3B \end{cases}$$

• Recall that $x = 1 \cdot x$.

Solving the system of equations for A and B, we have $A = -1$ and $B = 2$. Substituting -1 for A and 2 for B into the form of the partial fraction decomposition (1), we obtain

$$\frac{x + 11}{(x + 3)(x - 5)} = \frac{-1}{x + 3} + \frac{2}{x - 5}$$

You should add the two expressions to verify the equality.

▶ **TRY EXERCISE 14, PAGE 808**

EXAMPLE 2 **Find the Partial Fraction Decomposition Case 2: Repeated Linear Factors**

Find a partial fraction decomposition of $\dfrac{x^2 + 2x + 7}{x(x - 1)^2}$.

Solution

The denominator has one nonrepeated factor and one repeated factor. The partial fraction decomposition will have the form

$$\frac{x^2 + 2x + 7}{x(x - 1)^2} = \frac{A}{x} + \frac{B}{x - 1} + \frac{C}{(x - 1)^2}$$

Multiplying each side by the LCD $x(x - 1)^2$, we have

$$x^2 + 2x + 7 = A(x - 1)^2 + B(x - 1)x + Cx$$

Expanding the right side and combining like terms gives

$$x^2 + 2x + 7 = (A + B)x^2 + (-2A - B + C)x + A$$

Using the Equality of Polynomials Theorem, equate coefficients of like powers. This will result in the system of equations

$$\begin{cases} 1 = A + B \\ 2 = -2A - B + C \\ 7 = A \end{cases}$$

The solution is $A = 7$, $B = -6$, and $C = 10$. Thus the partial fraction decomposition is

$$\frac{x^2 + 2x + 7}{x(x - 1)^2} = \frac{7}{x} + \frac{-6}{x - 1} + \frac{10}{(x - 1)^2}$$

▶ **TRY EXERCISE 22, PAGE 808**

EXAMPLE 3	**Find the Partial Fraction Decomposition** **Case 3: Nonrepeated Quadratic Factor**

Find the partial fraction decomposition of $\dfrac{3x + 16}{(x - 2)(x^2 + 7)}$.

Solution

Because $(x - 2)$ is a nonrepeated linear factor and $x^2 + 7$ is an irreducible quadratic over the real numbers, the partial fraction decomposition will have the form

$$\frac{3x + 16}{(x - 2)(x^2 + 7)} = \frac{A}{x - 2} + \frac{Bx + C}{x^2 + 7}$$

Multiplying each side by the LCD $(x - 2)(x^2 + 7)$ yields

$$3x + 16 = A(x^2 + 7) + (Bx + C)(x - 2)$$

Expanding the right side and combining like terms, we have

$$3x + 16 = (A + B)x^2 + (-2B + C)x + (7A - 2C)$$

Using the Equality of Polynomials Theorem, equate coefficients of like powers. This will result in the system of equations

$$\begin{cases} 0 = A + B \\ 3 = -2B + C \\ 16 = 7A - 2C \end{cases}$$

• Think of $3x + 16$ as $0x^2 + 3x + 16$.

The solution is $A = 2$, $B = -2$, and $C = -1$. Thus the partial fraction decomposition is

$$\frac{3x + 16}{(x - 2)(x^2 + 7)} = \frac{2}{x - 2} + \frac{-2x - 1}{x^2 + 7}$$

▶ **TRY EXERCISE 24, PAGE 808**

EXAMPLE 4	**Find a Partial Fraction Decomposition** **Case 4: Repeated Quadratic Factors**

Find the partial fraction decomposition of $\dfrac{4x^3 + 5x^2 + 7x - 1}{(x^2 + x + 1)^2}$.

Solution

The quadratic factor $(x^2 + x + 1)$ is irreducible over the real numbers and is a repeated factor. The partial fraction decomposition will be of the form

$$\frac{4x^3 + 5x^2 + 7x - 1}{(x^2 + x + 1)^2} = \frac{Ax + B}{x^2 + x + 1} + \frac{Cx + D}{(x^2 + x + 1)^2}$$

Multiplying each side by the LCD $(x^2 + x + 1)^2$ and collecting like terms, we obtain

$$4x^3 + 5x^2 + 7x - 1 = (Ax + B)(x^2 + x + 1) + Cx + D$$
$$= Ax^3 + Ax^2 + Ax + Bx^2 + Bx + B + Cx + D$$
$$= Ax^3 + (A + B)x^2 + (A + B + C)x + (B + D)$$

Equating coefficients of like powers gives the system of equations

$$\begin{cases} 4 = A \\ 5 = A + B \\ 7 = A + B + C \\ -1 = B + D \end{cases}$$

Solving this system, we have $A = 4$, $B = 1$, $C = 2$, and $D = -2$. Thus the partial fraction decomposition is

$$\frac{4x^3 + 5x^2 + 7x - 1}{(x^2 + x + 1)^2} = \frac{4x + 1}{x^2 + x + 1} + \frac{2x - 2}{(x^2 + x + 1)^2}$$

▶ **TRY EXERCISE 30, PAGE 809**

The Partial Fraction Decomposition Theorem requires that the degree of the numerator be less than the degree of the denominator. If this is *not* the case, use long division to first write the rational expression as a polynomial plus a remainder over the denominator.

EXAMPLE 5   **Find a Partial Fraction Decomposition When the Degree of the Numerator Exceeds the Degree of the Denominator**

Find the partial fraction decomposition of $F(x) = \dfrac{x^3 - 4x^2 - 19x - 35}{x^2 - 7x}$.

Solution

Because the degree of the denominator is less than the degree of the numerator, use long division first to obtain

$$F(x) = x + 3 + \frac{2x - 35}{x^2 - 7x}$$

The partial fraction decomposition of $\dfrac{2x - 35}{x^2 - 7x}$ will have the form

$$\frac{2x - 35}{x^2 - 7x} = \frac{2x - 35}{x(x - 7)} = \frac{A}{x} + \frac{B}{x - 7}$$

Multiplying each side by $x(x - 7)$ and combining like terms, we have

$$2x - 35 = (A + B)x + (-7A)$$

Equating coefficients of like powers yields

$$\begin{cases} 2 = A + B \\ -35 = -7A \end{cases}$$

The solution of this system is $A = 5$ and $B = -3$. The partial fraction decomposition is

$$\frac{x^3 - 4x^2 - 19x - 35}{x^2 - 7x} = x + 3 + \frac{5}{x} + \frac{-3}{x - 7}$$

▶ **TRY EXERCISE 34, PAGE 809**

TOPICS FOR DISCUSSION

1. What is the purpose of a partial fraction decomposition?

2. Discuss how the factors of the denominator of a rational expression dictate how a partial fraction decomposition is determined.

3. Discuss the Equality of Polynomials Theorem and how it is used in a partial fraction decomposition.

4. For the rational expression $\dfrac{3x - 1}{x^3 - 2x^2 - x + 2}$, what is the first step you perform to find a partial fraction decomposition? What equation or equations do you solve to find the partial fraction decomposition?

EXERCISE SET 9.4

In Exercises 1 to 10, determine the constants A, B, C, and D.

1. $\dfrac{x + 15}{x(x - 5)} = \dfrac{A}{x} + \dfrac{B}{x - 5}$

2. $\dfrac{5x - 6}{x(x + 3)} = \dfrac{A}{x} + \dfrac{B}{x + 3}$

3. $\dfrac{1}{(2x + 3)(x - 1)} = \dfrac{A}{2x + 3} + \dfrac{B}{x - 1}$

4. $\dfrac{6x - 5}{(x + 4)(3x + 2)} = \dfrac{A}{x + 4} + \dfrac{B}{3x + 2}$

5. $\dfrac{x + 9}{x(x - 3)^2} = \dfrac{A}{x} + \dfrac{B}{x - 3} + \dfrac{C}{(x - 3)^2}$

6. $\dfrac{2x - 7}{(x + 1)(x - 2)^2} = \dfrac{A}{x + 1} + \dfrac{B}{x - 2} + \dfrac{C}{(x - 2)^2}$

7. $\dfrac{4x^2 + 3}{(x - 1)(x^2 + x + 5)} = \dfrac{A}{x - 1} + \dfrac{Bx + C}{x^2 + x + 5}$

8. $\dfrac{x^2 + x + 3}{(x^2 + 7)(x - 3)} = \dfrac{Ax + B}{x^2 + 7} + \dfrac{C}{x - 3}$

9. $\dfrac{x^3 + 2x}{(x^2 + 1)^2} = \dfrac{Ax + B}{x^2 + 1} + \dfrac{Cx + D}{(x^2 + 1)^2}$

10. $\dfrac{3x^3 + x^2 - x - 5}{(x^2 + 2x + 5)^2} = \dfrac{Ax + B}{x^2 + 2x + 5} + \dfrac{Cx + D}{(x^2 + 2x + 5)^2}$

In Exercises 11 to 36, find the partial fraction decomposition of the given rational expression.

11. $\dfrac{8x + 12}{x(x + 4)}$

12. $\dfrac{x - 14}{x(x - 7)}$

13. $\dfrac{3x + 50}{x^2 - 7x - 18}$

▶ 14. $\dfrac{7x + 44}{x^2 + 10x + 24}$

15. $\dfrac{16x + 34}{4x^2 + 16x + 15}$

16. $\dfrac{-15x + 37}{9x^2 - 12x - 5}$

17. $\dfrac{x - 5}{(3x + 5)(x - 2)}$

18. $\dfrac{1}{(x + 7)(2x - 5)}$

19. $\dfrac{x^3 + 3x^2 - 4x - 8}{x^2 - 4}$

20. $\dfrac{x^3 - 13x - 9}{x^2 - x - 12}$

21. $\dfrac{3x^2 + 49}{x(x + 7)^2}$

▶ 22. $\dfrac{x - 18}{x(x - 3)^2}$

23. $\dfrac{5x^2 - 7x + 2}{x^3 - 3x^2 + x}$

▶ 24. $\dfrac{9x^2 - 3x + 49}{x^3 - x^2 + 10x - 10}$

28. Minimize Cost A producer of animal feed makes two food products: F_1 and F_2. The products contain three major ingredients: M_1, M_2, and M_3. Each ton of F_1 requires 200 pounds of M_1, 100 pounds of M_2, and 100 pounds of M_3. Each ton of F_2 requires 100 pounds of M_1, 200 pounds of M_2, and 400 pounds of M_3. There are at least 5000 pounds of M_1 available, at least 7000 pounds of M_2 available, and at least 10,000 pounds of M_3 available. Each ton of F_1 costs \$450 to make, and each ton of F_2 costs \$300 to make. How many tons of each food product should the feed producer make to minimize cost? What is the minimum cost?

PROJECTS

1. 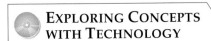 **History of Linear Programming** Linear programming has been used successfully to solve a wide range of problems in fields as diverse as providing health care and hardening nuclear silos. Write an essay on linear programming and some of the applications of this procedure in solving practical problems. Include in your essay the contributions of George Danzig, Narendra Karmarkar, and L. G. Khachian.

EXPLORING CONCEPTS WITH TECHNOLOGY

Ill-Conditioned Systems of Equations

Solving systems of equations algebraically as we did in this chapter is not practical for systems of equations that contain a large number of variables. In those cases, a computer solution is the only hope. Computer solutions are not without some problems, however.

Consider the system of equations

$$\begin{cases} 0.24567x + 0.49133y = 0.73700 \\ 0.84312x + 1.68623y = 2.52935 \end{cases}$$

It is easy to verify that the solution of this system of equations is $(1, 1)$. However, change the constant 0.73700 to 0.73701 (add 0.00001) and the constant 2.52935 to 2.52936 (add 0.00001), and the solution is now $(3, 0)$. Thus a very small change in the constant terms produces a dramatic change in the solution. A system of equations of this sort is said to be *ill-conditioned*.

These types of systems are important because computers generally cannot store numbers beyond a certain number of significant digits. Your calculator, for example, probably allows you to enter no more than 10 significant digits. If an exact number cannot be entered, then an approximation to that number is necessary. When a computer is solving an equation or a system of equations, the hope is that approximations of the coefficients it uses will give reasonable approximations to the solutions. For ill-conditioned systems of equations, this is not always true.

In the system of equations above, small changes in the constant terms caused a large change in the solution. It is possible that small changes in the coefficients of the variables will also cause large changes in the solution.

In the two systems of equations that follow, examine the effects of approximating the fractional coefficients on the solutions. Try approximating each fraction to the nearest hundredth, to the nearest thousandth, to the nearest ten-thousandth, and then to the limits of your calculator. The exact solution of the first system of equations is $(27, -192, 210)$. The exact solution of the second system of equations is $(-64, 900, -2520, 1820)$.

$$\begin{cases} x + \dfrac{1}{2}y + \dfrac{1}{3}z = 1 \\ \dfrac{1}{2}x + \dfrac{1}{3}y + \dfrac{1}{4}z = 2 \\ \dfrac{1}{3}x + \dfrac{1}{4}y + \dfrac{1}{5}z = 3 \end{cases} \qquad \begin{cases} x + \dfrac{1}{2}y + \dfrac{1}{3}z + \dfrac{1}{4}w = 1 \\ \dfrac{1}{2}x + \dfrac{1}{3}y + \dfrac{1}{4}z + \dfrac{1}{5}w = 2 \\ \dfrac{1}{3}x + \dfrac{1}{4}y + \dfrac{1}{5}z + \dfrac{1}{6}w = 3 \\ \dfrac{1}{4}x + \dfrac{1}{5}y + \dfrac{1}{6}z + \dfrac{1}{7}w = 4 \end{cases}$$

Note how the solutions change as the approximations change and thus how important it is to know whether a system of equations is ill-conditioned. For systems that are not ill-conditioned, approximations of the coefficients yield reasonable approximations of the solution. For ill-conditioned systems of equations, this is not always true.

CHAPTER 9 SUMMARY

9.1 Systems of Linear Equations in Two Variables

- A system of equations is two or more equations considered together. A solution of a system of equations in two variables is an ordered pair that satisfies each equation of the system. Equivalent systems of equations have the same solution set.

- A system of equations is consistent if it has one or more solutions. A system of linear equations is independent if it has exactly one solution. A system is dependent if it has infinitely many solutions. An inconsistent system of equations has no solution.

- **Operations That Produce Equivalent Systems of Equations**

   1. Interchange any two equations.
   2. Replace an equation with a nonzero multiple of that equation.
   3. Replace an equation with the sum of that equation and a nonzero constant multiple of another equation in the system.

9.2 Systems of Linear Equations in More Than Two Variables

- An equation of the form $ax + by + cz = d$, with a, b, and c not all zero, is a linear equation in three variables. A solution of a system of equations in three variables is an ordered triple that satisfies each equation of the system.

- The graph of a linear equation in three variables is a plane.

- A linear system of equations for which the constant term is zero for all equations of the system is called a homogeneous system of equations.

9.3 Nonlinear Systems of Equations

- A nonlinear system of equations is a system in which one or more equations of the system are nonlinear.

9.4 Partial Fractions

- A rational expression can be written as the sum of terms whose denominators are factors of the denominator of the rational expression. This is called a partial fraction decomposition.

9.5 Inequalities in Two Variables and Systems of Inequalities

- The graph of an inequality in two variables frequently separates the plane into two or more regions.

- The solution set of a system of inequalities is the intersection of the solution sets of the individual inequalities.

9.6 Linear Programming

- A linear programming problem consists of a linear objective function and a number of constraints, which are inequalities or equations that restrict the values of the variables.

- The Fundamental Linear Programming Theorem states that if an objective function has an optimal solution, then that solution will be at a vertex of the set of feasible solutions.

CHAPTER 9 TRUE/FALSE EXERCISES

In Exercises 1 to 10, answer true or false. If the statement is false, give an example or state a reason to show that the statement is false.

1. A system of equations will always have a solution as long as the number of equations is equal to the number of variables.

2. A system of two different quadratic equations can have at most four solutions.

3. A homogeneous system of equations is one in which all the variables have the same exponent.

4. In an xyz-coordinate system, the graph of the set of points formed by the intersection of two different planes is a straight line.

5. It is possible to find a partial fraction decomposition of a rational expression if the degree of the numerator is greater than the degree of the denominator.

6. Two systems of equations with the same solution set have the same equations in their respective systems.

7. The systems of equations
$$\begin{cases} x = 0 \\ y = 0 \end{cases} \text{ and } \begin{cases} y = x \\ y = -x \end{cases}$$
are equivalent systems of equations.

8. For a linear programming problem, one or more constraints are used to define the set of feasible solutions.

9. A system of three linear equations in three variables for which two of the planes are parallel and the third plane intersects the first two is a dependent system of equations.

10. The inequality $xy < 1$ and the inequality $y < \dfrac{1}{x}$ are equivalent inequalities.

CHAPTER 9 REVIEW EXERCISES

In Exercises 1 to 30, solve each system of equations.

1. $\begin{cases} 2x - 4y = -3 \\ 3x + 8y = -12 \end{cases}$

2. $\begin{cases} 4x - 3y = 15 \\ 2x + 5y = -12 \end{cases}$

3. $\begin{cases} 3x - 4y = -5 \\ y = \dfrac{2}{3}x + 1 \end{cases}$

4. $\begin{cases} 7x + 2y = -14 \\ y = -\dfrac{5}{2}x - 3 \end{cases}$

5. $\begin{cases} y = 2x - 5 \\ x = 4y - 1 \end{cases}$

6. $\begin{cases} y = 3x + 4 \\ x = 4y - 5 \end{cases}$

7. $\begin{cases} 6x + 9y = 15 \\ 10x + 15y = 25 \end{cases}$

8. $\begin{cases} 4x - 8y = 9 \\ 2x - 4y = 5 \end{cases}$

9. $\begin{cases} 2x - 3y + z = -9 \\ 2x + 5y - 2z = 18 \\ 4x - y + 3z = -4 \end{cases}$

10. $\begin{cases} x - 3y + 5z = 1 \\ 2x + 3y - 5z = 15 \\ 3x + 6y + 5z = 15 \end{cases}$

11. $\begin{cases} x + 3y - 5z = -12 \\ 3x - 2y + z = 7 \\ 5x + 4y - 9z = -17 \end{cases}$

12. $\begin{cases} 2x - y + 2z = 5 \\ x + 3y - 3z = 2 \\ 5x - 9y + 8z = 13 \end{cases}$

13. $\begin{cases} 3x + 4y - 6z = 10 \\ 2x + 2y - 3z = 6 \\ x - 6y + 9z = -4 \end{cases}$ **14.** $\begin{cases} x - 6y + 4z = 6 \\ 4x + 3y - 4z = 1 \\ 5x - 9y + 8z = 13 \end{cases}$

15. $\begin{cases} 2x + 3y - 2z = 0 \\ 3x - y - 4z = 0 \\ 5x + 13y - 4z = 0 \end{cases}$ **16.** $\begin{cases} 3x - 5y + z = 0 \\ x + 4y - 3z = 0 \\ 2x + y - 2z = 0 \end{cases}$

17. $\begin{cases} x - 2y + z = 1 \\ 3x + 2y - 3z = 1 \end{cases}$ **18.** $\begin{cases} 2x - 3y + z = 1 \\ 4x + 2y + 3z = 21 \end{cases}$

19. $\begin{cases} y = x^2 - 2x - 3 \\ y = 2x - 7 \end{cases}$ **20.** $\begin{cases} y = 2x^2 + x \\ y = 2x + 1 \end{cases}$

21. $\begin{cases} y = 3x^2 - x + 1 \\ y = x^2 + 2x - 1 \end{cases}$ **22.** $\begin{cases} y = 4x^2 - 2x - 3 \\ y = 2x^2 + 3x - 6 \end{cases}$

23. $\begin{cases} (x + 1)^2 + (y - 2)^2 = 4 \\ 2x + y = 4 \end{cases}$

24. $\begin{cases} (x - 1)^2 + (y + 1)^2 = 5 \\ y = 2x - 3 \end{cases}$

25. $\begin{cases} (x - 2)^2 + (y + 2)^2 = 4 \\ (x + 2)^2 + (y + 1)^2 = 17 \end{cases}$

26. $\begin{cases} (x + 1)^2 + (y - 2)^2 = 1 \\ (x - 2)^2 + (y + 2)^2 = 20 \end{cases}$

27. $\begin{cases} x^2 - 3xy + y^2 = -1 \\ 3x^2 - 5xy - 2y^2 = 0 \end{cases}$

28. $\begin{cases} 2x^2 + 2xy - y^2 = -1 \\ 6x^2 + xy - y^2 = 0 \end{cases}$

29. $\begin{cases} 2x^2 - 5xy + 2y^2 = 56 \\ 14x^2 - 3xy - 2y^2 = 56 \end{cases}$

30. $\begin{cases} 2x^2 + 7xy + 6y^2 = 1 \\ 6x^2 + 7xy + 2y^2 = 1 \end{cases}$

In Exercises 31 to 36, find the partial fraction decomposition.

31. $\dfrac{7x - 5}{x^2 - x - 2}$ **32.** $\dfrac{x + 1}{(x - 1)^2}$

33. $\dfrac{2x - 2}{(x^2 + 1)(x + 2)}$ **34.** $\dfrac{5x^2 - 10x + 9}{(x - 2)^2(x + 1)}$

35. $\dfrac{11x^2 - x - 2}{x^3 - x}$ **36.** $\dfrac{x^4 + x^3 + 4x^2 + x + 3}{(x^2 + 1)^2}$

In Exercises 37 to 48, graph the solution set of each inequality.

37. $4x - 5y < 20$ **38.** $2x + 7y \geq -14$

39. $y \geq 2x^2 - x - 1$ **40.** $y < x^2 - 5x - 6$

41. $(x - 2)^2 + (y - 1)^2 > 4$ **42.** $(x + 3)^2 + (y + 1)^2 \leq 9$

43. $\dfrac{(x - 3)^2}{16} - \dfrac{(y + 2)^2}{25} \leq 1$

44. $\dfrac{(x + 1)^2}{9} - \dfrac{(y - 3)^2}{4} < -1$

45. $(2x - y + 1)(x - 2y - 2) > 0$

46. $(2x - 3y - 6)(x + 2y - 4) < 0$

47. $x^2y^2 < 1$

48. $xy \geq 0$

In Exercises 49 to 60, graph the solution set of each system of inequalities.

49. $\begin{cases} 2x - 5y < 9 \\ 3x + 4y \geq 2 \end{cases}$ **50.** $\begin{cases} 3x + y > 7 \\ 2x + 5y < 9 \end{cases}$

51. $\begin{cases} 2x + 3y > 6 \\ 2x - y > -2 \\ x \leq 4 \end{cases}$ **52.** $\begin{cases} 2x + 5y > 10 \\ x - y > -2 \\ x \leq 4 \end{cases}$

53. $\begin{cases} 2x + 3y \leq 18 \\ x + y \leq 7 \\ x \geq 0, y \geq 0 \end{cases}$ **54.** $\begin{cases} 3x + 5y \geq 25 \\ 2x + 3y \geq 16 \\ x \geq 0, y \geq 0 \end{cases}$

55. $\begin{cases} 3x + y \geq 6 \\ x + 4y \geq 14 \\ 2x + 3y \geq 16 \\ x \geq 0, y \geq 0 \end{cases}$ **56.** $\begin{cases} 3x + 2y \geq 14 \\ x + y \geq 6 \\ 11x + 4y \leq 48 \\ x \geq 0, y \geq 0 \end{cases}$

57. $\begin{cases} y < x^2 - x - 2 \\ y \geq 2x - 4 \end{cases}$ **58.** $\begin{cases} y > 2x^2 + x - 1 \\ y > x + 3 \end{cases}$

59. $\begin{cases} x^2 + y^2 - 2x + 4y > 4 \\ y < 2x^2 - 1 \end{cases}$

60. $\begin{cases} x^2 - y^2 - 4x - 2y < -4 \\ x^2 + y^2 - 4x + 4y > 8 \end{cases}$

In Exercises 61 to 66, solve the linear programming problem. In each problem, assume $x \geq 0$ and $y \geq 0$.

61. Objective function: $P = 2x + 2y$
Constraints: $\begin{cases} x + 2y \leq 14 \\ 5x + 2y \leq 30 \end{cases}$
Maximize the objective function.

62. Objective function: $P = 4x + 5y$
Constraints: $\begin{cases} 2x + 3y \leq 24 \\ 4x + 3y \leq 36 \end{cases}$
Maximize the objective function.

63. Objective function: $P = 4x + y$
Constraints: $\begin{cases} 5x + 2y \geq 16 \\ x + 2y \geq 8 \\ x \leq 20, y \leq 20 \end{cases}$
Minimize the objective function.

64. Objective function: $P = 2x + 7y$
Constraints: $\begin{cases} 4x + 3y \geq 24 \\ 4x + 7y \geq 40 \\ x \leq 10, y \leq 10 \end{cases}$
Minimize the objective function.

65. Objective function: $P = 6x + 3y$
Constraints: $\begin{cases} 5x + 2y \geq 20 \\ x + y \geq 7 \\ x + 2y \geq 10 \\ x \leq 15, y \leq 15 \end{cases}$
Minimize the objective function.

66. Objective function: $P = 5x + 4y$
Constraints: $\begin{cases} x + y \leq 10 \\ 2x + y \leq 13 \\ 3x + y \leq 18 \end{cases}$
Maximize the objective function.

In Exercises 67 to 73, solve each exercise by solving a system of equations.

67. Find an equation of the form $y = ax^2 + bx + c$ whose graph passes through the points $(1, 0)$, $(-1, 5)$, and $(2, 3)$.

68. Find an equation of the circle that passes through the points $(4, 2)$, $(0, 1)$, and $(3, -1)$.

69. Find an equation of the plane that passes through the points $(2, 1, 2)$, $(3, 1, 0)$, and $(-2, -3, -2)$. Use the equation $z = ax + by + c$.

70. How many liters of a 20% acid solution should be mixed with 10 liters of a 10% acid solution so that the resulting solution is a 16% acid solution?

71. Flying with the wind, a small plane traveled 855 miles in 5 hours. Flying against the wind, the same plane traveled 575 miles in the same time. Find the rate of the wind and the rate of the plane in calm air.

72. A collection of 10 coins has a value of $1.25. The collection consists of only nickels, dimes, and quarters. How many of each coin are in the collection? (*Hint:* There is more than one solution.)

73. Consider the ordered triple (a, b, c). Find all real number values for a, b, and c so that the product of any two numbers equals the remaining number.

CHAPTER 9 TEST

In Exercises 1 to 8, solve each system of equations. If a system of equations is inconsistent, so state.

1. $\begin{cases} 3x + 2y = -5 \\ 2x - 5y = -16 \end{cases}$

2. $\begin{cases} x - \dfrac{1}{2}y = 3 \\ 2x - y = 6 \end{cases}$

3. $\begin{cases} x + 3y - z = 8 \\ 2x - 7y + 2z = 1 \\ 4x - y + 3z = 13 \end{cases}$

4. $\begin{cases} 3x - 2y + z = 2 \\ x + 2y - 2z = 1 \\ 4x - z = 3 \end{cases}$

5. $\begin{cases} 2x - 3y + z = -1 \\ x + 5y - 2z = 5 \end{cases}$

6. $\begin{cases} 4x + 2y + z = 0 \\ x - 3y - 2z = 0 \\ 3x + 5y + 3z = 0 \end{cases}$

7. $\begin{cases} y = x + 3 \\ y = x^2 + x - 1 \end{cases}$

8. $\begin{cases} y = x^2 - x - 3 \\ y = 2x^2 + 2x - 1 \end{cases}$

In Exercises 9 to 12, graph each inequality.

9. $3x - 4y > 8$

10. $y \le x^2 - 2x - 3$

11. $x^2 + 4y^2 \ge 16$

12. $x + y^2 < 0$

In Exercises 13 to 16, graph each system of inequalities. If the solution set is empty, so state.

13. $\begin{cases} 2x - 5y \le 16 \\ x + 3y \ge -3 \end{cases}$

14. $\begin{cases} x^2 + y^2 > 9 \\ x^2 + y^2 < 4 \end{cases}$

15. $\begin{cases} x + y \ge 8 \\ 2x + y \ge 11 \\ x \ge 0, y \ge 0 \end{cases}$

16. $\begin{cases} 2x + 3y \le 12 \\ x + y \le 5 \\ 3x + 2y \le 11 \\ x \ge 0, y \ge 0 \end{cases}$

In Exercises 17 and 18, find the partial fraction decomposition.

17. $\dfrac{3x - 5}{x^2 - 3x - 4}$

18. $\dfrac{2x + 1}{x(x^2 + 1)}$

19. A farmer has 160 acres available on which to plant oats and barley. It costs \$15 per acre for oat seed and \$13 per acre for barley seed. The labor cost is \$15 per acre for oats and \$20 per acre for barley. The farmer has \$2200 available to purchase seed and has set aside \$2600 for labor. The profit per acre for oats is \$120, and the profit per acre for barley is \$150. How many acres of oats should the farmer plant to maximize profit?

20. Find an equation of the circle that passes through the points $(3, 5)$, $(-3, -3)$, and $(4, 4)$. (*Hint:* Use $x^2 + y^2 + ax + by + c = 0$.)

CUMULATIVE REVIEW EXERCISES

1. Find the range of $f(x) = -x^2 + 2x - 4$.

2. Write $\log_6(x - 5) + 3\log_6(2x)$ as a single logarithm with a coefficient of 1.

3. Find the equation in standard form of the parabola that has vertex $(4, 2)$, axis of symmetry parallel to the y-axis, and passes through the point $(-1, 1)$.

4. Let $f(x) = \dfrac{x^2 - 1}{x^4}$. Is f an even function, an odd function, or neither?

5. Solve: $\log x - \log(2x - 3) = 2$

6. Find the equation in standard form of the hyperbola with vertices $(2, 2)$ and $(10, 2)$ and eccentricity 3.

7. Given $g(x) = \dfrac{x - 2}{x}$, find $g\left(-\dfrac{1}{2}\right)$.

8. Given $f(x) = x^2 - 1$ and $g(x) = x^2 - 4x - 2$, find $(f \cdot g)(-2)$.

9. Find a quadratic regression model for the data $\{(1, 1), (2, 3), (3, 10), (4, 17), (5, 26)\}$.

10. Find the polynomial of lowest degree that has zeros of -2, $3i$, and $-3i$.

11. Find the inverse function of $Q(r) = \dfrac{2}{1 - r}$.

12. Find the slant asymptote of the graph of $H(x) = \dfrac{2x^3 - x^2 - 2}{x^2 - x - 1}$.

13. Given that $f(x) = 2^x$ and $g(x) = 3^{2x}$, find $g[f(1)]$.

14. Sketch the graph of $F(x) = \dfrac{2^x - 2^{-x}}{3}$.

15. Find the exact value of $\cos\left(\dfrac{11\pi}{6}\right)$.

16. In triangle ABC, $a = 30$ feet, $b = 25$ feet, and $c = 35$ feet. Find B to the nearest degree.

17. Use a subtraction formula for $\sin(\alpha - \beta)$ to find the exact value of $\sin 15°$.

18. Find the measure of the smallest positive angle between the vectors $\mathbf{v} = 3\mathbf{i} - 2\mathbf{j}$ and $\mathbf{w} = \mathbf{i} + 4\mathbf{j}$. Round to the nearest tenth.

19. Solve: $\cos^{-1} x + \tan^{-1} x = \dfrac{\pi}{2}$

20. Find the rectangular coordinates of the point whose polar coordinates are $\left(10, \dfrac{5\pi}{4}\right)$.

MATRICES

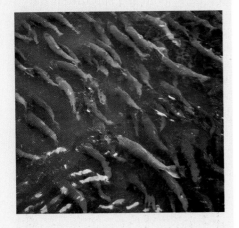

Matrices and Ecology

A *matrix* is a rectangular array of numbers such as the example below. This particular matrix is called a *Leslie* matrix and is used to model an ecological environment.

$$\begin{bmatrix} 0 & 7 & 28 & 84 \\ 0.1 & 0 & 0 & 0 \\ 0 & 0.4 & 0 & 0 \\ 0 & 0 & 0.6 & 0 \end{bmatrix}$$

To see how this matrix might be used, suppose that a biologist is studying a certain type of fish and is considering four stages of life: egg, yearling, 2-year-old, and 3-year-old. Then the numbers in the first row represent the number of eggs laid for each fish in that group. For instance, a 2-year-old fish lays 28 eggs.

The remaining numbers represent the survival rates for each stage. The 0.6 in the last row means that 60% of 2-year-old fish live to be 3 years old.

Various operations, such as addition and multiplication, can be applied to matrices. By applying these procedures, a biologist can predict the survival rates of succeeding generations of fish. If, for instance, a survey of the environment shows that today there are 600 eggs, 200 yearlings, 100 2-year-olds, and 25 3-year-olds, then the matrix product (you will study matrix multiplication in this chapter)

$$\begin{bmatrix} 0 & 7 & 28 & 84 \\ 0.1 & 0 & 0 & 0 \\ 0 & 0.4 & 0 & 0 \\ 0 & 0 & 0.6 & 0 \end{bmatrix}\begin{bmatrix} 600 \\ 200 \\ 100 \\ 25 \end{bmatrix} = \begin{bmatrix} 6300 \\ 60 \\ 80 \\ 60 \end{bmatrix}$$

tells the biologist that 1 year from now, there will be 6300 new eggs, 60 yearlings, 80 2-year-olds, and 60 3-year-olds. Subsequent products would enable the biologist to predict the number of fish in each stage for future years.

In addition to this application, there are applications of matrices to economics, physics, chemistry, and a host of other disciplines. To read about another application, see **Exercise 56, page 863.**

  VIDEO & DVD SSG WWW

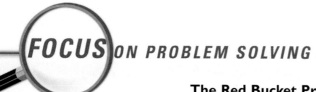

The Red Bucket Procedure

A chemistry laboratory has two buckets of sand that are available to put out any accidental fire that might occur. One bucket is red, and the other is blue.

One day a fire started as a student was working on an experiment. The student quickly put out the fire by pouring the sand from the red bucket onto the flames. Later, the *empty* red bucket was put back on the shelf alongside the blue bucket. Several weeks later another fire broke out. How would you put out this second fire? Most people would use the sand in the blue bucket to put out the second fire; however, another approach would be to pour the sand from the blue bucket into the red bucket, knowing from previous experience that sand from the red bucket can be used to put out a fire. This approach may seem a bit silly, but mathematicians *often* use this procedure of solving a problem by putting the problem in a form for which there is a known solution process.

Keep this red bucket story in mind as you study how to solve a system of equations using the Gaussian elimination method in Section 10.1. The goal of the Gaussian elimination method is to write the system in an equivalent form for which we have a known solution process. That is, to "put the problem in a red bucket."

SECTION 10.1   GAUSSIAN ELIMINATION METHOD

INTRODUCTION TO MATRICES

A **matrix** is a rectangular array of numbers. Each number in a matrix is called an **element** of the matrix. The matrix below, with three rows and four columns, is called a 3×4 (read "3 by 4") matrix.

$$\begin{bmatrix} 2 & 5 & -2 & 5 \\ -3 & 6 & 4 & 0 \\ 1 & 3 & 7 & 2 \end{bmatrix}$$

A matrix of m rows and n columns is said to be of **order $m \times n$** or **dimension $m \times n$**. A **square matrix of order n** is a matrix with n rows and n columns. The matrix above has order 3×4. We will use the notation a_{ij} to refer to the element of a matrix in the ith row and jth column. For the matrix given above, $a_{23} = 4$, $a_{31} = 1$, and $a_{13} = -2$.

The elements $a_{11}, a_{22}, a_{23}, \ldots, a_{mm}$ form the **main diagonal** of a matrix. The elements 2, 6, and 7 form the main diagonal of the matrix shown above.

A matrix can be created from a system of linear equations. Consider the system of linear equations

$$\begin{cases} 2x - 3y + z = 2 \\ x \quad\quad - 3z = 4 \\ 4x - y + 4z = 3 \end{cases}$$

take note

When a term is missing from one of the equations of the system (as in the second equation), the coefficient of that term is 0, and a 0 is entered in the matrix. A vertical bar that separates the coefficients of the variables from the constants is frequently drawn in the matrix.

Using only the coefficients and constants of this system, we can write the 3×4 matrix

$$\begin{bmatrix} 2 & -3 & 1 & | & 2 \\ 1 & 0 & -3 & | & 4 \\ 4 & -1 & 4 & | & 3 \end{bmatrix}$$

This matrix is called the **augmented matrix** of the system of equations. The matrix formed by the coefficients of the system is the **coefficient matrix.** The matrix formed from the constants is the **constant matrix** for the system. The coefficient matrix and constant matrix for the given system are

Coefficient matrix: $\begin{bmatrix} 2 & -3 & 1 \\ 1 & 0 & -3 \\ 4 & -1 & 4 \end{bmatrix}$ Constant matrix: $\begin{bmatrix} 2 \\ 4 \\ 3 \end{bmatrix}$

We can write a system of equations from an augmented matrix.

Augmented matrix: $\begin{bmatrix} 2 & -1 & 4 & | & 3 \\ 1 & 1 & 0 & | & 2 \\ 3 & -2 & -1 & | & 2 \end{bmatrix}$ System: $\longrightarrow$ $\begin{cases} 2x - y + 4z = 3 \\ x + y \quad\quad = 2 \\ 3x - 2y - z = 2 \end{cases}$

In certain cases, an augmented matrix represents a system of equations that we can solve by back substitution. Consider the following augmented matrix and the equivalent system of equations.

$$\begin{bmatrix} 1 & -3 & 4 & 5 \\ 0 & 1 & 2 & -4 \\ 0 & 0 & 1 & -1 \end{bmatrix} \xrightarrow{\text{equivalent system}} \begin{cases} x - 3y + 4z = 5 \\ y + 2z = -4 \\ z = -1 \end{cases}$$

Solving this system by using back substitution, we find that the solution is $(3, -2, -1)$. The matrix above is in *row echelon form*.

Row Echelon Form

A matrix is in **row echelon form** if all the following conditions are satisfied.

1. The first nonzero number in any row is a 1.

2. Rows are arranged so that the column containing the first nonzero number in any row is to the left of the column containing the first nonzero number of the next row.

3. All rows consisting entirely of zeros appear at the bottom of the matrix.

Following are three examples of matrices in row echelon form.

$$\begin{bmatrix} 1 & -3 & 4 & 2 \\ 0 & 1 & -2 & -1 \\ 0 & 0 & 0 & 0 \end{bmatrix} \quad \begin{bmatrix} 1 & 2 & -1 & 3 \\ 0 & 1 & 2 & -1 \end{bmatrix} \quad \begin{bmatrix} 1 & -1 & 3 & 2 \\ 0 & 1 & 2 & 5 \\ 0 & 0 & 1 & -2 \end{bmatrix}$$

❓ QUESTION Is the augmented matrix $\begin{bmatrix} 1 & -2 & 3 & 2 \\ 0 & 0 & 1 & 4 \\ 0 & 1 & -1 & 3 \end{bmatrix}$ in row echelon form?

● ELEMENTARY ROW OPERATIONS

We can write an augmented matrix in row echelon form by using **elementary row operations**. These operations are a rewording, in matrix terminology, of the operations that produce equivalent equations.

INTEGRATING TECHNOLOGY

Many graphing calculators have the elementary row operations as built-in functions. See the Project on page 845 for more details.

Elementary Row Operations

Given the augmented matrix for a system of linear equations, each of the following elementary row operations produces a matrix of an equivalent system of equations.

1. Interchanging any two rows

2. Multiplying all the elements in a row by the same nonzero number

3. Replacing a row by the sum of that row and a nonzero multiple of any other row

---

❓ ANSWER   No. The matrix does not satisfy condition (2) of row echelon form.

It is convenient to specify each operation symbolically as follows:

1. Interchanging the ith and jth rows: $R_i \longleftrightarrow R_j$

2. Multiplying the ith row by k, a nonzero constant: kR_i

3. Replacing the jth row by the sum of that row and a nonzero multiple of the ith row: $kR_i + R_j$

To demonstrate these operations, we will use the 3×3 matrix

$$\begin{bmatrix} 2 & 1 & -2 \\ 3 & -2 & 2 \\ 1 & -2 & 3 \end{bmatrix}$$

$$\begin{bmatrix} 2 & 1 & -2 \\ 3 & -2 & 2 \\ 1 & -2 & 3 \end{bmatrix} \xrightarrow{R_1 \longleftrightarrow R_3} \begin{bmatrix} 1 & -2 & 3 \\ 3 & -2 & 2 \\ 2 & 1 & -2 \end{bmatrix}$$ • Interchange row 1 and row 3.

$$\begin{bmatrix} 2 & 1 & -2 \\ 3 & -2 & 2 \\ 1 & -2 & 3 \end{bmatrix} \xrightarrow{-3R_2} \begin{bmatrix} 2 & 1 & -2 \\ -9 & 6 & -6 \\ 1 & -2 & 3 \end{bmatrix}$$ • Multiply row 2 by −3.

$$\begin{bmatrix} 2 & 1 & -2 \\ 3 & -2 & 2 \\ 1 & -2 & 3 \end{bmatrix} \xrightarrow{-2R_3 + R_1} \begin{bmatrix} 0 & 5 & -8 \\ 3 & -2 & 2 \\ 1 & -2 & 3 \end{bmatrix}$$ • Multiply row 3 by −2 and add to row 1. Replace row 1 by the sum.

In Example 1 we use elementary row operations to write a matrix in row echelon form. As we carry out this procedure, to conserve space, we will occasionally perform more than one elementary row operation in one step. For instance, the notation

$$\begin{array}{c} 3R_1 + R_2 \\ -5R_1 + R_3 \\ \hline \longrightarrow \end{array}$$

means that two elementary row operations were performed. First, multiply row 1 by 3 and add it to row 2. Replace row 2. Second, multiply row 1 by −5 and add it to row 3. Replace row 3.

EXAMPLE 1   Write a Matrix in Row Echelon Form

Write the matrix $\begin{bmatrix} -3 & 13 & -1 & -7 \\ 1 & -5 & 2 & 0 \\ 5 & -20 & -2 & 5 \end{bmatrix}$ in row echelon form.

Solution

Follow the procedure to write a matrix in row echelon form. Change a_{11} to 1.

$$\begin{bmatrix} -3 & 13 & -1 & -7 \\ 1 & -5 & 2 & 0 \\ 5 & -20 & -2 & 5 \end{bmatrix} \xrightarrow{R_1 \longleftrightarrow R_2} \begin{bmatrix} 1 & -5 & 2 & 0 \\ -3 & 13 & -1 & -7 \\ 5 & -20 & -2 & 5 \end{bmatrix}$$

Continued ▶

take note

The sequence of steps used to place a matrix in row echelon form is not unique. For instance, in Example 1 we could have started by multiplying row 1 by $-\frac{1}{3}$. The sequence of steps you use may result in a row echelon form that is different from the one we show. See the Integrating Technology following Example 1.

Change the remaining elements in the first column to 0.

$$\begin{bmatrix} 1 & -5 & 2 & 0 \\ -3 & 13 & -1 & -7 \\ 5 & -20 & -2 & 5 \end{bmatrix} \xrightarrow[\;-5R_1 + R_3\;]{3R_1 + R_2} \begin{bmatrix} 1 & -5 & 2 & 19 \\ 0 & -2 & 5 & -7 \\ 0 & 5 & -12 & 5 \end{bmatrix}$$

Change a_{22} to 1.

$$\begin{bmatrix} 1 & -5 & 2 & 0 \\ 0 & -2 & 5 & -7 \\ 0 & 5 & -12 & 5 \end{bmatrix} \xrightarrow{-\frac{1}{2}R_2} \begin{bmatrix} 1 & -5 & 2 & 0 \\ 0 & 1 & -\frac{5}{2} & \frac{7}{2} \\ 0 & 5 & -12 & 5 \end{bmatrix}$$

Change the remaining elements under a_{22} to 0.

$$\begin{bmatrix} 1 & -5 & 2 & 0 \\ 0 & 1 & -\frac{5}{2} & \frac{7}{2} \\ 0 & 5 & -12 & 5 \end{bmatrix} \xrightarrow{-5R_2 + R_3} \begin{bmatrix} 1 & -5 & 2 & 0 \\ 0 & 1 & -\frac{5}{2} & \frac{7}{2} \\ 0 & 0 & \frac{1}{2} & -\frac{25}{2} \end{bmatrix}$$

Change a_{33} to 1.

$$\begin{bmatrix} 1 & -5 & 2 & 0 \\ 0 & 1 & -\frac{5}{2} & \frac{7}{2} \\ 0 & 0 & \frac{1}{2} & -\frac{25}{2} \end{bmatrix} \xrightarrow{2R_3} \begin{bmatrix} 1 & -5 & 2 & 0 \\ 0 & 1 & -\frac{5}{2} & \frac{7}{2} \\ 0 & 0 & 1 & -25 \end{bmatrix}$$

A row echelon form for the matrix is $\begin{bmatrix} 1 & -5 & 2 & 0 \\ 0 & 1 & -\frac{5}{2} & \frac{7}{2} \\ 0 & 0 & 1 & -25 \end{bmatrix}$.

▶ **TRY EXERCISE 6, PAGE 843**

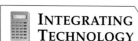

INTEGRATING TECHNOLOGY

A graphing calculator can be used to find a row echelon form for a matrix. The screens below, from a TI-83 Plus calculator, show a row echelon form for the matrix in Example 1. The abbreviation **ref(** stands for *row echelon form*. If you need assistance with keystrokes, please see the website for this text at college.hmco.com.

```
Names Math Edit
1: (A)   3x4
2: (B)
3: (C)
4: (D)
5: (E)
6: (F)
7↓(G)
```

```
Matrix(A)  3x4
[ -3    13    -1    ...
[ 1    -5     2    ...
[ 5    -20    -2    ...
```

```
Names Math Edit
0↑cumSum(
A: ref(
B: rref(
C: rowSwap(
D: row+(
E: *row(
F: *row+(
```

```
ref([A])
[ [1    -4    -.4    1...
 [0     1    -2.4    1...
 [0     0     1    -25...
```

Note that the row echelon form produced by the calculator is different from the one we produced. The two forms are equivalent and either can be used for any calculation for which a row echelon form of the matrix is required.

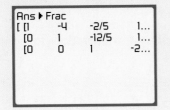
● GAUSSIAN ELIMINATION METHOD

The **Gaussian elimination method** is an algorithm[1] that uses elementary row operations to solve a system of linear equations. The goal of this method is to rewrite an augmented matrix in row echelon form.

We will now demonstrate how to solve a system of two equations in two variables by the Gaussian elimination method. Consider the system of equations

$$\begin{cases} 2x + 5y = -1 \\ 3x - 2y = 8 \end{cases} \quad (1)$$

The augmented matrix for this system is

$$\begin{bmatrix} 2 & 5 & | & -1 \\ 3 & -2 & | & 8 \end{bmatrix}$$

The goal of the Gaussian elimination method is to rewrite the augmented matrix in row echelon form by using elementary row operations. The row operations are chosen so that first, there is a 1 as a_{11}; second, there is a 0 as a_{21}; and third, there is a 1 as a_{22}.

Begin by multiplying row 1 by $\dfrac{1}{2}$. The result is a 1 as a_{11}.

$$\begin{bmatrix} 2 & 5 & | & -1 \\ 3 & -2 & | & 8 \end{bmatrix} \xrightarrow{\frac{1}{2}R_1} \begin{bmatrix} 1 & \frac{5}{2} & | & -\frac{1}{2} \\ 3 & -2 & | & 8 \end{bmatrix}$$

Now multiply row 1 by -3 and add the result to row 2. Replace row 2. The result is a 0 as a_{21}.

$$\begin{bmatrix} 1 & \frac{5}{2} & | & -\frac{1}{2} \\ 3 & -2 & | & 8 \end{bmatrix} \xrightarrow{-3R_1 + R_2} \begin{bmatrix} 1 & \frac{5}{2} & | & -\frac{1}{2} \\ 0 & -\frac{19}{2} & | & \frac{19}{2} \end{bmatrix}$$

Now multiply row 2 by $-\dfrac{2}{19}$. The result is a 1 as a_{22}. The matrix is now in row echelon form.

$$\begin{bmatrix} 1 & \frac{5}{2} & | & -\frac{1}{2} \\ 0 & -\frac{19}{2} & | & \frac{19}{2} \end{bmatrix} \xrightarrow{-\frac{2}{19}R_2} \begin{bmatrix} 1 & \frac{5}{2} & | & -\frac{1}{2} \\ 0 & 1 & | & -1 \end{bmatrix}$$

The system of equations written from the echelon form of the matrix is

$$\begin{cases} x + \dfrac{5}{2}y = -\dfrac{1}{2} \\ \phantom{x + \dfrac{5}{2}}y = -1 \end{cases} \quad (2)$$

To solve by back substitution, replace y in the first equation by -1 and solve for x.

$$x + \left(\dfrac{5}{2}\right)(-1) = -\dfrac{1}{2}$$

$$x = 2$$

The solution of system (1) is $(2, -1)$.

---

[1] An algorithm is a procedure used in calculations. The word is derived from *Al-Khwarizmi,* the name of the author of an Arabic algebra book written around A.D. 825.

EXAMPLE 2   Solve a System of Equations by the
Gaussian Elimination Method

Solve by using the Gaussian elimination method.

$$\begin{cases} 3t - 8u + 8v + 7w = 41 \\ t - 2u + 2v + w = 9 \\ 2t - 2u + 6v - 4w = -1 \\ 2t - 2u + 3v - 3w = 3 \end{cases}$$

Solution

Write the augmented matrix and then use elementary row operations to rewrite the matrix in row echelon form.

$$\begin{bmatrix} 3 & -8 & 8 & 7 & | & 41 \\ 1 & -2 & 2 & 1 & | & 9 \\ 2 & -2 & 6 & -4 & | & -1 \\ 2 & -2 & 3 & -3 & | & 3 \end{bmatrix} \xrightarrow{R_1 \longleftrightarrow R_2} \begin{bmatrix} 1 & -2 & 2 & 1 & | & 9 \\ 3 & -8 & 8 & 7 & | & 41 \\ 2 & -2 & 6 & -4 & | & -1 \\ 2 & -2 & 3 & -3 & | & 3 \end{bmatrix}$$

$$\begin{matrix} -3R_1 + R_2 \\ -2R_1 + R_3 \\ -2R_1 + R_4 \\ \xrightarrow{} \end{matrix} \begin{bmatrix} 1 & -2 & 2 & 1 & | & 9 \\ 0 & -2 & 2 & 4 & | & 14 \\ 0 & 2 & 2 & -6 & | & -19 \\ 0 & 2 & -1 & -5 & | & -15 \end{bmatrix} \xrightarrow{-\frac{1}{2}R_2} \begin{bmatrix} 1 & -2 & 2 & 1 & | & 9 \\ 0 & 1 & -1 & -2 & | & -7 \\ 0 & 2 & 2 & -6 & | & -19 \\ 0 & 2 & -1 & -5 & | & -15 \end{bmatrix}$$

$$\begin{matrix} -2R_2 + R_3 \\ -2R_2 + R_4 \\ \xrightarrow{} \end{matrix} \begin{bmatrix} 1 & -2 & 2 & 1 & | & 9 \\ 0 & 1 & -1 & -2 & | & -7 \\ 0 & 0 & 4 & -2 & | & -5 \\ 0 & 0 & 1 & -1 & | & -1 \end{bmatrix} \xrightarrow{R_4 \longleftrightarrow R_3} \begin{bmatrix} 1 & -2 & 2 & 1 & | & 9 \\ 0 & 1 & -1 & -2 & | & -7 \\ 0 & 0 & 1 & -1 & | & -1 \\ 0 & 0 & 4 & -2 & | & -5 \end{bmatrix}$$

$$\begin{matrix} -4R_3 + R_4 \\ \xrightarrow{} \end{matrix} \begin{bmatrix} 1 & -2 & 2 & 1 & | & 9 \\ 0 & 1 & -1 & -2 & | & -7 \\ 0 & 0 & 1 & -1 & | & -1 \\ 0 & 0 & 0 & 2 & | & -1 \end{bmatrix} \xrightarrow{\frac{1}{2}R_4} \begin{bmatrix} 1 & -2 & 2 & 1 & | & 9 \\ 0 & 1 & -1 & -2 & | & -7 \\ 0 & 0 & 1 & -1 & | & -1 \\ 0 & 0 & 0 & 1 & | & -\frac{1}{2} \end{bmatrix}$$

The last matrix is in row echelon form. The system of equations written from the matrix is

$$\begin{cases} t - 2u + 2v + w = 9 \\ u - v - 2w = -7 \\ v - w = -1 \\ w = -\dfrac{1}{2} \end{cases}$$

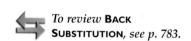

To review **BACK SUBSTITUTION,** *see p. 783.*

Solve by back substitution. The solution is $\left(-\dfrac{13}{2}, -\dfrac{19}{2}, -\dfrac{3}{2}, -\dfrac{1}{2}\right)$.

▶ **TRY EXERCISE 14, PAGE 843**

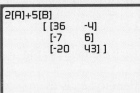

INTEGRATING TECHNOLOGY

A graphing calculator can be used to perform matrix operations. Once the matrices are entered into the calculator, you can use regular arithmetic operations keys and the variable names of the matrices to perform many operations. The screens below, from a TI-83 Plus calculator, show the operations performed in Examples 1 and 2 in this section. For assistance with keystrokes, see your calculator manual or the website for this text at **college.hmco.com.**

Example 1	Example 2a	Example 2b
2[A]+5[B]	[A] [B]	[A] [B]
[[36 -4]	[[-1 4 5 -6...	[[11 -14 2]
[-7 6]	[-4 -6 9 -13...	[4 5 1]
[-20 43]]	[4 2 -11 15...	[8 -15 1]]

Generally, matrix multiplication is not commutative. That is, given two matrices A and B, $AB \neq BA$. In some cases, if we reverse the order of the matrices, the product will not be defined. For instance, if

$$A = \begin{bmatrix} -2 & 3 \\ 1 & 4 \end{bmatrix} \quad \text{and} \quad B = \begin{bmatrix} 5 & 6 & -3 \\ 4 & 1 & 2 \end{bmatrix}, \quad \text{then}$$

$$AB = \begin{bmatrix} -2 & 3 \\ 1 & 4 \end{bmatrix}\begin{bmatrix} 5 & 6 & -3 \\ 4 & 1 & 2 \end{bmatrix} = \begin{bmatrix} 2 & -9 & 12 \\ 21 & 10 & 5 \end{bmatrix}$$

However, BA is undefined because the number of columns of B does not equal the number of rows of A.

$$BA = \begin{bmatrix} 5 & 6 & -3 \\ 4 & 1 & 2 \end{bmatrix}_{2\times3}\begin{bmatrix} -2 & 3 \\ 1 & 4 \end{bmatrix}_{2\times2}$$

$$\text{columns} \neq \text{rows}$$

Even in cases where multiplication is defined, the products AB and BA may not be equal. For instance, if

$$A = \begin{bmatrix} 2 & -3 & 1 \\ 0 & -4 & 3 \\ 5 & 1 & 7 \end{bmatrix} \quad \text{and} \quad B = \begin{bmatrix} 4 & 6 & 1 \\ 5 & 3 & -2 \\ 1 & 0 & 5 \end{bmatrix}, \quad \text{then}$$

$$AB = \begin{bmatrix} 2 & -3 & 1 \\ 0 & -4 & 3 \\ 5 & 1 & 7 \end{bmatrix}\begin{bmatrix} 4 & 6 & 1 \\ 5 & 3 & -2 \\ 1 & 0 & 5 \end{bmatrix} = \begin{bmatrix} -6 & 3 & 13 \\ -17 & -12 & 23 \\ 32 & 33 & 38 \end{bmatrix}$$

$$BA = \begin{bmatrix} 4 & 6 & 1 \\ 5 & 3 & -2 \\ 1 & 0 & 5 \end{bmatrix}\begin{bmatrix} 2 & -3 & 1 \\ 0 & -4 & 3 \\ 5 & 1 & 7 \end{bmatrix} = \begin{bmatrix} 13 & -35 & 29 \\ 0 & -29 & 0 \\ 27 & 2 & 36 \end{bmatrix}$$

Thus, in this example, $AB \neq BA$.

Although matrix multiplication is not a commutative operation, the associative property of multiplication and the distributive property do hold for matrices.

Properties of Matrix Multiplication

Associative property Given matrices A, B, and C of orders $m \times n$, $n \times p$, and $p \times q$, respectively, then

$$A(BC) = (AB)C$$

Distributive property Given matrices A_1 and A_2 of order $m \times n$ and matrices B_1 and B_2 of order $n \times p$, then

$$A_1(B_1 + B_2) = A_1B_1 + A_1B_2 \qquad \text{• Left distributive property}$$
$$(A_1 + A_2)B_1 = A_1B_1 + A_2B_1 \qquad \text{• Right distributive property}$$

A square matrix that has a 1 for each element on the main diagonal and zeros elsewhere is called an *identity matrix*.

The **identity matrix** of order n, denoted I_n, is the $n \times n$ matrix

$$I_n = \begin{bmatrix} 1 & 0 & 0 & \cdots & 0 \\ 0 & 1 & 0 & \cdots & 0 \\ 0 & 0 & 1 & \cdots & 0 \\ \vdots & \vdots & \vdots & \cdots & \vdots \\ 0 & 0 & 0 & \cdots & 1 \end{bmatrix}_{n \times n}$$

The identity matrix has properties similar to those of the real number 1. For example, the product of the matrix A below and I_3 is A.

$$\begin{bmatrix} 2 & -3 & 0 \\ 4 & 7 & -5 \\ 9 & 8 & -6 \end{bmatrix} \begin{bmatrix} 1 & 0 & 0 \\ 0 & 1 & 0 \\ 0 & 0 & 1 \end{bmatrix} = \begin{bmatrix} 2 & -3 & 0 \\ 4 & 7 & -5 \\ 9 & 8 & -6 \end{bmatrix}$$

Multiplicative Identity Property for Matrices

If A is a square matrix of order n, and I_n is the identity matrix of order n, then $AI_n = I_nA = A$.

● MATRIX PRODUCTS AND SYSTEMS OF EQUATIONS

Consider the system of equations

$$\begin{cases} 2x + 3y - z = 5 \\ x - 2y + 2z = 6 \\ 4x + y - 3z = 5 \end{cases}$$

This system can be expressed as a product of matrices, as follows.

$$\begin{bmatrix} 2x + 3y - z \\ x - 2y + 2z \\ 4x + y - 3z \end{bmatrix} = \begin{bmatrix} 5 \\ 6 \\ 5 \end{bmatrix} \qquad \text{• Equality of matrices}$$

$$\begin{bmatrix} 2 & 3 & -1 \\ 1 & -2 & 2 \\ 4 & 1 & -3 \end{bmatrix} \begin{bmatrix} x \\ y \\ z \end{bmatrix} = \begin{bmatrix} 5 \\ 6 \\ 5 \end{bmatrix}$$ • **Definition of matrix multiplication**

Reversing this procedure, certain matrix products can represent systems of equations. Consider the matrix equation

$$\begin{bmatrix} 4 & 3 & -2 \\ 1 & -2 & 3 \\ 1 & 0 & 5 \end{bmatrix}_{3\times3} \begin{bmatrix} x \\ y \\ z \end{bmatrix}_{3\times1} = \begin{bmatrix} 2 \\ -1 \\ 3 \end{bmatrix}_{3\times1}$$

$$\begin{bmatrix} 4x + 3y - 2z \\ x - 2y + 3z \\ x \quad\quad + 5z \end{bmatrix}_{3\times1} = \begin{bmatrix} 2 \\ -1 \\ 3 \end{bmatrix}_{3\times1}$$ • **Definition of matrix multiplication**

$$\begin{cases} 4x + 3y - 2z = 2 \\ x - 2y + 3z = -1 \\ x \quad\quad + 5z = 3 \end{cases}$$ • **Equality of matrices**

Performing operations on matrices that represent a system of equations is another method of solving systems of equations. This is discussed in the next section.

EXAMPLE 3   **Write a System of Equations from a Matrix Equation**

Write the matrix equation $\begin{bmatrix} 2 & 3 & 1 \\ 0 & -1 & 4 \\ 5 & -3 & 4 \end{bmatrix} \begin{bmatrix} x \\ y \\ z \end{bmatrix} = \begin{bmatrix} 0 \\ -2 \\ 8 \end{bmatrix}$ as a system of

equations.

Solution

$$\begin{bmatrix} 2 & 3 & 1 \\ 0 & -1 & 4 \\ 5 & -3 & 4 \end{bmatrix} \begin{bmatrix} x \\ y \\ z \end{bmatrix} = \begin{bmatrix} 0 \\ -2 \\ 8 \end{bmatrix}$$

$$\begin{bmatrix} 2x + 3y + z \\ -y + 4z \\ 5x - 3y + 4z \end{bmatrix} = \begin{bmatrix} 0 \\ -2 \\ 8 \end{bmatrix}$$

Using the equality of matrices, we have the system of equations

$$\begin{cases} 2x + 3y + z = 0 \\ -y + 4z = -2 \\ 5x - 3y + 4z = 8 \end{cases}$$

▶ **TRY EXERCISE 36, PAGE 860**

Matrices often can be used to solve applications in which a sequence of events repeats itself over a period of time. For instance, the following application is from the field of botany.

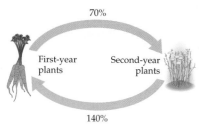

70%

First-year plants   Second-year plants

140%

Yearly transition diagram for biennial plants

FIGURE 10.2

A biennial plant matures 1 year after a seed is planted. In the second year the plant produces seeds that will become the new plants in the third year, and then the 2-year-old plant dies. Suppose that a wilderness area currently contains 500,000 of a certain biennial and that there are 225,000 plants in their first year and 275,000 in their second year. Suppose also that 70% of the 1-year-old plants survive to the second year and that each 1000 second-year plants give rise to 1400 first-year plants. See **Figure 10.2**.

With this information, a botanist can predict how many plants will be in the wilderness area after n years. The formula for the number of plants is given by

$$[225{,}000 \quad 275{,}000]\begin{bmatrix} 0 & 0.7 \\ 1.4 & 0 \end{bmatrix}^n = [\text{first-year plants} \quad \text{second-year plants}]$$

For instance, to find the number of each type of plant in 5 years, the botanist would calculate

$$[225{,}000 \quad 275{,}000]\begin{bmatrix} 0 & 0.7 \\ 1.4 & 0 \end{bmatrix}^5 = [369{,}754 \quad 151{,}263]$$

The resulting matrix indicates that after 5 years there would be 369,754 first-year plants and 151,263 second-year plants.

EXAMPLE 4 Solve an Application

A local trailer rental agency has offices in Tampa and St. Petersburg. To start with, the agency has 40% of its trailers in Tampa and the other 60% in St. Petersburg. The agency finds that each week:

- 86% of the Tampa rentals are returned to the Tampa office, and the other 14% are returned to the St. Petersburg office.

- 81% of the St. Petersburg rentals are returned to the St. Petersburg office, and the other 19% are returned to the Tampa office. See **Figure 10.3**.

The owner of the agency has determined that after n weeks the percent of the trailers that will be at the Tampa office T and the percent of the trailers that will be at the St. Petersburg office S is given by

$$[0.4 \quad 0.6]\begin{bmatrix} 0.86 & 0.14 \\ 0.19 & 0.81 \end{bmatrix}^n = [T \quad S]$$

Find the percent of the trailers that will be at the Tampa office after 3 weeks and after 8 weeks.

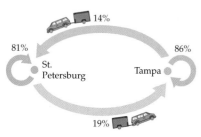

14%

81%   86%

St. Petersburg   Tampa

19%

A transition diagram for the trailer rentals

FIGURE 10.3

Solution

Using a calculator, we find

$$[0.4 \quad 0.6]\begin{bmatrix} 0.86 & 0.14 \\ 0.19 & 0.81 \end{bmatrix}^3 \approx [0.523 \quad 0.477]$$

and

$$[0.4 \quad 0.6]\begin{bmatrix} 0.86 & 0.14 \\ 0.19 & 0.81 \end{bmatrix}^8 \approx [0.569 \quad 0.431]$$

Thus, after 3 weeks, the Tampa office will have about 52.3% of the trailers, and after 8 weeks, the Tampa office will have about 56.9% of the trailers.

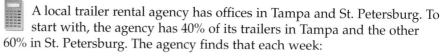

▶ **TRY EXERCISE 56, PAGE 863**

TOPICS FOR DISCUSSION

1. How are matrices related to spreadsheet programs?

2. Is it always possible to add two matrices? If so, explain why. If not, discuss what conditions must be met for two matrices to be added. Is matrix addition a commutative operation?

3. Is it always possible to multiply two matrices? If so, explain why. If not, discuss what conditions must be met for two matrices to be multiplied. Is matrix multiplication a commutative operation?

4. How does scalar multiplication differ from matrix multiplication?

5. Is it possible that two matrices could be added but not multiplied? Can two matrices be multiplied but not added? Discuss what types of conditions must be met for two matrices to be both added and multiplied.

EXERCISE SET 10.2

In Exercises 1 to 8, find *a.* **$A + B$,** *b.* **$A - B$,** *c.* **$2B$, and** *d.* **$2A - 3B$.**

1. $A = \begin{bmatrix} 2 & -1 \\ 3 & 3 \end{bmatrix}$ $B = \begin{bmatrix} -1 & 3 \\ 2 & 1 \end{bmatrix}$

2. $A = \begin{bmatrix} 0 & -2 \\ 2 & 3 \end{bmatrix}$ $B = \begin{bmatrix} 5 & -1 \\ 3 & 0 \end{bmatrix}$

3. $A = \begin{bmatrix} 0 & -1 & 3 \\ 1 & 0 & -2 \end{bmatrix}$ $B = \begin{bmatrix} -3 & 1 & 2 \\ 2 & 5 & -3 \end{bmatrix}$

4. $A = \begin{bmatrix} 2 & -2 & 4 \\ 0 & -3 & -4 \end{bmatrix}$ $B = \begin{bmatrix} 1 & -5 & 6 \\ 4 & -2 & -3 \end{bmatrix}$

5. $A = \begin{bmatrix} -3 & 4 \\ 2 & -3 \\ -1 & 0 \end{bmatrix}$ $B = \begin{bmatrix} 4 & 1 \\ 1 & -2 \\ 3 & -4 \end{bmatrix}$

▶**6.** $A = \begin{bmatrix} 2 & -2 \\ 3 & 4 \\ 1 & 0 \end{bmatrix}$ $B = \begin{bmatrix} -1 & 8 \\ 2 & -2 \\ -4 & 3 \end{bmatrix}$

7. $A = \begin{bmatrix} -2 & 3 & -1 \\ 0 & -1 & 2 \\ -4 & 3 & 3 \end{bmatrix}$ $B = \begin{bmatrix} 1 & -2 & 0 \\ 2 & 3 & -1 \\ 3 & -1 & 2 \end{bmatrix}$

8. $A = \begin{bmatrix} 0 & 2 & 0 \\ 1 & -3 & 3 \\ 5 & 4 & -2 \end{bmatrix}$ $B = \begin{bmatrix} -1 & 2 & 4 \\ 3 & 3 & -2 \\ -4 & 4 & 3 \end{bmatrix}$

In Exercises 9 to 16, find AB and BA if possible.

9. $A = \begin{bmatrix} 2 & -3 \\ 1 & 4 \end{bmatrix}$ $B = \begin{bmatrix} -2 & 4 \\ 2 & -3 \end{bmatrix}$

10. $A = \begin{bmatrix} 3 & -2 \\ 4 & 1 \end{bmatrix}$ $B = \begin{bmatrix} -1 & -1 \\ 0 & 4 \end{bmatrix}$

11. $A = \begin{bmatrix} 3 & -1 \\ 2 & 3 \end{bmatrix}$ $B = \begin{bmatrix} 4 & 1 \\ 2 & -3 \end{bmatrix}$

12. $A = \begin{bmatrix} -3 & 2 \\ 2 & -2 \end{bmatrix}$ $B = \begin{bmatrix} 0 & 2 \\ -2 & 4 \end{bmatrix}$

13. $A = \begin{bmatrix} 2 & -1 \\ 0 & 3 \\ 1 & -2 \end{bmatrix}$ $B = \begin{bmatrix} 1 & -2 & 3 \\ 2 & 0 & 1 \end{bmatrix}$

14. $A = \begin{bmatrix} -1 & 3 \\ 2 & 1 \\ -3 & -2 \end{bmatrix}$ $B = \begin{bmatrix} 0 & -1 & 2 \\ 1 & 2 & -4 \end{bmatrix}$

15. $A = \begin{bmatrix} 2 & -1 & 3 \\ 0 & 2 & -1 \\ 0 & 0 & 2 \end{bmatrix}$ $B = \begin{bmatrix} 2 & 0 & 0 \\ 1 & -1 & 0 \\ 2 & -1 & -2 \end{bmatrix}$

▶**16.** $A = \begin{bmatrix} -1 & 2 & 0 \\ 2 & -1 & 1 \\ -2 & 2 & -1 \end{bmatrix}$ $B = \begin{bmatrix} 2 & -1 & 0 \\ 1 & 5 & -1 \\ 0 & -1 & 3 \end{bmatrix}$

In Exercises 17 to 24, find AB if possible.

17. $A = [1 \quad -2 \quad 3]$ $B = \begin{bmatrix} 1 & 0 \\ 2 & -1 \\ 1 & 2 \end{bmatrix}$

18. $A = \begin{bmatrix} -2 & 3 \\ 1 & -2 \\ 0 & 2 \end{bmatrix}$ $B = \begin{bmatrix} 3 \\ -2 \end{bmatrix}$

19. $A = \begin{bmatrix} 2 & -1 \\ 3 & 3 \end{bmatrix}$ $B = \begin{bmatrix} 1 & -2 \\ 3 & 1 \\ 0 & -2 \end{bmatrix}$

20. $A = \begin{bmatrix} 2 & 0 & -1 \\ 3 & 4 & -3 \end{bmatrix}$ $B = \begin{bmatrix} 3 & -1 & 0 \\ 2 & 4 & 5 \end{bmatrix}$

21. $A = \begin{bmatrix} 2 & 3 \\ -4 & -6 \end{bmatrix}$ $B = \begin{bmatrix} 3 & 6 \\ -2 & -4 \end{bmatrix}$

22. $A = \begin{bmatrix} 2 & -1 & 3 \\ -1 & 2 & 1 \end{bmatrix}$ $B = \begin{bmatrix} 1 & 3 & 2 \\ 2 & -1 & 0 \\ 3 & 1 & 2 \end{bmatrix}$

23. $A = \begin{bmatrix} 1 & 2 & -2 & 3 \\ 0 & -2 & 1 & -3 \end{bmatrix}$ $B = \begin{bmatrix} -2 & 0 \\ 4 & -2 \end{bmatrix}$

24. $A = \begin{bmatrix} 2 & -2 & 4 \\ 1 & 0 & -1 \\ 2 & 1 & 3 \end{bmatrix}$ $B = \begin{bmatrix} 2 & 1 & -3 & 0 \\ 0 & -2 & 1 & -2 \\ 1 & -1 & 0 & 2 \end{bmatrix}$

In Exercises 25 to 28, given the matrices

$$A = \begin{bmatrix} -1 & 3 \\ 2 & -1 \\ 3 & 1 \end{bmatrix} \quad \text{and} \quad B = \begin{bmatrix} 0 & -2 \\ 1 & 3 \\ 4 & -3 \end{bmatrix}$$

find the 3 × 2 matrix X that is a solution of the equation.

25. $3X + A = B$

26. $2A - 3X = 5B$

27. $2X - A = X + B$

28. $3X + 2B = X - 2A$

In Exercises 29 to 32, use the matrices

$$A = \begin{bmatrix} 2 & -3 \\ 1 & -1 \end{bmatrix} \quad \text{and} \quad B = \begin{bmatrix} 3 & -1 & 0 \\ 2 & -2 & -1 \\ 1 & 0 & 2 \end{bmatrix}$$

If A is a square matrix, then $A^n = A \cdot A \cdot A \cdots A$, where the matrix A is repeated n times.

29. Find A^2.

30. Find A^3.

31. Find B^2.

32. Find B^3.

In Exercises 33 to 38, find the system of equations that is equivalent to the given matrix equation.

33. $\begin{bmatrix} 3 & -8 \\ 4 & 3 \end{bmatrix} \begin{bmatrix} x \\ y \end{bmatrix} = \begin{bmatrix} 11 \\ 1 \end{bmatrix}$

34. $\begin{bmatrix} 2 & 7 \\ 3 & -4 \end{bmatrix} \begin{bmatrix} x \\ y \end{bmatrix} = \begin{bmatrix} 1 \\ 16 \end{bmatrix}$

35. $\begin{bmatrix} 1 & -3 & -2 \\ 3 & 1 & 0 \\ 2 & -4 & 5 \end{bmatrix} \begin{bmatrix} x \\ y \\ z \end{bmatrix} = \begin{bmatrix} 6 \\ 2 \\ 1 \end{bmatrix}$

▶ **36.** $\begin{bmatrix} 2 & 0 & 5 \\ 3 & -5 & 1 \\ 4 & -7 & 6 \end{bmatrix} \begin{bmatrix} x \\ y \\ z \end{bmatrix} = \begin{bmatrix} 9 \\ 7 \\ 14 \end{bmatrix}$

37. $\begin{bmatrix} 2 & -1 & 0 & 2 \\ 4 & 1 & 2 & -3 \\ 6 & 0 & 1 & -2 \\ 5 & 2 & -1 & -4 \end{bmatrix} \begin{bmatrix} x_1 \\ x_2 \\ x_3 \\ x_4 \end{bmatrix} = \begin{bmatrix} 5 \\ 6 \\ 10 \\ 8 \end{bmatrix}$

38. $\begin{bmatrix} 5 & -1 & 2 & -3 \\ 4 & 0 & 2 & 0 \\ 2 & -2 & 5 & -4 \\ 3 & 1 & -3 & 4 \end{bmatrix} \begin{bmatrix} x_1 \\ x_2 \\ x_3 \\ x_4 \end{bmatrix} = \begin{bmatrix} -2 \\ 2 \\ -1 \\ 2 \end{bmatrix}$

39. LIFE SCIENCES Biologists use capture-recapture models to estimate how many animals live in a certain area. A sample of, say, fish are caught and tagged. When subsequent samples of fish are caught, a biologist can use a capture history matrix to record (with a 1) which, if any, of the fish in the original sample have been caught again. The rows of this matrix represent the particular fish (each has its own identification number), and the columns represent the number of the sample in which the fish was caught. Here is a small capture history matrix.

Samples

	1	2	3	4
Fish A	1	0	0	1
Fish B	0	1	1	1
Fish C	0	0	1	1

a. What is the dimension of this matrix? Write a sentence that explains the meaning of dimension in this case.

b. What is the meaning of the 1 in row A, column 4?

c. Which fish was captured the most times?

40. LIFE SCIENCE Biologists can use a predator-prey matrix to study the relationships among animals in an ecosystem. Each row and each column represents an animal in the system. A 1 as an element in the matrix indicates that the animal represented by that row preys on the animal represented by that column. A 0 indicates that the animal in that row does not prey on the animal in that column. A simple predator-prey matrix is shown below. The abbreviations are H = hawk, R = rabbit, S = snake, and C = coyote.

$$
\begin{array}{c}
\\
H \\
R \\
S \\
C
\end{array}
\begin{array}{cccc}
H & R & S & C \\
\left[\begin{array}{cccc}
0 & 1 & 1 & 0 \\
0 & 0 & 0 & 0 \\
1 & 1 & 0 & 0 \\
0 & 1 & 1 & 0
\end{array}\right]
\end{array}
$$

a. What is the dimension of this matrix? Write a sentence that explains the meaning of dimension in this case.

b. What is the meaning of the 0 in row R, column H?

c. What is the meaning of there being all zeros in column C?

d. What is the meaning of all zeros in row R?

41. BUSINESS The matrix below shows the sales revenues, in millions of dollars, that a pharmaceutical company received from various divisions in different parts of the country. The abbreviations are W = western states, N = northern states, S = southern states, and E = eastern states.

$$
\begin{array}{c}
\\
\text{Patented drugs} \\
\text{Generic drugs} \\
\text{Nonprescription drugs}
\end{array}
\begin{array}{cccc}
W & N & S & E \\
\left[\begin{array}{cccc}
2.0 & 1.4 & 3.0 & 1.4 \\
0.8 & 1.1 & 2.0 & 0.9 \\
3.6 & 1.2 & 4.5 & 1.5
\end{array}\right]
\end{array}
$$

Suppose the business plan for this company indicates that it anticipates a 2% decrease in sales (because of competition) for each of its drug divisions for each region of the country. Express, to the nearest ten thousand dollars, this matrix as a scalar product, and compute the anticipated sales matrix.

42. SALARY SCHEDULES The partial current-year salary matrix for an elementary school district is given below. Column A indicates a B.A. degree, column B a B.A. degree plus 15 graduate units, column C an M.A. degree, and column D an M.A. degree plus 30 additional graduate units. The rows give the numbers of years of teaching experience. Each entry is the annual salary in thousands of dollars.

$$
\begin{array}{c}
\\
0\text{ to }4 \\
\text{Years} \quad 5\text{ to }9 \\
10\text{ to }15
\end{array}
\begin{array}{cccc}
A & B & C & D \\
\left[\begin{array}{cccc}
18.0 & 18.9 & 20.0 & 21.5 \\
19.0 & 20.3 & 22.5 & 24.5 \\
20.0 & 21.4 & 24.0 & 27.0
\end{array}\right]
\end{array}
$$

Express, as a matrix scalar multiplication to the nearest hundred dollars, the result of the school board's approving a 6% salary increase for all teachers in this district, and compute the scalar product.

43. SPORTS The matrices for the numbers of wins and losses at home, H, and away, A, are shown for the top three finishers of the 2000 American League East division baseball teams.

$$
H = \begin{array}{c}
\\
\\
\\
\end{array}
\begin{array}{cc}
W & L \\
\left[\begin{array}{cc}
44 & 36 \\
42 & 39 \\
45 & 36
\end{array}\right]
\end{array}
\begin{array}{c}
\text{New York} \\
\text{Boston} \\
\text{Toronto}
\end{array}
\qquad
A = \begin{array}{cc}
W & L \\
\left[\begin{array}{cc}
43 & 38 \\
43 & 38 \\
38 & 43
\end{array}\right]
\end{array}
\begin{array}{c}
\text{New York} \\
\text{Boston} \\
\text{Toronto}
\end{array}
$$

a. Find $H + A$.

b. Write a sentence that explains the meaning of the sum of the two matrices.

c. Find $H - A$.

d. Write a sentence that explains the meaning of the difference of the two matrices.

44. BUSINESS Let A represent the number of televisions of various sizes in two stores of a company in one city, and let B represent the same situation for the company in a second city.

$$
A = \begin{array}{ccc}
\text{19-inch} & \text{25-inch} & \text{40-inch} \\
\left[\begin{array}{ccc}
23 & 35 & 49 \\
32 & 41 & 24
\end{array}\right]
\end{array}
\begin{array}{c}
\text{Store 1} \\
\text{Store 2}
\end{array}
$$

$$
B = \begin{array}{ccc}
\text{19-inch} & \text{25-inch} & \text{40-inch} \\
\left[\begin{array}{ccc}
19 & 28 & 36 \\
25 & 38 & 26
\end{array}\right]
\end{array}
\begin{array}{c}
\text{Store 1} \\
\text{Store 2}
\end{array}
$$

a. Find $A + B$.

b. Write a sentence that explains the meaning of the sum of the two matrices.

45. GEOMETRIC TRANSFORMATION Consider the rectangle shown below. Each pair of x- and y-coordinates of the points shown appears as a column of a matrix of dimension 2×4. This is matrix A below. Matrix T is called a *translation matrix*.

$$
A = \begin{bmatrix}
-2 & 4 & 2 & -4 \\
5 & 2 & -2 & 1
\end{bmatrix}
$$

$$
T = \begin{bmatrix}
2 & 2 & 2 & 2 \\
-1 & -1 & -1 & -1
\end{bmatrix}
$$

a. Find $A + T$.

b. Using the columns of $A + T$ as the x- and y-coordinates of four points, plot the points on the same coordinate grid as the original rectangle. Construct a polygon by connecting the points in the order of the columns. What polygon have you constructed?

c. Write a sentence that explains the relationship between the original polygon and the new one.

46. GEOMETRIC TRANSFORMATION Use the information in Exercise 45.

a. Find $A - T$.

b. Using the columns of $A - T$ as the x- and y-coordinates of four points, plot the points on the same coordinate grid as the original rectangle. Construct a polygon by connecting the points in the order of the columns. What polygon have you constructed?

c. Write a sentence that explains the relationship between the original polygon and the new one.

47. GEOMETRIC TRANSFORMATION Consider the information in Exercise 45 and the matrix $R = \begin{bmatrix} 0 & -1 \\ 1 & 0 \end{bmatrix}$.

a. Find $R \cdot A$.

b. Using the columns of $R \cdot A$ as the x- and y-coordinates of four points, plot the points on the same coordinate grid as the original rectangle. Construct a polygon by connecting the points in the order of the columns. What polygon have you constructed?

c. Write a sentence that explains the meaning of the product of these two matrices.

48. GEOMETRIC TRANSFORMATION Consider the information in Exercise 45 and the matrix $R = \begin{bmatrix} 0 & 1 \\ 1 & 0 \end{bmatrix}$.

a. Find $R \cdot A$.

b. Using the columns of $R \cdot A$ as the x- and y-coordinates of four points, plot the points on the same coordinate grid as the original rectangle. Construct a polygon by connecting the points in the order of the columns. What polygon have you constructed?

c. Write a sentence that explains the meaning of the product of these two matrices.

49. BUSINESS INVENTORY Matrix A gives the stock on hand of four products in a warehouse at the beginning of the week, and matrix B gives the stock on hand for the same

four items at the end of the week. Find and interpret $A - B$.

$$A = \begin{matrix} & \text{Blue} & \text{Green} & \text{Red} \\ \begin{bmatrix} & 530 & 650 & 815 \\ & 190 & 385 & 715 \\ & 485 & 600 & 610 \\ & 150 & 210 & 305 \end{bmatrix} & \begin{matrix} \text{Pens} \\ \text{Pencils} \\ \text{Ink} \\ \text{Colored lead} \end{matrix} \end{matrix}$$

$$B = \begin{matrix} & \text{Blue} & \text{Green} & \text{Red} \\ \begin{bmatrix} & 480 & 500 & 675 \\ & 175 & 215 & 345 \\ & 400 & 350 & 480 \\ & 70 & 95 & 280 \end{bmatrix} & \begin{matrix} \text{Pens} \\ \text{Pencils} \\ \text{Ink} \\ \text{Colored lead} \end{matrix} \end{matrix}$$

50. BUSINESS SERVICES Matrix A gives the numbers of employees in the divisions of a company in the west coast branch, and matrix B gives the same information for the east coast branch. Find and interpret $A + B$.

$$A = \begin{matrix} & \text{Engineering} & \begin{matrix}\text{Admini-}\\\text{stration}\end{matrix} & \begin{matrix}\text{Data}\\\text{Processing}\end{matrix} \\ \begin{bmatrix} & 315 & 200 & 415 \\ & 285 & 175 & 300 \\ & 275 & 195 & 250 \end{bmatrix} & \begin{matrix} \text{Division I} \\ \text{Division II} \\ \text{Division III} \end{matrix} \end{matrix}$$

$$B = \begin{matrix} & \text{Engineering} & \begin{matrix}\text{Admini-}\\\text{stration}\end{matrix} & \begin{matrix}\text{Data}\\\text{Processing}\end{matrix} \\ \begin{bmatrix} & 200 & 175 & 350 \\ & 150 & 90 & 180 \\ & 105 & 50 & 175 \end{bmatrix} & \begin{matrix} \text{Division I} \\ \text{Division II} \\ \text{Division III} \end{matrix} \end{matrix}$$

51. STOCK MARKET The commission rates that three companies T_1, T_2, and T_3 charge to sell one share of stock on the New York Stock Exchange (NYSE), NASDQ, and the American Stock Exchange (ASE) are shown in the matrix below.

$$C = \begin{matrix} & \text{NYSE} & \text{NASDQ} & \text{ASE} \\ \begin{bmatrix} & 0.04 & 0.06 & 0.05 \\ & 0.04 & 0.04 & 0.04 \\ & 0.03 & 0.07 & 0.06 \end{bmatrix} & \begin{matrix} T_1 \\ T_2 \\ T_3 \end{matrix} \end{matrix}$$

Two customers, S_1 and S_2, own stocks that are traded on the NYSE, NASDQ, and the ASE. The numbers of shares of stocks they own for each exchange are given in the following matrix.

$$S = \begin{matrix} & S_1 & S_2 \\ \begin{bmatrix} & 500 & 600 \\ & 250 & 450 \\ & 600 & 750 \end{bmatrix} & \begin{matrix} \text{NYSE} \\ \text{NASDQ} \\ \text{ASE} \end{matrix} \end{matrix}$$

Find CS. Which company should customer S_1 use to minimize commission costs?

52. YOUTH SPORTS The total unit sales matrix at three soccer games in a summer league for children is given by

$$S = \begin{array}{c} \\ \\ \end{array} \begin{array}{cccc} \text{Soft} & \text{Hot} & & \\ \text{Drinks} & \text{Dogs} & \text{Candy} & \text{Popcorn} \\ \left[\begin{array}{cccc} 52 & 50 & 75 & 20 \\ 45 & 48 & 80 & 20 \\ 62 & 70 & 78 & 25 \end{array}\right] & \begin{array}{l} \text{Game 1} \\ \text{Game 2} \\ \text{Game 3} \end{array} \end{array}$$

The unit pricing matrix in dollars for the wholesale cost of each item and the retail price of each item is given by

$$P = \begin{array}{c} \\ \end{array} \begin{array}{cc} \text{Wholesale} & \text{Retail} \\ \left[\begin{array}{cc} 0.25 & 0.50 \\ 0.30 & 0.75 \\ 0.15 & 0.45 \\ 0.10 & 0.50 \end{array}\right] & \begin{array}{l} \text{Soft drinks} \\ \text{Hot dogs} \\ \text{Candy} \\ \text{Popcorn} \end{array} \end{array}$$

Use matrix multiplication to find the total cost and total revenue at each game.

53. CONSUMER PREFERENCES A soft drink company has determined that every six months 1.1% of its customers switch from regular soda to diet soda and 0.7% of its customers switch from diet soda to regular soda. At the present time, 55% of its customers drink regular soda and 45% drink diet soda. After n 6-month periods, the percent of its customers who drink regular soda r and the percent of its customers who drink diet soda d are given by

$$[0.55 \quad 0.45]\begin{bmatrix} 0.989 & 0.011 \\ 0.007 & 0.993 \end{bmatrix}^n = [r \quad d]$$

Use a calculator and the above matrix equation to predict the percent, to the nearest 0.1%, of the customers who will be drinking diet soda

a. 1 year from now.

b. 3 years from now.

54. CONSUMER PREFERENCES Experiment with several different values of n in the matrix equation from Exercise 53 to determine how long, to the nearest year, it will be before the diet soda drinkers outnumber the regular soda drinkers.

55. CONSUMER PREFERENCES A video store has determined that every month 0.8% of its customers switch from VHS to DVD and only 0.1% of its customers switch from DVD to VHS. At the present time, 88% of its customers rent VHS movies, and

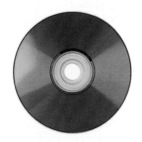

the other 12% rent DVD movies. After n months, the percent of its customers who rent VHS movies V and the percent of its customers who rent DVD movies D are given by

$$[0.88 \quad 0.12]\begin{bmatrix} 0.992 & 0.008 \\ 0.001 & 0.999 \end{bmatrix}^n = [V \quad D]$$

Use a calculator and the above matrix equation to predict the percent, to the nearest 0.1%, of the customers who will be renting DVD movies

a. 6 months from now.

b. 18 months from now.

▶ **56. CONSUMER PREFERENCES** A town has two grocery stores, A and B. Each month

• store A retains 98% of its customers and loses 2% to Store B.

• store B retains 95% of its customers and loses 5% to Store A.

To start with, store B has 75% of town's customers, and store A has the other 25%. After n months, the percent of the customers who shop at store A, denoted by a, and the percent of the customers who shop at store B, denoted by b, are given by

$$[0.25 \quad 0.75]\begin{bmatrix} 0.98 & 0.02 \\ 0.05 & 0.95 \end{bmatrix}^n = [a \quad b]$$

Find the percent, to the nearest 0.1%, of the customers who shop at store A after 5 months.

57. CONSUMER PREFERENCES Experiment with several different values of n in the matrix equation from Exercise 56 to determine how long, to the nearest month, it will be before store A has 50% of the town's customers.

58. PLANT REPRODUCTION A biennial plant matures 1 year after a seed is planted. In the second year, the plant produces seeds that will become the new plants in the third year, and then the 2-year-old plant dies. Suppose that an animal preserve currently contains 850,000 of a certain biennial and that there are 475,000 plants in their first year and 375,000 in their second year. Suppose also that 65% of the 1-year-old plants survive to the second year and that each 1000 second-year plants give rise to 1250 first-year plants. Write a matrix product that will predict the number of plants that will be in the animal preserve in n years. Use this product to determine the number of these plants in the preserve after 4 years. Round to the nearest thousand.

In Exercises 59 to 64, use a graphing calculator to perform the indicated operations on matrices A and B.

$$A = \begin{bmatrix} 2 & -1 & 3 & 5 & -1 \\ 2 & 0 & 2 & -1 & 1 \\ -1 & -3 & 2 & 3 & 3 \\ 5 & -4 & 1 & 0 & 3 \\ 0 & 2 & -1 & 4 & 3 \end{bmatrix}$$

$$B = \begin{bmatrix} 0 & -2 & 1 & 7 & 2 \\ -3 & 0 & 2 & 3 & 1 \\ -2 & 1 & 1 & 4 & 5 \\ 6 & 4 & -4 & 2 & -3 \\ 3 & -2 & -5 & 1 & 3 \end{bmatrix}$$

59. AB

60. BA

61. A^3

62. B^3

63. $A^2 + B^2$

64. $AB - BA$

CONNECTING CONCEPTS

The elements of a matrix can be complex numbers. In Exercises 65 to 74, let

$$A = \begin{bmatrix} 2 + 3i & 1 - 2i \\ 1 + i & 2 - i \end{bmatrix} \quad \text{and} \quad B = \begin{bmatrix} 1 - i & 2 + 3i \\ 3 + 2i & 4 - i \end{bmatrix}$$

Perform the indicated operations.

65. $3A$

66. $-2B$

67. $2iB$

68. $3iA$

69. $A + B$

70. $A - B$

71. AB

72. BA

73. A^2

74. B^2

Matrices with complex number elements play a role in the theory of the atom. The following three matrices, called Pauli spin matrices, were used by Wolfgang Pauli in his early study of the electron. Use these matrices in Exercises 75 to 77.

$$\sigma_1 = \begin{bmatrix} 0 & 1 \\ 1 & 0 \end{bmatrix} \quad \sigma_2 = \begin{bmatrix} 0 & -i \\ i & 0 \end{bmatrix} \quad \sigma_3 = \begin{bmatrix} 1 & 0 \\ 0 & -1 \end{bmatrix}$$

75. Show that $(\sigma_i)^2 = I_2$ for $i = 1, 2$, and 3.

76. Show that $\sigma_1 \cdot \sigma_2 = i\sigma_3$.

77. Show that $\sigma_1 \cdot \sigma_2 + \sigma_2 \cdot \sigma_1 = O$.

78. Given two real numbers a and b and a matrix A of order 2×2, prove that $(a + b)A = aA + bA$.

79. Given two real numbers a and b and a matrix A of order 2×2, prove that $a(bA) = (ab)A$.

PREPARE FOR SECTION 10.3

80. What is the multiplicative inverse of $-\dfrac{2}{3}$? [P.1]

81. Write the 3×3 multiplicative identity matrix. [10.2]

82. State the three elementary row operations for matrices. [10.1]

83. Complete the following:

$$\begin{bmatrix} 1 & -2 & 3 \\ 2 & -1 & 4 \\ -3 & 2 & 2 \end{bmatrix} \xrightarrow[\begin{array}{c} -2R_1 + R_2 \\ 3R_1 + R_3 \end{array}]{} \begin{bmatrix} & & \\ & ? & \\ & & \end{bmatrix}$$

[10.1]

84. Solve for X: $AX = B$. Write the answer using negative exponents. [P.2/1.1]

85. What system of equations is represented by the matrix equation

$$\begin{bmatrix} 2 & 3 \\ 4 & -5 \end{bmatrix} \begin{bmatrix} x \\ y \end{bmatrix} = \begin{bmatrix} 9 \\ 7 \end{bmatrix}? \quad [10.2]$$

PROJECTS

I.   **MATRICES IN GRAPHICS ART** Matrices can be used to translate and dilate geometric figures in the plane. These concepts are important in graphics design and art. Consider the kite shown below.

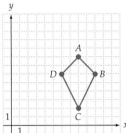

a. Find the coordinates of each vertex of the kite, and prepare a matrix K of dimension 2×4 in which the columns are the x- and y-coordinates, respectively, of the vertices of the kite.

b. Find the product $\frac{1}{2}K = M$. Plot the points of M using the columns as the x- and y-coordinates, respectively, of the vertices of a new kite.

c. Find the lengths of any two line segments of the original kite and the corresponding lengths for the new

kite. Show that the **scale factor**, which is the ratio $\dfrac{\text{length of new}}{\text{length of original}}$, is $\dfrac{1}{2}$ for each pair of line segments. Observe that this is the coefficient of K in the product in part **b**.

d. Find the product $2K = N$. Plot the points of N again, using the columns as the x- and y-coordinates, respectively, of the vertices of a new kite. Show that the new kite has a scale factor of 2.

e. Consider the vertices A, B, C, and D of the original kite and the corresponding vertices A', B', C', and D' of the kite from part **c**. Show that the lines through AA', BB', CC', and DD' pass through the origin of the coordinate system. This point of intersection is called the **center of dilatation**.

2.   See our website at **http://college.hmco.com** for a program that animates translation of geometric figures. *Note:* A TI-83 graphing calculator program is available to allow you to observe the effects of translation on a geometric figure. This program, TRANSLATE, can be found on our website at **http://college.hmco.com**.

SECTION 10.3

THE INVERSE OF A MATRIX

- **FINDING THE INVERSE OF A MATRIX**
- **SOLVING SYSTEMS OF EQUATIONS USING INVERSE MATRICES**
- **INPUT-OUTPUT ANALYSIS**

● FINDING THE INVERSE OF A MATRIX

Recall that the multiplicative inverse of a nonzero real number c is $\dfrac{1}{c}$, the number whose product with c is 1. For example, the multiplicative inverse of $\dfrac{2}{3}$ is $\dfrac{3}{2}$ because $\dfrac{2}{3} \cdot \dfrac{3}{2} = 1$.

For some square matrices we can define a multiplicative inverse.

Multiplicative Inverse of a Matrix

If A is a square matrix of order n, then the **inverse** of matrix A, denoted by A^{-1}, has the property that
$$A \cdot A^{-1} = A^{-1} \cdot A = I_n$$
where I_n is the identity matrix of order n.

As we will see shortly, not all square matrices have a multiplicative inverse.

? QUESTION Are there any real numbers that do not have a multiplicative inverse?

A procedure for finding the inverse (we will simply say *inverse* for *multiplicative inverse*) uses elementary row operations. The procedure will be illustrated by finding the inverse of a 2×2 matrix.

Let $A = \begin{bmatrix} 2 & 7 \\ 1 & 4 \end{bmatrix}$. To the matrix A we will merge the identity matrix I_2 to the right of A and denote this new matrix by $[A; I_2]$.

$$[A; I_2] = \begin{bmatrix} 2 & 7 & 1 & 0 \\ 1 & 4 & 0 & 1 \end{bmatrix}$$

$$A \longrightarrow \uparrow \qquad \uparrow \longrightarrow I_2$$

Now we use elementary row operations in a manner similar to that of the Gaussian elimination method. The goal is to produce

$$[I_2; A^{-1}] = \begin{bmatrix} 1 & 0 & b_{11} & b_{12} \\ 0 & 1 & b_{21} & b_{22} \end{bmatrix}$$

$$I_2 \longrightarrow \uparrow \qquad \uparrow \longrightarrow A^{-1}$$

In this form, the inverse matrix is the matrix that is to the right of the identity matrix. That is,

$$A^{-1} = \begin{bmatrix} b_{11} & b_{12} \\ b_{21} & b_{22} \end{bmatrix}$$

To find A^{-1}, we first use a series of elementary row operations that will result in a 1 in the first row and the first column.

$$\begin{bmatrix} 2 & 7 & 1 & 0 \\ 1 & 4 & 0 & 1 \end{bmatrix} \xrightarrow{\frac{1}{2}R_1} \begin{bmatrix} 1 & \frac{7}{2} & \frac{1}{2} & 0 \\ 1 & 4 & 0 & 1 \end{bmatrix} \xrightarrow{-1R_1 + R_2} \begin{bmatrix} 1 & \frac{7}{2} & \frac{1}{2} & 0 \\ 0 & \frac{1}{2} & -\frac{1}{2} & 1 \end{bmatrix}$$

$$\xrightarrow{2R_2} \begin{bmatrix} 1 & \frac{7}{2} & \frac{1}{2} & 0 \\ 0 & 1 & -1 & 2 \end{bmatrix} \xrightarrow{-\frac{7}{2}R_2 + R_1} \begin{bmatrix} 1 & 0 & 4 & -7 \\ 0 & 1 & -1 & 2 \end{bmatrix}$$

The inverse matrix is the matrix to the right of the identity matrix. Therefore,

$$A^{-1} = \begin{bmatrix} 4 & -7 \\ -1 & 2 \end{bmatrix}$$

Each elementary row operation is chosen to advance the process of transforming the original matrix into the identity matrix.

? ANSWER The real number zero does not have a multiplicative inverse.

EXAMPLE I   **Find the Inverse of a 3 × 3 Matrix**

Find the inverse of the matrix $A = \begin{bmatrix} 1 & -1 & 2 \\ 2 & 0 & 6 \\ 3 & -5 & 7 \end{bmatrix}$.

Solution

$$\begin{bmatrix} 1 & -1 & 2 & | & 1 & 0 & 0 \\ 2 & 0 & 6 & | & 0 & 1 & 0 \\ 3 & -5 & 7 & | & 0 & 0 & 1 \end{bmatrix}$$

• Merge the given matrix with the identity matrix I_3.

$\begin{matrix} -2R_1 + R_2 \\ -3R_1 + R_3 \\ \longrightarrow \end{matrix} \begin{bmatrix} 1 & -1 & 2 & | & 1 & 0 & 0 \\ 0 & 2 & 2 & | & -2 & 1 & 0 \\ 0 & -2 & 1 & | & -3 & 0 & 1 \end{bmatrix}$

• Because a_{11} is already 1, we next produce zeros in a_{21} and a_{31}.

$\begin{matrix} \frac{1}{2}R_2 \\ \longrightarrow \end{matrix} \begin{bmatrix} 1 & -1 & 2 & | & 1 & 0 & 0 \\ 0 & 1 & 1 & | & -1 & \frac{1}{2} & 0 \\ 0 & -2 & 1 & | & -3 & 0 & 1 \end{bmatrix}$

• Produce a 1 in a_{22}.

$\begin{matrix} 2R_2 + R_3 \\ \longrightarrow \end{matrix} \begin{bmatrix} 1 & -1 & 2 & | & 1 & 0 & 0 \\ 0 & 1 & 1 & | & -1 & \frac{1}{2} & 0 \\ 0 & 0 & 3 & | & -5 & 1 & 1 \end{bmatrix}$

• Produce a 0 in a_{32}.

$\begin{matrix} \frac{1}{3}R_3 \\ \longrightarrow \end{matrix} \begin{bmatrix} 1 & -1 & 2 & | & 1 & 0 & 0 \\ 0 & 1 & 1 & | & -1 & \frac{1}{2} & 0 \\ 0 & 0 & 1 & | & -\frac{5}{3} & \frac{1}{3} & \frac{1}{3} \end{bmatrix}$

• Produce a 1 in a_{33}.

$\begin{matrix} -1R_3 + R_2 \\ -2R_3 + R_1 \\ \longrightarrow \end{matrix} \begin{bmatrix} 1 & -1 & 0 & | & \frac{13}{3} & -\frac{2}{3} & -\frac{2}{3} \\ 0 & 1 & 0 & | & \frac{2}{3} & \frac{1}{6} & -\frac{1}{3} \\ 0 & 0 & 1 & | & -\frac{5}{3} & \frac{1}{3} & \frac{1}{3} \end{bmatrix}$

• Now work upward. Produce a 0 in a_{23} and a_{13}.

$\begin{matrix} R_2 + R_1 \\ \longrightarrow \end{matrix} \begin{bmatrix} 1 & 0 & 0 & | & 5 & -\frac{1}{2} & -1 \\ 0 & 1 & 0 & | & \frac{2}{3} & \frac{1}{6} & -\frac{1}{3} \\ 0 & 0 & 1 & | & -\frac{5}{3} & \frac{1}{3} & \frac{1}{3} \end{bmatrix}$

• Produce a 0 in a_{12}.

The inverse matrix is $A^{-1} = \begin{bmatrix} 5 & -\frac{1}{2} & -1 \\ \frac{2}{3} & \frac{1}{6} & -\frac{1}{3} \\ -\frac{5}{3} & \frac{1}{3} & \frac{1}{3} \end{bmatrix}$.

You should verify that this matrix satisfies the condition of an inverse matrix. That is, show that $A^{-1} \cdot A = A \cdot A^{-1} = I_3$.

▶ **TRY EXERCISE 6, PAGE 874**

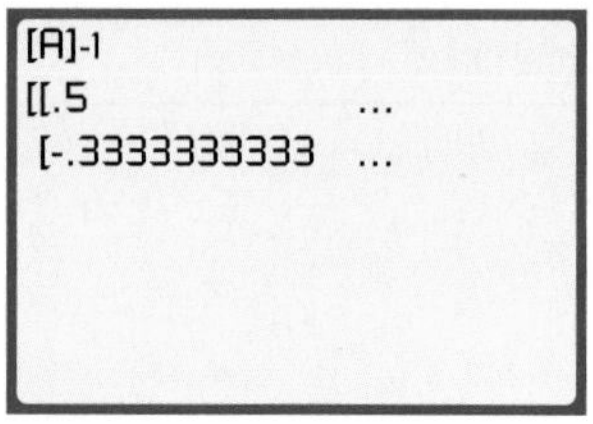

[A]⁻¹
[[.5   ...
 [-.3333333333 ...

FIGURE 10.4a

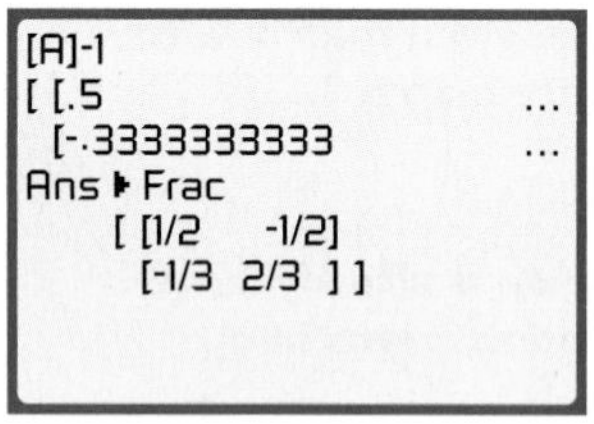

[A]⁻¹
[[.5   ...
 [-.3333333333 ...
Ans ▶ Frac
   [[1/2 -1/2]
    [-1/3 2/3]]

FIGURE 10.4b

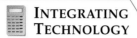

INTEGRATING TECHNOLOGY

The inverse of a matrix can be found by using a graphing calculator. Enter and store the matrix in, say, [A]. To compute the inverse of A, use the

$\boxed{x^{-1}}$ key. For instance, let $A = \begin{bmatrix} 4 & 3 \\ 2 & 3 \end{bmatrix}$. A typical calculator display of the

inverse of A is shown in **Figure 10.4a.** Because the elements of the matrix are decimals, it is possible to see only the first column of the inverse matrix. Use the arrow keys to see the remaining columns.

Another possibility for viewing the inverse of A is to use the function on your calculator that converts a decimal to a fraction. This will change the decimals to fractions, as in **Figure 10.4b.**

A **singular matrix** is a matrix that does not have a multiplicative inverse. A matrix that has a multiplicative inverse is a **nonsingular matrix.** As you apply elementary row operations to a singular matrix, there will come a point where there are all zeros in a row of the *original* matrix. When that condition exists, the original matrix does not have an inverse.

EXAMPLE 2 Identify a Singular Matrix

Show that the matrix $\begin{bmatrix} 1 & -1 & -1 \\ 2 & -3 & 0 \\ 1 & -2 & 1 \end{bmatrix}$ is a singular matrix.

Solution

$$\begin{bmatrix} 1 & -1 & -1 & | & 1 & 0 & 0 \\ 2 & -3 & 0 & | & 0 & 1 & 0 \\ 1 & -2 & 1 & | & 0 & 0 & 1 \end{bmatrix} \xrightarrow[-1R_1 + R_3]{-2R_1 + R_2} \begin{bmatrix} 1 & -1 & -1 & | & 1 & 0 & 0 \\ 0 & -1 & 2 & | & -2 & 1 & 0 \\ 0 & -1 & 2 & | & -1 & 0 & 1 \end{bmatrix}$$

$$\xrightarrow{-1 \cdot R_2} \begin{bmatrix} 1 & -1 & -1 & | & 1 & 0 & 0 \\ 0 & 1 & -2 & | & 2 & -1 & 0 \\ 0 & -1 & 2 & | & -1 & 0 & 1 \end{bmatrix} \xrightarrow{R_2 + R_3} \begin{bmatrix} 1 & -1 & -1 & | & -1 & 0 & 0 \\ 0 & 1 & -2 & | & 2 & -1 & 0 \\ 0 & 0 & 0 & | & 1 & -1 & 1 \end{bmatrix}$$

There are zeros in a row of the original matrix. The original matrix does not have an inverse.

▶ **TRY EXERCISE 10, PAGE 874**

● SOLVING SYSTEMS OF EQUATIONS USING INVERSE MATRICES

Systems of linear equations can be solved by finding the inverse of the coefficient matrix. Consider the following system of equations.

$$\begin{cases} 3x_1 + 4x_2 = -1 \\ 3x_1 + 5x_2 = 1 \end{cases} \quad (1)$$

Using matrix multiplication and the concept of equality of matrices, we can write this system as a matrix equation.

$$\begin{bmatrix} 3 & 4 \\ 3 & 5 \end{bmatrix}\begin{bmatrix} x_1 \\ x_2 \end{bmatrix} = \begin{bmatrix} -1 \\ 1 \end{bmatrix} \quad (2)$$

If we let

$$A = \begin{bmatrix} 3 & 4 \\ 3 & 5 \end{bmatrix} \qquad X = \begin{bmatrix} x_1 \\ x_2 \end{bmatrix} \qquad B = \begin{bmatrix} -1 \\ 1 \end{bmatrix}$$

then Equation (2) can be written as $AX = B$. The inverse of the coefficient matrix A is $A^{-1} = \begin{bmatrix} \frac{5}{3} & -\frac{4}{3} \\ -1 & 1 \end{bmatrix}$.

To solve the system of equations, multiply each side of the equation $AX = B$ by the inverse A^{-1}.

$$\begin{bmatrix} \frac{5}{3} & -\frac{4}{3} \\ -1 & 1 \end{bmatrix}\begin{bmatrix} 3 & 4 \\ 3 & 5 \end{bmatrix}\begin{bmatrix} x_1 \\ x_2 \end{bmatrix} = \begin{bmatrix} \frac{5}{3} & -\frac{4}{3} \\ -1 & 1 \end{bmatrix}\begin{bmatrix} -1 \\ 1 \end{bmatrix}$$

$$\begin{bmatrix} x_1 \\ x_2 \end{bmatrix} = \begin{bmatrix} -3 \\ 2 \end{bmatrix}$$

Thus $x_1 = -3$ and $x_2 = 2$. The solution to System (1) is $(-3, 2)$.

EXAMPLE 3 **Solve a System of Equations by Using the Inverse of the Coefficient Matrix**

Find the solution of the system of equations by using the inverse of the coefficient matrix.

$$\begin{cases} x_1 + 7x_3 = 20 \\ 2x_1 + x_2 - x_3 = -3 \\ 7x_1 + 3x_2 + x_3 = 2 \end{cases} \quad (1)$$

Solution

Write the system as a matrix equation.

$$\begin{bmatrix} 1 & 0 & 7 \\ 2 & 1 & -1 \\ 7 & 3 & 1 \end{bmatrix}\begin{bmatrix} x_1 \\ x_2 \\ x_3 \end{bmatrix} = \begin{bmatrix} 20 \\ -3 \\ 2 \end{bmatrix} \quad (2)$$

The inverse of the coefficient matrix is $\begin{bmatrix} -\frac{4}{3} & -7 & \frac{7}{3} \\ 3 & 16 & -5 \\ \frac{1}{3} & 1 & -\frac{1}{3} \end{bmatrix}$.

Continued ▶

Multiply each side of the matrix equation (2) by the inverse.

$$\begin{bmatrix} -\dfrac{4}{3} & -7 & \dfrac{7}{3} \\ 3 & 16 & -5 \\ \dfrac{1}{3} & 1 & -\dfrac{1}{3} \end{bmatrix} \begin{bmatrix} 1 & 0 & 7 \\ 2 & 1 & -1 \\ 7 & 3 & 1 \end{bmatrix} \begin{bmatrix} x_1 \\ x_2 \\ x_3 \end{bmatrix} = \begin{bmatrix} -\dfrac{4}{3} & -7 & \dfrac{7}{3} \\ 3 & 16 & -5 \\ \dfrac{1}{3} & 1 & -\dfrac{1}{3} \end{bmatrix} \begin{bmatrix} 20 \\ -3 \\ 2 \end{bmatrix}$$

$$\begin{bmatrix} x_1 \\ x_2 \\ x_3 \end{bmatrix} = \begin{bmatrix} -1 \\ 2 \\ 3 \end{bmatrix}$$

Thus $x_1 = -1$, $x_2 = 2$, and $x_3 = 3$. The solution to System (1) is $(-1, 2, 3)$.

▶ **TRY EXERCISE 20, PAGE 874**

The temperature distribution on a metal plate can be approximated by a system of equations. The idea is based on the assumption that the temperature of a point P on the plate is the average of the temperatures of the four points shown nearest P.

EXAMPLE 4   Temperature Distribution on a Metal Plate

The edges of a metal plate are kept at a constant temperature, as shown below. Find the temperatures at x_1, x_2, x_3, and x_4.

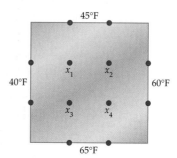

Solution

The temperature at x_1 is the average of the temperatures of the four points nearest x_1. That is,

$$x_1 = \frac{40 + 45 + x_2 + x_3}{4} = \frac{85 + x_2 + x_3}{4} \text{ or } 4x_1 - x_2 - x_3 = 85$$

Similarly, the temperatures at the remaining three points are

$$x_2 = \frac{60 + 45 + x_1 + x_4}{4} = \frac{105 + x_1 + x_4}{4} \text{ or } -x_1 + 4x_2 - x_4 = 105$$

$$x_3 = \frac{40 + 65 + x_1 + x_4}{4} = \frac{105 + x_1 + x_4}{4} \text{ or } -x_1 + 4x_3 - x_4 = 105$$

$$x_4 = \frac{60 + 65 + x_2 + x_3}{4} = \frac{125 + x_2 + x_3}{4} \text{ or } -x_2 - x_3 + 4x_4 = 125$$

The system of equations and the associated matrix equation are

$$\begin{cases} 4x_1 - x_2 - x_3 = 85 \\ -x_1 + 4x_2 - x_4 = 105 \\ -x_1 + 4x_3 - x_4 = 105 \\ -x_2 - x_3 + 4x_4 = 125 \end{cases} \qquad \begin{bmatrix} 4 & -1 & -1 & 0 \\ -1 & 4 & 0 & -1 \\ -1 & 0 & 4 & -1 \\ 0 & -1 & -1 & 4 \end{bmatrix} \begin{bmatrix} x_1 \\ x_2 \\ x_3 \\ x_4 \end{bmatrix} = \begin{bmatrix} 85 \\ 105 \\ 105 \\ 125 \end{bmatrix}$$

Solving the matrix equation by using an inverse matrix gives $\begin{bmatrix} x_1 \\ x_2 \\ x_3 \\ x_4 \end{bmatrix} = \begin{bmatrix} 47.5 \\ 52.5 \\ 52.5 \\ 57.5 \end{bmatrix}.$

The temperatures are $x_1 = 47.5°F$, $x_2 = 52.5°F$, $x_3 = 52.5°F$, and $x_4 = 57.5°F$.

▶ **TRY EXERCISE 26, PAGE 874**

● INPUT-OUTPUT ANALYSIS

The advantage of using the inverse matrix to solve a system of equations is not apparent unless it is necessary to solve repeatedly a system of equations with the same coefficient matrix but different constant matrices. *Input-output analysis* is one such application of this method.

In an economy, some of the output of an industry is used by the industry to produce its own product. For example, an electric company uses water and electricity to produce electricity, and a water company uses water and electricity to produce drinking water. **Input-output analysis** attempts to determine the necessary output of industries to satisfy each other's demands plus the demands of consumers. Wassily Leontief, a Harvard economist, was awarded the Nobel prize for his work in this field.

An **input-output matrix** is used to express the interdependence among industries in an economy. Each column of this matrix gives the dollar values of the inputs an industry needs to produce $1 worth of output.

To illustrate the concepts, we will assume an economy with only three industries: agriculture, transportation, and oil. Suppose that to produce $1 worth of agricultural products requires $.05 worth of agriculture, $.02 worth of transportation, and $.05 worth of oil. To produce $1 worth of transportation requires $.10 worth of agriculture, $.08 worth of transportation, and $.10 worth of oil. To produce $1 worth of oil requires $.10 worth of agriculture, $.15 worth of transportation, and $.13 worth of oil. The input-output matrix A is

Input requirements of

		Agriculture	Transportation	Oil
	Agriculture	0.05	0.10	0.10
from	Transportation	0.02	0.08	0.15
	Oil	0.05	0.10	0.13

Consumers (other than the industries themselves) want to purchase some of the output from these industries. The amount of output that the consumer will want is called the **final demand** on the economy. This is represented by a column matrix.

Suppose in our example that the final demand is \$3 billion worth of agriculture, \$1 billion worth of transportation, and \$2 billion worth of oil. The final demand matrix is

$$\begin{bmatrix} 3 \\ 1 \\ 2 \end{bmatrix} = D$$

We represent the total output of each industry (in billions of dollars) as follows:

$$x = \text{total output of agriculture}$$
$$y = \text{total output of transportation}$$
$$z = \text{total output of oil}$$

The object of input-output analysis is to determine the values of x, y, and z that will satisfy the amount the consumer demands. To find these values, consider agriculture. The amount of agriculture left for the consumer (demand d) is

$$d = x - \text{(amount of agriculture used by industries)} \qquad (1)$$

To find the amount of agriculture used by the three industries in our economy, refer to the input-output matrix. Production of x billion dollars worth of agriculture takes $0.05x$ of agriculture, production of y billion dollars worth of transportation takes $0.10y$ of agriculture, and production of z billion dollars worth of oil takes $0.10z$ of agriculture. Thus

$$\text{Amount of agriculture used by industries} = 0.05x + 0.10y + 0.10z \qquad (2)$$

Combining Equations (1) and (2), we have

$$d = x - (0.05x + 0.10y + 0.10z)$$
$$3 = 0.95x - 0.10y - 0.10z \qquad \bullet \; d \text{ is \$3 billion for agriculture.}$$

We could continue this way for each of the other industries. The result would be a system of equations. Instead, however, we will use a matrix approach.

If $X = $ total output of the three industries of the economy, then

$$X = \begin{bmatrix} x \\ y \\ z \end{bmatrix}$$

The product of A, the input-output matrix, and X is

$$AX = \begin{bmatrix} 0.05 & 0.10 & 0.10 \\ 0.02 & 0.08 & 0.15 \\ 0.05 & 0.10 & 0.13 \end{bmatrix} \begin{bmatrix} x \\ y \\ z \end{bmatrix}$$

This matrix represents the dollar amount of products used in production for all three industries. Thus the amount available for consumer demand is $X - AX$. As a matrix equation, we can write

$$X - AX = D$$

Solving this equation for X, we determine the output necessary to meet the needs of our industries and the consumer.

$$IX - AX = D \qquad \qquad \bullet \; I \text{ is the identity matrix. Thus } IX = X.$$
$$(I - A)X = D \qquad \qquad \bullet \; \textbf{Right distributive property}$$
$$X = (I - A)^{-1}D \qquad \qquad \bullet \; \textbf{Assuming the inverse of } (I - A) \textbf{ exists}$$

The last equation states that the solution to an input-output problem can be found by multiplying the demand matrix D by the inverse of $(I - A)$. In our example, we have

$$I - A = \begin{bmatrix} 1 & 0 & 0 \\ 0 & 1 & 0 \\ 0 & 0 & 1 \end{bmatrix} - \begin{bmatrix} 0.05 & 0.10 & 0.10 \\ 0.02 & 0.08 & 0.15 \\ 0.05 & 0.10 & 0.13 \end{bmatrix} = \begin{bmatrix} 0.95 & -0.10 & -0.10 \\ -0.02 & 0.92 & -0.15 \\ -0.05 & -0.10 & 0.87 \end{bmatrix}$$

$$(I - A)^{-1} \approx \begin{bmatrix} 1.063 & 0.131 & 0.145 \\ 0.034 & 1.112 & 0.196 \\ 0.065 & 0.135 & 1.180 \end{bmatrix}$$

The consumer demand is

$$X = (I - A)^{-1}D$$

$$X \approx \begin{bmatrix} 1.063 & 0.131 & 0.145 \\ 0.034 & 1.112 & 0.196 \\ 0.065 & 0.135 & 1.180 \end{bmatrix} \begin{bmatrix} 3 \\ 1 \\ 2 \end{bmatrix} \approx \begin{bmatrix} 3.61 \\ 1.61 \\ 2.69 \end{bmatrix}$$

This matrix indicates that $3.61 billion worth of agriculture, $1.61 billion worth of transportation, and $2.69 billion worth of oil must be produced by the industries to satisfy consumers' demands and the industries' internal requirements.

If we change the final demand matrix to,

$$D = \begin{bmatrix} 2 \\ 2 \\ 3 \end{bmatrix}$$

then the total output of the economy can be found as

$$X \approx \begin{bmatrix} 1.063 & 0.131 & 0.145 \\ 0.034 & 1.112 & 0.196 \\ 0.065 & 0.135 & 1.180 \end{bmatrix} \begin{bmatrix} 2 \\ 2 \\ 3 \end{bmatrix} \approx \begin{bmatrix} 2.82 \\ 2.88 \\ 3.94 \end{bmatrix}$$

Thus agriculture must produce output worth $2.82 billion, transportation must produce output worth $2.88 billion, and oil must produce output worth $3.94 billion to satisfy the given consumer demand and the industries' internal requirements.

TOPICS FOR DISCUSSION

1. Explain how to find the inverse of a matrix.

2. Explain the difference between a singular matrix and a nonsingular matrix.

3. Discuss the advantages and disadvantages of solving a system of equations by using an inverse matrix.

4. Do all square matrices have an inverse? If not, give an example of a square matrix that does not have an inverse.

EXERCISE SET 10.3

In Exercises 1 to 10, find the inverse of the given matrix.

1. $\begin{bmatrix} 1 & -3 \\ -2 & 5 \end{bmatrix}$

2. $\begin{bmatrix} 1 & 2 \\ -2 & -3 \end{bmatrix}$

3. $\begin{bmatrix} 1 & 4 \\ 2 & 10 \end{bmatrix}$

4. $\begin{bmatrix} -2 & 3 \\ -6 & -8 \end{bmatrix}$

5. $\begin{bmatrix} 1 & 2 & -1 \\ 2 & 5 & 1 \\ 3 & 6 & -2 \end{bmatrix}$

▶**6.** $\begin{bmatrix} 1 & 3 & -2 \\ -1 & -5 & 6 \\ 2 & 6 & -3 \end{bmatrix}$

7. $\begin{bmatrix} 1 & 2 & -1 \\ 2 & 6 & 1 \\ 3 & 6 & -4 \end{bmatrix}$

8. $\begin{bmatrix} 2 & 1 & -1 \\ 6 & 4 & -1 \\ 4 & 2 & -3 \end{bmatrix}$

9. $\begin{bmatrix} 2 & 4 & -4 \\ 1 & 3 & -4 \\ 2 & 4 & -3 \end{bmatrix}$

▶**10.** $\begin{bmatrix} 1 & -2 & 2 \\ 2 & -3 & 1 \\ 3 & -6 & 6 \end{bmatrix}$

In Exercises 11 to 14, use a graphing calculator to find the inverse of the given matrix.

11. $\begin{bmatrix} 1 & -1 & 2 & 1 \\ 2 & -1 & 5 & 1 \\ 3 & -3 & 7 & 5 \\ -2 & 3 & -4 & -1 \end{bmatrix}$

12. $\begin{bmatrix} 1 & 1 & -1 & 2 \\ 3 & 2 & -1 & 5 \\ 2 & 2 & -1 & 5 \\ 4 & 4 & -4 & 7 \end{bmatrix}$

13. $\begin{bmatrix} 1 & -1 & 1 & 3 \\ 2 & -1 & 4 & 8 \\ 1 & 1 & 6 & 10 \\ -1 & 5 & 5 & 4 \end{bmatrix}$

14. $\begin{bmatrix} 1 & -1 & 1 & 2 \\ 2 & -1 & 6 & 6 \\ 3 & -1 & 12 & 12 \\ -2 & -1 & -14 & -10 \end{bmatrix}$

In Exercises 15 to 24, solve each system of equations by using inverse matrix methods.

15. $\begin{cases} x + 4y = 6 \\ 2x + 7y = 11 \end{cases}$

16. $\begin{cases} 2x + 3y = 5 \\ x + 2y = 4 \end{cases}$

17. $\begin{cases} x - 2y = 8 \\ 3x + 2y = -1 \end{cases}$

18. $\begin{cases} 3x - 5y = -18 \\ 2x - 3y = -11 \end{cases}$

19. $\begin{cases} x + y + 2z = 4 \\ 2x + 3y + 3z = 5 \\ 3x + 3y + 7z = 14 \end{cases}$

▶**20.** $\begin{cases} x + 2y - z = 5 \\ 2x + 3y - z = 8 \\ 3x + 6y - 2z = 14 \end{cases}$

21. $\begin{cases} x + 2y + 2z = 5 \\ -2x - 5y - 2z = 8 \\ 2x + 4y + 7z = 19 \end{cases}$

22. $\begin{cases} x - y + 3z = 5 \\ 3x - y + 10z = 16 \\ 2x - 2y + 5z = 9 \end{cases}$

23. $\begin{cases} w + 2x + z = 6 \\ 2w + 5x + y + 2z = 10 \\ 2w + 4x + y + z = 8 \\ 3w + 6x + 4z = 16 \end{cases}$

24. $\begin{cases} w - x + 2y = 5 \\ 2w - x + 6y + 2z = 16 \\ 3w - 2x + 9y + 4z = 28 \\ w - 2x - z = 2 \end{cases}$

25. The edges of a metal plate are kept at a constant temperature, as shown below. Find the temperatures at x_1 and x_2. Round to the nearest tenth.

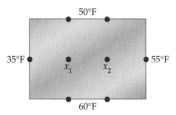

▶26. The edges of a metal plate are kept at a constant temperature, as shown below. Find the temperatures at x_1 and x_2. Round to the nearest tenth.

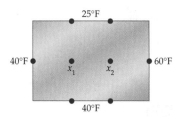

27. The edges of a metal plate are kept at a constant temperature, as shown below. Find the temperatures at x_1, x_2, x_3, and x_4.

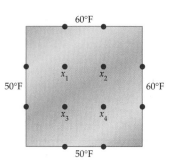

28. The edges of a metal plate are kept at a constant temperature, as shown below. Find the temperatures at x_1, x_2, x_3, and x_4.

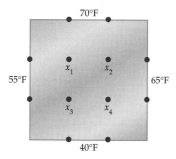

In Exercises 29–32, solve each application by writing a system of equations that models the conditions and then applying inverse matrix methods.

29. BUSINESS REVENUE A vacation resort offers a helicopter tour of an island. The price for an adult ticket is $20; the price for a children's ticket is $15. The records of the tour operator show that 100 people took the tour on Saturday and 120 people took the tour on Sunday. The total receipts for Saturday were $1900, and on Sunday the receipts were $2275. Find the number of adults and the number of children who took the tour on Saturday and on Sunday.

30. BUSINESS REVENUE A company sells a standard and a deluxe model tape recorder. Each standard tape recorder costs $45 to manufacture, and each deluxe model costs $60 to manufacture. The January manufacturing budget for 90 of these recorders was $4650; the February budget for 100 recorders was $5250. Find the number of each type of recorder manufactured in January and in February.

31. SOIL SCIENCE The following table shows the active chemical content of three different soil additives.

Additive	Grams per 100 Grams		
	Ammonium Nitrate	Phosphorus	Iron
1	30	10	10
2	40	15	10
3	50	5	5

A soil chemist wants to prepare two chemical samples. The first sample contains 380 grams of ammonium nitrate, 95 grams of phosphorus, and 85 grams of iron. The second sample requires 380 grams of ammonium nitrate, 110 grams of phosphorus, and 90 grams of iron. How many grams of each additive are required for sample 1, and how many grams of each additive are required for sample 2?

32. NUTRITION The following table shows the carbohydrate, fat, and protein content of three food types.

Food Type	Grams per 100 Grams		
	Carbohydrate	Fat	Protein
I	13	10	13
II	4	4	3
III	1	0	10

A nutritionist must prepare two diets from these three food groups. The first diet must contain 23 grams of carbohydrate, 18 grams of fat, and 39 grams of protein. The second diet must contain 35 grams of carbohydrate, 28 grams of fat, and 42 grams of protein. How many grams of each food type are required for the first diet, and how many grams of each food type are required for the second diet?

 In Exercises 33 to 36, use a graphing calculator to find the inverse of each matrix. Where necessary, round values to the nearest thousandth.

33. $\begin{bmatrix} 2 & -2 & 3 & 1 \\ 5 & 2 & -2 & 3 \\ 6 & -1 & 2 & 3 \\ 2 & 3 & -1 & 5 \end{bmatrix}$ **34.** $\begin{bmatrix} 3 & -1 & 0 & 1 \\ 2 & -2 & 3 & 0 \\ -1 & -3 & 5 & 3 \\ 5 & 3 & -2 & 1 \end{bmatrix}$

35. $\begin{bmatrix} -\frac{2}{7} & 4 & -\frac{1}{6} \\ -2 & \sqrt{2} & -3 \\ \sqrt{3} & 3 & -\sqrt{5} \end{bmatrix}$ **36.** $\begin{bmatrix} 6 & \pi & -\frac{4}{7} \\ -5 & \sqrt{7} & 2 \\ \frac{5}{6} & -\sqrt{3} & \sqrt{10} \end{bmatrix}$

37. INPUT-OUTPUT ANALYSIS A simplified economy has three industries: manufacturing, transportation, and service. The input-output matrix for this economy is

$$\begin{bmatrix} 0.20 & 0.15 & 0.10 \\ 0.10 & 0.30 & 0.25 \\ 0.20 & 0.10 & 0.10 \end{bmatrix}$$

Find the gross output needed to satisfy the consumer demand of $120 million worth of manufacturing, $60 million worth of transportation, and $55 million worth of service.

38. INPUT-OUTPUT ANALYSIS A four-sector economy consists of manufacturing, agriculture, service, and transportation. The input-output matrix for this economy is

$$\begin{bmatrix} 0.10 & 0.05 & 0.20 & 0.15 \\ 0.20 & 0.10 & 0.30 & 0.10 \\ 0.05 & 0.30 & 0.20 & 0.40 \\ 0.10 & 0.20 & 0.15 & 0.20 \end{bmatrix}$$

Find the gross output needed to satisfy the consumer demand of $80 million worth of manufacturing, $100 million worth of agriculture, $50 million worth of service, and $80 million worth of transportation.

39. INPUT-OUTPUT ANALYSIS A conglomerate is composed of three industries: coal, iron, and steel. To produce $1 worth of coal requires $.05 worth of coal, $.02 worth of iron, and $.10 worth of steel. To produce $1 worth of iron requires $.20 worth of coal, $.03 worth of iron, and $.12 worth of steel. To produce $1 worth of steel requires $.15 worth of coal, $.25 worth of iron, and $.05 worth of steel. How much should each industry produce to allow for a consumer demand of $30 million worth of coal, $5 million worth of iron, and $25 million worth of steel?

40. INPUT-OUTPUT ANALYSIS A conglomerate has three divisions: plastics, semiconductors, and computers. For each $1 worth of output, the plastics division needs $.01 worth of plastics, $.03 worth of semiconductors, and $.10 worth of computers. Each $1 worth of output from the semiconductor division requires $.08 worth from plastics, $.05 worth from semiconductors, and $.15 worth from computers. For each $1 worth of output, the computer division needs $.20 worth from plastics, $.20 worth from semiconductors, and $.10 worth from computers. The conglomerate estimates consumer demand of $100 million worth from the plastics division, $75 million worth from the semiconductor division, and $150 million worth from the computer division. At what level should each division produce to satisfy this demand?

CONNECTING CONCEPTS

41. Let $A = \begin{bmatrix} 2 & -3 \\ -6 & 9 \end{bmatrix}$ and $B = \begin{bmatrix} -3 & 15 \\ -2 & 10 \end{bmatrix}$. Show that $AB = O$, the 2×2 zero matrix. This illustrates that for matrices, if $AB = O$, it is not necessarily so that $A = O$ or $B = O$.

42. Show that if a matrix A has an inverse and $AB = O$, then $B = O$.

43. Let $A = \begin{bmatrix} 2 & -1 \\ -4 & 2 \end{bmatrix}$, $B = \begin{bmatrix} 3 & 4 \\ 1 & 5 \end{bmatrix}$, and $C = \begin{bmatrix} 4 & 7 \\ 3 & 11 \end{bmatrix}$. Show that $AB = AC$ but that $B \neq C$. This illustrates that the cancellation rule of real numbers may not apply to matrices.

44. (Continuation of Exercise 43.) Show that if A is a matrix that has an inverse and $AB = AC$, then $B = C$.

45. Show that if $A = \begin{bmatrix} a & b \\ c & d \end{bmatrix}$ and $ad - bc \neq 0$, then

$$A^{-1} = \frac{1}{ad - bc} \begin{bmatrix} d & -b \\ -c & a \end{bmatrix}$$

46. Use the result of Exercise 45 to show that a square matrix of order 2 has an inverse if and only if $ad - bc \neq 0$.

47. Use the result of Exercise 45 to find the inverse of each matrix.

a. $\begin{bmatrix} 2 & -3 \\ 4 & -5 \end{bmatrix}$ **b.** $\begin{bmatrix} 5 & 6 \\ 3 & 4 \end{bmatrix}$ **c.** $\begin{bmatrix} 0 & -1 \\ 4 & 4 \end{bmatrix}$

48. Let $A = \begin{bmatrix} 3 & -2 \\ 1 & 1 \end{bmatrix}$ and $B = \begin{bmatrix} 2 & -1 \\ 2 & 3 \end{bmatrix}$. Use Exercise 45 to show that

$$A^{-1} = \frac{1}{5} \begin{bmatrix} 1 & 2 \\ -1 & 3 \end{bmatrix} \quad \text{and} \quad B^{-1} = \frac{1}{8} \begin{bmatrix} 3 & 1 \\ -2 & 2 \end{bmatrix}$$

Now show that $(AB)^{-1} = B^{-1} \cdot A^{-1}$.

49. Generalize the last result in Exercise 48. That is, show that if A and B are square matrices of order n and each has an inverse matrix, then $(AB)^{-1} = B^{-1} \cdot A^{-1}$. (*Hint:* Begin with the equation $(AB)(AB)^{-1} = I$, where I is the identity matrix. Now multiply each side of the equation by A^{-1} and then by B^{-1}.)

PREPARE FOR SECTION 10.4

50. What is the order of the matrix $\begin{bmatrix} -3 & 5 \\ 0 & 1 \end{bmatrix}$? [10.1]

51. Evaluate $(-1)^{i+j}$ for $i = 2$ and $j = 6$. [P.2]

52. Evaluate $(-1)^{1+1}(-3) + (-1)^{1+2}(-2) + (-1)^{1+3}(5)$ [P.2]

53. If $A = \begin{bmatrix} 0 & 1 & -2 \\ 4 & -5 & 1 \\ 2 & -3 & 5 \end{bmatrix}$, what is a_{23}? [10.1]

54. Find $3\begin{bmatrix} -2 & 1 \\ 3 & -5 \end{bmatrix}$. [10.2]

55. Complete the following:

$$\begin{bmatrix} 1 & 3 & -2 \\ -2 & -1 & 1 \\ 4 & 0 & 1 \end{bmatrix} \xrightarrow[-4R_1 + R_3]{2R_1 + R_2} \begin{bmatrix} & & \\ & ? & \\ & & \end{bmatrix}$$

[10.1]

PROJECTS

1. Cryptography is the study of the techniques of concealing the meaning of a message. The message that is to be concealed is called **plaintext**. The concealed message is called **ciphertext**. One way to change plaintext to ciphertext is to give each letter of the alphabet a numerical equivalent. Then matrices are used to scramble the numbers so that it is difficult to determine which number is associated with which letter.

a. One way to assign each letter a number is to use the ASCII coding system. Check the Internet to determine how this system assigns a number to each letter and punctuation mark.

b. Now write a short sentence, such as "THE BUCK STOPS HERE." Group the letters of the sentence into packets of, say, 3, using 0 (zero) for a space. Our sentence would look like

(THE)(0BU)(CK0)(STO)(PS0)(HER)(E.0)

Replace each letter and punctuation mark by its numerical ASCII equivalent. For our message, the first three groups would be

(84 72 69)(48 66 85)(67 75 48)...

Place these numbers in a matrix, using the set of three numbers as a column. For our example, the first three columns are

$$W = \begin{bmatrix} 84 & 48 & 67 & \cdots \\ 72 & 66 & 75 & \cdots \\ 69 & 85 & 48 & \cdots \end{bmatrix}$$

c. Now construct a 3×3 matrix E with integer elements that has an inverse. You can use any 3×3 matrix as long as you can find the inverse. (A graphing calculator may be useful here.)

d. Find the product $E \cdot W = M$. The numbers in the matrix M would be sent as the coded message. Do this for your message.

e. The person who receives this message would multiply the matrix M by E^{-1} to restore the message to its original form. Do this for your message.

SECTION 10.4 DETERMINANTS

DETERMINANT OF A 2 × 2 MATRIX

Associated with each square matrix A is a number called the *determinant* of A. We will denote the determinant of the matrix A by $\det(A)$ or by $|A|$. For the remainder of this chapter, we assume that all matrices are square matrices.

The Determinant of a 2 × 2 Matrix

The **determinant** of the matrix $A = [a_{ij}]$ of order 2 is

$$|A| = \begin{vmatrix} a_{11} & a_{12} \\ a_{21} & a_{22} \end{vmatrix} = a_{11}a_{22} - a_{21}a_{12}$$

Caution Be careful not to confuse the notation for a matrix and that for a determinant. The symbol [ ] (brackets) is used for a matrix; the symbol | | (vertical bars) is used for the determinant of a matrix.

An easy way to remember the formula for the determinant of a 2 × 2 matrix is to recognize that the determinant is the difference between the products of the diagonal elements. That is,

$$\begin{vmatrix} a_{11} & a_{12} \\ a_{21} \diagdown a_{22} \end{vmatrix} = a_{11}a_{22} - a_{21}a_{12}$$

EXAMPLE 1 Find the Value of a Determinant

Find the value of the determinant of the matrix $A = \begin{bmatrix} 5 & 3 \\ 2 & -3 \end{bmatrix}$.

Solution

$$|A| = \begin{vmatrix} 5 & 3 \\ 2 & -3 \end{vmatrix} = 5(-3) - 2(3) = -15 - 6 = -21$$

▶ **TRY EXERCISE 2, PAGE 885**

MINORS AND COFACTORS

To define the determinant of a matrix of order greater than 2, we first need two other definitions.

The Minor of a Matrix

The **minor** M_{ij} of the element a_{ij} of a square matrix A of order $n \geq 3$ is the determinant of the matrix of order $n - 1$ obtained by deleting the ith row and the jth column of A.

Consider the matrix $A = \begin{bmatrix} 2 & -1 & 5 \\ 4 & 3 & -7 \\ 8 & -7 & 6 \end{bmatrix}$. The minor M_{23} is the determinant of the matrix A formed by deleting row 2 and column 3 from A.

$$M_{23} = \begin{vmatrix} 2 & -1 \\ 8 & -7 \end{vmatrix} \qquad \bullet \begin{vmatrix} 2 & -1 & 5 \\ 4 & 3 & -7 \\ 8 & -7 & 6 \end{vmatrix}$$
$$= 2(-7) - 8(-1) = -14 + 8 = -6$$

The minor M_{31} is the determinant of the matrix A formed by deleting row 3 and column 1 from A.

$$M_{31} = \begin{vmatrix} -1 & 5 \\ 3 & -7 \end{vmatrix} \qquad \bullet \begin{vmatrix} 2 & -1 & 5 \\ 4 & 3 & -7 \\ 8 & -7 & 6 \end{vmatrix}$$
$$= (-1)(-7) - 3(5) = 7 - 15 = -8$$

The second definition we need is that of the *cofactor* of a matrix.

Cofactor of a Matrix

The **cofactor** C_{ij} of the element a_{ij} of a square matrix A is given by $C_{ij} = (-1)^{i+j}M_{ij}$, where M_{ij} is the minor of a_{ij}.

When $i + j$ is an even integer, $(-1)^{i+j} = 1$. When $i + j$ is an odd integer, $(-1)^{i+j} = -1$. Thus

$$C_{ij} = \begin{cases} M_{ij}, & i + j \text{ is an even integer} \\ -M_{ij}, & i + j \text{ is an odd integer} \end{cases}$$

EXAMPLE 2   Find the Minor and Cofactor of a Matrix

Given $A = \begin{bmatrix} 4 & 3 & -2 \\ 5 & -2 & 4 \\ 3 & -2 & -6 \end{bmatrix}$, find M_{32} and C_{12}.

Solution

$$M_{32} = \begin{vmatrix} 4 & -2 \\ 5 & 4 \end{vmatrix} = 4(4) - 5(-2) = 16 - (-10) = 16 + 10 = 26$$

$$C_{12} = (-1)^{1+2}M_{12} = -M_{12} = -\begin{vmatrix} 5 & 4 \\ 3 & -6 \end{vmatrix} = -(-30 - 12) = 42$$

▶ **TRY EXERCISE 14, PAGE 885**

● EVALUATE A DETERMINANT USING EXPANDING BY COFACTORS

Cofactors are used to evaluate the determinant of a matrix of order 3 or greater. The technique used to evaluate a determinant by using cofactors is called *expanding by cofactors*.

Evaluating Determinants by Expanding by Cofactors

Given the square matrix A of order 3 or greater, the value of the determinant of A is the sum of the products of the elements of any row or column and their cofactors. For the rth row of A, the value of the determinant of A is

$$|A| = a_{r1}C_{r1} + a_{r2}C_{r2} + a_{r3}C_{r3} + \cdots + a_{rn}C_{rn}$$

For the cth column of A, the determinant of A is

$$|A| = a_{1c}C_{1c} + a_{2c}C_{2c} + a_{3c}C_{3c} + \cdots + a_{nc}C_{nc}$$

This theorem states that the value of a determinant can be found by expanding by cofactors of *any* row or column. The value of the determinant is the same in each case. To illustrate the method, consider the matrix $A = \begin{bmatrix} 2 & 3 & -1 \\ 4 & -2 & 3 \\ 1 & -3 & 4 \end{bmatrix}$.

Expanding the determinant of A by some row, say row 2, gives

$$|A| = \begin{vmatrix} 2 & 3 & -1 \\ 4 & -2 & 3 \\ 1 & -3 & 4 \end{vmatrix} = 4C_{21} + (-2)C_{22} + 3C_{23}$$

$$= 4(-1)^{2+1}M_{21} + (-2)(-1)^{2+2}M_{22} + 3(-1)^{2+3}M_{23}$$

$$= (-4)\begin{vmatrix} 3 & -1 \\ -3 & 4 \end{vmatrix} + (-2)\begin{vmatrix} 2 & -1 \\ 1 & 4 \end{vmatrix} + (-3)\begin{vmatrix} 2 & 3 \\ 1 & -3 \end{vmatrix}$$

$$= (-4)9 + (-2)9 + (-3)(-9) = -27$$

Expanding the determinant of A by some column, say, column 3, gives

$$|A| = \begin{vmatrix} 2 & 3 & -1 \\ 4 & -2 & 3 \\ 1 & -3 & 4 \end{vmatrix} = (-1)C_{13} + 3C_{23} + 4C_{33}$$

$$= (-1)(-1)^{1+3}M_{13} + 3(-1)^{2+3}M_{23} + 4(-1)^{3+3}M_{33}$$

$$= (-1)\begin{vmatrix} 4 & -2 \\ 1 & -3 \end{vmatrix} + (-3)\begin{vmatrix} 2 & 3 \\ 1 & -3 \end{vmatrix} + 4\begin{vmatrix} 2 & 3 \\ 4 & -2 \end{vmatrix}$$

$$= (-1)(-10) + (-3)(-9) + 4(-16) = -27$$

The value of the determinant of A is the same whether we expanded by cofactors of the elements of a row or by cofactors of the elements of a column. When evaluating a determinant, choose the most convenient row or column, which usually is the row or column containing the most zeros.

| EXAMPLE 3 | Evaluate a Determinant by Cofactors |

Evaluate the determinant of $A = \begin{bmatrix} 5 & -3 & -1 \\ -2 & 1 & -1 \\ 1 & 0 & 2 \end{bmatrix}$ by expanding by cofactors.

Solution

Because $a_{32} = 0$, expand using row 3 or column 2. Row 3 will be used here.

$$|A| = 1C_{31} + 0C_{32} + 2C_{33} = 1(-1)^{3+1}M_{31} + 0(-1)^{3+2}M_{32} + 2(-1)^{3+3}M_{33}$$

$$= 1\begin{vmatrix} -3 & -1 \\ 1 & -1 \end{vmatrix} + 0 + 2\begin{vmatrix} 5 & -3 \\ -2 & 1 \end{vmatrix} = 1[3 - (-1)] + 0 + 2[5 - 6]$$

$$= 4 - 2 = 2$$

▶ **TRY EXERCISE 20, PAGE 885**

take note

Example 3 illustrates that choosing a row or column with the most zeros and then expanding about that row or column will reduce the number of calculations you must perform. For Example 3 we have $0 \cdot C_{32} = 0$, and it is not necessary to compute C_{32}.

INTEGRATING TECHNOLOGY

The determinant of a matrix can be found by using a graphing calculator. Many of these calculators use *det* as the operation that produces the value of the determinant. For instance, if $A = \begin{bmatrix} -2 & 3 & 4 \\ 1 & 0 & -2 \\ 3 & 1 & 1 \end{bmatrix}$, then a typical calculator display of the determinant of A is shown in **Figure 10.5**.

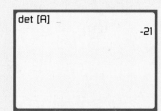

det [A]

-21

FIGURE 10.5

● EVALUATE A DETERMINANT USING ELEMENTARY ROW OPERATIONS

take note

The properties of determinants stated at the right remain true when the word *row* is replaced by *column*. In that case, we would have elementary *column* operations.

Effects of Elementary Row Operations on the Value of a Determinant of a Matrix

If A is a square matrix of order n, then the following elementary row operations produce the indicated changes in the determinant of A.

1. Interchanging any two rows of A changes the sign of $|A|$.

2. Multiplying a row of A by a constant k multiplies the determinant of A by k.

3. Adding a multiple of a row of A to another row does not change the value of the determinant of A.

To illustrate these properties, consider the matrix $A = \begin{bmatrix} 2 & 3 \\ 1 & -2 \end{bmatrix}$. The determinant of A is $|A| = 2(-2) - 1(3) = -7$. Now consider each of the elementary row operations.

Interchange the rows of A and evaluate the determinant.

$$\begin{vmatrix} 1 & -2 \\ 2 & 3 \end{vmatrix} = 1(3) - 2(-2) = 3 + 4 = 7 = -|A|$$

Multiply row 2 of A by -3 and evaluate the determinant.

$$\begin{vmatrix} 2 & 3 \\ -3 & 6 \end{vmatrix} = 2(6) - (-3)3 = 12 + 9 = 21 = -3|A|$$

Multiply row 1 of A by -2 and add it to row 2. Then evaluate the determinant.

$$\begin{vmatrix} 2 & 3 \\ -3 & -8 \end{vmatrix} = 2(-8) - (-3)(3) = -16 + 9 = -7 = |A|.$$

These elementary row operations are often used to rewrite a matrix in *triangular form*. A matrix is in **triangular form** if all elements below or above the main diagonal are zero. The matrices

$$A = \begin{bmatrix} 2 & -2 & 3 & 1 \\ 0 & -2 & 4 & 2 \\ 0 & 0 & 6 & 9 \\ 0 & 0 & 0 & -5 \end{bmatrix} \quad \text{and} \quad B = \begin{bmatrix} 3 & 0 & 0 & 0 \\ 2 & -3 & 0 & 0 \\ 6 & 4 & -2 & 0 \\ 8 & 3 & 4 & 2 \end{bmatrix}$$

are in triangular form.

Determinant of a Matrix in Triangular Form

Let A be a square matrix of order n in triangular form. The determinant of A is the product of the elements on the main diagonal.

$$|A| = a_{11}a_{22}a_{33} \cdots a_{nn}$$

For the matrices A and B given above,

$$|A| = 2(-2)(6)(-5) = 120$$
$$|B| = 3(-3)(-2)(2) = 36$$

EXAMPLE 4 **Evaluate a Determinant by Elementary Row Operations**

Evaluate the determinant by rewriting it in triangular form.

$$\begin{vmatrix} 2 & 1 & -1 & 3 \\ 2 & 2 & 0 & 1 \\ 4 & 5 & 4 & -3 \\ 2 & 2 & 7 & -3 \end{vmatrix}$$

Solution

Rewrite the determinant in triangular form by using elementary row operations.

$$\begin{vmatrix} 2 & 1 & -1 & 3 \\ 2 & 2 & 0 & 1 \\ 4 & 5 & 4 & -3 \\ 2 & 2 & 7 & -3 \end{vmatrix} \begin{matrix} -1R_1 + R_2 \\ -2R_1 + R_3 \\ -1R_1 + R_4 \\ = \end{matrix} \begin{vmatrix} 2 & 1 & -1 & 3 \\ 0 & 1 & 1 & -2 \\ 0 & 3 & 6 & -9 \\ 0 & 1 & 8 & -6 \end{vmatrix}$$

$$\begin{matrix} \text{Factor 3} \\ \text{from row 3.} \\ = \end{matrix} \; 3 \begin{vmatrix} 2 & 1 & -1 & 3 \\ 0 & 1 & 1 & -2 \\ 0 & 1 & 2 & -3 \\ 0 & 1 & 8 & -6 \end{vmatrix} \begin{matrix} -1R_2 + R_3 \\ -1R_2 + R_4 \\ = \end{matrix} \; 3 \begin{vmatrix} 2 & 1 & -1 & 3 \\ 0 & 1 & 1 & -2 \\ 0 & 0 & 1 & -1 \\ 0 & 0 & 7 & -4 \end{vmatrix}$$

$$\begin{matrix} -7R_3 + R_4 \\ = \end{matrix} \; 3 \begin{vmatrix} 2 & 1 & -1 & 3 \\ 0 & 1 & 1 & -2 \\ 0 & 0 & 1 & -1 \\ 0 & 0 & 0 & 3 \end{vmatrix} = 3(2)(1)(1)(3) = 18$$

▶ **TRY EXERCISE 42, PAGE 886**

❓ **QUESTION**   If I is the identity matrix of order n, what is the value of $|I|$?

In some cases it is possible to recognize when the determinant of a matrix is zero.

Conditions for a Zero Determinant

If A is a square matrix, then $|A| = 0$ when any one of the following is true.

1. A row (column) consists entirely of zeros.

2. Two rows (columns) are identical.

3. One row (column) is a constant multiple of a second row (column).

Proof To prove part 2 of this theorem, let A be the given matrix and let $D = |A|$. Now interchange the two identical rows. Then $|A| = -D$. Thus

$$D = -D$$

Zero is the only real number that is its own additive inverse, and hence $D = |A| = 0$.   ◆

The proofs of the other two properties are left as exercises.

The last property of determinants that we will discuss is a product property.

❓ **ANSWER**   The identity matrix is in diagonal form with 1s on the main diagonal. Thus $|I|$ is a product of 1s, or $|I| = 1$.

Product Property of Determinants

If A and B are square matrices of order n, then

$$|AB| = |A||B|$$

● CONDITION FOR A SQUARE MATRIX TO HAVE A MULTIPLICATIVE INVERSE

Recall that a singular matrix is one that does not have a multiplicative inverse. The Product Property of Determinants can be used to determine whether a matrix has an inverse.

Consider a matrix A with an inverse A^{-1}. Then, by the last theorem,

$$|A \cdot A^{-1}| = |A||A^{-1}|$$

But $A \cdot A^{-1} = I$, the identity matrix, and $|I| = 1$. Therefore,

$$1 = |A||A^{-1}|$$

From the last equation, $|A| \neq 0$. And, in particular,

$$|A^{-1}| = \frac{1}{|A|}$$

These results are summarized in the following theorem.

Existence of the Inverse of a Square Matrix

If A is a square matrix of order n, then A has a multiplicative inverse if and only if $|A| \neq 0$. Furthermore,

$$|A^{-1}| = \frac{1}{|A|}$$

We proved only part of this theorem. It remains to show that given $|A| \neq 0$, then A has an inverse. This proof can be found in most texts on linear algebra.

 ## TOPICS FOR DISCUSSION

1. Discuss the difference between a matrix and a determinant.

2. Explain the difference between the minor and the cofactor of an element of a matrix.

3. Discuss how determinants are used to discover whether a matrix has an inverse.

4. Explain how to calculate the value of a determinant by expanding by cofactors.

5. Discuss how elementary row operations are used to find the determinant of a matrix.

EXERCISE SET 10.4

In Exercises 1 to 8, evaluate the determinant.

1. $\begin{vmatrix} 2 & -1 \\ 3 & 5 \end{vmatrix}$

► 2. $\begin{vmatrix} 2 & 9 \\ -6 & 2 \end{vmatrix}$

3. $\begin{vmatrix} 5 & 0 \\ 2 & -3 \end{vmatrix}$

4. $\begin{vmatrix} 0 & -8 \\ 3 & 4 \end{vmatrix}$

5. $\begin{vmatrix} 4 & 6 \\ 2 & 3 \end{vmatrix}$

6. $\begin{vmatrix} -3 & 6 \\ 4 & -8 \end{vmatrix}$

7. $\begin{vmatrix} 0 & 9 \\ 0 & -2 \end{vmatrix}$

8. $\begin{vmatrix} -3 & 9 \\ 0 & 0 \end{vmatrix}$

In Exercises 9 to 12, evaluate the indicated minor and co-factor for the determinant

$$\begin{vmatrix} 5 & -2 & -3 \\ 2 & 4 & -1 \\ 4 & -5 & 6 \end{vmatrix}$$

9. M_{11}, C_{11}

10. M_{21}, C_{21}

11. M_{32}, C_{32}

12. M_{33}, C_{33}

In Exercises 13 to 16, evaluate the indicated minor and co-factor for the determinant

$$\begin{vmatrix} 3 & -2 & 3 \\ 1 & 3 & 0 \\ 6 & -2 & 3 \end{vmatrix}$$

13. M_{22}, C_{22}

► 14. M_{13}, C_{13}

15. M_{31}, C_{31}

16. M_{23}, C_{23}

In Exercises 17 to 26, evaluate the determinant by expanding by cofactors.

17. $\begin{vmatrix} 2 & -3 & 1 \\ 2 & 0 & 2 \\ 3 & -2 & 4 \end{vmatrix}$

18. $\begin{vmatrix} 3 & 1 & -2 \\ 2 & -5 & 4 \\ 3 & 2 & 1 \end{vmatrix}$

19. $\begin{vmatrix} -2 & 3 & 2 \\ 1 & 2 & -3 \\ -4 & -2 & 1 \end{vmatrix}$

► 20. $\begin{vmatrix} 3 & -2 & 0 \\ 2 & -3 & 2 \\ 8 & -2 & 5 \end{vmatrix}$

21. $\begin{vmatrix} 2 & -3 & 10 \\ 0 & 2 & -3 \\ 0 & 0 & 5 \end{vmatrix}$

22. $\begin{vmatrix} 6 & 0 & 0 \\ 2 & -3 & 0 \\ 7 & -8 & 2 \end{vmatrix}$

23. $\begin{vmatrix} 0 & -2 & 4 \\ 1 & 0 & -7 \\ 5 & -6 & 0 \end{vmatrix}$

24. $\begin{vmatrix} 5 & -8 & 0 \\ 2 & 0 & -7 \\ 0 & -2 & -1 \end{vmatrix}$

25. $\begin{vmatrix} 4 & -3 & 3 \\ 2 & 1 & -4 \\ 6 & -2 & -1 \end{vmatrix}$

26. $\begin{vmatrix} -2 & 3 & 9 \\ 4 & -2 & -6 \\ 0 & -8 & -24 \end{vmatrix}$

In Exercises 27 to 40, without expanding, give a reason for each equality.

27. $\begin{vmatrix} 2 & -1 & 3 \\ 0 & 0 & 0 \\ 3 & 4 & 1 \end{vmatrix} = 0$

28. $\begin{vmatrix} 2 & 3 & 0 \\ 1 & -2 & 0 \\ 4 & 1 & 0 \end{vmatrix} = 0$

29. $\begin{vmatrix} 1 & 4 & -1 \\ 2 & 4 & 12 \\ 3 & 1 & 4 \end{vmatrix} = 2\begin{vmatrix} 1 & 4 & -1 \\ 1 & 2 & 6 \\ 3 & 1 & 4 \end{vmatrix}$

30. $\begin{vmatrix} 1 & -3 & 4 \\ 4 & 6 & 1 \\ 0 & -9 & 3 \end{vmatrix} = -3\begin{vmatrix} 1 & 1 & 4 \\ 4 & -2 & 1 \\ 0 & 3 & 3 \end{vmatrix}$

31. $\begin{vmatrix} 1 & 5 & -2 \\ 2 & -1 & 4 \\ 3 & 0 & -2 \end{vmatrix} = \begin{vmatrix} 1 & 5 & -2 \\ 0 & -11 & 8 \\ 3 & 0 & -2 \end{vmatrix}$

32. $\begin{vmatrix} 1 & 1 & -3 \\ 2 & 2 & 5 \\ 1 & -2 & 4 \end{vmatrix} = \begin{vmatrix} 1 & 1 & -3 \\ 2 & 2 & 5 \\ 0 & -3 & 7 \end{vmatrix}$

33. $\begin{vmatrix} 4 & -3 & 2 \\ 6 & 2 & 1 \\ -2 & 2 & 4 \end{vmatrix} = 2\begin{vmatrix} 2 & -3 & 2 \\ 3 & 2 & 1 \\ -1 & 2 & 4 \end{vmatrix}$

34. $\begin{vmatrix} 2 & -1 & 3 \\ 3 & 0 & 1 \\ -4 & 2 & -6 \end{vmatrix} = 0$

35. $\begin{vmatrix} 2 & -4 & 5 \\ 0 & 3 & 4 \\ 0 & 0 & -2 \end{vmatrix} = -12$

36. $\begin{vmatrix} 3 & 0 & 0 \\ 2 & -1 & 0 \\ 3 & 4 & 5 \end{vmatrix} = -15$

37. $\begin{vmatrix} 3 & 5 & -2 \\ 2 & 1 & 0 \\ 9 & -2 & -3 \end{vmatrix} = -\begin{vmatrix} 9 & -2 & -3 \\ 2 & 1 & 0 \\ 3 & 5 & -2 \end{vmatrix}$

38. $\begin{vmatrix} 6 & 0 & -2 \\ 2 & -1 & -3 \\ 1 & 5 & -7 \end{vmatrix} = -\begin{vmatrix} 0 & 6 & -2 \\ -1 & 2 & -3 \\ 5 & 1 & -7 \end{vmatrix}$

39. $a^3 \begin{vmatrix} 1 & 1 & 1 \\ a & a & a \\ a^2 & a^2 & a^2 \end{vmatrix} = \begin{vmatrix} a & a & a \\ a^2 & a^2 & a^2 \\ a^3 & a^3 & a^3 \end{vmatrix}$

40. $\begin{vmatrix} 1 & 1 & 1 \\ 2 & 2 & 2 \\ 3 & 3 & 3 \end{vmatrix} = 0$

In Exercises 41 to 50, evaluate the determinant by first rewriting the determinant in triangular form.

41. $\begin{vmatrix} 2 & 4 & 1 \\ 1 & 2 & -1 \\ 1 & 2 & 2 \end{vmatrix}$

▶ **42.** $\begin{vmatrix} 3 & -2 & -1 \\ 1 & 2 & 4 \\ 2 & -2 & 3 \end{vmatrix}$

43. $\begin{vmatrix} 1 & 2 & -1 \\ 2 & 3 & 1 \\ 3 & 4 & 3 \end{vmatrix}$

44. $\begin{vmatrix} 1 & 2 & 5 \\ -1 & 1 & -2 \\ 3 & 1 & 10 \end{vmatrix}$

45. $\begin{vmatrix} 0 & -1 & 1 \\ 1 & 0 & -2 \\ 2 & 2 & 0 \end{vmatrix}$

46. $\begin{vmatrix} 2 & -1 & 3 \\ 1 & 1 & 1 \\ 3 & -4 & 5 \end{vmatrix}$

47. $\begin{vmatrix} 1 & 2 & -1 & 2 \\ 1 & -2 & 0 & 3 \\ 3 & 0 & 1 & 5 \\ -2 & -4 & 1 & 6 \end{vmatrix}$

48. $\begin{vmatrix} 1 & -1 & -1 & 2 \\ 0 & 2 & 4 & 6 \\ 1 & 1 & 4 & 12 \\ 1 & -1 & 0 & 8 \end{vmatrix}$

49. $\begin{vmatrix} 1 & 2 & 3 & -1 \\ 6 & 5 & 9 & 8 \\ 2 & 4 & 12 & -1 \\ 1 & 2 & 6 & -1 \end{vmatrix}$

50. $\begin{vmatrix} 1 & 2 & 0 & -2 \\ -1 & 1 & 3 & 5 \\ 2 & 1 & 4 & 0 \\ -2 & 5 & 2 & 6 \end{vmatrix}$

In Exercises 51 to 54, use a graphing calculator to find the value of the determinant of the matrix. Where necessary, round your answer to the nearest thousandth.

51. $\begin{bmatrix} 2 & -2 & 3 & 1 \\ 5 & 2 & -2 & 3 \\ 6 & -1 & 2 & 3 \\ 2 & 3 & -1 & 5 \end{bmatrix}$

52. $\begin{bmatrix} 3 & -1 & 0 & 1 \\ 2 & -2 & 3 & 0 \\ -1 & -3 & 5 & 3 \\ 5 & 3 & -2 & 1 \end{bmatrix}$

53. $\begin{bmatrix} -\frac{2}{7} & 4 & -\frac{1}{6} \\ -2 & \sqrt{2} & -3 \\ \sqrt{3} & 3 & -\sqrt{5} \end{bmatrix}$

54. $\begin{bmatrix} 6 & \pi & -\frac{4}{7} \\ -5 & \sqrt{7} & 2 \\ \frac{5}{6} & -\sqrt{3} & \sqrt{10} \end{bmatrix}$

CONNECTING CONCEPTS

The area of a triangle with vertices (x_1, y_1), (x_2, y_2), and (x_3, y_3) can be given as the absolute value of the determinant

$$\frac{1}{2} \begin{vmatrix} x_1 & y_1 & 1 \\ x_2 & y_2 & 1 \\ x_3 & y_3 & 1 \end{vmatrix}$$

Use this formula to find the area of each triangle whose coordinates are given in Exercises 55 to 58.

55. $(2, 3), (-1, 0), (4, 8)$

56. $(-3, 4), (1, 5), (5, -2)$

57. $(4, 9), (8, 2), (-3, -2)$

58. $(0, 4), (-5, 7), (2, 9)$

59. Given a square matrix of order 3 in which one row is a constant multiple of a second row, show that the determinant of the matrix is zero. (*Hint:* Use an elementary row operation and part 2 of the theorem for conditions for a zero determinant.)

60. Given a square matrix of order 3 with a zero as every element in a column, show that the determinant of the matrix is zero. (*Hint:* Expand the determinant by cofactors using the column of zeros.)

61. Show that the determinant $\begin{vmatrix} x & y & 1 \\ x_1 & y_1 & 1 \\ x_2 & y_2 & 1 \end{vmatrix} = 0$ is the equation of a line through the points (x_1, y_1) and (x_2, y_2).

62. Use Exercise 61 to find the equation of the line passing through the points $(2, 3)$ and $(-1, 4)$.

63. Use Exercise 61 to find the equation of the line passing through the points $(-3, 4)$ and $(2, -3)$.

64. Show that $\begin{vmatrix} a_1 & b_1 \\ a_2 & b_2 \end{vmatrix} = \begin{vmatrix} a_1 & b_1 \\ ka_1 + a_2 & kb_1 + b_2 \end{vmatrix}$.
What property of determinants does this illustrate?

65. Surveyors use a formula to find the area of a plot of land. *Surveyor's Area Formula:* If the vertices (x_1, y_1), (x_2, y_2), $(x_3, y_3), \ldots, (x_n, y_n)$ of a convex polygon are listed counterclockwise around the perimeter, the area of the polygon is

$$A = \frac{1}{2} \left\{ \begin{vmatrix} x_1 & x_2 \\ y_1 & y_2 \end{vmatrix} + \begin{vmatrix} x_2 & x_3 \\ y_2 & y_3 \end{vmatrix} + \begin{vmatrix} x_3 & x_4 \\ y_3 & y_4 \end{vmatrix} + \cdots + \begin{vmatrix} x_n & x_1 \\ y_n & y_1 \end{vmatrix} \right\}$$

Use the Surveyor's Area Formula to find the area of the polygon with vertices $(8, -4)$, $(25, 5)$, $(15, 9)$, $(17, 20)$, and $(0, 10)$.

11. Find the difference quotient of $f(x) = x^2 - 3x + 2$.

12. Solve: $125^x = \dfrac{1}{25}$

13. Find the partial fraction decomposition of $\dfrac{x - 2}{x^2 - 5x - 6}$.

14. Solve $10^x - 10^{-x} = 2$. Round to the nearest ten-thousandth.

15. Find the exact value of $\cos 30° \sin 30° + \sec 45° \tan 60°$.

16. Use a half-angle formula to find the exact value of $\sin 292\dfrac{1}{2}°$.

17. Given angle $A = 65°$, $b = 11$ meters, and $c = 4$ meters, find the area of triangle ABC. Round to the nearest square meter.

18. Simplify: $\tan \theta + \dfrac{\cos \theta}{1 + \sin \theta}$

19. Write the complex number $z = 2 - 2i\sqrt{3}$ in trigonometric form.

20. Solve $\sin \dfrac{x}{2} + \cos x = 1$ for $0 \le x < 2\pi$.

CHAPTER 11

SEQUENCES, SERIES, AND PROBABILITY

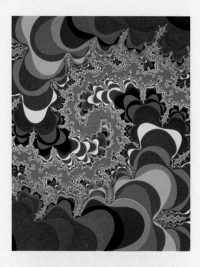

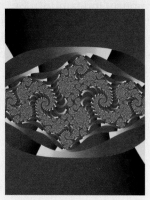

Fractals

One of the topics of this chapter is the concept of a *sequence.* The list of numbers $1, \frac{1}{2}, \frac{1}{3}, \frac{1}{4}, \frac{1}{5}, \ldots, \frac{1}{n}, \ldots$ is an example of a sequence. As the natural number n becomes very large, the numbers in the sequence become closer and closer to 0.

In addition to a sequence of numbers, we could have, for instance, a sequence of functions or a sequence of geometric figures. Although each of these sequences has important applications, we will focus here on a sequence of geometric figures.

Look at the sequence of figures below. Each succeeding figure is created by constructing right triangles using the line segments in the preceding figure as the hypotenuses.

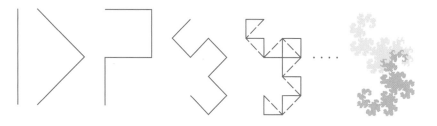

If this process is repeated over and over, the sequence of figures becomes like the last figure on the right, which is called a *fractal.* The computer-generated fractals at the left were created using a different procedure from that shown above. We will look at fractals again in **Project 1 of Section 11.3, page 926.**

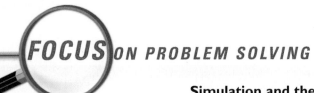
Simulation and the Monte Hall Problem

In the television game show "Let's Make a Deal," the grand prize was hidden behind one of three curtains. Behind each of the other two curtains was a less desirable prize. For instance, a box of candy might be behind one curtain and a goat behind the other.

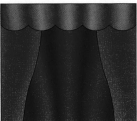
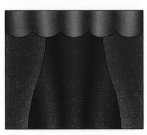
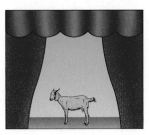

Curtain A   Curtain B   Curtain C

After a contestant chose a curtain, say, curtain *A*, the host, Monte Hall, would show the contestant one of the less desirable prizes behind one of the other two curtains. At this point the contestant could stay with her or his original choice of curtain or switch to the other unopened curtain.

Do you have a better chance of winning the grand prize by

- switching to the other closed curtain or

- staying with your original choice?

Of course there is also the possibility that it does not matter, if the chances of winning are the same with either strategy.

An analysis of the best strategy is based on calculating the probability of each alternative.

One way to decide the best strategy is to use a computer *simulation* and test each strategy. This simulation may give you some insight into a mathematical solution. One Internet site at which you can do simulations of this problem is http://www.shodor.org/interactivate/activities/monty/index.html. By selecting stay or switch and then clicking the Run Simulation button, you can see how many times you would win the grand prize by remaining with your original curtain and by switching. You may be surprised to learn that by switching, you double your chances of winning the grand prize.

An analytical solution of this problem is based on the principles of probability. For further examples of probability problems, see **Project 1 of Section 11.7, on page 958.**

INFINITE SEQUENCES AND SUMMATION NOTATION

• INFINITE SEQUENCES

The *ordered* list of numbers 2, 4, 8, 16, 32, ... is called an *infinite sequence*. The list is ordered simply because order makes a difference. The sequence 2, 8, 4, 16, 32, ... contains the same numbers but in a different order. Therefore, it is a different infinite sequence.

An infinite sequence can be thought of as a pairing between positive integers and real numbers. For example, 1, 4, 9, 16, 25, 36, ..., n^2, ... pairs a natural number with its square.

$$
\begin{array}{cccccccc}
1 & 2 & 3 & 4 & 5 & 6 & \cdots & n & \cdots \\
\downarrow & \downarrow & \downarrow & \downarrow & \downarrow & \downarrow & & \downarrow & \\
1 & 4 & 9 & 16 & 25 & 36 & \cdots & n^2 & \cdots
\end{array}
$$

This pairing of numbers enables us to define an infinite sequence as a function whose domain is the positive integers.

Infinite Sequence

An **infinite sequence** is a function whose domain is the positive integers and whose range is a set of real numbers.

Although the positive integers do not include zero, it is occasionally convenient to include zero in the domain of an infinite sequence. Also, we will frequently use the word *sequence* instead of the phrase *infinite sequence*.

As an example of a sequence, let $f(n) = 2n - 1$. The range of this function is

$$
f(1), f(2), f(3), f(4), \ldots, \quad f(n), \quad \cdots
$$
$$
1, \quad 3, \quad 5, \quad 7, \quad \ldots, \quad 2n - 1, \quad \cdots
$$

The elements in the range of a sequence are called the **terms** of the sequence. For our example, the terms are 1, 3, 5, 7, ..., $2n - 1$, ... The **first term** of the sequence is 1, the **second term** is 3, and so on. The **nth term**, or the **general term**, is $2n - 1$.

? QUESTION What is the fifth term of the sequence $f(n) = 2n - 1$ given above?

Rather than use functional notation for sequences, it is customary to use a subscript notation. Thus a_n represents the nth term of a sequence. Using this notation, we would write

$$
a_n = 2n - 1
$$

Thus $a_1 = 1$, $a_2 = 3$, $a_3 = 5$, $a_4 = 7$, and so on.

? ANSWER $f(5) = 2(5) - 1 = 9$

EXAMPLE 1   **Find the Terms of a Sequence**

a.   Find the first three terms of the sequence $a_n = \dfrac{1}{n(n+1)}$.

b.   Find the eighth term of the sequence $a_n = \dfrac{2^n}{n^2}$.

Solution

a. $a_1 = \dfrac{1}{1(1+1)} = \dfrac{1}{2}, a_2 = \dfrac{1}{2(2+1)} = \dfrac{1}{6}, a_3 = \dfrac{1}{3(3+1)} = \dfrac{1}{12}$

b. $a_8 = \dfrac{2^8}{8^2} = \dfrac{256}{64} = 4$

▶ **TRY EXERCISE 6, PAGE 908**

An **alternating sequence** is one in which the signs of the terms *alternate* between positive and negative values. The sequence defined by $a_n = (-1)^{n+1} \cdot \dfrac{1}{n}$ is an alternating sequence.

$$a_1 = (-1)^{1+1} \cdot \frac{1}{1} = 1 \qquad a_2 = (-1)^{2+1} \cdot \frac{1}{2} = -\frac{1}{2} \qquad a_3 = (-1)^{3+1} \cdot \frac{1}{3} = \frac{1}{3}$$

The first six terms of the sequence are

$$1, -\frac{1}{2}, \frac{1}{3}, -\frac{1}{4}, \frac{1}{5}, -\frac{1}{6}$$

A **recursively defined sequence** is one in which each succeeding term of the sequence is defined by using some of the preceding terms. For example, let $a_1 = 1$, $a_2 = 1$, and $a_{n+1} = a_{n-1} + a_n$.

$$\begin{array}{ll} a_3 = a_1 + a_2 = 1 + 1 = 2 & \bullet\, n = 2 \\ a_4 = a_2 + a_3 = 1 + 2 = 3 & \bullet\, n = 3 \\ a_5 = a_3 + a_4 = 2 + 3 = 5 & \bullet\, n = 4 \\ a_6 = a_4 + a_5 = 3 + 5 = 8 & \bullet\, n = 5 \end{array}$$

This recursive sequence 1, 1, 2, 3, 5, 8, ... is called the **Fibonacci sequence**, named after Leonardo Fibonacci (1180?–?1250), an Italian mathematician.

MATH MATTERS

Recursive subroutines are a part of many computer programming languages. One of the most influential people in the development of computer languages (especially COBOL) was Grace Murray Hooper. She was the first woman to receive a doctorate in mathematics from Yale University. She went on to become an admiral in the U.S. Navy.

EXAMPLE 2   **Find Terms of a Sequence Defined Recursively**

Let $a_1 = 1$ and $a_n = na_{n-1}$. Find a_2, a_3, and a_4.

Solution

$$a_2 = 2a_1 = 2 \cdot 1 = 2 \qquad a_3 = 3a_2 = 3 \cdot 2 = 6 \qquad a_4 = 4a_3 = 4 \cdot 6 = 24$$

▶ **TRY EXERCISE 28, PAGE 908**

● FACTORIALS

It is possible to find an nth term formula for the sequence defined recursively in Example 2 by

$$a_1 = 1 \qquad a_n = na_{n-1}$$

Consider the term a_5 of that sequence.

$$
\begin{aligned}
a_5 &= 5a_4 \\
&= 5 \cdot 4a_3 & &\bullet \; a_4 = 4a_3 \\
&= 5 \cdot 4 \cdot 3a_2 & &\bullet \; a_3 = 3a_2 \\
&= 5 \cdot 4 \cdot 3 \cdot 2a_1 & &\bullet \; a_2 = 2a_1 \\
&= 5 \cdot 4 \cdot 3 \cdot 2 \cdot 1 & &\bullet \; a_1 = 1
\end{aligned}
$$

Continuing in this manner for a_n, we have

$$
\begin{aligned}
a_n &= na_{n-1} \\
&= n(n-1)a_{n-2} \\
&= n(n-1)(n-2)a_{n-3} \\
&\;\;\vdots \\
&= n(n-1)(n-2)(n-3) \cdots 2 \cdot 1
\end{aligned}
$$

The number $n \cdot (n-1) \cdots 3 \cdot 2 \cdot 1$ is called **n factorial** and is written $n!$.

The Factorial of a Number

If n is a positive integer, then $n!$, which is read "n factorial," is

$$n! = n \cdot (n-1) \cdots 3 \cdot 2 \cdot 1$$

We also define

$$0! = 1$$

MATH MATTERS

The factorial symbol as an exclamation point was first introduced in 1808 so that printers setting the type in a book would not have to write out, for example, $5 \cdot 4 \cdot 3 \cdot 2 \cdot 1$ but instead could write 5!. This notation was not accepted by everyone, and alternative suggestions were made. One other notation was $\underline{|n}$. Although this angle bracket saved the printer from writing a product, it had to be constructed, whereas the exclamation point was already in the printer's tray.

It may seem strange to define $0! = 1$, but we shall see later that it is a reasonable definition.

Examples of factorials include

$$5! = 5 \cdot 4 \cdot 3 \cdot 2 \cdot 1 = 120$$
$$10! = 10 \cdot 9 \cdot 8 \cdot 7 \cdot 6 \cdot 5 \cdot 4 \cdot 3 \cdot 2 \cdot 1 = 3{,}628{,}800$$

Note that we can write 12! as

$$12! = 12 \cdot 11! = 12 \cdot 11 \cdot 10! = 12 \cdot 11 \cdot 10 \cdot 9!$$

In general,

$$n! = n \cdot (n-1)!$$

EXAMPLE 3   **Evaluate Factorial Expressions**

Evaluate each factorial expression.   **a.** $\dfrac{8!}{5!}$ **b.** $6! - 4!$

Solution

a. $\dfrac{8!}{5!} = \dfrac{8 \cdot 7 \cdot 6 \cdot 5!}{5!} = 8 \cdot 7 \cdot 6 = 336$

b. $6! - 4! = (6 \cdot 5 \cdot 4 \cdot 3 \cdot 2 \cdot 1) - (4 \cdot 3 \cdot 2 \cdot 1) = 720 - 24 = 696$

▶ **TRY EXERCISE 42, PAGE 908**

● **PARTIAL SUMS AND SUMMATION NOTATION**

Another important way of obtaining a sequence is by adding the terms of a given sequence. For example, consider the sequence whose general term is given by $a_n = \dfrac{1}{2^n}$. The terms of this sequence are

$$\frac{1}{2}, \frac{1}{4}, \frac{1}{8}, \frac{1}{16}, \frac{1}{32}, \cdots, \frac{1}{2^n}, \cdots$$

From this sequence we can generate a new sequence that is the sum of the terms of $\dfrac{1}{2^n}$.

$$S_1 = \frac{1}{2}$$

$$S_2 = \frac{1}{2} + \frac{1}{4} = \frac{3}{4}$$

$$S_3 = \frac{1}{2} + \frac{1}{4} + \frac{1}{8} = \frac{7}{8}$$

$$S_4 = \frac{1}{2} + \frac{1}{4} + \frac{1}{8} + \frac{1}{16} = \frac{15}{16}$$

MATH MATTERS

Leonhard Euler (1707–1783) found that the sequence of terms given by

$$S_n = 1 - \frac{1}{3} + \frac{1}{5} - \frac{1}{7} + \cdots$$
$$+ \frac{(-1)^{n-1}}{2n - 1}$$

became closer and closer to $\dfrac{\pi}{4}$ as n increased. In summation notation, we would write

$$S_n = \sum_{k=1}^{n} \frac{(-1)^{k-1}}{2k - 1}$$

and, in general, $S_n = \dfrac{1}{2} + \dfrac{1}{4} + \dfrac{1}{8} + \dfrac{1}{16} + \cdots + \dfrac{1}{2^n}$

The term S_n is called the **nth partial sum** of the infinite sequence, and the sequence $S_1, S_2, S_3, \ldots, S_n$ is called the **sequence of partial sums**.

A convenient notation used for partial sums is called **summation notation**. The sum of the first n terms of a sequence a_n is represented by using the Greek letter Σ (sigma).

$$\sum_{i=1}^{n} a_i = a_1 + a_2 + a_3 + \cdots + a_n$$

This sum is called a **series**. The letter i is called the **index of the summation**; n is the **upper limit** of the summation; 1 is the **lower limit** of the summation.

EXAMPLE 4	**Evaluate Series**

Evaluate each series.   **a.** $\displaystyle\sum_{i=1}^{4} \frac{i}{i+1}$ **b.** $\displaystyle\sum_{j=2}^{5} (-1)^j j^2$

Solution

a. $\displaystyle\sum_{i=1}^{4} \frac{i}{i+1} = \frac{1}{2} + \frac{2}{3} + \frac{3}{4} + \frac{4}{5} = \frac{163}{60}$

b. $\displaystyle\sum_{j=2}^{5} (-1)^j j^2 = (-1)^2 2^2 + (-1)^3 3^2 + (-1)^4 4^2 + (-1)^5 5^2$

$= 4 - 9 + 16 - 25 = -14$

▶ **TRY EXERCISE 52, PAGE 909**

take note

Example 4b illustrates that it is not necessary for a summation to begin at 1. The index of the summation can be any letter.

Properties of Summation Notation

If a_n and b_n are sequences and c is a real number, then

1. $\displaystyle\sum_{i=1}^{n} (a_i \pm b_i) = \sum_{i=1}^{n} a_i \pm \sum_{i=1}^{n} b_i$

2. $\displaystyle\sum_{i=1}^{n} ca_i = c \sum_{i=1}^{n} a_i$

3. $\displaystyle\sum_{i=1}^{n} c = nc$

The proof of property (1) depends on the commutative and associative properties of real numbers.

$$\sum_{i=1}^{n} (a_i \pm b_i) = (a_1 \pm b_1) + (a_2 \pm b_2) + \cdots + (a_n \pm b_n)$$

$$= (a_1 + a_2 + \cdots + a_n) \pm (b_1 + b_2 + \cdots + b_n)$$

$$= \sum_{i=1}^{n} a_i \pm \sum_{i=1}^{n} b_i$$

Property (2) is proved by using the distributive property; this is left as an exercise.

To prove property (3), let $a_n = c$. That is, each a_n is equal to the same constant c. (This is called a **constant sequence**.) Then

$$\sum_{i=1}^{n} a_n = a_1 + a_2 + \cdots + a_n = \underbrace{c + c + \cdots + c}_{n \text{ terms}} = nc$$

TOPICS FOR DISCUSSION

1. Discuss the difference between a finite sequence and an infinite sequence. Give an example of each type.

2. Discuss the difference between a sequence and a series.

3. What is a recursive sequence? Give an example of a recursive sequence.

4. What is an alternating sequence? Give an example of an alternating sequence.

EXERCISE SET 11.1

In Exercises 1 to 24, find the first three terms and the eighth term of the sequence that has the given nth term.

1. $a_n = n(n - 1)$

2. $a_n = 2n$

3. $a_n = 1 - \dfrac{1}{n}$

4. $a_n = \dfrac{n + 1}{n}$

5. $a_n = \dfrac{(-1)^{n+1}}{n^2}$

▶ **6.** $a_n = \dfrac{(-1)^{n+1}}{n(n + 1)}$

7. $a_n = \dfrac{(-1)^{2n-1}}{3n}$

8. $a_n = \dfrac{(-1)^n}{2n - 1}$

9. $a_n = \left(\dfrac{2}{3}\right)^n$

10. $a_n = \left(\dfrac{-1}{2}\right)^n$

11. $a_n = 1 + (-1)^n$

12. $a_n = 1 + (-0.1)^n$

13. $a_n = (1.1)^n$

14. $a_n = \dfrac{n}{n^2 + 1}$

15. $a_n = \dfrac{(-1)^{n+1}}{\sqrt{n}}$

16. $a_n = \dfrac{3^{n-1}}{2^n}$

17. $a_n = n!$

18. $a_n = \dfrac{n!}{(n - 1)!}$

19. $a_n = \log n$

20. $a_n = \ln n$ (natural logarithm)

21. a_n is the digit in the nth place in the decimal expansion of $\dfrac{1}{7}$.

22. a_n is the digit in the nth place in the decimal expansion of $\dfrac{1}{13}$.

23. $a_n = 3$

24. $a_n = -2$

In Exercises 25 to 34, find the first three terms of each recursively defined sequence.

25. $a_1 = 5, a_n = 2a_{n-1}$

26. $a_1 = 2, a_n = 3a_{n-1}$

27. $a_1 = 2, a_n = na_{n-1}$

▶ **28.** $a_1 = 1, a_n = n^2 a_{n-1}$

29. $a_1 = 2, a_n = (a_{n-1})^2$

30. $a_1 = 4, a_n = \dfrac{1}{a_{n-1}}$

31. $a_1 = 2, a_n = 2na_{n-1}$

32. $a_1 = 2, a_n = (-3)na_{n-1}$

33. $a_1 = 3, a_n = (a_{n-1})^{1/n}$

34. $a_1 = 2, a_n = (a_{n-1})^n$

35. $a_1 = 1, a_2 = 3, a_n = \dfrac{1}{2}(a_{n-1} + a_{n-2})$. Find $a_3, a_4,$ and a_5.

36. $a_1 = 1, a_2 = 4, a_n = (a_{n-1})(a_{n-2})$. Find $a_3, a_4,$ and a_5.

In Exercises 37 to 44, evaluate the factorial expression.

37. $7! - 6!$

38. $(4!)^2$

39. $\dfrac{9!}{7!}$

40. $\dfrac{10!}{5!}$

41. $\dfrac{8!}{3! \, 5!}$

▶ **42.** $\dfrac{12!}{4! \, 8!}$

43. $\dfrac{100!}{99!}$

44. $\dfrac{100!}{98! \, 2!}$

In Exercises 45 to 58, evaluate the series.

45. $\displaystyle\sum_{i=1}^{5} i$

46. $\displaystyle\sum_{i=1}^{4} i^2$

47. $\displaystyle\sum_{i=1}^{5} i(i - 1)$

48. $\displaystyle\sum_{i=1}^{7} (2i + 1)$

49. $\displaystyle\sum_{k=1}^{4} \dfrac{1}{k}$

50. $\displaystyle\sum_{k=1}^{6} \dfrac{1}{k(k + 1)}$

51. $\displaystyle\sum_{j=1}^{8} 2j$

▶ **52.** $\displaystyle\sum_{i=1}^{6} (2i + 1)(2i - 1)$

53. $\displaystyle\sum_{i=3}^{5} (-1)^i 2^i$

54. $\displaystyle\sum_{i=3}^{5} \frac{(-1)^i}{2^i}$

55. $\displaystyle\sum_{n=1}^{7} \log \frac{n+1}{n}$

56. $\displaystyle\sum_{n=2}^{8} \ln \frac{n}{n+1}$

57. $\displaystyle\sum_{k=0}^{8} \frac{8!}{k!\,(8-k)!}$

58. $\displaystyle\sum_{k=0}^{7} \frac{1}{k!}$

In Exercises 59 to 66, write the given series in summation notation.

59. $\dfrac{1}{1} + \dfrac{1}{4} + \dfrac{1}{9} + \dfrac{1}{16} + \dfrac{1}{25} + \dfrac{1}{36}$

60. $2 + 4 + 6 + 8 + 10 + 12 + 14$

61. $2 - 4 + 8 - 16 + 32 - 64 + 128$

62. $1 - 8 + 27 - 64 + 125$

63. $7 + 10 + 13 + 16 + 19$

64. $30 + 26 + 22 + 18 + 14 + 10$

65. $\dfrac{1}{2} + \dfrac{1}{4} + \dfrac{1}{8} + \dfrac{1}{16}$

66. $1 - \dfrac{2}{3} + \dfrac{4}{9} - \dfrac{8}{27} + \dfrac{16}{81} - \dfrac{32}{243}$

CONNECTING CONCEPTS

67. NEWTON'S METHOD Newton's approximation to the square root of a number, N, is given by the recursive sequence

$$a_1 = \frac{N}{2} \qquad a_n = \frac{1}{2}\left(a_{n-1} + \frac{N}{a_{n-1}}\right)$$

Approximate $\sqrt{7}$ by computing a_4. Compare this result with the calculator value of $\sqrt{7} \approx 2.6457513$.

68. Use the formula in Exercise 67 to approximate $\sqrt{10}$ by finding a_5.

69. Let $a_1 = N$ and $a_n = \sqrt{a_{n-1}}$. Find a_{20} when $N = 7$. (*Hint:* Enter 7 into your calculator and then press the $\boxed{\sqrt{}}$ key 19 times.) Make a conjecture as to the value of a_n as n increases without bound.

70. Let $a_n = i^n$, where i is the imaginary unit. Find the first eight terms of the sequence defined by a_n. Find a_{237}.

71. Let $a_n = \left[\dfrac{1}{2}\left(-1 + i\sqrt{3}\right)\right]^n$. Find the first six terms of the sequence defined by a_n. Find a_{99}.

72. **STIRLING'S FORMULA** By using a calculator, evaluate $\sqrt{2\pi n}(n/e)^n$, where e is the base of the natural logarithms for $n = 10$, 20, and 30. This formula is called Stirling's formula and is used as an approximation for $n!$. For $n > 20$, the error in the approximation is less than 0.1%.

73. Prove that $\displaystyle\sum_{i=1}^{n} ca_i = c\sum_{i=1}^{n} a_i$, where c is a constant.

PREPARE FOR SECTION 11.2

74. Solve $a = b + (n-1)d$ for d given $a = -3$, $b = 25$, and $n = 15$. [1.1]

75. Solve $a = b + (n-1)d$ for d given $a = 13$, $b = 3$, and $n = 5$. [1.1]

76. Evaluate $S = \dfrac{n[2a_1 + (n-1)d]}{2}$ when $a_1 = 2$, $d = \dfrac{5}{4}$, and $n = 50$. [1.2]

77. Find the fifth term of the sequence whose nth term is $a_n = 5 + (n-1)4$. [11.1]

78. Find the twentieth term of the sequence whose nth term is $a_n = 52 + (n-1)(-3)$. [11.1]

79. Given the sequence $2, 5, 8, \ldots, 3n - 1, \ldots$, are the differences between successive terms equal to the same constant? [11.1]

━━━━ *PROJECTS* ━━━

1. FORMULAS FOR INFINITE SEQUENCES It is not possible to define an infinite sequence by giving a finite number of terms of the sequence. For instance, the question "What is the next term in the sequence 2, 4, 6, 8, ...?" does not have a unique answer.

   a. Verify this statement by finding a formula for a_n such that the first four terms of the sequence are 2, 4, 6, 8 and the next term is 43. *Suggestion:* The formula

$$a_n = \frac{n(n-1)(n-2)(n-3)(n-4)}{4!} + 2n$$

generates the sequence 2, 4, 6, 8, 15 for $n = 1, 2, 3, 4, 5$.

   b. Extend the result in part **a.** by finding a formula for a_n that will give the first four terms as 2, 4, 6, 8 and the fifth term as x, where x is any real number.

SECTION **11.2**   **ARITHMETIC SEQUENCES AND SERIES**

● **ARITHMETIC SEQUENCES**
● **ARITHMETIC SERIES**
● **ARITHMETIC MEANS**

● **ARITHMETIC SEQUENCES**

Note that in the sequence

$$2, 5, 8, 11, 14, \ldots, 3n - 1, \ldots$$

the difference between successive terms is always 3. Such a sequence is an *arithmetic sequence* or an *arithmetic progression.* These sequences have the following property: The difference between successive terms is the same constant. This constant is called the *common difference.* For the sequence above, the common difference is 3.

   In general, an arithmetic sequence can be defined as follows:

Arithmetic Sequence

Let d be a real number. A sequence a_n is an **arithmetic sequence** if

$$a_{i+1} - a_i = d \quad \text{for all } i$$

The number d is the **common difference** for the sequence.

Further examples of arithmetic sequences include

$$3, 8, 13, 18, \ldots, 5n - 2, \ldots$$
$$11, 7, 3, -1, \ldots, -4n + 15, \ldots$$
$$1, 2, 3, 4, \ldots, n, \ldots$$

? QUESTION Is the sequence $2, 6, 10, 14, \ldots, 4n - 2, \ldots$ an arithmetic sequence?

? ANSWER Yes. The difference between any two successive terms is 4.

TOPICS FOR DISCUSSION

1. Discuss what distinguishes an arithmetic sequence from all other types of sequences.

2. Is $a_n = \dfrac{2}{n}$ a possible formula for the nth term of an arithmetic sequence? Why or why not?

3. Discuss the characteristics of an arithmetic series. Give an example of an arithmetic series.

4. Consider the series $\sum\limits_{k=1}^{n} f(k)$. Discuss how you can determine whether this is an arithmetic series.

EXERCISE SET 11.2

In Exercises 1 to 14, find the ninth, twenty-fourth, and nth terms of the arithmetic sequence.

1. $6, 10, 14, \ldots$

2. $7, 12, 17, \ldots$

3. $6, 4, 2, \ldots$

4. $11, 4, -3, \ldots$

5. $-8, -5, -2, \ldots$

6. $-15, -9, -3, \ldots$

7. $1, 4, 7, \ldots$

8. $-4, 1, 6, \ldots$

9. $a, a + 2, a + 4, \ldots$

10. $a - 3, a + 1, a + 5, \ldots$

11. $\log 7, \log 14, \log 28, \ldots$

12. $\ln 4, \ln 16, \ln 64, \ldots$

13. $\log a, \log a^2, \log a^3, \ldots$

14. $\log_2 5, \log_2 5a, \log_2 5a^2, \ldots$

15. The fourth and fifth terms of an arithmetic sequence are 13 and 15. Find the twentieth term.

▶ **16.** The sixth and eighth terms of an arithmetic sequence are -14 and -20. Find the fifteenth term.

17. The fifth and seventh terms of an arithmetic sequence are -19 and -29. Find the seventeenth term.

18. The fourth and seventh terms of an arithmetic sequence are 22 and 34. Find the twenty-third term.

In Exercises 19 to 32, find the nth partial sum of the arithmetic sequence.

19. $a_n = 3n + 2; n = 10$

20. $a_n = 4n - 3; n = 12$

21. $a_n = 3 - 5n; n = 15$

▶ **22.** $a_n = 1 - 2n; n = 20$

23. $a_n = 6n; n = 12$

24. $a_n = 7n; n = 14$

25. $a_n = n + 8; n = 25$

26. $a_n = n - 4; n = 25$

27. $a_n = -n; n = 30$

28. $a_n = 4 - n; n = 40$

29. $a_n = n + x; n = 12$

30. $a_n = 2n - x; n = 15$

31. $a_n = nx; n = 20$

32. $a_n = -nx; n = 14$

In Exercises 33 to 36, insert k arithmetic means between the given numbers.

33. -1 and $23; k = 5$

▶ **34.** 7 and $19; k = 5$

35. 3 and $\dfrac{1}{2}; k = 4$

36. $\dfrac{11}{3}$ and $6; k = 4$

37. Show that the sum of the first n positive odd integers is n^2.

38. Show that the sum of the first n positive even integers is $n^2 + n$.

39. STACKING LOGS Logs are stacked so that there are 25 logs in the bottom row, 24 logs in the second row, and so on, decreasing by 1 log each row. How many logs are stacked in the sixth row? How many logs are there in all six rows?

40. THEATER SEATING The seating section in a theater has 27 seats in the first row, 29 seats in the second row, and so on, increasing by 2 seats each row for a total of 10 rows. How many seats are in the tenth row, and how many seats are there in the section?

41. CONTEST PRIZES A contest offers 15 prizes. The first prize is $5000, and each successive prize is $250 less than the preceding prize. What is the value of the fifteenth prize? What is the total amount of money distributed in prizes?

42. PHYSICAL FITNESS An exercise program calls for walking 15 minutes each day for a week. Each week thereafter, the amount of time spent walking increases by 5 minutes per day. In how many weeks will a person be walking 60 minutes each day?

43. PHYSICS An object dropped from a cliff will fall 16 feet the first second, 48 feet the second second, 80 feet the third second, and so on, increasing by 32 feet each second. What is the total distance the object will fall in 7 seconds?

44. PHYSICS The distance a ball rolls down a ramp each second is given by the arithmetic sequence whose nth term is $2n - 1$ feet. Find the distance the ball rolls during the tenth second and the total distance the ball travels in 10 seconds.

——— CONNECTING CONCEPTS ———

45. If $f(x)$ is a linear polynomial, show that $f(n)$, where n is a positive integer, is an arithmetic sequence.

46. Find the formula for a_n in terms of a_1 and n for the sequence that is defined recursively by $a_1 = 3$, $a_n = a_{n-1} + 5$.

47. Find a formula for a_n in terms of a_1 and n for the sequence that is defined recursively by $a_1 = 4$, $a_n = a_{n-1} - 3$.

48. Suppose a_n and b_n are two sequences such that $a_1 = 4$, $a_n = b_{n-1} + 5$ and $b_1 = 2$, $b_n = a_{n-1} + 1$. Show that a_n and b_n are arithmetic sequences. Find a_{100}.

49. Suppose a_n and b_n are two sequences such that $a_1 = 1$, $a_n = b_{n-1} + 7$ and $b_1 = -2$, $b_n = a_{n-1} + 1$. Show that a_n and b_n are arithmetic sequences. Find a_{50}.

——— PREPARE FOR SECTION 11.3 ———

50. For the sequence 2, 4, 8, . . . , 2^n, . . . , what is the ratio of any two successive terms? [11.1]

51. Evaluate: $\displaystyle\sum_{n=1}^{4} \frac{1}{2^{n-1}}$ [11.1]

52. Evaluate $S = \dfrac{a(1 - r^n)}{1 - r}$ when $a = 3$, $r = -2$, and $n = 5$. [P.2]

53. Solve $S - rS = a - ar^2$, $r \neq 1$, for S. [1.1]

54. Write the first three terms of the sequence whose nth term is $a_n = 3\left(-\dfrac{1}{2}\right)^n$. [11.1]

55. Find the first three terms of the sequence of partial sums for the sequence 2, 4, 8, . . . , 2^n, . . . [11.1]

——— PROJECTS ———

1. ANGLES OF A TRIANGLE The sum of the interior angles of a triangle is 180°.

  a. Using this fact, what is the sum of the interior angles of a quadrilateral?

  b. What is the sum of the interior angles of a pentagon?

  c. What is the sum of the interior angles of a hexagon?

  d. On the basis of your previous results, what is the apparent formula for the sum of the interior angles of a polygon of n sides?

2. PROVE A FORMULA Prove the Alternative Formula for the Sum of an Arithmetic Sequence.

GEOMETRIC SEQUENCES AND SERIES

● GEOMETRIC SEQUENCES

Arithmetic sequences are characterized by a common *difference* between successive terms. A *geometric sequence* is characterized by a common *ratio* between successive terms.

The sequence

$$3, 6, 12, 24, \ldots, 3(2^{n-1}), \ldots$$

is a geometric sequence. Note that the ratio of any two successive terms is 2.

$$\frac{6}{3} = 2 \qquad \frac{12}{6} = 2 \qquad \frac{24}{12} = 2$$

Geometric Sequence

Let r be a nonzero constant real number. A sequence is a **geometric sequence** if

$$\frac{a_{i+1}}{a_i} = r \quad \text{for all positive integers } i.$$

The number r is called the **common ratio.**

EXAMPLE 1   Determine Whether a Sequence Is a Geometric Sequence

Which of the following are geometric sequences?

a. $4, -2, 1, \ldots, 4\left(-\frac{1}{2}\right)^{n-1}, \ldots$ b. $1, 4, 9, \ldots, n^2, \ldots$

Solution

To determine whether the sequence is a geometric sequence, calculate the ratio of successive terms.

a. $\dfrac{a_{i+1}}{a_i} = \dfrac{4\left(-\dfrac{1}{2}\right)^{i}}{4\left(-\dfrac{1}{2}\right)^{i-1}} = -\dfrac{1}{2}.$

Because the ratio of successive terms is a constant, the sequence is a geometric sequence.

b. $\dfrac{a_{i+1}}{a_i} = \dfrac{(i+1)^2}{i^2} = \left(1 + \dfrac{1}{i}\right)^2$

Because the ratio of successive terms is not a constant, the sequence is not a geometric sequence.

▶ **TRY EXERCISE 6, PAGE 923**

Consider a geometric sequence in which the first term is a_1 and the common ratio is r. By multiplying each successive term of the geometric sequence by the common ratio, we can derive a formula for the nth term.

$$a_1 = a_1$$
$$a_2 = a_1 r$$
$$a_3 = a_2 r = (a_1 r)r = a_1 r^2$$
$$a_4 = a_3 r = (a_1 r^2)r = a_1 r^3$$

Note the relationship between the number of the term and the number that is the exponent on r. The exponent on r is 1 less than the number of the term. With this observation, we can write a formula for the nth term of a geometric sequence.

The nth Term of a Geometric Sequence

The **nth term of a geometric sequence** with first term a_1 and common ratio r is

$$a_n = a_1 r^{n-1}$$

EXAMPLE 2   Find the nth Term of a Geometric Sequence

Find the nth term of the geometric sequence whose first three terms are

a. $4, \dfrac{8}{3}, \dfrac{16}{9}, \ldots$ b. $5, -10, 20, \ldots$

Solution

a. $r = \dfrac{8/3}{4} = \dfrac{2}{3}$ and $a_1 = 4$. Thus $a_n = 4\left(\dfrac{2}{3}\right)^{n-1}$.

b. $r = \dfrac{-10}{5} = -2$ and $a_1 = 5$. Thus $a_n = 5(-2)^{n-1}$.

▶ **TRY EXERCISE 18, PAGE 924**

● FINITE GEOMETRIC SERIES

Adding the terms of a geometric sequence, we can define the nth partial sum of a geometric sequence in a manner similar to that of an arithmetic sequence. Consider the geometric sequence $1, 2, 4, 8, \ldots, 2^{n-1}, \ldots$.

$$S_1 = 1$$
$$S_2 = 1 + 2 = 3$$
$$S_3 = 1 + 2 + 4 = 7$$
$$S_4 = 1 + 2 + 4 + 8 = 15$$
$$\vdots \qquad \vdots$$
$$S_n = 1 + 2 + 4 + 8 + \cdots + 2^{n-1}$$

The first four terms of the sequence of partial sums are 1, 3, 7, and 15.

This pattern suggests the conjecture that

$$S_n = \frac{1}{1 \cdot 2} + \frac{1}{2 \cdot 3} + \frac{1}{3 \cdot 4} + \cdots + \frac{1}{n(n + 1)} = \frac{n}{n + 1}$$

How can we be sure that the pattern does not break down when $n = 50$ or maybe when $n = 2000$ or some other large number? As we will show, this conjecture is true for all values of n.

As a second example, consider the conjecture that the expression $n^2 - n + 41$ is a prime number for all positive integers n. To test this conjecture, we will try various values of n. See **Table 11.3**. The results suggest that the conjecture is true. But again, how can we be sure? In fact, this conjecture is false when $n = 41$. In that case we have

$$n^2 - n + 41 = (41)^2 - 41 + 41 = (41)^2$$

and $(41)^2$ is not a prime.

The last example illustrates that just verifying a conjecture for a few values of n does not constitute a proof of the conjecture. To prove theorems about statements involving positive integers, a process called *mathematical induction* is used. This process is based on an axiom called the *induction axiom*.

TABLE 11.3

n	$n^2 - n + 41$	
1	41	Prime
2	43	Prime
3	47	Prime
4	53	Prime
5	61	Prime

Induction Axiom

Suppose S is a set of positive integers with the following two properties:

1. 1 is an element of S.

2. If the positive integer k is in S, then $k + 1$ is in S.

Then S contains all the positive integers.

Part 2 of this axiom states that if some positive integer, say, 8, is in S, then $8 + 1$, or 9, is in S. But because 9 is in S, part 2 says that $9 + 1$, or 10, is in S, and so on. Part 1 states that 1 is in S. Thus 2 is in S; thus 3 is in S; thus 4 is in S; .... Therefore all the positive integers are in S.

● PRINCIPLE OF MATHEMATICAL INDUCTION

The induction axiom is used to prove the *Principle of Mathematical Induction.*

Principle of Mathematical Induction

Let P_n be a statement about a positive integer n. If

1. P_1 is true, and

2. the truth of P_k implies the truth of P_{k+1},

then P_n is true for all positive integers.

Part 2 of the Principle of Mathematical Induction is referred to as the **induction hypothesis**. When applying this step, we assume that the statement P_k is true, and then we try to prove that P_{k+1} is also true.

As an example, we will prove that the first conjecture we made in this section is true for all positive integers. Every induction proof has the two distinct parts stated in the theorem. First, we must show that the result is true for $n = 1$. Second, we assume that the statement is true for some positive integer k, and using that assumption, we prove that the statement is true for $n = k + 1$.

Prove that

$$S_n = \frac{1}{1 \cdot 2} + \frac{1}{2 \cdot 3} + \frac{1}{3 \cdot 4} + \cdots + \frac{1}{n(n+1)} = \frac{n}{n+1}$$

for all positive integers n.

Proof

1. For $n = 1$,

$$S_1 = \frac{1}{1(1+1)} = \frac{1}{2}, \text{ and } \frac{n}{n+1} = \frac{1}{1+1} = \frac{1}{2}$$

The statement is true for $n = 1$.

2. Assume the statement is true for some positive integer k.

$$S_k = \frac{1}{1 \cdot 2} + \frac{1}{2 \cdot 3} + \frac{1}{3 \cdot 4} + \cdots + \frac{1}{k(k+1)} = \frac{k}{k+1} \qquad \text{• Induction hypothesis}$$

Now verify that the formula is true when $n = k + 1$. That is, verify that

$$S_{k+1} = \frac{k+1}{(k+1)+1} = \frac{k+1}{k+2} \qquad \text{• This is the goal of the induction proof.}$$

It is helpful, when proving a theorem about sums, to note that

$$S_{k+1} = S_k + a_{k+1}$$

Begin by noting that $a_k = \dfrac{1}{k(k+1)}$; thus, $a_{k+1} = \dfrac{1}{(k+1)(k+2)}$.

$$
\begin{aligned}
S_{k+1} &= S_k + a_{k+1} \\
&= \frac{k}{k+1} + \frac{1}{(k+1)(k+2)} \qquad \text{• By the induction hypothesis} \\
&\qquad\qquad\qquad\qquad\qquad\qquad \text{and substituting for } a_{k+1} \\
&= \frac{k(k+2)}{(k+1)(k+2)} + \frac{1}{(k+1)(k+2)} \\
&= \frac{k(k+2)+1}{(k+1)(k+2)} = \frac{k^2+2k+1}{(k+1)(k+2)} = \frac{(k+1)^2}{(k+1)(k+2)} \\
S_{k+1} &= \frac{k+1}{k+2}
\end{aligned}
$$

Because we have verified the two parts of the Principle of Mathematical Induction, we can conclude that the statement is true for all positive integers.   ◆

EXAMPLE 1   **Prove by Mathematical Induction**

Prove that $1^2 + 2^2 + 3^2 + \cdots + n^2 = \dfrac{n(n + 1)(2n + 1)}{6}$.

Solution

Verify the two parts of the Principle of Mathematical Induction.

1. Let $n = 1$.

$$S_1 = 1^2 = 1 = \frac{1(1 + 1)(2 \cdot 1 + 1)}{6}$$

2. Assume the statement is true for some positive integer k.

$$S_k = 1^2 + 2^2 + 3^2 + \cdots + k^2 = \frac{k(k + 1)(2k + 1)}{6} \qquad \bullet \text{ Induction hypothesis}$$

Verify that the statement is true when $n = k + 1$. Show that

$$S_{k+1} = \frac{(k + 1)(k + 2)(2k + 3)}{6}$$

Because $a_k = k^2$, $a_{k+1} = (k + 1)^2$.

$$S_{k+1} = \qquad S_k \qquad + a_{k+1}$$

$$= \frac{k(k + 1)(2k + 1)}{6} + (k + 1)^2$$

$$= \frac{k(k + 1)(2k + 1)}{6} + \frac{6(k + 1)^2}{6} = \frac{k(k + 1)(2k + 1) + 6(k + 1)^2}{6}$$

$$= \frac{(k + 1)[k(2k + 1) + 6(k + 1)]}{6} = \frac{(k + 1)(2k^2 + 7k + 6)}{6}$$

$$S_{k+1} = \frac{(k + 1)(k + 2)(2k + 3)}{6}$$

By the Principle of Mathematical Induction, the statement is true for all positive integers.

▶ **TRY EXERCISE 8, PAGE 932**

Mathematical induction can also be used to prove statements about sequences, products, and inequalities.

EXAMPLE 2   **Prove a Product Formula by Mathematical Induction**

Prove that

$$\left(1 + \frac{1}{1}\right)\left(1 + \frac{1}{2}\right)\left(1 + \frac{1}{3}\right)\cdots\left(1 + \frac{1}{n}\right) = n + 1$$

Continued ▶

Solution

1. Verify for $n = 1$.

$$\left(1 + \frac{1}{1}\right) = 2, \text{ and } 1 + 1 = 2$$

2. Assume the statement is true for some positive integer k.

$$P_k = \left(1 + \frac{1}{1}\right)\left(1 + \frac{1}{2}\right)\left(1 + \frac{1}{3}\right)\cdots\left(1 + \frac{1}{k}\right) = k + 1 \qquad \bullet \textbf{ Induction hypothesis}$$

Verify that the statement is true when $n = k + 1$. That is, prove that $P_{k+1} = k + 2$.

$$P_{k+1} = \left(1 + \frac{1}{1}\right)\left(1 + \frac{1}{2}\right)\left(1 + \frac{1}{3}\right)\cdots\left(1 + \frac{1}{k}\right)\left(1 + \frac{1}{k+1}\right)$$

$$= P_k\left(1 + \frac{1}{k+1}\right) = (k+1)\left(1 + \frac{1}{k+1}\right) = k + 1 + 1$$

$$P_{k+1} = k + 2$$

By the Principle of Mathematical Induction, the statement is true for all positive integers.

▶ **TRY EXERCISE 12, PAGE 932**

EXAMPLE 3 **Prove an Inequality by Mathematical Induction**

Prove that $1 + 2n \leq 3^n$ for all positive integers.

Solution

1. Let $n = 1$. Then $1 + 2(1) = 3 \leq 3^1$. The statement is true when n is 1.

2. Assume the statement is true for some positive integer k.

$$1 + 2k \leq 3^k \qquad \bullet \textbf{ Induction hypothesis}$$

Now prove that the statement is true for $n = k + 1$. That is, prove that $1 + 2(k + 1) \leq 3^{k+1}$.

$$3^{k+1} = 3^k(3)$$
$$\geq (1 + 2k)(3) \qquad \bullet \textbf{ Because, by the induction hypothesis,}$$
$$\qquad\qquad\qquad\qquad \mathbf{1 + 2k \leq 3^k.}$$
$$= 6k + 3$$
$$> 2k + 2 + 1 \qquad \bullet \mathbf{6k > 2k, and\ 3 = 2 + 1.}$$
$$= 2(k + 1) + 1$$

Thus $1 + 2(k + 1) \leq 3^{k+1}$.

By the Principle of Mathematical Induction, $1 + 2k \leq 3^k$ for all positive integers.

▶ **TRY EXERCISE 16, PAGE 932**

● EXTENDED PRINCIPLE OF MATHEMATICAL INDUCTION

The Principle of Mathematical Induction can be extended to cases in which the beginning index is greater than 1.

Extended Principle of Mathematical Induction

Let P_n be a statement about a positive integer n. If

1. P_j is true for some positive integer j, and

2. for $k \geq j$ the truth of P_k implies the truth of P_{k+1},

then P_n is true for all positive integers $n \geq j$.

EXAMPLE 4   **Prove an Inequality by Mathematical Induction**

For $n \geq 3$, prove that $n^2 > 2n + 1$.

Solution

1.   Let $n = 3$. Then $3^2 = 9$; $2(3) + 1 = 7$. Thus $n^2 > 2n + 1$ for $n = 3$.

2.   Assume the statement is true for some positive integer $k \geq 3$.

$$k^2 > 2k + 1 \qquad \text{• Induction hypothesis}$$

Verify that the statement is true when $n = k + 1$. That is, show that

$$(k + 1)^2 > 2(k + 1) + 1 = 2k + 3$$

$$
\begin{aligned}
(k + 1)^2 &= k^2 + 2k + 1 \\
&> (2k + 1) + 2k + 1 \qquad &\text{• Induction hypothesis} \\
&> 2k + 1 + 1 + 1 \qquad &\text{• } 2k > 1 \\
&= 2k + 3
\end{aligned}
$$

Thus $(k + 1)^2 > 2k + 3$.

By the Extended Principle of Mathematical Induction, $n^2 > 2n + 1$ for all $n \geq 3$.

▶ **TRY EXERCISE 20, PAGE 932**

TOPICS FOR DISCUSSION

1.   Discuss the purpose of mathematical induction.

2.   What is an induction hypothesis?

3.   What is the Extended Principle of Mathematical Induction and how is it used?

4.   Mathematical induction can be used to prove that $x^m \cdot x^n = x^{m+n}$ when m and n are natural numbers. Explain why mathematical induction cannot be used if we are attempting to prove the result when m and n are real numbers.

Exercise Set 11.4

In Exercises 1 to 12, use mathematical induction to prove each statement.

1. $\displaystyle\sum_{i=1}^{n} (3i - 2) = 1 + 4 + 7 + \cdots + 3n - 2 = \frac{n(3n - 1)}{2}$

2. $\displaystyle\sum_{i=1}^{n} 2i = 2 + 4 + 6 + \cdots + 2n = n(n + 1)$

3. $\displaystyle\sum_{i=1}^{n} i^3 = 1 + 8 + 27 + \cdots + n^3 = \frac{n^2(n + 1)^2}{4}$

4. $\displaystyle\sum_{i=1}^{n} 2^i = 2 + 4 + 8 + \cdots + 2^n = 2(2^n - 1)$

5. $\displaystyle\sum_{i=1}^{n} (4i - 1) = 3 + 7 + 11 + \cdots + 4n - 1 = n(2n + 1)$

6. $\displaystyle\sum_{i=1}^{n} 3^i = 3 + 9 + 27 + \cdots + 3^n = \frac{3(3^n - 1)}{2}$

7. $\displaystyle\sum_{i=1}^{n} (2i - 1)^3 = 1 + 27 + 125 + \cdots + (2n - 1)^3$

$\qquad = n^2(2n^2 - 1)$

▶ **8.** $\displaystyle\sum_{i=1}^{n} i(i + 1) = 2 + 6 + 12 + \cdots + n(n + 1)$

$\qquad = \dfrac{n(n + 1)(n + 2)}{3}$

9. $\displaystyle\sum_{i=1}^{n} \frac{1}{(2i - 1)(2i + 1)} = \frac{1}{1 \cdot 3} + \frac{1}{3 \cdot 5} + \frac{1}{5 \cdot 7} + \cdots$

$\qquad + \dfrac{1}{(2n - 1)(2n + 1)} = \dfrac{n}{2n + 1}$

10. $\displaystyle\sum_{i=1}^{n} \frac{1}{2i(2i + 2)} = \frac{1}{2 \cdot 4} + \frac{1}{4 \cdot 6} + \frac{1}{6 \cdot 8} + \cdots + \frac{1}{2n(2n + 2)}$

$\qquad = \dfrac{n}{4(n + 1)}$

11. $\displaystyle\sum_{i=1}^{n} i^4 = 1 + 16 + 81 + \cdots + n^4$

$\qquad = \dfrac{n(n + 1)(2n + 1)(3n^2 + 3n - 1)}{30}$

▶ **12.** $P_n = \left(1 - \dfrac{1}{2}\right)\left(1 - \dfrac{1}{3}\right)\left(1 - \dfrac{1}{4}\right) \cdots \left(1 - \dfrac{1}{n + 1}\right)$

$\qquad = \dfrac{1}{n + 1}$

In Exercises 13 to 20, use mathematical induction to prove each inequality.

13. $\left(\dfrac{3}{2}\right)^n > n + 1, n \geq 4$

14. $\left(\dfrac{4}{3}\right)^n > n, n \geq 7$

15. If $0 < a < 1$, show that $a^{n+1} < a^n$ for all positive integers n.

▶ **16.** If $a > 1$, show that $a^{n+1} > a^n$ for all positive integers n.

17. $1 \cdot 2 \cdot 3 \cdot \cdots \cdot n > 2^n, n \geq 4$

18. $\dfrac{1}{\sqrt{1}} + \dfrac{1}{\sqrt{2}} + \dfrac{1}{\sqrt{3}} + \cdots + \dfrac{1}{\sqrt{n}} \geq \sqrt{n}$

19. For $a > 0$, show that $(1 + a)^n \geq 1 + na$ for all positive integers.

▶ **20.** $\log_{10} n < n$ for all positive integers. (*Hint:* Because $\log_{10} x$ is an increasing function, $\log_{10} (n + 1) \leq \log_{10} (n + n)$.)

In Exercises 21 to 30, use mathematical induction to prove each statement.

21. 2 is a factor of $n^2 + n$ for all positive integers n.

22. 3 is a factor of $n^3 - n$ for all positive integers n.

23. 4 is a factor of $5^n - 1$ for all positive integers n. (*Hint:* $5^{k+1} - 1 = 5 \cdot 5^k - 5 + 4$.)

24. 5 is a factor of $6^n - 1$ for all positive integers n.

25. $(xy)^n = x^n y^n$ for all positive integers n.

26. $\left(\dfrac{x}{y}\right)^n = \dfrac{x^n}{y^n}$ for all positive integers n.

27. For $a \neq b$, show that $(a - b)$ is a factor of $a^n - b^n$, where n is a positive integer. *Hint:*

$\qquad a^{k+1} - b^{k+1} = (a \cdot a^k - ab^k) + (ab^k - b \cdot b^k)$

28. For $a \neq -b$, show that $(a + b)$ is a factor of $a^{2n+1} + b^{2n+1}$, where n is a positive integer. *Hint:*

$\qquad a^{2k+3} + b^{2k+3} = (a^{2k+2} + b^{2k+2})(a + b) - ab(a^{2k+1} + b^{2k+1})$

29. $\displaystyle\sum_{k=1}^{n} ar^{k-1} = \frac{a(1 - r^n)}{1 - r}$ for $r \neq 1$

30. $\displaystyle\sum_{k=1}^{n} (ak + b) = \frac{n[(n + 1)a + 2b]}{2}$

CONNECTING CONCEPTS

In Exercises 31 to 35, use mathematical induction to prove each statement.

31. Using a calculator, find the smallest integer N for which $\log N! > N$. Now prove that $\log n! > n$ for all $n > N$.

32. Let a_n be a sequence for which there is a number r and an integer N for which $\dfrac{a_{n+1}}{a_n} < r$ for $n \geq N$. Show that $a_{N+k} < a_N r^k$ for each positive integer k.

33. For constant positive integers m and n, show that $(x^m)^n = x^{mn}$.

34. Prove that $\displaystyle\sum_{i=0}^{n} \frac{1}{i!} \leq 3 - \frac{1}{n}$ for all positive integers n.

35. Prove that $\left(\dfrac{n+1}{n}\right)^n < 3$ for all integers $n \geq 3$.

PREPARE FOR SECTION 11.5

36. Expand $(a + b)^3$. [P.3]

37. Evaluate $5!$. [11.1]

38. Evaluate $0!$. [11.1]

39. Evaluate $\dfrac{n!}{k!\,(n-k)!}$ when $n = 6$ and $k = 2$. [11.1]

40. Evaluate $\dfrac{n!}{k!\,(n-k)!}$ when $n = 7$ and $k = 3$. [11.1]

41. Evaluate $\dfrac{n!}{k!\,(n-k)!}$ when $n = 10$ and $k = 10$. [11.1]

PROJECTS

1. **STEPS IN A MATHEMATICAL INDUCTION PROOF** In every proof by mathematical induction, it is important that both parts of the Principle of Mathematical Induction be verified. For instance, consider the formula

$$2 + 4 + 8 + \cdots + 2^n = 2^{n+1} + 1$$

a. Show that if we assume the formula is true for some positive integer k, then the formula is true for $k + 1$.

b. Show that the formula is not true for $n = 1$.

c. Show that the formula is not valid for any value of n by showing that the left side is always an even number and the right side is always an odd number.

d. Explain how this shows that both steps of the Principle of Mathematical Induction must be verified.

2. **THE TOWER OF HANOI** The Tower of Hanoi is a game that consists of three pegs and n disks of distinct diameter arranged on one of the pegs such that the largest disk is on the bottom, the next largest is on top of the largest, and so on. The object of the game is to move all the disks from one peg to a second peg. The rules require that only one

disk be moved at a time and that a larger disk may not be placed on a smaller disk. All pegs may be used.

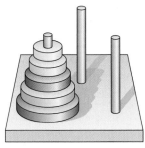

a. Show that it is possible to complete the game in $2^n - 1$ moves.

b. A legend says that in the center of the universe, high priests have the task of moving 64 golden disks from one of three diamond needles by using the rules of the Tower of Hanoi game. When they have completed the transfer, the universe will cease to exist. If one move is made every second, and the priests started 5 billion years ago (the approximate age of Earth), for how many more years does the legend predict the universe will continue to exist?

SECTION 11.5 — THE BINOMIAL THEOREM

In certain situations in mathematics it is necessary to write $(a + b)^n$ as the sum of its terms. Because $(a + b)$ is a binomial, this process is called **expanding the binomial.** For small values of n, it is relatively easy to write the expansion by using multiplication.

Earlier in the text we found that

$$(a + b)^1 = a + b$$

$$(a + b)^2 = a^2 + 2ab + b^2$$

$$(a + b)^3 = a^3 + 3a^2b + 3ab^2 + b^3$$

Building on these expansions, we can write a few more.

$$(a + b)^4 = a^4 + 4a^3b + 6a^2b^2 + 4ab^3 + b^4$$

$$(a + b)^5 = a^5 + 5a^4b + 10a^3b^2 + 10a^2b^3 + 5ab^4 + b^5$$

We could continue to build on previous expansions and eventually have quite a comprehensive list of binomial expansions. Instead, however, we will look for a theorem that will enable us to expand $(a + b)^n$ directly without multiplying.

Look at the variable parts of each expansion above. Note that for each $n = 1, 2, 3, 4, 5,$

- The first term is a^n. The exponent on a decreases by 1 for each successive term.

- The exponent on b increases by 1 for each successive term. The last term is b^n.

- The degree of each term is n.

❓ QUESTION What is the degree of the fourth term of the expansion of $(a + b)^{11}$?

To find a pattern for the coefficients in each expansion of $(a + b)^n$, first note that there are $n + 1$ terms and that the coefficient of the first and last term is 1. To find the remaining coefficients, consider the expansion of $(a + b)^5$.

$$(a + b)^5 = a^5 + 5a^4b + 10a^3b^2 + 10a^2b^3 + 5ab^4 + b^5$$

$$\frac{5}{1} = 5 \qquad \frac{5 \cdot 4}{2 \cdot 1} = 10 \qquad \frac{5 \cdot 4 \cdot 3}{3 \cdot 2 \cdot 1} = 10 \qquad \frac{5 \cdot 4 \cdot 3 \cdot 2}{4 \cdot 3 \cdot 2 \cdot 1} = 5$$

Observe from these patterns that there is a strong relationship to factorials. In fact, we can express each coefficient by using factorial notation.

$$\frac{5!}{1!\,4!} = 5 \qquad \frac{5!}{2!\,3!} = 10 \qquad \frac{5!}{3!\,2!} = 10 \qquad \frac{5!}{4!\,1!} = 5$$

In each denominator the first factorial is the exponent of b, and the second factorial is the exponent of a.

❓ ANSWER 11

In general, we will conjecture that the coefficient of the term $a^{n-k}b^k$ in the expansion of $(a + b)^n$ is $\dfrac{n!}{k!\,(n - k)!}$. Each coefficient of a term of a binomial expansion is called a **binomial coefficient** and is denoted by $\dbinom{n}{k}$.

Formula for a Binomial Coefficient

The coefficient of the term whose variable part is $a^{n-k}b^k$ in the expansion of $(a + b)^n$ is

$$\binom{n}{k} = \frac{n!}{k!\,(n - k)!}$$

The first term of the expansion of $(a + b)^n$ can be thought of as $a^n b^0$. In this case, we can calculate the coefficient of this term as

$$\binom{n}{0} = \frac{n!}{0!\,(n - 0)!} = \frac{n!}{1 \cdot n!} = 1$$

EXAMPLE 1   Evaluate a Binomial Coefficient

Evaluate each binomial coefficient.   **a.** $\dbinom{9}{6}$ **b.** $\dbinom{10}{10}$

Solution

a. $\dbinom{9}{6} = \dfrac{9!}{6!\,(9 - 6)!} = \dfrac{9!}{6!\,3!} = \dfrac{9 \cdot 8 \cdot 7 \cdot 6!}{6! \cdot 3 \cdot 2 \cdot 1} = 84$

b. $\dbinom{10}{10} = \dfrac{10!}{10!\,(10 - 10)!} = \dfrac{10!}{10!\,0!} = 1.$ • **Remember that 0! = 1.**

▶ **TRY EXERCISE 4, PAGE 938**

● BINOMIAL THEOREM

We are now ready to state the Binomial Theorem for positive integers.

Binomial Theorem for Positive Integers

If n is a positive integer, then

$$(a + b)^n = \sum_{i=0}^{n} \binom{n}{i} a^{n-i} b^i$$

$$= \binom{n}{0} a^n + \binom{n}{1} a^{n-1}b + \binom{n}{2} a^{n-2}b^2 + \cdots + \binom{n}{n} b^n$$

EXAMPLE 2   **Expand the Sum of Two Terms**

Expand: $(2x^2 + 3)^4$

Solution

$$(2x^2 + 3)^4 = \binom{4}{0}(2x^2)^4 + \binom{4}{1}(2x^2)^3(3) + \binom{4}{2}(2x^2)^2(3)^2$$

$$+ \binom{4}{3}(2x^2)(3)^3 + \binom{4}{4}(3)^4$$

$$= 16x^8 + 96x^6 + 216x^4 + 216x^2 + 81$$

▶ **TRY EXERCISE 18, PAGE 938**

EXAMPLE 3   **Expand a Difference of Two Terms**

Expand: $\left(\sqrt{x} - 2y\right)^5$

Solution

$$\left(\sqrt{x} - 2y\right)^5 = \binom{5}{0}(\sqrt{x})^5 + \binom{5}{1}(\sqrt{x})^4(-2y) + \binom{5}{2}(\sqrt{x})^3(-2y)^2$$

$$+ \binom{5}{3}(\sqrt{x})^2(-2y)^3 + \binom{5}{4}(\sqrt{x})(-2y)^4 + \binom{5}{5}(-2y)^5$$

$$= x^{5/2} - 10x^2y + 40x^{3/2}y^2 - 80xy^3 + 80x^{1/2}y^4 - 32y^5$$

▶ **TRY EXERCISE 20, PAGE 938**

> ***take note***
>
> If exactly one of the terms a or b in $(a + b)^n$ is negative, the terms of the expansion alternate in sign.

● ***i*th TERM OF A BINOMIAL EXPANSION**

The Binomial Theorem also can be used to find a specific term in the expansion of $(a + b)^n$.

Formula for the *i*th Term of a Binomial Expansion

The *i*th term of the expansion of $(a + b)^n$ is given by

$$\binom{n}{i-1}a^{n-i+1}b^{i-1}$$

> ***take note***
>
> The exponent on b is 1 less than the term number.

EXAMPLE 4   **Find the *i*th Term of a Binomial Expansion**

Find the fourth term in the expansion of $(2x^3 - 3y^2)^5$.

Solution

With $a = 2x^3$ and $b = -3y^2$, and using the preceding theorem with $i = 4$ and $n = 5$, we have

$$\binom{5}{3}(2x^3)^2(-3y^2)^3 = -1080x^6y^6$$

The fourth term is $-1080x^6y^6$.

▶ **TRY EXERCISE 34, PAGE 938**

● PASCAL'S TRIANGLE

A pattern for the coefficients of the terms of an expanded binomial can be found by writing the coefficients in a triangular array known as **Pascal's Triangle.** See **Figure 11.1.**

Each row begins and ends with the number 1. Any other number in a row is the sum of the two closest numbers above it. For example, $4 + 6 = 10$. Thus each succeeding row can be found from the preceding row.

$(a + b)^1$:					1		1				
$(a + b)^2$:				1		2		1			
$(a + b)^3$:			1		3		3		1		
$(a + b)^4$:		1		4		6		4		1	
$(a + b)^5$:	1		5		10		10		5		1
$(a + b)^6$:	1	6		15		20		15		6	1

FIGURE 11.1

Pascal's Triangle can be used to expand a binomial for small values of n. For instance, the seventh row of Pascal's Triangle is

1 7 21 35 35 21 7 1

Therefore,

$$(a + b)^7 = a^7 + 7a^6b + 21a^5b^2 + 35a^4b^3 + 35a^3b^4 + 21a^2b^5 + 7ab^6 + b^7$$

TOPICS FOR DISCUSSION

1. Discuss the use of the Binomial Theorem.

2. Can the Binomial Theorem be used to expand $(a + b)^n$, n a natural number, for any expressions a and b? Why or why not?

3. What is Pascal's Triangle and how is it related to expanding a binomial?

4. Explain how Pascal's Triangle suggests that $\binom{n-1}{k-1} + \binom{n-1}{k} = \binom{n}{k}$.

EXERCISE SET 11.5

In Exercises 1 to 8, evaluate the binomial coefficient.

1. $\dbinom{7}{4}$ **2.** $\dbinom{8}{6}$ **3.** $\dbinom{9}{2}$ ▶**4.** $\dbinom{10}{5}$

5. $\dbinom{12}{9}$ **6.** $\dbinom{6}{5}$ **7.** $\dbinom{11}{0}$ **8.** $\dbinom{14}{14}$

In Exercises 9 to 28, expand the binomial.

9. $(x - y)^6$ **10.** $(a - b)^5$ **11.** $(x + 3)^5$

12. $(x - 5)^4$ **13.** $(2x - 1)^7$ **14.** $(2x + y)^6$

15. $(x + 3y)^6$ **16.** $(x - 4y)^5$ **17.** $(2x - 5y)^4$

▶**18.** $(3x + 2y)^4$ **19.** $\left(x + \dfrac{1}{x}\right)^6$ ▶**20.** $\left(2x - \sqrt{y}\right)^7$

21. $(x^2 - 4)^7$ **22.** $(x - y^3)^6$ **23.** $(2x^2 + y^3)^5$

24. $(2x - y^3)^6$ **25.** $\left(\dfrac{2}{x} - \dfrac{x}{2}\right)^4$ **26.** $\left(\dfrac{a}{b} + \dfrac{b}{a}\right)^3$

27. $(s^{-2} + s^2)^6$ **28.** $(2r^{-1} + s^{-1})^5$

In Exercises 29 to 36, find the indicated term without expanding.

29. $(3x - y)^{10}$; eighth term

30. $(x + 2y)^{12}$; fourth term

31. $(x + 4y)^{12}$; third term

32. $(2x - 1)^{14}$; thirteenth term

33. $\left(\sqrt{x} - \sqrt{y}\right)^9$; fifth term

▶**34.** $(x^{-1/2} + x^{1/2})^{10}$; sixth term

35. $\left(\dfrac{a}{b} + \dfrac{b}{a}\right)^{11}$; ninth term

36. $\left(\dfrac{3}{x} - \dfrac{x}{3}\right)^{13}$; seventh term

37. Find the term that contains b^8 in the expansion of $(2a - b)^{10}$.

38. Find the term that contains s^7 in the expansion of $(3r + 2s)^9$.

39. Find the term that contains y^8 in the expansion of $(2x + y^2)^6$.

40. Find the term that contains b^9 in the expansion of $(a - b^3)^8$.

41. Find the middle term of $(3a - b)^{10}$.

42. Find the middle term of $(a + b^2)^8$.

43. Find the two middle terms of $(s^{-1} + s)^9$.

44. Find the two middle terms of $(x^{1/2} - y^{1/2})^7$.

In Exercises 45 to 50, use the Binomial Theorem to simplify the powers of the complex numbers.

45. $(2 - i)^4$ **46.** $(3 + 2i)^3$

47. $(1 + 2i)^5$ **48.** $(1 - 3i)^5$

49. $\left(\dfrac{\sqrt{2}}{2} + i\dfrac{\sqrt{2}}{2}\right)^8$ **50.** $\left(\dfrac{1}{2} + i\dfrac{\sqrt{3}}{2}\right)^6$

CONNECTING CONCEPTS

51. Let n be a positive integer. Expand and simplify $\dfrac{(x + h)^n - x^n}{h}$, where x is any real number and $h \neq 0$.

52. Show that $\dbinom{n}{k} = \dbinom{n}{n - k}$ for all positive integers n and k with $0 \le k \le n$.

53. Show that $\displaystyle\sum_{k=0}^{n} \dbinom{n}{k} = 2^n$. (*Hint:* Use the Binomial Theorem with $x = 1, y = 1$.)

54. Prove that $\dbinom{n}{k} + \dbinom{n}{k + 1} = \dbinom{n + 1}{k + 1}$, n and k integers, $0 \le k \le n$.

55. Prove that $\sum_{i=0}^{n} (-1)^i \binom{n}{i} = 0$.

56. Approximate $(0.98)^8$ by evaluating the first three terms of $(1 - 0.02)^8$.

57. Approximate $(1.02)^8$ by evaluating the first three terms of $(1 + 0.02)^8$.

There is an extension of the Binomial Theorem called the *Multinomial Theorem.* **This theorem is used in determining probabilities.** *Multinomial Theorem:* **If *n*, *r*, and *k* are positive integers, then the coefficient of $a^r b^k c^{n-r-k}$ in the expansion of $(a + b + c)^n$ is**

$$\frac{n!}{r!\, k!\, (n - r - k)!}$$

In Exercises 58 to 61, use the Multinomial Theorem to find the indicated coefficient.

58. Find the coefficient of $a^2 b^3 c^5$ in the expansion of $(a + b + c)^{10}$.

59. Find the coefficient of $a^5 b^2 c^2$ in the expansion of $(a + b + c)^9$.

60. Find the coefficient of $a^4 b^5$ in the expansion of $(a + b + c)^9$.

61. Find the coefficient of $a^3 c^5$ in the expansion of $(a + b + c)^8$.

PREPARE FOR SECTION 11.6

62. Evaluate: 7! [11.1]

63. Evaluate: 0! [11.1]

64. Evaluate $\binom{n}{k}$ when $n = 7$ and $k = 1$. [11.1]

65. Evaluate $\binom{n}{k}$ when $n = 8$ and $k = 5$. [11.1]

66. Evaluate $\dfrac{n!}{(n - k)!}$ when $n = 10$ and $k = 2$. [11.1]

67. Evaluate $\dfrac{n!}{(n - k)!}$ when $n = 6$ and $k = 6$. [11.1]

PROJECTS

1. PASCAL'S TRIANGLE Write an essay on Pascal's Triangle. Include some of the earliest known examples of the triangle and some of its applications.

2. SOME OTHER FUNCTIONS Do some research and determine a definition of positive integers for each of the following types of numbers. Give examples of calculations using each type of number.

   a. Pochammer (m, n) **b.** double factorial $(n!!)$

PERMUTATIONS AND COMBINATIONS

- FUNDAMENTAL COUNTING PRINCIPLE
- PERMUTATIONS
- COMBINATIONS

● FUNDAMENTAL COUNTING PRINCIPLE

Suppose that an electronics store offers a three-component stereo system for $250. A buyer must choose one amplifier, one tuner, and one pair of speakers. If the store has two models of amplifiers, three models of tuners, and two speaker models, how many different stereo systems could a consumer purchase?

This problem belongs to a class of problems called *counting problems.* The problem is to determine the number of ways in which the conditions of the problem can be satisfied. One way to do this is to make a tree diagram and then count the items on the list. We will organize the list in a table using A_1 and A_2 for the amplifiers; T_1, T_2, and T_3 for the tuners; and S_1 and S_2 for the speakers. See **Figure 11.2.**

By counting the possible systems that can be purchased, we find there are 12 different systems. Another way to arrive at this result is to find the product of the numbers of options available.

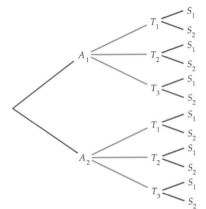

FIGURE 11.2

$$\underset{\text{amplifiers}}{\text{Number of}} \times \underset{\text{tuners}}{\text{number of}} \times \underset{\text{speakers}}{\text{number of}} = \underset{\text{systems}}{\text{number of}}$$

$$2 \quad \times \quad 3 \quad \times \quad 2 \quad = \quad 12$$

In some states, a standard car license plate consists of a nonzero digit, followed by three letters, followed by three more digits. What is the maximum number of car license plates of this type that could be issued? If we begin a list of the possible license plates, it soon becomes apparent that listing them all would be very time-consuming and impractical.

$$1AAA000, \quad 1AAA001, \quad 1AAA002, \quad 1AAA003, \ldots$$

Instead, the following counting principle is used. This principle forms the basis for all counting problems.

Fundamental Counting Principle

Let $T_1, T_2, T_3, \ldots, T_n$ be a sequence of n conditions. Suppose that T_1 can occur in m_1 ways, T_2 can occur in m_2 ways, T_3 can occur in m_3 ways, and so on until finally T_n can occur in m_n ways. Then the number of ways of satisfying the conditions $T_1, T_2, T_3, \ldots, T_n$ in succession is given by the product

$$m_1 m_2 m_3 \cdots m_n$$

TABLE 11.4

Condition	Number of ways
T_1: a nonzero digit	$m_1 = 9$
T_2: a letter	$m_2 = 26$
T_3: a letter	$m_3 = 26$
T_4: a letter	$m_4 = 26$
T_5: a digit	$m_5 = 10$
T_6: a digit	$m_6 = 10$
T_7: a digit	$m_7 = 10$

To apply the counting principle to the license plate problem, first find the number of ways each condition can be satisfied, as shown in **Table 11.4.** Thus, we have

$$\underset{\text{license plates}}{\text{Number of car}} = 9 \cdot 26 \cdot 26 \cdot 26 \cdot 10 \cdot 10 \cdot 10 = 158,184,000$$

? **QUESTION** Suppose that a license plate begins with two letters. How many different ways could the license plate begin?

EXAMPLE 1 \ Apply the Fundamental Counting Principle

An automobile dealer offers three mid-size cars. A customer selecting one of these cars must choose one of three different engines, one of five different colors, and one of four different interior packages. How many different selections can the customer make?

Solution

T_1: mid-size car $m_1 = 3$

T_2: engine $m_2 = 3$

T_3: color $m_3 = 5$

T_4: interior $m_4 = 4$

The number of different selections is $3 \cdot 3 \cdot 5 \cdot 4 = 180$.

▶ **TRY EXERCISE 12, PAGE 945**

● PERMUTATIONS

An application of the Fundamental Counting Principle is to determine the number of arrangements of distinct elements in a definite order.

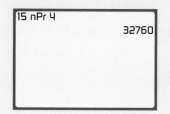

INTEGRATING TECHNOLOGY

Some graphing calculators use the notation nPr to represent the number of permutations of n objects taken r at a time. The calculation of the number of permutations of 15 objects taken 4 at a time is shown below.

```
15 nPr 4
               32760
```

> **Permutation**
>
> A **permutation** is an arrangement of distinct objects in a definite order.

For example, abc and bca are two of the possible permutations of the three elements a, b, and c.

Consider a race with 10 runners. In how many different orders can the runners finish first, second, and third (assuming no ties)?

Any one of the 10 runners could finish first: $m_1 = 10$

Any one of the remaining 9 runners could be second: $m_2 = 9$

Any one of the remaining 8 runners could be third: $m_3 = 8$

By the Fundamental Counting Principle, there are $10 \cdot 9 \cdot 8 = 720$ possible first-, second-, and third-place finishes for the 10 runners. Using the language of permutations, we would say, "There are 720 permutations of 10 objects (the runners) taken 3 (the possible finishes) at a time."

Permutations occur so frequently in counting problems that a formula, rather than the counting principle, is often used.

? **ANSWER** $26 \times 26 = 676$

Formula for the Number of Permutations of n Distinct Objects Taken r at a Time

The number of permutations of n distinct objects taken r at a time is

$$P(n, r) = \frac{n!}{(n - r)!}$$

E X A M P L E   2   Find the Number of Permutations

In how many ways can a president, a vice president, a secretary, and a treasurer be selected from a committee of 15 people?

Solution

There are 15 distinct people to place in four positions. Thus $n = 15$ and $r = 4$.

$$P(15, 4) = \frac{15!}{(15 - 4)!} = \frac{15!}{11!} = \frac{15 \cdot 14 \cdot 13 \cdot 12 \cdot 11!}{11!} = 32{,}760$$

▶ **TRY EXERCISE 16, PAGE 945**

E X A M P L E   3   Find the Number of Seating Permutations

Six people attend a movie and all sit in the same row containing six seats.

a. Find the number of ways the group can sit together.

b. Find the number of ways the group can sit together if two people in the group must sit side-by-side.

c. Find the number of ways the group can sit together if two people in the group refuse to sit side-by-side.

Solution

a. There are six distinct people to place in six distinct positions. Thus $n = 6$ and $r = 6$.

$$P(6, 6) = \frac{6!}{(6 - 6)!} = \frac{6!}{0!} = \frac{6!}{1} = 720$$

b. Think of the two people who must sit together as a single object and count the number of arrangements of the *five* objects (AB), C, D, E, and F.

Thus $n = 5$ and $r = 5$.

$$P(5, 5) = \frac{5!}{(5 - 5)!} = \frac{5!}{0!} = \frac{5!}{1} = 120$$

There are also 120 arrangements with A and B reversed [(BA), C, D, E, F]. Thus the total number of arrangements is $120 + 120 = 240$.

c. From part **a.**, there are 720 possible seating arrangements. From part **b.**, there are 240 arrangements with two specific people next to each other. Thus there are $720 - 240 = 480$ arrangements in which two specific people are not seated together.

▶ **TRY EXERCISE 22, PAGE 945**

• COMBINATIONS

Up to this point we have been counting the number of distinct arrangements of objects. In some cases we may be interested in determining the number of ways of selecting objects without regard to the order of the selection. For example, suppose we want to select a committee of three people from five candidates denoted by A, B, C, D, and E. One possible committee is A, C, D. If we select D, C, A, we still have the same committee because the order of the selection is not important. An arrangement of objects for which the order of the selection is not important is a **combination**.

take note

Recall that a binomial coefficient is given by $\binom{n}{r} = \frac{n!}{r!\,(n - r)!}$, which is the same as $C(n, r)$.

Formula for the Number of Combinations of n Objects Taken r at a Time

The number of combinations of n objects taken r at a time is

$$C(n, r) = \frac{n!}{r!\,(n - r)!}$$

EXAMPLE 4   Find the Number of Combinations

A standard deck of playing cards consists of 52 cards. How many five-card hands can be chosen from this deck?

INTEGRATING TECHNOLOGY

Some calculators use the notation nCr to represent a combination of n objects taken r at a time.

Solution

We have $n = 52$ and $r = 5$. Thus

$$C(52, 5) = \frac{52!}{5!\,(52 - 5)!} = \frac{52!}{5!\,47!} = \frac{52 \cdot 51 \cdot 50 \cdot 49 \cdot 48 \cdot 47!}{5 \cdot 4 \cdot 3 \cdot 2 \cdot 1 \cdot 47!} = 2{,}598{,}960$$

▶ **TRY EXERCISE 20, PAGE 945**

EXAMPLE 5   **Find the Number of Combinations**

A chemist has nine solution samples, of which four are type A and five are type B. If the chemist chooses three of the solutions at random, determine in how many ways the chemist can choose exactly one type A solution.

Solution

The chemist has chosen three solutions, one of which is type A. If one is type A, then two are type B. The number of ways of choosing one type A solution from four type A solutions is $C(4, 1)$.

$$C(4, 1) = \frac{4!}{1!\,(4 - 1)!} = \frac{4!}{1!\,3!} = 4$$

The number of ways of choosing two type B solutions from five type B solutions is $C(5, 2)$.

$$C(5, 2) = \frac{5!}{2!\,(5 - 2)!} = \frac{5!}{2!\,3!} = 10$$

By the counting principle, there are

$$C(4, 1) \cdot C(5, 2) = 4 \cdot 10 = 40$$

ways to choose one type A and two type B solutions.

▶ **TRY EXERCISE 30, PAGE 946**

The difficult part of counting is determining whether to use the counting principle, the permutation formula, or the combination formula. Following is a summary of guidelines.

Guidelines for Solving Counting Problems

1. The counting principle will always work but is not always the easiest method to apply.

2. When reading a problem, ask yourself, "Is the order of the selection process important?" If the answer is yes, the arrangements are permutations. If the answer is no, the arrangements are combinations.

 TOPICS FOR DISCUSSION

1. Discuss the Fundamental Counting Principle and how it is used.

2. Discuss the difference between a permutation and a combination.

3. Explain why $\binom{n}{k}$ occurs as a coefficient in the binomial formula.

Exercise Set 11.6

In Exercises 1 to 10, evaluate each quantity.

1. $P(6, 2)$ **2.** $P(8, 7)$ **3.** $C(8, 4)$ **4.** $C(9, 2)$

5. $P(8, 0)$ **6.** $P(9, 9)$ **7.** $C(7, 7)$ **8.** $C(6, 0)$

9. $C(10, 4)$ **10.** $P(10, 4)$

11. COMPUTER SYSTEMS A computer manufacturer offers a computer system with three different disk drives, two different monitors, and two different keyboards. How many different computer systems could a consumer purchase from this manufacturer?

▶ **12. COLOR MONITORS** A computer monitor produces color by blending colors on *palettes*. If a computer monitor has four palettes and each palette has four colors, how many blended colors can be formed? Assume each palette must be used each time.

13. COMPUTER SCIENCE A 4-digit binary number is a sequence of four digits consisting of 0 or 1. For instance, 0011 and 1011 are binary numbers. How many four-digit binary numbers are possible?

14. COMPUTER MEMORY An integer is stored in a computer's memory as a series of zeros and ones. Each memory unit contains eight spaces for a zero or a one. The first space is used for the sign of the number, and the remaining seven spaces are used for the integer. How many positive integers can be stored in one memory unit of this computer?

15. SCHEDULING In how many different ways can six employees be assigned to six different jobs?

▶ **16. CONTEST WINNERS** First-, second-, and third-place prizes are to be awarded in a dance contest in which 12 contestants are entered. In how many ways can the prizes be awarded?

17. MAILBOXES There are five mailboxes outside a post office. In how many ways can three letters be deposited into the five boxes?

18. COMMITTEE MEMBERSHIP How many different committees of three people can be selected from nine people?

19. TEST QUESTIONS A professor provides to a class 25 possible essay questions for an upcoming test. Of the 25 questions, the professor will ask 5 of the questions on the exam. How many different tests can the professor prepare?

▶ **20. TENNIS MATCHES** Twenty-six people enter a tennis tournament. How many different first-round matches are possible if each player can be matched with any other player?

21. EMPLOYEE INITIALS A company has more than 676 employees. Explain why there must be at least two employees who have the same first and last initials.

▶ **22. SEATING ARRANGEMENTS** A car holds six passengers, three in the front seat and three in the back seat. How many different seating arrangements of six people are possible if one person refuses to sit in front and one person refuses to sit in back?

23. COMMITTEE MEMBERSHIP A committee of six people is chosen from six senators and eight representatives. How many committees are possible if there are to be three senators and three representatives on the committee?

24. ARRANGING NUMBERS The numbers 1, 2, 3, 4, 5, 6 are to be arranged. How many different arrangements are possible under each of the following conditions?

a. All the even numbers come first.

b. The arrangements are such that the numbers alternate between even and odd.

25. TEST QUESTIONS A true-false examination contains 10 questions. In how many ways can a person answer the questions on this test by just guessing? Assume that all questions are answered.

26. TEST QUESTIONS A twenty-question, four-option multiple-choice examination is given as a pre-employment test. In how many ways could a prospective employee answer the questions on this test by just guessing? Assume that all questions are answered.

27. STATE LOTTERY A state lottery game requires a person to select six different numbers from forty numbers. The order of the selection is not important. In how many ways can this be done?

28. TEST QUESTIONS A student must answer eight of ten questions on an exam. How many different choices can the student make?

29. ACCEPTANCE SAMPLING A warehouse receives a shipment of ten computers, of which three are defective. Five computers are then randomly selected from the ten and delivered to a store.

a. In how many ways can the store receive no defective computers?

b. In how many ways can the store receive one defective computer?

c. In how many ways can the store receive all three defective computers?

▶ **30.** **CONTEST** Fifteen students, of whom seven are seniors, are selected as semifinalists for a literary award. Of the fifteen students, ten finalists will be selected.

a. In how many ways can ten finalists be selected from the fifteen students?

b. In how many ways can the ten finalists contain three seniors?

c. In how many ways can the ten finalists contain at least five seniors?

31. **SERIAL NUMBERS** A television manufacturer uses a code for the serial number of a television set. The first symbol is the letter *A*, *B*, or *C* and represents the location of the manufacturing plant. The next two symbols (01, 02, . . . , 12) represent the month in which the set was manufactured. The next symbol is a 5, a 6, a 7, an 8, or a 9 and represents the year the set was manufactured. The last seven symbols are digits. How many serial numbers are possible?

32. **CARD GAMES** Five cards are chosen at random from an ordinary deck of playing cards. In how many ways can the cards be chosen under each of the following conditions?

a. All are hearts.   **b.** All are the same suit.

c. Exactly three are kings.

d. Two or more are aces.

33. **ACCEPTANCE SAMPLING** A quality control inspector receives a shipment of ten computer disk drives and randomly selects three of the drives for testing. If two of the disk drives in the shipment are defective, find the number of ways in which the inspector could select at most one defective drive.

34. **BASKETBALL TEAMS** A basketball team has twelve members. In how many ways can five players be chosen under each of the following conditions?

a. The selection is random.

b. The two tallest players are always among the five selected.

35. **ARRANGING NUMBERS** The numbers 1, 2, 3, 4, 5, and 6 are arranged in random order. In how many ways can the numbers 1 and 2 appear next to one another in the order 1, 2?

36. **OCCUPANCY PROBLEM** Seven identical balls are randomly placed in seven available containers in such a way that two balls are in one container. Of the remaining six containers, each receives at most one ball. Find the number of ways in which this can be accomplished.

37. **LINES IN A PLANE** Seven points lie in a plane in such a way that no three points lie on the same line. How many lines are determined by the seven points?

38. **CHESS MATCHES** A chess tournament has 12 participants. How many games must be scheduled if every player must play every other player exactly once?

39. **CONTEST WINNERS** Eight couples attend a benefit at which two prizes are given. In how many ways can two names be randomly drawn so that the prizes are not awarded to the same couple?

40. **GEOMETRY** Suppose there are 12 distinct points on a circle. How many different triangles can be formed with vertices at the given points?

41. **TEST QUESTIONS** In how many ways can a student answer a 20-question true-false test if the student marks 10 of the questions true and 10 of the questions false?

42. **COMMITTEE MEMBERSHIP** From a group of fifteen people, a committee of eight is formed. From the committee a president, a secretary, and a treasurer are selected. Find the number of ways in which the two consecutive operations can be carried out.

43. **COMMITTEE MEMBERSHIP** From a group of twenty people, a committee of twelve is formed. From the committee of twelve, a subcommittee of four people is chosen. Find the number of ways in which the two consecutive operations can be carried out.

44. **CHECKERBOARDS** A checkerboard consists of eight rows and eight columns of squares. Starting at the top left square of a checkerboard, how many possible paths will end at the bottom right square if the only way a player can legally move is right one square or down one square from the current position?

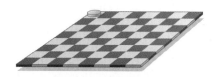

45. ICE CREAM CONES An ice cream store offers 31 flavors of ice cream. How many different triple-decker cones are possible? *Note:* Assume that different orders of the same flavors are *not* different cones. Thus a scoop of rocky road followed by two scoops of mint chocolate is the same as one scoop of mint chocolate followed by one scoop of rocky road followed by a second scoop of mint chocolate.

46. COMPUTER SCREENS A typical computer monitor consists of pixels, each of which can, in some cases, be assigned any one of 2^{16} different colors. If a computer screen has a resolution of 1024 pixels by 768 pixels, how many different images can be displayed on the screen? *Suggestion:* Write your answer as a power of 2.

47. DART BOARDS How many different arrangements are there of the integers 1 through 20 on a typical dart board, assuming that 20 is always at the top?

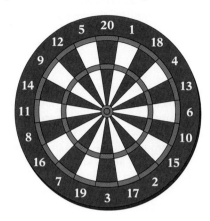

CONNECTING CONCEPTS

48. LINES IN A PLANE Generalize Exercise 37. That is, given n points in a plane, no three of which lie on the same line, how many lines are determined by the n points?

49. BIRTHDAYS Seven people are asked the month of their birth. In how many ways can each of the following conditions exist?

   a. No two people have a birthday in the same month.

   b. At least two people have a birthday in the same month.

50. SUMS OF COINS From a penny, a nickel, a dime, and a quarter, how many different sums of money can be formed using one or more of the coins?

51. BIOLOGY Five sticks of equal length are broken into a short piece and a long piece. The ten pieces are randomly arranged in five pairs. In how many ways will each pair consist of a long stick and a short stick? (This exercise actually has a practical side. When cells are exposed to harmful radiation, some chromosomes break. If two long sides unite or two short sides unite, the cell dies.)

52. ARRANGING NUMBERS Four random digits are drawn (repetitions are allowed). Among the four digits, in how many ways can two or more repetitions occur?

53. RANDOM WALK An aimless tourist, standing on a street corner, tosses a coin. If the result is heads, the tourist walks one block north. If the result is tails, the tourist walks one block south. At the new corner, the coin is tossed again and the same rule applied. If the coin is tossed ten times, in how many ways will the tourist be back at the original corner? This problem is an elementary example of what is called a *random walk*. Random walk problems have many applications in physics, chemistry, and economics.

PREPARE FOR SECTION 11.7

54. What is the Fundamental Counting Principle? [11.6]

55. How many ways can a two-digit number be formed from the digits 1, 2, 3, and 4 if no digit can be repeated in the number? [11.6]

56. Evaluate $P(7, 2)$. [11.6]

57. Evaluate $C(7, 2)$. [11.6]

58. Evaluate $\binom{n}{k} p^k q^{n-k}$ when $n = 8$, $k = 5$, $p = \dfrac{1}{4}$, and $q = \dfrac{3}{4}$. [11.1]

59. A light switch panel has four switches that control the lights in four different segments of a room. In how many on/off configurations can the four switches be placed? [11.6]

PROJECTS

1. ✎ **EXPLAIN PERMUTATIONS AND COMBINATIONS**
   Write an outline of a lesson that you could use to teach permutations and combinations. Include at least five examples of permutations and five examples of combinations.

2. **APPLICATION OF COUNTING** Calculating the number of ways in which balls can be distributed in boxes has a variety of applications. For instance, a traffic engineer may want to know how traffic accidents (the balls) are distributed throughout the days of the week (the boxes). Or a physicist may want to know how electrons (the balls) can be distributed in the energy orbits (the boxes) of an atom. The formula for counting the number of ways in which n distinguishable balls can be placed in k distinguishable boxes, where each box must have at least one ball, is given by

$$\binom{k}{0}k^n - \binom{k}{1}(k-1)^n + \binom{k}{2}(k-2)^n + \cdots + (-1)^{k-1}\binom{k}{k-1}$$

The word *distinguishable* is important. This formula refers to counting under circumstances similar to the situation depicted in the first figure at the top of the next column, where the boxes are numbered and the balls are numbered. The second figure shows a situation that is not covered by this formula. Although the boxes are numbered, the balls are not and are therefore *indistinguishable*.

Box 1   Box 2   Box 3

Box 1   Box 2   Box 3

a. A computer network consists of five computers and three printers. How many possible connections can be made if each computer must be hooked to a printer and all printers are used?

b. A supermarket has four checkout lanes. Assuming shoppers are efficient and will not leave a checkout lane empty, in how many ways can 10 shoppers line up for the checkout lanes?

SECTION 11.7	# INTRODUCTION TO PROBABILITY

- **SAMPLE SPACES**
- **EVENTS**
- **PROBABILITY OF AN EVENT**
- **BINOMIAL PROBABILITIES**

Many events in the world around us have random character, such as the chances of an accident occurring on a certain freeway, the chances of winning a state lottery, and the chances that the nucleus of an atom will undergo fission. By repeatedly observing such events, it is often possible to recognize certain patterns. **Probability** is the mathematical study of random patterns.

When a weather reporter predicts a 30% chance of rain, the forecaster is saying that similar weather conditions have led to rain 30 times out of 100. When a fair coin is tossed, we expect heads to occur $\frac{1}{2}$, or 50%, of the time. The numbers 30% (or 0.3) and $\frac{1}{2}$ are the probabilities of the events.

MATH MATTERS

The beginning of probability theory is frequently associated with letters sent between Pascal (of Pascal's Triangle) and Fermat (Fermat's Last Theorem) in which they discuss the solution of a problem posed to them by Antoine Gombaud, Chevalier de Mere, a French aristocrat who liked to gamble. The basic question was "How many tosses of two dice are necessary to have a better than 50–50 chance of rolling two sixes?" Although the correct analysis of this problem by Fermat and Pascal pre-saged probability theory, historical records indicate that commerce and the need to insure ships, cargo, and lives was another motivating factor to calculating probabilities. The first merchant insurance companies were established in the fourteenth century to insure ships. Life insurance companies were established at the end of the seventeenth century. Lloyd's of London was established sometime before 1690.

● SAMPLE SPACES

An activity with an observable outcome is called an **experiment**. Examples of experiments include

1. Flipping a coin and observing the side facing upward

2. Observing the incidence of a disease in a certain population

3. Observing the length of time a person waits in a checkout line in a grocery store

The **sample space** of an experiment is the set of *all possible* outcomes of that experiment.

Consider the experiment of tossing one coin three times and recording the number of occurrences of the upward side of the coin. The sample space is

$$S = \{HHH, HHT, HTH, THH, HTT, THT, TTH, TTT\}$$

EXAMPLE 1   List the Elements of a Sample Space

Suppose that among five batteries, two are defective. Two batteries are randomly drawn from the five and tested for defects. List the elements in the sample space.

Solution

Label the nondefective batteries N_1, N_2, and N_3 and the defective batteries D_1 and D_2. The sample space is

$$S = \{ N_1D_1, N_2D_1, N_3D_1, N_1D_2, N_2D_2, N_3D_2, N_1N_2, N_1N_3, N_2N_3, D_1D_2\}$$

▶ **TRY EXERCISE 6, PAGE 956**

● EVENTS

An **event** E is any subset of a sample space. For the sample space defined in Example 1, several of the events we could define are

E_1: There are no defective batteries.

E_2: At least one battery is defective.

E_3: Both batteries are defective.

Because an event is a subset of the sample space, each of these events can be expressed as a set.

$$E_1 = \{N_1N_2, N_1N_3, N_2N_3\}$$
$$E_2 = \{N_1D_1, N_2D_1, N_3D_1, N_1D_2, N_2D_2, N_3D_2, D_1D_2\}$$
$$E_3 = \{D_1D_2\}$$

There are two methods by which elements are drawn from a sample space: with replacement and without replacement. *With replacement* means that after the element is drawn, it is returned to the sample space. The same element could be selected on the next drawing. When elements are drawn *without replacement*, an element drawn is not returned to the sample space and therefore is not available for any subsequent drawing.

EXAMPLE 2   List the Elements of an Event

A two-digit number is formed by choosing from the digits 1, 2, 3, and 4, both with replacement and without replacement. Express each event as a set.

a. E_1:   The second digit is greater than or equal to the first digit.

b. E_2:   Both digits are less than zero.

Solution

a. With replacement: $E_1 = \{11, 12, 13, 14, 22, 23, 24, 33, 34, 44\}$

Without replacement: $E_1 = \{12, 13, 14, 23, 24, 34\}$

b. $E_2 = \varnothing$
Choosing from the digits 1, 2, 3, and 4, this event is impossible. The impossible event is denoted by the empty set or null set.

▶ **TRY EXERCISE 14, PAGE 956**

● PROBABILITY OF AN EVENT

The probability of an event is defined in terms of the concepts of sample space and event.

Probability of an Event

Let $n(S)$ and $n(E)$ represent the number of elements in the sample space S and the number of elements in the event E, respectively. The probability of event E, $P(E)$, is

$$P(E) = \frac{n(E)}{n(S)}$$

Because E is a subset of S, $n(E) \leq n(S)$. Thus $P(E) \leq 1$. If E is an impossible event, then $E = \varnothing$ and $n(E) = 0$. Thus $P(E) = 0$. If E is an event that *always* occurs, then $E = S$ and $n(E) = n(S)$. Thus $P(E) = 1$. Thus we have, for any event E,

$$0 \leq P(E) \leq 1$$

❓ QUESTION Is it possible for the probability of an event to equal 1.25?

EXAMPLE 3   Calculate the Probability of an Event

A coin is tossed three times. What is the probability of each outcome?

a. E_1:   Two or more heads will appear.

b. E_2:   At least one tail will appear.

❓ ANSWER No. All probabilities must be between 0 and 1, inclusive.

Solution

First determine the number of elements in the sample space. The sample space for this experiment is

$$S = \{HHH, HHT, HTH, THH, HTT, THT, TTH, TTT\}$$

Therefore $n(S) = 8$. Now determine the number of elements in each event. Then calculate the probability of the event by using $P(E) = n(E)/n(S)$.

a. $E_1 = \{HHH, HHT, HTH, THH\}$

$$P(E_1) = \frac{n(E_1)}{n(S)} = \frac{4}{8} = \frac{1}{2}$$

b. $E_2 = \{HHT, HTH, THH, HTT, THT, TTH, TTT\}$

$$P(E_2) = \frac{n(E_2)}{n(S)} = \frac{7}{8}$$

▶ **TRY EXERCISE 22, PAGE 956**

Calculating probabilities by listing and then counting the elements of a sample space is not always practical. Instead, we will use the counting principles developed in the last section to determine the number of elements in the sample space and in an event.

EXAMPLE 4   **Use the Counting Principles to Calculate a Probability**

A state lottery game allows a person to choose five numbers from the integers 1 to 40. Repetitions of numbers are not allowed. If three or more numbers match the numbers chosen by the lottery, the player wins a prize. Find the probability that a player will match

a.   exactly three numbers

b.   exactly four numbers

Solution

The sample space S is the number of ways in which five numbers can be chosen from forty numbers. This is a combination because the order of the drawing is not important.

$$n(S) = C(40, 5) = \frac{40!}{5!\,35!} = 658{,}008$$

We will call the five numbers chosen by the state lottery "lucky" and the remaining thirty-five numbers "unlucky."

a.   Let E_1 be the event a player has three lucky and therefore two unlucky numbers. The three lucky numbers are chosen from the five lucky numbers. There are $C(5, 3)$ ways to do this. The two unlucky numbers

Continued ▶

are chosen from the thirty-five unlucky numbers. There are $C(35, 2)$ ways to do this. By the counting principle, the number of ways the event E_1 can occur is

$$N(E_1) = C(5, 3) \cdot C(35, 2) = 10 \cdot 595 = 5950$$

$$P(E_1) = \frac{n(E_1)}{n(S)} = \frac{C(5, 3) \cdot C(35, 2)}{C(40, 5)} = \frac{5950}{658,008} \approx 0.009042$$

b. Let E_2 be the event a player has four lucky numbers and one unlucky number. The number of ways a person can select four lucky numbers and one unlucky number is $C(5, 4) \cdot C(35, 1)$.

$$P(E_2) = \frac{n(E_2)}{n(S)} = \frac{C(5, 4) \cdot C(35, 1)}{C(40, 5)} = \frac{175}{658,008} \approx 0.000266$$

▶ **TRY EXERCISE 32, PAGE 957**

The expression "one or the other of two events occurs" is written as the union of the two sets. For example, suppose an experiment leads to the sample space $S = \{1, 2, 3, 4, 5, 6\}$ and the events are

Draw a number less than four, $E_1 = \{1, 2, 3\}$

Draw an even number, $E_2 = \{2, 4, 6\}$

Then the event $E_1 \cup E_2$ is described by drawing a number less than four *or* an even number. Thus

$$E_1 \cup E_2 = \{1, 2, 3\} \cup \{2, 4, 6\} = \{1, 2, 3, 4, 6\}$$

Two events E_1 and E_2 that cannot occur at the same time are **mutually exclusive** events. Using set notation, if $E_1 \cap E_2 = \emptyset$, then E_1 and E_2 are mutually exclusive.

For example, using the same sample space $\{1, 2, 3, 4, 5, 6\}$, a third event is

Draw an odd number, $E_3 = \{1, 3, 5\}$

Then $E_2 \cap E_3 = \emptyset$ and the events E_2 and E_3 are mutually exclusive. On the other hand,

$$E_1 \cap E_2 = \{2\}$$

so the events E_1 and E_2 are not mutually exclusive.

One of the axioms of probability involves the union of mutually exclusive events.

A Probability Axiom

If E_1 and E_2 are mutually exclusive events, then

$$P(E_1 \cup E_2) = P(E_1) + P(E_2)$$

If the events are not mutually exclusive, the addition rule for probabilities can be used.

78. **ARRANGEMENTS OF CARDS** A deck of ten cards contains five red and five black cards. If four cards are drawn from the deck, what is the probability that two are red and two are black?

79. **NUMBER THEORY** For the 1000 numbers from 000 to 999, what is the probability that the middle digit of a number chosen at random is greater than the other two digits?

80. **NUMBER THEORY** Two numbers are chosen, with replacement, from the digits 1, 2, 3, 4, 5, and 6, and their sum is recorded. Now two more digits are selected and their sum noted. This process continues until the sum is 7 or the original sum is obtained. If the original sum was 9, what is the probability of obtaining another sum of 9 before obtaining a sum of 7? (*Hint:* Assume the events are independent. The probability can be found by summing an infinite geometric series.)

81. **CARD GAMES** Which of the following has the greater probability: drawing an ace and a ten-card (ten, jack, queen, or king) from a regular deck of fifty-two playing cards, or drawing an ace and a ten-card from two decks of regular playing cards?

82. **NUMBER THEORY** From the digits 1, 2, 3, 4, and 5, two numbers are chosen without replacement. What is the probability that the second number is greater than the first number?

83. **EMPLOYEE BADGES** A room contains twelve people who are wearing badges numbered 1 to 12. If three people are randomly selected, what is the probability that the person wearing badge 6 will be included?

CHAPTER 11 TEST

In Exercises 1 to 3, find the third and fifth terms of the sequence defined by a_n.

1. $a_n = \dfrac{2^n}{n!}$

2. $a_n = \dfrac{(-1)^{n+1}}{2n}$

3. $a_1 = 3, a_n = 2a_{n-1}$

In Exercises 4 to 6, classify each sequence as an arithmetic sequence, a geometric sequence, or neither.

4. $a_n = -2n + 3$

5. $a_n = 2n^2$

6. $a_n = \dfrac{(-1)^{n-1}}{3^n}$

In Exercises 7 to 9, find the indicated sum of the series.

7. $\displaystyle\sum_{i=1}^{6} \dfrac{1}{i}$

8. $\displaystyle\sum_{j=1}^{10} \dfrac{1}{2^j}$

9. $\displaystyle\sum_{k=1}^{20} (3k - 2)$

10. The third term of an arithmetic sequence is 7 and the eighth term is 22. Find the twentieth term.

11. Find the sum of the infinite geometric series given by $\displaystyle\sum_{k=1}^{\infty} \left(\dfrac{3}{8}\right)^k$.

12. Write $0.\overline{15}$ as the quotient of integers in simplest form.

In Exercises 13 and 14, prove the statement by mathematical induction.

13. $\displaystyle\sum_{i=1}^{n} (2 - 3i) = \dfrac{n(1 - 3n)}{2}$

14. $n! > 3^n, \quad n \geq 7$

15. Write the binomial expansion of $(x - 2y)^5$.

16. Write the binomial expansion of $\left(x + \dfrac{1}{x}\right)^6$.

17. Find the sixth term in the expansion of $(3x + 2y)^8$.

18. Three cards are randomly chosen from a regular deck of playing cards. In how many ways can the cards be chosen?

19. A serial number consists of seven characters. The first three characters are upper-case letters of the alphabet. The next two characters are selected from the digits 1 through 9. The last two characters are upper-case letters of the alphabet. How many serial numbers are possible if no letter or number can be used twice in the same serial number?

20. Five cards are randomly selected from a deck of cards containing eight black cards and ten red cards. What is the probability that three black cards and two red cards are selected?

CUMULATIVE REVIEW EXERCISES

1. Find the linear regression equation for the set $\{(1, 5), (3, 8),$ $(4, 11), (6, 15), (8, 16)\}$. Round values to the nearest tenth.

2. Find the value of x in the domain of $F(x) = 5 + \dfrac{x}{3}$ for which $F(x) = -3$.

3. Solve: $2x^2 - 3x = 4$

4. Write $\log_b\left(\dfrac{xy^2}{z^3}\right)$ in terms of the logarithms of x, y, and z.

5. Find the eccentricity of the graph of $16x^2 + 25y^2 - 96x + 100y - 156 = 0$.

6. Solve: $\begin{cases} 2x - 3y = 8 \\ x + 4y = -7 \end{cases}$

7. Given $A = \begin{bmatrix} -1 & 2 \\ 5 & 3 \\ 0 & 3 \end{bmatrix}$ and $B = \begin{bmatrix} 7 & -3 \\ 6 & 5 \\ 1 & -2 \end{bmatrix}$, find $3A - 2B$.

8. Let $g(x) = x^2 - x + 4$ and $h(x) = x - 2$. Find $\left(\dfrac{h}{g}\right)(-3)$.

9. Find the horizontal asymptote of the graph of $F(x) = \dfrac{x^3 - 8}{x^5}$.

10. Evaluate: $\log_{\frac{1}{2}} 64$

11. Solve $4^{2x+1} = 3^{x-2}$. Round to the nearest tenth.

12. Solve: $\begin{cases} x^2 + y^2 + xy = 10 \\ x - y = 1 \end{cases}$

13. Find the product of $\begin{bmatrix} 3 & 2 \\ -2 & 1 \\ 1 & -4 \end{bmatrix} \begin{bmatrix} 2 & 3 & 1 & 1 \\ -2 & 0 & 4 & -3 \end{bmatrix}$.

14. The time t in seconds required for a sky diver to reach a velocity of v feet per second is given by $t = -\dfrac{175}{32} \ln\left(1 - \dfrac{v}{175}\right)$. Determine the velocity of the sky diver after 5 seconds. Round to the nearest mile per hour.

15. Given $\sin \theta = -\dfrac{1}{2}$ and $\sec \theta = \dfrac{2\sqrt{3}}{3}$, find $\cot \theta$.

16. Express $\dfrac{\sin x}{1 + \cos x} + \dfrac{1 + \cos x}{\sin x}$ in terms of $\csc x$.

17. Solve the triangle ABC if $A = 40°$, $B = 65°$, and $c = 20$ centimeters.

18. Find the angle of rotation α that eliminates the xy term of $9x^2 + 4xy + 6y^2 + 12x + 36y + 44 = 0$. Round to the nearest degree.

19. Evaluate: $\sin\left(\dfrac{1}{2} \cos^{-1} \dfrac{4}{5}\right)$

20. Given $\mathbf{v} = 2\mathbf{i} + 5\mathbf{j}$ and $\mathbf{w} = 3\mathbf{i} - 6\mathbf{j}$, find $2\mathbf{v} - 3\mathbf{w}$.

SOLUTIONS
TO THE TRY EXERCISES

2. a. Integers: 31, 51

  b. Rational numbers: $\dfrac{5}{7}$, 31, $-2\dfrac{1}{2}$, 4.235653907493, 51,

    0.888...

  c. Irrational number: $\dfrac{5}{\sqrt{7}}$

  d. Prime number: 31

  e. Real numbers: All the numbers are real numbers.

6. In absolute value, the four smallest integers are 0, 1, 2, and 3. Replacing x in $x^2 - 1$ by these values, we obtain $\{-1, 0, 3, 8\}$.

16. $A \cap B = \{-2, 0, 2\}$ and $A \cap C = \{0, 1, 2, 3\}$. Therefore, $(A \cap B) \cup (A \cap C) = \{-2, 0, 1, 2, 3\}$.

48. $d(z, 5) = |z - 5|$; therefore $|z - 5| > 7$.

54. The interval $(-\infty, 3]$ includes all real numbers from $-\infty$ to 3, including 3. The interval $(2, 6)$ includes all real numbers between 2 and 6, not including 2 and not including 6. Therefore, $(-\infty, 3] \cap (2, 6) = (2, 3]$. The graph is

$$\begin{array}{c}\text{-5 -4 -3 -2 -1 \ 0 \ 1 \ 2 \ 3 \ 4 \ 5}\end{array}$$

64. $\{x \mid -3 \le x < 0\} \cup \{x \mid x \ge 2\}$ is the set of all real numbers between -3 and 0, including -3 but excluding 0, together with (union) all real numbers greater than or equal to 2. The graph is

$$\begin{array}{c}\text{-5 -4 -3 -2 -1 \ 0 \ 1 \ 2 \ 3 \ 4 \ 5}\end{array}$$

74. $(z - 2y)^2 - 3z^3$

$$[(-1) - 2(-2)]^2 - 3(-1)^3 = [-1 + 4]^2 - 3(-1)$$
$$= 3^2 + 3$$
$$= 9 + 3 = 12$$

86. Commutative property of addition

90. Substitution

106. $6 + 3[2x - 4(3x - 2)] = 6 + 3[2x - 12x + 8]$
$$= 6 + 3[-10x + 8]$$
$$= 6 - 30x + 24 = -30x + 30$$

10. $\dfrac{4^{-2}}{2^{-3}} = \dfrac{2^3}{4^2} = \dfrac{8}{16} = \dfrac{1}{2}$

30. $\dfrac{(-3a^2b^3)^2}{(-2ab^4)^3} = \dfrac{(-3)^{1\cdot2}a^{2\cdot2}b^{3\cdot2}}{(-2)^{1\cdot3}a^{1\cdot3}b^{4\cdot3}}$

$$= \dfrac{9a^4b^6}{-8a^3b^{12}} = -\dfrac{9a}{8b^6}$$

46. $\dfrac{(6.9 \times 10^{27})(8.2 \times 10^{-13})}{4.1 \times 10^{15}} = \dfrac{(6.9)(8.2) \times 10^{27-13}}{4.1 \times 10^{15}}$

$$= \dfrac{56.58 \times 10^{14}}{4.1 \times 10^{15}}$$

$$= 13.8 \times 10^{-1} = 1.38$$

62. $(-5x^{1/3})(-4x^{1/2}) = (-5)(-4)x^{1/3+1/2}$

$$= 20x^{2/6+3/6} = 20x^{5/6}$$

78. $\sqrt{18x^2y^5} = \sqrt{9x^2y^4}\sqrt{2y} = 3|x|y^2\sqrt{2y}$

86. $-3x\sqrt[3]{54x^4} + 2\sqrt[3]{16x^7} = -3x\sqrt[3]{3^3 \cdot 2x^4} + 2\sqrt[3]{2^4x^7}$

$$= -3x\sqrt[3]{3^3x^3}\sqrt[3]{2x} + 2\sqrt[3]{2^3x^6}\sqrt[3]{2x}$$
$$= -3x(3x\sqrt[3]{2x}) + 2(2x^2\sqrt[3]{2x})$$
$$= -9x^2\sqrt[3]{2x} + 4x^2\sqrt[3]{2x}$$
$$= -5x^2\sqrt[3]{2x}$$

96. $\left(3\sqrt{5y} - 4\right)^2 = \left(3\sqrt{5y} - 4\right)\left(3\sqrt{5y} - 4\right)$

$$= 9 \cdot 5y - 12\sqrt{5y} - 12\sqrt{5y} + 16$$
$$= 45y - 24\sqrt{5y} + 16$$

106. $\dfrac{2}{\sqrt[4]{4y}} = \dfrac{2}{\sqrt[4]{4y}} \cdot \dfrac{\sqrt[4]{4y^3}}{\sqrt[4]{4y^3}} = \dfrac{2\sqrt[4]{4y^3}}{2y} = \dfrac{\sqrt[4]{4y^3}}{y}$

110. $-\dfrac{7}{3\sqrt{2} - 5} = -\dfrac{7}{3\sqrt{2} - 5} \cdot \dfrac{3\sqrt{2} + 5}{3\sqrt{2} + 5}$

$$= \dfrac{-21\sqrt{2} - 35}{18 - 25}$$

$$= \dfrac{-21\sqrt{2} - 35}{-7} = 3\sqrt{2} + 5$$

12. a. $-12x^4 - 3x^2 - 11$

  b. 4

  c. $-12, -3, -11$

  d. -12

  e. $-12x^4, -3x^2, -11$

24. $(5y^2 - 7y + 3) + (2y^2 + 8y + 1) = 7y^2 + y + 4$

32.
$$
\begin{array}{r}
3x^2 - 8x - 5 \\
5x - 7 \\
\hline
-21x^2 + 56x + 35 \\
15x^3 - 40x^2 - 25x \\
\hline
15x^3 - 61x^2 + 31x + 35
\end{array}
$$

56. $(4x^2 - 3y)(4x^2 + 3y) = (4x^2)^2 - (3y)^2 = 16x^4 - 9y^2$

66. $-x^2 - 5x + 4$

     $-(-5)^2 - 5(-5) + 4$ • Replace x by -5.

     $= -25 + 25 + 4$ • Simplify.

     $= 4$

76. $\dfrac{1}{6}n^3 - \dfrac{1}{2}n^2 + \dfrac{1}{3}n$

     $\dfrac{1}{6}(21)^3 - \dfrac{1}{2}(21)^2 + \dfrac{1}{3}(21) = 1330$

     It is possible to form 1330 different committees.

78. a. $4.3 \times 10^{-6}(1000)^2 - 2.1 \times 10^{-4}(1000)$

       $= 4.09$ seconds

     b. $4.3 \times 10^{-6}(5000)^2 - 2.1 \times 10^{-4}(5000)$

       $= 106.45$ seconds

     c. $4.3 \times 10^{-6}(10,000)^2 - 2.1 \times 10^{-4}(10,000)$

       $= 427.9$ seconds

Exercise Set P.4, page 53

6. $6a^3b^2 - 12a^2b + 72ab^3 = 6ab(a^2b - 2a + 12b^2)$

12. $b^2 + 12b - 28 = (b + 14)(b - 2)$

16. $57y^2 + y - 6 = (19y - 6)(3y + 1)$

24. $b^2 - 4ac = 8^2 - 4(16)(-35) = 2304 = 48^2$

     The trinomial is factorable over the integers.

36. $8x^6 - 10x^3 - 3 = (4x^3 + 1)(2x^3 - 3)$

40. $81b^2 - 16c^2 = (9b - 4c)(9b + 4c)$

50. $b^2 - 24b + 144 = (b - 12)^2$

56. $b^3 + 64 = (b + 4)(b^2 - 4b + 16)$

66. $a^2y^2 - ay^3 + ac - cy = ay^2(a - y) + c(a - y)$

       $= (a - y)(ay^2 + c)$

72. $81y^4 - 16 = (9y^2 - 4)(9y^2 + 4)$

       $= (3y - 2)(3y + 2)(9y^2 + 4)$

Exercise Set P.5, page 62

2. $\dfrac{2x^2 - 5x - 12}{2x^2 + 5x + 3} = \dfrac{(2x + 3)(x - 4)}{(2x + 3)(x + 1)} = \dfrac{x - 4}{x + 1}$

16. $\dfrac{x^2 - 16}{x^2 + 7x + 12} \cdot \dfrac{x^2 - 4x - 21}{x^2 - 4x}$

     $= \dfrac{(x - 4)(x + 4)(x + 3)(x - 7)}{(x + 3)(x + 4)x(x - 4)} = \dfrac{x - 7}{x}$

30. $\dfrac{3y - 1}{3y + 1} - \dfrac{2y - 5}{y - 3} = \dfrac{(3y - 1)(y - 3)}{(3y + 1)(y - 3)} - \dfrac{(2y - 5)(3y + 1)}{(y - 3)(3y + 1)}$

     $= \dfrac{(3y^2 - 10y + 3) - (6y^2 - 13y - 5)}{(3y + 1)(y - 3)}$

     $= \dfrac{-3y^2 + 3y + 8}{(3y + 1)(y - 3)}$

42. $\dfrac{3 - \dfrac{2}{a}}{5 + \dfrac{3}{a}} = \dfrac{\left(3 - \dfrac{2}{a}\right)a}{\left(5 + \dfrac{3}{a}\right)a} = \dfrac{3a - 2}{5a + 3}$

60. $\dfrac{e^{-2} - f^{-1}}{ef} = \dfrac{\dfrac{1}{e^2} - \dfrac{1}{f}}{ef} = \dfrac{f - e^2}{e^2 f} \div \dfrac{ef}{1}$

     $= \dfrac{f - e^2}{e^2 f} \cdot \dfrac{1}{ef} = \dfrac{f - e^2}{e^3 f^2}$

64. a. $\dfrac{v_1 + v_2}{1 + \dfrac{v_1 v_2}{c^2}} = \dfrac{1.2 \times 10^8 + 2.4 \times 10^8}{1 + \dfrac{(1.2 \times 10^8)(2.4 \times 10^8)}{(6.7 \times 10^8)^2}} \approx 3.4 \times 10^8$ mph

     b. $\dfrac{v_1 + v_2}{1 + \dfrac{v_1 \cdot v_2}{c^2}} = \dfrac{c^2(v_1 + v_2)}{c^2\left(1 + \dfrac{v_1 \cdot v_2}{c^2}\right)} = \dfrac{c^2(v_1 + v_2)}{c^2 + v_1 \cdot v_2}$

Exercise Set P.6, page 71

8. $6 - \sqrt{-1} = 6 - i$

18. $(5 - 3i) - (2 + 9i) = 5 - 3i - 2 - 9i = 3 - 12i$

34. $\left(5 + 2\sqrt{-16}\right)\left(1 - \sqrt{-25}\right) = [5 + 2(4i)](1 - 5i)$

       $= (5 + 8i)(1 - 5i)$

       $= 5 - 25i + 8i - 40i^2$

       $= 5 - 25i + 8i - 40(-1)$

       $= 5 - 25i + 8i + 40$

       $= 45 - 17i$

48. $\dfrac{8 - i}{2 + 3i} = \dfrac{8 - i}{2 + 3i} \cdot \dfrac{2 - 3i}{2 - 3i}$

     $= \dfrac{16 - 24i - 2i + 3i^2}{2^2 + 3^2}$

     $= \dfrac{16 - 24i - 2i + 3(-1)}{4 + 9}$

     $= \dfrac{16 - 26i - 3}{13}$

     $= \dfrac{13 - 26i}{13} = \dfrac{13(1 - 2i)}{13}$

     $= 1 - 2i$

60. $\dfrac{1}{i^{83}} = \dfrac{1}{i^{80} \cdot i^3} = \dfrac{1}{i^3} = \dfrac{1}{-i}$

$= \dfrac{1}{-i} \cdot \dfrac{i}{i} = \dfrac{i}{-i^2} = \dfrac{i}{-(-1)} = \dfrac{i}{1} = i$

Exercise Set 1.1, page 88

2. $-3y + 20 = 2$

$\qquad -3y = -18$ • Subtract 20 from each side.

$\qquad\quad y = 6$ • Divide each side by -3.

12. $\dfrac{1}{2}x + 7 - \dfrac{1}{4}x = \dfrac{19}{2}$

$4\left(\dfrac{1}{2}x + 7 - \dfrac{1}{4}x\right) = 4\left(\dfrac{19}{2}\right)$ • Multiply each side by 4.

$\qquad\quad 2x + 28 - x = 38$

$\qquad\qquad\qquad x = 38 - 28$ • Collect like terms.

$\qquad\qquad\qquad x = 10$

18. $5(x + 4)(x - 4) = (x - 3)(5x + 4)$

$\qquad 5(x^2 - 16) = 5x^2 - 11x - 12$

$\qquad 5x^2 - 80 = 5x^2 - 11x - 12$

$\qquad -80 + 12 = -11x$

$\qquad\quad -68 = -11x$

$\qquad\quad \dfrac{68}{11} = x$

24. $2x + \dfrac{1}{3} = \dfrac{6x + 1}{3}$ • Rewrite the left side.

$\qquad \dfrac{6x + 1}{3} = \dfrac{6x + 1}{3}$

The left side of this equation is now identical to the right side. Thus the original equation is an identity. The solution set of the original equation consists of all real numbers.

38. $\qquad\qquad |2x - 3| = 21$

$2x - 3 = 21 \qquad \text{or} \qquad 2x - 3 = -21$

$\qquad 2x = 24 \qquad\qquad\qquad 2x = -18$

$\qquad\quad x = 12 \qquad\qquad\qquad\quad x = -9$

The solutions of $|2x - 3| = 21$ are -9 and 12.

50. Substitute 175 for P, the number of patents measured in thousands, and solve for x.

$\qquad P = 5.4x + 110$

$\quad 175 = 5.4x + 110$

$\qquad 65 = 5.4x$ • Subtract 110 from each side.

$\qquad\quad x = \dfrac{65}{5.4}$ • Solve for x.

$\qquad\quad x \approx 12.04$

Adding 12.04 to 1993 yields 2005.04. Thus, according to the model, the number of patents will first exceed 175,000 in the year 2005.

52. Substitute 22 for m in the given equation and solve for s.

$22 = -\dfrac{1}{2}|s - 55| + 25$ • Substitute 22 for m.

$-44 = |s - 55| - 50$ • Multiply each side by -2 to clear the equation of fractions.

$\quad 6 = |s - 55|$ • Add 50 to each side.

$s - 55 = 6 \qquad \text{or} \qquad s - 55 = -6$

$\qquad s = 61 \qquad\qquad\qquad s = 49$

Kate should drive her car at either 61 miles per hour or 49 miles per hour to obtain a gas mileage of 22 miles per gallon.

Exercise Set 1.2, page 98

4. $A = P + Prt$

$A = P(1 + rt)$ • Factor.

$P = \dfrac{A}{(1 + rt)}$ • Solve for P.

14. Substitute 105 for w.

SMOG reading grade level $= \sqrt{105} + 3$

$\approx 10.2 + 3$

$= 13.2$

According to the SMOG formula, the estimated reading grade level required to fully understand *A Tale of Two Cities* is 13.2. (*Note*: A different sample of 30 sentences likely would produce a different result. It is for this reason that reading grade levels are often estimated by using several different samples and then computing an average of the results.)

20. $P = 2l + 2w, \qquad w = \dfrac{1}{2}l + 1$

$110 = 2l + 2\left(\dfrac{1}{2}l + 1\right)$ • Substitute for w.

$110 = 2l + l + 2$ • Simplify.

$108 = 3l$

$\quad 36 = l$

$\qquad l = 36$ meters

$w = \dfrac{1}{2}l + 1 = \dfrac{1}{2}(36) + 1 = 19$ meters

24. Let $t_1 =$ the time it takes to travel to the island.
Let $t_2 =$ the time it takes to make the return trip.

$t_1 + t_2 = 7.5$

$\qquad t_2 = 7.5 - t_1$

(continued)

$$15t_1 = 10t_2$$
$$15t_1 = 10(7.5 - t_1) \quad \bullet \text{ Substitute for } t_2.$$
$$15t_1 = 75 - 10t_1$$
$$25t_1 = 75$$
$$t_1 = 3 \text{ hours}$$
$$D = 15t_1 = 15(3) = 45 \text{ nautical miles}$$

32. Let x = the number of glasses of orange juice.

Profit = revenue − cost
$$\$2337 = 0.75x - 0.18x$$
$$2337 = 0.57x$$
$$x = \frac{2337}{0.57}$$
$$x = 4100$$

36. Let x = the amount of money invested at 5%.
Then $7500 - x$ is the amount of money invested at 7%.

5%	x
7%	$7500 - x$

$$0.05x + 0.07(7500 - x) = 405$$
$$0.05x + 525 - 0.07x = 405$$
$$-0.02x = -120$$
$$x = 6000$$
$$7500 - x = 1500$$

$6000 was invested at 5%. $1500 was invested at 7%.

40. Let x = the number of liters of the 40% solution to be mixed with the 24% solution.

0.40	x
0.24	4
0.30	$4 + x$

$$0.40x + 0.24(4) = 0.30(4 + x)$$
$$0.40x + 0.96 = 1.2 + 0.30x$$
$$0.10x = 0.24$$
$$x = 2.4$$

Thus 2.4 liters of 40% sulfuric acid should be mixed with 4 liters of a 24% sulfuric acid solution to produce the 30% solution.

50. Let x = the number of hours needed to print the report if both the printers are used.

Printer A prints $\dfrac{1}{3}$ of the report every hour.

Printer B prints $\dfrac{1}{4}$ of the report every hour.

Thus

$$\frac{1}{3}x + \frac{1}{4}x = 1$$
$$4x + 3x = 12 \cdot 1$$
$$7x = 12$$
$$x = \frac{12}{7} \approx 1.71$$

It would take approximately 1.71 hours to print the report.

Exercise Set 1.3, page 113

6. $12x^2 - 41x + 24 = 0$
$$(4x - 3)(3x - 8) = 0 \qquad \bullet \text{ Factor.}$$
$$4x - 3 = 0 \quad \text{ or } \quad 3x - 8 = 0 \qquad \bullet \text{ Apply the zero product property.}$$
$$x = \frac{3}{4} \qquad\qquad x = \frac{8}{3}$$

A check shows that $\dfrac{3}{4}$ and $\dfrac{8}{3}$ are both solutions of
$$12x^2 - 41x + 24 = 0.$$

20. $(x + 2)^2 + 28 = 0$
$$(x + 2)^2 = -28$$
$$\sqrt{(x + 2)^2} = \sqrt{-28}$$
$$x + 2 = \pm i\sqrt{28} = \pm 2i\sqrt{7}$$
$$x = -2 \pm 2i\sqrt{7}$$

The solutions are $-2 - 2i\sqrt{7}$ and $-2 + 2i\sqrt{7}$.

26. $x^2 - 6x + 10 = 0$
$$x^2 - 6x = -10$$
$$x^2 - 6x + 9 = -10 + 9 \quad \bullet \text{ Add } \left[\frac{1}{2}(-6)\right]^2 \text{ to each side.}$$
$$(x - 3)^2 = -1 \qquad \bullet \text{ Factor the left side.}$$
$$\sqrt{(x - 3)^2} = \pm\sqrt{-1} \qquad \bullet \text{ The square root procedure.}$$
$$x - 3 = \pm i$$
$$x = 3 \pm i$$

The solutions are $3 - i$ and $3 + i$.

30. $2x^2 + 10x - 3 = 0$
$$2x^2 + 10x = 3$$
$$2(x^2 + 5x) = 3$$
$$x^2 + 5x = \frac{3}{2} \qquad \bullet \text{ Divide each side by 2.}$$
$$x^2 + 5x + \frac{25}{4} = \frac{3}{2} + \frac{25}{4} \qquad \bullet \text{ Complete the square.}$$
$$\left(x + \frac{5}{2}\right)^2 = \frac{31}{4}$$
$$x + \frac{5}{2} = \pm\sqrt{\frac{31}{4}}$$
$$x = -\frac{5}{2} \pm \frac{\sqrt{31}}{2}$$
$$x = \frac{-5 + \sqrt{31}}{2} \quad \text{ or } \quad x = \frac{-5 - \sqrt{31}}{2}$$

38. $2x^2 + 4x - 1 = 0$
$$x = \frac{-4 \pm \sqrt{4^2 - 4(2)(-1)}}{4}$$
$$x = \frac{-4 \pm \sqrt{16 + 8}}{4} = \frac{-4 \pm \sqrt{24}}{4}$$

$$x = \frac{-4 \pm 2\sqrt{6}}{4} = \frac{-2 \pm \sqrt{6}}{2}$$

$$x = \frac{-2 + \sqrt{6}}{2} \quad \text{or} \quad x = \frac{-2 - \sqrt{6}}{2}$$

48. $x^2 + 3x - 11 = 0$

$b^2 - 4ac = 3^2 - 4(1)(-11) = 9 + 44 = 53 > 0$

Thus the equation has two distinct real roots.

58. Home plate, first base, and second base form a right triangle. The legs of this right triangle each measure 90 ft. The distance from home plate to second base is the length of the hypotenuse c of this right triangle.

$c^2 = 90^2 + 90^2$

$c^2 = 16{,}200$

$c = \sqrt{16{,}200}$

$c \approx 127.3$

To the nearest tenth of a foot, the distance from home plate to second base is 127.3 feet. (*Note:* We have not considered $-\sqrt{16{,}200}$ as a solution because we know that the distance must be positive.)

70. Let w be the width, in inches, of the new candy bar. Then the length, in inches, of the new candy bar is $2.5w$, and the height is 0.5 inch. The volume of the original candy bar is $5 \cdot 2 \cdot 0.5 = 5$ cubic inches. Thus the volume of the new candy bar is $0.80(5) = 4$ cubic inches. Substitute in the formula for the volume of a rectangular solid to produce

$$lwh = V$$

$$(2.5w)(w)(0.5) = 4$$

$$1.25w^2 = 4$$

$$w^2 = 3.2$$

$$w = \sqrt{3.2}$$

$$\approx 1.8$$

The width of the new candy bar should be about 1.8 inches and the length $2.5(1.8) \approx 4.5$ inches.

72. When the ball hits the ground, $h = 0$. Thus we need to solve $0 = -16t^2 + 52t + 4.5$ for t.

$$0 = -16t^2 + 52t + 4.5$$

$$t = \frac{-(52) \pm \sqrt{(52)^2 - 4(-16)(4.5)}}{2(-16)}$$

$$= \frac{-52 \pm \sqrt{2992}}{-32}$$

$$\approx 3.3$$

The ball will hit the ground in about 3.3 seconds. Disregard the negative solution because the time must be positive.

Exercise Set 1.4, page 125

6.
$$x^4 - 36x^2 = 0$$
$$x^2(x^2 - 36) = 0$$
$$x^2(x - 6)(x + 6) = 0$$
$$x = 0, x = 6, x = -6$$

14. Multiply each side of the equation by $(y + 2)(y - 4)$ to clear the equation of fractions.

$$\frac{4}{y + 2} = \frac{7}{y - 4} \qquad y \neq -2, y \neq 4$$

$$(y + 2)(y - 4)\left(\frac{4}{y + 2}\right) = (y + 2)(y - 4)\left(\frac{7}{y - 4}\right)$$

$$(y - 4)4 = (y + 2)7$$

$$4y - 16 = 7y + 14$$

$$4y - 7y = 14 + 16$$

$$-3y = 30$$

$$y = -10$$

Check to verify that -10 is the solution.

28. $\sqrt{10 - x} = 4 \qquad$ *Check:* $\sqrt{10 - (-6)} = 4$

$10 - x = 16 \qquad\qquad\qquad \sqrt{16} = 4$

$-x = 6 \qquad\qquad\qquad\quad 4 = 4$

$x = -6$

The solution is -6.

30.
$$x = \sqrt{5 - x} + 5$$
$$(x - 5)^2 = \left(\sqrt{5 - x}\right)^2$$
$$x^2 - 10x + 25 = 5 - x$$
$$x^2 - 9x + 20 = 0$$
$$(x - 5)(x - 4) = 0$$
$$x = 5 \quad \text{or} \quad x = 4$$

Check: $5 = \sqrt{5 - 5} + 5 \qquad 4 = \sqrt{5 - 4} + 5$

$5 = 0 + 5 \qquad\qquad\quad 4 = 1 + 5$

$5 = 5 \qquad\qquad\qquad\quad 4 = 6 \quad$ False

The solution is 5.

34.
$$\sqrt{x + 7} - 2 = \sqrt{x - 9}$$
$$\left(\sqrt{x + 7} - 2\right)^2 = \left(\sqrt{x - 9}\right)^2$$
$$x + 7 - 4\sqrt{x + 7} + 4 = x - 9$$
$$-4\sqrt{x + 7} = -20$$
$$\left(\sqrt{x + 7}\right)^2 = (5)^2$$
$$x + 7 = 25$$
$$x = 18$$

Check: $\sqrt{18 + 7} - 2 = \sqrt{18 - 9}$

$\sqrt{25} - 2 = \sqrt{9}$ (continued)

$$5 - 2 = 3$$
$$3 = 3$$

The solution is 18.

42. $x^4 - 10x^2 + 9 = 0$ • **Let $u = x^2$.**

$u^2 - 10u + 9 = 0$

$(u - 9)(u - 1) = 0$

$u = 9$ or $u = 1$

$x^2 = 9$ $x^2 = 1$

$x = \pm 3$ $x = \pm 1$

The solutions are 3, −3, 1, and −1.

50. $6x^{2/3} - 7x^{1/3} - 20 = 0$ • **Let $u = x^{1/3}$.**

$6u^2 - 7u - 20 = 0$

$(3u + 4)(2u - 5) = 0$

$u = -\dfrac{4}{3}$ or $u = \dfrac{5}{2}$

$x^{1/3} = -\dfrac{4}{3}$ $x^{1/3} = \dfrac{5}{2}$

$(x^{1/3})^3 = \left(-\dfrac{4}{3}\right)^3$ $(x^{1/3})^3 = \left(\dfrac{5}{2}\right)^3$

$x = -\dfrac{64}{27}$ $x = \dfrac{125}{8}$

The solutions are $-\dfrac{64}{27}$ and $\dfrac{125}{8}$.

60. Sandy is to receive $\dfrac{1}{2}$ of an adult dosage of a particular medication. Substituting into Young's rule yields:

$$\frac{1}{2} = \frac{x}{x + 12}$$

$$2(x + 12)\frac{1}{2} = 2(x + 12)\left(\frac{x}{x + 12}\right)$$

$$x + 12 = 2x$$

$$12 = x$$

Sandy is 12 years old.

62. Substitute 4 for the reading grade level in the SMOG formula to produce

$4 = \sqrt{w} + 3$

$1 = \sqrt{w}$ • **Subtract 3 from each side.**

$1 = w$ • **Square each side.**

The writer should strive for a maximum of one word with three or more syllables in any sample of 30 sentences.

Exercise Set 1.5, page 140

6. $-4(x - 5) \geq 2x + 15$

$-4x + 20 \geq 2x + 15$

$-6x \geq -5$

$x \leq \dfrac{5}{6}$

The solution set is $\left\{x \,\middle|\, x \leq \dfrac{5}{6}\right\}$.

10. $2x + 5 > -16$ and $2x + 5 < 9$

$2x > -21$ and $2x < 4$

$x > -\dfrac{21}{2}$ and $x < 2$

$\left\{x \,\middle|\, x > -\dfrac{21}{2}\right\} \cap \{x \,|\, x < 2\} = \left\{x \,\middle|\, -\dfrac{21}{2} < x < 2\right\}$

The solution set is $\left\{x \,\middle|\, -\dfrac{21}{2} < x < 2\right\}$.

18. $|2x - 9| < 7$

$-7 < 2x - 9 < 7$

$2 < 2x < 16$

$1 < x < 8$

In interval notation, the solution set is $(1, 8)$.

34. $x^2 + 5x + 6 < 0$

$(x + 2)(x + 3) = 0$

$x = -2$ and $x = -3$ • **Critical values**

Use a test number from each of the intervals $(-\infty, -3)$, $(-3, -2)$, and $(-2, \infty)$ to determine where $x^2 + 5x + 6$ is negative.

In interval notation, the solution set is $(-3, -2)$.

46.
$$\frac{3x + 1}{x - 2} \geq 4$$

$$\frac{3x + 1}{x - 2} - 4 \geq 0$$

$$\frac{3x + 1 - 4(x - 2)}{x - 2} \geq 0$$

$$\frac{-x + 9}{x - 2} \geq 0$$

$x = 2$ and $x = 9$ • **Critical values**

Use a test number from each of the intervals $(-\infty, 2)$, $(2, 9)$, and $(9, \infty)$ to determine where $\frac{-x + 9}{x - 2}$ is positive. The solution set is $(2, 9]$.

52. Let $m =$ the number of miles driven.

Company A: $29 + 0.12m$
Company B: $22 + 0.21m$

$$29 + 0.12m < 22 + 0.21m$$
$$77.\overline{7} < m$$

Company A is less expensive if you drive at least 78 miles.

54. Substitute 6.50 for P. Then solve the following inequality.

$$0.218t + 4.02 > 6.50$$
$$0.218t > 2.48$$
$$t > \frac{2.48}{0.218} \approx 11.38$$

Adding 11.38 to 1994 yields 2005.38. According to the given mathematical model, we should first expect to see the average price of a movie ticket exceed $6.50 in the year 2005.

58. $41 \leq$ F ≤ 68

$$41 \leq \frac{9}{5}C + 32 \leq 68$$

$$9 \leq \frac{9}{5}C \leq 36$$

$$\frac{5}{9}(9) \leq \left(\frac{5}{9}\right)\left(\frac{9}{5}\right)C \leq \frac{5}{9}(36)$$

$$5 \leq \quad C \quad \leq 20$$

The Celsius temperature is between 5°C and 20°C, inclusive.

66. We need to solve:

$$\frac{0.00014x^2 + 12x + 400,000}{x} < 30 \qquad \text{(I)}$$

Because $x > 0$, we can multiply both sides by x to produce

$$0.00014x^2 + 12x + 400,000 < 30x$$
$$0.00014x^2 - 18x + 400,000 < 0$$

Use the quadratic formula to find the critical values.

$$x = \frac{-(-18) \pm \sqrt{(-18)^2 - 4(0.00014)(400,000)}}{2(0.00014)}$$

$$= \frac{18 \pm \sqrt{100}}{0.00028}$$

Use a calculator to find that the critical values are approximately 28,571.4 and 100,000. A test value can be

used to show that our original inequality (I) is true on the interval (28,571.4, 100,000). Thus the company should manufacture from 28,572 to 99,999 pairs of running shoes if it wishes to bring the average cost below $30 per pair.

Exercise Set 1.6, page 150

22. $d = kw$

$$6 = k \cdot 80$$
$$\frac{6}{80} = k$$
$$k = \frac{3}{40}$$

Thus $d = \frac{3}{40} \cdot 100 = 7.5$ inches.

26. $r = kv^2$

$$140 = k \cdot 60^2$$
$$\frac{140}{60^2} = k$$
$$\frac{7}{180} = k$$

Thus $r = \frac{7}{180} \cdot 65^2 \approx 164.3$ feet.

30. The general variation is

$$f = \frac{k}{l}$$

where f is the frequency, in vibrations per second, of the vibrating string and l is the length of the string in inches. We are given that $f = 144$ when $l = 20$ inches.

Solving $144 = \frac{k}{20}$ for k yields $k = 2880$. Thus the specific variation is $f = \frac{2880}{l}$. When $l = 18$ inches, we find

$f = \frac{2880}{18} = 160$. The frequency of a guitar string with a length of 18 inches is 160 vibrations per second. (*Note:* We have assumed that the tension is the same for both strings.)

32. $I = \frac{k}{d^2}$

$$50 = \frac{k}{10^2}$$
$$5000 = k$$

Thus $I = \frac{5000}{d^2} = \frac{5000}{15^2} = \frac{5000}{225} \approx 22.2$ footcandles.

34. $L = kwd^2$

$200 = k \cdot 2 \cdot 6^2$

$k = \dfrac{200}{2 \cdot 6^2} = \dfrac{25}{9}$

Thus $L = \dfrac{25}{9} \cdot 4 \cdot 4^2 = \dfrac{1600}{9} \approx 178$ pounds.

38. $L = k\dfrac{wd^2}{l}$

$800 = k\dfrac{4 \cdot 8^2}{12}$

$\dfrac{12 \cdot 800}{4 \cdot 8^2} = k$

$37.5 = k$

Thus $L = 37.5\dfrac{3.5 \cdot 6^2}{16} = 295.3125 \approx 295$ pounds.

Exercise Set 2.1, page 174

6. $d = \sqrt{(x_2 - x_1)^2 + (y_2 - y_1)^2}$

$d = \sqrt{[-10 - (-5)]^2 + (14 - 8)^2}$

$= \sqrt{(-5)^2 + 6^2} = \sqrt{25 + 36}$

$= \sqrt{61}$

26.
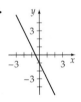

30.

32.
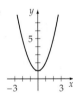

40. y-intercept: $\left(0, -\dfrac{15}{4}\right)$

x-intercept: $(5, 0)$

64. $r = \sqrt{(1 - (-2))^2 + (7 - 5)^2}$

$= \sqrt{9 + 4} = \sqrt{13}$

Using the standard form

$(x - h)^2 + (y - k)^2 = r^2$

with $h = -2, k = 5$, and $r = \sqrt{13}$ yields

$(x + 2)^2 + (y - 5)^2 = \left(\sqrt{13}\right)^2$

66. $x^2 + y^2 - 6x - 4y + 12 = 0$

$x^2 - 6x + y^2 - 4y = -12$

$x^2 - 6x + 9 + y^2 - 4y + 4 = -12 + 9 + 4$

$(x - 3)^2 + (y - 2)^2 = 1^2$

center $(3, 2)$, radius 1

Exercise Set 2.2, page 190

2. Given $g(x) = 2x^2 + 3$

   a. $g(3) = 2(3)^2 + 3 = 18 + 3 = 21$

   b. $g(-1) = 2(-1)^2 + 3 = 2 + 3 = 5$

   c. $g(0) = 2(0)^2 + 3 = 0 + 3 = 3$

   d. $g\left(\dfrac{1}{2}\right) = 2\left(\dfrac{1}{2}\right)^2 + 3 = \dfrac{1}{2} + 3 = \dfrac{7}{2}$

   e. $g(c) = 2(c)^2 + 3 = 2c^2 + 3$

   f. $g(c + 5) = 2(c + 5)^2 + 3 = 2c^2 + 20c + 50 + 3$

$= 2c^2 + 20c + 53$

10. a. Because $0 \le 0 \le 5, Q(0) = 4$.

   b. Because $6 < e < 7, Q(e) = -e + 9$.

   c. Because $1 < n < 2, Q(n) = 4$.

   d. Because $1 < m \le 2, 8 < m^2 + 7 \le 11$. Thus

$Q(m^2 + 7) = \sqrt{(m^2 + 7) - 7} = \sqrt{m^2} = m$

14. $x^2 - 2y = 2$ • Solve for y.

$-2y = -x^2 + 2$

$y = \dfrac{1}{2}x^2 - 1$

y is a function of x because each x value will yield one and only one y value.

28. Domain is the set of all real numbers.

40. Domain is the set of all real numbers.

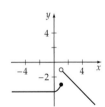

48. a. $[0, \infty]$

   b. Since \$31,250 is between \$27,950 and \$67,700 use $T(x) = 0.27(x - 27,950) + 3892.50$. Then, $T(31,250) = 0.27(31,250 - 27,950) + 3892.50 = \$4783.50$.

   c. Since \$72,000 is between \$67,700 and \$141,250 use $T(x) = 0.30(x - 67,700) + 14,625$. Then, $T(72,000) = 0.30(72,000 - 67,700) + 14,625 = \$15,915$.

50. a. This is the graph of a function. Every vertical line intersects the graph in at most one point.

   b. This is not the graph of a function. Some vertical lines intersect the graph at two points.

   c. This is not the graph of a function. The vertical line at $x = -2$ intersects the graph at more than one point.

   d. This is the graph of a function. Every vertical line intersects the graph at exactly one point.

66. $v(t) = 44,000 - 4200t, 0 \le t \le 8$

68. a. $V(x) = (30 - 2x)^2 x$

$$= (900 - 120x + 4x^2)x$$

$$= 900x - 120x^2 + 4x^3$$

b. Domain: $\{x \mid 0 < x < 15\}$

72. $d(A, B) = \sqrt{1 + x^2}$. The time required to swim from A to B at 2 mph is $\dfrac{\sqrt{1 + x^2}}{2}$ hours.

$d(B, C) = 3 - x$. The time required to run from B to C at 8 mph is $\dfrac{3 - x}{8}$ hours.

Thus the total time to reach point C is

$$t = \frac{\sqrt{1 + x^2}}{2} + \frac{3 - x}{8} \text{ hours}$$

Exercise Set 2.3, page 207

2. $m = \dfrac{1 - 4}{5 - (-2)} = -\dfrac{3}{7}$

16. $m = -1$
$b = 1$

28. $y - 5 = -2(x - 0)$

$y = -2x + 5$

42. $f(x) = \dfrac{2x}{3} + 2$

$4 = \dfrac{2x}{3} + 2$ • Replace $f(x)$ by 4 and solve for x.

$2 = \dfrac{2x}{3}$

$3 = x$

When $x = 3, f(x) = 4$.

46. $f(x) = 0$

$-2x - 4 = 0$

$-2x = 4$

$x = -2$

50. $f_1(x) = f_2(x)$

$-2x - 11 = 3x + 7$

$-5x - 11 = 7$

$-5x = 18$

$x = -\dfrac{18}{5} = -3.6$

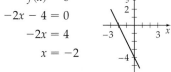

56. a. Using the data for 1997 and 2003, two ordered pairs on the line are $(1997, 531.0)$ and $(2003, 725.0)$. Find the slope of the line.

$$m = \frac{725.0 - 531.0}{2003 - 1997} \approx 32.3$$

Use the point-slope formula to find the equation of the line between the given points.

$$y - y_1 = m(x - x_1)$$

$$y - 531.0 = 32.3(x - 1997)$$

$$y - 531.0 = 32.3x - 64{,}503.1$$

$$y = 32.3x - 64{,}038.7$$

Using functional notation, the linear function is $C(t) = 35.6t - 70{,}585.4$.

b. To find the year when consumer debt first exceeds \$850 billion, let $C(t) = 850$ and solve for t.

$$C(t) = 32.3t - 64{,}038.7$$

$$850 = 32.3t - 64{,}038.7$$

$$64{,}888.7 = 32.3t$$

$$2008 \approx t$$

According to the model, revolving consumer debt will first exceed \$850 billion in 2008.

66. $P(x) = R(x) - C(x)$

$P(x) = 124x - (78.5x + 5005)$

$P(x) = 45.5x - 5005$

$45.5x - 5005 = 0$

$45.5x = 5005$

$x = 110$ • **The break-even point**

78. a. The slope of the radius from $(0, 0)$ to $(\sqrt{15}, 1)$ is $\dfrac{1}{\sqrt{15}}$.

The slope of the linear path of the rock is $-\sqrt{15}$. The path of the rock is given by

$$y - 1 = -\sqrt{15}(x - \sqrt{15})$$

$$y - 1 = -\sqrt{15}x + 15$$

$$y = -\sqrt{15}x + 16$$

Every point on the wall has a y value of 14. Thus

$$14 = -\sqrt{15}x + 16$$

$$-2 = -\sqrt{15}x$$

$$x = \frac{2}{\sqrt{15}} \approx 0.52$$

The rock hits the wall at $(0.52, 14)$.

Exercise Set 2.4, page 222

10. $f(x) = x^2 + 6x - 1$

$\quad = x^2 + 6x + 9 + (-1 - 9)$

$\quad = (x + 3)^2 - 10$

vertex $(-3, -10)$

axis of symmetry $x = -3$

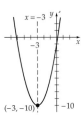

20. $h = -\dfrac{b}{2a} = -\dfrac{-6}{2(1)} = 3$

$k = f(3) = 3^2 - 6(3) = -9$

vertex $(3, -9)$

$f(x) = (x - 3)^2 - 9$

32. Determine the y-coordinate of the vertex of the graph of $f(x) = 2x^2 + 6x - 5$.

$f(x) = 2x^2 + 6x - 5$ • $a = 2, b = 6, c = -5$.

$h = -\dfrac{b}{2a} = -\dfrac{6}{2(2)} = -\dfrac{3}{2}$ • Find the x-coordinate of the vertex.

$k = f\left(-\dfrac{3}{2}\right) = 2\left(-\dfrac{3}{2}\right)^2 + 6\left(-\dfrac{3}{2}\right) - 5 = -\dfrac{19}{2}$

• Find the y-coordinate of the vertex.

The vertex is $\left(-\dfrac{3}{2}, -\dfrac{19}{2}\right)$. Because the parabola opens up, $-\dfrac{19}{2}$ is the minimum value of f. Therefore, the range of f is $\left\{y \,|\, y \geq -\dfrac{19}{2}\right\}$. To determine the values of x for which $f(x) = 15$, replace $f(x)$ by $2x^2 + 6x - 5$ and solve for x.

$\qquad\qquad f(x) = 15$

$\quad 2x^2 + 6x - 5 = 15$ • Replace $f(x)$ by $2x^2 + 6x - 5$.

$\quad 2x^2 + 6x - 20 = 0$ • Solve for x.

$\quad 2(x - 2)(x + 5) = 0$ • Factor.

$x - 2 = 0 \qquad x + 5 = 0$ • Use the Principle of Zero Products to solve for x.

$\quad x = 2 \qquad\qquad x = -5$

The values of x for which $f(x) = 15$ are 2 and -5.

36. $f(x) = -x^2 - 6x$

$\quad = -(x^2 + 6x)$

$\quad = -(x^2 + 6x + 9) + 9$

$\quad = -(x + 3)^2 + 9$

Maximum value of f is 9 when $x = -3$.

46. a. $l + w = 240$, so $w = 240 - l$.

b. $A = lw = l(240 - l) = 240l - l^2$

c. The l value of the vertex point of the graph of $A = 240l - l^2$ is

$-\dfrac{b}{2a} = -\dfrac{240}{2(-1)} = 120$

Thus $l = 120$ meters and $w = 240 - 120 = 120$ meters are the dimensions that produce the greatest area.

68. Let $x =$ the number of parcels.

a. $R(x) = xp = x(22 - 0.01x) = -0.01x^2 + 22x$

b. $P(x) = R(x) - C(x)$

$\quad = (-0.01x^2 + 22x) - (2025 + 7x)$

$\quad = -0.01x^2 + 15x - 2025$

c. $-\dfrac{b}{2a} = -\dfrac{15}{2(-0.01)} = 750$

The maximum profit is

$P(750) = -0.01(750)^2 + 15(750) - 2025 = \$3600$

d. The price per parcel that yields the maximum profit is

$p(750) = 22 - 0.01(750) = \$14.50$

e. The break-even point(s) occur when $R(x) = C(x)$.

$\quad -0.01x^2 + 22x = 2025 + 7x$

$\quad 0 = 0.01x^2 - 15x + 2025$

$\quad x = \dfrac{-(-15) \pm \sqrt{(-15)^2 - 4(0.01)(2025)}}{2(0.01)}$

$x = 150$ and $x = 1350$ are the break-even points.

Thus the minimum number of parcels the air freight company must ship to break even is 150.

70. $h(t) = -16t^2 + 64t + 80$

$t = -\dfrac{b}{2a} = -\dfrac{64}{2(-16)} = 2$

$h(2) = -16(2)^2 + 64(2) + 80$

$\quad = -64 + 128 + 80 = 144$

a. The vertex $(2, 144)$ gives us the maximum height of 144 feet.

b. The vertex of the graph of h is $(2, 144)$, so the time when the projectile achieves this maximum height is at time $t = 2$ seconds.

c. $-16t^2 + 64t + 80 = 0$ • Solve for t with $h = 0$.

$\quad -16(t^2 - 4t - 5) = 0$

$\quad -16(t + 1)(t - 5) = 0$

$\quad t = -1 \qquad\quad t - 5 = 0$

$\quad$ no $\qquad\qquad t = 5$

The projectile will have a height of 0 feet at time $t = 5$ seconds.

Exercise Set 2.5, page 238

14. The graph is symmetric with respect to the x-axis, because replacing y with $-y$ leaves the equation unaltered. The graph is not symmetric with respect to the y-axis, because replacing x with $-x$ alters the equation.

24. The graph is symmetric with respect to the origin because $(-y) = (-x)^3 - (-x)$ simplifies to $-y = -x^3 + x$, which is equivalent to the original equation $y = x^3 - x$.

44. Even, because $h(-x) = (-x)^2 + 1 = x^2 + 1 = h(x)$.

58. 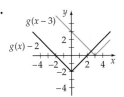   **68.**

70.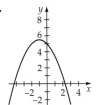

72. a.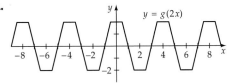

b.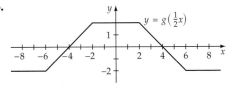

Exercise Set 2.6, page 251

10. $f(x) + g(x) = \sqrt{x - 4} - x$ Domain: $\{x \mid x \geq 4\}$
$f(x) - g(x) = \sqrt{x - 4} + x$ Domain: $\{x \mid x \geq 4\}$
$f(x)g(x) = -x\sqrt{x - 4}$ Domain: $\{x \mid x \geq 4\}$
$\dfrac{f(x)}{g(x)} = -\dfrac{\sqrt{x - 4}}{x}$ Domain: $\{x \mid x \geq 4\}$

14. $(f + g)(x) = (x^2 - 3x + 2) + (2x - 4) = x^2 - x - 2$
$(f + g)(-7) = (-7)^2 - (-7) - 2 = 49 + 7 - 2 = 54$

30. $\dfrac{f(x + h) - f(x)}{h} = \dfrac{[4(x + h) - 5] - (4x - 5)}{h}$
$= \dfrac{4x + 4(h) - 5 - 4x + 5}{h}$

$= \dfrac{4(h)}{h} = 4$

38. $(g \circ f)(x) = g[f(x)] = g[2x - 7]$
$= 3[2x - 7] + 2 = 6x - 19$
$(f \circ g)(x) = f[g(x)] = f[3x + 2]$
$= 2[3x + 2] - 7 = 6x - 3$

50. $(f \circ g)(4) = f[g(4)]$
$= f[4^2 - 5(4)]$
$= f[-4] = 2(-4) + 3 = -5$

66. a. $l = 3 - 0.5t$ for $0 \leq t \leq 6$. $l = -3 + 0.5t$ for $t > 6$. In either case, $l = |3 - 0.5t|$. $w = |2 - 0.2t|$ as in Example 7.

b. $A(t) = |3 - 0.5t||2 - 0.2t|$

c. A is decreasing on $[0, 6]$ and on $[8, 10]$. A is increasing on $[6, 8]$ and on $[10, 14]$.

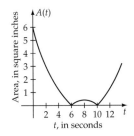

d. The highest point on the graph of A for $0 \leq t \leq 14$ occurs when $t = 0$ seconds.

72. a. On $[2, 3]$,

$a = 2$
$\Delta t = 3 - 2 = 1$
$s(a + \Delta t) = s(3) = 6 \cdot 3^2 = 54$
$s(a) = s(2) = 6 \cdot 2^2 = 24$

$\text{Average velocity} = \dfrac{s(a + \Delta t) - s(a)}{\Delta t}$

$= \dfrac{s(3) - s(2)}{1}$

$= 54 - 24 = 30$ feet per second

This is identical to the slope of the line through $(2, f(2))$ and $(3, f(3))$ because

$m = \dfrac{s(3) - s(2)}{3 - 2} = s(3) - s(2) = 54 - 24 = 30$

b. On $[2, 2.5]$,

$a = 2$
$\Delta t = 2.5 - 2 = 0.5$
$s(a + \Delta t) = s(2.5) = 6(2.5)^2 = 37.5$

$$\text{Average velocity} = \frac{s(2.5) - s(2)}{0.5}$$

$$= \frac{37.5 - 24}{0.5}$$

$$= \frac{13.5}{0.5} = 27 \text{ feet per second}$$

c. On $[2, 2.1]$,

$a = 2$

$\Delta t = 2.1 - 2 = 0.1$

$s(a + \Delta t) = s(2.1) = 6(2.1)^2 = 26.46$

$$\text{Average velocity} = \frac{s(2.1) - s(2)}{0.1}$$

$$= \frac{26.46 - 24}{0.1}$$

$$= \frac{2.46}{0.1} = 24.6 \text{ feet per second}$$

d. On $[2, 2.01]$,

$a = 2$

$\Delta t = 2.01 - 2 = 0.01$

$s(a + \Delta t) = s(2.01) = 6(2.01)^2 = 24.2406$

$$\text{Average velocity} = \frac{s(2.01) - s(2)}{0.01}$$

$$= \frac{24.2406 - 24}{0.01}$$

$$= \frac{0.2406}{0.01} = 24.06 \text{ feet per second}$$

e. On $[2, 2.001]$,

$a = 2$

$\Delta t = 2.001 - 2 = 0.001$

$s(a + \Delta t) = s(2.001) = 6(2.001)^2 = 24.024006$

$$\text{Average velocity} = \frac{s(2.001) - s(2)}{0.001}$$

$$= \frac{24.024006 - 24}{0.001}$$

$$= \frac{0.024006}{0.001} = 24.006 \text{ feet per second}$$

f. On $[2, 2 + \Delta t]$,

$$\frac{s(2 + \Delta t) - s(2)}{\Delta t} = \frac{6(2 + \Delta t)^2 - 24}{\Delta t}$$

$$= \frac{6(4 + 4(\Delta t) + (\Delta t)^2) - 24}{\Delta t}$$

$$= \frac{24 + 24(\Delta t) + 6(\Delta t)^2 - 24}{\Delta t}$$

$$= \frac{24\Delta t + 6(\Delta t)^2}{\Delta t} = 24 + 6(\Delta t)$$

As Δt approaches zero, the average velocity approaches 24 feet per second.

Exercise Set 2.7, page 263

18. Enter the data in the table. Then use your calculator to find the linear regression equation.

    a. The linear regression equation is

$$y = -72.06131724x + 14926.16191$$

    b. Evaluate the linear regression equation when $x = 55$.

$$y = -72.06131724(55) + 14926.16191$$

$$\approx 10{,}962.79$$

    The approximate trade-in value of the car is $10,963.

32. Enter the data in the table. Then use your calculator to find the quadratic regression model.

    a. $y = 0.05208x^2 - 3.56026x + 82.32999$

    b. The speed at which the bird has minimum oxygen consumption is the x-coordinate of the vertex of the graph of the regression equation. Recall that the x-coordinate of the vertex is given by $x = -\dfrac{b}{2a}$.

$$x = -\frac{b}{2a} = -\frac{-3.56026}{2(0.05208)}$$

$$\approx 34$$

The speed that minimizes oxygen consumption is approximately 34 kilometers per hour.

Exercise Set 3.1, page 287

2.

$$\begin{array}{r} 6x^2 - 9x + 28 \\ x + 4 \overline{)6x^3 + 15x^2 - 8x + 2} \\ \underline{6x^3 + 24x^2} \\ -9x^2 - 8x \\ \underline{-9x^2 - 36x} \\ 28x + 2 \\ \underline{28x + 112} \\ -110 \end{array}$$

$$\frac{6x^3 + 15x^2 - 8x + 2}{x + 4} = 6x^2 - 9x + 28 - \frac{110}{x + 4}$$

12.

$$\begin{array}{r|rrrr} 5 & 5 & 6 & -8 & 1 \\ & & 25 & 155 & 735 \\ \hline & 5 & 31 & 147 & 736 \end{array}$$

$$\frac{5x^3 + 6x^2 - 8x + 1}{x - 5} = 5x^2 + 31x + 147 + \frac{736}{x - 5}$$

Exercise Set 5.6, page 526

22. $y = \dfrac{1}{3} \tan x$

period $= \dfrac{\pi}{b} = \pi$

30. $y = -3 \tan 3x$

period $= \dfrac{\pi}{b} = \dfrac{\pi}{3}$

32. $y = \dfrac{1}{2} \cot 2x$

period $= \dfrac{\pi}{b} = \dfrac{\pi}{2}$

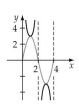

38. $y = 3 \csc \dfrac{\pi x}{2}$

period $= \dfrac{2\pi}{b} = \dfrac{2\pi}{\pi/2} = 4$

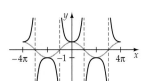

42. $y = \sec \dfrac{x}{2}$

period $= \dfrac{2\pi}{b} = \dfrac{2\pi}{1/2} = 4\pi$

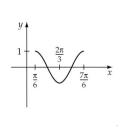

Exercise Set 5.7, page 535

20. $y = \cos\left(2x - \dfrac{\pi}{3}\right)$

$a = 1$

period $= \pi$

phase shift $= -\dfrac{c}{b}$

$= -\dfrac{-\pi/3}{2} = \dfrac{\pi}{6}$

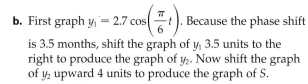

22. $y = \tan(x - \pi)$

period $= \pi$

phase shift $= -\dfrac{c}{b}$

$= -\dfrac{-\pi}{1} = \pi$

40. $y = 2 \sin\left(\dfrac{\pi x}{2} + 1\right) - 2$

$a = 2$

period $= 4$

phase shift $= -\dfrac{c}{b}$

$= -\dfrac{1}{2/\pi} = -\dfrac{2}{\pi}$

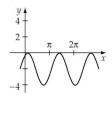

42. $y = -3 \cos(2\pi x - 3) + 1$

$a = 3$

period $= 1$

phase shift $= -\dfrac{c}{b} = \dfrac{3}{2\pi}$

48. $y = \csc \dfrac{x}{3} + 4$

period $= 6\pi$

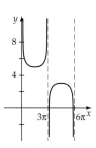

52. a. Phase shift $= -\dfrac{c}{b} = -\dfrac{\left(-\dfrac{7}{12}\pi\right)}{\left(\dfrac{\pi}{6}\right)} = 3.5$ months,

period $= \dfrac{2\pi}{b} = \dfrac{2\pi}{\pi/6} = 12$ months

b. First graph $y_1 = 2.7 \cos\left(\dfrac{\pi}{6}t\right)$. Because the phase shift is 3.5 months, shift the graph of y_1 3.5 units to the right to produce the graph of y_2. Now shift the graph of y_2 upward 4 units to produce the graph of S.

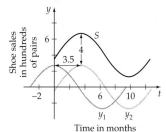

c. 3.5 months after January 1 is the middle of April.

54. $y = \dfrac{x}{2} + \cos x$

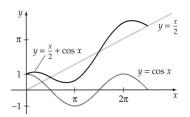

58. $y = -\sin x + \cos x$

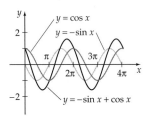

78. $y = x \cos x$

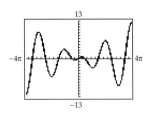

Exercise Set 5.8, page 544

20. Amplitude $= 3$, frequency $= \dfrac{1}{\pi}$, period $= \pi$.

Because $\dfrac{2\pi}{b} = \pi$, we have $b = 2$. Thus $y = 3 \cos 2t$.

28. Amplitude $= |-1.5| = 1.5$

$f = \dfrac{1}{2\pi}\sqrt{\dfrac{k}{m}} = \dfrac{1}{2\pi}\sqrt{\dfrac{3}{27}} = \dfrac{1}{2\pi}\cdot\dfrac{1}{3} = \dfrac{1}{6\pi}$, period $= 6\pi$

$y = a \cos 2\pi f t = -1.5 \cos\left[2\pi\left(\dfrac{1}{6\pi}\right)t\right] = -1.5 \cos\dfrac{1}{3}t$

30.

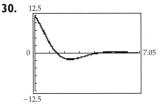

a. f has pseudoperiod $\dfrac{2\pi}{1} = 2\pi$.

$10 \div (2\pi) \approx 1.59$

Thus f completes only one full oscillation on $0 \le t \le 10$.

b. The following graph of f shows that $|f(t)| < 0.01$ for $t > 10.5$.

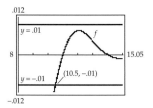

Exercise Set 6.1, page 559

2. The equation $\tan 2x = 2 \tan x$ is not an identity. To verify that it is not an identity, let $x = \dfrac{\pi}{6}$. Then

$$\tan 2x = \tan 2\left(\dfrac{\pi}{6}\right) = \tan\dfrac{\pi}{3} = \sqrt{3}$$

whereas $2 \tan x = 2 \tan\dfrac{\pi}{6} = \dfrac{2\sqrt{3}}{3}$. Because $\sqrt{3} \neq \dfrac{2\sqrt{3}}{3}$, we have shown that the equation is not an identity.

22. $\sin^4 x - \cos^4 x = (\sin^2 x + \cos^2 x)(\sin^2 x - \cos^2 x)$
$\qquad\qquad = 1(\sin^2 x - \cos^2 x) = \sin^2 x - \cos^2 x$

34. $\dfrac{2 \sin x \cot x + \sin x - 4 \cot x - 2}{2 \cot x + 1}$

$= \dfrac{(\sin x)(2 \cot x + 1) - 2(2 \cot x + 1)}{2 \cot x + 1}$

$= \dfrac{(2 \cot x + 1)(\sin x - 2)}{2 \cot x + 1} = \sin x - 2$

44. $\dfrac{\dfrac{1}{\sin x} + \dfrac{1}{\cos x}}{\dfrac{1}{\sin x} - \dfrac{1}{\cos x}} = \dfrac{\dfrac{1}{\sin x} + \dfrac{1}{\cos x}}{\dfrac{1}{\sin x} - \dfrac{1}{\cos x}} \cdot \dfrac{\sin x \cos x}{\sin x \cos x}$

$= \dfrac{\cos x + \sin x}{\cos x - \sin x}$

$= \dfrac{\cos x + \sin x}{\cos x - \sin x} \cdot \dfrac{\cos x - \sin x}{\cos x - \sin x}$

$= \dfrac{\cos^2 x - \sin^2 x}{\cos^2 x - 2 \sin x \cos x + \sin^2 x}$

$= \dfrac{\cos^2 x - \sin^2 x}{1 - 2 \sin x \cos x}$

54. $\dfrac{\dfrac{1}{\tan x} + \cot x}{\dfrac{1}{\tan x} + \tan x} = \dfrac{\dfrac{1}{\tan x} + \cot x}{\dfrac{1}{\tan x} + \tan x} \cdot \dfrac{\tan x}{\tan x}$

$= \dfrac{1 + 1}{1 + \tan^2 x} = \dfrac{2}{\sec^2 x}$

Exercise Set 6.2, page 570

4. Use the identity $\cos(\alpha - \beta) = \cos \alpha \cos \beta + \sin \alpha \sin \beta$ with $\alpha = 120°$ and $\beta = 45°$.

$\cos(120° - 45°) = \cos 120° \cos 45° + \sin 120° \sin 45°$

$= \left(-\dfrac{1}{2}\right)\left(\dfrac{\sqrt{2}}{2}\right) + \left(\dfrac{\sqrt{3}}{2}\right)\left(\dfrac{\sqrt{2}}{2}\right)$

$= -\dfrac{\sqrt{2}}{4} + \dfrac{\sqrt{6}}{4}$

$= \dfrac{\sqrt{6} - \sqrt{2}}{4}$

20. The value of a given trigonometric function of θ, measured in degrees, is equal to its cofunction of $90° - \theta$. Thus

$\cos 80° = \sin(90° - 80°)$

$= \sin 10°$

26. $\sin x \cos 3x + \cos x \sin 3x = \sin(x + 3x) = \sin 4x$

38. $\tan \alpha = \dfrac{24}{7}$, with $0° < \alpha < 90°$, $\sin \alpha = \dfrac{24}{25}$, $\cos \alpha = \dfrac{7}{25}$

$\sin \beta = -\dfrac{8}{17}$, with $180° < \beta < 270°$

$\cos \beta = -\dfrac{15}{17}$, $\tan \beta = \dfrac{8}{15}$

a. $\sin(\alpha + \beta) = \sin \alpha \cos \beta + \cos \alpha \sin \beta$

$= \left(\dfrac{24}{25}\right)\left(-\dfrac{15}{17}\right) + \left(\dfrac{7}{25}\right)\left(-\dfrac{8}{17}\right)$

$= -\dfrac{360}{425} - \dfrac{56}{425} = -\dfrac{416}{425}$

b. $\cos(\alpha + \beta) = \cos \alpha \cos \beta - \sin \alpha \sin \beta$

$= \left(\dfrac{7}{25}\right)\left(-\dfrac{15}{17}\right) - \left(\dfrac{24}{25}\right)\left(-\dfrac{8}{17}\right)$

$= -\dfrac{105}{425} + \dfrac{192}{425} = \dfrac{87}{425}$

c. $\tan(\alpha - \beta) = \dfrac{\tan \alpha - \tan \beta}{1 + \tan \alpha \tan \beta}$

$= \dfrac{\dfrac{24}{7} - \dfrac{8}{15}}{1 + \left(\dfrac{24}{7}\right)\left(\dfrac{8}{15}\right)} = \dfrac{\dfrac{24}{7} - \dfrac{8}{15}}{1 + \dfrac{192}{105}} \cdot \dfrac{105}{105}$

$= \dfrac{360 - 56}{105 + 192} = \dfrac{304}{297}$

50. $\cos(\theta + \pi) = \cos \theta \cos \pi - \sin \theta \sin \pi$

$= (\cos \theta)(-1) - (\sin \theta)(0) = -\cos \theta$

62. $\cos 5x \cos 3x + \sin 5x \sin 3x = \cos(5x - 3x) = \cos 2x$

$= \cos(x + x) = \cos x \cos x - \sin x \sin x$

$= \cos^2 x - \sin^2 x$

74. $\sin(\theta + 2\pi) = \sin \theta \cos 2\pi + \cos \theta \sin 2\pi$

$= (\sin \theta)(1) + (\cos \theta)(0) = \sin \theta$

Exercise Set 6.3, page 578

2. $2 \sin 3\theta \cos 3\theta = \sin[2(3\theta)] = \sin 6\theta$

26. $\cos \theta = \dfrac{24}{25}$ with $270° < \theta < 360°$

$\sin \theta = -\sqrt{1 - \left(\dfrac{24}{25}\right)^2}$ ・ $\tan \theta = \dfrac{-7/25}{24/25}$

$= -\dfrac{7}{25}$ ・ $= -\dfrac{7}{24}$

$\sin 2\theta = 2 \sin \theta \cos \theta$ ・ $\cos 2\theta = \cos^2 \theta - \sin^2 \theta$

$= 2\left(-\dfrac{7}{25}\right)\left(\dfrac{24}{25}\right)$ ・ $= \left(\dfrac{24}{25}\right)^2 - \left(-\dfrac{7}{25}\right)^2$

$= -\dfrac{336}{625}$ ・ $= \dfrac{527}{625}$

$\tan 2\theta = \dfrac{2 \tan \theta}{1 - \tan^2 \theta}$

$= \dfrac{2\left(-\dfrac{7}{24}\right)}{1 - \left(-\dfrac{7}{24}\right)^2}$

$= \dfrac{-\dfrac{7}{12}}{1 - \dfrac{49}{576}} \cdot \dfrac{576}{576} = -\dfrac{336}{527}$

54. $\dfrac{1}{1 - \cos 2x} = \dfrac{1}{1 - 1 + 2 \sin^2 x}$

$= \dfrac{1}{2 \sin^2 x} = \dfrac{1}{2} \csc^2 x$

72. $\cos^2 \dfrac{x}{2} = \left[\pm\sqrt{\dfrac{1 + \cos x}{2}}\right]^2 = \dfrac{1 + \cos x}{2}$

$= \dfrac{1 + \cos x}{2} \cdot \dfrac{\sec x}{\sec x} = \dfrac{\sec x + 1}{2 \sec x}$

78. $\tan^2 \dfrac{x}{2} = \left(\dfrac{1 - \cos x}{\sin x}\right)^2 = \dfrac{(1 - \cos x)^2}{\sin^2 x} = \dfrac{(1 - \cos x)^2}{1 - \cos^2 x}$

$\qquad = \dfrac{(1 - \cos x)^2}{(1 - \cos x)(1 + \cos x)} = \dfrac{1 - \cos x}{1 + \cos x}$

$\qquad = \dfrac{\dfrac{1}{\cos x} - \dfrac{\cos x}{\cos x}}{\dfrac{1}{\cos x} + \dfrac{\cos x}{\cos x}} = \dfrac{\sec x - 1}{\sec x + 1}$

Exercise Set 6.4, page 587

22. $\cos 3\theta + \cos 5\theta = 2 \cos \dfrac{3\theta + 5\theta}{2} \cos \dfrac{3\theta - 5\theta}{2}$

$\qquad = 2 \cos 4\theta \cos(-\theta) = 2 \cos 4\theta \cos \theta$

36. $\sin 5x \cos 3x = \dfrac{1}{2}[\sin(5x + 3x) + \sin(5x - 3x)]$

$\qquad = \dfrac{1}{2}(\sin 8x + \sin 2x)$

$\qquad = \dfrac{1}{2}(2 \sin 4x \cos 4x + 2 \sin x \cos x)$

$\qquad = \sin 4x \cos 4x + \sin x \cos x$

44. $\dfrac{\cos 5x - \cos 3x}{\sin 5x + \sin 3x} = \dfrac{-2 \sin \dfrac{5x + 3x}{2} \sin \dfrac{5x - 3x}{2}}{2 \sin \dfrac{5x + 3x}{2} \cos \dfrac{5x - 3x}{2}}$

$\qquad = -\dfrac{\sin 4x \sin x}{\sin 4x \cos x} = -\tan x$

62. $a = 1, b = \sqrt{3}, k = \sqrt{(\sqrt{3})^2 + (1)^2} = 2.$ Thus α is a first-quadrant angle.

$\sin \alpha = \dfrac{\sqrt{3}}{2}$ and $\cos \alpha = \dfrac{1}{2}$

Thus $\alpha = \dfrac{\pi}{3}$.

$y = k \sin(x + \alpha)$

$y = 2 \sin\left(x + \dfrac{\pi}{3}\right)$

70. From Exercise 62, we know that

$y = \sin x + \sqrt{3} \cos x = 2 \sin\left(x + \dfrac{\pi}{3}\right)$

Thus the graph of $y = \sin x + \sqrt{3} \cos x$ is the graph of $y = 2 \sin x$ shifted $\dfrac{\pi}{3}$ units to the left.

Exercise Set 6.5, page 601

2. $y = \sin^{-1} \dfrac{\sqrt{2}}{2}$ implies

$\sin y = \dfrac{\sqrt{2}}{2}$ for $-\dfrac{\pi}{2} \le y \le \dfrac{\pi}{2}$

Thus $y = \dfrac{\pi}{4}$.

24. Because $\tan(\tan^{-1}x) = x$ for all real numbers x, we have

$\tan\left[\tan^{-1}\left(\dfrac{1}{2}\right)\right] = \dfrac{1}{2}$.

46. Let $x = \cos^{-1} \dfrac{3}{5}$. Thus

$\cos x = \dfrac{3}{5}$ and $\sin x = \sqrt{1 - \left(\dfrac{3}{5}\right)^2} = \dfrac{4}{5}$

$y = \tan\left(\cos^{-1} \dfrac{3}{5}\right) = \tan x = \dfrac{\sin x}{\cos x} = \dfrac{4/5}{3/5} = \dfrac{4}{3}$

54. $y = \cos\left(\sin^{-1} \dfrac{3}{4} + \cos^{-1} \dfrac{5}{13}\right)$

Let $\alpha = \sin^{-1} \dfrac{3}{4}, \sin \alpha = \dfrac{3}{4}, \cos \alpha = \sqrt{1 - \left(\dfrac{3}{4}\right)^2} = \dfrac{\sqrt{7}}{4}$.

$\beta = \cos^{-1} \dfrac{5}{13}, \cos \beta = \dfrac{5}{13}, \sin \beta = \sqrt{1 - \left(\dfrac{5}{13}\right)^2} = \dfrac{12}{13}$.

$y = \cos(\alpha + \beta)$

$\quad = \cos \alpha \cos \beta - \sin \alpha \sin \beta$

$\quad = \dfrac{\sqrt{7}}{4} \cdot \dfrac{5}{13} - \dfrac{3}{4} \cdot \dfrac{12}{13} = \dfrac{5\sqrt{7}}{52} - \dfrac{36}{52} = \dfrac{5\sqrt{7} - 36}{52}$

64. $\sin^{-1}x + \cos^{-1} \dfrac{4}{5} = \dfrac{\pi}{6}$

$\sin^{-1}x = \dfrac{\pi}{6} - \cos^{-1} \dfrac{4}{5}$

$\sin(\sin^{-1}x) = \sin\left(\dfrac{\pi}{6} - \cos^{-1} \dfrac{4}{5}\right)$

$x = \sin \dfrac{\pi}{6} \cos\left(\cos^{-1} \dfrac{4}{5}\right) - \cos \dfrac{\pi}{6} \sin\left(\cos^{-1} \dfrac{4}{5}\right)$

$\quad = \dfrac{1}{2} \cdot \dfrac{4}{5} - \dfrac{\sqrt{3}}{2} \cdot \dfrac{3}{5} = \dfrac{4 - 3\sqrt{3}}{10}$

72. Let $\alpha = \cos^{-1}x$ and $\beta = \cos^{-1}(-x)$. Thus $\cos \alpha = x$ and $\cos \beta = -x$. We know $\sin \alpha = \sqrt{1 - x^2}$ and $\sin \beta = \sqrt{1 - x^2}$ because α is in Quadrant I and β is in Quadrant II.

$\cos^{-1}x + \cos^{-1}(-x)$

$\quad = \alpha + \beta$

$\quad = \cos^{-1}[\cos(\alpha + \beta)]$

$= \cos^{-1}(\cos \alpha \cos \beta - \sin \alpha \sin \beta)$

$= \cos^{-1}\left[x(-x) - \sqrt{1 - x^2} \cdot \sqrt{1 - x^2}\right]$

$= \cos^{-1}(-x^2 - 1 + x^2)$

$= \cos^{-1}(-1) = \pi$

76. Recall that the graph of $y = f(x - a)$ is a horizontal shift of the graph of $y = f(x)$. Therefore, the graph of $y = \cos^{-1}(x - 1)$ is the graph of $y = \cos^{-1}x$ shifted 1 unit to the right.

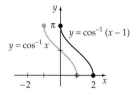

84. a.

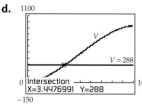

b. Although the water rises 0.1 foot in each case, there is more surface area (and thus more volume of water) at the 4.9- to 5.0-foot level near the diameter of the cylinder than at the 0.1- to 0.2-foot level near the bottom.

c.

$$V(4) = 12\left[25\cos^{-1}\left(\frac{5 - (4)}{5}\right) - [5 - (4)]\sqrt{10(4) - (4)^2}\right]$$

$$= 12\left[25\cos^{-1}\left(\frac{1}{5}\right) - \sqrt{24}\right]$$

$$\approx 352.04 \text{ cubic feet}$$

d.

When $V = 288$ cubic feet, $x \approx 3.45$ feet.

Exercise Set 6.6, page 614

14.
$$2\cos^2 x + 1 = -3\cos x$$

$$2\cos^2 x + 3\cos x + 1 = 0$$

$$(2\cos x + 1)(\cos x + 1) = 0$$

$2\cos x + 1 = 0$ or $\cos x + 1 = 0$

$\cos x = -\dfrac{1}{2}$ $\cos x = -1$

$x = \dfrac{2\pi}{3}, \dfrac{4\pi}{3}$ $x = \pi$

The solutions in the interval $0 \le x < 2\pi$ are $\dfrac{2\pi}{3}$, π, and $\dfrac{4\pi}{3}$.

52. $\sin x + 2\cos x = 1$

$$\sin x = 1 - 2\cos x$$

$$(\sin x)^2 = (1 - 2\cos x)^2$$

$$\sin^2 x = 1 - 4\cos x + 4\cos^2 x$$

$$1 - \cos^2 x = 1 - 4\cos x + 4\cos^2 x$$

$$0 = \cos x(5\cos x - 4)$$

$\cos x = 0$ or $5\cos x - 4 = 0$

$x = 90°, 270°$ $\cos x = \dfrac{4}{5}$

$x \approx 36.9°, 323.1°$

The solutions in the interval $0 \le x < 360°$ are $90°$ and $323.1°$. (*Note:* $x = 270°$ and $x = 36.9°$ are extraneous solutions. Neither of these values satisfy the original equation.)

56. $2\cos^2 x - 5\cos x - 5 = 0$

$$\cos x = \frac{5 \pm \sqrt{(-5)^2 - 4(2)(-5)}}{2(2)} = \frac{5 \pm \sqrt{65}}{4}$$

$\cos x \approx 3.27$ or $\cos x \approx -0.7656$

no solution $x \approx 140.0°, 220.0°$

The solutions in the interval $0° \le x < 360°$ are $140.0°$ and $220.0°$.

66. $\cos 2x = -\dfrac{\sqrt{3}}{2}$

$2x = \dfrac{5\pi}{6} + 2k\pi$ or $2x = \dfrac{7\pi}{6} + 2k\pi$, k an integer

$x = \dfrac{5\pi}{12} + k\pi$ or $x = \dfrac{7\pi}{12} + k\pi$, k an integer

84. $2\sin x \cos x - 2\sqrt{2}\sin x - \sqrt{3}\cos x + \sqrt{6} = 0$

$2\sin x(\cos x - \sqrt{2}) - \sqrt{3}(\cos x - \sqrt{2}) = 0$

$(\cos x - \sqrt{2})(2\sin x - \sqrt{3}) = 0$

$\cos x = \sqrt{2}$ or $\sin x = \dfrac{\sqrt{3}}{2}$

no solution $x = \dfrac{\pi}{3}, \dfrac{2\pi}{3}$

The solutions in the interval $0 \le x < 2\pi$ are $\dfrac{\pi}{3}$ and $\dfrac{2\pi}{3}$.

86. The following graph shows that the solutions in the interval $[0, 2\pi)$ are $x = 0$ and $x = 1.895$.

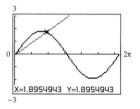

X=1.8954943 Y=1.8954943

92. When $\theta = 45°$, d attains its maximum of 4394.5 feet.

94. a. The following work shows the sine regression procedure for a TI-83 calculator. Note that each time (given in hours:minutes) must be converted to hours. Thus 17:03 is $17 + \dfrac{3}{60} = 17.05$ hours.

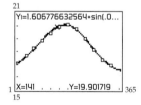

The sine regression function is
$$Y_1 \approx 1.6068 \sin(0.01622x - 1.1384) + 18.436$$

b. Using the "value" command in the CALC menu gives

$$Y_1(141) \approx 19.901719$$
$$\approx 19{:}54 \text{ (to the nearest minute)}$$

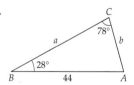

Exercise Set 7.1, page 634

4.

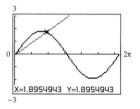

$$A = 180° - 78° - 28° = 74°$$

$$\frac{b}{\sin B} = \frac{c}{\sin C} \qquad\qquad \frac{a}{\sin A} = \frac{c}{\sin C}$$

$$\frac{b}{\sin 28°} = \frac{44}{\sin 78°} \qquad\qquad \frac{a}{\sin 74°} = \frac{44}{\sin 78°}$$

$$b = \frac{44 \sin 28°}{\sin 78°} \approx 21 \qquad\qquad a = \frac{44 \sin 74°}{\sin 78°} \approx 43$$

18.
$$\frac{a}{\sin A} = \frac{b}{\sin B}$$

$$\frac{13.8}{\sin A} = \frac{5.55}{\sin 22.6}$$

$$\sin A = 0.9555$$

$$A \approx 72.9° \quad \text{or} \quad 107.1°$$

If $A = 72.9°$, $C \approx 180° - 72.9° - 22.6° = 84.5°$.

$$\frac{c}{\sin 84.5°} = \frac{5.55}{\sin 22.6°}$$

$$c = \frac{5.55 \sin 84.5°}{\sin 22.6°} \approx 14.4$$

If $A = 107.1°$, $C \approx 180° - 107.1° - 22.6° = 50.3°$.

$$\frac{c}{\sin 50.3°} = \frac{5.55}{\sin 22.6°}$$

$$c = \frac{5.55 \sin 50.3°}{\sin 22.6°} \approx 11.1$$

Case 1: $A = 72.9°$, $C = 84.5°$, and $c = 14.4$
Case 2: $A = 107.1°$, $C = 50.3°$, and $c = 11.1$

26. The angle with its vertex at the position of the helicopter measures $180° - (59.0° + 77.2°) = 43.8°$. Let the distance from the helicopter to the carrier be x. Using the Law of Sines, we have

$$\frac{x}{\sin 77.2°} = \frac{7620}{\sin 43.8°}$$

$$x = \frac{7620 \sin 77.2°}{\sin 43.8°}$$

$$\approx 10{,}700 \text{ feet} \qquad \text{(three significant digits)}$$

36.

$$\alpha = 65°$$
$$B = 65° + 8° = 73°$$
$$A = 180° - 50° - 65° = 65°$$
$$C = 180° - 65° - 73° = 42°$$

26. $16x^2 - 9y^2 - 32x - 54y + 79 = 0$

$16(x^2 - 2x + 1) - 9(y^2 + 6y + 9) = -79 + 16 - 81$

$$= -144$$

$$\frac{(y + 3)^2}{16} - \frac{(x - 1)^2}{9} = 1$$

Transverse axis is parallel to y-axis because y^2 term is positive. Center is at $(1, -3)$; $a^2 = 16$ so $a = 4$.

vertices $(h, k + a) = (1, 1)$

$(h, k - a) = (1, -7)$

$c^2 = a^2 + b^2 = 16 + 9 = 25$

$c = \sqrt{25} = 5$

foci $(h, k + c) = (1, 2)$

$(h, k - c) = (1, -8)$

Because $b^2 = 9$ and $b = 3$, the asymptotes are

$y + 3 = \dfrac{4}{3}(x - 1)$ and

$y + 3 = -\dfrac{4}{3}(x - 1)$.

48. Because the vertices are $(2, 3)$ and $(-2, 3)$, $a = 2$ and the center is $(0, 3)$.

$e = \dfrac{c}{a}$ $c^2 = a^2 + b^2$

$\dfrac{5}{2} = \dfrac{c}{2}$ $5^2 = 2^2 + b^2$

$b^2 = 25 - 4 = 21$

$c = 5$

Substituting into the standard equation yields

$$\frac{x^2}{4} - \frac{(y - 3)^2}{21} = 1.$$

54. a. Because the transmitters are 300 miles apart, $2c = 300$ and $c = 150$.

$2a = \text{rate} \times \text{time}$

$2a = 0.186 \times 800 = 148.8$ miles

Thus $a = 74.4$ miles.

$b = \sqrt{c^2 - a^2}$

$= \sqrt{150^2 - 74.4^2} \approx 130.25$ miles

The ship is located on the hyperbola given by

$$\frac{x^2}{74.4^2} - \frac{y^2}{130.25^2} = 1$$

b. The ship will reach the coastline when $x < 0$ and $y = 0$. Thus

$$\frac{x^2}{74.4^2} - \frac{0^2}{130.25^2} = 1$$

$$\frac{x^2}{74.4^2} = 1$$

$x^2 = 74.4^2$

$x = -74.4$

The ship reaches the coastline 74.4 miles to the left of the origin at the point $(-74.4, 0)$.

Exercise Set 8.4, page 730

8. $xy = -10$

$xy + 10 = 0$

$A = 0, B = 1, C = 0, F = 10$

$\cot 2\alpha = \dfrac{A - C}{B} = \dfrac{0 - 0}{1} = 0$

Thus $2\alpha = 90°$ and $\alpha = 45°$.

$A' = A \cos^2 \alpha + B \cos \alpha \sin \alpha + C \sin^2 \alpha$

$= 0\left(\dfrac{\sqrt{2}}{2}\right)^2 + 1\left(\dfrac{\sqrt{2}}{2}\right)\left(\dfrac{\sqrt{2}}{2}\right) + 0\left(\dfrac{\sqrt{2}}{2}\right)^2 = \dfrac{1}{2}$

$C' = A \sin^2 \alpha - B \cos \alpha \sin \alpha + C \cos^2 \alpha$

$= 0\left(\dfrac{\sqrt{2}}{2}\right)^2 - 1\left(\dfrac{\sqrt{2}}{2}\right)\left(\dfrac{\sqrt{2}}{2}\right) + 0\left(\dfrac{\sqrt{2}}{2}\right)^2 = -\dfrac{1}{2}$

$F' = F = 10$

$\dfrac{1}{2}x'^2 - \dfrac{1}{2}y'^2 + 10 = 0$ or $\dfrac{y'^2}{20} - \dfrac{x'^2}{20} = 1$

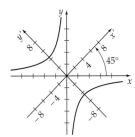

18. $x^2 + 4xy + 4y^2 - 2\sqrt{5}x + \sqrt{5}y = 0$

$A = 1, B = 4, C = 4, D = -2\sqrt{5}, E = \sqrt{5}, F = 0$

$\cot 2\alpha = \dfrac{A - C}{B} = \dfrac{1 - 4}{4} = -\dfrac{3}{4}$

$\csc^2 2\alpha = \cot^2 2\alpha + 1$

$\csc^2 2\alpha = \left(-\dfrac{3}{4}\right)^2 + 1 = \dfrac{25}{16}$

$\csc 2\alpha = +\sqrt{\dfrac{25}{16}} = \dfrac{5}{4}$ (2α is in Quadrant II)

$\sin 2\alpha = \dfrac{1}{\csc 2\alpha} = \dfrac{4}{5}$

$\sin^2 2\alpha + \cos^2 2\alpha = 1$

$\cos^2 2\alpha = 1 - \sin^2 2\alpha$

$\cos^2 2\alpha = 1 - \left(\dfrac{4}{5}\right)^2 = \dfrac{9}{25}$

(continued)

$$\cos 2\alpha = -\sqrt{\frac{9}{25}} = -\frac{3}{5} \quad (2\alpha \text{ is in Quadrant II})$$

$$\sin \alpha = \sqrt{\frac{1 - \left(-\dfrac{3}{5}\right)}{2}} = \frac{2\sqrt{5}}{5}$$

$$\cos \alpha = \sqrt{\frac{1 + \left(-\dfrac{3}{5}\right)}{2}} = \frac{\sqrt{5}}{5}$$

$\alpha \approx 63.4°$

$A' = A \cos^2 \alpha + B \cos \alpha \sin \alpha + C \sin^2 \alpha$

$$= 1\left(\frac{\sqrt{5}}{5}\right)^2 + 4\left(\frac{\sqrt{5}}{5}\right)\left(\frac{2\sqrt{5}}{5}\right) + 4\left(\frac{2\sqrt{5}}{5}\right)^2 = 5$$

$C' = A \sin^2 \alpha - B \cos \alpha \sin \alpha + C \cos^2 \alpha$

$$= 1\left(\frac{2\sqrt{5}}{5}\right)^2 - 4\left(\frac{\sqrt{5}}{5}\right)\left(\frac{2\sqrt{5}}{5}\right) + 4\left(\frac{\sqrt{5}}{5}\right)^2 = 0$$

$D' = D \cos \alpha + E \sin \alpha$

$$= -2\sqrt{5}\left(\frac{\sqrt{5}}{5}\right) + \sqrt{5}\left(\frac{2\sqrt{5}}{5}\right) = 0$$

$E' = -D \sin \alpha + E \cos \alpha$

$$= 2\sqrt{5}\left(\frac{2\sqrt{5}}{5}\right) + \sqrt{5}\left(\frac{\sqrt{5}}{5}\right) = 5$$

$5x'^2 + 5y' = 0$ or $y' = -x'^2$

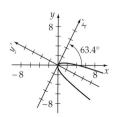

24. Use y_1 and y_2 as in Equations (11) and (12) that precede Example 4. Store the following constants.

$A = 2, B = -8, C = 8, D = 20, E = -24, F = -3$

Graph y_1 and y_2 on the same screen to produce the following parabola.

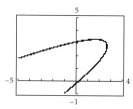

30. Because

$$B^2 - 4AC = \left(-10\sqrt{3}\right)^2 - 4(11)(1)$$
$$= 300 - 44 > 0$$

the graph is a hyperbola.

Exercise Set 8.5, page 743

14.

16. $r = 5 \cos 3\theta$

Because 3 is odd, this is a rose with three petals.

20. Because $|a| = |b| = 2$, the graph of $r = 2 - 2 \csc \theta$, $-\pi \le \theta \le \pi$, is a cardioid.

26. $r = 4 - 4 \sin \theta$

$\theta \min = 0$
$\theta \max = 2\pi$
$\theta \text{ step} = 0.1$

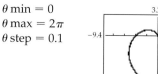

32. $r = -4 \sec \theta$

$\theta \min = 0$
$\theta \max = 2\pi$
$\theta \text{ step} = 0.1$

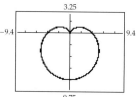

44.

$\begin{aligned} x &= r \cos \theta \\ &= (2)\left[\cos\left(-\frac{\pi}{3}\right)\right] \\ &= (2)\left(\frac{1}{2}\right) = 1 \end{aligned}$

$\begin{aligned} y &= r \sin \theta \\ &= (2)\left[\sin\left(-\frac{\pi}{3}\right)\right] \\ &= (2)\left(-\frac{\sqrt{3}}{2}\right) = -\sqrt{3} \end{aligned}$

The rectangular coordinates of the point are $\left(1, -\sqrt{3}\right)$.

48. $r = \sqrt{x^2 + y^2}$ $\alpha = \tan^{-1}\left|\dfrac{y}{x}\right|$ • α is the reference
      $= \sqrt{(12)^2 + (-5)^2}$ angle for θ.
      $= \sqrt{144 + 25}$
      $= \sqrt{169}$ $= \tan^{-1}\left|\dfrac{-5}{12}\right| \approx 22.6°$
      $= 13$

θ is a Quadrant IV angle. Thus
$\theta \approx 360° - 22.6° = 337.4°$.

The approximate polar coordinates of the point are
$(13, 337.4°)$.

56. $r = \cot \theta$

$r = \dfrac{\cos \theta}{\sin \theta}$ • $\cot \theta = \dfrac{\cos \theta}{\sin \theta}$

$r \sin \theta = \cos \theta$

$r(r \sin \theta) = r \cos \theta$ • Multiply both sides by r.

$\left(\sqrt{x^2 + y^2}\right)y = x$ • $y = r \sin \theta; x = r \cos \theta$

$(x^2 + y^2)y^2 = x^2$ • Square each side.

$y^4 + x^2y^2 - x^2 = 0$

64. $2x - 3y = 6$

$2r \cos \theta - 3r \sin \theta = 6$ • $x = r \cos \theta; y = r \sin \theta$

$r(2 \cos \theta - 3 \sin \theta) = 6$

$r = \dfrac{6}{2 \cos \theta - 3 \sin \theta}$

Exercise Set 8.6, page 750

2. $r = \dfrac{8}{2 - 4 \cos \theta}$ • Divide numerator and
                                denominator by 2.

$r = \dfrac{4}{1 - 2 \cos \theta}$

$e = 2$, so the graph is a hyperbola. The transverse axis is
on the polar axis because the equation involves $\cos \theta$. Let
$\theta = 0$.

$r = \dfrac{8}{2 - 4 \cos 0} = \dfrac{8}{2 - 4} = -4$

Let $\theta = \pi$.

$r = \dfrac{8}{2 - 4 \cos \pi} = \dfrac{8}{2 + 4} = \dfrac{4}{3}$

The vertices are $(-4, 0)$ and $\left(\dfrac{4}{3}, \pi\right)$.

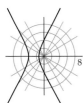

4. $r = \dfrac{6}{3 + 2 \cos \theta}$

$r = \dfrac{2}{1 + \left(\dfrac{2}{3}\right) \cos \theta}$ • Divide numerator and
                                      denominator by 3.

$e = \dfrac{2}{3}$, so the graph is an ellipse. The major axis is on
the polar axis because the equation involves $\cos \theta$. Let
$\theta = 0$.

$r = \dfrac{6}{3 + 2 \cos 0} = \dfrac{6}{3 + 2} = \dfrac{6}{5}$

Let $\theta = \pi$.

$r = \dfrac{6}{3 + 2 \cos \pi} = \dfrac{6}{3 - 2} = 6$

The vertices on the major axis are $\left(\dfrac{6}{5}, 0\right)$ and $(6, \pi)$. Let

$\theta = \dfrac{\pi}{2}$.

$r = \dfrac{6}{3 + 2 \cos\left(\dfrac{\pi}{2}\right)} = \dfrac{6}{3 + 0} = 2$

Let $\theta = \dfrac{3\pi}{2}$.

$r = \dfrac{6}{3 + 2 \cos\left(\dfrac{3\pi}{2}\right)} = \dfrac{6}{3 + 0} = 2$

Two additional points on the ellipse are $\left(2, \dfrac{\pi}{2}\right)$ and

$\left(2, \dfrac{3\pi}{2}\right)$.

24. $e = 1, r \cos \theta = -2$

Thus the directrix is $x = -2$. The distance from the focus
(pole) to the directrix is $d = 2$.

$r = \dfrac{ed}{1 - e \cos \theta}$ • Use Eq (2) since the vertical
                              directrix is to the left of the pole.

$r = \dfrac{(1)(2)}{1 - (1) \cos \theta}$ • $e = 1, d = 2$

$r = \dfrac{2}{1 - \cos \theta}$

Exercise Set 8.7, page 758

6. Plotting points for several values of t yields the following graph.

12. $x = 3 + 2\cos t, y = -1 - 3\sin t, 0 \le t < 2\pi$

$$\cos t = \frac{x-3}{2}, \sin t = -\frac{y+1}{3}$$

$$\cos^2 t + \sin^2 t = 1$$

$$\left(\frac{x-3}{2}\right)^2 + \left(-\frac{y+1}{3}\right)^2 = 1$$

$$\frac{(x-3)^2}{4} + \frac{(y+1)^2}{9} = 1$$

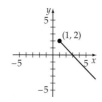

14. $x = 1 + t^2, y = 2 - t^2, t \in R$

$$x = 1 + t^2$$

$$t^2 = x - 1$$

$$y = 2 - (x - 1)$$

$$y = -x + 3$$

Because $x = 1 + t^2$ and $t^2 \ge 0$ for all real numbers t, $x \ge 1$ for all t. Similarly, $y \le 2$ for all t.

26. The maximum height of the cycloid is $2a = 2(3) = 6$. The cycloid intersects the x-axis every $2\pi a = 2\pi(3) = 6\pi$ units.

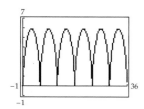

32.

The maximum height is about 195 feet when $t \approx 3.50$ seconds. The range is 1117 feet when $t \approx 6.99$ seconds.

Exercise Set 9.1, page 777

6. $\begin{cases} 8x + 3y = -7 & (1) \\ \quad\quad x = 3y + 15 & (2) \end{cases}$

$8(3y + 15) + 3y = -7$ • Replace x in Eq. (1).

$24y + 120 + 3y = -7$ • Simplify.

$27y = -7$

$$y = -\frac{127}{27}$$

$x = 3\left(-\frac{127}{27}\right) + 15 = \frac{8}{9}$ • Substitute $-\frac{127}{27}$ for y in Eq. (2).

The solution is $\left(\frac{8}{9}, -\frac{127}{27}\right)$.

18. $\begin{cases} 3x - 4y = 8 & (1) \\ 6x - 8y = 9 & (2) \end{cases}$

$8y = 6x - 9$ • Solve Eq. (2) for y.

$$y = \frac{3}{4}x - \frac{9}{8}$$

$3x - 4\left(\frac{3}{4}x - \frac{9}{8}\right) = 8$ • Replace y in Eq. (1).

$3x - 3x + \frac{9}{2} = 8$ • Simplify.

$$\frac{9}{2} = 8$$

This is a false equation. Therefore, the system of equations is inconsistent and has no solution.

20. $\begin{cases} 5x + 2y = 2 & (1) \\ \quad\quad y = -\frac{5}{2}x + 1 & (2) \end{cases}$

$5x + 2\left(-\frac{5}{2}x + 1\right) = 2$ • Replace y in Eq. (1).

$5x - 5x + 2 = 2$ • Simplify.

$2 = 2$

This is a true statement; therefore, the system of equations is dependent. Let $x = c$. Then $y = -\frac{5}{2}c + 1$. Thus the solutions are $\left(c, -\frac{5}{2}c + 1\right)$.

24. $\begin{cases} 3x - 8y = -6 & (1) \\ -5x + 4y = 10 & (2) \end{cases}$

$3x - 8y = -6$

$-10x + 8y = 20$ • 2 times Eq. (2)

$-7x \quad\quad = 14$

$x = -2$

$3(-2) - 8y = -6$ • Substitute -2 for x in Eq. (1).
Solve for y.

$-8y = 0$

$y = 0$

The solution is $(-2, 0)$.

28. $\begin{cases} 4x + 5y = 2 & (1) \\ 8x - 15y = 9 & (2) \end{cases}$

$12x + 15y = 6$ • 3 times Eq. (1)

$\underline{8x - 15y = 9}$

$20x \qquad = 15$

$x = \dfrac{3}{4}$

$4\left(\dfrac{3}{4}\right) + 5y = 2$ • Substitute $\dfrac{3}{4}$ for x in Eq. (1).

$3 + 5y = 2$ • Solve for y.

$y = -\dfrac{1}{5}$

The solution is $\left(\dfrac{3}{4}, -\dfrac{1}{5}\right)$.

42. Solve the system of equations $\begin{cases} x = 25p - 500 \\ x = -7p + 1100 \end{cases}$
by the substitution method.

$25p - 500 = -7p + 1100$

$32p = 1600$

$p = 50$

The equilibrium price is $50.

46. Let r = the rate of the canoeist.
Let w = the rate of the current.
Rate of canoeist with the current: $r + w$
Rate of canoeist against the current: $r - w$

$r \cdot t = d$

$(r + w) \cdot 2 = 12$ (1)

$(r - w) \cdot 4 = 12$ (2)

$r + w = 6$ • Divide Eq. (1) by 2.

$\underline{r - w = 3}$ • Divide Eq. (2) by 4.

$2r \qquad = 9$

$r = 4.5$

$4.5 + w = 6$

$w = 1.5$

Rate of canoeist = 4.5 miles per hour
Rate of current = 1.5 miles per hour

Exercise Set 9.2, page 791

12. $\begin{cases} 3x + 2y - 5z = 6 & (1) \\ 5x - 4y + 3z = -12 & (2) \\ 4x + 5y - 2z = 15 & (3) \end{cases}$

$15x + 10y - 25z = 30$ • 5 times Eq. (1)

$\underline{-15x + 12y - 9z = 36}$ • -3 times Eq. (2)

$22y - 34z = 66$ • Divide by 2.

$11y - 17z = 33$ (4)

$12x + 8y - 20z = 24$ • 4 times Eq. (1)

$\underline{-12x - 15y + 6z = -45}$ • -3 times Eq. (3)

$-7y - 14z = -21$ • Divide by -7.

$y + 2z = 3$ (5)

$11y - 17z = 33$ (4)

$\underline{-11y - 22z = -33}$ • -11 times Eq. (5)

$-39z = 0$

$z = 0$ (6)

$11y - 17(0) = 33$

$y = 3$

$3x + 2(3) - 5(0) = 6$

$x = 0$

The solution is $(0, 3, 0)$.

16. $\begin{cases} 2x + 3y + 2z = 14 & (1) \\ x - 3y + 4z = 4 & (2) \\ -x + 12y - 6z = 2 & (3) \end{cases}$

$2x + 3y + 2z = 14$ (1)

$\underline{-2x + 6y - 8z = -8}$ • -2 times Eq. (2)

$9y - 6z = 6$ • Divide by 3.

$3y - 2z = 2$ (4)

$2x + 3y + 2z = 14$ (1)

$\underline{-2x + 24y - 12z = 4}$ • 2 times Eq. (3)

$27y - 10z = 18$ (5)

$-27y + 18z = -18$ • -9 times Eq. (4)

$\underline{27y - 10z = 18}$ (5)

$8z = 0$

$z = 0$ (6)

$3y - 2(0) = 2$ • Substitute $z = 0$ in Eq. (4).

$y = \dfrac{2}{3}$

$2x + 3\left(\dfrac{2}{3}\right) + 2(0) = 14$ • Substitute $y = \dfrac{2}{3}$ and $z = 0$ in Eq. (1).

$x = 6$

The solution is $\left(6, \dfrac{2}{3}, 0\right)$.

18. $\begin{cases} 2x + 3y - 6z = 4 \quad (1) \\ 3x - 2y - 9z = -7 \quad (2) \\ 2x + 5y - 6z = 8 \quad (3) \end{cases}$

$6x + 9y - 18z = 12$ • 3 times Eq. (1)

$\underline{-6x + 4y + 18z = 14}$ • −2 times Eq. (2)

$13y = 26$

$y = 2 \quad (4)$

$2x + 3y - 6z = 4 \quad (1)$

$\underline{-2x - 5y + 6z = -8}$ • −1 times Eq. (3)

$-2y = -4$

$y = 2 \quad (5)$

$y = 2 \quad (4)$

$\underline{-y = -2}$ • −1 times Eq. (5)

$0 = 0 \quad (6)$

The equations are dependent. Let $z = c$.

$2x + 3(2) - 6c = 4$ • Substitute $y = 2$ and $z = c$ in Eq. (1).

$x = 3c - 1$

The solutions are $(3c - 1, 2, c)$.

20. $\begin{cases} x - 3y + 4z = 9 \quad (1) \\ 3x - 8y - 2z = 4 \quad (2) \end{cases}$

$-3x + 9y - 12z = -27$ • −3 times Eq. (1)

$\underline{3x - 8y - 2z = 4} \quad (2)$

$y - 14z = -23 \quad (3)$

$y = 14z - 23$ • Solve Eq. (3) for y.

$x - 3(14z - 23) + 4z = 9$ • Substitute $14z - 23$ for y in Eq. (1).

$x = 38z - 60$ • Solve for x.

Let $z = c$. The solutions are $(38c - 60, 14c - 23, c)$.

32. $\begin{cases} 5x + 2y + 3z = 0 \quad (1) \\ 3x + y - 2z = 0 \quad (2) \\ 4x - 7y + 5z = 0 \quad (3) \end{cases}$

$15x + 6y + 9z = 0$ • 3 times Eq. (1)

$\underline{-15x - 5y + 10z = 0}$ • −5 times Eq. (2)

$y + 19z = 0 \quad (4)$

$20x + 8y + 12z = 0$ • 4 times Eq. (1)

$\underline{-20x + 35y - 25z = 0}$ • −5 times Eq. (3)

$43y - 13z = 0 \quad (5)$

$-43y - 817z = 0$ • −43 times Eq. (4)

$\underline{43y - 13z = 0} \quad (5)$

$-830z = 0$

$z = 0 \quad (6)$

Solving by back substitution, the only solution is $(0, 0, 0)$.

36. $x^2 + y^2 + ax + by + c = 0$

$\begin{cases} 0 + 36 + a(0) + b(6) + c = 0 \\ 1 + 25 + a(1) + b(5) + c = 0 \\ 49 + 1 + a(-7) + b(-1) + c = 0 \end{cases}$

• Let $x = 0$, $y = 6$.
• Let $x = 1$, $y = 5$.
• Let $x = -7$, $y = -1$.

$\begin{cases} 6b + c = -36 \quad (1) \\ a + 5b + c = -26 \quad (2) \\ -7a - b + c = -50 \quad (3) \end{cases}$

$7a + 35b + 7c = -182$ • 7 times Eq. (2)

$\underline{-7a - b + c = -50} \quad (3)$

$34b + 8c = -232$

$17b + 4c = -116 \quad (4)$

$-24b - 4c = 144$ • −4 times Eq. (1)

$\underline{17b + 4c = -116} \quad (4)$

$-7b = 28$

$b = -4$

$17(-4) + 4c = -116$ • Substitute −4 for b in Eq. (4).

$c = -12$

$-7a - (-4) - 12 = -50$ • Substitute −4 for b and −12 for c in Eq. (3).

$a = 6$

An equation of the circle whose graph passes through the three given points is $x^2 + y^2 + 6x - 4y - 12 = 0$.

42. Let x_1, x_2, x_3, and x_4 represent the numbers of cars per hour that travel AB, BC, CD, and DA, respectively. Using the principle that the number of cars entering an intersection must equal the number of cars leaving the intersection, we can write the following equations.

$A: 75 + x_4 = x_1 + 60$

$B: x_1 + 50 = x_2 + 100$

$C: x_2 + 45 = x_3 + 50$

$D: x_3 + 80 = x_4 + 40$

The equations for the traffic intersections result in the following system of equations.

$\begin{cases} x_1 - x_4 = 15 \quad (1) \\ x_1 - x_2 = 50 \quad (2) \\ x_2 - x_3 = 5 \quad (3) \\ x_3 - x_4 = -40 \quad (4) \end{cases}$

Subtracting Equation (2) from Equation (1) gives

$x_1 - x_4 = 15$

$\underline{x_1 - x_2 = 50}$

$x_2 - x_4 = -35 \quad (5)$

Adding Equation (3) and Equation (4) gives

$x_2 - x_3 = 5$

$\underline{x_3 - x_4 = -40}$

$x_2 - x_4 = -35 \quad (6)$

Because Equation (5) and Equation (6) are the same, the system of equations is dependent. Because we want to know the cars per hour between B and C, solve the system in terms of x_2.

$x_1 = x_2 + 50$

$x_3 = x_2 - 5$

$x_4 = x_2 + 35$

Because there cannot be a negative number of cars per hour between two intersections, to ensure that $x_3 \geq 0$, we must have $x_2 \geq 5$. The minimum number of cars traveling between B and C is 5 cars per hour.

Exercise Set 9.3, page 800

8. $\begin{cases} x - 2y = 3 & (1) \\ xy = -1 & (2) \end{cases}$

$x = 2y + 3$ • Solve Eq. (1) for x.

$(2y + 3)y = -1$ • Replace x by
           2y + 3 in Eq. (2).

$2y^2 + 3y + 1 = 0$ • Solve for y.

$(2y + 1)(y + 1) = 0$

$y = -\dfrac{1}{2}$ or $y = -1$

$x - 2\left(-\dfrac{1}{2}\right) = 3$ $x - 2(-1) = 3$ • Substitute for y
                                     in Eq. (1).

$x = 2$ $x = 1$

The solutions are $\left(2, -\dfrac{1}{2}\right)$ and $(1, -1)$.

16. $\begin{cases} 3x^2 - 2y^2 = 19 & (1) \\ x^2 - y^2 = 5 & (2) \end{cases}$

$3x^2 - 2y^2 = 19$ (1)

$\underline{-3x^2 + 3y^2 = -15}$ • Multiply Eq. (2) by −3.

$y^2 = 4$ • Add the equations.

$y = \pm 2$ • Solve for y.

$x^2 - (-2)^2 = 5$ • Substitute −2 for y in Eq. (2).

$x^2 - 4 = 5$

$x^2 = 9$

$x = \pm 3$

$x^2 - 2^2 = 5$ • Substitute 2 for y in Eq. (2).

$x^2 - 4 = 5$

$x^2 = 9$

$x = \pm 3$

The solutions are $(3, -2), (-3, -2), (3, 2), (-3, 2)$.

20. $\begin{cases} 2x^2 + 3y^2 = 11 & (1) \\ 3x^2 + 2y^2 = 19 & (2) \end{cases}$

Use the elimination method to eliminate y^2.

$4x^2 + 6y^2 = 22$ • 2 times Eq. (1)

$\dfrac{-9x^2 - 6y^2 = -57}{}$ • −3 times Eq. (2)

$-5x^2 = -35$

$x^2 = 7$

$2(7) + 3y^2 = 11$ • Substitute for x in Eq. (1).

$3y^2 = -3$

$y^2 = -1$

$y^2 = -1$ has no real number solutions. The graphs of the equations do not intersect. The system is inconsistent and has no solution.

28. $\begin{cases} (x + 2)^2 + (y - 3)^2 = 10 \\ (x - 3)^2 + (y + 1)^2 = 13 \end{cases}$

$x^2 + 4x + 4 + y^2 - 6y + 9 = 10$ (1)

$\underline{x^2 - 6x + 9 + y^2 + 2y + 1 = 13}$ (2)

$10x - 5 - 8y + 8 = -3$ • Subtract.

$10x - 8y = -6$

$y = \dfrac{5x + 3}{4}$ (3)   • Solve for y.

$(x + 2)^2 + \left(\dfrac{5x - 9}{4}\right)^2 = 10$ • Substitute for y.

$x^2 + 4x + 4 + \dfrac{25x^2 - 90x + 81}{16} = 10$ • Solve for x.

$16x^2 + 64x + 64 + 25x^2 - 90x + 81 = 160$

$41x^2 - 26x - 15 = 0$

$(41x + 15)(x - 1) = 0$

$x = -\dfrac{15}{41}$ or $x = 1$

$y = \dfrac{5}{4}\left(-\dfrac{15}{41}\right) + \dfrac{3}{4}$ or $y = \dfrac{5(1) + 3}{4}$ • Substitute for x into Eq. (3).

$y = \dfrac{12}{41}$ or $y = 2$

The solutions are $\left(-\dfrac{15}{41}, \dfrac{12}{41}\right)$ and $(1, 2)$.

Exercise Set 9.4, page 808

14. $\dfrac{7x + 44}{x^2 + 10x + 24} = \dfrac{7x + 44}{(x + 4)(x + 6)} = \dfrac{A}{x + 4} + \dfrac{B}{x + 6}$

$7x + 44 = A(x + 6) + B(x + 4)$

$7x + 44 = (A + B)x + (6A + 4B)$

$\begin{cases} 7 = A + B \\ 44 = 6A + 4B \end{cases}$

The solution is $A = 8, B = -1$.

$\dfrac{7x + 44}{x^2 + 10x + 24} = \dfrac{8}{x + 4} + \dfrac{-1}{x + 6}$

22. $\dfrac{x-18}{x(x-3)^2} = \dfrac{A}{x} + \dfrac{B}{x-3} + \dfrac{C}{(x-3)^2}$

$x - 18 = A(x-3)^2 + Bx(x-3) + Cx$

$x - 18 = Ax^2 - 6Ax + 9A + Bx^2 - 3Bx + Cx$

$x - 18 = (A+B)x^2 + (-6A - 3B + C)x + 9A$

$$\begin{cases} 0 = A + B \\ 1 = -6A - 3B + C \\ -18 = 9A \end{cases}$$

The solution is $A = -2$, $B = 2$, $C = -5$.

$\dfrac{x-18}{x(x-3)^2} = \dfrac{-2}{x} + \dfrac{2}{x-3} + \dfrac{-5}{(x-3)^2}$

24. $x^3 - x^2 + 10x - 10 = (x-1)(x^2+10)$

$\dfrac{9x^2 - 3x + 49}{(x-1)(x^2+10)} = \dfrac{A}{x-1} + \dfrac{Bx+C}{x^2+10}$

$9x^2 - 3x + 49 = A(x^2+10) + (Bx+C)(x-1)$

$9x^2 - 3x + 49 = (A+B)x^2 + (-B+C)x + (10A - C)$

$$\begin{cases} 9 = A + B \\ -3 = -B + C \\ 49 = 10A - C \end{cases}$$

The solution is $A = 5$, $B = 4$, $C = 1$.

$\dfrac{9x^2 - 3x + 49}{x^3 - x^2 + 10x - 10} = \dfrac{5}{x-1} + \dfrac{4x+1}{x^2+10}$

30. $\dfrac{2x^3 + 9x + 1}{(x^2+7)^2} = \dfrac{Ax+B}{x^2+7} + \dfrac{Cx+D}{(x^2+7)^2}$

$2x^3 + 9x + 1 = (Ax+B)(x^2+7) + Cx + D$

$2x^3 + 9x + 1 = Ax^3 + Bx^2 + (7A+C)x + (7B+D)$

$$\begin{cases} 2 = A \\ 0 = B \\ 9 = 7A + C \\ 1 = 7B + D \end{cases}$$

The solutions are $A = 2$, $B = 0$, $C = -5$, $D = 1$.

$\dfrac{2x^3 + 9x + 1}{x^4 + 14x^2 + 49} = \dfrac{2x}{x^2+7} + \dfrac{-5x+1}{(x^2+7)^2}$

34.
$$\begin{array}{r}
x + 1 \\
2x^2 + 3x - 2 \overline{\smash{\big)}\, 2x^3 + 5x^2 + 3x - 8} \\
\underline{2x^3 + 3x^2 - 2x} \\
2x^2 + 5x - 8 \\
\underline{2x^2 + 3x - 2} \\
2x - 6
\end{array}$$

$\dfrac{2x^3 + 5x^2 + 3x - 8}{2x^2 + 3x - 2} = x + 1 + \dfrac{2x - 6}{2x^2 + 3x - 2}$

$\dfrac{2x-6}{(2x-1)(x+2)} = \dfrac{A}{2x-1} + \dfrac{B}{x+2}$

$2x - 6 = A(x+2) + B(2x-1)$

$2x - 6 = Ax + 2A + 2Bx - B$

$2x - 6 = (A + 2B)x + (2A - B)$

$$\begin{cases} 2 = A + 2B \\ -6 = 2A - B \end{cases}$$

The solutions are $A = -2$, $B = 2$.

$\dfrac{2x^3 + 5x^2 + 3x - 8}{2x^2 + 3x - 2} = x + 1 + \dfrac{-2}{2x-1} + \dfrac{2}{x+2}$

Exercise Set 9.5, page 815

6.
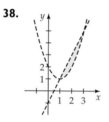

12.
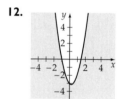

20.
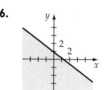

28.
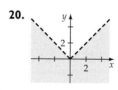

38.
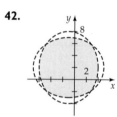

42.
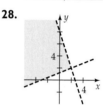

44.
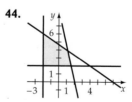

Exercise Set 9.6, page 823

12. $C = 4x + 3y$

(x, y)	C
(0, 8)	24
(2, 4)	20
(5, 2)	26
(11, 0)	44
(20, 0)	80
(20, 20)	140
(0, 20)	60

• Minimum

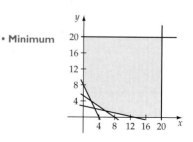

34. $a = 7, c_1, c_2, c_3, c_4, c_5, b = 19$

$a_n = a_1 + (n - 1)d$

$19 = 7 + (7 - 1)d$ • **There are seven terms, so $n = 7$.**

$19 = 7 + 6d$

$d = 2$

$c_1 = a_1 + d = 7 + 2 = 9$

$c_2 = a_1 + 2d = 7 + 4 = 11$

$c_3 = a_1 + 3d = 7 + 6 = 13$

$c_4 = a_1 + 4d = 7 + 8 = 15$

$c_5 = a_1 + 5d = 7 + 10 = 17$

Exercise Set 11.3, page 923

6. $\dfrac{a_{i+1}}{a_i} = \dfrac{(-1)^i e^{(i+1)x}}{(-1)^{i-1} e^{ix}} = -e^{(i+1)x - ix} = -e^x$

Because x is a constant, $-e^x$ is a constant and the sequence is a geometric sequence.

18. $\dfrac{a_2}{a_1} = \dfrac{6}{8} = \dfrac{3}{4} = r$

$a_n = a_1 r^{n-1}$

$a_n = 8\left(\dfrac{3}{4}\right)^{n-1}$

40. $r = \dfrac{4}{3}, a_1 = \dfrac{4}{3}, n = 14$

$S_n = \dfrac{a_1(1 - r^n)}{1 - r}$

$S_{14} = \dfrac{\dfrac{4}{3}\left[1 - \left(\dfrac{4}{3}\right)^{14}\right]}{1 - \dfrac{4}{3}} = \dfrac{\dfrac{4}{3}\left[\dfrac{-263,652,487}{4,782,969}\right]}{-\dfrac{1}{3}} \approx 220.49$

62. $0.3\overline{95} = \dfrac{3}{10} + \dfrac{95}{1000} + \dfrac{95}{100,000} + \cdots = \dfrac{3}{10} + \dfrac{\dfrac{95}{1000}}{1 - \dfrac{1}{100}}$

$= \dfrac{3}{10} + \dfrac{95}{990} = \dfrac{392}{990} = \dfrac{196}{495}$

70. $A = \dfrac{P[(1 + r)^m - 1]}{r}; \quad P = 250, r = \dfrac{0.08}{12}, m = 12(4)$

$A = \dfrac{250\left[\left(1 + \dfrac{0.08}{12}\right)^{48} - 1\right]}{\dfrac{0.08}{12}} \approx 14,087.48$

Exercise Set 11.4, page 932

8. $S_n = 2 + 6 + 12 + \cdots + n(n + 1) = \dfrac{n(n + 1)(n + 2)}{3}$

1. When $n = 1$, $S_1 = 1(1 + 1) = 2$; $\dfrac{1(1 + 1)(1 + 2)}{3} = 2$.

   Therefore, the statement is true for $n = 1$.

2. Assume the statement is true for $n = k$.

   $S_k = 2 + 6 + 12 + \cdots + k(k + 1)$

   $= \dfrac{k(k + 1)(k + 2)}{3}$ • **Induction hypothesis**

   Prove the statement is true for $n = k + 1$. That is, prove

   $S_{k+1} = \dfrac{(k + 1)(k + 2)(k + 3)}{3}$.

   Because $a_k = k(k + 1)$ and $a_{k+1} = (k + 1)(k + 2)$,

   $S_{k+1} = S_k + a_{k+1} = \dfrac{k(k + 1)(k + 2)}{3} + (k + 1)(k + 2)$

   $= \dfrac{k(k + 1)(k + 2) + 3(k + 1)(k + 2)}{3}$

   $= \dfrac{(k + 1)(k + 2)(k + 3)}{3}$ • **Factor out $(k + 1)$ and $(k + 2)$ from each term.**

   By the Principle of Mathematical Induction, the statement is true for all positive integers n.

12. $P_n = \left(1 - \dfrac{1}{2}\right)\left(1 - \dfrac{1}{3}\right)\cdots\left(1 - \dfrac{1}{n + 1}\right) = \dfrac{1}{n + 1}$

1. Let $n = 1$; then $P_1 = \left(1 - \dfrac{1}{2}\right) = \dfrac{1}{2}$; $\dfrac{1}{1 + 1} = \dfrac{1}{2}$.

   The statement is true for $n = 1$.

2. Assume the statement is true for $n = k$.

   $P_k = \left(1 - \dfrac{1}{2}\right)\left(1 - \dfrac{1}{3}\right)\cdots\left(1 - \dfrac{1}{k + 1}\right) = \dfrac{1}{k + 1}$

   Prove the statement is true for $n = k + 1$. That is, prove

   $P_{k+1} = \left(1 - \dfrac{1}{2}\right)\left(1 - \dfrac{1}{3}\right)\cdots\left(1 - \dfrac{1}{k + 1}\right)\left(1 - \dfrac{1}{k + 2}\right)$

   $= \dfrac{1}{k + 2}$

   Because $a_k = \left(1 - \dfrac{1}{k + 1}\right)$ and $a_{k+1} = \left(1 - \dfrac{1}{k + 2}\right)$,

   $P_{k+1} = P_k \cdot a_{k+1} = \dfrac{1}{k + 1} \cdot \left(1 - \dfrac{1}{k + 2}\right)$

   $= \dfrac{1}{k + 1} \cdot \dfrac{k + 1}{k + 2} = \dfrac{1}{k + 2}$

   By the Principle of Mathematical Induction, the statement is true for all positive integers n.

16. If $a > 1$, show that $a^{n+1} > a^n$ for all positive integers n.

1. Because $a > 1$, $a \cdot a > a \cdot 1$ or $a^2 > a$. Thus the statement is true when $n = 1$.

2. Assume the statement is true for $n = k$.

$a^{k+1} > a^k$ • **Induction hypothesis**

Prove the statement is true for $n = k + 1$. That is, prove $a^{k+2} > a^{k+1}$.

Because $a^{k+1} > a^k$ and $a > 0$,

$a(a^{k+1}) > a(a^k)$
$a^{k+2} > a^{k+1}$

By the Principle of Mathematical Induction, the statement is true for all positive integers n.

20. 1. Let $n = 1$. Because $\log_{10} 1 = 0$,

$\log_{10} 1 < 1$

The inequality is true for $n = 1$.

2. Assume $\log_{10} k < k$ is true for some positive integer k (induction hypothesis). Prove the inequality is true for $n = k + 1$. That is, prove $\log_{10}(k + 1) < k + 1$ is true when $n = k + 1$.

$\log_{10}(k + 1) \le \log_{10}(k + k)$
$= \log_{10} 2k = \log_{10} 2 + \log_{10} k < 1 + k$

Thus $\log_{10}(k + 1) < k + 1$. By the Principle of Mathematical Induction, $\log_{10} n < n$ for all positive integers n.

Exercise Set 11.5, page 938

4. $\dbinom{10}{5} = \dfrac{10!}{5! \, 5!} = \dfrac{10 \cdot 9 \cdot 8 \cdot 7 \cdot 6 \cdot 5!}{5! \, 5!} = \dfrac{10 \cdot 9 \cdot 8 \cdot 7 \cdot 6}{5 \cdot 4 \cdot 3 \cdot 2 \cdot 1}$

$= 252$

18. $(3x + 2y)^4$

$= (3x)^4 + 4(3x)^3(2y) + 6(3x)^2(2y)^2 + 4(3x)(2y)^3 + (2y)^4$

$= 81x^4 + 216x^3y + 216x^2y^2 + 96xy^3 + 16y^4$

20. $(2x - \sqrt{y})^7 = \dbinom{7}{0}(2x)^7 + \dbinom{7}{1}(2x)^6(-\sqrt{y})$

$+ \dbinom{7}{2}(2x)^5(-\sqrt{y})^2 + \dbinom{7}{3}(2x)^4(-\sqrt{y})^3$

$+ \dbinom{7}{4}(2x)^3(-\sqrt{y})^4 + \dbinom{7}{5}(2x)^2(-\sqrt{y})^5$

$+ \dbinom{7}{6}(2x)(-\sqrt{y})^6 + \dbinom{7}{7}(-\sqrt{y})^7$

$= 128x^7 - 448x^6\sqrt{y} + 672x^5y - 560x^4y\sqrt{y}$
$+ 280x^3y^2 - 84x^2y^2\sqrt{y}$
$+ 14xy^3 - y^3\sqrt{y}$

34. $\dbinom{10}{6-1}(x^{-1/2})^{10-6+1}(x^{1/2})^{6-1} = \dbinom{10}{5}(x^{-1/2})^5(x^{1/2})^5 = 252$

Exercise Set 11.6, page 945

12. Because there are four palettes and each palette contains four colors, by the counting principle there are $4 \cdot 4 \cdot 4 \cdot 4 = 256$ possible colors.

16. There are three possible finishes (first, second, and third) for the 12 contestants. Because the order of finish is important, these are the permutations of the 12 contestants selected 3 at a time.

$P(12, 3) = \dfrac{12!}{(12 - 3)!} = \dfrac{12!}{9!} = 12 \cdot 11 \cdot 10 = 1320$

There are 1320 possible finishes.

20. Player A matched against Player B is the same tennis match as Player B matched against Player A. Therefore, this is a combination of 26 players selected 2 at a time.

$C(26, 2) = \dfrac{26!}{2!(26 - 2)!} = \dfrac{26!}{2! \, 24!} = \dfrac{26 \cdot 25 \cdot 24!}{2 \cdot 1 \cdot 24!} = 325$

There are 325 possible first-round matches.

22. The person who refuses to sit in the back seat can be placed in any one of the three front seats. Similarly, the person who refuses to sit in the front can be placed in any of the three back seats. The remaining four people can sit in any of the remaining seats. The number of seating arrangements is

$3 \cdot 3 \cdot 4 \cdot 3 \cdot 2 \cdot 1 = 216$

30. a. The number of ways in which 10 finalists can be selected from 15 semifinalists is the combination of 15 students selected 10 at a time.

$C(15, 10) = 3003$

There are 3003 ways in which the finalists can be chosen.

b. The number of ways in which the 10 finalists can include 3 seniors is the product of the combination of 7 seniors selected 3 at a time and the combination of 8 remaining students selected 7 at a time.

$C(7, 3)C(8, 7) = 35 \cdot 8 = 280$

There are 280 ways in which the finalists can include 3 seniors.

c. "At least five seniors" means 5 or 6 or 7 seniors are finalists (there are only 7 seniors). Because the events are related by "or," sum the number of ways each event can occur.

$C(7, 5)C(8, 5) + C(7, 6)C(8, 4) + C(7, 7)C(8, 3)$
$= 21 \cdot 56 + 7 \cdot 70 + 1 \cdot 56 = 1176 + 490 + 56 = 1722$

There are 1722 ways in which the finalists can include at least 5 seniors.

Exercise Set 11.7, page 956

6. Let R represent the Republican, D the Democrat, and I the Independent. The sample space is

$\{(R, D), (R, I), (D, I)\}$

14. {HHHT, HHTH, HTHH, THHH, HHHH}

22. Let $E = \{2, 4, 6\}$, $T = \{3, 6\}$, and $S = \{1, 2, 3, 4, 5, 6\}$.

$E \cup T = \{2, 3, 4, 6\}$

$P(E \cup T) = \dfrac{N(E \cup T)}{N(S)} = \dfrac{4}{6} = \dfrac{2}{3}$

32. $\dfrac{C(3, 2) \cdot C(5, 2)}{C(8, 4)} = \dfrac{3 \cdot 10}{70} = \dfrac{3}{7}$

34. Yes, because the card was replaced. The probability of an ace on each draw is $\dfrac{4}{52} = \dfrac{1}{13}$.

$P(2 \text{ aces}) = \dfrac{1}{13} \cdot \dfrac{1}{13} = \dfrac{1}{169}$

40. This is a binomial experiment; $p = \dfrac{1}{4}$, $q = \dfrac{3}{4}$, $n = 8$, and $k = 3$.

$\dbinom{8}{3}\left(\dfrac{1}{4}\right)^3\left(\dfrac{3}{4}\right)^5 = 56\left(\dfrac{1}{64}\right)\left(\dfrac{243}{1024}\right) \approx 0.2076$

ANSWERS TO SELECTED EXERCISES

Exercise Set P.1, page 15

1. Integers: $0, -44, \sqrt{81}, 53$; Rational numbers: $-\frac{1}{5}, 0, -44, 3.14, \sqrt{81}, 53$; Irrational numbers: $\pi, 5.05005000500005\ldots$; Prime number: 53;
Real numbers: All the numbers are real numbers.   **3.** 2, 4, 6, 8   **5.** 3, 5, 7, 9   **7.** 0, 1, 2, 3   **9.** $\{-3, -2, -1, 0, 1, 2, 3, 4, 6\}$
11. $\{0, 1, 2, 3\}$ **13.** $\varnothing$ **15.** $\{1, 3\}$ **17.** $\{-2, 0, 1, 2, 3, 4, 6\}$ **19.** $\{x \mid -2 < x < 3\}$,

21. $\{x \mid -5 \le x \le -1\}$,   **23.** $\{x \mid x \ge 2\}$,

25. $(3, 5)$,   **27.** $[-2, \infty)$,   **29.** $[0, 1]$,

31. -5 **33.** 12   **35.** $\pi^2 + 10$ **37.** 9   **39.** $3x - 1$ **41.** $|x - 3|$ **43.** $|x + 2| = 4$ **45.** $|m - n|$ **47.** $|a - 4| < 5$

49. $|x + 2| > 4$ **51.**   **53.**   **55.**

57.   **59.**   **61.**

63.   **65.**   **67.** 8   **69.** 12   **71.** -72 **73.** 19   **75.** 13   **77.** -3

79. Associative property of multiplication   **81.** Distributive property   **83.** Commutative property of multiplication   **85.** Identity
property of multiplication   **87.** Reflexive property of equality   **89.** Transitive property of equality   **91.** Inverse property of multiplication
93. $6x$ **95.** $3x + 6$ **97.** $\frac{3}{2}a$ **99.** $6x - 13$ **101.** $-12x + 6y + 5$ **103.** $2a$ **105.** $21a + 6$ **107.** 6 square inches
109. \$5150   **111.** 66 beats per minute   **113.** 100 feet   **115.** A **117.** $\varnothing$,   **119.** B is a subset of A.
121. No. $(8 \div 4) \div 2 = 2 \div 2 = 1, 8 \div (4 \div 2) = 8 \div 2 = 4$ **123.** All but the multiplicative inverse property   **125.** $x + 7$
127. $|x - 2| < |x - 6|$ **129.** $|x - 3| > |x + 7|$ **131.** $2 < |x - 4| < 7$

Prepare for Section P.2, page 18

133. 32   **134.** $\frac{1}{16}$ **135.** 64   **136.** 314,000   **137.** False   **138.** False

Exercise Set P.2, page 31

1. -125 **3.** 1   **5.** $\frac{1}{16}$ **7.** 32   **9.** 27   **11.** -2 **13.** $\frac{2}{x^4}$ **15.** $6a^3b^8$ **17.** $\frac{3}{4a^4}$ **19.** $\frac{2y^2}{3x^2}$ **21.** $\frac{12}{a^3b}$ **23.** $-18m^5n^6$ **25.** $\frac{1}{x^6}$
27. $\frac{1}{4a^4b^2}$ **29.** $2x$ **31.** $\frac{b^{10}}{a^{10}}$ **33.** 2.011×10^{12} **35.** 5.62×10^{-10} **37.** 31,400,000   **39.** -0.0000023 **41.** 2.7×10^8 **43.** 1.5×10^{-11}
45. 7.2×10^{12} **47.** 8×10^{-16} **49.** 8   **51.** -16 **53.** $\frac{1}{27}$ **55.** $\frac{2}{3}$ **57.** 16   **59.** $8ab^2$ **61.** $-12x^{11/12}$ **63.** $3x^2y^3$ **65.** $\frac{4z^{2/5}}{3}$
67. $6x^{5/6}y^{5/6}$ **69.** $\frac{3a^{1/12}}{b}$ **71.** $3\sqrt{5}$ **73.** $2\sqrt[3]{3}$ **75.** $-3\sqrt[3]{5}$ **77.** $2|x|y\sqrt{6y}$ **79.** $2ay^2\sqrt[3]{2y}$ **81.** $-13\sqrt{2}$ **83.** $-10\sqrt[4]{3}$
85. $17y\sqrt[3]{4y}$ **87.** $-14x^2y\sqrt[3]{y}$ **89.** $17 + 7\sqrt{5}$ **91.** -7 **93.** $12z + \sqrt{z} - 6$ **95.** $x + 4\sqrt{x} + 4$ **97.** $x + 4\sqrt{x - 3} + 1$ **99.** $\sqrt{2}$
101. $\frac{\sqrt{10}}{6}$ **103.** $\frac{3\sqrt[3]{4}}{2}$ **105.** $\frac{2\sqrt[3]{x}}{x}$ **107.** $-\frac{3\sqrt{3} - 12}{13}$ **109.** $\frac{3\sqrt{5} - 3}{4}$ **111.** $\frac{3\sqrt{5} - 3\sqrt{x}}{5 - x}$ **113.** $\approx 2.21 \times 10^4$ dollars
115. $\approx 3.13 \times 10^7$ **117.** $\approx 1.38 \times 10^{-2}$ **119.** 8 minutes   **121.** \$22,688   **123.** ≈ 8.91 billion   **125. a.** 56%   **b.** 24%
127. No. $2 < 3$, but $\frac{1}{2} > \frac{1}{3}$.   **129.** $\frac{8}{5}$ **131.** $-\frac{19}{12}$ **133.** $\frac{1}{\sqrt{4 + h} + 2}$ **135.** $\frac{1}{\sqrt{n^2 + 1} + n}$ **137.** 2

Prepare for Section P.3, page 34

138. $-6a + 12b$ **139.** $-4x + 19$ **140.** $3x^2 - 3x - 6$ **141.** $-x^2 - 5x - 1$ **142.** False   **143.** False

Exercise Set P.3, page 40

1. D   **3.** H   **5.** G   **7.** B   **9.** J   **11. a.** $x^2 + 2x - 7$ **b.** 2   **c.** 1, 2, -7 **d.** 1   **e.** $x^2, 2x, -7$ **13. a.** $x^3 - 1$ **b.** 3   **c.** 1, -1
d. 1   **e.** $x^3, -1$ **15. a.** $2x^4 + 3x^3 + 4x^2 + 5$ **b.** 4   **c.** 2, 3, 4, 5   **d.** 2   **e.** $2x^4, 3x^3, 4x^2, 5$ **17.** 3   **19.** 5   **21.** 2
23. $5x^2 + 11x + 3$ **25.** $9w^3 + 8w^2 - 2w + 6$ **27.** $-2r^2 + 3r^2 - 12$ **29.** $-3u^2 - 2u + 4$ **31.** $8x^3 + 18x^2 - 67x + 40$
33. $6x^4 - 19x^3 + 26x^2 - 29x + 10$ **35.** $10x^2 + 22x + 4$ **37.** $y^2 + 3y + 2$ **39.** $4z^2 - 19z + 12$ **41.** $a^2 + 3a - 18$
43. $10x^2 - 57xy + 77y^2$ **45.** $18x^2 + 55xy + 25y^2$ **47.** $6p^2 - 11pq - 35q^2$ **49.** $12d^2 + 4d - 8$ **51.** $r^3 + s^3$ **53.** $60c^3 - 49c^2 + 4$
55. $9x^2 - 25$ **57.** $9x^4 - 6x^2y + y^2$ **59.** $16w^2 + 8wz + z^2$ **61.** $x^2 + 10x + 25 - y^2$ **63.** 29   **65.** -17 **67.** -1 **69.** 33
71. a. 1.6 pounds   **b.** 3.6 pounds   **73. a.** 72π cubic inches   **b.** 300π cubic centimeters   **75. a.** 0.076 second   **b.** 0.085 second
77. 11,175 matches   **79.** 14.8 seconds; 90.4 seconds   **81.** Yes. The ball is approximately 4.4 feet high when it crosses home plate.   **83.** 15
85. $a^3 - 3a^2b + 3ab^2 - b^3$ **87.** $y^3 + 6y^2 + 12y + 8$ **89.** $27x^3 + 135x^2y + 225xy^2 + 125y^3$

Prepare for Section P.4, page 43

90. $3x^2$ **91.** $-36x^6$ **92. a.** $(x^2)^3$ **b.** $(x^3)^2$ **93.** $3b^3$ **94.** 7   **95.** 1

Exercise Set P.4, page 53

1. $5(x + 4)$ **3.** $-3x(5x + 4)$ **5.** $2xy(5x + 3 - 7y)$ **7.** $(x - 3)(2a + 3b)$ **9.** $(x + 3)(x + 4)$ **11.** $(a - 12)(a + 2)$ **13.** $(6x + 1)(x + 4)$
15. $(17x + 4)(3x - 1)$ **17.** $(3x + 8y)(2x - 5y)$ **19.** $(x^2 + 5)(x^2 + 1)$ **21.** $(6x^2 + 5)(x^2 + 3)$ **23.** factorable over the integers
25. not factorable over the integers   **27.** not factorable over the integers   **29.** $(x^2 - 3)(x^2 + 2)$ **31.** $(xy - 4)(xy + 2)$
33. $(3x^2 - 1)(x^2 + 4)$ **35.** $(3x^3 - 4)(x^3 + 2)$ **37.** $(x - 3)(x + 3)$ **39.** $(2a - 7)(2a + 7)$ **41.** $(1 - 10x)(1 + 10x)$ **43.** $(x^2 - 3)(x^2 + 3)$
45. $(x + 3)(x + 7)$ **47.** $(x + 5)^2$ **49.** $(a - 7)^2$ **51.** $(2x + 3)^2$ **53.** $(z^2 + 2w^2)^2$ **55.** $(x - 2)(x^2 + 2x + 4)$
57. $(2x - 3y)(4x^2 + 6xy + 9y^2)$ **59.** $(2 - x^2)(4 + 2x^2 + x^4)$ **61.** $(x - 3)(x^2 - 3x + 3)$ **63.** $(3x + 1)(x^2 + 2)$ **65.** $(x - 1)(ax + b)$
67. $(3w + 2)(2w^2 - 5)$ **69.** $2(3x - 1)(3x + 1)$ **71.** $(2x - 1)(2x + 1)(4x^2 + 1)$ **73.** $a(3x - 2y)(4x - 5y)$ **75.** $b(3x + 4)(x - 1)(x + 1)$
77. $2b(6x + y)^2$ **79.** $(w - 3)(w^2 - 12w + 39)$ **81.** $(x + 3y - 1)(x + 3y + 1)$ **83.** not factorable over the integers   **85.** $(2x - 5)^2(3x + 5)$
87. $(2x - y)(2x + y + 1)$ **89.** 8   **91.** 64   **93.** $(x^n - 1)(x^n + 1)(x^{2n} + 1)$ **95.** $\pi(R - r)(R + r)$ **97.** $r^2(4 - \pi)$

Prepare for Section P.5, page 55

99. $\dfrac{8}{5}$ **100.** $\dfrac{xz}{wy}$ **101.** $x + 3$ **102.** $2x(2x - 3)$ **103.** $(x - 6)(x + 1)$ **104.** $(x - 4)(x^2 + 4x + 16)$

Exercise Set P.5, page 62

1. $\dfrac{x + 4}{3}$ **3.** $\dfrac{x - 3}{x - 2}$ **5.** $\dfrac{a^2 - 2a + 4}{a - 2}$ **7.** $-\dfrac{x + 8}{x + 2}$ **9.** $-\dfrac{4y^2 + 7}{y + 7}$ **11.** $-\dfrac{8}{a^3b}$ **13.** $\dfrac{10}{27q^2}$ **15.** $\dfrac{x(3x + 7)}{2x + 3}$ **17.** $\dfrac{x + 3}{2x + 3}$
19. $\dfrac{(2y + 3)(3y - 4)}{(2y - 3)(y + 1)}$ **21.** $\dfrac{1}{a - 8}$ **23.** $\dfrac{3p - 2}{r}$ **25.** $\dfrac{8x(x - 4)}{(x - 5)(x + 3)}$ **27.** $\dfrac{3y - 4}{y + 4}$ **29.** $\dfrac{7z(2z - 5)}{(2z - 3)(z - 5)}$ **31.** $\dfrac{-2x^2 + 14x - 3}{(x - 3)(x + 3)(x + 4)}$
33. $\dfrac{(2x - 1)(x + 5)}{x(x - 5)}$ **35.** $\dfrac{-q^2 + 12q + 5}{(q - 3)(q + 5)}$ **37.** $\dfrac{3x^2 - 7x - 13}{(x + 3)(x + 4)(x - 3)(x - 4)}$ **39.** $\dfrac{(x + 2)(3x - 1)}{x^2}$ **41.** $\dfrac{4x + 1}{x - 1}$ **43.** $\dfrac{x - 2y}{y(y - x)}$
45. $\dfrac{(5x + 9)(x + 3)}{(x + 2)(4x + 3)}$ **47.** $\dfrac{(b + 3)(b - 1)}{(b - 2)(b + 2)}$ **49.** $\dfrac{x - 1}{x}$ **51.** $2 - m^2$ **53.** $\dfrac{-x^2 + 5x + 1}{x^2}$ **55.** $\dfrac{-x - 7}{x^2 + 6x - 3}$ **57.** $\dfrac{2x - 3}{x + 3}$ **59.** $\dfrac{a + b}{ab(a - b)}$
61. $\dfrac{(b - a)(b + a)}{ab(a^2 + b^2)}$ **63. a.** ≈ 136.55 miles per hour   **b.** $\dfrac{2v_1v_2}{v_1 + v_2}$ **65.** 0.040   **67.** 9446 kilometers per second   **69.** $\dfrac{2x + 1}{x(x + 1)}$
71. $\dfrac{3x^2 - 4}{x(x - 2)(x + 2)}$ **73.** $\dfrac{x^2 + 9x + 25}{(x + 5)^2}$ **75.** $\dfrac{x(1 - 4xy)}{(1 - 2xy)(1 + 2xy)}$ **77.** $R\left[\dfrac{(1 + i)^n - 1}{i(1 + i)^n}\right]$

Prepare for Section P.6, page 65

79. $15x^2 - 22x + 8$ **80.** $25x^2 - 20x + 4$ **81.** $4\sqrt{6}$ **82.** $-54 + \sqrt{5}$ **83.** $\dfrac{17 + 8\sqrt{2}}{7}$ **84.** b

Exercise Set P.6, page 71

1. $9i$ **3.** $7i\sqrt{2}$ **5.** $4 + 9i$ **7.** $5 + 7i$ **9.** $8 - 3i\sqrt{2}$ **11.** $11 - 5i$ **13.** $-7 + 4i$ **15.** $8 - 5i$ **17.** -10 **19.** $-2 + 16i$ **21.** -40
23. -10 **25.** $19i$ **27.** $20 - 10i$ **29.** $22 - 29i$ **31.** 41   **33.** $12 - 5i$ **35.** $-114 + 42i\sqrt{2}$ **37.** $-6i$ **39.** $3 - 6i$ **41.** $\dfrac{7}{53} - \dfrac{2}{53}i$

43. $1 + i$ **45.** $\dfrac{15}{41} - \dfrac{29}{41}i$ **47.** $\dfrac{5}{13} + \dfrac{12}{13}i$ **49.** $2 + 5i$ **51.** $-16 - 30i$ **53.** $-11 - 2i$ **55.** $-i$ **57.** -1 **59.** $-i$ **61.** -1

63. $\dfrac{1}{2} + \dfrac{\sqrt{3}}{2}i$ **65.** $-\dfrac{3}{2} + \dfrac{\sqrt{3}}{2}i$ **67.** $\dfrac{1}{2} + \dfrac{1}{2}i$ **69.** $(x + 4i)(x - 4i)$ **71.** $(z + 5i)(z - 5i)$ **73.** $(2x + 9i)(2x - 9i)$ **79.** 0

Chapter P True/False Exercises, page 76

1. True   **2.** False; if $a = \dfrac{1}{2}$, then $\left(\dfrac{1}{2}\right)^2 = \dfrac{1}{4} < \dfrac{1}{2}$.   **3.** True   **4.** False; $\sqrt{2} + \left(-\sqrt{2}\right) = 0$, which is a rational number.
5. False. $\sqrt{-2}\sqrt{-6} = -4$ $(2 \oplus 4) \oplus 6 \neq 2 \oplus (4 \oplus 6)$.   **6.** False; $x > a$ is written as (a, ∞).   **7.** False; $\sqrt{(-2)^2} \neq -2$.   **8.** False. Let $a = 2$
and $b = 3$.   **9.** False. Let $a = 1$ and $b = 2$.   **10.** False. $\sqrt{-2} \cdot \sqrt{-8} = i\sqrt{2} \cdot i\sqrt{8} = i^2\sqrt{16} = -4$.

Chapter P Review Exercises, page 76

1. integer, rational number, real number, prime number [P.1]   **2.** irrational number, real number [P.1]   **3.** rational number, real number [P.1]
4. rational number, real number [P.1]   **5.** $\{1, 2, 3, 5, 7, 11\}$ [P.1]   **6.** $\{5\}$ [P.1]   **7.** Distributive property [P.1]   **8.** Commutative property of
addition [P.1]   **9.** Associative property of multiplication [P.1]   **10.** Closure property of addition [P.1]   **11.** Identity property of addition [P.1]
12. Identity property of multiplication [P.1]   **13.** Symmetric property of equality [P.1]   **14.** Transitive property of equality [P.1]
15. ⟵┼┼┤━━━━━┫┼┼┼⟶ $(-4, 2]$ [P.1]   **16.** ⟵┼┼┼┼┼┫┼┼┤━━━⟶ $(-\infty, -1] \cup (3, \infty)$ [P.1]
　　$-5\ -4\ -3\ -2\ -1\ \ 0\ \ 1\ \ 2\ \ 3\ \ 4\ \ 5$　　　　　　　　　　　$-5\ -4\ -3\ -2\ -1\ \ 0\ \ 1\ \ 2\ \ 3\ \ 4\ \ 5$
17. ⟵┼┤━━━━━━┫┼┼┼⟶ $-3 \le x < 2$ [P.1]   **18.** ⟵┼┼┼┼┼┼┤━━━━⟶ $x > -1$ [P.1]   **19.** 7 [P.1]   **20.** $\pi - 2$ [P.1]
　　$-5\ -4\ -3\ -2\ -1\ \ 0\ \ 1\ \ 2\ \ 3\ \ 4\ \ 5$　　　　　　　　　$-5\ -4\ -3\ -2\ -1\ \ 0\ \ 1\ \ 2\ \ 3\ \ 4\ \ 5$

21. $4 - \pi$ [P.1]   **22.** 11 [P.1]   **23.** 17 [P.1]   **24.** $\sqrt{5} + \sqrt{2}$ [P.1]   **25.** -36 [P.1]   **26.** $\dfrac{2}{27}$ [P.2]   **27.** $12x^8y^3$ [P.2]   **28.** $\dfrac{4a^2b^8}{9c^4}$ [P.2]

29. 5 [P.2]   **30.** -9 [P.2]   **31.** $x^{17/12}$ [P.2]   **32.** $4x^{1/2}$ [P.2]   **33.** $x^{3/4}y^2$ [P.2]   **34.** $x - y$ [P.2]   **35.** $4ab^3\sqrt{3b}$ [P.2]   **36.** $2a\sqrt{3ab}$ [P.2]

37. $6x\sqrt{2y}$ [P.2]   **38.** $3xy^2\sqrt{2xy}$ [P.2]   **39.** $\dfrac{3y\sqrt{15y}}{5}$ [P.2]   **40.** $-\dfrac{2\sqrt{10xyz}}{5z^2}$ [P.2]   **41.** $\dfrac{7\sqrt[3]{4x}}{2}$ [P.2]   **42.** $\dfrac{5\sqrt[3]{3y^2}}{3}$ [P.2]   **43.** $-3y^2\sqrt[3]{5x^2y}$ [P.2]

44. $-5y^2\sqrt[3]{2x}$ [P.2]   **45.** 6.2×10^5 [P.2]   **46.** 1.7×10^{-6} [P.2]   **47.** $35{,}000$ [P.2]   **48.** 0.000000431 [P.2]   **49.** $-a^2 - 2a - 1$ [P.3]
50. $2b^2 + 8b - 8$ [P.3]   **51.** $6x^4 + 5x^3 - 13x^2 + 22x - 20$ [P.3]   **52.** $27y^3 - 135y^2 + 225y - 125$ [P.3]   **53.** $3(x + 5)^2$ [P.4]
54. $(5x - 3y)^2$ [P.4]   **55.** $4(5a^2 - b^2)$ [P.4]   **56.** $2(2a + 5)(4a^2 - 10a + 25)$ [P.4]   **57.** $\dfrac{3x - 2}{x + 4}$ [P.5]   **58.** $\dfrac{2x - 5}{4x^2 - 10x + 25}$ [P.5]

59. $\dfrac{2x + 3}{2x - 5}$ [P.5]   **60.** $\dfrac{2x + 1}{x + 3}$ [P.5]   **61.** $\dfrac{x(3x + 10)}{(x + 3)(x - 3)(x + 4)}$ [P.5]   **62.** $\dfrac{x(5x - 7)}{(x + 3)(x + 4)(2x - 1)}$ [P.5]   **63.** $\dfrac{2x - 9}{3x - 17}$ [P.5]   **64.** $\dfrac{x + 4}{5x + 8}$ [P.5]
65. $5 + 8i$ [P.6]   **66.** $2 - 3i\sqrt{2}$ [P.6]   **67.** $6 - i$ [P.6]   **68.** $-2 + 10i$ [P.6]   **69.** $8 + 6i$ [P.6]   **70.** $29 + 22i$ [P.6]   **71.** $8 + 6i$ [P.6]

72. i [P.6]   **73.** $-3 - 2i$ [P.6]   **74.** $-\dfrac{14}{25} - \dfrac{23}{25}i$ [P.6]

Chapter P Test, page 78

1. Distributive property [P.1]   **2.** $\{0, 1, 2, 3, 4, 5, 6, 7, 8, 9\}$ [P.1]   **3.** 7 [P.1]   **4.** $\dfrac{4}{9x^4y^2}$ [P.2]   **5.** $\dfrac{96bc^2}{a^5}$ [P.2]   **6.** 1.37×10^{-3} [P.2]   **7.** $\dfrac{x^{5/6}}{y^{9/4}}$ [P.2]

8. $7xy\sqrt[3]{3xy}$ [P.2]   **9.** $\dfrac{\sqrt[4]{8x}}{2}$ [P.2]   **10.** $\dfrac{3\sqrt{x} - 6}{x - 4}$ [P.2]   **11.** $x^3 - 2x^2 + 5xy - 2x^2y - 2y^2$ [P.3]   **12.** -94 [P.3]   **13.** $(7x - 1)(x + 5)$ [P.4]

14. $(a - 4b)(3x - 2)$ [P.4]   **15.** $2x(2x - y)(4x^2 + 2xy + y^2)$ [P.4]   **16.** $-\dfrac{x + 3}{x + 5}$ [P.5]   **17.** $\dfrac{(x - 6)(x + 1)}{(x + 3)(x - 2)(x - 3)}$ [P.5]   **18.** $\dfrac{x(x + 2)}{x - 3}$ [P.5]

19. $\dfrac{3a^2 - 3ab - 10a + 5b}{a(2a - b)}$ [P.5]   **20.** $\dfrac{x(2x - 1)}{2x + 1}$ [P.5]   **21.** $7 + 2i\sqrt{5}$ [P.6]   **22.** $2 + 2i$ [P.6]   **23.** $22 - 3i$ [P.6]   **24.** $\dfrac{11}{26} + \dfrac{23}{26}i$ [P.6]   **25.** i [P.6]

Exercise Set 1.1, page 88

1. 15 **3.** -4 **5.** $\dfrac{9}{2}$ **7.** $\dfrac{108}{23}$ **9.** $\dfrac{2}{9}$ **11.** 12 **13.** 16 **15.** 9 **17.** $\dfrac{1}{2}$ **19.** $\dfrac{22}{13}$ **21.** $\dfrac{95}{18}$ **23.** identity   **25.** conditional

equation   **27.** contradiction   **29.** identity   **31.** conditional equation   **33.** $-4, 4$ **35.** $7, 3$ **37.** $8, -3$ **39.** $2, -8$ **41.** $20, -12$

43. no solution   **45.** $12, -18$ **47.** $\dfrac{a + b}{2}, \dfrac{a - b}{2}$ **49.** 2008 **51.** after 3 hours and after 5 hours 24 minutes   **53.** 72 square yards

55. 15 minutes   **57.** maximum 166 beats per minute, minimum 127 beats per minute   **61.** $\{x \,|\, x \ge -4\}$ **63.** $\{x \,|\, x \le -7\}$ **65.** $\left\{x \,\middle|\, x \ge -\dfrac{7}{2}\right\}$

Prepare for Section 1.2, page 90

67. $23\frac{1}{2}$ **68.** $\frac{4}{15}$ **69.** the distributive property   **70.** the associative property of multiplication   **71.** $\frac{11}{15}x$ **72.** $\frac{ab}{a+b}$

Exercise Set 1.2, page 98

1. $h = \frac{3V}{\pi r^2}$ **3.** $t = \frac{I}{Pr}$ **5.** $m_1 = \frac{Fd^2}{Gm_2}$ **7.** $d = \frac{a_n - a_1}{n-1}$ **9.** $r = \frac{S - a_1}{S}$ **11.** 88.8   **13.** 9.5   **15.** 11.2   **17.** 100   **19.** 30 feet by 57 feet
21. 12 centimeters, 36 centimeters, 36 centimeters   **23.** 240 meters   **25.** 2 hours   **27.** 3 miles   **29.** 98   **31.** 850   **33.** $937.50
35. $7600 invested at 8%, $6400 invested at 6.5%   **37.** $3750   **39.** $18\frac{2}{11}$ grams   **41.** 64 liters   **43.** 1200 at $14 and 1800 at $25
45. $6\frac{2}{3}$ pounds of the $12 coffee and $13\frac{1}{3}$ pounds of the $9 coffee   **47.** 10 grams   **49.** 7.875 hours   **51.** $10.05 for book, $0.05 for
bookmark   **53.** 6.25 feet   **55.** 40 pounds   **57.** 1384 feet   **59.** 84 years old

Prepare for Section 1.3, page 102
61. $(x + 6)(x - 7)$ **62.** $(2x + 3)(3x - 5)$ **63.** $3 + 4i$ **64.** 1   **65.** 1   **66.** 0

Exercise Set 1.3, page 113

1. $-3, 5$ **3.** $-\frac{1}{2}, 1$ **5.** $-24, \frac{3}{8}$ **7.** $0, \frac{7}{3}$ **9.** $2, 8$ **11.** ± 9 **13.** $\pm 2\sqrt{6}$ **15.** $\pm 2i$ **17.** $-1, 11$ **19.** $3 \pm 4i$ **21.** $-3 \pm 2\sqrt{2}$
23. $-3, 5$ **25.** $-2 \pm i$ **27.** $\frac{-3 \pm \sqrt{13}}{2}$ **29.** $\frac{-2 \pm \sqrt{6}}{2}$ **31.** $\frac{4 \pm \sqrt{13}}{3}$ **33.** $-3, 5$ **35.** $\frac{-1 \pm \sqrt{5}}{2}$ **37.** $\frac{-2 \pm \sqrt{2}}{2}$ **39.** $\frac{5}{6} \pm \frac{\sqrt{11}}{6}i$
41. $\frac{-3 \pm \sqrt{41}}{4}$ **43.** $-\frac{5}{6}, \frac{7}{4}$ **45.** $-2, \frac{4}{5}$ **47.** 81; two real solutions   **49.** -116; no real solutions   **51.** 0; one real solution   **53.** 2116;
two real solutions   **55.** -111; no real solutions   **57.** 26.8 centimeters   **59.** width 43.2 inches; height 32.4 inches   **61.** 1996   **63.** 5800 or
11,000 racquets   **65.** 12 feet by 48 feet or 32 feet by 18 feet   **67.** 0.3 miles and 3.9 miles   **69.** 1.7 seconds   **71.** 1.8 seconds and
11.9 seconds   **73.** No   **75.** 9 people   **77.** 2006   **79. a.** 44.8 million pounds   **b.** 2007   **81. a.** $l = \left(\frac{1 + \sqrt{5}}{2}\right)w$ **b.** 163.4 feet
c. Answers will vary.   **83.** Yes   **85.** Yes

Prepare for Section 1.4, page 117
87. $x(x + 4)(x - 4)$ **88.** $x^2(x + 6)(x - 6)$ **89.** 4   **90.** 64   **91.** $x + 2\sqrt{x - 5} - 4$ **92.** $x - 4\sqrt{x + 3} + 7$

Exercise Set 1.4, page 125

1. $0, \pm 5$ **3.** $2, \pm 1$ **5.** $0, \pm 3$ **7.** $0, -5, 8$ **9.** $0, \pm 4$ **11.** $2, -1 \pm i\sqrt{3}$ **13.** 31   **15.** 2   **17.** no solution   **19.** no solution
21. $\frac{7}{2}$ **23.** -12 **25.** 1   **27.** 40   **29.** 3   **31.** 7   **33.** 7   **35.** 9   **37.** $\frac{5}{2}$ **39.** $1, -6$ **41.** $\pm\sqrt{7}, \pm\sqrt{2}$ **43.** $\pm 2, \pm\frac{\sqrt{6}}{2}$
45. $\sqrt[3]{2}, -\sqrt[3]{3}$ **47.** 1, 16   **49.** $-\frac{1}{27}, 64$ **51.** $\pm\frac{\sqrt{15}}{3}$ **53.** ± 1 **55.** $\frac{256}{81}, 16$ **57.** $13\frac{1}{3}$ hours   **59.** 8 games   **61.** 9 words with
three or more syllables   **63.** 3 inches   **65.** 10.5 millimeters   **67.** 87 feet   **69. a.** 8.93 inches   **b.** $5\sqrt{3}$ inches
71. $s = \left(\frac{-275 + 5\sqrt{3025 + 176T}}{2}\right)^2$

Prepare for Section 1.5, page 128
73. $\{x \mid x > 5\}$ **74.** 38   **75.** 2   **76.** $(2x + 3)(5x - 3)$ **77.** $\frac{7}{2}$ **78.** $\frac{5}{2}, 3$

Exercise Set 1.5, page 140

1. $\{x \mid x < 4\}$,

3. $\{x \mid x < -6\}$,

5. $\left\{x \mid x \geq -\frac{13}{8}\right\}$,

7. $\{x \mid x < 2\}$,

9. $\left\{x \mid -\frac{3}{4} < x \leq 4\right\}$,

11. $\left\{x \mid \frac{1}{3} \leq x \leq \frac{11}{3}\right\}$,

13. $\{x \mid x < -3 \text{ or } x \ge -1\}$, [number line: -5 to 5] **15.** $\{x \mid x < 1\}$, [number line: -5 to 5] **17.** $\left(-\infty, -\dfrac{3}{2}\right) \cup \left(\dfrac{5}{2}, \infty\right)$

19. $(-\infty, -8] \cup [2, \infty)$ **21.** $\left[-\dfrac{4}{3}, 8\right]$ **23.** $(-\infty, -4] \cup \left[\dfrac{28}{5}, \infty\right)$ **25.** $(-\infty, \infty)$ **27.** $\{4\}$ **29.** $(-\infty, -7) \cup (0, \infty)$ **31.** $[-4, 4]$

33. $(-5, -2)$ **35.** $(-\infty, -4] \cup [7, \infty)$ **37.** $(-4, 1)$ **39.** $\left[-\dfrac{29}{2}, -8\right)$ **41.** $\left[-4, -\dfrac{7}{2}\right)$ **43.** $(-\infty, -1) \cup (2, 4)$

45. $(-\infty, 5) \cup [12, \infty)$ **47.** $\left(-\dfrac{2}{3}, 0\right) \cup \left(\dfrac{5}{2}, \infty\right)$ **49.** $(-\infty, 5)$ **51.** if you write more than 57 checks a month **53.** $0 < h \le 26$ inches
55. at least 34 sales **57.** $20° \le C \le 40°$ **59.** $\{12, 14, 16\}, \{14, 16, 18\}$ **61.** 130.0 to 137.5 centimeters **63.** $(0, 210)$ **65.** at least 9791 books **67.** maximum radius 4.480 inches, minimum radius 4.432 inches **69.** $(-\infty, 3) \cup (3, 6) \cup (6, \infty)$ **71.** $(-3, \infty)$
73. $(-5, -1) \cup (1, 5)$ **75.** $(-7, -3] \cup [3, 7)$ **77.** $(a - \delta, a) \cup (a, a + \delta)$ **79.** more than 1 second but less than 3 seconds

Prepare for Section 1.6, page 143

81. 65 **82.** 45 **83.** 27 **84.** 28.125 **85.** The area becomes 4 times as large. **86.** No. The volume becomes 9 times as large.

Exercise Set 1.6, page 150

1. $d = kt$ **3.** $y = \dfrac{k}{x}$ **5.** $m = knp$ **7.** $V = klwh$ **9.** $A = ks^2$ **11.** $F = \dfrac{km_1m_2}{d^2}$ **13.** $y = kx, k = \dfrac{4}{3}$ **15.** $r = kt^2, k = \dfrac{1}{81}$

17. $T = krs^2, k = \dfrac{7}{25}$ **19.** $V = klwh, k = 1$ **21.** 1.02 liters **23.** 62 semester hours **25.** 11.7 fluid ounces **27. a.** 3.3 seconds
b. 3.7 feet **29.** 40 revolutions per minute **31.** 112 decibels **33. a.** 9 times larger **b.** 3 times larger **c.** 27 times larger
35. 6 times larger **37.** 2.97 **39.** 3950 pounds **41.** 142 million miles

Chapter 1 True/False Exercises, page 155

1. False; $(-3)^2 = 9$. **2.** False; one has solution set $\{3\}$, and the other has solution set $\{3, -4\}$. **3.** True **4.** True **5.** False; $100 > 1$ but $\dfrac{1}{100} \not> \dfrac{1}{1}$. **6.** False; the discriminant is $b^2 - 4ac$. **7.** False; $\sqrt{1} + \sqrt{1} = 1 + 1 = 2$ but $1 + 1 = 2 \ne 2^2$. **8.** True **9.** False; $3x^2 - 48 = 0$ has roots of 4 and -4. **10.** True

Chapter 1 Review Exercises, page 155

1. $\dfrac{3}{2}$ [1.1] **2.** $\dfrac{11}{3}$ [1.1] **3.** $\dfrac{1}{2}$ [1.1] **4.** $\dfrac{11}{4}$ [1.1] **5.** $-\dfrac{38}{15}$ [1.4] **6.** $-\dfrac{1}{2}$ [1.4] **7.** 3, 2 [1.3] **8.** $\dfrac{4}{3}, -\dfrac{3}{2}$ [1.3] **9.** $\dfrac{1 \pm \sqrt{13}}{6}$ [1.3]

10. $\dfrac{1}{2} \pm \dfrac{\sqrt{3}}{2}i$ [1.3] **11.** $0, \dfrac{5}{3}$ [1.4] **12.** $0, \pm 2$ [1.4] **13.** $\pm\dfrac{2\sqrt{3}}{3}, \pm\dfrac{\sqrt{10}}{2}$ [1.4] **14.** $\dfrac{4}{9}$ [1.4] **15.** $-5, 3$ [1.4] **16.** $6, -4$ [1.4]

17. 4 [1.4] **18.** 7 [1.4] **19.** -4 [1.4] **20.** 0 [1.4] **21.** $-2, -4$ [1.4] **22.** $\dfrac{11}{4}, \dfrac{9}{4}$ [1.4] **23.** 5, 1 [1.1] **24.** $-1, -9$ [1.1] **25.** 2, -3 [1.1]

26. $5, -\dfrac{1}{3}$ [1.1] **27.** $-2, -1$ [1.4] **28.** $1, -4, \dfrac{4}{3}$ [1.4] **29.** $(-\infty, 2]$ [1.5] **30.** $\left[\dfrac{6}{7}, \infty\right)$ [1.5] **31.** $[-5, 2]$ [1.5] **32.** $(-\infty, -1) \cup (3, \infty)$ [1.5]

33. $\left[\dfrac{145}{9}, 35\right]$ [1.5] **34.** $(86, 149)$ [1.5] **35.** $(-\infty, 0] \cup [3, 4]$ [1.5] **36.** $(-7, 0) \cup (3, \infty)$ [1.5] **37.** $(-\infty, -3) \cup (4, \infty)$ [1.5]

38. $(-\infty, -7) \cup [0, 5]$ [1.5] **39.** $\left(-\infty, \dfrac{5}{2}\right) \cup (3, \infty)$ [1.5] **40.** $\left[\dfrac{5}{2}, 5\right)$ [1.5] **41.** $\left(\dfrac{2}{3}, 2\right)$ [1.5] **42.** $(-\infty, 1] \cup [2, \infty)$ [1.5]

43. $(1, 2) \cup (2, 3)$ [1.5] **44.** $(a - b, a) \cup (a, a + b)$ [1.5] **45.** $h = \dfrac{V}{\pi r^2}$ [1.2] **46.** $t = \dfrac{A - P}{Pr}$ [1.2] **47.** $b_1 = \dfrac{2A - hb_2}{h}$ [1.2]

48. $w = \dfrac{P - 2l}{2}$ [1.2] **49.** $m = \dfrac{e}{c^2}$ [1.2] **50.** $m_1 = \dfrac{Fs^2}{Gm_2}$ [1.2] **51.** 80 [1.2] **52.** width = 12 feet by length = 15 feet [1.2]
53. 24 nautical miles [1.2] **54.** $20.00 [1.2] **55.** $1750 in the 4% account, $3750 in the 6% account [1.2] **56.** Price of calculator is $20.50. Price of battery is $0.50. [1.2] **57.** $864 [1.2] **58.** length = 12 inches by width = 8 inches, or length = 8 inches by width = 12 inches [1.2]
59. 18 hours [1.4] **60.** 4024 adult tickets, 502 student tickets [1.2] **61.** ≈ 13 feet [1.2] **62.** $(12, 24)$ The revenue is greater than $576 when the price is between $12 and $24. [1.5] **63. a.** $|B - 218| > 48$ [1.5] **b.** $(0, 170) \cup (266, \infty)$ **64.** more than $425 but less than $725 [1.5]
65. $0 < h \le 23.5$ inches [1.5] **66.** $[27, 82]$ [1.5] **67.** 9.39 to 9.55 inches [1.5] **68.** more than 0.6 miles but less than 3.6 miles from the city center [1.5] **69.** 1.64 meters per second squared [1.6] **70.** 89.9 tons [1.6]

Chapter 1 Test, page 158

1. 3 [1.1]   **2.** −5, 11 [1.1]   **3.** $-\dfrac{1}{2}, \dfrac{8}{3}$ [1.3]   **4.** $\dfrac{4 \pm \sqrt{14}}{2}$ [1.3]   **5.** $\dfrac{5 \pm \sqrt{37}}{6}$ [1.3]   **6.** discriminant: 1; two real solutions [1.3]

7. $x = \dfrac{c - cd}{a - c}, a \neq c$ [1.2]   **8.** 3 [1.4]   **9.** $\dfrac{8}{27}, -64$ [1.4]   **10.** $-\dfrac{14}{3}$ [1.4]   **11. a.** $\{x \,|\, x \le 8\}$ [1.5]   **b.** $[-2, 5)$ [1.5]   **12.** $[-4, -1) \cup [3, \infty)$ [1.5]

13. from $10\dfrac{7}{8}$ inches to $11\dfrac{7}{16}$ inches [1.5]   **14.** 2 miles per hour [1.2]   **15.** 2.25 liters [1.2]   **16.** 15 hours [1.4]   **17.** more than 100 miles [1.5]

18. more than 14.7 feet but less than 145.0 feet from a side line [1.5]   **19.** more than 1.25 miles but less than 10 miles from the city center [1.5]
20. 4.4 miles per second [1.6]

Cumulative Review Exercises, page 159

1. −11 [P.1]   **2.** 1.7×10^{-4} [P.2]   **3.** $8x^2 - 30x + 41$ [P.3]   **4.** $(8x - 5)(x + 3)$ [P.4]   **5.** $\dfrac{2x + 17}{x - 4}$ [P.5]   **6.** $a^{11/12}$ [P.2]   **7.** 29 [P.6]   **8.** $\dfrac{10}{3}$ [1.1]

9. $\dfrac{2 \pm \sqrt{10}}{2}$ [1.3]   **10.** 1, 5 [1.1]   **11.** 5 [1.4]   **12.** −6, 0, 6 [1.4]   **13.** $\pm\sqrt{3}, \pm\dfrac{\sqrt{10}}{2}$ [1.4]   **14.** $\{x \,|\, x \le -1 \text{ or } x > 1\}$ [1.5]

15. $(-\infty, 4] \cup [8, \infty)$ [1.5]   **16.** $\left\{x \,\middle|\, \dfrac{10}{7} \le x < \dfrac{3}{2}\right\}$ [1.5]   **17.** length 58 feet, width 42 feet [1.2]   **18.** 9475 to 24,275 printers [1.5]
19. 68 to 100 [1.5]   **20.** between 14.3% and 23.1% [1.5]

Exercise Set 2.1, page 174

1. 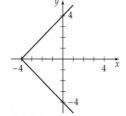   **3. a.** 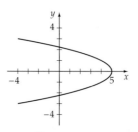   **b.** 23.4 beats per minute   **5.** $7\sqrt{5}$ **7.** $\sqrt{1261}$ **9.** $\sqrt{89}$ **11.** $\sqrt{38 - 12\sqrt{6}}$
13. $2\sqrt{a^2 + b^2}$ **15.** $-x\sqrt{10}$ **17.** $(12, 0), (-4, 0)$ **19.** $(3, 2)$ **21.** $(6, 4)$
23. $(-0.875, 3.91)$ **25.** 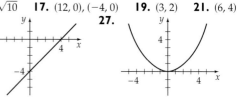   **27.**

29.   **31.**   **33.**   **35.**   **37.**

39. $(6, 0), \left(0, \dfrac{12}{5}\right)$ **41.** $(5, 0); \left(0, \sqrt{5}\right), \left(0, -\sqrt{5}\right)$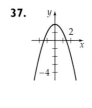

43. $(-4, 0); (0, 4), (0, -4)$ **45.** $(\pm 2, 0), (0, \pm 2)$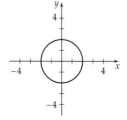

47. $(\pm 4, 0)$, $(0, \pm 4)$

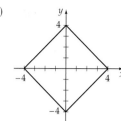

49. center $(0, 0)$, radius 6   **51.** center $(1, 3)$, radius 7   **53.** center $(-2, -5)$, radius 5

55. center $(8, 0)$, radius $\dfrac{1}{2}$ **57.** $(x - 4)^2 + (y - 1)^2 = 2^2$

59. $\left(x - \dfrac{1}{2}\right)^2 + \left(y - \dfrac{1}{4}\right)^2 = (\sqrt{5})^2$ **61.** $(x - 0)^2 + (y - 0)^2 = 5^2$

63. $(x - 1)^2 + (y - 3)^2 = 5^2$ **65.** center $(3, 0)$, radius 2   **67.** center $(7, -4)$, radius 3

69. center $\left(-\dfrac{1}{2}, 0\right)$, radius 4   **71.** center $\left(\dfrac{1}{2}, -\dfrac{3}{2}\right)$, radius $\dfrac{5}{2}$

73. $(x + 1)^2 + (y - 7)^2 = 25$ **75.** $(x - 7)^2 + (y - 11)^2 = 121$ **77.**

79.

81.

83.

85.

87. $(13, 5)$ **89.** $(7, -6)$ **91.** $x^2 - 6x + y^2 - 8y = 0$ **93.** $9x^2 + 25y^2 = 225$

95. $(x + 3)^2 + (y - 3)^2 = 3^2$

Prepare for Section 2.2, page 176

97. -4 **98.** $D = \{-3, -2, -1, 0, 2\}$; $R = \{1, 2, 4, 5\}$ **99.** $\sqrt{58}$ **100.** $x \geq 3$ **101.** $-2, 3$ **102.** 13

Exercise Set 2.2, page 190

1. a. 5   **b.** -4 **c.** -1 **d.** 1   **e.** $3k - 1$ **f.** $3k + 5$ **3. a.** $\sqrt{5}$ **b.** 3   **c.** 3   **d.** $\sqrt{21}$ **e.** $\sqrt{r^2 + 2r + 6}$ **f.** $\sqrt{c^2 + 5}$

5. a. $\dfrac{1}{2}$ **b.** $\dfrac{1}{2}$ **c.** $\dfrac{5}{3}$ **d.** 1   **e.** $\dfrac{1}{c^2 + 4}$ **f.** $\dfrac{1}{|2 + h|}$ **7. a.** 1   **b.** 1   **c.** -1 **d.** -1 **e.** 1   **f.** -1 **9. a.** -11 **b.** 6

c. $3c + 1$ **d.** $-k^2 - 2k + 10$ **11.** Yes   **13.** No   **15.** No   **17.** Yes   **19.** No   **21.** Yes   **23.** Yes   **25.** Yes   **27.** all real numbers

29. all real numbers   **31.** $\{x \mid x \neq -2\}$ **33.** $\{x \mid x \geq -7\}$ **35.** $\{x \mid -2 \leq x \leq 2\}$ **37.** $\{x \mid x > -4\}$ **39.**

41.

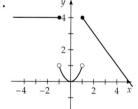

43.

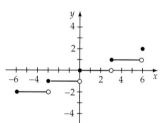

45.

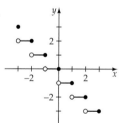

47. a. $1.05   **b.**

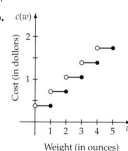

Cost (in dollars)

Weight (in ounces)

49. a, b, and d.   **51.** decreasing on $(-\infty, 0]$; increasing on $[0, \infty)$

53. increasing on $(-\infty, \infty)$ **55.** decreasing on $(-\infty, -3]$; increasing on $[-3, 0]$; decreasing on $[0, 3]$; increasing on $[3, \infty)$ **57.** constant on $(-\infty, 0]$; increasing on $[0, \infty)$

59. decreasing on $(-\infty, 0]$; constant on $[0, 1]$; increasing on $[1, \infty)$ **61.** g and F

63. a. $w = 25 - l$ **b.** $A = 25l - l^2$ **65.** $v(t) = 80{,}000 - 6500t$, $0 \leq t \leq 10$

67. a. $C(x) = 2000 + 22.80x$ **b.** $R(x) = 37.00x$ **c.** $P(x) = 14.20x - 2000$

69. $h = 15 - 5r$ **71.** $d = \sqrt{(3t)^2 + 50^2}$ **73.** $d = \sqrt{(45 - 8t)^2 + (6t)^2}$

75. a. $L(x) = \left(\dfrac{1}{4\pi} + \dfrac{1}{16}\right)x^2 - \dfrac{5}{2}x + 25$ **b.** $25, 17.27, 14.09, 15.46, 21.37, 31.83$

c. $[0, 20]$ **77. a.** $A(x) = \sqrt{900 + x^2} + \sqrt{400 + (40 - x)^2}$

b. $74.72, 67.68, 64.34, 64.79, 70$ **c.** $[0, 40]$ **79.** $275, 375, 385, 390, 394$

81. $c = -2$ or $c = 3$ **83.** 1 is not in the range of f.

85.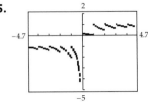

87.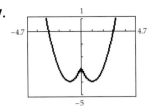

89.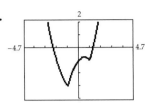

91. 4   **93.** 2   **95. a.** 36
b. 13   **c.** 12   **d.** 30
e. $13k - 2$ **f.** $8k - 11$
97. $4\sqrt{21}$ **99.** $1, -3$

101.

Prepare for Section 2.3, page 197

103. 7   **104.** -1 **105.** $-\dfrac{8}{5}$ **106.** $y = -2x + 9$ **107.** $y = \dfrac{3}{5}x - 3$ **108.** 2

Exercise Set 2.3, page 207

1. $-\dfrac{3}{2}$ **3.** $-\dfrac{1}{2}$ **5.** The line does not have slope.   **7.** 6   **9.** $\dfrac{9}{19}$ **11.** $\dfrac{f(3 + h) - f(3)}{h}$ **13.** $\dfrac{f(h) - f(0)}{h}$ **15.**

17.   **19.**   **21.**   **23.**   **25.**   **27.** $y = x + 3$
29. $y = \dfrac{3}{4}x + \dfrac{1}{2}$
31. $y = (0)x + 4 = 4$
33. $y = -4x - 10$

35. $y = -\dfrac{3}{4}x + \dfrac{13}{4}$ **37.** $y = \dfrac{12}{5}x - \dfrac{29}{5}$ **39.** -2 **41.** $-\dfrac{1}{2}$ **43.** -4 **45.** 4   **47.** -20 **49.** $\dfrac{1}{3}$ **51.** $\dfrac{16}{3}$ **53.** $m = 2.875$. The
value of the slope indicates that the speed of sound in water increases 2.875 feet per second for a 1-degree increase in temperature.
55. a. $H(c) = 1.45c$ **b.** 26 miles per gallon   **57. a.** $N(t) = 2500t - 4,962,000$ **b.** 2008   **59. a.** $B(d) = 30d - 300$ **b.** The value of
the slope means that a 1-inch increase in the diameter of a log 32 feet long results in an increase of 30 board-feet of lumber that can be obtained
from the log.   **c.** 270 board-feet   **61.** line A, Michelle; line B, Amanda; line C, distance between Michelle and Amanda
63. a. $y = 1.842x - 18.947$ **b.** 147   **65.** $P(x) = 40.50x - 1782, x = 44$, the break-even point   **67.** $P(x) = 79x - 10,270, x = 130$, the break-
even point   **69. a.** \$275   **b.** \$283   **c.** \$355   **d.** \$8   **71. a.** $C(t) = 19,500.00 + 6.75t$ **b.** $R(t) = 55.00t$ **c.** $P(t) = 48.25t - 19,500.00$
d. approximately 405 days   **73.** $y = -\dfrac{3}{4}x + \dfrac{15}{4}$ **75.** $y = x + 1$ **77.** -5 ft   **79. a.** $Q = (3, 10), m = 5$ **b.** $Q = (2.1, 5.41), m = 4.1$
c. $Q = (2.01, 5.0401), m = 4.01$ **d.** 4   **85.** $y = -2x + 11$ **87.** $5x + 3y = 15$ **89.** $3x + y = 17$ **93.** $\left(\dfrac{9}{2}, \dfrac{81}{4}\right)$

Prepare for Section 2.4, page 213

95. $(3x - 2)(x + 4)$ **96.** $x^2 - 8x + 16 = (x - 4)^2$ **97.** 26   **98.** $-\dfrac{1}{2}, 1$ **99.** $\dfrac{-3 \pm \sqrt{17}}{2}$ **100.** $1, 3$

Exercise Set 2.4, page 222

1. d **3.** b **5.** g **7.** c **9.** $f(x) = (x + 2)^2 - 3$

11. $f(x) = (x - 4)^2 - 11$ **13.** $f(x) = \left(x - \left(-\dfrac{3}{2}\right)\right)^2 - \dfrac{5}{4}$

vertex: $(-2, -3)$ vertex: $(4, -11)$ vertex: $\left(-\dfrac{3}{2}, -\dfrac{5}{4}\right)$

axis of symmetry: $x = -2$ axis of symmetry: $x = 4$ axis of symmetry: $x = -\dfrac{3}{2}$

15. $f(x) = -(x - 2)^2 + 6$ **17.** $f(x) = -3\left(x - \dfrac{1}{2}\right)^2 + \dfrac{31}{4}$

vertex: $(2, 6)$ vertex: $\left(\dfrac{1}{2}, \dfrac{31}{4}\right)$

axis of symmetry: $x = 2$ axis of symmetry: $x = \dfrac{1}{2}$

19. vertex: $(5, -25)$, $f(x) = (x - 5)^2 - 25$
21. vertex: $(0, -10)$, $f(x) = x^2 - 10$
23. vertex: $(3, 10)$, $f(x) = -(x - 3)^2 + 10$
25. vertex: $\left(\dfrac{3}{4}, \dfrac{47}{8}\right)$, $f(x) = 2\left(x - \dfrac{3}{4}\right)^2 + \dfrac{47}{8}$
27. vertex: $\left(\dfrac{1}{8}, \dfrac{17}{16}\right)$, $f(x) = -4\left(x - \dfrac{1}{8}\right)^2 + \dfrac{17}{16}$

29. $\{y \,|\, y \geq -2\}$, -1 and 3 **31.** $\left\{y \,|\, y \leq \dfrac{17}{8}\right\}$, 1 and $\dfrac{3}{2}$

33. No, $3 \notin \left\{y \,|\, y \geq \dfrac{15}{4}\right\}$ **35.** -16, minimum   **37.** 11, maximum

39. $-\dfrac{1}{8}$, minimum   **41.** -11, minimum   **43.** 35, maximum

45. a. 27 feet   **b.** $22\dfrac{5}{16}$ feet   **c.** 20.1 feet from the center   **47. a.** $w = \dfrac{600 - 2l}{3}$ **b.** $A = 200l - \dfrac{2}{3}l^2$ **c.** $w = 100$ feet, $l = 150$ feet
49. a. 12:43 P.M.   **b.** 91°F   **51.** 1993, 2500 homes   **53.** Yes   **55. a.** 41 miles per gallon   **b.** 34 miles per gallon   **57.** y-intercept $(0, 0)$;
x-intercepts $(0, 0)$ and $(-6, 0)$ **59.** y-intercept $(0, -6)$; no x-intercepts   **61.** 740 units yield a maximum revenue of \$109,520.   **63.** 85 units
yield a maximum profit of \$24.25.   **65.** $P(x) = -0.1x^2 + 50x - 1840$, break-even points: $x = 40$ and $x = 460$ **67. a.** $R(x) = -0.25x^2 + 30.00x$
b. $P(x) = -0.25x^2 + 27.50x - 180$ **c.** \$576.25   **d.** 55   **69. a.** $t = 4$ seconds   **b.** 256 feet   **c.** $t = 8$ seconds   **71.** 30 feet
73. $r = \dfrac{48}{4 + \pi} \approx 6.72$ feet, $h = r \approx 6.72$ feet   **77.** $f(x) = \dfrac{3}{4}x^2 - 3x + 4$ **79. a.** $w = 16 - x$ **b.** $A = 16x - x^2$ **81.** The discriminant is
$b^2 - 4(1)(-1) = b^2 + 4$, which is positive for all b.   **83.** increases the height of each point on the graph by c units   **85.** 4, 4

Prepare for Section 2.5, page 227
89. $x = -2$ **92.** $3, -1, -3, -3, -1$ **93.** $(0, b)$ **94.** $(0, 0)$

Exercise Set 2.5, page 238

1. 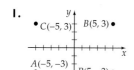   **3.** 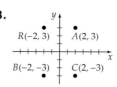   **5.**   **7.**   **9.**   **11.**

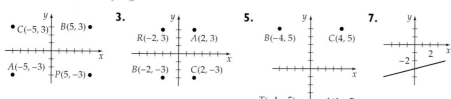

13. a. No   **b.** Yes   **15. a.** No   **b.** No   **17. a.** Yes   **b.** Yes   **19. a.** Yes   **b.** Yes   **21. a.** Yes   **b.** Yes   **23.** No   **25.** Yes

27. Yes   **29.** Yes   **31.**

33.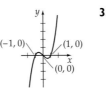

35.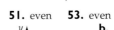

37.

39.

41.

43. even   **45.** odd   **47.** even   **49.** even   **51.** even   **53.** even   **55.** neither

57. a., b.

59. a. 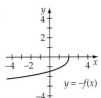   **b.**

61. a. $(-5, 5), (-3, -2), (-2, 0)$ **b.** $(-2, 6), (0, -1), (1, 1)$ **63. a.** 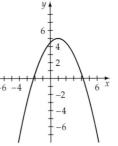   **b.**

65. a. $(1, 3), (-2, -4)$ **b.** $(-1, -3), (2, 4)$ **67. a., b.** 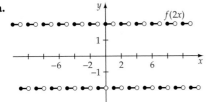   **69.**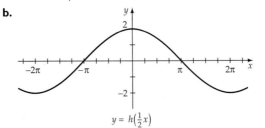

71. a. 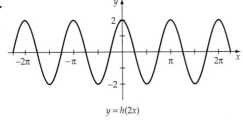   **b.**

73. a. 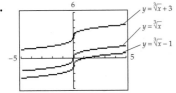 $y = h(2x)$ **b.** 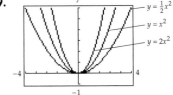 $y = h\left(\frac{1}{2}x\right)$

75. 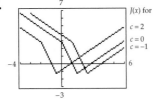 $y = \sqrt[3]{x} + 3$, $y = \sqrt[3]{x}$, $y = \sqrt[3]{x} - 1$

77. $J(x)$ for $c = 2$, $c = 0$, $c = -1$

79. $y = \frac{1}{2}x^2$, $y = x^2$, $y = 2x^2$

81.

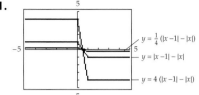

83. a.

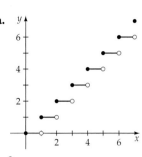

b.

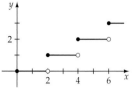

c.

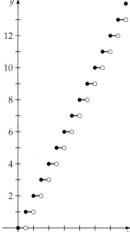

85. a. $f(x) = \dfrac{2}{(x+1)^2 + 1} + 1$ **b.** $f(x) = -\dfrac{2}{(x-2)^2 + 1}$

Prepare for Section 2.6, page 242

87. $x^2 + 1$ **88.** $6x^3 - 11x^2 + 7x - 6$ **89.** $18a^2 - 15a + 2$ **90.** $2h^2 + 3h$ **91.** all real numbers except $x = 1$ **92.** $[4, \infty)$

Exercise Set 2.6, page 251

1. $f(x) + g(x) = x^2 - x - 12$,   Domain is the set of all real numbers.
$f(x) - g(x) = x^2 - 3x - 18$,   Domain is the set of all real numbers.
$f(x) \cdot g(x) = x^3 + x^2 - 21x - 45$,   Domain is the set of all real numbers.
$\dfrac{f(x)}{g(x)} = x - 5$,   Domain $\{x \,|\, x \ne -3\}$

3. $f(x) + g(x) = 3x + 12$,   Domain is the set of all real numbers.
$f(x) - g(x) = x + 4$,   Domain is the set of all real numbers.
$f(x) \cdot g(x) = 2x^2 + 16x + 32$,   Domain is the set of all real numbers.
$\dfrac{f(x)}{g(x)} = 2$,   Domain $\{x \,|\, x \ne -4\}$

5. $f(x) + g(x) = x^3 - 2x^2 + 8x$,   Domain is the set of all real numbers.
$f(x) - g(x) = x^3 - 2x^2 + 6x$,   Domain is the set of all real numbers
$f(x) \cdot g(x) = x^4 - 2x^3 + 7x^2$,   Domain is the set of all real numbers.
$\dfrac{f(x)}{g(x)} = x^2 - 2x + 7$,   Domain $\{x \,|\, x \ne 0\}$

7. $f(x) + g(x) = 4x^2 + 7x - 12$,   Domain is the set of all real numbers.
$f(x) - g(x) = x - 2$,   Domain is the set of all real numbers.
$f(x) \cdot g(x) = 4x^4 + 14x^3 - 12x^2 - 41x + 35$,   Domain is the set of all real numbers.
$\dfrac{f(x)}{g(x)} = 1 + \dfrac{x - 2}{2x^2 + 3x - 5}$,   Domain $\left\{x \,\middle|\, x \ne 1, x \ne -\dfrac{5}{2}\right\}$

9. $f(x) + g(x) = \sqrt{x - 3} + x$,   Domain $\{x \,|\, x \ge 3\}$
$f(x) - g(x) = \sqrt{x - 3} - x$,   Domain $\{x \,|\, x \ge 3\}$
$f(x) \cdot g(x) = x\sqrt{x - 3}$,   Domain $\{x \,|\, x \ge 3\}$
$\dfrac{f(x)}{g(x)} = \dfrac{\sqrt{x - 3}}{x}$,   Domain $\{x \,|\, x \ge 3\}$

11. $f(x) + g(x) = \sqrt{4 - x^2} + 2 + x$,   Domain $\{x \,|\, -2 \le x \le 2\}$
$f(x) - g(x) = \sqrt{4 - x^2} - 2 - x$,   Domain $\{x \,|\, -2 \le x \le 2\}$
$f(x) \cdot g(x) = \left(\sqrt{4 - x^2}\right)(2 + x)$,   Domain $\{x \,|\, -2 \le x \le 2\}$
$\dfrac{f(x)}{g(x)} = \dfrac{\sqrt{4 - x^2}}{2 + x}$ Domain $\{x \,|\, -2 < x \le 2\}$

13. 18   **15.** $-\dfrac{9}{4}$ **17.** 30   **19.** 12   **21.** 300   **23.** $-\dfrac{384}{125}$ **25.** $-\dfrac{5}{2}$ **27.** $-\dfrac{1}{4}$ **29.** 2   **31.** $2x + h$ **33.** $4x + 2h + 4$

35. $-8x - 4h$ **37.** $(g \circ f)(x) = 6x + 3$ **39.** $(g \circ f)(x) = x^2 + 4x + 1$ **41.** $(g \circ f)(x) = -5x^3 - 10x$ **43.** $(g \circ f)(x) = \dfrac{1 - 5x}{x + 1}$
$(f \circ g)(x) = 6x - 16$ $(f \circ g)(x) = x^2 + 8x + 11$ $(f \circ g)(x) = -125x^3 - 10x$
$(f \circ g)(x) = \dfrac{2}{3x - 4}$

45. $(g \circ f)(x) = \dfrac{\sqrt{1 - x^2}}{|x|}$ **47.** $(g \circ f)(x) = -\dfrac{2|5 - x|}{3}$ **49.** 66   **51.** 51   **53.** -4 **55.** 41   **57.** $-\dfrac{3848}{625}$ **59.** $6 + 2\sqrt{3}$
$(f \circ g)(x) = \dfrac{1}{x - 1}$ $(f \circ g)(x) = \dfrac{3|x|}{|5x + 2|}$

61. $16c^2 + 4c - 6$ **63.** $9k^4 + 36k^3 + 45k^2 + 18k - 4$ **65. a.** $A(t) = \pi(1.5t)^2$, $A(2) = 9\pi$ square feet ≈ 28.27 square feet   **b.** $V(t) = 2.25\pi t^3$, $V(3) = 60.75\pi$ cubic feet ≈ 190.85 cubic feet   **67. a.** $d(t) = \sqrt{(48 - t)^2 - 4^2}$ **b.** $s(35) = 13$ feet, $d(35) \approx 12.37$ feet   **69.** $(Y \circ F)(x)$ converts x inches to yards.   **71. a.** 99.8; this is identical to the slope of the line through $(0, C(0))$ and $(1, C(1))$.   **b.** 156.2   **c.** -49.7 **d.** -30.8 **e.** -16.4 **f.** 0

Prepare for Section 2.7, page 254

83. slope: $-\dfrac{1}{3}$; y-intercept: $(0, 4)$ **84.** slope: $\dfrac{3}{4}$; y-intercept: $(0, -3)$ **85.** $y = -0.45x + 2.3$ **86.** $y = -\dfrac{2}{3}x - 2$ **87.** 19   **88.** 3

Exercise Set 2.7, page 263

1. no linear relationship   **3.** linear   **5.** Figure A   **7.** $y = 2.00862069x + 0.5603448276$ **9.** $y = -0.7231182796x + 9.233870968$
11. $y = 2.222641509x - 7.364150943$ **13.** $y = 1.095779221x^2 - 2.69642857x + 1.136363636$
15. $y = -0.2987274717x^2 - 3.20998141x + 3.416463667$ **17. a.** $y = 23.55706665x - 24.4271215$ **b.** 1247.7 centimeters
19. a. $y = 1.671510024x + 16.32830605$ **b.** 46.4 centimeters   **21. a.** $y = 0.1628623408x - 0.6875682232$ **b.** 25   **23.** No, because the linear correlation coefficient is close to 0.   **25. a.** Yes, there is a strong linear correlation.   **b.** $y = -0.9033088235x + 78.62573529$
c. 56 years   **27.** $r^2 \approx 0.667$. The coefficient of determination means that approximately 66.7% of the variation in EPA mileage estimates can be attributed to the horsepower of the car.   **29.** $y = -0.6328671329x^2 + 33.6160839x - 379.4405594$
31. a. $y = -0.0165034965x^2 + 1.366713287x + 5.685314685$ **b.** 32.8 miles per gallon
33. a.   5 pound: $s = 0.6130952381t^2 - 0.0714285714t + 0.1071428571$
10 pound: $s = 0.6091269841t^2 - 0.0011904762t - 0.3$
15 pound: $s = 0.5922619048t^2 + 0.3571428571t - 1.520833333$
b. All the regression equations are approximately the same. Therefore, the equations of motion of the three masses are the same.
35. quadratic; r^2 is closer to 1 for the quadratic model.

Chapter 2 True/False Exercises, page 271

1. False. Let $f(x) = x^2$. Then $f(3) = f(-3) = 9$, but $3 \neq -3$.   **2.** False. Consider $f(x) = x + 1$ and $g(x) = x^2 - 2$.   **3.** True   **4.** True
5. False. Let $f(x) = 3x$. $[f(x)]^2 = 9x^2$, whereas $f[f(x)] = f(3x) = 3(3x) = 9x$.   **6.** False. Let $f(x) = x^2$. Then $f(1) = 1$, $f(2) = 4$. Thus
$\dfrac{f(2)}{f(1)} = 4 \neq \dfrac{2}{1}$.   **7.** True   **8.** False. Let $f(x) = |x|$. Then $f(-1 + 3) = f(2) = 2$. $f(-1) + f(3) = 1 + 3 = 4$.   **9.** True   **10.** True   **11.** True
12. True   **13.** True   **14.** False. The coefficient of determination is r^2 and therefore nonnegative.

Chapter 2 Review Exercises, page 272

1. $\sqrt{181}$ [2.1]   **2.** $\sqrt{80} = 4\sqrt{5}$ [2.1]   **3.** $\left(-\dfrac{1}{2}, 10\right)$ [2.1]   **4.** $(2, -2)$ [2.1]   **5.** center $(3, -4)$, radius 9 [2.1]   **6.** center $(-5, -2)$,
radius 3 [2.1]   **7.** $(x - 2)^2 + (y + 3)^2 = 5^2$ [2.1]   **8.** $(x + 5)^2 + (y - 1)^2 = 8^2$, radius $= |-5 - (3)| = 8$ [2.1]   **9. a.** 2   **b.** 10   **c.** $3t^2 + 4t - 5$
d. $3x^2 + 6xh + 3h^2 + 4x + 4h - 5$ **e.** $9t^2 + 12t - 15$ **f.** $27t^2 + 12t - 5$ [2.2]   **10. a.** $\sqrt{55}$ **b.** $\sqrt{39}$ **c.** 0   **d.** $\sqrt{64 - x^2}$
e. $2\sqrt{64 - t^2}$ **f.** $2\sqrt{16 - t^2}$ [2.2]   **11. a.** 5   **b.** -11 **c.** $x^2 - 12x + 32$ **d.** $x^2 + 4x - 8$ [2.6]   **12. a.** 79   **b.** 56   **c.** $2x^2 - 4x + 9$
d. $2x^2 + 6$ [2.6]   **13.** $8x + 4h - 3$ [2.6]   **14.** $3x^2 + 3xh + h^2 - 1$ [2.6]   **15.** [2.2]   **16.** [2.2]

increasing on $[3, \infty)$
decreasing on $(-\infty, 3]$

f is increasing on $[0, \infty)$
f is decreasing on $(-\infty, 0]$

17. 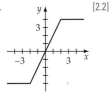 [2.2]

increasing on $[-2, 2]$
constant on $(-\infty, -2] \cup [2, \infty)$

18. 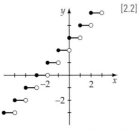 [2.2]

f is constant on..., $[-6, -5)$,
$[-5, -4), [-4, -3), [-3, -2)$,
$[-2, -1), [-1, 0), [0, 1), \ldots$

19. 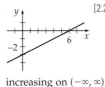 [2.2]

increasing on $(-\infty, \infty)$

20. 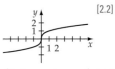 [2.2]

f is increasing on $(-\infty, \infty)$

21. Domain: $\{x \,|\, x \text{ is a real number}\}$ [2.2]   **22.** Domain: $\{x \,|\, x \le 6\}$ [2.2]   **23.** Domain: $\{x \,|\, -5 \le x \le 5\}$ [2.2]   **24.** Domain: $\{x \,|\, x \ne -3, x \ne 5\}$ [2.2]

25. $y = -2x + 1$ [2.3]   **26.** $y = \dfrac{11}{7}x$ [2.3]   **27.** $y = \dfrac{3}{4}x + \dfrac{19}{2}$ [2.3]   **28.** $y = \dfrac{5}{2}x + \dfrac{1}{2}$ [2.3]   **29.** $f(x) = (x + 3)^2 + 1$ [2.4]

30. $f(x) = 2(x + 1)^2 + 3$ [2.4]   **31.** $f(x) = -(x + 4)^2 + 19$ [2.4]   **32.** $f(x) = 4\left(x - \dfrac{3}{4}\right)^2 - \dfrac{5}{4}$ [2.4]   **33.** $f(x) = -3\left(x - \dfrac{2}{3}\right)^2 - \dfrac{11}{3}$ [2.4]

34. $f(x) = (x - 3)^2 + 0$ [2.4]   **35.** $(1, 8)$ [2.4]   **36.** $(0, -10)$ [2.4]   **37.** $(5, 161)$ [2.4]   **38.** $(-4, 30)$ [2.4]   **39.** $\dfrac{4\sqrt{5}}{5}$ [2.3]

40. a. $R = 13x$ **b.** $P = 12.5x - 1050$ **c.** $x = 84$ [2.3]   **41.** 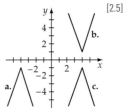 [2.5]   **42.** 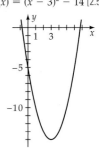 [2.5]

43. symmetric to the y-axis [2.5]   **44.** symmetric to the x-axis [2.5]   **45.** symmetric to the origin [2.5]   **46.** symmetric to the x-axis, the y-axis, and the origin [2.5]   **47.** symmetric to the x-axis, the y-axis, and the origin [2.5]   **48.** symmetric to the origin [2.5]   **49.** symmetric to the x-axis, the y-axis, and the origin [2.5]   **50.** symmetric to the origin [2.5]

51.

a. Domain is the set of all real numbers.
Range: $\{y \,|\, y \le 4\}$
b. even [2.5]

52.

a. Domain is the set of all real numbers.
Range is the set of all real numbers.
b. g is neither even nor odd [2.5]

53.

a. Domain is the set of all real numbers.
Range: $\{y \,|\, y \ge 4\}$
b. even [2.5]

54.

a. Domain: $\{x \,|\, -4 \le x \le 4\}$
Range: $\{y \,|\, 0 \le y \le 4\}$
b. even [2.5]

55.

a. Domain is the set of all real numbers.
Range is the set of all real numbers.
b. odd [2.5]

56.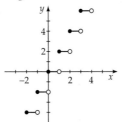

a. Domain: $\{x \,|\, x \text{ is a real number}\}$
Range: $\{y \,|\, y \text{ is an even integer}\}$
b. g is neither even nor odd [2.5]

57. $F(x) = (x + 2)^2 - 11$ [2.5]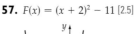

58. $A(x) = (x - 3)^2 - 14$ [2.5]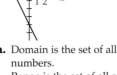

59. $P(x) = 3(x - 0)^2 - 4$ [2.5]   **60.** $G(x) = 2(x - 2)^2 - 5$ [2.5]   **61.** $W(x) = -4\left(x + \dfrac{3}{4}\right)^2 + \dfrac{33}{4}$ [2.5]   **62.** $T(x) = -2\left(x + \dfrac{5}{2}\right)^2 + \dfrac{25}{2}$

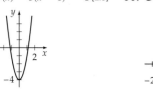
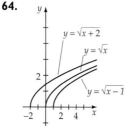
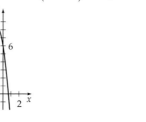
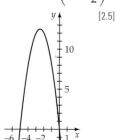
[2.5]

63. [2.5]   **64.** [2.5]   **65.** [2.5]

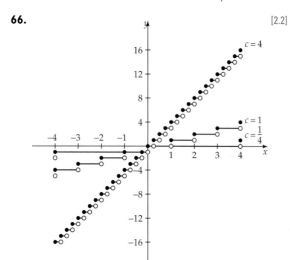
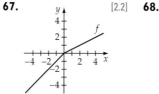
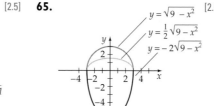

66. [2.2]   **67.** [2.2]   **68.** [2.2]

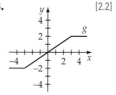

69. $f(x) + g(x) = x^2 + x - 6$, Domain is the set of all real numbers.
$f(x) - g(x) = x^2 - x - 12$, Domain is the set of all real numbers.
$f(x) \cdot g(x) = x^3 + 3x^2 - 9x - 27$, Domain is the set of all real numbers.
$\dfrac{f(x)}{g(x)} = x - 3$, Domain $\{x \mid x \neq -3\}$ [2.6]

70. $(f + g)(x) = x^3 + x^2 - 2x + 12$, Domain is the set of all real numbers.
$(f - g)(x) = x^3 - x^2 + 2x + 4$, Domain is the set of all real numbers.
$(fg)(x) = x^5 - 2x^4 + 4x^3 + 8x^2 - 16x + 32$, Domain is the set of all real numbers.
$\left(\dfrac{f}{g}\right)(x) = x + 2$, Domain is the set of all real numbers. [2.6]

71. 25, 25 [2.4]   **72.** -5 and 5 [2.4]   **73. a.** 18 feet per second   **b.** 15 feet per second   **c.** 13.5 feet per second   **d.** 12.03 feet per second
e. 12 feet per second [2.4]   **74. a.** 17 feet per second   **b.** 15 feet per second   **c.** 14 feet per second   **d.** 13.02 feet per second
e. 13 feet per second [2.4]   **75. a.** $y = 0.0180247x + 0.0005005$ **b.** Yes. $r = 0.999$, which is very close to 1.   **c.** 1.8 seconds [2.7]
76. a. $y = 0.0047952048x^2 - 1.756843157x + 180.4065934$ **b.** No   **c.** The regression line is a model of the data and is not based on physical principles. [2.7]

Chapter 2 Test, page 274

1. midpoint $(1, 1)$; length $2\sqrt{13}$ [2.1]   **2.** $(-4, 0)$; $\left(0, \sqrt{2}\right), \left(0, -\sqrt{2}\right)$ [2.1]   **3.** [2.1]

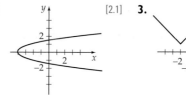
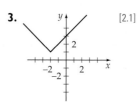

4. center $(2, -1)$; radius 3 [2.1]   **5.** domain $\{x \mid x \geq 4 \text{ or } x \leq -4\}$ [2.2]   **6.**

[2.2]   **7. a.** $R = 12.00x$ **b.** $P = 11.25x - 875$
c. $x = 78$ [2.4]

a. increasing on $(-\infty, 2]$
b. not consistent on any interval
c. decreasing on $[2, \infty)$

8.

[2.5]   **9. a.** even   **b.** odd   **c.** neither [2.5]   **10.** $y = -\dfrac{2}{3}x + \dfrac{2}{3}$ [2.3]   **11.** -12, minimum [2.4]

12. $x^2 + x - 3; \dfrac{x^2 - 1}{x - 2}, x \neq 2$ [2.6]   **13.** $2x + h$ [2.6]   **14.** $4x^2 + 16x + 15$ [2.6]

15. a. 25 feet per second   **b.** 22.5 feet per second   **c.** 20.05 feet per second [2.6]
16. a. $y = -7.98245614x + 767.122807$ **b.** 57 calories [2.7]

Cumulative Review Exercises, page 275

1. Commutative Property of Addition [P.1]   **2.** $\dfrac{6}{\pi}, \sqrt{2}$ [P.1]   **3.** $8x - 33$ [P.1]   **4.** $128x^5y^{10}$ [P.2]   **5.** $\dfrac{4}{3b^2}$ [P.2]   **6.** $6x^2 - 5x - 21$ [P.3]

7. $\dfrac{x + 9}{x + 3}$ [P.5]   **8.** $\dfrac{-2}{(2x - 1)(x - 1)}$ [P.5]   **9.** 0 [1.1]   **10.** $\dfrac{1 \pm \sqrt{5}}{2}$ [1.3]   **11.** $-\dfrac{7}{2}, 1$ [1.3]   **12.** $x = -\dfrac{2}{3}y + 5$ [1.1]   **13.** $\pm\sqrt{2}, \pm i$ [1.4]

14. $x > -4$ [1.5]   **15.** $\sqrt{17}$ [2.1]   **16.** -15 [2.2]   **17.** $y = -\dfrac{1}{2}x - 2$ [2.3]   **18.** 100 ounces [1.1]   **19.** Yes [2.4]   **20.** 0.04°F per minute [2.3]

Exercise Set 3.1, page 287

1. $5x^2 - 9x + 10 - \dfrac{10}{x + 3}$ **3.** $x^3 + 2x^2 - x + 1 + \dfrac{1}{x - 2}$ **5.** $x^2 + 4x + 10 + \dfrac{25}{x - 3}$ **7.** $x^3 + 7x^2 + 31x + 119 + \dfrac{475}{x - 4}$

9. $x^4 + 2x^3 + 2x - 1 - \dfrac{8}{x - 1}$ **11.** $4x^2 + 3x + 12 + \dfrac{17}{x - 2}$ **13.** $4x^2 - 4x + 2 + \dfrac{1}{x + 1}$ **15.** $x^4 + 4x^3 + 6x^2 + 24x + 101 + \dfrac{403}{x - 4}$

17. $x^4 + x^3 + x^2 + x + 1$ **19.** $8x^2 + 6$ **21.** $x^7 + 2x^6 + 5x^5 + 10x^4 + 21x^3 + 42x^2 + 85x + 170 + \dfrac{344}{x - 2}$

23. $x^5 - 3x^4 + 9x^3 - 27x^2 + 81x - 242 + \dfrac{716}{x + 3}$ **25.** 25   **27.** 45   **29.** -2230 **31.** -80 **33.** -187 **35.** Yes   **37.** No

39. Yes   **41.** Yes   **43.** No   **55.** $(x - 2)(x^2 + 3x + 7)$ **57.** $(x - 4)(x^3 + 3x^2 + 3x + 1)$ **59. a.** \$19,968   **b.** \$23,007   **61. a.** 336
b. 336; They are the same.   **63. a.** 100 cards   **b.** 610 cards   **65. a.** 400 people per square mile   **b.** 240 people per square mile
67. a. 304 cubic inches   **b.** 892 cubic inches   **69.** 13   **71.** Yes

Prepare for Section 3.2, page 289

73. 2   **74.** $\dfrac{9}{8}$ **75.** $[-1, \infty)$ **76.** $[1, \infty)$ **77.** $(x + 1)(x - 1)(x + 2)(x - 2)$ **78.** $\left(\dfrac{2}{3}, 0\right), \left(-\dfrac{1}{2}, 0\right)$

Exercise Set 3.2, page 301

1. up to the far left, up to the far right   **3.** down to the far left, up to the far right   **5.** down to the far left, down to the far right
7. down to the far left, up to the far right   **9.** $a < 0$ **11.** Vertex is $(-2, -5)$, minimum is -5.   **13.** Vertex is $(-4, 17)$, maximum is 17.
15. relative maximum $y \approx 5.0$ at   **17.** relative maximum $y \approx 31.0$ at   **19.** relative maximum $y \approx 2.0$ at $x \approx 1.0$,
$x \approx -2.1$, relative minimum $x \approx -2.0$, relative minimum   relative minima $y \approx -14.0$ at $x \approx -1.0$
$y \approx -16.9$ at $x \approx 1.4$ $y \approx -77.0$ at $x \approx 4.0$ and $y \approx -14.0$ at $x \approx 3.0$

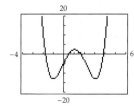

21. $-3, 0, 5$ **23.** $-3, -2, 2, 3$ **25.** $-2, -1, 0, 1, 2$ **33.** crosses the x-axis at $(-1, 0)$, $(1, 0)$, and $(3, 0)$ **35.** crosses the x-axis at $(7, 0)$;

intersects but does not cross at $(3, 0)$ **37.** crosses the x-axis at $(1, 0)$; intersects but does not cross at $\left(\dfrac{3}{2}, 0\right)$ **39.** crosses the x-axis at $(0, 0)$; intersects but does not cross at $(3, 0)$

41. 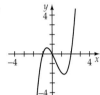   **43.** 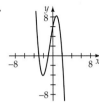   **45.**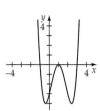

47. a. $V(x) = x(15 - 2x)(10 - 2x) = 4x^3 - 50x^2 + 150x$ **b.** 1.96 inches   **49.** 2.137 inches   **51.** $464,000   **53. a.** 1918   **b.** 9.5 marriages per thousand population   **55. a.** 20.69 milligrams   **b.** 118 minutes   **57. a.** 3.24 inches   **b.** 4 feet from either end; 3.84 inches   **c.** 3.24 inches   **59.** between 3 and 4   **61.** $(5, 0)$ **63.** Shift the graph of $y = x^3$ horizontally 2 units to the right and vertically upward 1 unit.

Prepare for Section 3.3, page 305

65. $\dfrac{2}{3}, \dfrac{7}{2}$ **66.** $2x^2 - x + 6 - \dfrac{19}{x + 2}$ **67.** $3x^3 + 9x^2 + 6x + 15 + \dfrac{40}{x - 3}$ **68.** $1, 2, 3, 4, 6, 12$ **69.** $\pm 1, \pm 3, \pm 9, \pm 27$
70. $P(-x) = -4x^3 - 3x^2 + 2x + 5$

Exercise Set 3.3, page 316

1. 3 (multiplicity 2), -5 (multiplicity 1)   **3.** 0 (multiplicity 2), $-\dfrac{5}{3}$ (multiplicity 2)   **5.** 2 (multiplicity 1), -2 (multiplicity 1), -3 (multiplicity 2)

7. $\pm 1, \pm 2, \pm 4, \pm 8$ **9.** $\pm 1, \pm 2, \pm 3, \pm 4, \pm 6, \pm 12, \pm \dfrac{1}{2}, \pm \dfrac{3}{2}$ **11.** $\pm 1, \pm 2, \pm 4, \pm \dfrac{1}{2}, \pm \dfrac{1}{3}, \pm \dfrac{2}{3}, \pm \dfrac{4}{3}, \pm \dfrac{1}{6}$ **13.** $\pm 1, \pm 7, \pm \dfrac{1}{2}, \pm \dfrac{7}{2},$

$\pm \dfrac{1}{4}, \pm \dfrac{7}{4}$ **15.** $\pm 1, \pm 2, \pm 4, \pm 8, \pm 16, \pm 32$ **17.** upper bound 2, lower bound -5 **19.** upper bound 4, lower bound -4

21. upper bound 1, lower bound -4 **23.** upper bound 4, lower bound -2 **25.** upper bound 2, lower bound -1 **27.** one positive zero, two or no negative zeros   **29.** two or no positive zeros, one negative zero   **31.** one positive zero, three or one negative zeros

33. three or one positive zeros, one negative zero   **35.** one positive zero, no negative zeros   **37.** $2, -1, -4$ **39.** $3, -4, \dfrac{1}{2}$

41. $\dfrac{1}{2}, -\dfrac{1}{3}, -2$ (multiplicity 2)   **43.** $\dfrac{1}{2}, 4, \sqrt{3}, -\sqrt{3}$ **45.** $6, 1 + \sqrt{5}, 1 - \sqrt{5}$ **47.** $5, \dfrac{1}{2}, 2 + \sqrt{3}, 2 - \sqrt{3}$

49. $1, -1, -2, -\dfrac{2}{3}, 3 + \sqrt{3}, 3 - \sqrt{3}$ **51.** $2, -1$ (multiplicity 2)   **53.** $0, -2, 1 + \sqrt{2}, 1 - \sqrt{2}$ **55.** -1 (multiplicity 3), 2

57. $-\dfrac{3}{2}, 1$ (multiplicity 2), 8   **59.** $n = 9$ inches   **61.** $x = 4$ inches   **63. a.** 26 pieces   **b.** 7 cuts   **65.** 7 rows   **67.** $x = 0.084$ inch

69. 1977 and 1986   **71.** 16.9 feet   **73. a.** 73 seconds   **b.** 93,000 digits   **75.** $B = 15$. The absolute value of each of the given zeros is less than B.   **77.** $B = 11$. The absolute value of each of the zeros is less than B.

Prepare for Section 3.4, page 320

79. $3 + 2i$ **80.** $2 - i\sqrt{5}$ **81.** $x^3 - 8x^2 + 19x - 12$ **82.** $x^2 - 4x + 5$ **83.** $-3i, 3i$ **84.** $\dfrac{1}{2} - \dfrac{1}{2}i\sqrt{19}, \dfrac{1}{2} + \dfrac{1}{2}i\sqrt{19}$

Exercise Set 3.4, page 327

1. $2, -3, 2i, -2i; P(x) = (x - 2)(x + 3)(x - 2i)(x + 2i)$ **3.** $\dfrac{1}{2}, -3, 1 + 5i, 1 - 5i; P(x) = \left(x - \dfrac{1}{2}\right)(x + 3)(x - 1 - 5i)(x - 1 + 5i)$

5. 1 (multiplicity 3), $3 + 2i, 3 - 2i; P(x) = (x - 1)^3(x - 3 - 2i)(x - 3 + 2i)$

7. $-3, -\dfrac{1}{2}, 2 + i, 2 - i; P(x) = (x + 3)\left(x + \dfrac{1}{2}\right)(x - 2 - i)(x - 2 + i)$

9. $4, 2, \dfrac{1}{2} + \dfrac{3}{2}i, \dfrac{1}{2} - \dfrac{3}{2}i; P(x) = (x - 4)(x - 2)\left(x - \dfrac{1}{2} - \dfrac{3}{2}i\right)\left(x - \dfrac{1}{2} + \dfrac{3}{2}i\right)$ **11.** $1 - i, \dfrac{1}{2}$ **13.** $i, -3$ **15.** $2 + 3i, i, -i$

17. $1 - 3i, 1 + 2i, 1 - 2i$ **19.** $2i, 1$ (multiplicity 3)   **21.** $5 - 2i, \dfrac{7}{2} + \dfrac{\sqrt{3}}{2}i, \dfrac{7}{2} - \dfrac{\sqrt{3}}{2}i$ **23.** $\dfrac{3}{2}, -\dfrac{1}{2} + \dfrac{\sqrt{7}}{2}i, -\dfrac{1}{2} - \dfrac{\sqrt{7}}{2}i$ **25.** $-\dfrac{2}{3}, \dfrac{3}{4}, \dfrac{5}{2}$

27. $-i, i, 2$ (multiplicity 2)   **29.** -3 (multiplicity 2), 1 (multiplicity 2)   **31.** $P(x) = x^3 - 3x^2 - 10x + 24$ **33.** $P(x) = x^3 - 3x^2 + 4x - 12$

35. $P(x) = x^4 - 10x^3 + 63x^2 - 214x + 290$ **37.** $P(x) = x^5 - 22x^4 + 212x^3 - 1012x^2 + 2251x - 1830$ **39.** $P(x) = 4x^3 - 19x^2 + 224x - 159$

41. $P(x) = x^3 + 13x + 116$ **43.** $P(x) = x^4 - 18x^3 + 131x^2 - 458x + 650$ **45.** $P(x) = x^5 - 4x^4 + 16x^3 - 18x^2 - 97x + 102$

47. $P(x) = 3x^3 - 12x^2 + 3x + 18$ **49.** $P(x) = -2x^4 + 4x^3 + 36x^2 - 140x + 150$ **51.** The Conjugate Pair Theorem does not apply because some of the coefficients of the polynomial are not real numbers.

Prepare for Section 3.5, page 328

53. $\dfrac{x - 3}{x - 5}$ **54.** $-\dfrac{3}{2}$ **55.** $\dfrac{1}{3}$ **56.** $x = 0, -3, \dfrac{5}{2}$ **57.** The degree of the numerator is 3. The degree of the denominator is 2.

58. $x + 4 + \dfrac{7x - 11}{x^2 - 2x}$

Exercise Set 3.5, page 341

1. $x = 0, x = -3$ **3.** $x = -\dfrac{1}{2}, x = \dfrac{4}{3}$ **5.** $y = 4$ **7.** $y = 30$

9. $x = -4, y = 0$ **11.** $x = 3, y = 0$ **13.** $x = 0, y = 0$ **15.** $x = -4, y = 1$ **17.** $x = 2, y = -1$

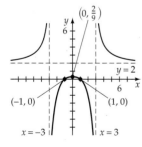 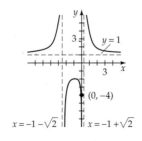 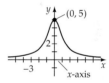 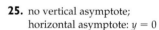 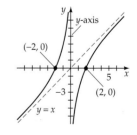

19. $x = 3, x = -3, y = 0$ **21.** $x = -3, x = 1, y = 0$ **23.** $x = -2, y = 1$ **25.** no vertical asymptote; horizontal asymptote: $y = 0$

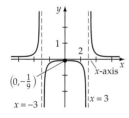 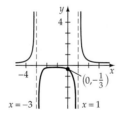 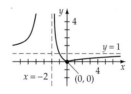 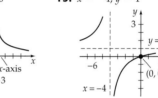

27. $x = 3, x = -3, y = 2$ **29.** $x = -1 + \sqrt{2}, x = -1 - \sqrt{2}, y = 1$ **31.** $y = 3x - 7$ **33.** $y = x$ **35.** $x = 0, y = x$

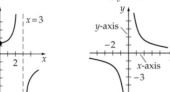 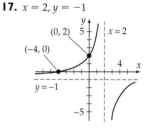

37. $x = -3, y = x - 6$ **39.** $x = 4, y = 2x + 13$ **41.** $x = -2, y = x - 3$ **43.** $x = 2, x = -2, y = x$

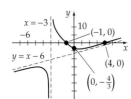 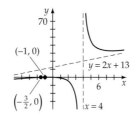 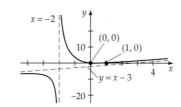 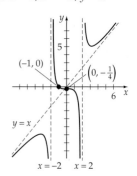

45.   **47.**   **49.**   **51.**

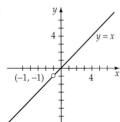 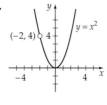 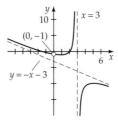 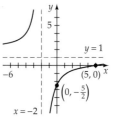

53. a. $76.43, $8.03, $1.19   **b.** $y = 0.43$. As the number of golf balls produced increases, the average cost per golf ball approaches $.43.

55. a. $1333.33   **b.** $8000   **c.**   **57. a.** $R(0) \approx 38.8\%$, $R(7) \approx 39.9\%$, $R(12) \approx 40.9\%$ **b.** $\approx 44.7\%$

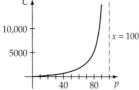

59. a. 26,923, 68,293, 56,000   **b.** 2001   **c.** The population will approach 0.   **61. a.** 3.8 centimeters

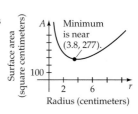

b. No   **c.** As the radius r increases without bound, the surface area approaches twice the area of a circle with radius r.   **63.** $(-2, 2)$
65. $(0, 1)$ and $(-4, 1)$

Chapter 3 True/False Exercises, page 347

1. False; $P(x) = x - i$ has a zero of i, but it does not have a zero of $-i$.   **2.** False; Descartes' Rule of Signs indicates that $P(x) = x^3 - x^2 + x - 1$

has three or one positive zeros. In fact, P has only one positive zero.   **3.** True   **4.** True   **5.** False; $F(x) = \dfrac{x}{x^2 + 1}$ does not have a vertical

asymptote.   **6.** False; $F(x) = \dfrac{(x - 2)^2}{(x - 3)(x - 2)} = \dfrac{x - 2}{x - 3}, x \neq 2$. The graph of F has a hole at $x = 2$.   **7.** True   **8.** True   **9.** True   **10.** True

11. True   **12.** False; $P(x) = x^2 + 1$ does not have a real zero.

Chapter 3 Review Exercises, page 348

1. $4x^2 + x + 8 + \dfrac{22}{x - 3}$ [3.1]   **2.** $5x^2 + 5x - 13 - \dfrac{11}{x - 1}$ [3.1]   **3.** $3x^2 - 6x + 7 - \dfrac{13}{x + 2}$ [3.1]   **4.** $2x^2 + 8x + 20$ [3.1]   **5.** $3x^2 + 5x - 11$ [3.1]

6. $x^3 + 2x^2 - 8x - 9$ [3.1]   **7.** 77 [3.1]   **8.** 22 [3.1]   **9.** 33 [3.1]   **10.** 558 [3.1]

The verifications in Exercises 11–14 make use of the concepts from Section 3.1.

15. [3.2] **16.** [3.2] **17.** [3.2] **18.** 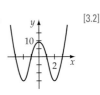 [3.2] **19.** 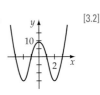 [3.2]

20. [3.2] **21.** $\pm 1, \pm 2, \pm 3, \pm 6$ [3.3] **22.** $\pm 1, \pm 2, \pm 3, \pm 5, \pm 6, \pm 10, \pm 15, \pm 30, \pm \dfrac{1}{2}, \pm \dfrac{3}{2}, \pm \dfrac{5}{2}, \pm \dfrac{15}{2}$ [3.3]

23. $\pm 1, \pm 2, \pm 3, \pm 4, \pm 6, \pm 12, \pm \dfrac{1}{3}, \pm \dfrac{2}{3}, \pm \dfrac{4}{3}, \pm \dfrac{1}{5}, \pm \dfrac{2}{5}, \pm \dfrac{3}{5}, \pm \dfrac{4}{5}, \pm \dfrac{6}{5}, \pm \dfrac{12}{5}, \pm \dfrac{1}{15}, \pm \dfrac{2}{15}, \pm \dfrac{4}{15}$ [3.3] **24.** $\pm 1, \pm 2, \pm 4, \pm 8, \pm 16, \pm 32, \pm 64$

[3.3] **25.** ± 1 [3.3] **26.** $\pm 1, \pm 2, \pm \dfrac{1}{6}, \pm \dfrac{1}{3}, \pm \dfrac{1}{2}, \pm \dfrac{2}{3}$ [3.3] **27.** no positive real zeros and three or one negative real zeros [3.3]

28. three or one positive real zeros, one negative real zero [3.3] **29.** one positive real zero and one negative real zero [3.3]

30. five, three, or one positive real zeros, no negative real zeros [3.3] **31.** $1, -2, -5$ [3.3] **32.** $2, 5, 3$ [3.3] **33.** -2 (multiplicity 2), $-\dfrac{1}{2}, -\dfrac{4}{3}$

[3.3] **34.** $-\dfrac{1}{2}, -3, i, -i$ [3.4] **35.** 1 (multiplicity 4) [3.3] **36.** $-\dfrac{1}{2}, 2 + 3i, 2 - 3i$ [3.4] **37.** $-1, 3, 1 + 2i$ [3.4] **38.** $-5, 2, 2 - i$ [3.4]

39. $P(x) = 2x^3 - 3x^2 - 23x + 12$ [3.4] **40.** $P(x) = x^4 + x^3 - 5x^2 + x - 6$ [3.4] **41.** $P(x) = x^4 - 3x^3 + 27x^2 - 75x + 50$ [3.4]

42. $P(x) = x^4 + 2x^3 + 6x^2 + 32x + 40$ [3.4] **43.** vertical asymptote: $x = -2$, horizontal asymptote: $y = 3$ [3.5]

44. vertical asymptotes: $x = -3, x = 1$, horizontal asymptote: $y = 2$ [3.5] **45.** vertical asymptote: $x = -1$, slant asymptote: $y = 2x + 3$ [3.5]

46. no vertical asymptote, horizontal asymptote: $y = 3$ [3.5]

47. [3.5] **48.** [3.5] **49.** [3.5] **50.** 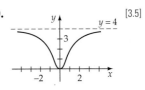 [3.5]

51. 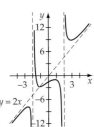 [3.5] **52.** 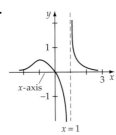 [3.5] **53.** 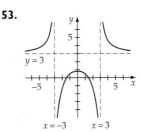 [3.5] **54.** 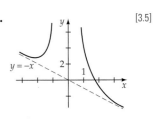 [3.5]

55. a. $12.59, $6.43 **b.** $y = 5.75$. As the number of skateboards produced increases, the average cost per skateboard approaches $5.75. [3.5]

56. a. 15°F **b.** 2.4°F **c.** 0°F [3.5] **57. a.** As the radius of the blood vessel approaches 0, the resistance gets larger. **b.** As the radius of the blood vessel gets larger, the resistance approaches zero. [3.5]

Chapter 3 Test, page 349

1. $3x^2 - x + 6 - \dfrac{13}{x + 2}$ [3.1] **2.** 43 [3.1] **3.** The verification for Exercise 3 makes use of the concepts from Section 3.1. **4.** up to the far left

and down to the far right [3.2] **5.** $0, \dfrac{2}{3}, -3$ [3.2] **6.** $P(1) < 0, P(2) > 0$. Therefore, by the Zero Location Theorem, the continuous polynomial function P has a zero between 1 and 2. [3.2]   **7.** 2 (multiplicity 2), -2 (multiplicity 2), $\dfrac{3}{2}$ (multiplicity 1), -1 (multiplicity 3) [3.3]

8. $\pm 1, \pm 3, \pm\dfrac{1}{2}, \pm\dfrac{3}{2}, \pm\dfrac{1}{3}, \pm\dfrac{1}{6}$ [3.3]   **9.** upper bound 4, lower bound -5 [3.3]   **10.** four, two, or zero positive zeros, no negative zero [3.3]

11. $\dfrac{1}{2}, 3, -2$ [3.3]   **12.** $2 - 3i, -\dfrac{2}{3}, -\dfrac{5}{2}$ [3.4]   **13.** 0, 1 (multiplicity 2), $2 + i, 2 - i$ [3.4]   **14.** $P(x) = x^4 - 5x^3 + 8x^2 - 6x$ [3.4]   **15.** vertical

asymptotes: $x = 3, x = 2$ [3.5]   **16.** horizontal asymptote: $y = \dfrac{3}{2}$ [3.5]   **17.**   [3.5]   **18.**   [3.5]

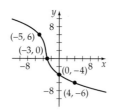

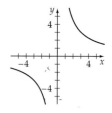

19. a. 5 words per minute, 16 words per minute, 25 words per minute   **b.** 70 words per minute [3.5]   **20.** 2.42 inches, 487.9 cubic inches [3.3]

Cumulative Review Exercises, page 351

1. $-1 + 2i$ [P.6]   **2.** $\dfrac{1 \pm \sqrt{5}}{2}$ [1.3]   **3.** 2, 10 [1.4]   **4.** $\{x \mid -8 \le x \le 14\}$ [1.5]   **5.** $\sqrt{281}$ [2.1]   **6.** Translate the graph of $y = x^2$ to the right

2 units and 4 units up. [2.5]   **7.** $2x + h - 2$ [2.6]   **8.** $32x^2 - 92x + 60$ [2.6]   **9.** $x^3 - x^2 + x + 11$ [2.6]   **10.** $4x^3 - 8x^2 + 14x - 32 + \dfrac{59}{x + 2}$
[3.1]   **11.** 141 [3.1]   **12.** The graph goes down. [3.2]   **13.** 0.3997 [3.2]   **14.** $\pm 1, \pm 2, \pm 4, \pm\dfrac{1}{3}, \pm\dfrac{2}{3}, \pm\dfrac{4}{3}$
[3.3]   **15.** zero positive real zeros, three or one negative real zeros [3.3]   **16.** $-2, 1 + 2i, 1 - 2i$ [3.4]   **17.** $P(x) = x^3 - 4x^2 - 2x + 20$ [3.4]
18. $(x - 2)(x + 3i)(x - 3i)$ [3.4]

19. vertical asymptotes: $x = -3, x = 2$; horizontal asymptote: $y = 4$ [3.5]   **20.** $y = x + 4$ [3.5]

Exercise Set 4.1, page 364
1. 3   **3.** -3 **5.** 3   **7.** range   **9.** Yes   **11.** Yes   **13.** Yes

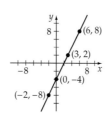

15. No   **17.** Yes   **19.** Yes   **21.** No   **23.** $\{(1, -3), (2, -2), (5, 1), (-7, 4)\}$ **25.** $\{(1, 0), (2, 1), (4, 2), (8, 3), (16, 4)\}$

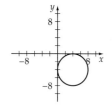

27. $f^{-1}(x) = \dfrac{1}{2}x - 2$ **29.** $f^{-1}(x) = \dfrac{1}{3}x + \dfrac{7}{3}$ **31.** $f^{-1}(x) = -\dfrac{1}{2}x + \dfrac{5}{2}$ **33.** $f^{-1}(x) = \dfrac{x}{x - 2}, x \ne 2$ **35.** $f^{-1}(x) = \dfrac{x + 1}{1 - x}, x \ne 1$
37. $f^{-1}(x) = \sqrt{x - 1}, x \ge 1$ **39.** $f^{-1}(x) = x^2 + 2, x \ge 0$ **41.** $f^{-1}(x) = \sqrt{x + 4} - 2, x \ge -4$ **43.** $f^{-1}(x) = -\sqrt{x + 5} - 2, x \ge -5$
45. $V^{-1}(x) = \sqrt[3]{x}$. V^{-1} finds the length of a side of a cube given the volume.
47. Yes. Yes. A conversion function is a nonconstant linear function. All nonconstant linear functions have inverses that are also functions.

49. $s^{-1}(x) = \dfrac{1}{2}x - 12$ **51.** $E^{-1}(s) = 20s - 50{,}000$. From the monthly earnings s the executive can find $E^{-1}(s)$, the value of the software sold.

53. $44205833; f^{-1}(x) = \dfrac{1}{2}x + \dfrac{1}{2}, f^{-1}(44205833) = 22102917$

55. Because the function is increasing and 4 is between 2 and 5, c must be between 7 and 12.   **57.** between 2 and 5

59. between 3 and 7   **61.** $f^{-1}(x) = \dfrac{x - b}{a}, a \neq 0$ **63.** The reflection of f across the line given by $y = x$ yields f. Thus f is its own inverse.

65. Yes   **67.** No

Prepare for Section 4.2, page 367

69. 8   **70.** $\dfrac{1}{81}$ **71.** $\dfrac{17}{8}$ **72.** $\dfrac{40}{9}$ **73.** $\dfrac{1}{10}$, 1, 10, and 100   **74.** 2, 1, $\dfrac{1}{2}$, and $\dfrac{1}{4}$

Exercise Set 4.2, page 376

1. $f(0) = 1; f(4) = 81$ **3.** $g(-2) = \dfrac{1}{100}; g(3) = 1000$ **5.** $h(2) = \dfrac{9}{4}; h(-3) = \dfrac{8}{27}$ **7.** $j(-2) = 4; j(4) = \dfrac{1}{16}$ **9.** 9.19   **11.** 9.03   **13.** 9.74

15. a. $k(x)$ **b.** $g(x)$ **c.** $h(x)$ **d.** $f(x)$

17.   **19.** 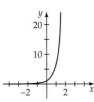   **21.**   **23.**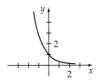

25. Shift the graph of f vertically upward 2 units.   **27.** Shift the graph of f horizontally to the right 2 units.
29. Reflect the graph of f across the y-axis.   **31.** Stretch the graph of f vertically away from the x-axis by a factor of 2.
33. Reflect the graph of f across the y-axis and then shift this graph vertically upward 2 units.

35. no horizontal asymptote   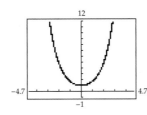   **37.** no horizontal asymptote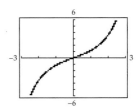

39. horizontal asymptote: $y = 0$ 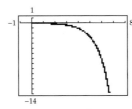   **41.** horizontal asymptote: $y = 10$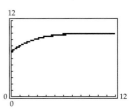

43. a. 122 million connections   **b.** 2006   **45. a.** 233 items per month; 59 items per month   **b.** The demand will approach
25 items per month.   **47. a.** 0.53   **b.** 0.89   **c.** 5.2 minutes   **d.** There is a 98% probability that at least one customer will arrive between
10:00 A.M. and 10:05.2 A.M.   **49. a.** 8.7%   **b.** 2.6%   **51. a.** 6400; 409,600   **b.** 11.6 hours   **53. a.** 515,000 people   **b.** 1997
55. a. 363 beneficiaries; 88,572 beneficiaries   **b.** 13 rounds   **57. a.** 141°F   **b.** after 28.3 minutes   **59. a.** 261.63 vibrations per second
b. No. The function $f(n)$ is not a linear function. Therefore, the graph of $f(n)$ does not increase at a constant rate.

63.   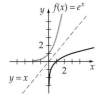   **65.** $(-\infty, \infty)$ **67.** $[0, \infty)$

Prepare for Section 4.3, page 381

68. 4   **69.** 3   **70.** 5   **71.** $f^{-1}(x) = \dfrac{3x}{2-x}$ **72.** $\{x | x \geq 2\}$ **73.** the set of all positive real numbers

Exercise Set 4.3, page 391

1. $10^1 = 10$ **3.** $8^2 = 64$ **5.** $7^0 = x$ **7.** $e^4 = x$ **9.** $e^0 = 1$ **11.** $\log_3 9 = 2$ **13.** $\log_4 \dfrac{1}{16} = -2$ **15.** $\log_b y = x$ **17.** $\ln y = x$

19. $\log 100 = 2$ **21.** 2   **23.** -5 **25.** 3   **27.** -2 **29.** -4 **31.**

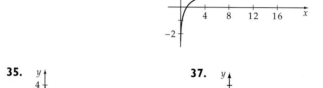

33.   **35.**   **37.**   **39.** $(3, \infty)$

41. $(-\infty, 11)$ **43.** $(-\infty, -2) \cup (2, \infty)$ **45.** $(4, \infty)$ **47.** $(-1, 0) \cup (1, \infty)$ **49.**

51.   **53.**   **55.**

57. a. $k(x)$ **b.** $f(x)$ **c.** $g(x)$ **d.** $h(x)$ **59.**   **61.**

63.   **65.**   **67.**

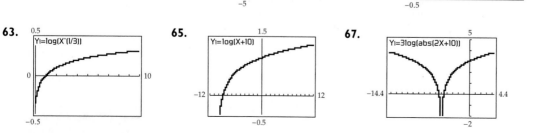

69. a. 2.0%   **b.** 45 months   **71. a.** 3298 units; 3418 units; 3490 units   **b.** 2750 units   **73.** 2.05 square meters
75. a. Answers will vary.   **b.** 96 digits   **c.** 3385 digits   **d.** 6,320,430 digits   **77.** f and g are inverse functions.
79. range of f: $\{y|-1 < y < 1\}$; range of g: all real numbers

Prepare for Section 4.4, page 394

81. ≈ 0.77815 for each expression   **82.** ≈ 0.98083 for each expression   **83.** ≈ 1.80618 for each expression
84. ≈ 3.21888 for each expression   **85.** ≈ 1.60944 for each expression   **86.** ≈ 0.90309 for each expression

Exercise Set 4.4, page 403

1. $\log_b x + \log_b y + \log_b z$ **3.** $\ln x - 4 \ln z$ **5.** $\frac{1}{2}\log_2 x - 3\log_2 y$ **7.** $\frac{1}{2}\log_7 x + \frac{1}{2}\log_7 z - 2\log_7 y$ **9.** $\log[x^2(x + 5)]$ **11.** $\ln(x + y)$
13. $\log\left[x^3 \cdot \sqrt[3]{y}(x + 1)\right]$ **15.** 1.5395   **17.** 0.8672   **19.** -0.6131 **21.** 0.6447   **23.**

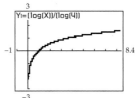

25. 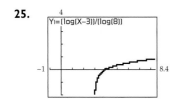   **27.** 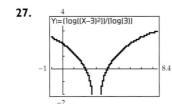   **29.**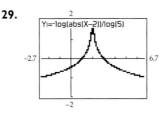

31. False; $\log 10 + \log 10 = 2$ but $\log(10 + 10) = \log 20 \neq 2$.   **33.** True   **35.** False; $\log 100 - \log 10 = 1$ but $\log(100 - 10) = \log 90 \neq 1$.
37. False; $\dfrac{\log 100}{\log 10} = \dfrac{2}{1} = 2$ but $\log 100 - \log 10 = 1$.   **39.** False; $(\log 10)^2 = 1$ but $2 \log 10 = 2$.   **41.** 2   **43.** 500^{501}
45. $1:870{,}551$; $1:757{,}858$; $1:659{,}754$; $1:574{,}349$; $1:500{,}000$ **47.** 10.4; base   **49.** 3.16×10^{-10} mole per liter
51. a. 82.0 decibels   **b.** 40.3 decibels   **c.** 115.0 decibels   **d.** 152.0 decibels   **53.** 10 times more intense   **55.** 5
57. $10^{6.5}I_0$ or about $3{,}162{,}277.7I_0$ **59.** 100 to 1   **61.** $10^{1.8}$ to 1 or about 63 to 1   **63.** 5.5   **65. a.** $M \approx 6$ **b.** $M \approx 4$ **c.** The results are
close to the magnitudes produced by the amplitude-time-difference formula.

Prepare for Section 4.5, page 406

66. $\log_3 729 = 6$ **67.** $5^4 = 625$ **68.** $\log_a b = x + 2$ **69.** $x = \dfrac{4a}{7b + 2c}$ **70.** $x = \dfrac{3}{44}$ **71.** $x = \dfrac{100(A - 1)}{A + 1}$

Exercise Set 4.5, page 415

1. 6   **3.** $-\dfrac{3}{2}$ **5.** $-\dfrac{6}{5}$ **7.** 3   **9.** $\dfrac{\log 70}{\log 5}$ **11.** $-\dfrac{\log 120}{\log 3}$ **13.** $\dfrac{\log 315 - 3}{2}$ **15.** $\ln 10$ **17.** $\dfrac{\ln 2 - \ln 3}{\ln 6}$ **19.** $\dfrac{3\log 2 - \log 5}{2\log 2 + \log 5}$
21. 7   **23.** 4   **25.** $2 + 2\sqrt{2}$ **27.** $\dfrac{199}{95}$ **29.** -1 **31.** 3   **33.** 10^{10} **35.** 2   **37.** 5   **39.** $\log\left(20 + \sqrt{401}\right)$ **41.** $\dfrac{1}{2}\log\left(\dfrac{3}{2}\right)$
43. $\ln\left(15 \pm 4\sqrt{14}\right)$ **45.** $\ln\left(1 + \sqrt{65}\right) - \ln 8$ **47.** 1.61   **49.** 0.96   **51.** 2.20   **53.** -1.93 **55.** -1.34

57. a. 8500, 10,285   **b.** in 6 years   **59. a.** 60°F   **b.** 27 minutes   **61. a.**

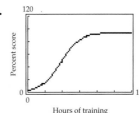

Hours of training

b. 48 hours
c. $P = 100$
d. As the number of hours of
training increases, the test
scores approach 100%.

63. a.

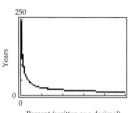

b. in 27 years or the year 2026   **c.** $B = 1000$ **d.** As the number of years increases, the bison population approaches but never reaches or exceeds 1000.

65. a.

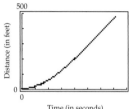

b. 78 years   **c.** 1.9%

67. a. 116 feet per second   **b.** $v = 150$ **c.** The velocity of the package approaches but never reaches or exceeds 150 feet per second.
69. a. 1.72 seconds   **b.** $v = 100$ **c.** The object cannot fall faster than 100 feet per second.

71. a.

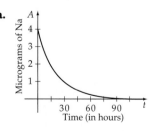

b. 2.6 seconds   **73.** 138

75. The second step; because $\log 0.5 < 0$, the inequality sign must be reversed.   **77.** $x = \dfrac{y}{y - 1}$ **79.** $e^{0.336} \approx 1.4$

Prepare for Section 4.6, page 419
81. 1220.39   **82.** 824.96   **83.** -0.0495 **84.** 1340   **85.** 0.025   **86.** 12.8

Exercise Set 4.6, page 430

1. a. \$9724.05   **b.** \$11,256.80   **3. a.** \$48,885.72   **b.** \$49,282.20   **c.** \$49,283.30   **5.** \$24,730.82   **7.** 8.8 years   **9.** $t = \dfrac{\ln 3}{r}$
11. 14 years   **13. a.** 2200 bacteria   **b.** 17,600 bacteria   **15. a.** $N(t) \approx 22,600e^{0.01368t}$ **b.** 27,700   **17. a.** 10,755,000   **b.** 2042

19. a.

b. 3.18 micrograms   **c.** ≈ 15.07 hours   **d.** ≈ 30.14 hours   **21.** ≈ 6601 years ago

23. ≈ 2378 years old   **25. a.** 0.056   **b.** 42°F   **c.** 54 minutes   **27. a.** 211 hours   **b.** 1386 hours   **29.** 3.1 years

23. $-\dfrac{\sqrt{3}}{3}$ **25.** $20°$ **27.** $9°$ **29.** $\dfrac{\pi}{5}$ **31.** $\pi - \dfrac{8}{3}$ **33.** $34°$ **35.** $65°$ **37.** $-\dfrac{\sqrt{2}}{2}$ **39.** 1 **41.** $-\dfrac{2\sqrt{3}}{3}$ **43.** $\dfrac{\sqrt{2}}{2}$ **45.** $\sqrt{2}$

47. $\cot 540°$ is undefined. **49.** 0.798636 **51.** -0.438371 **53.** -1.26902 **55.** -0.587785 **57.** -1.70130 **59.** -3.85522 **61.** 0

63. 1 **65.** $-\dfrac{3}{2}$ **67.** 1 **69.** $30°, 150°$ **71.** $150°, 210°$ **73.** $225°, 315°$ **75.** $\dfrac{3\pi}{4}, \dfrac{7\pi}{4}$ **77.** $\dfrac{5\pi}{6}, \dfrac{11\pi}{6}$ **79.** $\dfrac{\pi}{3}, \dfrac{2\pi}{3}$

Prepare for Section 5.4, page 498

91. Yes **92.** Yes **93.** No **94.** 2π **95.** even function **96.** neither

Exercise Set 5.4, page 508

1. $\left(\dfrac{\sqrt{3}}{2}, \dfrac{1}{2}\right)$ **3.** $\left(-\dfrac{\sqrt{3}}{2}, -\dfrac{1}{2}\right)$ **5.** $\left(\dfrac{1}{2}, -\dfrac{\sqrt{3}}{2}\right)$ **7.** $\left(\dfrac{\sqrt{3}}{2}, -\dfrac{1}{2}\right)$ **9.** $(-1, 0)$ **11.** $\left(-\dfrac{1}{2}, -\dfrac{\sqrt{3}}{2}\right)$ **13.** $-\dfrac{\sqrt{3}}{3}$ **15.** $-\dfrac{1}{2}$

17. $-\dfrac{2\sqrt{3}}{3}$ **19.** -1 **21.** $-\dfrac{2\sqrt{3}}{3}$ **23.** 0.9391 **25.** -1.1528 **27.** -0.2679 **29.** 0.8090 **31.** 48.0889 **33. a.** 0.9 **b.** -0.4

35. a. -0.8 **b.** 0.6 **37.** $0.4, 2.7$ **39.** $3.4, 6.0$ **41.** odd **43.** neither **45.** even **47.** odd **57.** $\sin t$ **59.** $\sec t$ **61.** $-\tan^2 t$

63. $-\cot t$ **65.** $\cos^2 t$ **67.** $2 \csc^2 t$ **69.** $\csc^2 t$ **71.** 1 **73.** $\sqrt{1 - \cos^2 t}$ **75.** $\sqrt{1 + \cot^2 t}$ **77.** 750 miles **79.** $-\dfrac{\sin^2 t}{\cos t}$

81. $\csc t \sec t$ **83.** $1 - 2 \sin t + \sin^2 t$ **85.** $1 - 2 \sin t \cos t$ **87.** $\cos^2 t$ **89.** $2 \csc t$ **91.** $(\cos t - \sin t)(\cos t + \sin t)$

93. $(\tan t + 2)(\tan t - 3)$ **95.** $(2 \sin t + 1)(\sin t - 1)$ **97.** $(\cos t - \sin t)(\cos t + \sin t)$ **99.** $\dfrac{\sqrt{2}}{2}$ **101.** $-\dfrac{\sqrt{3}}{3}$

Prepare for Section 5.5, page 510

107. 0.7 **108.** -0.7 **109.** Reflect the graph of $y = f(x)$ across the x-axis. **110.** Contract each point on the graph of $y = f(x)$ toward the y-axis by a factor of $\dfrac{1}{2}$. **111.** 6π **112.** 5π

Exercise Set 5.5, page 518

1. $2, 2\pi$ **3.** $1, \pi$ **5.** $\dfrac{1}{2}, 1$ **7.** $2, 4\pi$ **9.** $\dfrac{1}{2}, 2\pi$ **11.** $1, 8\pi$ **13.** $2, 6$ **15.** $3, 3\pi$ **17.** 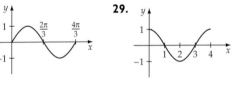 **19.**

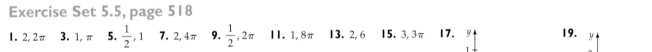

21. **23.** **25.** **27.** **29.**

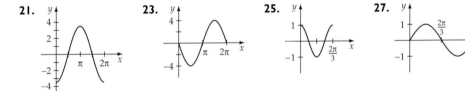

31. **33.** **35.** **37.** **39.**

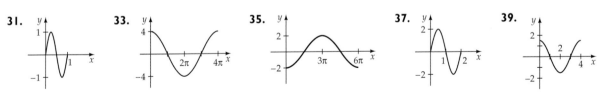

41. **43.** **45.** **47.** **49.**

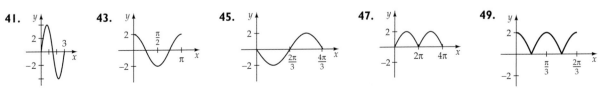

51.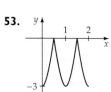

53.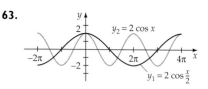

55. $y = \cos 2x$ **57.** $y = 2 \sin \dfrac{2}{3} x$ **59.** $y = -2 \cos \pi x$

61.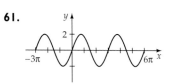

63.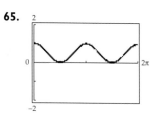

65.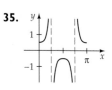

67.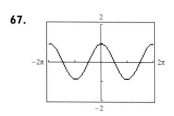

69.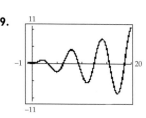

71.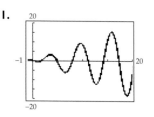

73.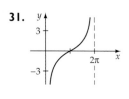

75. $f(t) = 60 \cos \dfrac{\pi}{10} t$ **77.** $y = 2 \sin \dfrac{2}{3} x$ **79.** $y = 4 \sin \pi x$ **81.** $y = 3 \cos 4x$

83. $y = 3 \cos \dfrac{4\pi}{5} x$

$\text{maximum} = e, \text{minimum} = \dfrac{1}{e} \approx 0.3679, \text{period} = 2\pi$

Prepare for Section 5.6, page 520

85. 1.7   **86.** 0.6   **87.** Stretch each point on the graph of $y = f(x)$ away from the x-axis by a factor of 2.

88. Shift the graph of $y = f(x)$ 2 units to the right and up 3 units.   **89.** 2π **90.** $\dfrac{4}{3} \pi$

Exercise Set 5.6, page 526

1. $\dfrac{\pi}{2} + k\pi$, k an integer   **3.** $\dfrac{\pi}{2} + k\pi$, k an integer   **5.** 2π **7.** π **9.** 2π **11.** $\dfrac{2\pi}{3}$ **13.** $\dfrac{\pi}{3}$ **15.** 8π **17.** 1   **19.** 4

21.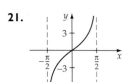

23.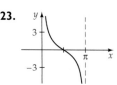

25.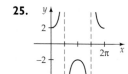

27.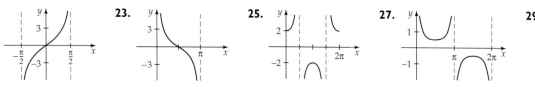

29.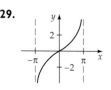

31.

33.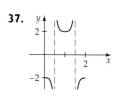

35.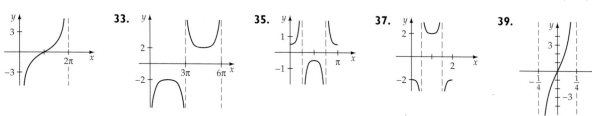

37.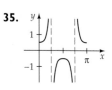

39.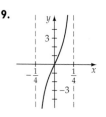

55. 46 feet [5.2]   **56.** 2.5, $\dfrac{\pi}{25}, \dfrac{25}{\pi}$ [5.8]   **57.** amplitude = 0.5, $f = \dfrac{1}{\pi}, p = \pi, y = -0.5 \cos 2t$ [5.8]   **58.** 7.2 seconds [5.8]

Chapter 5 Test, page 550

1. $\dfrac{5\pi}{6}$ [5.1]   **2.** $\dfrac{\pi}{12}$ [5.1]   **3.** 13.1 centimeters [5.1]   **4.** 12π radians/second [5.1]   **5.** 80 centimeters/second [5.1]   **6.** $\dfrac{\sqrt{58}}{7}$ [5.2]

7. 1.0864 [5.2]   **8.** $\dfrac{\sqrt{3} - 6}{6}$ [5.3]   **9.** $\left(\dfrac{\sqrt{3}}{2}, -\dfrac{1}{2}\right)$ [5.4]   **10.** $\sin^2 t$ [5.4]   **11.** $\dfrac{\pi}{3}$ [5.6]   **12.** amplitude 3, period π, phase shift $-\dfrac{\pi}{4}$ [5.7]

13. period 3, phase shift $-\dfrac{1}{2}$ [5.7]   **14.**   [5.5]   **15.**   [5.6]

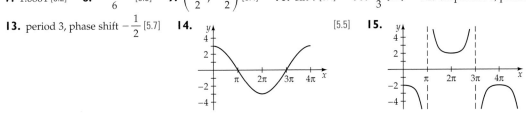

16. Shift the graph [of $y = 2 \sin(2x)$] $\dfrac{\pi}{4}$ units to the right and down 1 unit. [5.7]   **17.**   [5.7]

18.   [5.7]   **19.** 25.5 meters [5.2]   **20.** $y = 13 \sin \dfrac{2\pi}{5} t$ [5.8]

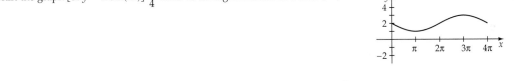

Cumulative Review Exercises, page 551

1. $(x + y)(x - y)$ [P.4]   **2.** $\sqrt{3}$ [P.5]   **3.** 12 square inches [1.2]   **4.** odd function [2.5]   **5.** $f^{-1}(x) = \dfrac{3x}{2x - 1}$ [4.1]   **6.** $(-\infty, 4) \cup (4, \infty)$ [2.2/3.5]

7. $[0, 2]$ [2.2]   **8.** Shift the graph of $y = f(x)$ horizontally 3 units to the right. [2.5]   **9.** Reflect the graph of $y = f(x)$ across the y-axis. [2.5]

10. $\dfrac{5\pi}{3}$ [5.1]   **11.** $225°$ [5.1]   **12.** 1 [5.3]   **13.** $\dfrac{\sqrt{3} + 1}{2}$ [5.2]   **14.** $\dfrac{5}{4}$ [5.2]   **15.** negative [5.3]   **16.** $30°$ [5.4]   **17.** $\dfrac{\pi}{3}$ [5.4]

18. $(-\infty, \infty)$ [5.4]   **19.** $[-1, 1]$ [5.4]   **20.** $\dfrac{3}{5}$ [5.2]

Exercise Set 6.1, page 559

1. If $x = \dfrac{\pi}{4}$, the left side is 2 and the right side is 1.   **3.** If $x = 0°$, the left side is $\dfrac{\sqrt{3}}{2}$ and the right side is $\dfrac{2 + \sqrt{3}}{2}$.   **5.** If $x = 0$, the left side

is -1 and the right side is 1.   **7.** If $x = 0$, the left side is -1 and the right side is 1.   **9.** If $x = \dfrac{\pi}{4}$, the left side is 2 and the right side is 1.

67. identity   **69.** identity   **71.** identity   **73.** not an identity

Prepare for Section 6.2, page 562

83. Both functional values equal $\dfrac{1}{2}$.   **84.** Both functional values equal $\dfrac{1}{2}$.   **85.** For each of the given values of θ, the functional values are

equal.   **86.** For each of the given values of θ, the functional values are equal.   **87.** Both functional values equal $\dfrac{\sqrt{3}}{3}$.   **88.** 0

Exercise Set 6.2, page 570

1. $\dfrac{\sqrt{6} + \sqrt{2}}{4}$ **3.** $\dfrac{\sqrt{6} + \sqrt{2}}{4}$ **5.** $2 - \sqrt{3}$ **7.** $\dfrac{-\sqrt{6} + \sqrt{2}}{4}$ **9.** $-\dfrac{\sqrt{6} + \sqrt{2}}{4}$ **11.** $2 + \sqrt{3}$ **13.** 0 **15.** $\dfrac{1}{2}$ **17.** $\sqrt{3}$ **19.** $\cos 48°$

21. $\cot 75°$ **23.** $\csc 65°$ **25.** $\sin 5x$ **27.** $\cos x$ **29.** $\sin 4x$ **31.** $\cos 2x$ **33.** $\sin x$ **35.** $\tan 7x$ **37. a.** $-\dfrac{77}{85}$ **b.** $\dfrac{84}{85}$ **c.** $\dfrac{77}{36}$

39. a. $-\dfrac{63}{65}$ **b.** $-\dfrac{56}{65}$ **c.** $-\dfrac{63}{16}$ **41. a.** $\dfrac{63}{65}$ **b.** $\dfrac{56}{65}$ **c.** $\dfrac{33}{56}$ **43. a.** $-\dfrac{77}{85}$ **b.** $-\dfrac{84}{85}$ **c.** $-\dfrac{13}{84}$ **45. a.** $-\dfrac{33}{65}$ **b.** $-\dfrac{16}{65}$ **c.** $\dfrac{63}{16}$

47. a. $-\dfrac{56}{65}$ **b.** $-\dfrac{63}{65}$ **c.** $\dfrac{16}{63}$ **73.** $-\cos\theta$ **75.** $\tan\theta$ **77.** $\sin\theta$ **79.** identity   **81.** identity

Prepare for Section 6.3, page 572

91. $2\sin\alpha\cos\alpha$ [6.2]   **92.** $\cos^2\alpha - \sin^2\alpha$ [6.2]   **93.** $\dfrac{2\tan\alpha}{1 - \tan^2\alpha}$ [6.2]   **94.** For each of the given values of α, the functional values are equal.

95. Let $\alpha = 45°$; then the left side of the equation is 1, and the right side of the equation is $\sqrt{2}$.   **96.** Let $\alpha = 60°$; then the left side of the equation is $\dfrac{\sqrt{3}}{2}$, and the right side of the equation is $\dfrac{1}{4}$.

Exercise Set 6.3, page 578

1. $\sin 4\alpha$ **3.** $\cos 10\beta$ **5.** $\cos 6\alpha$ **7.** $\tan 6\alpha$ **9.** $\dfrac{\sqrt{2 + \sqrt{3}}}{2}$ **11.** $\sqrt{2} + 1$ **13.** $-\dfrac{\sqrt{2 + \sqrt{2}}}{2}$ **15.** $\dfrac{\sqrt{2 - \sqrt{2}}}{2}$ **17.** $\dfrac{\sqrt{2 - \sqrt{2}}}{2}$

19. $\dfrac{\sqrt{2 - \sqrt{3}}}{2}$ **21.** $-2 - \sqrt{3}$ **23.** $\dfrac{\sqrt{2 + \sqrt{3}}}{2}$ **25.** $\sin 2\theta = -\dfrac{24}{25}$, $\cos 2\theta = \dfrac{7}{25}$, $\tan 2\theta = -\dfrac{24}{7}$

27. $\sin 2\theta = -\dfrac{240}{289}$, $\cos 2\theta = \dfrac{161}{289}$, $\tan 2\theta = -\dfrac{240}{161}$ **29.** $\sin 2\theta = -\dfrac{336}{625}$, $\cos 2\theta = -\dfrac{527}{625}$, $\tan 2\theta = \dfrac{336}{527}$

31. $\sin 2\theta = \dfrac{240}{289}$, $\cos 2\theta = -\dfrac{161}{289}$, $\tan 2\theta = -\dfrac{240}{161}$ **33.** $\sin 2\theta = -\dfrac{720}{1681}$, $\cos 2\theta = \dfrac{1519}{1681}$, $\tan 2\theta = -\dfrac{720}{1519}$

35. $\sin 2\theta = \dfrac{240}{289}$, $\cos 2\theta = -\dfrac{161}{289}$, $\tan 2\theta = -\dfrac{240}{161}$ **37.** $\sin\dfrac{\alpha}{2} = \dfrac{5\sqrt{26}}{26}$, $\cos\dfrac{\alpha}{2} = \dfrac{\sqrt{26}}{26}$, $\tan\dfrac{\alpha}{2} = 5$

39. $\sin\dfrac{\alpha}{2} = \dfrac{5\sqrt{34}}{34}$, $\cos\dfrac{\alpha}{2} = -\dfrac{3\sqrt{34}}{34}$, $\tan\dfrac{\alpha}{2} = -\dfrac{5}{3}$ **41.** $\sin\dfrac{\alpha}{2} = \dfrac{\sqrt{5}}{5}$, $\cos\dfrac{\alpha}{2} = \dfrac{2\sqrt{5}}{5}$, $\tan\dfrac{\alpha}{2} = \dfrac{1}{2}$

43. $\sin\dfrac{\alpha}{2} = \dfrac{\sqrt{2}}{10}$, $\cos\dfrac{\alpha}{2} = -\dfrac{7\sqrt{2}}{10}$, $\tan\dfrac{\alpha}{2} = -\dfrac{1}{7}$ **45.** $\sin\dfrac{\alpha}{2} = \dfrac{\sqrt{17}}{17}$, $\cos\dfrac{\alpha}{2} = \dfrac{4\sqrt{17}}{17}$, $\tan\dfrac{\alpha}{2} = \dfrac{1}{4}$

47. $\sin\dfrac{\alpha}{2} = \dfrac{5\sqrt{34}}{34}$, $\cos\dfrac{\alpha}{2} = -\dfrac{3\sqrt{34}}{34}$, $\tan\dfrac{\alpha}{2} = -\dfrac{5}{3}$ **95.** identity   **97.** not an identity

Prepare for Section 6.4, page 581

107. $\sin\alpha\cos\beta$ **108.** $\cos\alpha\cos\beta$ **109.** Both functional values equal $-\dfrac{1}{2}$.   **110.** $\sin x + \cos x$ **111.** Answers will vary.   **112.** 2

Exercise Set 6.4, page 587

1. $\sin 3x - \sin x$ **3.** $\dfrac{1}{2}(\sin 8x - \sin 4x)$ **5.** $\sin 8x + \sin 2x$ **7.** $\dfrac{1}{2}(\cos 4x - \cos 6x)$ **9.** $\dfrac{1}{4}$ **11.** $-\dfrac{\sqrt{2}}{4}$ **13.** $-\dfrac{1}{4}$ **15.** $\dfrac{\sqrt{3} - 2}{4}$

17. $2\sin 3\theta\cos\theta$ **19.** $2\cos 2\theta\cos\theta$ **21.** $-2\sin 4\theta\sin 2\theta$ **23.** $2\cos 4\theta\cos 3\theta$ **25.** $2\sin 7\theta\cos 2\theta$ **27.** $-2\sin\dfrac{3}{2}\theta\sin\dfrac{1}{2}\theta$

29. $2\sin\dfrac{3}{4}\theta\sin\dfrac{\theta}{4}$ **31.** $2\cos\dfrac{5}{12}\theta\sin\dfrac{1}{12}\theta$ **49.** $y = \sqrt{2}\sin(x - 135°)$ **51.** $y = \sin(x - 60°)$ **53.** $y = \dfrac{\sqrt{2}}{2}\sin(x - 45°)$

55. $y = 3\sqrt{2}\sin(x + 135°)$ **57.** $y = \pi\sqrt{2}\sin(x - 45°)$ **59.** $y = \sqrt{2}\sin\left(x + \dfrac{3\pi}{4}\right)$ **61.** $y = \sin\left(x + \dfrac{\pi}{6}\right)$ **63.** $y = 20\sin\left(x + \dfrac{2\pi}{3}\right)$

65. $y = 5\sqrt{2}\sin\left(x + \dfrac{3\pi}{4}\right)$

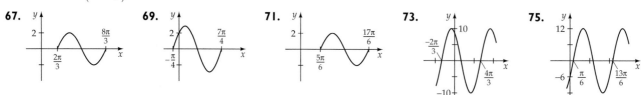

67.   **69.**   **71.**   **73.**   **75.**

77. a. $p(t) = \cos(2\pi \cdot 1336t) + \cos(2\pi \cdot 770t)$ **b.** $p(t) = 2\cos(2106\pi t)\cos(566\pi t)$ **c.** 1053 cycles per second   **79.** identity   **81.** identity
83. identity

Prepare for Section 6.5, page 589

99. A one-to-one function is a function for which each range value (y-value) is paired with one and only one domain value (x-value).
100. If every horizontal line intersects the graph of a function at most once, then the function is a one-to-one function.
101. $f[g(x)] = x$ **102.** $f[f^{-1}(x)] = x$ **103.** The graph of f^{-1} is the reflection of the graph of f across the line given by $y = x$.   **104.** No

Exercise Set 6.5, page 601

1. $\dfrac{\pi}{2}$ **3.** $\dfrac{5\pi}{6}$ **5.** $-\dfrac{\pi}{4}$ **7.** $\dfrac{\pi}{3}$ **9.** $\dfrac{\pi}{3}$ **11.** $-\dfrac{\pi}{4}$ **13.** $-\dfrac{\pi}{3}$ **15.** $\dfrac{2\pi}{3}$ **17.** $\dfrac{\pi}{6}$ **19.** $\dfrac{\pi}{6}$ **21.** $\dfrac{1}{2}$ **23.** 2   **25.** $\dfrac{3}{5}$ **27.** 1

29. $\dfrac{1}{2}$ **31.** $\dfrac{\pi}{6}$ **33.** $\dfrac{\pi}{4}$ **35.** not defined   **37.** 0.4636   **39.** $-\dfrac{\pi}{6}$ **41.** $\dfrac{\sqrt{3}}{3}$ **43.** $\dfrac{4\sqrt{15}}{15}$ **45.** $\dfrac{24}{25}$ **47.** $\dfrac{13}{5}$ **49.** 0   **51.** $\dfrac{24}{25}$

53. $\dfrac{2 + \sqrt{15}}{6}$ **55.** $\dfrac{1}{5}(3\sqrt{7} - 4\sqrt{3})$ **57.** $\dfrac{12}{13}$ **59.** 2   **61.** $\dfrac{2 - \sqrt{2}}{2}$ **63.** $\dfrac{7\sqrt{2}}{10}$ **65.** $\cos\dfrac{5\pi}{12} \approx 0.2588$ **67.** $\sqrt{1 - x^2}$ **69.** $\dfrac{\sqrt{x^2 - 1}}{|x|}$

75.   **77.**   **79.**   **81.**   **83. a.** 0.1014   **b.** 0.1552

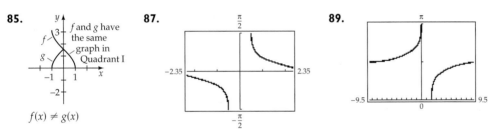

85.   **87.**   **89.**

f and g have the same graph in Quadrant I

$f(x) \neq g(x)$

91.   **93.**   **99.** $y = \dfrac{1}{3}\tan 5x$ **101.** $y = 3 + \cos\left(x - \dfrac{\pi}{3}\right)$

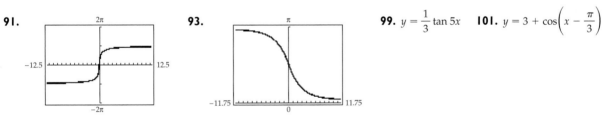

Prepare for Section 6.6, page 603

103. $x = \dfrac{5 \pm \sqrt{73}}{6}$ **104.** $1 - \cos^2 x$ **105.** $\dfrac{5}{2}\pi, \dfrac{9}{2}\pi$, and $\dfrac{13}{2}\pi$ **106.** $(x + 1)\left(x - \dfrac{\sqrt{3}}{2}\right)$ **107.**   **108.** 0, 1

Exercise Set 6.6, page 614

1. $\dfrac{\pi}{4}, \dfrac{7\pi}{4}$ **3.** $\dfrac{\pi}{3}, \dfrac{4\pi}{3}$ **5.** $\dfrac{\pi}{4}, \dfrac{\pi}{2}, \dfrac{3\pi}{4}, \dfrac{3\pi}{2}$ **7.** $\dfrac{\pi}{2}, \dfrac{3\pi}{2}$ **9.** $\dfrac{\pi}{6}, \dfrac{\pi}{4}, \dfrac{3\pi}{4}, \dfrac{11\pi}{6}$ **11.** $\dfrac{\pi}{4}, \dfrac{3\pi}{4}$ **13.** $\dfrac{\pi}{6}, \dfrac{\pi}{2}, \dfrac{5\pi}{6}$ **15.** $\dfrac{\pi}{6}, \dfrac{5\pi}{6}, \dfrac{7\pi}{6}, \dfrac{11\pi}{6}$

17. $0, \dfrac{\pi}{4}, \dfrac{3\pi}{4}, \pi, \dfrac{5\pi}{4}, \dfrac{7\pi}{4}$ **19.** $\dfrac{\pi}{6}, \dfrac{5\pi}{6}, \dfrac{4\pi}{3}, \dfrac{5\pi}{3}$ **21.** $0, \dfrac{\pi}{2}, \pi, \dfrac{3\pi}{2}$ **23.** $41.4°, 318.6°$ **25.** no solution **27.** $68.0°, 292.0°$ **29.** no solution

31. $12.8°, 167.2°$ **33.** $15.5°, 164.5°$ **35.** $0°, 33.7°, 180°, 213.7°$ **37.** no solution **39.** no solution **41.** $0°, 120°, 240°$ **43.** $70.5°, 289.5°$
45. $68.2°, 116.6°, 248.2°, 296.6°$ **47.** $19.5°, 90°, 160.5°, 270°$ **49.** $60°, 90°, 300°$ **51.** $53.1°, 180°$ **53.** $72.4°, 220.2°$ **55.** $50.1°, 129.9°, 205.7°,$

$334.3°$ **57.** no solution **59.** $22.5°, 157.5°$ **61.** $\dfrac{\pi}{8} + \dfrac{k\pi}{2}$, where k is an integer **63.** $\dfrac{\pi}{10} + \dfrac{2k\pi}{5}$, where k is an integer **65.** $0 + 2k\pi,$

$\dfrac{\pi}{3} + 2k\pi, \pi + 2k\pi, \dfrac{5\pi}{3} + 2k\pi$, where k is an integer **67.** $\dfrac{\pi}{2} + k\pi, \dfrac{5\pi}{6} + k\pi$, where k is an integer **69.** $0 + 2k\pi$, where k is an integer

71. $0, \pi$ **73.** $0, \dfrac{\pi}{6}, \dfrac{\pi}{2}, \dfrac{5\pi}{6}, \pi, \dfrac{7\pi}{6}, \dfrac{3\pi}{2}, \dfrac{11\pi}{6}$ **75.** $0, \dfrac{\pi}{2}, \dfrac{3\pi}{2}$ **77.** $0, \dfrac{\pi}{3}, \dfrac{2\pi}{3}, \pi, \dfrac{4\pi}{3}, \dfrac{5\pi}{3}$ **79.** $\dfrac{4\pi}{3}, \dfrac{5\pi}{3}$ **81.** $0, \dfrac{\pi}{4}, \dfrac{3\pi}{4}, \pi, \dfrac{5\pi}{4}, \dfrac{7\pi}{4}$

83. $\dfrac{\pi}{6}, \dfrac{5\pi}{6}, \pi$ **85.** 0.7391 **87.** $-3.2957, 3.2957$ **89.** 1.16 **91.** $14.99°$ and $75.01°$

The sine regression functions in Exercises 93, 95, and 97 were obtained on a TI-83 calculator by using an iteration factor of 16. The use of a different iteration factor may produce a sine regression function that varies from the regression functions listed below.
93. a. $y \approx 1.1213 \sin(0.01595x + 1.8362) + 6.6257$ **b.** 6:49 **95. a.** $y \approx 49.572 \sin(0.21261x - 2.1922) + 49.490$ **b.** 2%
97. a. $y \approx 35.185 \sin(0.30395x - 2.1630) + 2.1515$ **b.** $24.8°$ **99.** day 74 to day 268 = 195 days **101. b.** $42°$ and $79°$ **c.** $60°$

103. $\dfrac{\pi}{6}, \dfrac{\pi}{2}$ **105.** $\dfrac{5\pi}{3}, 0$ **107.** $0, \dfrac{\pi}{4}, \dfrac{\pi}{2}, \dfrac{3\pi}{4}, \pi, \dfrac{5\pi}{4}, \dfrac{3\pi}{2}, \dfrac{7\pi}{4}$ **109.** $0, \dfrac{\pi}{2}, \pi, \dfrac{3\pi}{2}$ **111.** $\dfrac{\pi}{6}, \dfrac{\pi}{2}, \dfrac{5\pi}{6}, \dfrac{7\pi}{6}, \dfrac{3\pi}{2}, \dfrac{11\pi}{6}$

113. 0.93 foot, 1.39 feet

Chapter 6 True/False Exercises, page 622

1. False; $\dfrac{\tan 45°}{\tan 60°} \neq \dfrac{45°}{60°}$. **2.** False; if $y = 0$, $\tan \dfrac{x}{y}$ is undefined. **3.** False. Let $x = 1$. Then $\sin^{-1} 1 = \dfrac{\pi}{2}$ and $\csc 1^{-1} \approx 1.18$. **4.** False; if

$\alpha = \dfrac{\pi}{2}$, then $\sin 2\alpha = \sin \pi = 0$ but $2 \sin \dfrac{\pi}{2} = 2$. **5.** False; $\sin(30° + 60°) \neq \sin 30° + \sin 60°$. **6.** False; $\sin x = 0$ has an infinite number of

solutions $x = k\pi$, but $\sin x = 0$ is not an identity. **7.** False; $\tan 45° = \tan 225°$ but $45° \neq 225°$.

8. False; $\cos^{-1}\left(\cos \dfrac{3\pi}{2}\right) = \cos^{-1}(0) = \dfrac{\pi}{2} \neq \dfrac{3\pi}{2}$. **9.** False; $\cos(\cos^{-1} 2) \neq 2$, because $\cos^{-1} 2$ is undefined. **10.** False; if $\alpha = 1$, then

$\csc^{-1} \dfrac{1}{\alpha} = \csc^{-1} 1 = \dfrac{\pi}{2} \neq \dfrac{1}{\csc 1} \approx 0.8415$. **11.** False; $\sin(180° - \theta) = \sin \theta$. **12.** False because $\sin^2 \theta \geq 0$ for all θ but $\sin \theta^2$ can be < 0.

Chapter 6 Review Exercises, page 622

1. $\dfrac{\sqrt{6} - \sqrt{2}}{4}$ [6.2] **2.** $\sqrt{3} - 2$ [6.2] **3.** $\dfrac{\sqrt{6} - \sqrt{2}}{4}$ [6.2] **4.** $\sqrt{2} - \sqrt{6}$ [6.2] **5.** $-\dfrac{\sqrt{6} + \sqrt{2}}{4}$ [6.2] **6.** $\dfrac{\sqrt{2} + \sqrt{6}}{4}$ [6.2] **7.** $\dfrac{\sqrt{2} - \sqrt{2}}{2}$ [6.3]

8. $-\dfrac{\sqrt{2 - \sqrt{3}}}{2}$ [6.3] **9.** $\sqrt{2} + 1$ [6.3] **10.** $\dfrac{\sqrt{2 + \sqrt{2}}}{2}$ [6.3] **11. a.** 0 **b.** $\sqrt{3}$ **c.** $\dfrac{1}{2}$ [6.2/6.3] **12. a.** 0 **b.** -2 **c.** $\dfrac{1}{2}$ [6.2/6.3]

13. a. $\dfrac{\sqrt{3}}{2}$ **b.** $-\sqrt{3}$ **c.** $-\dfrac{\sqrt{2 - \sqrt{3}}}{2}$ [6.2/6.3] **14. a.** $\dfrac{\sqrt{6} - \sqrt{2}}{4}$ **b.** $-\sqrt{3}$ **c.** 1 [6.2/6.3] **15.** $\sin 6x$ [6.3] **16.** $\tan 3x$ [6.2]

17. $\sin 3x$ [6.2] **18.** $\cos 4\theta$ [6.3] **19.** $\tan 2\theta$ [6.3] **20.** $\tan \theta$ [6.3] **21.** $2 \sin 3\theta \sin \theta$ [6.4] **22.** $-2 \cos 4\theta \sin \theta$ [6.4]

23. $2 \sin 4\theta \cos 2\theta$ [6.4] **24.** $2 \cos 3\theta \sin 2\theta$ [6.4] **43.** $\dfrac{13}{5}$ [6.5] **44.** $\dfrac{4}{5}$ [6.5] **45.** $\dfrac{56}{65}$ [6.5] **46.** $\dfrac{7}{25}$ [6.5] **47.** $\dfrac{3}{2}$ [6.6] **48.** $\dfrac{4}{5}$ [6.6]

49. $30°, 150°, 240°, 300°$ [6.6] **50.** $0°, 45°, 135°$ [6.6] **51.** $\dfrac{\pi}{2} + 2k\pi, 3.8713 + 2k\pi, 5.553 + 2k\pi$, where k is an integer. [6.6] **52.** $-\dfrac{\pi}{4} + k\pi,$

$1.2490 + k\pi$, where k is an integer. [6.6] **53.** $\dfrac{\pi}{12}, \dfrac{5\pi}{12}, \dfrac{13\pi}{12}, \dfrac{17\pi}{12}$ [6.6] **54.** $\dfrac{7\pi}{12}, \dfrac{19\pi}{12}, \dfrac{3\pi}{4}, \dfrac{7\pi}{4}$ [6.6]

55. $y = 2 \sin\left(x + \dfrac{\pi}{6}\right)$ [6.4]   **56.** $y = 2\sqrt{2} \sin\left(x + \dfrac{5\pi}{4}\right)$ [6.4]   **57.** $y = 2 \sin\left(x + \dfrac{4\pi}{3}\right)$ [6.4]   **58.** $y = \sin\left(x - \dfrac{\pi}{6}\right)$ [6.4]

 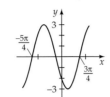

59. [6.5]   **60.** [6.5]   **61.** [6.5]   **62.** [6.5]

63. a. $y \approx 1.5698 \sin(0.01632x + 1.8145) + 5.9360$ **b.** 5:22 [6.6]

Chapter 6 Test, page 625

5. $\dfrac{-\sqrt{6} + \sqrt{2}}{4}$ [6.2]   **6.** $-\dfrac{\sqrt{2}}{10}$ [6.2]   **8.** $\sin 9x$ [6.2]   **9.** $-\dfrac{7}{25}$ [6.3]   **12.** $\dfrac{2 - \sqrt{3}}{4}$ [6.4]   **13.** $y = \sin\left(x + \dfrac{5\pi}{6}\right)$ [6.4]   **14.** 0.701 [6.5]

15. $\dfrac{5}{13}$ [6.5]   **16.** [6.5]

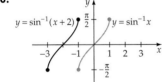

17. 41.8°, 138.2° [6.6]   **18.** $0, \dfrac{\pi}{6}, \pi, \dfrac{11\pi}{6}$ [6.6]   **19.** $\dfrac{\pi}{2}, \dfrac{2\pi}{3}, \dfrac{4\pi}{3}$ [6.6]

20. a. $y \approx 38.965 \sin(0.40089x + 2.9143) + 33.197$ **b.** 63.9° [6.6]

Cumulative Review Exercises, page 626

1. $(x - y)(x^2 + xy + y^2)$ [P.4]   **2.** 2, 8 [1.1]   **3.** Shift the graph of $y = f(x)$ to the left 1 unit and up 2 units. [2.5]   **4.** Reflect the graph of $y = f(x)$ across the x-axis. [2.5]   **5.** $x = 2$ [3.5]   **6.** odd function [2.5/5.5]   **7.** $f^{-1}(x) = \dfrac{x}{x - 5}$ [4.1]   **8.** $\log_2 x = 5$ [4.3]   **9.** 3 [4.3]

10. $\dfrac{4\pi}{3}$ [5.1]   **11.** 300° [5.1]   **12.** $\dfrac{2\sqrt{5}}{5}$ [5.2]   **13.** positive [5.3]   **14.** 50° [5.4]   **15.** $\dfrac{\pi}{3}$ [5.4]   **16.** $x = \dfrac{1}{2}, y = \dfrac{\sqrt{3}}{2}$ [5.4]   **17.** 0.43, π, $\dfrac{\pi}{12}$ [5.7]

18. $\dfrac{\pi}{6}$ [6.5]   **19.** $[-1, 1]$ [6.5]   **20.** $\left(-\dfrac{\pi}{2}, \dfrac{\pi}{2}\right)$ [6.5]

Exercise Set 7.1, page 634

1. $C = 77°, b \approx 16, c \approx 17$ **3.** $B = 38°, a \approx 18, c \approx 10$ **5.** $C \approx 15°, B \approx 33°, c \approx 7.8$ **7.** $C = 32.6°, c \approx 21.6, a \approx 39.8$
9. $B = 47.7°, a \approx 57.4, b \approx 76.3$ **11.** $A \approx 58.5°, B \approx 7.3°, a \approx 81.5$ **13.** $C = 59°, B = 84°, b \approx 46$ or $C = 121°, B = 22°, b \approx 17$
15. No triangle is formed.   **17.** No triangle is formed.   **19.** $C = 19.8°, B = 145.4°, b \approx 10.7$ or $C = 160.2°, B = 5.0°, b \approx 1.64$
21. No triangle is formed.   **23.** $C = 51.21°, A = 11.47°, c \approx 59.00$ **25.** ≈ 68.8 miles   **27.** 231 yards   **29.** ≈ 110 feet   **31.** ≈ 130 yards

33. ≈ 96 feet   **35.** ≈ 33 feet   **37.** ≈ 8.1 miles   **39.** ≈ 1200 miles   **41.** ≈ 260 meters   **45.**

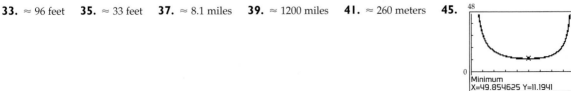

minimum value of $L \approx 11.19$ meters

Prepare for Section 7.2, page 637

46. 20.7   **47.** 25.5 square inches   **48.** $C = \cos^{-1}\left(\dfrac{a^2 + b^2 - c^2}{2ab}\right)$ **49.** 12.5 meters   **50.** 6   **51.** $c^2 = a^2 + b^2$

Exercise Set 7.2, page 644

1. ≈ 13   **3.** ≈ 150   **5.** ≈ 29   **7.** ≈ 9.5   **9.** ≈ 10   **11.** ≈ 40.1   **13.** ≈ 90.7   **15.** ≈ 39°   **17.** ≈ 90°   **19.** ≈ 47.9°   **21.** ≈ 116.67°
23. ≈ 80.3°   **25.** ≈ 140 square units   **27.** ≈ 53 square units   **29.** ≈ 81 square units   **31.** ≈ 299 square units   **33.** ≈ 36 square units
35. ≈ 7.3 square units   **37.** ≈ 710 miles   **39.** ≈ 74 feet   **41.** ≈ 60.9°   **43.** ≈ 350 miles   **45.** 40 centimeters
47. ≈ 9.7 inches, 25 inches   **49.** ≈ 55 centimeters   **51.** ≈ 2800 feet   **53.** ≈ 47,500 square meters   **55.** 162 square inches
57. ≈ $41,000   **59.** ≈ 6.23 acres   **61.** Triangle *DEF* has an incorrect dimension.   **63. a.** 64 square units   **b.** 55 square units
65. ≈ 12.5°   **67.** ≈ 52.0 centimeters   **73.** ≈ 140 cubic inches

Prepare for Section 7.3, page 648

75. 1   **76.** −6.691   **77.** 30°   **78.** 157.6°   **79.** $\dfrac{\sqrt{5}}{5}$ **80.** $\dfrac{14\sqrt{17}}{17}$

Exercise Set 7.3, page 661

1. $a = 7, b = -1; \langle 7, -1 \rangle$ **3.** $a = -7, b = -5; \langle -7, -5 \rangle$ **5.** $a = 0, b = 8; \langle 0, 8 \rangle$ **7.** $5, ≈ 126.9°, \left\langle -\dfrac{3}{5}, \dfrac{4}{5} \right\rangle$

9. $≈ 44.7, ≈ 296.6°, \left\langle \dfrac{\sqrt{5}}{5}, \dfrac{-2\sqrt{5}}{5} \right\rangle$ **11.** $≈ 4.5, ≈ 296.6°, \left\langle \dfrac{\sqrt{5}}{5}, \dfrac{-2\sqrt{5}}{5} \right\rangle$ **13.** $≈ 45.7, ≈ 336.8°, \left\langle \dfrac{7\sqrt{58}}{58}, \dfrac{-3\sqrt{58}}{58} \right\rangle$ **15.** $\langle -6, 12 \rangle$

17. $\langle -1, 10 \rangle$ **19.** $\left\langle -\dfrac{11}{6}, \dfrac{7}{3} \right\rangle$ **21.** $2\sqrt{5}$ **23.** $2\sqrt{109}$ **25.** $-8i + 12j$ **27.** $14i - 6j$ **29.** $\dfrac{11}{12}i + \dfrac{1}{2}j$ **31.** $\sqrt{113}$

33. $a_1 ≈ 4.5, a_2 ≈ 2.3, 4.5i + 2.3j$ **35.** $a_1 ≈ 2.8, a_2 ≈ 2.8, 2.8i + 2.8j$ **37.** ≈ 380 miles per hour
39. ≈ 250 miles per hour at a heading of 86°   **41.** ≈ 293 pounds   **43.** ≈ 24.7 pounds   **45.** −3   **47.** 0   **49.** 1   **51.** 0   **53.** ≈ 79.7°

55. 45°   **57.** 90°, orthogonal   **59.** 180°   **61.** $\dfrac{46}{5}$ **63.** $\dfrac{14\sqrt{29}}{29} ≈ 2.6$ **65.** $\sqrt{5} ≈ 2.2$ **67.** $-\dfrac{11\sqrt{5}}{5} ≈ -4.9$ **69.** ≈ 954 foot-pounds

71. ≈ 779 foot-pounds   **73.**

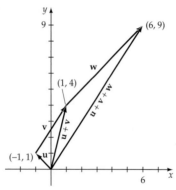

$\langle 6, 9 \rangle$ **75.** The vector from $P_1(3, -1)$ to $P_2(5, -4)$ is equivalent to $2i - 3j$.

77. $\mathbf{v} \cdot \mathbf{w} = 0$; the vectors are perpendicular.   **79.** $\langle 7, 2 \rangle$ is one example.   **81.** 6.4 pounds   **83.** No
87. The same amount of work is done.

Prepare for Section 7.4, page 663

88. $1 + 3i$ **89.** $\dfrac{1}{2} + \dfrac{1}{2}i$ **90.** $2 - 3i$ **91.** $3 + 5i$ **92.** $-\dfrac{1}{2} \pm \dfrac{\sqrt{3}}{2}i$ **93.** $-3i, 3i$

Exercise Set 7.4, page 669

1–7.   **9.** $\sqrt{2}\operatorname{cis} 315°$ **11.** $2\operatorname{cis} 330°$ **13.** $3\operatorname{cis} 90°$ **15.** $5\operatorname{cis} 180°$ **17.** $\sqrt{2} + i\sqrt{2}$ **19.** $\dfrac{\sqrt{2}}{2} - \dfrac{\sqrt{2}}{2}i$

Im ↑
Re →
−2 − 2i •
• $\sqrt{3} - i$
−2i
• 3 − 5i

$|-2 - 2i| = 2\sqrt{2}$
$|\sqrt{3} - i| = 2$
$|-2i| = 2$
$|3 - 5i| = \sqrt{34}$

21. $-3\sqrt{2} + 3i\sqrt{2}$ **23.** 8   **25.** $-\sqrt{3} + i$ **27.** $-3i$ **29.** $-4\sqrt{2} + 4i\sqrt{2}$ **31.** $\dfrac{9\sqrt{3}}{2} - \dfrac{9}{2}i$ **33.** $≈ -0.832 + 1.819i$ **35.** $6\operatorname{cis} 255°$

Prepare for Section 8.3, page 708

83. midpoint: $(1, -1)$; length: $2\sqrt{13}$ **84.** $-4, 2$ **85.** $\sqrt{2}$ **86.** $4(x^2 + 6x + 9) = 4(x + 3)^2$ **87.** $y = \pm\dfrac{3}{2}\sqrt{x^2 - 4}$

88.

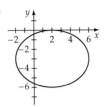

Exercise Set 8.3, page 719

1. center: $(0, 0)$

vertices: $(\pm 4, 0)$

foci: $\left(\pm\sqrt{41}, 0\right)$

asymptotes: $y = \pm\dfrac{5}{4}x$

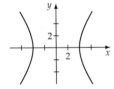

3. center: $(0, 0)$

vertices: $(0, \pm 2)$

foci: $\left(0, \pm\sqrt{29}\right)$

asymptotes: $y = \pm\dfrac{2}{5}x$

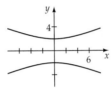

5. center: $(0, 0)$

vertices: $\left(\pm\sqrt{7}, 0\right)$

foci: $(\pm 4, 0)$

asymptotes: $y = \pm\dfrac{3\sqrt{7}}{7}x$

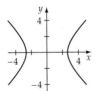

7. center: $(0, 0)$

vertices: $\left(\pm\dfrac{3}{2}, 0\right)$

foci: $\left(\pm\dfrac{\sqrt{73}}{2}, 0\right)$

asymptotes: $y = \pm\dfrac{8}{3}x$

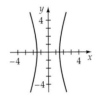

9. center: $(3, -4)$
vertices: $(7, -4), (-1, -4)$
foci: $(8, -4), (-2, -4)$

asymptotes: $y + 4 = \pm\dfrac{3}{4}(x - 3)$

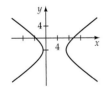

11. center: $(1, -2)$
vertices: $(1, 0), (1, -4)$
foci: $\left(1, -2 \pm 2\sqrt{5}\right)$

asymptotes: $y + 2 = \pm\dfrac{1}{2}(x - 1)$

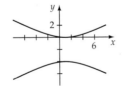

13. center: $(-2, 0)$
vertices: $(1, 0), (-5, 0)$
foci: $\left(-2 \pm \sqrt{34}, 0\right)$

asymptotes: $y = \pm\dfrac{5}{3}(x + 2)$

15. center: $(1, -1)$

vertices: $\left(\dfrac{7}{3}, -1\right), \left(-\dfrac{1}{3}, -1\right)$

foci: $\left(1 \pm \dfrac{\sqrt{97}}{3}, -1\right)$

asymptotes: $y + 1 = \pm\dfrac{9}{4}(x - 1)$

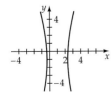

17. center: $(0, 0)$

vertices: $(\pm 3, 0)$

foci: $\left(\pm 3\sqrt{2}, 0\right)$

asymptotes: $y = \pm x$

19. center: $(0, 0)$

vertices: $(0, \pm 3)$

foci: $(0, \pm 5)$

asymptotes: $y = \pm\dfrac{3}{4}x$

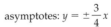

21. center: $(0, 0)$

vertices: $\left(0, \pm\dfrac{2}{3}\right)$

foci: $\left(0, \pm\dfrac{\sqrt{5}}{3}\right)$

asymptotes: $y = \pm 2x$

23. center: $(3, 4)$
vertices: $(3, 6)$, $(3, 2)$
foci: $\left(3, 4 \pm 2\sqrt{2}\right)$

asymptotes: $y - 4 = \pm(x - 3)$

25. center: $(-2, -1)$
vertices: $(-2, 2)$, $(-2, -4)$
foci: $\left(-2, -1 \pm \sqrt{13}\right)$

asymptotes: $y + 1 = \pm\dfrac{3}{2}(x + 2)$

27. $y = \dfrac{-6 \pm \sqrt{36 + 4(4x^2 + 32x + 39)}}{-2}$

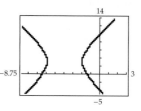

29. $y = \dfrac{64 \pm \sqrt{4096 + 64(9x^2 - 36x + 116)}}{-32}$

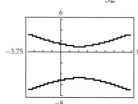

31. $y = \dfrac{18 \pm \sqrt{324 + 36(4x^2 + 8x - 6)}}{-18}$

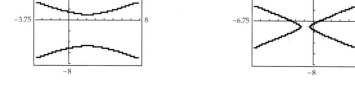

33. $\dfrac{x^2}{9} - \dfrac{y^2}{7} = 1$ **35.** $\dfrac{y^2}{20} - \dfrac{x^2}{5} = 1$

37. $\dfrac{y^2}{9} - \dfrac{x^2}{36/7} = 1$ **39.** $\dfrac{y^2}{16} - \dfrac{x^2}{64} = 1$ **41.** $\dfrac{(x - 4)^2}{4} - \dfrac{(y - 3)^2}{5} = 1$ **43.** $\dfrac{(x - 4)^2}{144/41} - \dfrac{(y + 2)^2}{225/41} = 1$ **45.** $\dfrac{(y - 2)^2}{3} - \dfrac{(x - 7)^2}{12} = 1$

47. $\dfrac{(y - 7)^2}{1} - \dfrac{(x - 1)^2}{3} = 1$ **49.** $\dfrac{x^2}{4} - \dfrac{y^2}{12} = 1$ **51.** $\dfrac{(x - 4)^2}{36/7} - \dfrac{(y - 1)^2}{4} = 1$ and $\dfrac{(y - 1)^2}{36/7} - \dfrac{(x - 4)^2}{4} = 1$ **53. a.** $\dfrac{x^2}{2162.25} - \dfrac{y^2}{13,462.75} = 1$

b. 221 miles   **55.** $y^2 - x^2 = 10{,}000^2$, hyperbola   **57.** ellipse   **59.** parabola   **61.** parabola

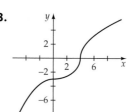

63. ellipse   **65.** $\dfrac{x^2}{1} - \dfrac{y^2}{3} = 1$ **67.** $\dfrac{y^2}{9} - \dfrac{x^2}{7} = 1$ **69.** $x = \pm\dfrac{16\sqrt{41}}{41}$ **73.**

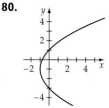

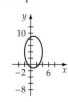

Prepare for Section 8.4, page 722

75. $\cos \alpha \cos \beta - \sin \alpha \sin \beta$ **76.** $\sin \alpha \cos \beta + \cos \alpha \sin \beta$ **77.** $\dfrac{\pi}{6}$ **78.** 150°   **79.** hyperbola

80.

Exercise Set 8.4, page 730

1. 45°   **3.** 36.9°   **5.** 73.5°   **7.** 45°, $\dfrac{x'^2}{8} - \dfrac{y'^2}{8} = 1$ **9.** 18.4°, $\dfrac{x'^2}{9} + \dfrac{y'^2}{3} = 1$ **11.** 26.6°, $\dfrac{x'^2}{1/2} - \dfrac{y'^2}{1/3} = 1$

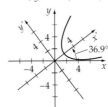   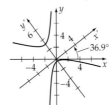   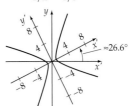

13. 30°, $y' = x'^2 + 4$ **15.** 36.9°, $y'^2 = 2(x' - 2)$ **17.** 36.9°, $15x'^2 - 10y'^2 + 6x' + 28y' + 11 = 0$

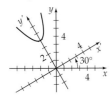   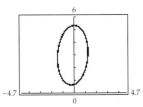   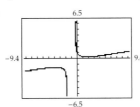

19.   **21.** 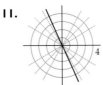   **23.**

25. $y = \dfrac{\sqrt{3} + 2\sqrt{2}}{2\sqrt{3} - \sqrt{2}}x$ and $y = \dfrac{\sqrt{3} - 2\sqrt{2}}{2\sqrt{3} + \sqrt{2}}x$ **27.** $\left(\dfrac{3\sqrt{15}}{5}, \dfrac{\sqrt{15}}{5}\right)$ and $\left(-\dfrac{3\sqrt{15}}{5}, -\dfrac{\sqrt{15}}{5}\right)$ **29.** hyperbola   **31.** parabola   **33.** parabola

35. hyperbola   **39.** $9x^2 - 4xy + 6y^2 = 100$

Prepare for Section 8.5, page 731

45. odd   **46.** even   **47.** $\dfrac{2\pi}{3}, \dfrac{5\pi}{3}$ **48.** 240°   **49.** r^2 **50.** (4.2, 2.6)

Exercise Set 8.5, page 743

1–7.   **9.**   **11.** 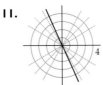   **13.**   **15.**

$0 \le \theta \le 2\pi$ $0 \le \theta \le \pi$ $0 \le \theta \le 2\pi$

17.   **19.**   **21.**   **23.**   **25.** 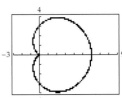 $0 \le \theta \le 2\pi$

$0 \le \theta \le \pi$ $0 \le \theta \le 2\pi$ $0 \le \theta \le 2\pi$ $0 \le \theta \le 2\pi$

27. 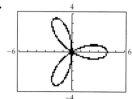 $0 \le \theta \le \pi$ **29.** 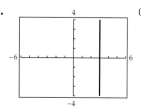 $0 \le \theta \le \pi$ **31.** 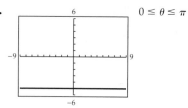 $0 \le \theta \le \pi$

33. 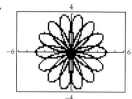 $0 \le \theta \le 4\pi$ **35.** 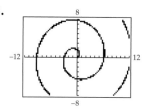 $0 \le \theta \le 6\pi$ **37.** 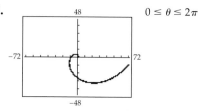 $0 \le \theta \le 2\pi$

39. 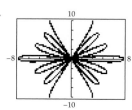 $0 \le \theta \le 2\pi$ **41.** $(2, -60°)$ **43.** $\left(\dfrac{3}{2}, -\dfrac{3\sqrt{3}}{2}\right)$ **45.** $(0, 0)$ **47.** $(5, 53.1°)$ **49.** $x^2 + y^2 - 3x = 0$

51. $x = 3$ **53.** $x^2 + y^2 = 16$ **55.** $x^4 - y^2 + x^2 y^2 = 0$ **57.** $y^2 + 4x - 4 = 0$ **59.** $y = 2x + 6$ **61.** $r = 2 \csc \theta$ **63.** $r = 2$

65. $r \cos^2 \theta = 8 \sin \theta$ **67.** $r^2 (\cos 2\theta) = 25$ **69.** 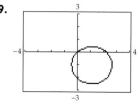   **71.**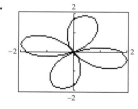

73. 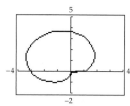   **75.** 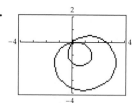   **79.** 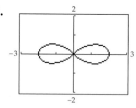 $0 \le \theta \le 4\pi$ **81.** 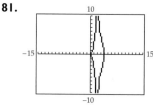 $0 \le \theta \le 2\pi$

83. 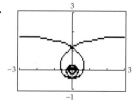 $-4\pi \le \theta \le 4\pi$ **85.** 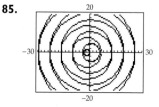 $-30 \le \theta \le 30$ **87. a.** 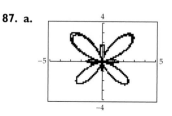 $0 \le \theta \le 5\pi$ **b.** 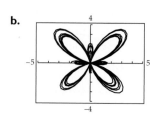 $0 \le \theta \le 20\pi$

Chapter 8 Test, page 764

1. focus: $(0, 2)$
vertex: $(0, 0)$
directrix: $y = -2$ [8.1]

2.   [8.2]

3. vertices: $(3, 4), (3, -6)$
foci: $(3, 3), (3, -5)$ [8.2]

4. $\dfrac{x^2}{45} + \dfrac{(y + 3)^2}{9} = 1$ [8.2]

5. 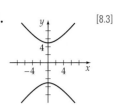 [8.3]

6. vertices: $(6, 0), (-6, 0)$
foci: $(-10, 0), (10, 0)$
asymptotes: $y = \pm\dfrac{4x}{3}$ [8.3]

7. 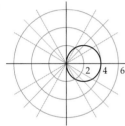 [8.3]

8. $73.15°$ [8.4]

9. ellipse [8.4]

10. $P(2, 300°)$ [8.5]

11.   [8.5]

12. [8.5]

13. [8.5]

14. $\left(\dfrac{5}{2}, \dfrac{5\sqrt{3}}{2}\right)$ [8.5]

15. $y^2 - 8x - 16 = 0$ [8.5]

16. $x^2 + 8y - 16 = 0$ [8.6]

17. $(x + 3)^2 = \dfrac{1}{2}y$

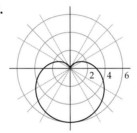 [8.7]

18. $\dfrac{x^2}{16} + \dfrac{(y - 2)^2}{1} = 1$

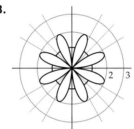 [8.7]

19. 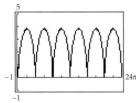 [8.7]

$\text{Xscl} = 2\pi$

20. $256\sqrt{3}$ feet ≈ 443 feet [8.7]

Cumulative Review Exercises, page 765

1. $\pm 2, \pm i\sqrt{2}$ [1.4]

2. $\dfrac{-x + 7}{(x - 1)(x + 2)}$ [P.5]

3. $-4 - h$ [2.6]

4. -19 [2.6]

5. 6 [3.4]

6. $y = -\dfrac{3}{2}x - \dfrac{5}{2}$ [2.3]

7. $x = -3, y = 2$ [3.5]

8. $\sqrt{29}$ [2.1]

9. 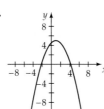   [4.2]   **10.** 1 [4.5]   **11.**   [2.5]   **12.** $f^{-1}(x) = \frac{1}{2}x + 4$ [4.1]   **13.** $-4, -2i$ [3.4]

14. odd [2.5]   **15.** $\frac{2\pi}{3}$ [5.1]   **16.** 12.6 centimeters [5.2]   **17.** period: $\frac{\pi}{2}$; amplitude: 3 [5.5]   **19.** $\frac{\pi}{6}, \frac{\pi}{2}, \frac{5\pi}{6}, \frac{3\pi}{2}$ [6.6]   **20.** 8 [7.3]

Exercise Set 9.1, page 777

1. $(2, -4)$ **3.** $\left(-\frac{6}{5}, \frac{27}{5}\right)$ **5.** $(3, 4)$ **7.** $(1, -1)$ **9.** $(3, -4)$ **11.** $(2, 5)$ **13.** $(-1, -1)$ **15.** $\left(\frac{62}{25}, \frac{34}{25}\right)$ **17.** no solution

19. $\left(c, -\frac{4}{3}c + 2\right)$ **21.** $(2, -4)$ **23.** $(0, 3)$ **25.** $\left(\frac{3}{5}c, c\right)$ **27.** $\left(-\frac{1}{2}, \frac{2}{3}\right)$ **29.** no solution   **31.** $(-6, 3)$ **33.** $\left(2, -\frac{3}{2}\right)$

35. $(2\sqrt{3}, 3)$ **37.** $\left(\frac{38}{17\pi}, \frac{3}{17}\right)$ **39.** $(\sqrt{2}, \sqrt{3})$ **41.** \$125   **43.** plane: 120 mph, wind: 30 mph

45. boat: 25 mph, current: 5 mph   **47.** \$12 per kilogram for iron, \$16 per kilogram for lead

49. 8 grams of 40% gold, 12 grams of 60% gold   **51.** $\frac{9}{5}$ square units   **53.** 8   **55.** 42, 56, 70; 40, 42, 58; 42, 144, 150; 42, 440, 442

57. 90 people   **59.** (63,800, 12,760) The point indicates that a person with an adjusted gross income of \$63,800 would pay the same tax using either method.   **61.** \$14,000 at 6%, \$11,000 at 6.5%   **63.** $x = -\frac{58}{17}, y = \frac{52}{17}$ **65.** $x = -2, y = -1$ **67.** $x = \frac{153}{26}, y = \frac{151}{26}$

69. $x = 2 + 3i, y = 1 - 2i$ **71.** $x = 3 - 5i, y = 4i$

Prepare for Section 9.2, page 780

72. $y = \frac{2}{5}x - 3$ **73.** $z = -c + 13$ **74.** $\left(\frac{18}{5}, 4\right)$ **75.** $(2, -5)$ **76.** $(-2, -10)$ **77.** $(c, -4c + 9)$

Exercise Set 9.2, page 791

1. $(2, -1, 3)$ **3.** $(2, 0, -3)$ **5.** $(2, -3, 1)$ **7.** $(-5, 1, -1)$ **9.** $(3, -5, 0)$ **11.** $(0, 2, 3)$ **13.** $(5c - 25, 48 - 9c, c)$ **15.** $(3, -1, 0)$

17. no solution   **19.** $\left(\frac{1}{11}(50 - 11c), \frac{1}{11}(11c - 18), c\right)$ **21.** no solution   **23.** $\left(\frac{1}{29}(25 + 4c), \frac{1}{29}(55 - 26c), c\right)$ **25.** $(0, 0, 0)$

27. $\left(\frac{5}{14}c, \frac{4}{7}c, c\right)$ **29.** $(-11c, -6c, c)$ **31.** $(0, 0, 0)$ **33.** $y = 2x^2 - x - 3$ **35.** $x^2 + y^2 - 4x + 2y - 20 = 0$

37. center $(-7, -2)$, radius 13   **39.** 500 cars per hour   **41.** AC: 258 to 308, CD: 209 to 259, BD: 262 to 312   **43.** $d_1 = 9$ inches, $d_2 = 3$ inches, $d_3 = 4$ inches   **45.** $(3, 5, 2, -3)$ **47.** $(1, -2, -1, 3)$ **49.** $(14a - 7b - 8, -6a + 2b + 5, a, b)$ **51.** $A = -\frac{13}{2}$

53. $A \neq -3, A \neq 1$ **55.** $A = -3$ **57.** $3x - 5y - 2z = -2$

Prepare for Section 9.3, page 794

58. $-1 \pm \sqrt{3}$ **59.** $(1, -3)$ **60.** parabola   **61.** hyperbola   **62.** 2   **63.** 4

Exercise Set 9.3, page 800

1. $(1, 0), (2, 2)$ **3.** $\left(\frac{2 + \sqrt{2}}{2}, \frac{-6 + \sqrt{2}}{2}\right), \left(\frac{2 - \sqrt{2}}{2}, \frac{-6 - \sqrt{2}}{2}\right)$ **5.** $(5, 18)$ **7.** $(4, 6), (6, 4)$ **9.** $\left(-\frac{3}{2}, -4\right), (2, 3)$

11. $\left(\frac{19}{29}, -\frac{11}{29}\right), (1, 1)$ **13.** $(-2, 9), (1, -3), (-1, 1)$ **15.** $(-2, 1), (-2, -1), (2, 1), (2, -1)$ **17.** $(4, 2), (-4, 2), (4, -2), (-4, -2)$

19. no real number solution   **21.** $\left(\frac{12}{5}, \frac{1}{5}\right), (2, 1)$ **23.** $\left(\frac{26}{5}, -\frac{3}{5}\right), (1, -2)$ **25.** $\left(\frac{39}{10}, -\frac{7}{10}\right), (3, 2)$

27. $\left(\frac{-3 + \sqrt{3}}{2}, \frac{1 + \sqrt{3}}{2}\right), \left(\frac{-3 - \sqrt{3}}{2}, \frac{1 - \sqrt{3}}{2}\right)$ **29.** $\left(\frac{19}{13}, \frac{22}{13}\right), (1, 4)$ **31.** no real number solution   **33.** 82 units

35. $r \geq \sqrt{\frac{1}{5}}$ or $\frac{\sqrt{5}}{5}$ **37.** \$45   **39.** $(0, 1), (1, 2)$ **41.** $(0.7035, 0.4949)$ **43.** $(1.7549, 1.3247)$ **45.** $(-0.7071, 0.7071), (0.7071, 0.7071)$

47. $(1, 5)$ **49.** $(-1, 1), (1, -1)$ **51.** $(1, -2), (-1, 2)$

Prepare for Section 9.4, page 801

54. $(x^2 + 7)^2$ **55.** $\dfrac{6x + 9}{(x - 1)(x + 2)}$ **56.** $\dfrac{x^2 + 2x + 7}{x(x - 1)^2}$ **57.** $(-1, 2)$ **58.** $(2, -2, -1)$ **59.** $x + 3 + \dfrac{2x - 35}{x^2 - 7x}$

Exercise Set 9.4, page 808

1. $A = -3, B = 4$ **3.** $A = -\dfrac{2}{5}, B = \dfrac{1}{5}$ **5.** $A = 1, B = -1, C = 4$ **7.** $A = 1, B = 3, C = 2$ **9.** $A = 1, B = 0, C = 1, D = 0$

11. $\dfrac{3}{x} + \dfrac{5}{x + 4}$ **13.** $\dfrac{7}{x - 9} + \dfrac{-4}{x + 2}$ **15.** $\dfrac{5}{2x + 3} + \dfrac{3}{2x + 5}$ **17.** $\dfrac{20}{11(3x + 5)} + \dfrac{-3}{11(x - 2)}$ **19.** $x + 3 + \dfrac{1}{x - 2} + \dfrac{-1}{x + 2}$

21. $\dfrac{1}{x} + \dfrac{2}{x + 7} + \dfrac{-28}{(x + 7)^2}$ **23.** $\dfrac{2}{x} + \dfrac{3x - 1}{x^2 - 3x + 1}$ **25.** $\dfrac{2}{x + 3} + \dfrac{-1}{(x + 3)^2} + \dfrac{4}{x^2 + 1}$ **27.** $\dfrac{3}{x - 4} + \dfrac{5}{(x - 4)^2}$ **29.** $\dfrac{3x - 1}{x^2 + 10} + \dfrac{4x}{(x^2 + 10)^2}$

31. $\dfrac{1}{2k(k - x)} + \dfrac{1}{2k(k + x)}$ **33.** $x + \dfrac{1}{x} + \dfrac{-2}{x - 1}$ **35.** $2x - 2 + \dfrac{3}{x^2 - x - 1}$ **37.** $\dfrac{1}{5(x + 2)} + \dfrac{4}{5(x - 3)}$ **39.** $\dfrac{1}{x} + \dfrac{2}{x^2} + \dfrac{3}{x^4} + \dfrac{-2}{x - 2}$

41. $\dfrac{4}{3(x - 1)} + \dfrac{2x + 7}{3(x^2 + x + 1)}$

Prepare for Section 9.5, page 809

45. 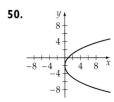 **46.**   **47.** 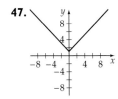 **48.**   **49.**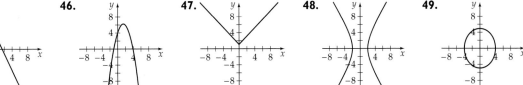

50.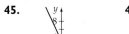

Exercise Set 9.5, page 815

1. 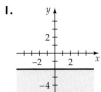 **3.**   **5.** 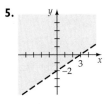 **7.** 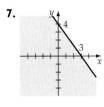 **9.**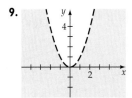

11. 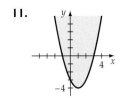 **13.** 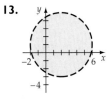 **15.** 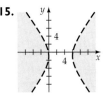 **17.** 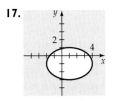 **19.**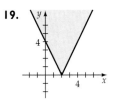

21. 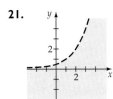 **23.** 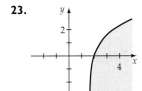 **25.** 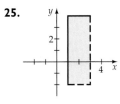 **27.** 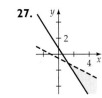 **29.** no solution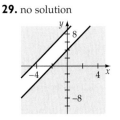

31. 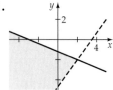   **33.**   **35.** 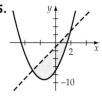   **37.**   **39.**

41. 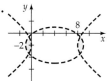   **43.** 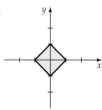   **45.**   **47.**   **49.**

51. 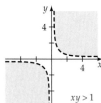   **53.** 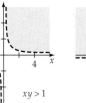   **55.**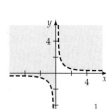

57.   If x is a negative number, then the inequality is reversed when both sides of the inequality are divided by a negative number.

$xy > 1$ $y > \dfrac{1}{x}$

Prepare for Section 9.6, page 816

59. 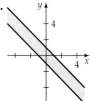   **60.**   **61.**   **62.**   **63.** $(1, 3)$ **64.** $(1, 6)$

Exercise Set 9.6, page 823

1. minimum at $(0, 8)$: 16   **3.** maximum at $(6, 5)$: 71   **5.** minimum at $\left(0, \dfrac{10}{3}\right)$: 20   **7.** maximum at $(0, 12)$: 72

9. minimum at $(0, 32)$: 32   **11.** maximum at $(0, 8)$: 56   **13.** minimum at $(2, 6)$: 18   **15.** maximum at $(3, 4)$: 25
17. minimum at $(2, 3)$: 12   **19.** maximum at $(100, 400)$: 3400   **21.** 20 acres of wheat and 40 acres of barley
23. 0 starter sets and 18 pro sets   **25.** 24 ounces of group B and 0 ounces of group A yields a minimum cost of $2.40.
27. Two 4-cylinder engines and seven 6-cylinder engines yields a maximum profit of $2050.

Chapter 9 True/False Exercises, page 827

1. False; $\begin{cases} x + y = 1 \\ x + y = 3 \end{cases}$ has no solution.   **2.** True   **3.** False; a homogeneous system is one in which the constant term in each equation is zero.

4. True   **5.** True   **6.** False; $\begin{cases} x + y = 2 \\ x + 2y = 3 \end{cases}$ and $\begin{cases} 2x + 3y = 5 \\ 2x - 2y = 0 \end{cases}$ are two systems with the same solution but without common equations.   **7.** True

8. True   **9.** False; it is inconsistent.   **10.** False; $(-1, 1)$ satisfies the first inequality but not the second, and $(-2, -1)$ satisfies the second but not the first.

Chapter 9 Review Exercises, page 827

1. $\left(-\dfrac{18}{7}, -\dfrac{15}{28}\right)$ [9.1]   **2.** $\left(\dfrac{3}{2}, -3\right)$ [9.1]   **3.** $(-3, -1)$ [9.1]   **4.** $(-4, 7)$ [9.1]   **5.** $(3, 1)$ [9.1]   **6.** $(-1, 1)$ [9.1]   **7.** $\left(\dfrac{1}{2}(5 - 3c), c\right)$ [9.1]

8. no solution [9.1]   **9.** $\left(\dfrac{1}{2}, 3, -1\right)$ [9.2]   **10.** $\left(\dfrac{16}{3}, \dfrac{10}{27}, -\dfrac{29}{45}\right)$ [9.2]   **11.** $\left(\dfrac{1}{11}(7c - 3), \dfrac{1}{11}(16c - 43), c\right)$ [9.2]   **12.** $\left(\dfrac{74}{31}, -\dfrac{1}{31}, \dfrac{3}{31}\right)$ [9.2]

13. $\left(2, \dfrac{1}{2}(3c + 2), c\right)$ [9.2]   **14.** $\left(1, -\dfrac{2}{3}, \dfrac{1}{4}\right)$ [9.2]   **15.** $\left(\dfrac{14}{11}c, -\dfrac{2}{11}c, c\right)$ [9.2]   **16.** $(0, 0, 0)$ [9.2]   **17.** $\left(\dfrac{1}{2}(c + 1), \dfrac{1}{4}(3c - 1), c\right)$ [9.2]

18. $\left(\dfrac{65 - 11c}{16}, \dfrac{19 - c}{8}, c\right)$ [9.2]   **19.** $(2, -3)$ [9.3]   **20.** $\left(-\dfrac{1}{2}, 0\right)$ and $(1, 3)$ [9.3]   **21.** no real solution [9.3]   **22.** $\left(\dfrac{3}{2}, 3\right)$ and $(1, -1)$ [9.3]

23. $\left(\dfrac{1}{5}, \dfrac{18}{5}\right)$, $(1, 2)$ [9.3]   **24.** $(0, -3)$ and $(2, 1)$ [9.3]   **25.** $(2, 0)$, $\left(\dfrac{18}{17}, -\dfrac{64}{17}\right)$ [9.3]   **26.** $(0, 2)$ and $\left(-\dfrac{32}{25}, \dfrac{26}{25}\right)$ [9.3]

27. $(2, 1)$, $(-2, -1)$ [9.3]   **28.** $(1, 3)$, $(-1, -3)$, $\left(\dfrac{\sqrt{6}}{6}, -\dfrac{\sqrt{6}}{3}\right)$, and $\left(-\dfrac{\sqrt{6}}{6}, \dfrac{\sqrt{6}}{3}\right)$ [9.3]   **29.** $(2, -3)$, $(-2, 3)$ [9.3]   **30.** $\left(\dfrac{\sqrt{15}}{15}, \dfrac{\sqrt{15}}{15}\right)$,

$\left(-\dfrac{\sqrt{15}}{15}, -\dfrac{\sqrt{15}}{15}\right)$, $(1, -1)$, $(-1, 1)$ [9.3]   **31.** $\dfrac{3}{x - 2} + \dfrac{4}{x + 1}$ [9.4]   **32.** $\dfrac{1}{x - 1} + \dfrac{2}{(x - 1)^2}$ [9.4]   **33.** $\dfrac{6x - 2}{5(x^2 + 1)} + \dfrac{-6}{5(x + 2)}$ [9.4]

34. $\dfrac{7}{3(x - 2)} + \dfrac{3}{(x - 2)^2} + \dfrac{8}{3(x + 1)}$ [9.4]   **35.** $\dfrac{2}{x} + \dfrac{4}{x - 1} + \dfrac{5}{x + 1}$ [9.4]   **36.** $1 + \dfrac{x + 2}{x^2 + 1}$ [9.4]   **37.** [9.5]

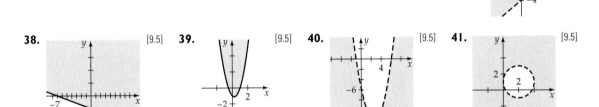

38. [9.5]   **39.** [9.5]   **40.** [9.5]   **41.** [9.5]

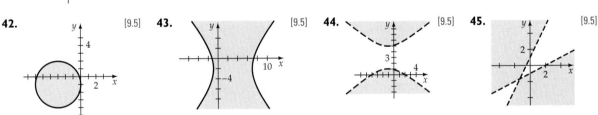

42. [9.5]   **43.** [9.5]   **44.** [9.5]   **45.** [9.5]

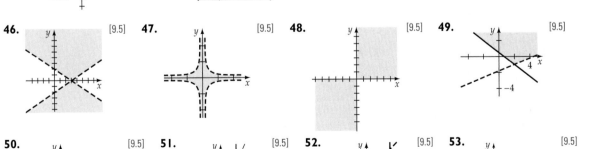

46. [9.5]   **47.** [9.5]   **48.** [9.5]   **49.** [9.5]

50. [9.5]   **51.** [9.5]   **52.** [9.5]   **53.** [9.5]

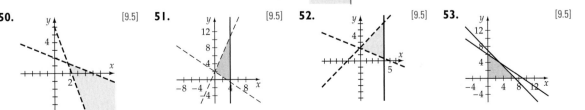

54. 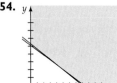 [9.5]   **55.** 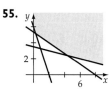 [9.5]   **56.** [9.5]   **57.** 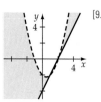 [9.5]

58. [9.5]   **59.** 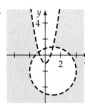 [9.5]   **60.** [9.5]

61. The maximum is 18 at $(4, 5)$. [9.6]   **62.** The maximum is 44 at $(6, 4)$. [9.6]   **63.** The minimum is 8 at $(0, 8)$. [9.6]   **64.** The minimum is 20 at $(10, 0)$. [9.6]   **65.** The minimum is 27 at $(2, 5)$. [9.6]   **66.** The maximum is 43 at $(3, 7)$. [9.6]   **67.** $y = \dfrac{11}{6}x^2 - \dfrac{5}{2}x + \dfrac{2}{3}$ [9.2]

68. $x^2 + y^2 - \dfrac{47}{11}x - \dfrac{21}{11}y + \dfrac{10}{11} = 0$ [9.2]   **69.** $z = -2x + 3y + 3$ [9.2]   **70.** $x = 15$ liters [9.1]   **71.** wind: 28 mph, plane: 143 mph [9.1]
72. 4 nickels, 3 dimes, 3 quarters; 1 nickel, 7 dimes, 2 quarters [9.2]   **73.** $(0, 0, 0), (1, 1, 1), (1, -1, -1), (-1, -1, 1), (-1, 1, -1)$ [9.2]

Chapter 9 Test, page 829

1. $(-3, 2)$ [9.1]   **2.** $\left(\dfrac{1}{2}(6 + c), c \right)$ [9.1]   **3.** $\left(\dfrac{173}{39}, \dfrac{29}{39}, -\dfrac{4}{3} \right)$ [9.2]   **4.** $\left(\dfrac{1}{4}(c + 3), \dfrac{1}{8}(7c + 1), c \right)$ [9.2]   **5.** $\left(\dfrac{1}{13}(c + 10), \dfrac{1}{13}(5c + 11), c \right)$ [9.2]

6. $\left(\dfrac{1}{14}c, -\dfrac{9}{14}c, c \right)$ [9.2]   **7.** $(2, 5), (-2, 1)$ [9.3]   **8.** $(-2, 3), (-1, -1)$ [9.3]   **9.** 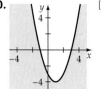 [9.5]   **10.** [9.5]

11. [9.5]   **12.** 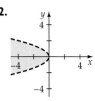 [9.5]   **13.** [9.5]   **14.** No graph; the solution set is the empty set. [9.5]

15. [9.5]   **16.** 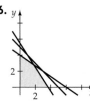 [9.5]   **17.** $\dfrac{7}{5(x - 4)} + \dfrac{8}{5(x + 1)}$ [9.4]   **18.** $\dfrac{1}{x} + \dfrac{-x + 2}{x^2 + 1}$ [9.4]

19. $\dfrac{680}{7}$ acres of oats and $\dfrac{400}{7}$ acres of barley [9.6]   **20.** $x^2 + y^2 - 2y - 24 = 0$ [9.2]

Cumulative Review Exercises, page 830

1. $\{y \mid y \le -3\}$ [2.4]   **2.** $\log_6[8x^3(x - 5)]$ [4.4]   **3.** $(x - 4)^2 = -25(y - 2)$ [8.1]   **4.** even [2.5]   **5.** $\dfrac{300}{199}$ [4.5]   **6.** $\dfrac{(x - 6)^2}{16} - \dfrac{(y - 2)^2}{128} = 1$ [8.3]

7. 5 [2.2]   **8.** 30 [2.6]   **9.** $x^2 + 0.4x - 0.8$ [2.7]   **10.** $x^3 + 2x^2 + 9x + 18$ [3.4]   **11.** $Q^{-1}(r) = \dfrac{r-2}{r}$ [4.1]   **12.** $y = 2x + 1$ [3.5]   **13.** 81 [4.2]

14.   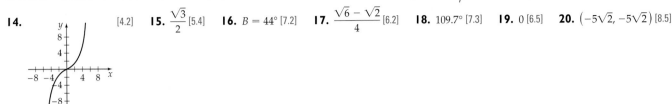   [4.2]   **15.** $\dfrac{\sqrt{3}}{2}$ [5.4]   **16.** $B = 44°$ [7.2]   **17.** $\dfrac{\sqrt{6} - \sqrt{2}}{4}$ [6.2]   **18.** $109.7°$ [7.3]   **19.** 0 [6.5]   **20.** $\left(-5\sqrt{2},\, -5\sqrt{2}\right)$ [8.5]

Exercise Set 10.1, page 842

1. $\begin{bmatrix} 2 & -3 & 1 & 1 \\ 3 & -2 & 3 & 0 \\ 1 & 0 & 5 & 4 \end{bmatrix}, \begin{bmatrix} 2 & -3 & 1 \\ 3 & -2 & 3 \\ 1 & 0 & 5 \end{bmatrix}, \begin{bmatrix} 1 \\ 0 \\ 4 \end{bmatrix}$ **3.** $\begin{bmatrix} 2 & -3 & -4 & 1 & 2 \\ 0 & 2 & 1 & 0 & 2 \\ 1 & -1 & 2 & 0 & 4 \\ 3 & -3 & -2 & 0 & 1 \end{bmatrix}, \begin{bmatrix} 2 & -3 & -4 & 1 \\ 0 & 2 & 1 & 0 \\ 1 & -1 & 2 & 0 \\ 3 & -3 & -2 & 0 \end{bmatrix}, \begin{bmatrix} 2 \\ 2 \\ 4 \\ 1 \end{bmatrix}$ **5.** $\begin{bmatrix} 1 & -1 & 2 & 2 \\ 0 & 1 & -1 & -6 \\ 0 & 0 & 1 & -\frac{27}{2} \end{bmatrix}$

7. $\begin{bmatrix} 1 & -2 & -1 & 3 \\ 0 & 1 & 2 & -\frac{11}{2} \\ 0 & 0 & 1 & -\frac{13}{6} \end{bmatrix}$ **9.** $\begin{bmatrix} 1 & -2 & 3 & -4 \\ 0 & 1 & 2 & -\frac{1}{2} \\ 0 & 0 & 1 & -2 \\ 0 & 0 & 0 & 0 \end{bmatrix}$ **11.** $\begin{bmatrix} 1 & -3 & 4 & 2 & 1 \\ 0 & 1 & -1 & -2 & -1 \\ 0 & 0 & 0 & 1 & 3 \end{bmatrix}$ **13.** $(2, -1, 1)$ **15.** $(1, -2, -1)$

17. $\left(2 - 2c,\, 2c + \dfrac{1}{2},\, c\right)$ **19.** $\left(\dfrac{1}{2}, \dfrac{1}{2}, \dfrac{3}{2}\right)$ **21.** $(16c, 6c, c)$ **23.** $(7c + 6, -11c - 8, c)$ **25.** $(c + 2, c, c)$ **27.** no solution

29. $(2, -2, 3, 4)$ **31.** $\left(\dfrac{21}{10}, -\dfrac{8}{5}, \dfrac{2}{5}, -\dfrac{5}{2}\right)$ **33.** $\left(3, -\dfrac{3}{2}, 1, -1\right)$ **35.** $\left(\dfrac{27}{2}c + 39,\, \dfrac{5}{2}c + 10,\, -4c - 10,\, c\right)$

37. $\left(c_1 - \dfrac{12}{7}c_2 + \dfrac{6}{7},\, c_1 - \dfrac{9}{7}c_2 + \dfrac{1}{7},\, c_1, c_2\right)$ **39.** $p(x) = 2x - 3$ **41.** $p(x) = x^2 - 2x + 3$ **43.** $p(x) = x^3 - 2x^2 - x + 2$

45. $p(x) = 2x + 5$ **47.** $z = 2x + 3y - 2$ **49.** $x^2 + y^2 + 2x - 4y - 20 = 0$ **51.** $(1, 0, -2, 1, 2)$

53. $\left(\dfrac{77c + 151}{3},\, \dfrac{-25c - 50}{3},\, \dfrac{14c + 34}{3},\, -3c - 7, c\right)$ **55.** all values of a except $a = 1$ and $a = -6$ **57.** $a = -6$

Prepare for Section 10.2, page 845

58. The additive inverse of c is $-c$. The additive inverse of zero is zero.   **59.** 1 **60.** No   **61.** $a = 5, b = -1$ **62.** 3×1

63. $\begin{cases} 3x - 5y = 16 \\ 2x + 7y = -10 \end{cases}$

Exercise Set 10.2, page 859

1. a. $\begin{bmatrix} 1 & 2 \\ 5 & 4 \end{bmatrix}$ **b.** $\begin{bmatrix} 3 & -4 \\ 1 & 2 \end{bmatrix}$ **c.** $\begin{bmatrix} -2 & 6 \\ 4 & 2 \end{bmatrix}$ **d.** $\begin{bmatrix} 7 & -11 \\ 0 & 3 \end{bmatrix}$ **3. a.** $\begin{bmatrix} -3 & 0 & 5 \\ 3 & 5 & -5 \end{bmatrix}$ **b.** $\begin{bmatrix} 3 & -2 & 1 \\ -1 & -5 & 1 \end{bmatrix}$ **c.** $\begin{bmatrix} -6 & 2 & 4 \\ 4 & 10 & -6 \end{bmatrix}$

d. $\begin{bmatrix} 9 & -5 & 0 \\ -4 & -15 & 5 \end{bmatrix}$ **5. a.** $\begin{bmatrix} 1 & 5 \\ 3 & -5 \\ 2 & -4 \end{bmatrix}$ **b.** $\begin{bmatrix} -7 & 3 \\ 1 & -1 \\ -4 & 4 \end{bmatrix}$ **c.** $\begin{bmatrix} 8 & 2 \\ 2 & -4 \\ 6 & -8 \end{bmatrix}$ **d.** $\begin{bmatrix} -18 & 5 \\ 1 & 0 \\ -11 & 12 \end{bmatrix}$ **7. a.** $\begin{bmatrix} -1 & 1 & -1 \\ 2 & 2 & 1 \\ -1 & 2 & 5 \end{bmatrix}$ **b.** $\begin{bmatrix} -3 & 5 & -1 \\ -2 & -4 & 3 \\ -7 & 4 & 1 \end{bmatrix}$

c. $\begin{bmatrix} 2 & -4 & 0 \\ 4 & 6 & -2 \\ 6 & -2 & 4 \end{bmatrix}$ **d.** $\begin{bmatrix} -7 & 12 & -2 \\ -6 & -11 & 7 \\ -17 & 9 & 0 \end{bmatrix}$ **9.** $\begin{bmatrix} -10 & 17 \\ 6 & -8 \end{bmatrix}\begin{bmatrix} 0 & 22 \\ 1 & -18 \end{bmatrix}$ **11.** $\begin{bmatrix} 10 & 6 \\ 14 & -7 \end{bmatrix}\begin{bmatrix} 14 & -1 \\ 0 & -11 \end{bmatrix}$ **13.** $\begin{bmatrix} 0 & -4 & 5 \\ 6 & 0 & 3 \\ -3 & -2 & 1 \end{bmatrix}\begin{bmatrix} 5 & -13 \\ 5 & -4 \end{bmatrix}$

15. $\begin{bmatrix} 9 & -2 & -6 \\ 0 & -1 & 2 \\ 4 & -2 & -4 \end{bmatrix}\begin{bmatrix} 4 & -2 & 6 \\ 2 & -3 & 4 \\ 4 & -4 & 3 \end{bmatrix}$ **17.** $[0, 8]$ **19.** The product is not possible. **21.** $\begin{bmatrix} 0 & 0 \\ 0 & 0 \end{bmatrix}$ **23.** The product is not possible.

25. $\begin{bmatrix} \frac{1}{3} & -\frac{5}{3} \\ -\frac{1}{3} & \frac{4}{3} \\ \frac{1}{3} & -\frac{4}{3} \end{bmatrix}$ **27.** $\begin{bmatrix} -1 & 1 \\ 3 & 2 \\ 7 & -2 \end{bmatrix}$ **29.** $\begin{bmatrix} 1 & -3 \\ 1 & -2 \end{bmatrix}$ **31.** $\begin{bmatrix} 7 & -1 & 1 \\ 1 & 2 & 0 \\ 5 & -1 & 4 \end{bmatrix}$ **33.** $\begin{cases} 3x - 8y = 11 \\ 4x + 3y = 1 \end{cases}$ **35.** $\begin{cases} x - 3y - 2z = 6 \\ 3x + y = 2 \\ 2x - 4y + 5z = 1 \end{cases}$

37. $\begin{cases} 2x_1 - x_2 \quad\quad + 2x_4 = 5 \\ 4x_1 + x_2 + 2x_3 - 3x_4 = 6 \\ 6x_1 \quad\quad + x_3 - 2x_4 = 10 \\ 5x_1 + 2x_2 - x_3 - 4x_4 = 8 \end{cases}$ **39. a.** 3×4. Three different fish were caught in four different samples.   **b.** Fish A was caught in sample 4.

c. Fish B   **41.** $\begin{bmatrix} 1.96 & 1.37 & 2.94 & 1.37 \\ 0.78 & 1.08 & 1.96 & 0.88 \\ 3.53 & 1.18 & 4.41 & 1.47 \end{bmatrix}$ **43. a.** $\begin{bmatrix} 87 & 74 \\ 85 & 77 \\ 83 & 79 \end{bmatrix}$ **b.** The matrix represents the total number of wins and losses for each team.

c. $\begin{bmatrix} 1 & -2 \\ -1 & 1 \\ 7 & -7 \end{bmatrix}$ **d.** The matrix represents the difference between performance at home and performance away.   **45. a.** $\begin{bmatrix} 0 & 6 & 4 & -2 \\ 4 & 1 & -3 & 0 \end{bmatrix}$

b. 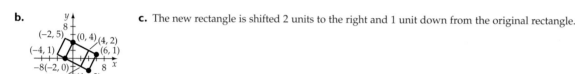   **c.** The new rectangle is shifted 2 units to the right and 1 unit down from the original rectangle.

47. a. $\begin{bmatrix} -5 & -2 & 2 & -1 \\ -2 & 4 & 2 & -4 \end{bmatrix}$ **b.**   **c.** The second rectangle is obtained by reflecting the first about $y = x$ and then reflecting the result about $x = 0$.

49. $A - B = \begin{bmatrix} 50 & 150 & 140 \\ 15 & 170 & 370 \\ 85 & 250 & 130 \\ 80 & 115 & 25 \end{bmatrix}$ $A - B$ is the number of each item sold during the week.   **51.** $\begin{bmatrix} 65 & 88.5 \\ 54 & 72 \\ 68.5 & 94.5 \end{bmatrix}; T_2$ **53. a.** 45.6%   **b.** 46.7%

55. a. 16.1%   **b.** 23.5%   **57.** 11 months   **59.** $\begin{bmatrix} 24 & 21 & -12 & 32 & 0 \\ -7 & -8 & 3 & 21 & 20 \\ 32 & 10 & -32 & 1 & 5 \\ 19 & -15 & -17 & 30 & 20 \\ 29 & 9 & -28 & 13 & -6 \end{bmatrix}$ **61.** $\begin{bmatrix} 46 & -100 & 36 & 273 & 93 \\ 82 & -93 & 19 & 27 & 97 \\ 73 & -10 & -23 & 109 & 83 \\ 212 & -189 & 52 & 37 & 156 \\ 68 & -22 & 54 & 221 & 58 \end{bmatrix}$

63. $\begin{bmatrix} 76 & -8 & -25 & 30 & 6 \\ 14 & 16 & -10 & 14 & 2 \\ 39 & 0 & -45 & 22 & 27 \\ 0 & -4 & 23 & 83 & -16 \\ 56 & -20 & -22 & 7 & 5 \end{bmatrix}$ **65.** $\begin{bmatrix} 6 + 9i & 3 - 6i \\ 3 + 3i & 6 - 3i \end{bmatrix}$ **67.** $\begin{bmatrix} 2 + 2i & -6 + 4i \\ -4 + 6i & 2 + 8i \end{bmatrix}$ **69.** $\begin{bmatrix} 3 + 2i & 3 + i \\ 4 + 3i & 6 - 2i \end{bmatrix}$ **71.** $\begin{bmatrix} 12 - 3i & -3 + 3i \\ 10 + i & 6 - i \end{bmatrix}$

73. $\begin{bmatrix} -2 + 11i & 8 - 6i \\ 2 + 6i & 6 - 5i \end{bmatrix}$

Prepare for Section 10.3, page 864

80. $-\dfrac{3}{2}$ **81.** $\begin{bmatrix} 1 & 0 & 0 \\ 0 & 1 & 0 \\ 0 & 0 & 1 \end{bmatrix}$ **82.** See Section 10.1.   **83.** $\begin{bmatrix} 1 & -2 & 3 \\ 0 & 3 & -2 \\ 0 & -4 & 11 \end{bmatrix}$ **84.** $X = A^{-1}B$ **85.** $\begin{cases} 2x + 3y = 9 \\ 4x - 5y = 7 \end{cases}$

Exercise Set 10.3, page 874

1. $\begin{bmatrix} -5 & -3 \\ -2 & -1 \end{bmatrix}$ **3.** $\begin{bmatrix} 5 & -2 \\ -1 & \frac{1}{2} \end{bmatrix}$ **5.** $\begin{bmatrix} -16 & -2 & 7 \\ 7 & 1 & -3 \\ -3 & 0 & 1 \end{bmatrix}$ **7.** $\begin{bmatrix} 15 & -1 & -4 \\ -\frac{11}{2} & \frac{1}{2} & \frac{3}{2} \\ 3 & 0 & -1 \end{bmatrix}$ **9.** $\begin{bmatrix} \frac{7}{2} & -2 & -2 \\ -\frac{5}{2} & 1 & 2 \\ -1 & 0 & 1 \end{bmatrix}$

A Library of Functions

Identity function

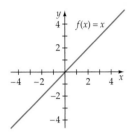

$f(x) = x$

Linear function

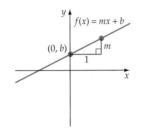

$f(x) = mx + b$

$(0, b)$

m

1

Constant function

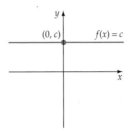

$(0, c)$ $f(x) = c$

Absolute value function

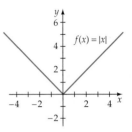

$f(x) = |x|$

Squaring function

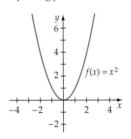

$f(x) = x^2$

Cubing function

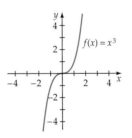

$f(x) = x^3$

Square root function

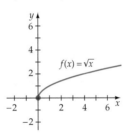

$f(x) = \sqrt{x}$

Cube root function

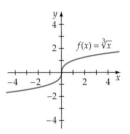

$f(x) = \sqrt[3]{x}$

Exponential function

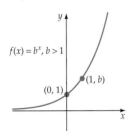

$f(x) = b^x, b > 1$

$(1, b)$

$(0, 1)$

Exponential function

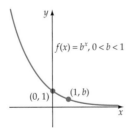

$f(x) = b^x, 0 < b < 1$

$(0, 1)$ $(1, b)$

Logarithmic function

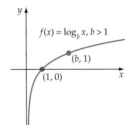

$f(x) = \log_b x, b > 1$

$(b, 1)$

$(1, 0)$

Logarithmic function

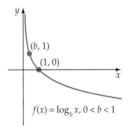

$(b, 1)$

$(1, 0)$

$f(x) = \log_b x, 0 < b < 1$

Logistic function

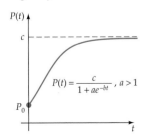

c

$P(t) = \dfrac{c}{1 + ae^{-bt}}, a > 1$

P_0

Logistic function

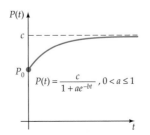

c

P_0

$P(t) = \dfrac{c}{1 + ae^{-bt}}, 0 < a \leq 1$

Reciprocal function

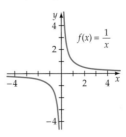

$f(x) = \dfrac{1}{x}$

A rational function

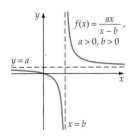

$f(x) = \dfrac{ax}{x - b}$,

$a > 0, b > 0$

$y = a$

$x = b$

A Library of Functions (cont'd)

Sine function

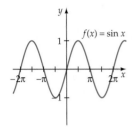

Cosine function

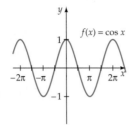

Tangent function

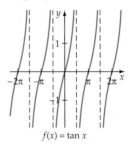

Cosecant function

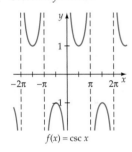

Secant function

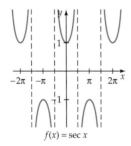

Cotangent function

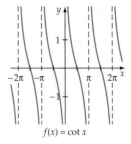

Properties of Exponents

$$a^m a^n = a^{m+n} \qquad \frac{a^m}{a^n} = a^{m-n} \qquad (a^m)^n = a^{mn}$$

$$(a^m b^n)^p = a^{mp} b^{np} \qquad \left(\frac{a^m}{b^n}\right)^p = \frac{a^{mp}}{b^{np}} \qquad b^{-p} = \frac{1}{b^p}$$

Properties of Logarithms

$y = \log_b x$ if and only if $b^y = x$

$$\log_b b = 1 \qquad \log_b 1 = 0 \qquad \log_b (b)^p = p$$

$$b^{\log_b p} = p \qquad \log x = \log_{10} x \qquad \ln x = \log_e x$$

$$\log_b (MN) = \log_b M + \log_b N$$

$$\log_b (M/N) = \log_b M - \log_b N$$

$$\log_b M^p = p \log_b M$$

Properties of Radicals

$$(\sqrt[n]{b})^m = \sqrt[n]{b^m} = b^{m/n} \qquad \frac{\sqrt[n]{a}}{\sqrt[n]{b}} = \sqrt[n]{\frac{a}{b}}$$

$$\sqrt[n]{a}\,\sqrt[n]{b} = \sqrt[n]{ab} \qquad \sqrt[m]{\sqrt[n]{b}} = \sqrt[mn]{b}$$

Properties of Absolute Value Inequalities

$|x| < c$ ($c \geq 0$) if and only if $-c < x < c$.

$|x| > c$ ($c \geq 0$) if and only if either $x > c$ or $x < -c$.

Important Theorems

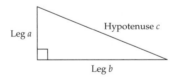
Leg a Hypotenuse c Leg b

Pythagorean Theorem
$c^2 = a^2 + b^2$

Remainder Theorem
If a polynomial $P(x)$ is divided by $x - c$, then the remainder is $P(c)$.

Factor Theorem
A polynomial $P(x)$ has a factor $(x - c)$ if and only if $P(c) = 0$.

Fundamental Theorem of Algebra
If P is a polynomial of degree $n \geq 1$ with complex coefficients, then P has at least one complex zero.

Binomial Theorem
$$(a + b)^n = a^n + \binom{n}{1} a^{n-1}b + \binom{n}{2} a^{n-2}b^2$$
$$+ \cdots + \binom{n}{k} a^{n-k}b^k + \cdots + b^n$$

Important Formulas

The *distance* between $P_1(x_1, y_1)$ and $P_2(x_2, y_2)$ is
$$d(P_1, P_2) = \sqrt{(x_1 - x_2)^2 + (y_1 - y_2)^2}$$

The *slope* m of a line through $P_1(x_1, y_1)$ and $P_2(x_2, y_2)$ is
$$m = \frac{y_2 - y_1}{x_2 - x_1}, \quad x_1 \neq x_2$$

The *slope-intercept form* of a line with slope m and y-intercept b is $y = mx + b$

The *point-slope formula* for a line with slope m passing through $P_1(x_1, y_1)$ is
$$y - y_1 = m(x - x_1)$$

Quadratic Formula
If $a \neq 0$, the solutions of $ax^2 + bx + c = 0$ are
$$x = \frac{-b \pm \sqrt{b^2 - 4ac}}{2a}$$

Properties of Functions

A *function* is a set of ordered pairs in which no two ordered pairs that have the same first coordinate have different second coordinates.

If a and b are elements of an interval I that is a subset of the domain of a function f, then

- f is an *increasing* function on I if $f(a) < f(b)$ whenever $a < b$.
- f is a *decreasing* function on I if $f(a) > f(b)$ whenever $a < b$.
- f is a *constant* function on I if $f(a) = f(b)$ for all a and b.

A *one-to-one* function satisfies the additional condition that given any y, there is one and only one x that can be paired with that given y.

Graphing Concepts

Odd Functions
A function f is an odd function if $f(-x) = -f(x)$ for all x in the domain of f. The graph of an odd function is symmetric with respect to the origin.

Even Functions
A function is an even function if $f(-x) = f(x)$ for all x in the domain of f. The graph of an even function is symmetric with respect to the y-axis.

Vertical and Horizontal Translations

If f is a function and c is a positive constant, then the graph of

- $y = f(x) + c$ is the graph of $y = f(x)$ shifted up *vertically* c units.
- $y = f(x) - c$ is the graph of $y = f(x)$ shifted down *vertically* c units.
- $y = f(x + c)$ is the graph of $y = f(x)$ shifted left *horizontally* c units.
- $y = f(x - c)$ is the graph of $y = f(x)$ shifted right *horizontally* c units.

Reflections

If f is a function then the graph of

- $y = -f(x)$ is the graph of $y = f(x)$ reflected across the x-axis.
- $y = f(-x)$ is the graph of $y = f(x)$ reflected across the y-axis.

Vertical Shrinking and Stretching

- If $c > 0$ and the graph of $y = f(x)$ contains the point (x, y), then the graph of $y = c \cdot f(x)$ contains the point (x, cy).
- If $c > 1$, the graph of $y = c \cdot f(x)$ is obtained by stretching the graph of $y = f(x)$ away from the x-axis by a factor of c.
- If $0 < c < 1$, the graph of $y = c \cdot f(x)$ is obtained by shrinking the graph of $y = f(x)$ toward the x-axis by a factor of c.

Horizontal Shrinking and Stretching

- If $a > 0$ and the graph of $y = f(x)$ contains the point (x, y), then the graph of $y = f(ax)$ contains the point $\left(\dfrac{1}{a}x, y\right)$.
- If $a > 1$, the graph of $y = f(ax)$ is a *horizontal shrinking* of the graph of $y = f(x)$.
- If $0 < a < 1$, the graph of $y = f(ax)$ is a *horizontal stretching* of the graph of $y = f(x)$.

Definitions of Trigonometric Functions

$$\sin \theta = \frac{b}{r} \qquad \csc \theta = \frac{r}{b}$$

$$\cos \theta = \frac{a}{r} \qquad \sec \theta = \frac{r}{a}$$

$$\tan \theta = \frac{b}{a} \qquad \cot \theta = \frac{a}{b}$$

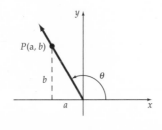

where $r = \sqrt{a^2 + b^2}$

Definitions of Circular Functions

$$\sin t = y \qquad \csc t = \frac{1}{y}$$

$$\cos t = x \qquad \sec t = \frac{1}{x}$$

$$\tan t = \frac{y}{x} \qquad \cot t = \frac{x}{y}$$

Formulas for Triangles

For any triangle ABC, the following formula can be used.

Law of Sines

$$\frac{a}{\sin A} = \frac{b}{\sin B} = \frac{c}{\sin C}$$

Law of Cosines

$$c^2 = a^2 + b^2 - 2ab \cos C$$

Area of a Triangle

$$K = \frac{1}{2}ab \sin C \qquad K = \frac{a^2 \sin B \sin C}{2 \sin A}$$

$$K = \sqrt{s(s - a)(s - b)(s - c)}, \text{ where } s = \frac{a + b + c}{2}$$

Fundamental Identities

$$\tan \theta = \frac{\sin \theta}{\cos \theta} \qquad \cot \theta = \frac{\cos \theta}{\sin \theta}$$

$$\sin^2 \theta + \cos^2 \theta = 1 \qquad 1 + \tan^2 \theta = \sec^2 \theta$$

$$1 + \cot^2 \theta = \csc^2 \theta$$

Formulas for Negatives

$$\sin(-\theta) = -\sin \theta \qquad \cos(-\theta) = \cos \theta$$

$$\tan(-\theta) = -\tan \theta$$

Reciprocal Identities

$$\csc \theta = \frac{1}{\sin \theta} \qquad \sec \theta = \frac{1}{\cos \theta} \qquad \cot \theta = \frac{1}{\tan \theta}$$